## Fundamental Constants

| Quantity | Symbol | Approximate Value | Current Best Value[†] |
|---|---|---|---|
| Speed of light in vacuum | $c$ | $3.00 \times 10^8$ m/s | $2.99792458 \times 10^8$ m/s |
| Gravitational constant | $G$ | $6.67 \times 10^{-11}$ N·m²/kg² | $6.67384(80) \times 10^{-11}$ N·m²/kg² |
| Avogadro's number | $N_A$ | $6.02 \times 10^{23}$ mol⁻¹ | $6.02214129(27) \times 10^{23}$ mol⁻¹ |
| Gas constant | $R$ | $8.314$ J/mol·K $= 1.99$ cal/mol·K $= 0.0821$ L·atm/mol·K | $8.3144621(75)$ J/mol·K |
| Boltzmann's constant | $k$ | $1.38 \times 10^{-23}$ J/K | $1.3806488(13) \times 10^{-23}$ J/K |
| Charge on electron | $e$ | $1.60 \times 10^{-19}$ C | $1.602176565(35) \times 10^{-19}$ C |
| Stefan-Boltzmann constant | $\sigma$ | $5.67 \times 10^{-8}$ W/m²·K⁴ | $5.670373(21) \times 10^{-8}$ W/m²·K⁴ |
| Permittivity of free space | $\epsilon_0 = (1/c^2\mu_0)$ | $8.85 \times 10^{-12}$ C²/N·m² | $8.854187817\ldots \times 10^{-12}$ C²/N·m² |
| Permeability of free space | $\mu_0$ | $4\pi \times 10^{-7}$ T·m/A | $1.2566370614\ldots \times 10^{-6}$ T·m/A |
| Planck's constant | $h$ | $6.63 \times 10^{-34}$ J·s | $6.62606957(29) \times 10^{-34}$ J·s |
| Electron mass | $m_e$ | $9.11 \times 10^{-31}$ kg $= 0.000549$ u $= 0.511$ MeV/$c^2$ | $9.10938291(40) \times 10^{-31}$ kg $= 5.48579909(25) \times 10^{-4}$ u |
| Proton mass | $m_p$ | $1.6726 \times 10^{-27}$ kg $= 1.00728$ u $= 938.27$ MeV/$c^2$ | $1.672621777(74) \times 10^{-27}$ kg $= 1.007276467(45)$ u |
| Neutron mass | $m_n$ | $1.6749 \times 10^{-27}$ kg $= 1.008665$ u $= 939.57$ MeV/$c^2$ | $1.674927351(74) \times 10^{-27}$ kg $= 1.008664916(45)$ u |
| Atomic mass unit (1 u) | | $1.6605 \times 10^{-27}$ kg $= 931.49$ MeV/$c^2$ | $1.660538921(73) \times 10^{-27}$ kg $= 931.494061(21)$ MeV/$c^2$ |

[†] Numbers in parentheses indicate one-standard-deviation experimental uncertainties in final digits.
Values without parentheses are exact (i.e., defined quantities).

## Other Useful Data

| | |
|---|---|
| Joule equivalent (1 cal) | 4.186 J |
| Absolute zero (0 K) | $-273.15$°C |
| Acceleration due to gravity at Earth's surface (avg.) | $9.80$ m/s² ($= g$) |
| Speed of sound in air (20°C) | 343 m/s |
| Density of air (dry) | 1.29 kg/m³ |
| Earth: Mass | $5.98 \times 10^{24}$ kg |
|     Radius (mean) | $6.38 \times 10^3$ km |
| Moon: Mass | $7.35 \times 10^{22}$ kg |
|     Radius (mean) | $1.74 \times 10^3$ km |
| Sun: Mass | $1.99 \times 10^{30}$ kg |
|     Radius (mean) | $6.96 \times 10^5$ km |
| Earth–Sun distance (mean) | $149.60 \times 10^6$ km |
| Earth–Moon distance (mean) | $384 \times 10^3$ km |

## The Greek Alphabet

| | | | | | |
|---|---|---|---|---|---|
| Alpha | A | $\alpha$ | Nu | N | $\nu$ |
| Beta | B | $\beta$ | Xi | $\Xi$ | $\xi$ |
| Gamma | $\Gamma$ | $\gamma$ | Omicron | O | $o$ |
| Delta | $\Delta$ | $\delta$ | Pi | $\Pi$ | $\pi$ |
| Epsilon | E | $\epsilon, \varepsilon$ | Rho | P | $\rho$ |
| Zeta | Z | $\zeta$ | Sigma | $\Sigma$ | $\sigma$ |
| Eta | H | $\eta$ | Tau | T | $\tau$ |
| Theta | $\Theta$ | $\theta$ | Upsilon | Y | $\upsilon$ |
| Iota | I | $\iota$ | Phi | $\Phi$ | $\phi, \varphi$ |
| Kappa | K | $\kappa$ | Chi | X | $\chi$ |
| Lambda | $\Lambda$ | $\lambda$ | Psi | $\Psi$ | $\psi$ |
| Mu | M | $\mu$ | Omega | $\Omega$ | $\omega$ |

## Values of Some Numbers

| | | | |
|---|---|---|---|
| $\pi = 3.1415927$ | $\sqrt{2} = 1.4142136$ | $\ln 2 = 0.6931472$ | $\log_{10} e = 0.4342945$ |
| $e = 2.7182818$ | $\sqrt{3} = 1.7320508$ | $\ln 10 = 2.3025851$ | $1$ rad $= 57.2957795$° |

## Mathematical Signs and Symbols

| | | | |
|---|---|---|---|
| $\propto$ | is proportional to | $\leq$ | is less than or equal to |
| $=$ | is equal to | $\geq$ | is greater than or equal to |
| $\approx$ | is approximately equal to | $\Sigma$ | sum of |
| $\neq$ | is not equal to | $\bar{x}$ | average value of $x$ |
| $>$ | is greater than | $\Delta x$ | change in $x$ |
| $\gg$ | is much greater than | $\Delta x \to 0$ | $\Delta x$ approaches zero |
| $<$ | is less than | $n!$ | $n(n-1)(n-2)\ldots(1)$ |
| $\ll$ | is much less than | | |

## Properties of Water

| | |
|---|---|
| Density (4°C) | $1.000 \times 10^3$ kg/m³ |
| Heat of fusion (0°C) | 334 kJ/kg (79.8 kcal/kg) |
| Heat of vaporization (100°C) | 2260 kJ/kg (539.9 kcal/kg) |
| Specific heat (15°C) | 4186 J/kg·C° (1.00 kcal/kg·C°) |
| Index of refraction | 1.33 |

## Unit Conversions (Equivalents)

### Length

1 in. = 2.54 cm (defined)
1 cm = 0.3937 in.
1 ft = 30.48 cm
1 m = 39.37 in. = 3.281 ft
1 mi = 5280 ft = 1.609 km
1 km = 0.6214 mi
1 nautical mile (U.S.) = 1.151 mi = 6076 ft = 1.852 km
1 fermi = 1 femtometer (fm) = $10^{-15}$ m
1 angstrom (Å) = $10^{-10}$ m = 0.1 nm
1 light-year (ly) = $9.461 \times 10^{15}$ m
1 parsec = 3.26 ly = $3.09 \times 10^{16}$ m

### Volume

1 liter (L) = 1000 mL = 1000 $cm^3$ = $1.0 \times 10^{-3}$ $m^3$ = 1.057 qt (U.S.) = 61.02 $in.^3$
1 gal (U.S.) = 4 qt (U.S.) = 231 $in.^3$ = 3.785 L = 0.8327 gal (British)
1 quart (U.S.) = 2 pints (U.S.) = 946 mL
1 pint (British) = 1.20 pints (U.S.) = 568 mL
1 $m^3$ = 35.31 $ft^3$

### Speed

1 mi/h = 1.4667 ft/s = 1.6093 km/h = 0.4470 m/s
1 km/h = 0.2778 m/s = 0.6214 mi/h
1 ft/s = 0.3048 m/s = 0.6818 mi/h = 1.0973 km/h
1 m/s = 3.281 ft/s = 3.600 km/h = 2.237 mi/h
1 knot = 1.151 mi/h = 0.5144 m/s

### Angle

1 radian (rad) = 57.30° = 57°18′
1° = 0.01745 rad
1 rev/min (rpm) = 0.1047 rad/s

### Time

1 day = $8.640 \times 10^4$ s
1 year = 365.242 days = $3.156 \times 10^7$ s

### Mass

1 atomic mass unit (u) = $1.6605 \times 10^{-27}$ kg
1 kg = 0.06852 slug
[1 kg has a weight of 2.20 lb where $g$ = 9.80 $m/s^2$.]

### Force

1 lb = 4.44822 N
1 N = $10^5$ dyne = 0.2248 lb

### Energy and Work

1 J = $10^7$ ergs = 0.7376 ft·lb
1 ft·lb = 1.356 J = $1.29 \times 10^{-3}$ Btu = $3.24 \times 10^{-4}$ kcal
1 kcal = $4.19 \times 10^3$ J = 3.97 Btu
1 eV = $1.6022 \times 10^{-19}$ J
1 kWh = $3.600 \times 10^6$ J = 860 kcal
1 Btu = $1.056 \times 10^3$ J

### Power

1 W = 1 J/s = 0.7376 ft·lb/s = 3.41 Btu/h
1 hp = 550 ft·lb/s = 746 W

### Pressure

1 atm = 1.01325 bar = $1.01325 \times 10^5$ $N/m^2$
         = 14.7 $lb/in.^2$ = 760 torr
1 $lb/in.^2$ = $6.895 \times 10^3$ $N/m^2$
1 Pa = 1 $N/m^2$ = $1.450 \times 10^{-4}$ $lb/in.^2$

## SI Derived Units and Their Abbreviations

| Quantity | Unit | Abbreviation | In Terms of Base Units† |
|---|---|---|---|
| Force | newton | N | $kg \cdot m/s^2$ |
| Energy and work | joule | J | $kg \cdot m^2/s^2$ |
| Power | watt | W | $kg \cdot m^2/s^3$ |
| Pressure | pascal | Pa | $kg/(m \cdot s^2)$ |
| Frequency | hertz | Hz | $s^{-1}$ |
| Electric charge | coulomb | C | $A \cdot s$ |
| Electric potential | volt | V | $kg \cdot m^2/(A \cdot s^3)$ |
| Electric resistance | ohm | Ω | $kg \cdot m^2/(A^2 \cdot s^3)$ |
| Capacitance | farad | F | $A^2 \cdot s^4/(kg \cdot m^2)$ |
| Magnetic field | tesla | T | $kg/(A \cdot s^2)$ |
| Magnetic flux | weber | Wb | $kg \cdot m^2/(A \cdot s^2)$ |
| Inductance | henry | H | $kg \cdot m^2/(A^2 \cdot s^2)$ |

†kg = kilogram (mass), m = meter (length), s = second (time), A = ampere (electric current).

## Metric (SI) Multipliers

| Prefix | Abbreviation | Value |
|---|---|---|
| yotta | Y | $10^{24}$ |
| zeta | Z | $10^{21}$ |
| exa | E | $10^{18}$ |
| peta | P | $10^{15}$ |
| tera | T | $10^{12}$ |
| giga | G | $10^9$ |
| mega | M | $10^6$ |
| kilo | k | $10^3$ |
| hecto | h | $10^2$ |
| deka | da | $10^1$ |
| deci | d | $10^{-1}$ |
| centi | c | $10^{-2}$ |
| milli | m | $10^{-3}$ |
| micro | $\mu$ | $10^{-6}$ |
| nano | n | $10^{-9}$ |
| pico | p | $10^{-12}$ |
| femto | f | $10^{-15}$ |
| atto | a | $10^{-18}$ |
| zepto | z | $10^{-21}$ |
| yocto | y | $10^{-24}$ |

VOLUME I

# PHYSICS

SEVENTH EDITION

## PRINCIPLES WITH APPLICATIONS

# DOUGLAS C. GIANCOLI

PEARSON

Boston   Columbus   Indianapolis   New York   San Francisco   Upper Saddle River
Amsterdam   Cape Town   Dubai   London   Madrid   Milan   Munich   Paris   Montréal   Toronto
Delhi   Mexico City   São Paulo   Sydney   Hong Kong   Seoul   Singapore   Taipei   Tokyo

President, Science, Business and Technology: Paul Corey
Publisher: Jim Smith
Executive Development Editor: Karen Karlin
Production Project Manager: Elisa Mandelbaum / Laura Ross
Marketing Manager: Will Moore
Senior Managing Editor: Corinne Benson
Managing Development Editor: Cathy Murphy
Copyeditor: Joanna Dinsmore
Proofreaders: Susan Fisher, Donna Young
Interior Designer: Mark Ong
Cover Designer: Derek Bacchus
Photo Permissions Management: Maya Melenchuk
Photo Research Manager: Eric Schrader
Photo Researcher: Mary Teresa Giancoli
Senior Administrative Assistant: Cathy Glenn
Senior Administrative Coordinator: Trisha Tarricone
Text Permissions Project Manager: Joseph Croscup
Editorial Media Producer: Kelly Reed
Manufacturing Buyer: Jeffrey Sargent
Indexer: Carol Reitz
Compositor: Preparé, Inc.
Illustrations: Precision Graphics

Cover Photo Credit: North Peak, California (D. Giancoli); Insets: left, analog to digital (page 488); right, electron microscope image—retina of human eye with cones artificially colored green, rods beige (page 785).
Back Cover Photo Credit: D. Giancoli

Credits and acknowledgments for materials borrowed from other sources and reproduced, with permission, in this textbook appear on page A-61.

**Library of Congress Cataloging-in-Publication Data on file**

ISBN-10: 0-321-76242-8
ISBN-13: 978-0-321-76242-9

3 4 5 6 7 8 9 10—**V011**—17 16 15

www.pearsonhighered.com

# Contents, Volume 1

Force    Displacement

# Contents, Volume 2

# Applications to Biology and Medicine (Selected)

# Applications to Other Fields and Everyday Life (Selected)

# Student Supplements

- **MasteringPhysics**™ (www.masteringphysics.com) is a homework, tutorial, and assessment system based on years of research into how students work physics problems and precisely where they need help. Studies show that students who use MasteringPhysics significantly increase their final scores compared to hand-written homework. Mastering-Physics achieves this improvement by providing students with instantaneous feedback specific to their wrong answers, simpler sub-problems upon request when they get stuck, and partial credit for their method(s) used. This individualized, 24/7 Socratic tutoring is recommended by nine out of ten students to their peers as the most effective and time-efficient way to study.

- The **Student Study Guide with Selected Solutions**, **Volume I** (Chapters 1–15, ISBN 978-0-321-76240-5) and **Volume II** (Chapters 16–33, ISBN 978-0-321-76808-7), written by Joseph Boyle (Miami-Dade Community College), contains overviews, key terms and phrases, key equations, self-study exams, problems for review, problem solving skills, and answers and solutions to selected end-of-chapter questions and problems for each chapter of this textbook.

- **Pearson eText** is available through MasteringPhysics, either automatically when MasteringPhysics is packaged with new books, or available as a purchased upgrade online. Allowing students access to the text wherever they have access to the Internet, Pearson eText comprises the full text, including figures that can be enlarged for better viewing. Within eText, students are also able to pop up definitions and terms to help with vocabulary and the reading of the material. Students can also take notes in eText using the annotation feature at the top of each page.

- **Pearson Tutor Services** (www.pearsontutorservices.com): Each student's subscription to MasteringPhysics also contains complimentary access to Pearson Tutor Services, powered by Smarthinking, Inc. By logging in with their MasteringPhysics ID and password, they will be connected to highly qualified e-instructors™ who provide additional, interactive online tutoring on the major concepts of physics.

- **ActivPhysics OnLine**™ (accessed through the Self Study area within www.masteringphysics.com) provides students with a group of highly regarded applet-based tutorials (see above). The following workbooks help students work though complex concepts and understand them more clearly.

- **ActivPhysics OnLine Workbook Volume 1: Mechanics • Thermal Physics • Oscillations & Waves** (ISBN 978-0-805-39060-5)

- **ActivPhysics OnLine Workbook Volume 2: Electricity & Magnetism • Optics • Modern Physics** (ISBN 978-0-805-39061-2)

# Preface

## What's New?

Lots! Much is new and unseen before. Here are the big four:

1. Multiple-choice Questions added to the end of each Chapter. They are not the usual type. These are called **MisConceptual Questions** because the responses (*a*, *b*, *c*, *d*, etc.) are intended to include common student misconceptions. Thus they are as much, or more, a learning experience than simply a testing experience.

2. **Search and Learn Problems** at the very end of each Chapter, after the other Problems. Some are pretty hard, others are fairly easy. They are intended to encourage students to go back and reread some part or parts of the text, and in this search for an answer they will hopefully learn more—if only because they have to read some material again.

3. **Chapter-Opening Questions** (COQ) that start each Chapter, a sort of "stimulant." Each is multiple choice, with responses including common misconceptions—to get preconceived notions out on the table right at the start. Where the relevant material is covered in the text, students find an Exercise asking them to return to the COQ to rethink and answer again.

4. **Digital.** Biggest of all. Crucial new applications. Today we are surrounded by digital electronics. How does it work? If you try to find out, say on the Internet, you won't find much physics: you may find shallow hand-waving with no real content, or some heavy jargon whose basis might take months or years to understand. So, for the first time, I have tried to explain

   - The basis of digital in bits and bytes, how analog gets transformed into digital, sampling rate, bit depth, quantization error, compression, noise (Section 17–10).

   - How digital TV works, including how each pixel is addressed for each frame, data stream, refresh rate (Section 17–11).

   - Semiconductor computer memory, DRAM, and flash (Section 21–8).

   - Digital cameras and sensors—revised and expanded Section 25–1.

   - New semiconductor physics, some of which is used in digital devices, including LED and OLED—how they work and what their uses are—plus more on transistors (MOSFET), chips, and technology generation as in 22-nm technology (Sections 29–9, 10, 11).

Besides those above, this new seventh edition includes

5. *New topics, new applications, principal revisions.*

   - *You can measure the Earth's radius* (Section 1–7).

   - Improved graphical analysis of linear motion (Section 2–8).

   - Planets (how first seen), heliocentric, geocentric (Section 5–8).

   - The Moon's orbit around the Earth: its phases and periods with diagram (Section 5–9).

   - Explanation of lake level change when large rock thrown from boat (Example 10–11).

- Biology and medicine, including:
  - Blood measurements (flow, sugar)—Chapters 10, 12, 14, 19, 20, 21;
  - Trees help offset $CO_2$ buildup—Chapter 15;
  - Pulse oximeter—Chapter 29;
  - Proton therapy—Chapter 31;
  - Radon exposure calculation—Chapter 31;
  - Cell phone use and brain—Chapter 31.
- Colors as seen underwater (Section 24–4).
- Soap film sequence of colors explained (Section 24–8).
- Solar sails (Section 22–6).
- Lots on sports.
- Symmetry—more emphasis and using italics or boldface to make visible.
- Flat screens (Sections 17–11, 24–11).
- Free-electron theory of metals, Fermi gas, Fermi level. New Section 29–6.
- Semiconductor devices—new details on diodes, LEDs, OLEDs, solar cells, compound semiconductors, diode lasers, MOSFET transistors, chips, 22-nm technology (Sections 29–9, 10, 11).
- Cross section (Chapter 31).
- Length of an object is a script $\ell$ rather than normal $l$, which looks like 1 or $I$ (moment of inertia, current), as in $F = I\ell B$. Capital $L$ is for angular momentum, latent heat, inductance, dimensions of length $[L]$.

6. *New photographs* taken by students and instructors (we asked).

7. *Page layout*: More than in previous editions, serious attention to how each page is formatted. Important derivations and Examples are on facing pages: no turning a page back in the middle of a derivation or Example. Throughout, readers see, on two facing pages, an important slice of physics.

8. *Greater clarity*: No topic, no paragraph in this book was overlooked in the search to improve the clarity and conciseness of the presentation. Phrases and sentences that may slow down the principal argument have been eliminated: keep to the essentials at first, give the elaborations later.

9. Much use has been made of physics education research. See the new powerful pedagogic features listed first.

10. *Examples modified*: More math steps are spelled out, and many new Examples added. About 10% of all Examples are Estimation Examples.

11. *This Book is Shorter* than other complete full-service books at this level. Shorter explanations are easier to understand and more likely to be read.

12. *Cosmological Revolution*: With generous help from top experts in the field, readers have the latest results.

## See the World through Eyes that Know Physics

I was motivated from the beginning to write a textbook different from the others which present physics as a sequence of facts, like a catalog: "Here are the facts and you better learn them." Instead of beginning formally and dogmatically, I have sought to begin each topic with concrete observations and experiences students can relate to: start with specifics, and after go to the great generalizations and the more formal aspects of a topic, showing *why* we believe what we believe. This approach reflects how science is actually practiced.

The ultimate aim is to give students a thorough understanding of the basic concepts of physics in all its aspects, from mechanics to modern physics. A second objective is to show students how useful physics is in their own everyday lives and in their future professions by means of interesting applications to biology, medicine, architecture, and more.

Also, much effort has gone into techniques and approaches for solving problems: worked-out Examples, Problem Solving sections (Sections 2–6, 3–6, 4–7, 4–8, 6–7, 6–9, 8–6, 9–2, 13–7, 14–4, and 16–6), and Problem Solving Strategies (pages 30, 57, 60, 88, 115, 141, 158, 184, 211, 234, 399, 436, 456, 534, 568, 594, 655, 666, and 697).

This textbook is especially suited for students taking a one-year introductory course in physics that uses algebra and trigonometry but not calculus.† Many of these students are majoring in biology or premed, as well as architecture, technology, and the earth and environmental sciences. Many applications to these fields are intended to answer that common student query: "Why must I study physics?" The answer is that physics is fundamental to a full understanding of these fields, and here they can see how. Physics is everywhere around us in the everyday world. It is the goal of this book to help students "see the world through eyes that know physics."

A major effort has been made to not throw too much material at students reading the first few chapters. The basics have to be learned first. Many aspects can come later, when students are less overloaded and more prepared. If we don't overwhelm students with too much detail, especially at the start, maybe they can find physics interesting, fun, and helpful—and those who were afraid may lose their fear.

Chapter 1 is *not* a throwaway. It is fundamental to physics to realize that every measurement has an *uncertainty*, and how significant figures are used. Converting units and being able to make rapid *estimates* are also basic.

*Mathematics* can be an obstacle to students. I have aimed at including all steps in a derivation. Important mathematical tools, such as addition of vectors and trigonometry, are incorporated in the text where first needed, so they come with a context rather than in a scary introductory Chapter. Appendices contain a review of algebra and geometry (plus a few advanced topics).

*Color* is used pedagogically to bring out the physics. Different types of vectors are given different colors (see the chart on page xix).

Sections marked with a star * are considered optional. These contain slightly more advanced physics material, or material not usually covered in typical courses and/or interesting applications; they contain no material needed in later Chapters (except perhaps in later optional Sections).

For a brief course, all optional material could be dropped as well as significant parts of Chapters 1, 10, 12, 22, 28, 29, 32, and selected parts of Chapters 7, 8, 9, 15, 21, 24, 25, 31. Topics not covered in class can be a valuable resource for later study by students. Indeed, this text can serve as a useful reference for years because of its wide range of coverage.

---

†It is fine to take a calculus course. But mixing calculus with physics for these students may often mean not learning the physics because of stumbling over the calculus.

# Thanks

Many physics professors provided input or direct feedback on every aspect of this textbook. They are listed below, and I owe each a debt of gratitude.

Edward Adelson, The Ohio State University
Lorraine Allen, United States Coast Guard Academy
Zaven Altounian, McGill University
Leon Amstutz, Taylor University
David T. Bannon, Oregon State University
Bruce Barnett, Johns Hopkins University
Michael Barnett, Lawrence Berkeley Lab
Anand Batra, Howard University
Cornelius Bennhold, George Washington University
Bruce Birkett, University of California Berkeley
Steven Boggs, University of California Berkeley
Robert Boivin, Auburn University
Subir Bose, University of Central Florida
David Branning, Trinity College
Meade Brooks, Collin County Community College
Bruce Bunker, University of Notre Dame
Grant Bunker, Illinois Institute of Technology
Wayne Carr, Stevens Institute of Technology
Charles Chiu, University of Texas Austin
Roger N. Clark, U. S. Geological Survey
Russell Clark, University of Pittsburgh
Robert Coakley, University of Southern Maine
David Curott, University of North Alabama
Biman Das, SUNY Potsdam
Bob Davis, Taylor University
Kaushik De, University of Texas Arlington
Michael Dennin, University of California Irvine
Karim Diff, Santa Fe College
Kathy Dimiduk, Cornell University
John DiNardo, Drexel University
Scott Dudley, United States Air Force Academy
Paul Dyke
John Essick, Reed College
Kim Farah, Lasell College
Cassandra Fesen, Dartmouth College
Leonard Finegold, Drexel University
Alex Filippenko, University of California Berkeley
Richard Firestone, Lawrence Berkeley Lab
Allen Flora, Hood College
Mike Fortner, Northern Illinois University
Tom Furtak, Colorado School of Mines
Edward Gibson, California State University Sacramento
John Hardy, Texas A&M
Thomas Hemmick, State University of New York Stonybrook
J. Erik Hendrickson, University of Wisconsin Eau Claire
Laurent Hodges, Iowa State University
David Hogg, New York University
Mark Hollabaugh, Normandale Community College
Andy Hollerman, University of Louisiana at Lafayette
Russell Holmes, University of Minnesota Twin Cities
William Holzapfel, University of California Berkeley
Chenming Hu, University of California Berkeley
Bob Jacobsen, University of California Berkeley
Arthur W. John, Northeastern University
Teruki Kamon, Texas A&M
Daryao Khatri, University of the District of Columbia
Tsu-Jae King Liu, University of California Berkeley
Richard Kronenfeld, South Mountain Community College
Jay Kunze, Idaho State University
Jim LaBelle, Dartmouth College
Amer Lahamer, Berea College
David Lamp, Texas Tech University
Kevin Lear, SpatialGraphics.com
Ran Li, Kent State University
Andreí Linde, Stanford University
M.A.K. Lodhi, Texas Tech
Lisa Madewell, University of Wisconsin

Bruce Mason, University of Oklahoma
Mark Mattson, James Madison University
Dan Mazilu, Washington and Lee University
Linda McDonald, North Park College
Bill McNairy, Duke University
Jo Ann Merrell, Saddleback College
Raj Mohanty, Boston University
Giuseppe Molesini, Istituto Nazionale di Ottica Florence
Wouter Montfrooij, University of Missouri
Eric Moore, Frostburg State University
Lisa K. Morris, Washington State University
Richard Muller, University of California Berkeley
Blaine Norum, University of Virginia
Lauren Novatne, Reedley College
Alexandria Oakes, Eastern Michigan University
Ralph Oberly, Marshall University
Michael Ottinger, Missouri Western State University
Lyman Page, Princeton and WMAP
Laurence Palmer, University of Maryland
Bruce Partridge, Haverford College
R. Daryl Pedigo, University of Washington
Robert Pelcovitz, Brown University
Saul Perlmutter, University of California Berkeley
Vahe Peroomian, UCLA
Harvey Picker, Trinity College
Amy Pope, Clemson University
James Rabchuk, Western Illinois University
Michele Rallis, Ohio State University
Paul Richards, University of California Berkeley
Peter Riley, University of Texas Austin
Dennis Rioux, University of Wisconsin Oshkosh
John Rollino, Rutgers University
Larry Rowan, University of North Carolina Chapel Hill
Arthur Schmidt, Northwestern University
Cindy Schwarz-Rachmilowitz, Vassar College
Peter Sheldon, Randolph-Macon Woman's College
Natalia A. Sidorovskaia, University of Louisiana at Lafayette
James Siegrist, University of California Berkeley
Christopher Sirola, University of Southern Mississippi
Earl Skelton, Georgetown University
George Smoot, University of California Berkeley
David Snoke, University of Pittsburgh
Stanley Sobolewski, Indiana University of Pennsylvania
Mark Sprague, East Carolina University
Michael Strauss, University of Oklahoma
Laszlo Takac, University of Maryland Baltimore Co.
Leo Takahashi, Pennsylvania State University
Richard Taylor, University of Oregon
Oswald Tekyi-Mensah, Alabama State University
Franklin D. Trumpy, Des Moines Area Community College
Ray Turner, Clemson University
Som Tyagi, Drexel University
David Vakil, El Camino College
Trina VanAusdal, Salt Lake Community College
John Vasut, Baylor University
Robert Webb, Texas A&M
Robert Weidman, Michigan Technological University
Edward A. Whittaker, Stevens Institute of Technology
Lisa M. Will, San Diego City College
Suzanne Willis, Northern Illinois University
John Wolbeck, Orange County Community College
Stanley George Wojcicki, Stanford University
Mark Worthy, Mississippi State University
Edward Wright, UCLA and WMAP
Todd Young, Wayne State College
William Younger, College of the Albemarle
Hsiao-Ling Zhou, Georgia State University
Michael Ziegler, The Ohio State University
Ulrich Zurcher, Cleveland State University

New photographs were offered by Professors Vickie Frohne (Holy Cross Coll.), Guillermo Gonzales (Grove City Coll.), Martin Hackworth (Idaho State U.), Walter H. G. Lewin (MIT), Nicholas Murgo (NEIT), Melissa Vigil (Marquette U.), Brian Woodahl (Indiana U. at Indianapolis), and Gary Wysin (Kansas State U.). New photographs shot by students are from the AAPT photo contest: Matt Buck, (John Burroughs School), Matthew Claspill (Helias H. S.), Greg Gentile (West Forsyth H. S.), Shilpa Hampole (Notre Dame H. S.), Sarah Lampen (John Burroughs School), Mrinalini Modak (Fayetteville–Manlius H. S.), Joey Moro (Ithaca H. S.), and Anna Russell and Annacy Wilson (both Tamalpais H. S.).

I owe special thanks to Prof. Bob Davis for much valuable input, and especially for working out all the Problems and producing the Solutions Manual for all Problems, as well as for providing the answers to odd-numbered Problems at the back of the book. Many thanks also to J. Erik Hendrickson who collaborated with Bob Davis on the solutions, and to the team they managed (Profs. Karim Diff, Thomas Hemmick, Lauren Novatne, Michael Ottinger, and Trina VanAusdal).

I am grateful to Profs. Lorraine Allen, David Bannon, Robert Coakley, Kathy Dimiduk, John Essick, Dan Mazilu, John Rollino, Cindy Schwarz, Earl Skelton, Michael Strauss, Ray Turner, Suzanne Willis, and Todd Young, who helped with developing the new MisConceptual Questions and Search and Learn Problems, and offered other significant clarifications.

Crucial for rooting out errors, as well as providing excellent suggestions, were Profs. Lorraine Allen, Kathy Dimiduk, Michael Strauss, Ray Turner, and David Vakil. A huge thank you to them and to Prof. Giuseppe Molesini for his suggestions and his exceptional photographs for optics.

For Chapters 32 and 33 on Particle Physics and Cosmology and Astrophysics, I was fortunate to receive generous input from some of the top experts in the field, to whom I owe a debt of gratitude: Saul Perlmutter, George Smoot, Richard Muller, Steven Boggs, Alex Filippenko, Paul Richards, James Siegrist, and William Holzapfel (UC Berkeley), Andreí Linde (Stanford U.), Lyman Page (Princeton and WMAP), Edward Wright (UCLA and WMAP), Michael Strauss (University of Oklahoma), Michael Barnett (LBNL), and Bob Jacobsen (UC Berkeley; so helpful in many areas, including digital and pedagogy).

I also wish to thank Profs. Howard Shugart, Chair Frances Hellman, and many others at the University of California, Berkeley, Physics Department for helpful discussions, and for hospitality. Thanks also to Profs. Tito Arecchi, Giuseppe Molesini, and Riccardo Meucci at the Istituto Nazionale di Ottica, Florence, Italy.

Finally, I am grateful to the many people at Pearson Education with whom I worked on this project, especially Paul Corey and the ever-perspicacious Karen Karlin.

The final responsibility for all errors lies with me. I welcome comments, corrections, and suggestions as soon as possible to benefit students for the next reprint.

D.C.G.

email: Jim.Smith@Pearson.com
Post: Jim Smith
1301 Sansome Street
San Francisco, CA 94111

## About the Author

Douglas C. Giancoli obtained his BA in physics (summa cum laude) from UC Berkeley, his MS in physics at MIT, and his PhD in elementary particle physics back at UC Berkeley. He spent 2 years as a post-doctoral fellow at UC Berkeley's Virus lab developing skills in molecular biology and biophysics. His mentors include Nobel winners Emilio Segrè and Donald Glaser.

He has taught a wide range of undergraduate courses, traditional as well as innovative ones, and continues to update his textbooks meticulously, seeking ways to better provide an understanding of physics for students.

Doug's favorite spare-time activity is the outdoors, especially climbing peaks. He says climbing peaks is like learning physics: it takes effort and the rewards are great.

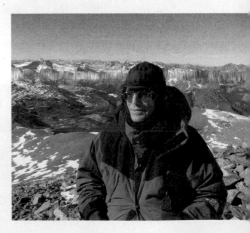

## To Students

### HOW TO STUDY

1. **Read the Chapter.** Learn new vocabulary and notation. Try to respond to questions and exercises as they occur.

2. **Attend all class meetings.** Listen. Take notes, especially about aspects you do not remember seeing in the book. Ask questions (everyone wants to, but maybe you will have the courage). You will get more out of class if you read the Chapter first.

3. **Read the Chapter again,** paying attention to details. Follow derivations and worked-out Examples. Absorb their logic. Answer Exercises and as many of the end-of-Chapter Questions as you can, and all MisConceptual Questions.

4. **Solve at least 10 to 20 end of Chapter Problems,** especially those assigned. In doing Problems you find out what you learned and what you didn't. Discuss them with other students. Problem solving is one of the great learning tools. Don't just look for a formula—it might be the wrong one.

### NOTES ON THE FORMAT AND PROBLEM SOLVING

1. Sections marked with a star (*) are considered **optional**. They can be omitted without interrupting the main flow of topics. No later material depends on them except possibly later starred Sections. They may be fun to read, though.

2. The customary **conventions** are used: symbols for quantities (such as $m$ for mass) are italicized, whereas units (such as m for meter) are not italicized. Symbols for vectors are shown in boldface with a small arrow above: $\vec{\mathbf{F}}$.

3. Few equations are valid in all situations. Where practical, the **limitations** of important equations are stated in square brackets next to the equation. The equations that represent the great laws of physics are displayed with a tan background, as are a few other indispensable equations.

4. At the end of each Chapter is a set of **Questions** you should try to answer. Attempt all the multiple-choice **MisConceptual Questions**. Most important are **Problems** which are ranked as Level I, II, or III, according to estimated difficulty. Level I Problems are easiest, Level II are standard Problems, and Level III are "challenge problems." These ranked Problems are arranged by Section, but Problems for a given Section may depend on earlier material too. There follows a group of **General Problems**, not arranged by Section or ranked. Problems that relate to optional Sections are starred (*). Answers to odd-numbered Problems are given at the end of the book. **Search and Learn Problems** at the end are meant to encourage you to return to parts of the text to find needed detail, and at the same time help you to learn.

5. Being able to solve **Problems** is a crucial part of learning physics, and provides a powerful means for understanding the concepts and principles. This book contains many aids to problem solving: (a) worked-out **Examples**, including an Approach and Solution, which should be studied as an integral part of the text; (b) some of the worked-out Examples are **Estimation Examples**, which show how rough or approximate results can be obtained even if the given data are sparse (see Section 1–7); (c) **Problem Solving Strategies** placed throughout the text to suggest a step-by-step approach to problem solving for a particular topic—but remember that the basics remain the same; most of these "Strategies" are followed by an Example that is solved by explicitly following the suggested steps; (d) special problem-solving Sections; (e) "Problem Solving" marginal notes which refer to hints within the text for solving Problems; (f) **Exercises** within the text that you should work out immediately, and then check your response against the answer given at the bottom of the last page of that Chapter; (g) the Problems themselves at the end of each Chapter (point 4 above).

6. **Conceptual Examples** pose a question which hopefully starts you to think and come up with a response. Give yourself a little time to come up with your own response before reading the Response given.

7. **Math** review, plus additional topics, are found in Appendices. Useful **data, conversion factors**, and math **formulas** are found inside the front and back covers.

# USE OF COLOR

## Vectors

| | |
|---|---|
| A general vector | |
| resultant vector (sum) is slightly thicker | |
| components of any vector are dashed | |
| Displacement ($\vec{\mathbf{D}}, \vec{\mathbf{r}}$) | |
| Velocity ($\vec{\mathbf{v}}$) | |
| Acceleration ($\vec{\mathbf{a}}$) | |
| Force ($\vec{\mathbf{F}}$) | |
| Force on second object | |
| or third object in same figure | |
| Momentum ($\vec{\mathbf{p}}$ or $m\vec{\mathbf{v}}$) | |
| Angular momentum ($\vec{\mathbf{L}}$) | |
| Angular velocity ($\vec{\boldsymbol{\omega}}$) | |
| Torque ($\vec{\boldsymbol{\tau}}$) | |
| Electric field ($\vec{\mathbf{E}}$) | |
| Magnetic field ($\vec{\mathbf{B}}$) | |

## Electricity and magnetism

| | |
|---|---|
| Electric field lines | |
| Equipotential lines | |
| Magnetic field lines | |
| Electric charge (+) | + or ● + |
| Electric charge (−) | − or ● − |

## Electric circuit symbols

| | |
|---|---|
| Wire, with switch S | S |
| Resistor | |
| Capacitor | |
| Inductor | |
| Battery | |
| Ground | |

## Optics

| | |
|---|---|
| Light rays | |
| Object | |
| Real image (dashed) | |
| Virtual image (dashed and paler) | |

## Other

| | |
|---|---|
| Energy level (atom, etc.) | |
| Measurement lines | ├—1.0 m—┤ |
| Path of a moving object | |
| Direction of motion or current | |

Image of the Earth from a NASA satellite. The sky appears black from out in space because there are so few molecules to reflect light. (Why the sky appears blue to us on Earth has to do with scattering of light by molecules of the atmosphere, as discussed in Chapter 24.) Note the storm off the coast of Mexico.

# Introduction, Measurement, Estimating

## CHAPTER 1

**CHAPTER-OPENING QUESTIONS—Guess now!**

**1.** How many $cm^3$ are in $1.0\ m^3$?

**(a)** 10. **(b)** 100. **(c)** 1000. **(d)** 10,000. **(e)** 100,000. **(f)** 1,000,000.

**2.** Suppose you wanted to actually measure the radius of the Earth, at least roughly, rather than taking other people's word for what it is. Which response below describes the best approach?

**(a)** Use an extremely long measuring tape.
**(b)** It is only possible by flying high enough to see the actual curvature of the Earth.
**(c)** Use a standard measuring tape, a step ladder, and a large smooth lake.
**(d)** Use a laser and a mirror on the Moon or on a satellite.
**(e)** Give up; it is impossible using ordinary means.

*[We start each Chapter with a Question—sometimes two. Try to answer right away. Don't worry about getting the right answer now—the idea is to get your preconceived notions out on the table. If they are misconceptions, we expect them to be cleared up as you read the Chapter. You will usually get another chance at the Question(s) later in the Chapter when the appropriate material has been covered. These Chapter-Opening Questions will also help you see the power and usefulness of physics.]*

CONTENTS

Physics is the most basic of the sciences. It deals with the behavior and structure of matter. The field of physics is usually divided into *classical physics* which includes motion, fluids, heat, sound, light, electricity, and magnetism; and *modern physics* which includes the topics of relativity, atomic structure, quantum theory, condensed matter, nuclear physics, elementary particles, and cosmology and astrophysics. We will cover all these topics in this book, beginning with motion (or mechanics, as it is often called) and ending with the most recent results in fundamental particles and the cosmos. But before we begin on the physics itself, we take a brief look at how this overall activity called "science," including physics, is actually practiced.

# 1–1 The Nature of Science

The principal aim of all sciences, including physics, is generally considered to be the search for order in our observations of the world around us. Many people think that science is a mechanical process of collecting facts and devising theories. But it is not so simple. Science is a creative activity that in many respects resembles other creative activities of the human mind.

One important aspect of science is **observation** of events, which includes the design and carrying out of experiments. But observation and experiments require imagination, because scientists can never include everything in a description of what they observe. Hence, scientists must make judgments about what is relevant in their observations and experiments.

Consider, for example, how two great minds, Aristotle (384–322 B.C.; Fig. 1–1) and Galileo (1564–1642; Fig. 2–18), interpreted motion along a horizontal surface. Aristotle noted that objects given an initial push along the ground (or on a tabletop) always slow down and stop. Consequently, Aristotle argued, the natural state of an object is to be at rest. Galileo, the first true experimentalist, reexamined horizontal motion in the 1600s. He imagined that if friction could be eliminated, an object given an initial push along a horizontal surface would continue to move indefinitely without stopping. He concluded that for an object to be in motion was just as natural as for it to be at rest. By inventing a new way of thinking about the same data, Galileo founded our modern view of motion (Chapters 2, 3, and 4), and he did so with a leap of the imagination. Galileo made this leap conceptually, without actually eliminating friction.

**FIGURE 1–1** Aristotle is the central figure (dressed in blue) at the top of the stairs (the figure next to him is Plato) in this famous Renaissance portrayal of *The School of Athens*, painted by Raphael around 1510. Also in this painting, considered one of the great masterpieces in art, are Euclid (drawing a circle at the lower right), Ptolemy (extreme right with globe), Pythagoras, Socrates, and Diogenes.

Observation, with careful experimentation and measurement, is one side of the scientific process. The other side is the invention or creation of **theories** to explain and order the observations. Theories are never derived directly from observations. Observations may help inspire a theory, and theories are accepted or rejected based on the results of observation and experiment.

Theories are inspirations that come from the minds of human beings. For example, the idea that matter is made up of atoms (the atomic theory) was not arrived at by direct observation of atoms—we can't see atoms directly. Rather, the idea sprang from creative minds. The theory of relativity, the electromagnetic theory of light, and Newton's law of universal gravitation were likewise the result of human imagination.

The great theories of science may be compared, as creative achievements, with great works of art or literature. But how does science differ from these other creative activities? One important difference is that science requires **testing** of its ideas or theories to see if their predictions are borne out by experiment. But theories are not "proved" by testing. First of all, no measuring instrument is perfect, so exact confirmation is not possible. Furthermore, it is not possible to test a theory for every possible set of circumstances. Hence a theory cannot be absolutely verified. Indeed, the history of science tells us that long-held theories can sometimes be replaced by new ones, particularly when new experimental techniques provide new or contradictory data.

A new theory is accepted by scientists in some cases because its predictions are quantitatively in better agreement with experiment than those of the older theory. But in many cases, a new theory is accepted only if it explains a greater *range* of phenomena than does the older one. Copernicus's Sun-centered theory of the universe (Fig. 1–2b), for example, was originally no more accurate than Ptolemy's Earth-centered theory (Fig. 1–2a) for predicting the motion of heavenly bodies (Sun, Moon, planets). But Copernicus's theory had consequences that Ptolemy's did not, such as predicting the moonlike phases of Venus. A simpler and richer theory, one which unifies and explains a greater variety of phenomena, is more useful and beautiful to a scientist. And this aspect, as well as quantitative agreement, plays a major role in the acceptance of a theory.

**FIGURE 1–2** (a) Ptolemy's geocentric view of the universe. Note at the center the four elements of the ancients: Earth, water, air (clouds around the Earth), and fire; then the circles, with symbols, for the Moon, Mercury, Venus, Sun, Mars, Jupiter, Saturn, the fixed stars, and the signs of the zodiac. (b) An early representation of Copernicus's heliocentric view of the universe with the Sun at the center. (See Chapter 5.)

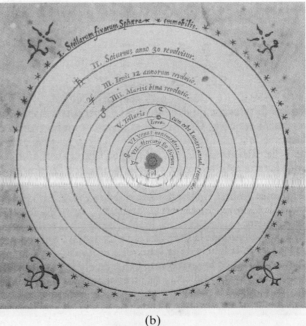

(a)　　　　　　　　　　　　　　　　　(b)

An important aspect of any theory is how well it can quantitatively predict phenomena, and from this point of view a new theory may often seem to be only a minor advance over the old one. For example, Einstein's theory of relativity gives predictions that differ very little from the older theories of Galileo and Newton in nearly all everyday situations. Its predictions are better mainly in the extreme case of very high speeds close to the speed of light. But quantitative prediction is not the only important outcome of a theory. Our view of the world is affected as well. As a result of Einstein's theory of relativity, for example, our concepts of space and time have been completely altered, and we have come to see mass and energy as a single entity (via the famous equation $E = mc^2$).

# 1–2 Physics and its Relation to Other Fields

For a long time science was more or less a united whole known as natural philosophy. Not until a century or two ago did the distinctions between physics and chemistry and even the life sciences become prominent. Indeed, the sharp distinction we now see between the arts and the sciences is itself only a few centuries old. It is no wonder then that the development of physics has both influenced and been influenced by other fields. For example, the notebooks (Fig. 1–3) of Leonardo da Vinci, the great Renaissance artist, researcher, and engineer, contain the first references to the forces acting within a structure, a subject we consider as physics today; but then, as now, it has great relevance to architecture and building.

Early work in electricity that led to the discovery of the electric battery and electric current was done by an eighteenth-century physiologist, Luigi Galvani (1737–1798). He noticed the twitching of frogs' legs in response to an electric spark and later that the muscles twitched when in contact with two dissimilar metals (Chapter 18). At first this phenomenon was known as "animal electricity," but it shortly became clear that electric current itself could exist in the absence of an animal.

Physics is used in many fields. A zoologist, for example, may find physics useful in understanding how prairie dogs and other animals can live underground without suffocating. A physical therapist will be more effective if aware of the principles of center of gravity and the action of forces within the human body. A knowledge of the operating principles of optical and electronic equipment is helpful in a variety of fields. Life scientists and architects alike will be interested in the nature of heat loss and gain in human beings and the resulting comfort or discomfort. Architects may have to calculate the dimensions of the pipes in a heating system or the forces involved in a given structure to determine if it will remain standing (Fig. 1–4). They must know physics principles in order to make realistic designs and to communicate effectively with engineering consultants and other specialists.

**FIGURE 1–3** Studies on the forces in structures by Leonardo da Vinci (1452–1519).

**FIGURE 1–4** (a) This bridge over the River Tiber in Rome was built 2000 years ago and still stands. (b) The 2007 collapse of a Mississippi River highway bridge built only 40 years before.

(a)

(b)

From the aesthetic or psychological point of view, too, architects must be aware of the forces involved in a structure—for example instability, even if only illusory, can be discomforting to those who must live or work in the structure.

The list of ways in which physics relates to other fields is extensive. In the Chapters that follow we will discuss many such applications as we carry out our principal aim of explaining basic physics.

# 1–3 Models, Theories, and Laws

When scientists are trying to understand a particular set of phenomena, they often make use of a **model**. A model, in the scientific sense, is a kind of analogy or mental image of the phenomena in terms of something else we are already familiar with. One example is the wave model of light. We cannot see waves of light as we can water waves. But it is valuable to think of light as made up of waves, because experiments indicate that light behaves in many respects as water waves do.

The purpose of a model is to give us an approximate mental or visual picture—something to hold on to—when we cannot see what actually is happening. Models often give us a deeper understanding: the analogy to a known system (for instance, the water waves above) can suggest new experiments to perform and can provide ideas about what other related phenomena might occur.

You may wonder what the difference is between a theory and a model. Usually a model is relatively simple and provides a structural similarity to the phenomena being studied. A **theory** is broader, more detailed, and can give quantitatively testable predictions, often with great precision. It is important, however, not to confuse a model or a theory with the real system or the phenomena themselves.

Scientists have given the title **law** to certain concise but general statements about how nature behaves (that electric charge is conserved, for example). Often the statement takes the form of a relationship or equation between quantities (such as Newton's second law, $F = ma$).

Statements that we call laws are usually experimentally valid over a wide range of observed phenomena. For less general statements, the term **principle** is often used (such as Archimedes' principle). We use "theory" for a more general picture of the phenomena dealt with.

Scientific laws are different from political laws in that the latter are *prescriptive*: they tell us how we ought to behave. Scientific laws are *descriptive*: they do not say how nature *should* behave, but rather are meant to describe how nature *does* behave. As with theories, laws cannot be tested in the infinite variety of cases possible. So we cannot be sure that any law is absolutely true. We use the term "law" when its validity has been tested over a wide range of cases, and when any limitations and the range of validity are clearly understood.

Scientists normally do their research as if the accepted laws and theories were true. But they are obliged to keep an open mind in case new information should alter the validity of any given law or theory.

# 1–4 Measurement and Uncertainty; Significant Figures

In the quest to understand the world around us, scientists seek to find relationships among physical quantities that can be measured.

## Uncertainty

Reliable measurements are an important part of physics. But no measurement is absolutely precise. There is an uncertainty associated with every measurement.

**FIGURE 1–5** Measuring the width of a board with a centimeter ruler. Accuracy is about ± 1 mm.

Among the most important sources of uncertainty, other than blunders, are the limited accuracy of every measuring instrument and the inability to read an instrument beyond some fraction of the smallest division shown. For example, if you were to use a centimeter ruler to measure the width of a board (Fig. 1–5), the result could be claimed to be precise to about 0.1 cm (1 mm), the smallest division on the ruler, although half of this value might be a valid claim as well. The reason is that it is difficult for the observer to estimate (or "interpolate") between the smallest divisions. Furthermore, the ruler itself may not have been manufactured to an accuracy very much better than this.

When giving the result of a measurement, it is important to state the **estimated uncertainty** in the measurement. For example, the width of a board might be written as 8.8 ± 0.1 cm. The ± 0.1 cm ("plus or minus 0.1 cm") represents the estimated uncertainty in the measurement, so that the actual width most likely lies between 8.7 and 8.9 cm. The **percent uncertainty** is the ratio of the uncertainty to the measured value, multiplied by 100. For example, if the measurement is 8.8 cm and the uncertainty about 0.1 cm, the percent uncertainty is

$$\frac{0.1}{8.8} \times 100\% \approx 1\%,$$

where ≈ means "is approximately equal to."

Often the uncertainty in a measured value is not specified explicitly. In such cases, the

**uncertainty in a numerical value is assumed to be *one or a few units* in the last digit specified.**

For example, if a length is given as 8.8 cm, the uncertainty is assumed to be about 0.1 cm or 0.2 cm, or possibly even 0.3 cm. It is important in this case that you do not write 8.80 cm, because this implies an uncertainty on the order of 0.01 cm; it assumes that the length is probably between about 8.79 cm and 8.81 cm, when actually you believe it is between about 8.7 and 8.9 cm.

---

**CONCEPTUAL EXAMPLE 1–1** | **Is the diamond yours?** A friend asks to borrow your precious diamond for a day to show her family. You are a bit worried, so you carefully have your diamond weighed on a scale which reads 8.17 grams. The scale's accuracy is claimed to be ± 0.05 gram. The next day you weigh the returned diamond again, getting 8.09 grams. Is this your diamond?

**RESPONSE** The scale readings are measurements and are not perfect. They do not necessarily give the "true" value of the mass. Each measurement could have been high or low by up to 0.05 gram or so. The actual mass of your diamond lies most likely between 8.12 grams and 8.22 grams. The actual mass of the returned diamond is most likely between 8.04 grams and 8.14 grams. These two ranges overlap, so the data do not give you a strong reason to doubt that the returned diamond is yours.

---

## Significant Figures

The number of reliably known digits in a number is called the number of **significant figures**. Thus there are four significant figures in the number 23.21 cm and two in the number 0.062 cm (the zeros in the latter are merely place holders that show where the decimal point goes). The number of significant figures may not always be clear. Take, for example, the number 80. Are there one or two significant figures? We need words here: If we say it is *roughly* 80 km between two cities, there is only one significant figure (the 8) since the zero is merely a place holder. If there is no suggestion that the 80 is a rough approximation, then we can often assume (as we will in this book) that it has 2 significant figures: so it is 80 km within an accuracy of about 1 or 2 km. If it is precisely 80 km, to within ± 0.1 or ± 0.2 km, then we write 80.0 km (three significant figures).

When specifying numerical results, you should avoid the temptation to keep more digits in the final answer than is justified: see boldface statement on previous page. For example, to calculate the area of a rectangle 11.3 cm by 6.8 cm, the result of multiplication would be 76.84 cm². But this answer can not be accurate to the implied 0.01 cm² uncertainty, because (using the outer limits of the assumed uncertainty for each measurement) the result could be between 11.2 cm × 6.7 cm = 75.04 cm² and 11.4 cm × 6.9 cm = 78.66 cm². At best, we can quote the answer as 77 cm², which implies an uncertainty of about 1 or 2 cm². The other two digits (in the number 76.84 cm²) must be dropped (rounded off) because they are not significant. As a rough general "significant figure" rule

> *the final result of a multiplication or division should have no more digits than the numerical value with the fewest significant figures.*

In our example, 6.8 cm has the least number of significant figures, namely two. Thus the result 76.84 cm² needs to be rounded off to 77 cm².

**EXERCISE A** The area of a rectangle 4.5 cm by 3.25 cm is correctly given by (*a*) 14.625 cm²; (*b*) 14.63 cm²; (*c*) 14.6 cm²; (*d*) 15 cm².

When adding or subtracting numbers, the final result should contain no more decimal places than the number with the fewest decimal places. For example, the result of subtracting 0.57 from 3.6 is 3.0 (not 3.03). Similarly 36 + 8.2 = 44, not 44.2.

Be careful not to confuse significant figures with the number of decimal places.

**EXERCISE B** For each of the following numbers, state the number of significant figures and the number of decimal places: (*a*) 1.23; (*b*) 0.123; (*c*) 0.0123.

Keep in mind when you use a calculator that all the digits it produces may not be significant. When you divide 2.0 by 3.0, the proper answer is 0.67, and not 0.666666666 as calculators give (Fig. 1–6a). Digits should not be quoted in a result unless they are truly significant figures. However, to obtain the most accurate result, you should normally *keep one or more extra significant figures throughout a calculation, and round off only in the final result.* (With a calculator, you can keep all its digits in intermediate results.) Note also that calculators sometimes give too few significant figures. For example, when you multiply 2.5 × 3.2, a calculator may give the answer as simply 8. But the answer is accurate to two significant figures, so the proper answer is 8.0. See Fig. 1–6b.

(a)

(b)

**FIGURE 1–6** These two calculators show the wrong number of significant figures. In (a), 2.0 was divided by 3.0. The correct final result should be 0.67. In (b), 2.5 was multiplied by 3.2. The correct result is 8.0.

📝 **PROBLEM SOLVING**

*Report only the proper number of significant figures in the final result. But keep extra digits during the calculation*

---

| CONCEPTUAL EXAMPLE 1–2 | **Significant figures.** Using a protractor (Fig. 1–7), you measure an angle to be 30°. (*a*) How many significant figures should you quote in this measurement? (*b*) Use a calculator to find the cosine of the angle you measured.

**RESPONSE** (*a*) If you look at a protractor, you will see that the precision with which you can measure an angle is about one degree (certainly not 0.1°). So you can quote two significant figures, namely 30° (not 30.0°). (*b*) If you enter cos 30° in your calculator, you will get a number like 0.866025403. But the angle you entered is known only to two significant figures, so its cosine is correctly given by 0.87; you must round your answer to two significant figures.

**NOTE** Trigonometric functions, like cosine, are reviewed in Chapter 3 and Appendix A.

**FIGURE 1–7** Example 1–2. A protractor used to measure an angle.

## Scientific Notation

We commonly write numbers in "powers of ten," or "scientific" notation—for instance 36,900 as 3.69 × 10⁴, or 0.0021 as 2.1 × 10⁻³. One advantage of scientific notation (reviewed in Appendix A) is that it allows the number of significant figures to be clearly expressed. For example, it is not clear whether 36,900 has three, four, or five significant figures. With powers of 10 notation the ambiguity can be avoided: if the number is known to three significant figures, we write 3.69 × 10⁴, but if it is known to four, we write 3.690 × 10⁴.

**EXERCISE C** Write each of the following in scientific notation and state the number of significant figures for each: (*a*) 0.0258; (*b*) 42,300; (*c*) 344.50.

## * Percent Uncertainty vs. Significant Figures

The significant figures rule is only approximate, and in some cases may under-estimate the accuracy (or uncertainty) of the answer. Suppose for example we divide 97 by 92:

$$\frac{97}{92} = 1.05 \approx 1.1.$$

Both 97 and 92 have two significant figures, so the rule says to give the answer as 1.1. Yet the numbers 97 and 92 both imply an uncertainty of $\pm 1$ if no other uncertainty is stated. Both $92 \pm 1$ and $97 \pm 1$ imply an uncertainty of about 1% ($1/92 \approx 0.01 = 1\%$). But the final result to two significant figures is 1.1, with an implied uncertainty of $\pm 0.1$, which is an uncertainty of about 10% ($0.1/1.1 \approx 0.1 \approx 10\%$). It is better in this case to give the answer as 1.05 (which is three significant figures). Why? Because 1.05 implies an uncertainty of $\pm 0.01$ which is $0.01/1.05 \approx 0.01 \approx 1\%$, just like the uncertainty in the original numbers 92 and 97.

SUGGESTION: Use the significant figures rule, but consider the % uncertainty too, and add an extra digit if it gives a more realistic estimate of uncertainty.

## Approximations

Much of physics involves approximations, often because we do not have the means to solve a problem precisely. For example, we may choose to ignore air resistance or friction in doing a Problem even though they are present in the real world, and then our calculation is only an approximation. In doing Problems, we should be aware of what approximations we are making, and be aware that the precision of our answer may not be nearly as good as the number of significant figures given in the result.

## Accuracy vs. Precision

There is a technical difference between "precision" and "accuracy." **Precision** in a strict sense refers to the repeatability of the measurement using a given instrument. For example, if you measure the width of a board many times, getting results like 8.81 cm, 8.85 cm, 8.78 cm, 8.82 cm (interpolating between the 0.1 cm marks as best as possible each time), you could say the measurements give a *precision* a bit better than 0.1 cm. **Accuracy** refers to how close a measurement is to the true value. For example, if the ruler shown in Fig. 1–5 was manufac-tured with a 2% error, the accuracy of its measurement of the board's width (about 8.8 cm) would be about 2% of 8.8 cm or about $\pm 0.2$ cm. Estimated uncertainty is meant to take both accuracy and precision into account.

# 1–5 Units, Standards, and the SI System

The measurement of any quantity is made relative to a particular standard or **unit**, and this unit must be specified along with the numerical value of the quantity. For example, we can measure length in British units such as inches, feet, or miles, or in the metric system in centimeters, meters, or kilometers. To specify that the length of a particular object is 18.6 is insufficient. The unit *must* be given, because 18.6 meters is very different from 18.6 inches or 18.6 millimeters.

For any unit we use, such as the meter for distance or the second for time, we need to define a **standard** which defines exactly how long one meter or one second is. It is important that standards be chosen that are readily reproducible so that anyone needing to make a very accurate measurement can refer to the standard in the laboratory and communicate with other people.

## Length

The first truly international standard was the **meter** (abbreviated m) established as the standard of **length** by the French Academy of Sciences in the 1790s. The standard meter was originally chosen to be one ten-millionth of the distance from the Earth's equator to either pole,[†] and a platinum rod to represent this length was made. (One meter is, very roughly, the distance from the tip of your nose to the tip of your finger, with arm and hand stretched out horizontally.) In 1889, the meter was defined more precisely as the distance between two finely engraved marks on a particular bar of platinum–iridium alloy. In 1960, to provide even greater precision and reproducibility, the meter was redefined as 1,650,763.73 wavelengths of a particular orange light emitted by the gas krypton-86. In 1983 the meter was again redefined, this time in terms of the speed of light (whose best measured value in terms of the older definition of the meter was 299,792,458 m/s, with an uncertainty of 1 m/s). The new definition reads: "The meter is the length of path traveled by light in vacuum during a time interval of 1/299,792,458 of a second."[‡]

British units of length (inch, foot, mile) are now defined in terms of the meter. The inch (in.) is defined as exactly 2.54 centimeters (cm; 1 cm = 0.01 m). Other conversion factors are given in the Table on the inside of the front cover of this book. Table 1–1 presents some typical lengths, from very small to very large, rounded off to the nearest power of 10. See also Fig. 1–8. [Note that the abbreviation for inches (in.) is the only one with a period, to distinguish it from the word "in".]

## Time

The standard unit of **time** is the **second** (s). For many years, the second was defined as 1/86,400 of a mean solar day (24 h/day × 60 min/h × 60 s/min = 86,400 s/day). The standard second is now defined more precisely in terms of the frequency of radiation emitted by cesium atoms when they pass between two particular states. [Specifically, one second is defined as the time required for 9,192,631,770 oscillations of this radiation.] There are, by definition, 60 s in one minute (min) and 60 minutes in one hour (h). Table 1–2 presents a range of measured time intervals, rounded off to the nearest power of 10.

[†]Modern measurements of the Earth's circumference reveal that the intended length is off by about one-fiftieth of 1%. Not bad!

[‡]The new definition of the meter has the effect of giving the speed of light the exact value of 299,792,458 m/s.

**FIGURE 1–8** Some lengths: (a) viruses (about $10^{-7}$ m long) attacking a cell; (b) Mt. Everest's height is on the order of $10^4$ m (8850 m above sea level, to be precise).

(a)

(b)

| TABLE 1–1 Some Typical Lengths or Distances (order of magnitude) | |
|---|---|
| **Length (or Distance)** | **Meters (approximate)** |
| Neutron or proton (diameter) | $10^{-15}$ m |
| Atom (diameter) | $10^{-10}$ m |
| Virus [see Fig. 1–8a] | $10^{-7}$ m |
| Sheet of paper (thickness) | $10^{-4}$ m |
| Finger width | $10^{-2}$ m |
| Football field length | $10^2$ m |
| Height of Mt. Everest [see Fig. 1–8b] | $10^4$ m |
| Earth diameter | $10^7$ m |
| Earth to Sun | $10^{11}$ m |
| Earth to nearest star | $10^{16}$ m |
| Earth to nearest galaxy | $10^{22}$ m |
| Earth to farthest galaxy visible | $10^{26}$ m |

| TABLE 1–2 Some Typical Time Intervals (order of magnitude) | |
|---|---|
| **Time Interval** | **Seconds (approximate)** |
| Lifetime of very unstable subatomic particle | $10^{-23}$ s |
| Lifetime of radioactive elements | $10^{-22}$ s to $10^{28}$ s |
| Lifetime of muon | $10^{-6}$ s |
| Time between human heartbeats | $10^0$ s ($\approx 1$ s) |
| One day | $10^5$ s |
| One year | $3 \times 10^7$ s |
| Human life span | $2 \times 10^9$ s |
| Length of recorded history | $10^{11}$ s |
| Humans on Earth | $10^{13}$ s |
| Age of Earth | $10^{17}$ s |
| Age of Universe | $4 \times 10^{17}$ s |

## TABLE 1-3 Some Masses

| Object | Kilograms (approximate) |
|---|---|
| Electron | $10^{-30}$ kg |
| Proton, neutron | $10^{-27}$ kg |
| DNA molecule | $10^{-17}$ kg |
| Bacterium | $10^{-15}$ kg |
| Mosquito | $10^{-5}$ kg |
| Plum | $10^{-1}$ kg |
| Human | $10^{2}$ kg |
| Ship | $10^{8}$ kg |
| Earth | $6 \times 10^{24}$ kg |
| Sun | $2 \times 10^{30}$ kg |
| Galaxy | $10^{41}$ kg |

**PROBLEM SOLVING**

*Always use a consistent set of units*

## TABLE 1-4 Metric (SI) Prefixes

| Prefix | Abbreviation | Value |
|---|---|---|
| yotta | Y | $10^{24}$ |
| zetta | Z | $10^{21}$ |
| exa | E | $10^{18}$ |
| peta | P | $10^{15}$ |
| tera | T | $10^{12}$ |
| giga | G | $10^{9}$ |
| mega | M | $10^{6}$ |
| kilo | k | $10^{3}$ |
| hecto | h | $10^{2}$ |
| deka | da | $10^{1}$ |
| deci | d | $10^{-1}$ |
| centi | c | $10^{-2}$ |
| milli | m | $10^{-3}$ |
| micro† | $\mu$ | $10^{-6}$ |
| nano | n | $10^{-9}$ |
| pico | p | $10^{-12}$ |
| femto | f | $10^{-15}$ |
| atto | a | $10^{-18}$ |
| zepto | z | $10^{-21}$ |
| yocto | y | $10^{-24}$ |

†$\mu$ is the Greek letter "mu."

## Mass

The standard unit of **mass** is the **kilogram** (kg). The standard mass is a particular platinum–iridium cylinder, kept at the International Bureau of Weights and Measures near Paris, France, whose mass is defined as exactly 1 kg. A range of masses is presented in Table 1–3. [For practical purposes, 1 kg weighs about 2.2 pounds on Earth.]

When dealing with atoms and molecules, we usually use the **unified atomic mass unit** (u or amu). In terms of the kilogram,

$$1 \text{ u } = 1.6605 \times 10^{-27} \text{ kg}.$$

The definitions of other standard units for other quantities will be given as we encounter them in later Chapters. (Precise values of this and other useful numbers are given inside the front cover.)

## Unit Prefixes

In the metric system, the larger and smaller units are defined in multiples of 10 from the standard unit, and this makes calculation particularly easy. Thus 1 kilometer (km) is 1000 m, 1 centimeter is $\frac{1}{100}$ m, 1 millimeter (mm) is $\frac{1}{1000}$ m or $\frac{1}{10}$ cm, and so on. The prefixes "centi-," "kilo-," and others are listed in Table 1–4 and can be applied not only to units of length but to units of volume, mass, or any other unit. For example, a centiliter (cL) is $\frac{1}{100}$ liter (L), and a kilogram (kg) is 1000 grams (g). An 8.2-megapixel camera has a detector with 8,200,000 pixels (individual "picture elements").

In common usage, $1 \mu m$ $(= 10^{-6} m)$ is called 1 **micron**.

## Systems of Units

When dealing with the laws and equations of physics it is very important to use a consistent set of units. Several systems of units have been in use over the years. Today the most important is the **Système International** (French for International System), which is abbreviated SI. In SI units, the standard of length is the meter, the standard for time is the second, and the standard for mass is the kilogram. This system used to be called the MKS (meter-kilogram-second) system.

A second metric system is the **cgs system**, in which the centimeter, gram, and second are the standard units of length, mass, and time, as abbreviated in the title. The **British engineering system** (although more used in the U.S. than Britain) has as its standards the foot for length, the pound for force, and the second for time.

We use SI units almost exclusively in this book.

## *Base vs. Derived Quantities

Physical quantities can be divided into two categories: *base quantities* and *derived quantities*. The corresponding units for these quantities are called *base units* and *derived units*. A **base quantity** must be defined in terms of a standard. Scientists, in the interest of simplicity, want the smallest number of base quantities possible consistent with a full description of the physical world. This number turns out to be seven, and those used in the SI are given in Table 1–5.

## TABLE 1-5 SI Base Quantities and Units

| Quantity | Unit | Unit Abbreviation |
|---|---|---|
| Length | meter | m |
| Time | second | s |
| Mass | kilogram | kg |
| Electric current | ampere | A |
| Temperature | kelvin | K |
| Amount of substance | mole | mol |
| Luminous intensity | candela | cd |

All other quantities can be defined in terms of these seven base quantities,[†] and hence are referred to as **derived quantities**. An example of a derived quantity is speed, which is defined as distance divided by the time it takes to travel that distance. A Table inside the front cover lists many derived quantities and their units in terms of base units. To define any quantity, whether base or derived, we can specify a rule or procedure, and this is called an **operational definition**.

# 1–6 Converting Units

Any quantity we measure, such as a length, a speed, or an electric current, consists of a number *and* a unit. Often we are given a quantity in one set of units, but we want it expressed in another set of units. For example, suppose we measure that a shelf is 21.5 inches wide, and we want to express this in centimeters. We must use a **conversion factor**, which in this case is, *by definition*, exactly

$$1 \text{ in.} = 2.54 \text{ cm}$$

or, written another way,

$$1 = 2.54 \text{ cm/in.}$$

Since multiplying by the number one does not change anything, the width of our shelf, in cm, is

$$21.5 \text{ inches} = (21.5 \text{ in.}) \times \left(2.54 \frac{\text{cm}}{\text{in.}}\right) = 54.6 \text{ cm.}$$

Note how the units (inches in this case) cancelled out (thin red lines). A Table containing many unit conversions is found inside the front cover of this book. Let's consider some Examples.

**EXAMPLE 1–3** **The 8000-m peaks.** There are only 14 peaks whose summits are over 8000 m above sea level. They are the tallest peaks in the world (Fig. 1–9 and Table 1–6) and are referred to as "eight-thousanders." What is the elevation, in feet, of an elevation of 8000 m?

**APPROACH** We need to convert meters to feet, and we can start with the conversion factor 1 in. = 2.54 cm, which is exact. That is, 1 in. = 2.5400 cm to any number of significant figures, because it is *defined* to be.

**SOLUTION** One foot is 12 in., so we can write

$$1 \text{ ft} = (12 \text{ in.})\left(2.54 \frac{\text{cm}}{\text{in.}}\right) = 30.48 \text{ cm} = 0.3048 \text{ m,}$$

which is exact. Note how the units cancel (colored slashes). We can rewrite this equation to find the number of feet in 1 meter:

$$1 \text{ m} = \frac{1 \text{ ft}}{0.3048} = 3.28084 \text{ ft.}$$

(We could carry the result to 6 significant figures because 0.3048 is exact, 0.304800....) We multiply this equation by 8000.0 (to have five significant figures):

$$8000.0 \text{ m} = (8000.0 \text{ m})\left(3.28084 \frac{\text{ft}}{\text{m}}\right) = 26{,}247 \text{ ft.}$$

An elevation of 8000 m is 26,247 ft above sea level.

**NOTE** We could have done the conversion all in one line:

$$8000.0 \text{ m} = (8000.0 \text{ m})\left(\frac{100 \text{ cm}}{1 \text{ m}}\right)\left(\frac{1 \text{ in.}}{2.54 \text{ cm}}\right)\left(\frac{1 \text{ ft}}{12 \text{ in.}}\right) = 26{,}247 \text{ ft.}$$

The key is to multiply conversion factors, each equal to one (= 1.0000), and to make sure which units cancel.

[†]Some exceptions are for angle (radians—see Chapter 8), solid angle (steradian), and sound level (bel or decibel, Chapter 12). No general agreement has been reached as to whether these are base or derived quantities.

**FIGURE 1–9** The world's second highest peak, K2, whose summit is considered the most difficult of the "8000-ers." K2 is seen here from the south (Pakistan). Example 1–3.

**PHYSICS APPLIED**
*The world's tallest peaks*

| TABLE 1–6 The 8000-m Peaks | |
|---|---|
| **Peak** | **Height (m)** |
| Mt. Everest | 8850 |
| K2 | 8611 |
| Kangchenjunga | 8586 |
| Lhotse | 8516 |
| Makalu | 8462 |
| Cho Oyu | 8201 |
| Dhaulagiri | 8167 |
| Manaslu | 8156 |
| Nanga Parbat | 8125 |
| Annapurna | 8091 |
| Gasherbrum I | 8068 |
| Broad Peak | 8047 |
| Gasherbrum II | 8035 |
| Shisha Pangma | 8013 |

**EXAMPLE 1–4** **Apartment area.** You have seen a nice apartment whose floor area is 880 square feet $(ft^2)$. What is its area in square meters?

**APPROACH** We use the same conversion factor, 1 in. = 2.54 cm, but this time we have to use it twice.

**SOLUTION** Because 1 in. = 2.54 cm = 0.0254 m, then

$$1 \, ft^2 = (12 \, in.)^2 (0.0254 \, m/in.)^2 = 0.0929 \, m^2.$$

So

$$880 \, ft^2 = (880 \, ft^2)(0.0929 \, m^2/ft^2) \approx 82 \, m^2.$$

**NOTE** As a rule of thumb, an area given in $ft^2$ is roughly 10 times the number of square meters (more precisely, about 10.8×).

**EXAMPLE 1–5** **Speeds.** Where the posted speed limit is 55 miles per hour (mi/h or mph), what is this speed (*a*) in meters per second (m/s) and (*b*) in kilometers per hour (km/h)?

**APPROACH** We again use the conversion factor 1 in. = 2.54 cm, and we recall that there are 5280 ft in a mile and 12 inches in a foot; also, one hour contains (60 min/h) × (60 s/min) = 3600 s/h.

**SOLUTION** (*a*) We can write 1 mile as

$$1 \, mi = (5280 \, \cancel{ft}) \left( 12 \, \frac{\cancel{in.}}{\cancel{ft}} \right) \left( 2.54 \, \frac{\cancel{cm}}{\cancel{in.}} \right) \left( \frac{1 \, m}{100 \, \cancel{cm}} \right)$$

$$= 1609 \, m.$$

We also know that 1 hour contains 3600 s, so

$$55 \, \frac{mi}{h} = \left( 55 \, \frac{\cancel{mi}}{\cancel{h}} \right) \left( 1609 \, \frac{m}{\cancel{mi}} \right) \left( \frac{1 \, \cancel{h}}{3600 \, s} \right)$$

$$= 25 \, \frac{m}{s},$$

where we rounded off to two significant figures.

(*b*) Now we use 1 mi = 1609 m = 1.609 km; then

$$55 \, \frac{mi}{h} = \left( 55 \, \frac{\cancel{mi}}{h} \right) \left( 1.609 \, \frac{km}{\cancel{mi}} \right)$$

$$= 88 \, \frac{km}{h}.$$

**NOTE** Each conversion factor is equal to one. You can look up most conversion factors in the Table inside the front cover.

**EXERCISE D** Return to the first Chapter-Opening Question, page 1, and answer it again now. Try to explain why you may have answered differently the first time.

**EXERCISE E** Would a driver traveling at 15 m/s in a 35 mi/h zone be exceeding the speed limit? Why or why not?

When changing units, you can avoid making an error in the use of conversion factors by checking that units cancel out properly. For example, in our conversion of 1 mi to 1609 m in Example 1–5(*a*), if we had incorrectly used the factor $\left( \frac{100 \, cm}{1 \, m} \right)$ instead of $\left( \frac{1 \, m}{100 \, cm} \right)$, the centimeter units would not have cancelled out; we would not have ended up with meters.

# 1–7 Order of Magnitude: Rapid Estimating

We are sometimes interested only in an approximate value for a quantity. This might be because an accurate calculation would take more time than it is worth or would require additional data that are not available. In other cases, we may want to make a rough estimate in order to check a calculation made on a calculator, to make sure that no blunders were made when the numbers were entered.

A rough estimate can be made by rounding off all numbers to one significant figure and its power of 10, and after the calculation is made, again keeping only one significant figure. Such an estimate is called an **order-of-magnitude estimate** and can be accurate within a factor of 10, and often better. In fact, the phrase "order of magnitude" is sometimes used to refer simply to the power of 10.

Let's do some Examples.

**PROBLEM SOLVING**
*How to make a rough estimate*

**FIGURE 1–10** Example 1–6. (a) How much water is in this lake? (Photo is one of the Rae Lakes in the Sierra Nevada of California.) (b) Model of the lake as a cylinder. [We could go one step further and estimate the mass or weight of this lake. We will see later that water has a density of $1000\ kg/m^3$, so this lake has a mass of about $(10^3\ kg/m^3)(10^7\ m^3) \approx 10^{10}\ kg$, which is about 10 billion kg or 10 million metric tons. (A metric ton is 1000 kg, about 2200 lb, slightly larger than a British ton, 2000 lb.)]

**EXAMPLE 1–6** | **ESTIMATE** | **Volume of a lake.** Estimate how much water there is in a particular lake, Fig. 1–10a, which is roughly circular, about 1 km across, and you guess it has an average depth of about 10 m.

**APPROACH** No lake is a perfect circle, nor can lakes be expected to have a perfectly flat bottom. We are only estimating here. To estimate the volume, we can use a simple model of the lake as a cylinder: we multiply the average depth of the lake times its roughly circular surface area, as if the lake were a cylinder (Fig. 1–10b).

**SOLUTION** The volume $V$ of a cylinder is the product of its height $h$ times the area of its base: $V = h\pi r^2$, where $r$ is the radius of the circular base.† The radius $r$ is $\frac{1}{2}$ km = 500 m, so the volume is approximately

$$V = h\pi r^2 \approx (10\ m) \times (3) \times (5 \times 10^2\ m)^2 \approx 8 \times 10^6\ m^3 \approx 10^7\ m^3,$$

where $\pi$ was rounded off to 3. So the volume is on the order of $10^7\ m^3$, ten million cubic meters. Because of all the estimates that went into this calculation, the order-of-magnitude estimate $(10^7\ m^3)$ is probably better to quote than the $8 \times 10^6\ m^3$ figure.

**NOTE** To express our result in U.S. gallons, we see in the Table on the inside front cover that 1 liter = $10^{-3}\ m^3 \approx \frac{1}{4}$ gallon. Hence, the lake contains $(8 \times 10^6\ m^3)(1\ gallon/4 \times 10^{-3}\ m^3) \approx 2 \times 10^9$ gallons of water.

**PHYSICS APPLIED**
*Estimating the volume (or mass) of a lake; see also Fig. 1–10*

†Formulas like this for volume, area, etc., are found inside the back cover of this book.

**FIGURE 1–11** Example 1–7. Micrometer used for measuring small thicknesses.

**FIGURE 1–12** Example 1–8. Diagrams are really useful!

(a)

1.5 m    3 m
2 m

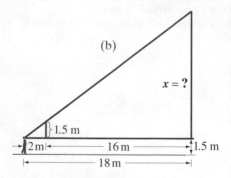

(b)

$x = ?$

1.5 m
2 m ← 16 m → 1.5 m
18 m

**FIGURE 1–13** Enrico Fermi. Fermi contributed significantly to both theoretical and experimental physics, a feat almost unique in modern times.

**EXAMPLE 1–7** | **ESTIMATE** | **Thickness of a sheet of paper.** Estimate the thickness of a page of this book.

**APPROACH** At first you might think that a special measuring device, a micrometer (Fig. 1–11), is needed to measure the thickness of one page since an ordinary ruler can not be read so finely. But we can use a trick or, to put it in physics terms, make use of a *symmetry*: we can make the reasonable assumption that all the pages of this book are equal in thickness.

**SOLUTION** We can use a ruler to measure hundreds of pages at once. If you measure the thickness of the first 500 pages of this book (page 1 to page 500), you might get something like 1.5 cm. Note that 500 numbered pages, counted front and back, is 250 separate pieces of paper. So one sheet must have a thickness of about

$$\frac{1.5 \text{ cm}}{250 \text{ sheets}} \approx 6 \times 10^{-3} \text{ cm} = 6 \times 10^{-2} \text{ mm},$$

or less than a tenth of a millimeter (0.1 mm).

It cannot be emphasized enough how important it is to draw a diagram when solving a physics Problem, as the next Example shows.

**EXAMPLE 1–8** | **ESTIMATE** | **Height by triangulation.** Estimate the height of the building shown in Fig. 1–12, by "triangulation," with the help of a bus-stop pole and a friend.

**APPROACH** By standing your friend next to the pole, you estimate the height of the pole to be 3 m. You next step away from the pole until the top of the pole is in line with the top of the building, Fig. 1–12a. You are 5 ft 6 in. tall, so your eyes are about 1.5 m above the ground. Your friend is taller, and when she stretches out her arms, one hand touches you, and the other touches the pole, so you estimate that distance as 2 m (Fig. 1–12a). You then pace off the distance from the pole to the base of the building with big, 1-m-long steps, and you get a total of 16 steps or 16 m.

**SOLUTION** Now you draw, to scale, the diagram shown in Fig. 1–12b using these measurements. You can measure, right on the diagram, the last side of the triangle to be about $x = 13$ m. Alternatively, you can use similar triangles to obtain the height $x$:

$$\frac{1.5 \text{ m}}{2 \text{ m}} = \frac{x}{18 \text{ m}},$$

so

$$x \approx 13\tfrac{1}{2} \text{ m}.$$

Finally you add in your eye height of 1.5 m above the ground to get your final result: the building is about 15 m tall.

Another approach, this one made famous by Enrico Fermi (1901–1954, Fig. 1–13), was to show his students how to estimate the number of piano tuners in a city, say, Chicago or San Francisco. To get a rough order-of-magnitude estimate of the number of piano tuners today in San Francisco, a city of about 800,000 inhabitants, we can proceed by estimating the number of functioning pianos, how often each piano is tuned, and how many pianos each tuner can tune. To estimate the number of pianos in San Francisco, we note that certainly not everyone has a piano. A guess of 1 family in 3 having a piano would correspond to 1 piano per 12 persons, assuming an average family of 4 persons.

As an order of magnitude, let's say 1 piano per 10 people. This is certainly more reasonable than 1 per 100 people, or 1 per every person, so let's proceed with the estimate that 1 person in 10 has a piano, or about 80,000 pianos in San Francisco. Now a piano tuner needs an hour or two to tune a piano. So let's estimate that a tuner can tune 4 or 5 pianos a day. A piano ought to be tuned every 6 months or a year—let's say once each year. A piano tuner tuning 4 pianos a day, 5 days a week, 50 weeks a year can tune about 1000 pianos a year. So San Francisco, with its (very) roughly 80,000 pianos, needs about 80 piano tuners. This is, of course, only a rough estimate.[†] It tells us that there must be many more than 10 piano tuners, and surely not as many as 1000.

**PROBLEM SOLVING**

*Estimating how many piano tuners there are in a city*

## A Harder Example—But Powerful

**EXAMPLE 1–9 | ESTIMATE | Estimating the radius of Earth.** Believe it or not, you can estimate the radius of the Earth without having to go into space (see the photograph on page 1). If you have ever been on the shore of a large lake, you may have noticed that you cannot see the beaches, piers, or rocks at water level across the lake on the opposite shore. The lake seems to bulge out between you and the opposite shore—a good clue that the Earth is round. Suppose you climb a stepladder and discover that when your eyes are 10 ft (3.0 m) above the water, you can just see the rocks at water level on the opposite shore. From a map, you estimate the distance to the opposite shore as $d \approx 6.1$ km. Use Fig. 1–14 with $h = 3.0$ m to estimate the radius $R$ of the Earth.

**APPROACH** We use simple geometry, including the theorem of Pythagoras,

$$c^2 = a^2 + b^2,$$

where $c$ is the length of the hypotenuse of any right triangle, and $a$ and $b$ are the lengths of the other two sides.

**SOLUTION** For the right triangle of Fig. 1–14, the two sides are the radius of the Earth $R$ and the distance $d = 6.1$ km $= 6100$ m. The hypotenuse is approximately the length $R + h$, where $h = 3.0$ m. By the Pythagorean theorem,

$$R^2 + d^2 \approx (R + h)^2$$
$$\approx R^2 + 2hR + h^2.$$

We solve algebraically for $R$, after cancelling $R^2$ on both sides:

$$R \approx \frac{d^2 - h^2}{2h} = \frac{(6100\,\text{m})^2 - (3.0\,\text{m})^2}{6.0\,\text{m}}$$
$$= 6.2 \times 10^6\,\text{m}$$
$$= 6200\,\text{km}.$$

**NOTE** Precise measurements give 6380 km. But look at your achievement! With a few simple rough measurements and simple geometry, you made a good estimate of the Earth's radius. You did not need to go out in space, nor did you need a very long measuring tape.

**FIGURE 1–14** Example 1–9, but not to scale. You can just barely see rocks at water level on the opposite shore of a lake 6.1 km wide if you stand on a stepladder.

**EXERCISE F** Return to the second Chapter-Opening Question, page 1, and answer it again now. Try to explain why you may have answered differently the first time.

[†]A check of the San Francisco Yellow Pages (done after this calculation) reveals about 60 listings. Each of these listings may employ more than one tuner, but on the other hand, each may also do repairs as well as tuning. In any case, our estimate is reasonable.

# *1–8 Dimensions and Dimensional Analysis

When we speak of the **dimensions** of a quantity, we are referring to the type of base units or base quantities that make it up. The dimensions of area, for example, are always length squared, abbreviated $[L^2]$, using square brackets; the units can be square meters, square feet, cm², and so on. Velocity, on the other hand, can be measured in units of km/h, m/s, or mi/h, but the dimensions are always a length $[L]$ divided by a time $[T]$: that is, $[L/T]$.

The formula for a quantity may be different in different cases, but the dimensions remain the same. For example, the area of a triangle of base $b$ and height $h$ is $A = \frac{1}{2}bh$, whereas the area of a circle of radius $r$ is $A = \pi r^2$. The formulas are different in the two cases, but the dimensions of area are always $[L^2]$.

Dimensions can be used as a help in working out relationships, a procedure referred to as **dimensional analysis**. One useful technique is the use of dimensions to check if a relationship is *incorrect*. Note that we add or subtract quantities only if they have the same dimensions (we don't add centimeters and hours); and the quantities on each side of an equals sign must have the same dimensions. (In numerical calculations, the units must also be the same on both sides of an equation.)

For example, suppose you derived the equation $v = v_0 + \frac{1}{2}at^2$, where $v$ is the speed of an object after a time $t$, $v_0$ is the object's initial speed, and the object undergoes an acceleration $a$. Let's do a dimensional check to see if this equation could be correct or is surely incorrect. Note that numerical factors, like the $\frac{1}{2}$ here, do not affect dimensional checks. We write a dimensional equation as follows, remembering that the dimensions of speed are $[L/T]$ and (as we shall see in Chapter 2) the dimensions of acceleration are $[L/T^2]$:

$$\left[\frac{L}{T}\right] \stackrel{?}{=} \left[\frac{L}{T}\right] + \left[\frac{L}{T^2}\right][T^2]$$

$$\stackrel{?}{=} \left[\frac{L}{T}\right] + [L].$$

The dimensions are incorrect: on the right side, we have the sum of quantities whose dimensions are not the same. Thus we conclude that an error was made in the derivation of the original equation.

A dimensional check can only tell you when a relationship is wrong. It can't tell you if it is completely right. For example, a dimensionless numerical factor (such as $\frac{1}{2}$ or $2\pi$) could be missing.

Dimensional analysis can also be used as a quick check on an equation you are not sure about. For example, consider a simple pendulum of length $\ell$. Suppose that you can't remember whether the equation for the period $T$ (the time to make one back-and-forth swing) is $T = 2\pi\sqrt{\ell/g}$ or $T = 2\pi\sqrt{g/\ell}$, where $g$ is the acceleration due to gravity and, like all accelerations, has dimensions $[L/T^2]$. (Do not worry about these formulas—the correct one will be derived in Chapter 11; what we are concerned about here is a person's recalling whether it contains $\ell/g$ or $g/\ell$.) A dimensional check shows that the former ($\ell/g$) is correct:

$$[T] = \sqrt{\frac{[L]}{[L/T^2]}} = \sqrt{[T^2]} = [T],$$

whereas the latter ($g/\ell$) is not:

$$[T] \neq \sqrt{\frac{[L/T^2]}{[L]}} = \sqrt{\frac{1}{[T^2]}} = \frac{1}{[T]}.$$

The constant $2\pi$ has no dimensions and so can't be checked using dimensions.

---

*Some Sections of this book, such as this one, may be considered *optional* at the discretion of the instructor, and they are marked with an asterisk (*). See the Preface for more details.

# Summary

[*The Summary that appears at the end of each Chapter in this book gives a brief overview of the main ideas of the Chapter. The Summary cannot serve to give an understanding of the material, which can be accomplished only by a detailed reading of the Chapter.*]

Physics, like other sciences, is a creative endeavor. It is not simply a collection of facts. Important **theories** are created with the idea of explaining **observations**. To be accepted, theories are "tested" by comparing their predictions with the results of actual experiments. Note that, in general, a theory cannot be "proved" in an absolute sense.

Scientists often devise models of physical phenomena. A **model** is a kind of picture or analogy that helps to describe the phenomena in terms of something we already know. A **theory**, often developed from a model, is usually deeper and more complex than a simple model.

A scientific **law** is a concise statement, often expressed in the form of an equation, which quantitatively describes a wide range of phenomena.

**Measurements** play a crucial role in physics, but can never be perfectly precise. It is important to specify the **uncertainty** of a measurement either by stating it directly using the ± notation, and/or by keeping only the correct number of **significant figures**.

Physical quantities are always specified relative to a particular standard or **unit**, and the unit used should always be stated. The commonly accepted set of units today is the **Système International** (SI), in which the standard units of length, mass, and time are the **meter**, **kilogram**, and **second**.

When converting units, check all **conversion factors** for correct cancellation of units.

Making rough, **order-of-magnitude estimates** is a very useful technique in science as well as in everyday life.

[*The **dimensions** of a quantity refer to the combination of base quantities that comprise it. Velocity, for example, has dimensions of [length/time] or $[L/T]$. Working with only the dimensions of the various quantities in a given relationship (this technique is called **dimensional analysis**) makes it possible to check a relationship for correct form.]

# Questions

1. What are the merits and drawbacks of using a person's foot as a standard? Consider both (*a*) a particular person's foot, and (*b*) any person's foot. Keep in mind that it is advantageous that fundamental standards be accessible (easy to compare to), invariable (do not change), indestructible, and reproducible.

2. What is wrong with this road sign:

   Memphis 7 mi (11.263 km)?

3. Why is it incorrect to think that the more digits you include in your answer, the more accurate it is?

4. For an answer to be complete, the units need to be specified. Why?

5. You measure the radius of a wheel to be 4.16 cm. If you multiply by 2 to get the diameter, should you write the result as 8 cm or as 8.32 cm? Justify your answer.

6. Express the sine of 30.0° with the correct number of significant figures.

7. List assumptions useful to estimate the number of car mechanics in (*a*) San Francisco, (*b*) your hometown, and then make the estimates.

# MisConceptual Questions

[List all answers that are valid.]

1. A student weighs herself on a digital bathroom scale as 117.4 lb. If all the digits displayed reflect the true precision of the scale, then probably her weight is
   (*a*) within 1% of 117.4 lb.
   (*b*) exactly 117.4 lb.
   (*c*) somewhere between 117.38 and 117.42 lb.
   (*d*) roughly between 117.2 and 117.6 lb.

2. Four students use different instruments to measure the length of the same pen. Which measurement implies the greatest precision?
   (*a*) 160.0 mm.   (*b*) 16.0 cm.   (*c*) 0.160 m.   (*d*) 0.00016 km.
   (*e*) Need more information.

3. The number 0.0078 has how many significant figures?
   (*a*) 1.        (*b*) 2.        (*c*) 3.        (*d*) 4.

4. How many significant figures does 1.362 + 25.2 have?
   (*a*) 2.        (*b*) 3.        (*c*) 4.        (*d*) 5.

5. Accuracy represents
   (*a*) repeatability of a measurement, using a given instrument.
   (*b*) how close a measurement is to the true value.
   (*c*) an ideal number of measurements to make.
   (*d*) how poorly an instrument is operating.

6. To convert from $ft^2$ to $yd^2$, you should
   (*a*) multiply by 3.
   (*b*) multiply by 1/3.
   (*c*) multiply by 9.
   (*d*) multiply by 1/9.
   (*e*) multiply by 6.
   (*f*) multiply by 1/6.

7. Which is *not* true about an order-of-magnitude estimation?
   (*a*) It gives you a rough idea of the answer.
   (*b*) It can be done by keeping only one significant figure.
   (*c*) It can be used to check if an exact calculation is reasonable.
   (*d*) It may require making some reasonable assumptions in order to calculate the answer.
   (*e*) It will always be accurate to at least two significant figures.

*8. $[L^2]$ represents the dimensions for which of the following?
   (*a*) $cm^2$.
   (*b*) square feet.
   (*c*) $m^2$.
   (*d*) All of the above.

# Problems

[*The Problems at the end of each Chapter are ranked* I, II, *or* III *according to estimated difficulty, with* (I) *Problems being easiest. Level* III *are meant as challenges for the best students. The Problems are arranged by Section, meaning that the reader should have read up to and including that Section, but not only that Section—Problems often depend on earlier material. Next is a set of "General Problems" not arranged by Section and not ranked. Finally, there are "Search and Learn" Problems that require rereading parts of the Chapter.*]

## 1–4 Measurement, Uncertainty, Significant Figures

(Note: *In Problems, assume a number like* 6.4 *is accurate to* ± 0.1; *and* 950 *is* ± 10 *unless* 950 *is said to be "precisely" or "very nearly"* 950, *in which case assume* 950 ± 1.)

1. (I) How many significant figures do each of the following numbers have: (a) 214, (b) 81.60, (c) 7.03, (d) 0.03, (e) 0.0086, (f) 3236, and (g) 8700?

2. (I) Write the following numbers in powers of 10 notation: (a) 1.156, (b) 21.8, (c) 0.0068, (d) 328.65, (e) 0.219, and (f) 444.

3. (I) Write out the following numbers in full with the correct number of zeros: (a) $8.69 \times 10^4$, (b) $9.1 \times 10^3$, (c) $8.8 \times 10^{-1}$, (d) $4.76 \times 10^2$, and (e) $3.62 \times 10^{-5}$.

4. (II) The age of the universe is thought to be about 14 billion years. Assuming two significant figures, write this in powers of 10 in (a) years, (b) seconds.

5. (II) What is the percent uncertainty in the measurement $5.48 \pm 0.25$ m?

6. (II) Time intervals measured with a stopwatch typically have an uncertainty of about 0.2 s, due to human reaction time at the start and stop moments. What is the percent uncertainty of a hand-timed measurement of (a) 5.5 s, (b) 55 s, (c) 5.5 min?

7. (II) Add $(9.2 \times 10^3 \text{ s}) + (8.3 \times 10^4 \text{ s}) + (0.008 \times 10^6 \text{ s})$.

8. (II) Multiply $3.079 \times 10^2$ m by $0.068 \times 10^{-1}$ m, taking into account significant figures.

9. (II) What, approximately, is the percent uncertainty for a measurement given as $1.57 \text{ m}^2$?

10. (III) What, roughly, is the percent uncertainty in the volume of a spherical beach ball of radius $r = 0.84 \pm 0.04$ m?

11. (III) What is the area, and its approximate uncertainty, of a circle of radius $3.1 \times 10^4$ cm?

## 1–5 and 1–6 Units, Standards, SI, Converting Units

12. (I) Write the following as full (decimal) numbers without prefixes on the units: (a) 286.6 mm, (b) 85 μV, (c) 760 mg, (d) 62.1 ps, (e) 22.5 nm, (f) 2.50 gigavolts.

13. (I) Express the following using the prefixes of Table 1–4: (a) $1 \times 10^6$ volts, (b) $2 \times 10^{-6}$ meters, (c) $6 \times 10^3$ days, (d) $18 \times 10^2$ bucks, and (e) $7 \times 10^{-7}$ seconds.

14. (I) One hectare is defined as $1.000 \times 10^4 \text{ m}^2$. One acre is $4.356 \times 10^4 \text{ ft}^2$. How many acres are in one hectare?

15. (II) The Sun, on average, is 93 million miles from Earth. How many meters is this? Express (a) using powers of 10, and (b) using a metric prefix (km).

16. (II) Express the following sum with the correct number of significant figures: $1.80 \text{ m} + 142.5 \text{ cm} + 5.34 \times 10^5 \text{ μm}$.

17. (II) A typical atom has a diameter of about $1.0 \times 10^{-10}$ m. (a) What is this in inches? (b) Approximately how many atoms are along a 1.0-cm line, assuming they just touch?

18. (II) Determine the conversion factor between (a) km/h and mi/h, (b) m/s and ft/s, and (c) km/h and m/s.

19. (II) A **light-year** is the distance light travels in one year (at speed $= 2.998 \times 10^8$ m/s). (a) How many meters are there in 1.00 light-year? (b) An astronomical unit (AU) is the average distance from the Sun to Earth, $1.50 \times 10^8$ km. How many AU are there in 1.00 light-year?

20. (II) How much longer (percentage) is a one-mile race than a 1500-m race ("the metric mile")?

21. (II) American football uses a field that is 100.0 yd long, whereas a soccer field is 100.0 m long. Which field is longer, and by how much (give yards, meters, and percent)?

22. (II) (a) How many seconds are there in 1.00 year? (b) How many nanoseconds are there in 1.00 year? (c) How many years are there in 1.00 second?

23. (II) Use Table 1–3 to estimate the total number of protons or neutrons in (a) a bacterium, (b) a DNA molecule, (c) the human body, (d) our Galaxy.

24. (III) A standard baseball has a circumference of approximately 23 cm. If a baseball had the same mass per unit volume (see Tables in Section 1–5) as a neutron or a proton, about what would its mass be?

## 1–7 Order-of-Magnitude Estimating

(Note: *Remember that for rough estimates, only round numbers are needed both as input to calculations and as final results.*)

25. (I) Estimate the order of magnitude (power of 10) of: (a) 2800, (b) $86.30 \times 10^3$, (c) 0.0076, and (d) $15.0 \times 10^8$.

26. (II) Estimate how many books can be shelved in a college library with 3500 $\text{m}^2$ of floor space. Assume 8 shelves high, having books on both sides, with corridors 1.5 m wide. Assume books are about the size of this one, on average.

27. (II) Estimate how many hours it would take to run (at 10 km/h) across the U.S. from New York to California.

28. (II) Estimate the number of liters of water a human drinks in a lifetime.

29. (II) Estimate how long it would take one person to mow a football field using an ordinary home lawn mower (Fig. 1–15). (State your assumption, such as the mower moves with a 1-km/h speed, and has a 0.5-m width.)

**FIGURE 1–15**
Problem 29.

30. (II) Estimate the number of gallons of gasoline consumed by the total of all automobile drivers in the U.S., per year.

31. (II) Estimate the number of dentists (a) in San Francisco and (b) in your town or city.

32. (III) You are in a hot air balloon, 200 m above the flat Texas plains. You look out toward the horizon. How far out can you see—that is, how far is your horizon? The Earth's radius is about 6400 km.

**33.** (III) I agree to hire you for 30 days. You can decide between two methods of payment: either (1) $1000 a day, or (2) one penny on the first day, two pennies on the second day and continue to double your daily pay each day up to day 30. Use quick estimation to make your decision, and justify it.

**34.** (III) Many sailboats are docked at a marina 4.4 km away on the opposite side of a lake. You stare at one of the sailboats because, when you are lying flat at the water's edge, you can just see its deck but none of the side of the sailboat. You then go to that sailboat on the other side of the lake and measure that the deck is 1.5 m above the level of the water. Using Fig. 1–16, where $h = 1.5$ m, estimate the radius $R$ of the Earth.

**FIGURE 1–16** Problem 34. You see a sailboat across a lake (not to scale). $R$ is the radius of the Earth. Because of the curvature of the Earth, the water "bulges out" between you and the boat.

**35.** (III) You are lying on a beach, your eyes 20 cm above the sand. Just as the Sun sets, fully disappearing over the horizon, you immediately jump up, your eyes now 150 cm above the sand, and you can again just see the top of the Sun. If you count the number of seconds $(= t)$ until the Sun fully disappears again, you can estimate the Earth's radius. But for this Problem, use the known radius of the Earth to calculate the time $t$.

## *1–8 Dimensions

*36.** (I) What are the dimensions of density, which is mass per volume?

*37.** (II) The speed $v$ of an object is given by the equation $v = At^3 - Bt$, where $t$ refers to time. (a) What are the dimensions of $A$ and $B$? (b) What are the SI units for the constants $A$ and $B$?

*38.** (II) Three students derive the following equations in which $x$ refers to distance traveled, $v$ the speed, $a$ the acceleration $(m/s^2)$, $t$ the time, and the subscript zero $(_0)$ means a quantity at time $t = 0$. Here are their equations: (a) $x = vt^2 + 2at$, (b) $x = v_0t + \frac{1}{2}at^2$, and (c) $x = v_0t + 2at^2$. Which of these could possibly be correct according to a dimensional check, and why?

*39.** (III) The smallest meaningful measure of length is called the **Planck length**, and is defined in terms of three fundamental constants in nature: the speed of light $c = 3.00 \times 10^8$ m/s, the gravitational constant $G = 6.67 \times 10^{-11}$ m$^3$/kg·s$^2$, and Planck's constant $h = 6.63 \times 10^{-34}$ kg·m$^2$/s. The Planck length $\ell_P$ is given by the following combination of these three constants:

$$\ell_P = \sqrt{\frac{Gh}{c^3}}.$$

Show that the dimensions of $\ell_P$ are length $[L]$, and find the order of magnitude of $\ell_P$. [Recent theories (Chapters 32 and 33) suggest that the smallest particles (quarks, leptons) are "strings" with lengths on the order of the Planck length, $10^{-35}$ m. These theories also suggest that the "Big Bang," with which the universe is believed to have begun, started from an initial size on the order of the Planck length.]

# General Problems

**40. Global positioning satellites (GPS)** can be used to determine your position with great accuracy. If one of the satellites is 20,000 km from you, and you want to know your position to $\pm 2$ m, what percent uncertainty in the distance is required? How many significant figures are needed in the distance?

**41. Computer chips** (Fig. 1–17) are etched on circular silicon wafers of thickness 0.300 mm that are sliced from a solid cylindrical silicon crystal of length 25 cm. If each wafer can hold 400 chips, what is the maximum number of chips that can be produced from one entire cylinder?

**FIGURE 1–17** Problem 41. The wafer held by the hand is shown below, enlarged and illuminated by colored light. Visible are rows of integrated circuits (chips).

**42.** A typical adult human lung contains about 300 million tiny cavities called alveoli. Estimate the average diameter of a single alveolus.

**43.** If you used only a keyboard to enter data, how many years would it take to fill up the hard drive in a computer that can store 1.0 terabytes $(1.0 \times 10^{12}$ bytes) of data? Assume 40-hour work weeks, and that you can type 180 characters per minute, and that one byte is one keyboard character.

**44.** An average family of four uses roughly 1200 L (about 300 gallons) of water per day $(1$ L $= 1000$ cm$^3)$. How much depth would a lake lose per year if it covered an area of 50 km$^2$ with uniform depth and supplied a local town with a population of 40,000 people? Consider only population uses, and neglect evaporation, rain, creeks and rivers.

**45.** Estimate the number of jelly beans in the jar of Fig. 1–18.

**FIGURE 1–18** Problem 45. Estimate the number of jelly beans in the jar.

**46.** How big is a ton? That is, what is the volume of something that weighs a ton? To be specific, estimate the diameter of a 1-ton rock, but first make a wild guess: will it be 1 ft across, 3 ft, or the size of a car? [*Hint*: Rock has mass per volume about 3 times that of water, which is 1 kg per liter $(10^3 \, \text{cm}^3)$ or 62 lb per cubic foot.]

**47.** A certain compact disc (CD) contains 783.216 megabytes of digital information. Each byte consists of exactly 8 bits. When played, a CD player reads the CD's information at a constant rate of 1.4 megabits per second. How many minutes does it take the player to read the entire CD?

**48.** Hold a pencil in front of your eye at a position where its blunt end just blocks out the Moon (Fig. 1–19). Make appropriate measurements to estimate the diameter of the Moon, given that the Earth–Moon distance is $3.8 \times 10^5$ km.

**FIGURE 1–19**
Problem 48. How big is the Moon?

**49.** A storm dumps 1.0 cm of rain on a city 6 km wide and 8 km long in a 2-h period. How many metric tons $(1 \text{ metric ton} = 10^3 \, \text{kg})$ of water fell on the city? $(1 \, \text{cm}^3$ of water has a mass of $1 \, \text{g} = 10^{-3}$ kg.) How many gallons of water was this?

**50.** Estimate how many days it would take to walk around the Earth, assuming 12 h walking per day at 4 km/h.

**51.** One liter $(1000 \, \text{cm}^3)$ of oil is spilled onto a smooth lake. If the oil spreads out uniformly until it makes an oil slick just one molecule thick, with adjacent molecules just touching, estimate the diameter of the oil slick. Assume the oil molecules have a diameter of $2 \times 10^{-10}$ m.

**52.** A watch manufacturer claims that its watches gain or lose no more than 8 seconds in a year. How accurate are these watches, expressed as a percentage?

**53.** An angstrom (symbol Å) is a unit of length, defined as $10^{-10}$ m, which is on the order of the diameter of an atom. (*a*) How many nanometers are in 1.0 angstrom? (*b*) How many femtometers or fermis (the common unit of length in nuclear physics) are in 1.0 angstrom? (*c*) How many angstroms are in 1.0 m? (*d*) How many angstroms are in 1.0 light-year (see Problem 19)?

**54.** Jim stands beside a wide river and wonders how wide it is. He spots a large rock on the bank directly across from him. He then walks upstream 65 strides and judges that the angle between him and the rock, which he can still see, is now at an angle of 30° downstream (Fig. 1–20). Jim measures his stride to be about 0.8 m long. Estimate the width of the river.

**FIGURE 1–20**
Problem 54.

65 Strides

**55.** Determine the percent uncertainty in $\theta$, and in $\sin \theta$, when (*a*) $\theta = 15.0° \pm 0.5°$, (*b*) $\theta = 75.0° \pm 0.5°$.

**56.** If you walked north along one of Earth's lines of longitude until you had changed latitude by 1 minute of arc (there are 60 minutes per degree), how far would you have walked (in miles)? This distance is a **nautical mile**.

**57.** Make a rough estimate of the volume of your body (in $\text{m}^3$).

**58.** The following formula estimates an average person's lung capacity $V$ (in liters, where $1 \, \text{L} = 10^3 \, \text{cm}^3$):

$$V = 4.1H - 0.018A - 2.7,$$

where $H$ and $A$ are the person's height (in meters) and age (in years), respectively. In this formula, what are the units of the numbers 4.1, 0.018, and 2.7?

**59.** One mole of atoms consists of $6.02 \times 10^{23}$ individual atoms. If a mole of atoms were spread uniformly over the Earth's surface, how many atoms would there be per square meter?

**60.** The density of an object is defined as its mass divided by its volume. Suppose a rock's mass and volume are measured to be 6 g and 2.8325 $\text{cm}^3$. To the correct number of significant figures, determine the rock's density (mass/volume).

**61.** Recent findings in astrophysics suggest that the observable universe can be modeled as a sphere of radius $R = 13.7 \times 10^9$ light-years $= 13.0 \times 10^{25}$ m with an average total mass density of about $1 \times 10^{-26} \, \text{kg/m}^3$. Only about 4% of total mass is due to "ordinary" matter (such as protons, neutrons, and electrons). Estimate how much ordinary matter (in kg) there is in the observable universe. (For the light-year, see Problem 19.)

# Search and Learn

**1.** Galileo is to Aristotle as Copernicus is to Ptolemy. See Section 1–1 and explain this analogy.

**2.** How many wavelengths of orange krypton-86 light (Section 1–5) would fit into the thickness of one page of this book?

**3.** Using the French Academy of Sciences' original definition of the meter, determine Earth's circumference and radius in those meters.

**4.** Estimate the ratio (order of magnitude) of the mass of a human to the mass of a DNA molecule.

**5.** To the correct number of significant figures, use the information inside the front cover of this book to determine the ratio of (*a*) the surface area of Earth compared to the surface area of the Moon; (*b*) the volume of Earth compared to the volume of the Moon.

## ANSWERS TO EXERCISES

**A:** (*d*).
**B:** All three have three significant figures; the number of decimal places is (*a*) 2, (*b*) 3, (*c*) 4.
**C:** (*a*) $2.58 \times 10^{-2}$, 3; (*b*) $4.23 \times 10^4$, 3 (probably); (*c*) $3.4450 \times 10^2$, 5.

**D:** (*f*).
**E:** No: 15 m/s $\approx$ 34 mi/h.
**F:** (*c*).

The space shuttle has released a parachute to reduce its speed quickly. The directions of the shuttle's velocity and acceleration are shown by the green ($\vec{v}$) and gold ($\vec{a}$) arrows.

Motion is described using the concepts of velocity and acceleration. In the case shown here, the velocity $\vec{v}$ is to the right, in the direction of motion. The acceleration $\vec{a}$ is in the opposite direction from the velocity $\vec{v}$, which means the object is slowing down.

We examine in detail motion with constant acceleration, including the vertical motion of objects falling under gravity.

# Describing Motion: Kinematics in One Dimension

## CHAPTER 2

**CHAPTER-OPENING QUESTION—Guess now!**
[*Don't worry about getting the right answer now—you will get another chance later in the Chapter. See also p. 1 of Chapter 1 for more explanation.*]
Two small heavy balls have the same diameter but one weighs twice as much as the other. The balls are dropped from a second-story balcony at the exact same time. The time to reach the ground below will be:
  **(a)** twice as long for the lighter ball as for the heavier one.
  **(b)** longer for the lighter ball, but not twice as long.
  **(c)** twice as long for the heavier ball as for the lighter one.
  **(d)** longer for the heavier ball, but not twice as long.
  **(e)** nearly the same for both balls.

## CONTENTS

The motion of objects—baseballs, automobiles, joggers, and even the Sun and Moon—is an obvious part of everyday life. It was not until the sixteenth and seventeenth centuries that our modern understanding of motion was established. Many individuals contributed to this understanding, particularly Galileo Galilei (1564–1642) and Isaac Newton (1642–1727).

The study of the motion of objects, and the related concepts of force and energy, form the field called **mechanics**. Mechanics is customarily divided into two parts: **kinematics**, which is the description of how objects move, and **dynamics**, which deals with force and why objects move as they do. This Chapter and the next deal with kinematics.

FIGURE 2–1 A falling pinecone undergoes (a) pure translation; (b) it is rotating as well as translating.

(a)          (b)

For now we only discuss objects that move without rotating (Fig. 2–1a). Such motion is called **translational motion**. In this Chapter we will be concerned with describing an object that moves along a straight-line path, which is one-dimensional translational motion. In Chapter 3 we will describe translational motion in two (or three) dimensions along paths that are not straight. (Rotation, shown in Fig. 2–1b, is discussed in Chapter 8.)

We will often use the concept, or *model*, of an idealized **particle** which is considered to be a mathematical **point** with no spatial extent (no size). A point particle can undergo only translational motion. The particle model is useful in many real situations where we are interested only in translational motion and the object's size is not significant. For example, we might consider a billiard ball, or even a spacecraft traveling toward the Moon, as a particle for many purposes.

# 2–1 Reference Frames and Displacement

Any measurement of position, distance, or speed must be made with respect to a **reference frame**, or **frame of reference**. For example, while you are on a train traveling at 80 km/h, suppose a person walks past you toward the front of the train at a speed of, say, 5 km/h (Fig. 2–2). This 5 km/h is the person's speed with respect to the train as frame of reference. With respect to the ground, that person is moving at a speed of 80 km/h + 5 km/h = 85 km/h. It is always important to specify the frame of reference when stating a speed. In everyday life, we usually mean "with respect to the Earth" without even thinking about it, but the reference frame must be specified whenever there might be confusion.

FIGURE 2–2 A person walks toward the front of a train at 5 km/h. The train is moving 80 km/h with respect to the ground, so the walking person's speed, relative to the ground, is 85 km/h.

When specifying the motion of an object, it is important to specify not only the speed but also the direction of motion. Often we can specify a direction by using north, east, south, and west, and by "up" and "down." In physics, we often draw a set of **coordinate axes**, as shown in Fig. 2–3, to represent a frame of reference. We can always place the origin 0, and the directions of the $x$ and $y$ axes, as we like for convenience. The $x$ and $y$ axes are always perpendicular to each other. The **origin** is where $x = 0$, $y = 0$. Objects positioned to the right of the origin of coordinates (0) on the $x$ axis have an $x$ coordinate which we almost always choose to be positive; then points to the left of 0 have a negative $x$ coordinate. The position along the $y$ axis is usually considered positive when above 0, and negative when below 0, although the reverse convention can be used if convenient. Any point on the plane can be specified by giving its $x$ and $y$ coordinates. In three dimensions, a $z$ axis perpendicular to the $x$ and $y$ axes is added.

For one-dimensional motion, we often choose the $x$ axis as the line along which the motion takes place. Then the **position** of an object at any moment is given by its $x$ coordinate. If the motion is vertical, as for a dropped object, we usually use the $y$ axis.

FIGURE 2–3 Standard set of $xy$ coordinate axes, sometimes called "rectangular coordinates."

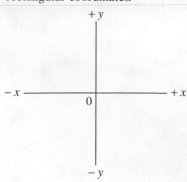

We need to make a distinction between the *distance* an object has traveled and its **displacement**, which is defined as the *change in position* of the object. That is, *displacement is how far the object is from its starting point.* To see the distinction between total distance and displacement, imagine a person walking 70 m to the east and then turning around and walking back (west) a distance of 30 m (see Fig. 2–4). The total *distance* traveled is 100 m, but the *displacement* is only 40 m since the person is now only 40 m from the starting point.

Displacement is a quantity that has both magnitude and direction. Such quantities are called **vectors**, and are represented by arrows in diagrams. For example, in Fig. 2–4, the blue arrow represents the displacement whose magnitude is 40 m and whose direction is to the right (east).

We will deal with vectors more fully in Chapter 3. For now, we deal only with motion in one dimension, along a line. In this case, vectors which point in one direction will be positive (typically to the right along the $x$ axis). Vectors that point in the opposite direction will have a negative sign in front of their magnitude.

Consider the motion of an object over a particular time interval. Suppose that at some initial time, call it $t_1$, the object is on the $x$ axis at the position $x_1$ in the coordinate system shown in Fig. 2–5. At some later time, $t_2$, suppose the object has moved to position $x_2$. The displacement of our object is $x_2 - x_1$, and is represented by the arrow pointing to the right in Fig. 2–5. It is convenient to write

$$\Delta x = x_2 - x_1,$$

where the symbol $\Delta$ (Greek letter delta) means "change in." Then $\Delta x$ means "the change in $x$," or "change in position," which is the displacement. The **change in** any quantity means *the final value of that quantity, minus the initial value.* Suppose $x_1 = 10.0$ m and $x_2 = 30.0$ m, as in Fig. 2–5. Then

$$\Delta x = x_2 - x_1 = 30.0 \text{ m} - 10.0 \text{ m} = 20.0 \text{ m},$$

so the displacement is 20.0 m in the positive direction, Fig. 2–5.

Now consider an object moving to the left as shown in Fig. 2–6. Here the object, a person, starts at $x_1 = 30.0$ m and walks to the left to the point $x_2 = 10.0$ m. In this case her displacement is

$$\Delta x = x_2 - x_1 = 10.0 \text{ m} - 30.0 \text{ m} = -20.0 \text{ m},$$

and the blue arrow representing the vector displacement points to the left. For one-dimensional motion along the $x$ axis, a vector pointing to the right is positive, whereas a vector pointing to the left has a negative sign.

**EXERCISE A** An ant starts at $x = 20$ cm on a piece of graph paper and walks along the $x$ axis to $x = -20$ cm. It then turns around and walks back to $x = -10$ cm. Determine (a) the ant's displacement and (b) the total distance traveled.

# 2–2 Average Velocity

An important aspect of the motion of a moving object is how *fast* it is moving—its speed or velocity.

The term "speed" refers to how far an object travels in a given time interval, regardless of direction. If a car travels 240 kilometers (km) in 3 hours (h), we say its average speed was 80 km/h. In general, the **average speed** of an object is defined as *the total distance traveled along its path divided by the time it takes to travel this distance:*

$$\text{average speed} = \frac{\text{distance traveled}}{\text{time elapsed}}. \quad \textbf{(2–1)}$$

The terms "velocity" and "speed" are often used interchangeably in ordinary language. But in physics we make a distinction between the two. Speed is simply a positive number, with units. **Velocity**, on the other hand, is used to signify both the *magnitude* (numerical value) of how fast an object is moving and also the *direction* in which it is moving. Velocity is therefore a *vector*.

*The displacement may not equal the total distance traveled*

**FIGURE 2–4** A person walks 70 m east, then 30 m west. The total distance traveled is 100 m (path is shown dashed in black); but the displacement, shown as a solid blue arrow, is 40 m to the east.

**FIGURE 2–5** The arrow represents the displacement $x_2 - x_1$. Distances are in meters.

**FIGURE 2–6** For the displacement $\Delta x = x_2 - x_1 = 10.0$ m $- 30.0$ m, the displacement vector points left.

There is a second difference between speed and velocity: namely, the *average velocity* is defined in terms of *displacement*, rather than total distance traveled:

$$\text{average velocity} = \frac{\text{displacement}}{\text{time elapsed}} = \frac{\text{final position} - \text{initial position}}{\text{time elapsed}}.$$

Average speed and average velocity have the same magnitude when the motion is all in one direction. In other cases, they may differ: recall the walk we described earlier, in Fig. 2–4, where a person walked 70 m east and then 30 m west. The total distance traveled was 70 m + 30 m = 100 m, but the displacement was 40 m. Suppose this walk took 70 s to complete. Then the average speed was:

$$\frac{\text{distance}}{\text{time elapsed}} = \frac{100 \text{ m}}{70 \text{ s}} = 1.4 \text{ m/s}.$$

The magnitude of the average velocity, on the other hand, was:

$$\frac{\text{displacement}}{\text{time elapsed}} = \frac{40 \text{ m}}{70 \text{ s}} = 0.57 \text{ m/s}.$$

To discuss one-dimensional motion of an object in general, suppose that at some moment in time, call it $t_1$, the object is on the $x$ axis at position $x_1$ in a coordinate system, and at some later time, $t_2$, suppose it is at position $x_2$. The **elapsed time** (= change in time) is $\Delta t = t_2 - t_1$; during this time interval the displacement of our object is $\Delta x = x_2 - x_1$. Then the **average velocity**, defined as *the displacement divided by the elapsed time*, can be written

$$\bar{v} = \frac{x_2 - x_1}{t_2 - t_1} = \frac{\Delta x}{\Delta t}, \qquad \text{[average velocity]} \quad (2–2)$$

where $v$ stands for velocity and the bar ($^-$) over the $v$ is a standard symbol meaning "average."

For one-dimensional motion in the usual case of the $+x$ axis to the right, note that if $x_2$ is less than $x_1$, the object is moving to the left, and then $\Delta x = x_2 - x_1$ is less than zero. The sign of the displacement, and thus of the average velocity, indicates the direction: the average velocity is positive for an object moving to the right along the $x$ axis and negative when the object moves to the left. The direction of the average velocity is always the same as the direction of the displacement.

It is always important to choose (and state) the *elapsed time*, or **time interval**, $t_2 - t_1$, the time that passes during our chosen period of observation.

**EXAMPLE 2–1** **Runner's average velocity.** The position of a runner is plotted as moving along the $x$ axis of a coordinate system. During a 3.00-s time interval, the runner's position changes from $x_1 = 50.0$ m to $x_2 = 30.5$ m, as shown in Fig. 2–7. What is the runner's average velocity?

**APPROACH** We want to find the average velocity, which is the displacement divided by the elapsed time.

**SOLUTION** The displacement is

$$\Delta x = x_2 - x_1$$
$$= 30.5 \text{ m} - 50.0 \text{ m} = -19.5 \text{ m}.$$

The elapsed time, or time interval, is given as $\Delta t = 3.00$ s. The average velocity (Eq. 2–2) is

$$\bar{v} = \frac{\Delta x}{\Delta t} = \frac{-19.5 \text{ m}}{3.00 \text{ s}} = -6.50 \text{ m/s}.$$

The displacement and average velocity are negative, which tells us that the runner is moving to the left along the $x$ axis, as indicated by the arrow in Fig. 2–7. The runner's average velocity is 6.50 m/s to the left.

**FIGURE 2–7** Example 2–1. A person runs from $x_1 = 50.0$ m to $x_2 = 30.5$ m. The displacement is −19.5 m.

> **EXAMPLE 2-2** **Distance a cyclist travels.** How far can a cyclist travel in 2.5 h along a straight road if her average velocity is 18 km/h?
>
> **APPROACH** We want to find the distance traveled, so we solve Eq. 2-2 for $\Delta x$.
>
> **SOLUTION** In Eq. 2-2, $\bar{v} = \Delta x / \Delta t$, we multiply both sides by $\Delta t$ and obtain
>
> $$\Delta x = \bar{v}\,\Delta t = (18\,\text{km/h})(2.5\,\text{h}) = 45\,\text{km}.$$

> **EXAMPLE 2-3** **Car changes speed.** A car travels at a constant 50 km/h for 100 km. It then speeds up to 100 km/h and is driven another 100 km. What is the car's average speed for the 200-km trip?
>
> **APPROACH** At 50 km/h, the car takes 2.0 h to travel 100 km. At 100 km/h it takes only 1.0 h to travel 100 km. We use the defintion of average velocity, Eq. 2-2.
>
> **SOLUTION** Average velocity (Eq. 2-2) is
>
> $$\bar{v} = \frac{\Delta x}{\Delta t} = \frac{100\,\text{km} + 100\,\text{km}}{2.0\,\text{h} + 1.0\,\text{h}} = 67\,\text{km/h}.$$
>
> **NOTE** Averaging the two speeds, (50 km/h + 100 km/h)/2 = 75 km/h, gives a wrong answer. Can you see why? You must use the definition of $\bar{v}$, Eq. 2-2.

# 2-3 Instantaneous Velocity

If you drive a car along a straight road for 150 km in 2.0 h, the magnitude of your average velocity is 75 km/h. It is unlikely, though, that you were moving at precisely 75 km/h at every instant. To describe this situation we need the concept of *instantaneous velocity*, which is the velocity at any instant of time. (Its magnitude is the number, with units, indicated by a speedometer, Fig. 2-8.) More precisely, the **instantaneous velocity** at any moment is defined as *the average velocity over an infinitesimally short time interval*. That is, Eq. 2-2 is to be evaluated in the limit of $\Delta t$ becoming extremely small, approaching zero. We can write the definition of instantaneous velocity, $v$, for one-dimensional motion as

$$v = \lim_{\Delta t \to 0} \frac{\Delta x}{\Delta t}. \qquad \text{[instantaneous velocity]} \quad (2\text{-}3)$$

The notation $\lim_{\Delta t \to 0}$ means the ratio $\Delta x / \Delta t$ is to be evaluated in the limit of $\Delta t$ approaching zero.[†]

For instantaneous velocity we use the symbol $v$, whereas for average velocity we use $\bar{v}$, with a bar above. In the rest of this book, when we use the term "velocity" it will refer to instantaneous velocity. When we want to speak of the average velocity, we will make this clear by including the word "average."

Note that the *instantaneous speed* always equals the magnitude of the instantaneous velocity. Why? Because distance traveled and the magnitude of the displacement become the same when they become infinitesimally small.

If an object moves at a uniform (that is, constant) velocity during a particular time interval, then its instantaneous velocity at any instant is the same as its average velocity (see Fig. 2-9a). But in many situations this is not the case. For example, a car may start from rest, speed up to 50 km/h, remain at that velocity for a time, then slow down to 20 km/h in a traffic jam, and finally stop at its destination after traveling a total of 15 km in 30 min. This trip is plotted on the graph of Fig. 2-9b. Also shown on the graph is the average velocity (dashed line), which is $\bar{v} = \Delta x / \Delta t = 15\,\text{km}/0.50\,\text{h} = 30\,\text{km/h}$.

Graphs are often useful for analysis of motion; we discuss additional insights graphs can provide as we go along, especially in Section 2-8.

> **EXERCISE B** What is your instantaneous speed at the instant you turn around to move in the opposite direction? (*a*) Depends on how quickly you turn around; (*b*) always zero; (*c*) always negative; (*d*) none of the above.

[†]We do not simply set $\Delta t = 0$ in this definition, for then $\Delta x$ would also be zero, and we would have an undetermined number. Rather, we consider the *ratio* $\Delta x / \Delta t$, as a whole. As we let $\Delta t$ approach zero, $\Delta x$ approaches zero as well. But the ratio $\Delta x / \Delta t$ approaches some definite value, which is the instantaneous velocity at a given instant.

**FIGURE 2-8** Car speedometer showing mi/h in white, and km/h in orange.

**FIGURE 2-9** Velocity of a car as a function of time: (a) at constant velocity; (b) with velocity varying in time.

# 2–4 Acceleration

An object whose velocity is changing is said to be accelerating. For instance, a car whose velocity increases in magnitude from zero to 80 km/h is accelerating. Acceleration specifies how *rapidly* the velocity of an object is changing.

*Average acceleration* is defined as the change in velocity divided by the time taken to make this change:

$$\text{average acceleration} = \frac{\text{change of velocity}}{\text{time elapsed}}.$$

In symbols, the **average acceleration**, $\bar{a}$, over a time interval $\Delta t = t_2 - t_1$, during which the velocity changes by $\Delta v = v_2 - v_1$, is defined as

$$\bar{a} = \frac{v_2 - v_1}{t_2 - t_1} = \frac{\Delta v}{\Delta t}. \qquad \text{[average acceleration]} \quad \textbf{(2–4)}$$

We saw that velocity is a vector (it has magnitude and direction), so acceleration is a vector too. But for one dimensional motion, we need only use a plus or minus sign to indicate acceleration direction relative to a chosen coordinate axis. (Usually, right is +, left is −.)

The **instantaneous acceleration**, $a$, can be defined in analogy to instantaneous velocity as the average acceleration over an infinitesimally short time interval at a given instant:

$$a = \lim_{\Delta t \to 0} \frac{\Delta v}{\Delta t}. \qquad \text{[instantaneous acceleration]} \quad \textbf{(2–5)}$$

Here $\Delta v$ is the very small change in velocity during the very short time interval $\Delta t$.

---

**EXAMPLE 2–4** **Average acceleration.** A car accelerates on a straight road from rest to 75 km/h in 5.0 s, Fig. 2–10. What is the magnitude of its average acceleration?

**APPROACH** Average acceleration is the change in velocity divided by the elapsed time, 5.0 s. The car starts from rest, so $v_1 = 0$. The final velocity is $v_2 = 75$ km/h.

**SOLUTION** From Eq. 2–4, the average acceleration is

$$\bar{a} = \frac{v_2 - v_1}{t_2 - t_1} = \frac{75 \text{ km/h} - 0 \text{ km/h}}{5.0 \text{ s}} = 15 \frac{\text{km/h}}{\text{s}}.$$

This is read as "fifteen kilometers per hour per second" and means that, on average, the velocity changed by 15 km/h during each second. That is, assuming the acceleration was constant, during the first second the car's velocity increased from zero to 15 km/h. During the next second its velocity increased by another 15 km/h, reaching a velocity of 30 km/h at $t = 2.0$ s, and so on. See Fig. 2–10.

**FIGURE 2–10** Example 2–4. The car is shown at the start with $v_1 = 0$ at $t_1 = 0$. The car is shown three more times, at $t = 1.0$ s, $t = 2.0$ s, and at the end of our time interval, $t_2 = 5.0$ s. The green arrows represent the velocity vectors, whose length represents the magnitude of the velocity at that moment. The acceleration vector is the orange arrow, whose magnitude is constant and equals 15 km/h/s or 4.2 m/s² (see top of next page). Distances are not to scale.

Our result in Example 2–4 contains two different time units: hours and seconds. We usually prefer to use only seconds. To do so we can change km/h to m/s (see Section 1–6, and Example 1–5):

$$75 \text{ km/h} = \left(75 \frac{\text{km}}{\text{h}}\right)\left(\frac{1000 \text{ m}}{1 \text{ km}}\right)\left(\frac{1 \text{ h}}{3600 \text{ s}}\right) = 21 \text{ m/s}.$$

Then

$$\bar{a} = \frac{21 \text{ m/s} - 0.0 \text{ m/s}}{5.0 \text{ s}} = 4.2 \frac{\text{m/s}}{\text{s}} = 4.2 \frac{\text{m}}{\text{s}^2}.$$

We almost always write the units for acceleration as $\text{m/s}^2$ (meters per second squared) instead of m/s/s. This is possible because:

$$\frac{\text{m/s}}{\text{s}} = \frac{\text{m}}{\text{s} \cdot \text{s}} = \frac{\text{m}}{\text{s}^2}.$$

Note that *acceleration tells us how quickly the velocity changes*, whereas *velocity tells us how quickly the position changes*.

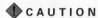

**◆ CAUTION**
*Distinguish velocity from acceleration*

**◆ CAUTION**
*If v or a is zero, is the other zero too?*

---

**CONCEPTUAL EXAMPLE 2–5** | **Velocity and acceleration.** (*a*) If the velocity of an object is zero, does it mean that the acceleration is zero? (*b*) If the acceleration is zero, does it mean that the velocity is zero? Think of some examples.

**RESPONSE** A zero velocity does not necessarily mean that the acceleration is zero, nor does a zero acceleration mean that the velocity is zero. (*a*) For example, when you put your foot on the gas pedal of your car which is at rest, the velocity starts from zero but the acceleration is not zero since the velocity of the car changes. (How else could your car start forward if its velocity weren't changing—that is, accelerating?) (*b*) As you cruise along a straight highway at a constant velocity of 100 km/h, your acceleration is zero: $a = 0$, $v \neq 0$.

---

**EXAMPLE 2–6** | **Car slowing down.** An automobile is moving to the right along a straight highway, which we choose to be the positive *x* axis (Fig. 2–11). Then the driver steps on the brakes. If the initial velocity (when the driver hits the brakes) is $v_1 = 15.0 \text{ m/s}$, and it takes 5.0 s to slow down to $v_2 = 5.0 \text{ m/s}$, what was the car's average acceleration?

**APPROACH** We put the given initial and final velocities, and the elapsed time, into Eq. 2–4 for $\bar{a}$.

**SOLUTION** In Eq. 2–4, we call the initial time $t_1 = 0$, and set $t_2 = 5.0 \text{ s}$:

$$\bar{a} = \frac{5.0 \text{ m/s} - 15.0 \text{ m/s}}{5.0 \text{ s}} = -2.0 \text{ m/s}^2.$$

The negative sign appears because the final velocity is less than the initial velocity. In this case the direction of the acceleration is to the left (in the negative *x* direction)—even though the velocity is always pointing to the right. We say that the acceleration is 2.0 m/s² to the left, and it is shown in Fig. 2–11 as an orange arrow.

**FIGURE 2–11** Example 2–6, showing the position of the car at times $t_1$ and $t_2$, as well as the car's velocity represented by the green arrows. The acceleration vector (orange) points to the left because the car slows down as it moves to the right.

## Deceleration

When an object is slowing down, we can say it is **decelerating**. But be careful: deceleration does *not* mean that the acceleration is necessarily negative. The velocity of an object moving to the right along the positive *x* axis is positive; if the object is slowing down (as in Fig. 2–11), the acceleration *is* negative. But the same car moving to the left (decreasing *x*), and slowing down, has positive acceleration that points to the right, as shown in Fig. 2–12. We have a deceleration whenever the magnitude of the velocity is decreasing; thus the *velocity and acceleration point in opposite directions* when there is deceleration.

**FIGURE 2–12** The car of Example 2–6, now moving to the *left* and decelerating. The acceleration is $a = (v_2 - v_1)/\Delta t$, or

$$a = \frac{(-5.0 \text{ m/s}) - (-15.0 \text{ m/s})}{5.0 \text{ s}}$$

$$= \frac{-5.0 \text{ m/s} + 15.0 \text{ m/s}}{5.0 \text{ s}} = +2.0 \text{ m/s}^2.$$

$v_2 = -5.0 \text{ m/s}$      $v_1 = -15.0 \text{ m/s}$

$a$

**EXERCISE C** A car moves along the *x* axis. What is the sign of the car's acceleration if it is moving in the positive *x* direction with (*a*) increasing speed or (*b*) decreasing speed? What is the sign of the acceleration if the car moves in the negative *x* direction with (*c*) increasing speed or (*d*) decreasing speed?

# 2–5 Motion at Constant Acceleration

We now examine motion in a straight line when the magnitude of the acceleration is constant. In this case, the instantaneous and average accelerations are equal. We use the definitions of average velocity and acceleration to derive a set of valuable equations that relate $x$, $v$, $a$, and $t$ when $a$ is constant, allowing us to determine any one of these variables if we know the others. We can then solve many interesting Problems.

Notation in physics varies from book to book; and different instructors use different notation. We are now going to change our notation, to simplify it a bit for our discussion here of motion at **constant acceleration**. First we choose the initial time in any discussion to be zero, and we call it $t_0$. That is, $t_1 = t_0 = 0$. (This is effectively starting a stopwatch at $t_0$.) We can then let $t_2 = t$ be the elapsed time. The initial position $(x_1)$ and the initial velocity $(v_1)$ of an object will now be represented by $x_0$ and $v_0$, since they represent $x$ and $v$ at $t = 0$. At time $t$ the position and velocity will be called $x$ and $v$ (rather than $x_2$ and $v_2$). The average velocity during the time interval $t - t_0$ will be (Eq. 2–2)

$$\bar{v} = \frac{\Delta x}{\Delta t} = \frac{x - x_0}{t - t_0} = \frac{x - x_0}{t}$$

since we chose $t_0 = 0$. The acceleration, assumed constant in time, is $a = \Delta v / \Delta t$ (Eq. 2–4), so

$$a = \frac{v - v_0}{t}.$$

A common problem is to determine the velocity of an object after any elapsed time $t$, when we are given the object's constant acceleration. We can solve such problems[†] by solving for $v$ in the last equation: first we multiply both sides by $t$,

$$at = v - v_0 \qquad \text{or} \qquad v - v_0 = at.$$

Then, adding $v_0$ to both sides, we obtain

$$v = v_0 + at. \qquad \text{[constant acceleration]} \quad \textbf{(2–6)}$$

If an object, such as a motorcycle (Fig. 2–13), starts from rest $(v_0 = 0)$ and accelerates at $4.0 \, \text{m/s}^2$, after an elapsed time $t = 6.0 \, \text{s}$ its velocity will be $v = 0 + at = (4.0 \, \text{m/s}^2)(6.0 \, \text{s}) = 24 \, \text{m/s}$.

Next, let us see how to calculate the position $x$ of an object after a time $t$ when it undergoes constant acceleration. The definition of average velocity (Eq. 2–2) is $\bar{v} = (x - x_0)/t$, which we can rewrite by multiplying both sides by $t$:

$$x = x_0 + \bar{v}t. \qquad \textbf{(2–7)}$$

Because the velocity increases at a uniform rate, the average velocity, $\bar{v}$, will be midway between the initial and final velocities:

$$\bar{v} = \frac{v_0 + v}{2}. \qquad \text{[constant acceleration]} \quad \textbf{(2–8)}$$

(Careful: Equation 2–8 is not necessarily valid if the acceleration is not constant.) We combine the last two Equations with Eq. 2–6 and find, starting with Eq. 2–7,

$$x = x_0 + \bar{v}t$$

$$= x_0 + \left( \frac{v_0 + v}{2} \right) t$$

$$= x_0 + \left( \frac{v_0 + v_0 + at}{2} \right) t$$

or

$$x = x_0 + v_0 t + \tfrac{1}{2} a t^2. \qquad \text{[constant acceleration]} \quad \textbf{(2–9)}$$

Equations 2–6, 2–8, and 2–9 are three of the four most useful equations for motion at constant acceleration. We now derive the fourth equation, which is useful

**FIGURE 2–13** An accelerating motorcycle.

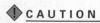

⚠ **CAUTION**

*Average velocity, but only if $a = $ constant*

[†]Appendix A–4 summarizes simple algebraic manipulations.

in situations where the time $t$ is not known. We substitute Eq. 2–8 into Eq. 2–7:

$$x = x_0 + \bar{v}t = x_0 + \left(\frac{v + v_0}{2}\right)t.$$

Next we solve Eq. 2–6 for $t$, obtaining (see Appendix A–4 for a quick review)

$$t = \frac{v - v_0}{a},$$

and substituting this into the previous equation we have

$$x = x_0 + \left(\frac{v + v_0}{2}\right)\left(\frac{v - v_0}{a}\right) = x_0 + \frac{v^2 - v_0^2}{2a}.$$

We solve this for $v^2$ and obtain

$$v^2 = v_0^2 + 2a(x - x_0), \qquad \text{[constant acceleration]} \quad \textbf{(2–10)}$$

which is the other useful equation we sought.

We now have four equations relating position, velocity, acceleration, and time, when the acceleration $a$ is constant. We collect these *kinematic equations for constant acceleration* here in one place for future reference (the tan background screen emphasizes their usefulness):

$$
\begin{aligned}
v &= v_0 + at & \text{[$a$ = constant]} \quad \textbf{(2–11a)} \\
x &= x_0 + v_0 t + \tfrac{1}{2}at^2 & \text{[$a$ = constant]} \quad \textbf{(2–11b)} \\
v^2 &= v_0^2 + 2a(x - x_0) & \text{[$a$ = constant]} \quad \textbf{(2–11c)} \\
\bar{v} &= \frac{v + v_0}{2}. & \text{[$a$ = constant]} \quad \textbf{(2–11d)}
\end{aligned}
$$

*Kinematic equations for constant acceleration (we'll use them a lot)*

These useful equations are not valid unless $a$ is a constant. In many cases we can set $x_0 = 0$, and this simplifies the above equations a bit. Note that $x$ represents position (not distance), also that $x - x_0$ is the displacement, and that $t$ is the elapsed time. Equations 2–11 are useful also when $a$ is approximately constant to obtain reasonable estimates.

---

**EXAMPLE 2–7** **Runway design.** You are designing an airport for small planes. One kind of airplane that might use this airfield must reach a speed before takeoff of at least 27.8 m/s (100 km/h), and can accelerate at 2.00 m/s². (*a*) If the runway is 150 m long, can this airplane reach the required speed for takeoff? (*b*) If not, what minimum length must the runway have?

**APPROACH** Assuming the plane's acceleration is constant, we use the kinematic equations for constant acceleration. In (*a*), we want to find $v$, and what we are given is shown in the Table in the margin.

**SOLUTION** (*a*) Of the above four equations, Eq. 2–11c will give us $v$ when we know $v_0$, $a$, $x$, and $x_0$:

$$
\begin{aligned}
v^2 &= v_0^2 + 2a(x - x_0) \\
&= 0 + 2(2.00 \text{ m/s}^2)(150 \text{ m}) = 600 \text{ m}^2/\text{s}^2 \\
v &= \sqrt{600 \text{ m}^2/\text{s}^2} = 24.5 \text{ m/s}.
\end{aligned}
$$

This runway length is *not* sufficient, because the minimum speed is not reached. (*b*) Now we want to find the minimum runway length, $x - x_0$, for a plane to reach $v = 27.8$ m/s, given $a = 2.00$ m/s². We again use Eq. 2–11c, but rewritten as

$$(x - x_0) = \frac{v^2 - v_0^2}{2a} = \frac{(27.8 \text{ m/s})^2 - 0}{2(2.00 \text{ m/s}^2)} = 193 \text{ m}.$$

A 200-m runway is more appropriate for this plane.

**NOTE** We did this Example as if the plane were a particle, so we round off our answer to 200 m.

**PHYSICS APPLIED**

*Airport design*

| Known | Wanted |
|---|---|
| $x_0 = 0$ | $v$ |
| $v_0 = 0$ | |
| $x = 150$ m | |
| $a = 2.00$ m/s² | |

**PROBLEM SOLVING**

*Equations 2–11 are valid only when the acceleration is constant, which we assume in this Example*

**EXERCISE D** A car starts from rest and accelerates at a constant $10 \, \text{m/s}^2$ during a $\frac{1}{4}$-mile (402 m) race. How fast is the car going at the finish line? (a) 8040 m/s; (b) 90 m/s; (c) 81 m/s; (d) 804 m/s.

# 2–6 Solving Problems

Before doing more worked-out Examples, let us look at how to approach problem solving. First, it is important to note that physics is *not* a collection of equations to be memorized. Simply searching for an equation that might work can lead you to a wrong result and will not help you understand physics (Fig. 2–14). A better approach is to use the following (rough) procedure, which we present as a special "Problem Solving Strategy." (Other such Problem Solving Strategies will be found throughout the book.)

**FIGURE 2–14** Read the book, study carefully, and work the Problems using your reasoning abilities.

1. Read and **reread** the whole problem carefully before trying to solve it.

2. Decide what **object** (or objects) you are going to study, and for what **time interval**. You can often choose the initial time to be $t = 0$.

3. **Draw** a **diagram** or picture of the situation, with coordinate axes wherever applicable. [You can place the origin of coordinates and the axes wherever you like to make your calculations easier. You also choose which direction is positive and which is negative. Usually we choose the $x$ axis to the right as positive.]

4. Write down what quantities are "**known**" or "given," and then what you *want* to know. Consider quantities both at the beginning and at the end of the chosen time interval. You may need to "translate" language into physical terms, such as "starts from rest" means $v_0 = 0$.

5. Think about which **principles of physics** apply in this problem. Use common sense and your own experiences. Then plan an approach.

6. Consider which **equations** (and/or definitions) relate the quantities involved. Before using them, be sure their **range of validity** includes your problem (for example, Eqs. 2–11 are valid only when the acceleration is constant). If you find an applicable equation that involves only known quantities and one desired unknown, **solve** the equation algebraically for the unknown. Sometimes several sequential calculations, or a combination of equations, may be needed. It is often preferable to solve algebraically for the desired unknown before putting in numerical values.

7. Carry out the **calculation** if it is a numerical problem. Keep one or two extra digits during the calculations, but round off the final answer(s) to the correct number of significant figures (Section 1–4).

8. Think carefully about the result you obtain: Is it **reasonable**? Does it make sense according to your own intuition and experience? A good check is to do a rough **estimate** using only powers of 10, as discussed in Section 1–7. Often it is preferable to do a rough estimate at the *start* of a numerical problem because it can help you focus your attention on finding a path toward a solution.

9. A very important aspect of doing problems is keeping track of **units**. An equals sign implies the units on each side must be the same, just as the numbers must. If the units do not balance, a mistake has been made. This can serve as a **check** on your solution (but it only tells you if you're wrong, not if you're right). Always use a consistent set of units.

**EXAMPLE 2–8** **Acceleration of a car.** How long does it take a car to cross a 30.0-m-wide intersection after the light turns green, if the car accelerates from rest at a constant 2.00 m/s²?

**APPROACH** We follow the Problem Solving Strategy on the previous page, step by step.

**SOLUTION**

1. **Reread** the problem. Be sure you understand what it asks for (here, a time interval: "how long does it take").

2. The **object** under study is the car. We need to choose the **time interval** during which we look at the car's motion: we choose $t = 0$, the initial time, to be the moment the car starts to accelerate from rest $(v_0 = 0)$; the time $t$ is the instant the car has traveled the full 30.0-m width of the intersection.

3. **Draw** a **diagram**: the situation is shown in Fig. 2–15, where the car is shown moving along the positive $x$ axis. We choose $x_0 = 0$ at the front bumper of the car before it starts to move.

4. The "**knowns**" and the "wanted" information are shown in the Table in the margin. Note that "starting from rest" means $v = 0$ at $t = 0$; that is, $v_0 = 0$. The wanted time $t$ is how long it takes the car to travel 30.0 m.

5. The **physics**: the car, starting from rest $(\text{at } t_0 = 0)$, increases in speed as it covers more distance. The acceleration is constant, so we can use the kinematic equations, Eqs. 2–11.

6. **Equations**: we want to find the time, given the distance and acceleration; Eq. 2–11b is perfect since the only unknown quantity is $t$. Setting $v_0 = 0$ and $x_0 = 0$ in Eq. 2–11b $(x = x_0 + v_0 t + \frac{1}{2}at^2)$, we have

$$x = \tfrac{1}{2}at^2.$$

We solve for $t$ by multiplying both sides by $\frac{2}{a}$:

$$\frac{2x}{a} = t^2.$$

Taking the square root, we get $t$:

$$t = \sqrt{\frac{2x}{a}}.$$

7. The **calculation**:

$$t = \sqrt{\frac{2x}{a}} = \sqrt{\frac{2(30.0\,\text{m})}{2.00\,\text{m/s}^2}} = 5.48\,\text{s}.$$

This is our answer. Note that the units come out correctly.

8. We can check the **reasonableness** of the answer by doing an alternate calculation: we first find the final velocity

$$v = at = (2.00\,\text{m/s}^2)(5.48\,\text{s}) = 10.96\,\text{m/s},$$

and then find the distance traveled

$$x = x_0 + \bar{v}t = 0 + \tfrac{1}{2}(10.96\,\text{m/s} + 0)(5.48\,\text{s}) = 30.0\,\text{m},$$

which checks with our given distance.

9. We checked the **units** in step 7, and they came out correctly (seconds).

**NOTE** In steps 6 and 7, when we took the square root, we should have written $t = \pm\sqrt{2x/a} = \pm 5.48\,\text{s}$. Mathematically there are two solutions. But the second solution, $t = -5.48\,\text{s}$, is a time *before* our chosen time interval and makes no sense physically. We say it is "unphysical" and ignore it.

We explicitly followed the steps of the Problem Solving Strategy in Example 2–8. In upcoming Examples, we will use our usual "Approach" and "Solution" to avoid being wordy.

**PROBLEM SOLVING**
*"Starting from rest" means*
$v = 0$ at $t = 0$ [*i.e.*, $v_0 = 0$]

$a = 2.00\,\text{m/s}^2$      $a = 2.00\,\text{m/s}^2$

$x_0 = 0$      $x =$
$v_0 = 0$      30.0 m

**FIGURE 2–15** Example 2–8.

| Known | Wanted |
|---|---|
| $x_0 = 0$ | $t$ |
| $x = 30.0\,\text{m}$ | |
| $a = 2.00\,\text{m/s}^2$ | |
| $v_0 = 0$ | |

**PROBLEM SOLVING**
*Check your answer*

**PROBLEM SOLVING**
*"Unphysical" solutions*

FIGURE 2–16 Example 2–9:
stopping distance for a
braking car.

**FIGURE 2–16** Example 2–9:
stopping distance for a
braking car.

$v = $ constant $= 14$ m/s
$t = 0.50$ s
$a = 0$

$v$ decreases from 14 m/s to zero
$a = -6.0$ m/s$^2$

**PHYSICS APPLIED**

*Car stopping distances*

**EXAMPLE 2–9 | ESTIMATE | Braking distances.** Estimate the minimum
stopping distance for a car, which is important for traffic safety and traffic design.
The problem is best dealt with in two parts, two separate time intervals. (1) The
first time interval begins when the driver decides to hit the brakes, and ends
when the foot touches the brake pedal. This is the "reaction time" during which
the speed is constant, so $a = 0$. (2) The second time interval is the actual
braking period when the vehicle slows down ($a \neq 0$) and comes to a stop. The
stopping distance depends on the reaction time of the driver, the initial speed of
the car (the final speed is zero), and the deceleration of the car. For a dry road
and good tires, good brakes can decelerate a car at a rate of about 5 m/s$^2$ to
8 m/s$^2$. Calculate the total stopping distance for an initial velocity of 50 km/h
($= 14$ m/s $\approx 31$ mi/h) and assume the acceleration of the car is $-6.0$ m/s$^2$
(the minus sign appears because the velocity is taken to be in the positive
$x$ direction and its magnitude is decreasing). Reaction time for normal drivers
varies from perhaps 0.3 s to about 1.0 s; take it to be 0.50 s.

**APPROACH** During the "reaction time," part (1), the car moves at constant
speed of 14 m/s, so $a = 0$. Once the brakes are applied, part (2), the acceleration is $a = -6.0$ m/s$^2$ and is constant over this time interval. For both parts
$a$ is constant, so we can use Eqs. 2–11.

*Part 1: Reaction time*

| Known | Wanted |
|---|---|
| $t = 0.50$ s | $x$ |
| $v_0 = 14$ m/s | |
| $v = 14$ m/s | |
| $a = 0$ | |
| $x_0 = 0$ | |

**SOLUTION** Part (1). We take $x_0 = 0$ for the first time interval, when the driver
is reacting (0.50 s): the car travels at a constant speed of 14 m/s so $a = 0$.
See Fig. 2–16 and the Table in the margin. To find $x$, the position of the car
at $t = 0.50$ s (when the brakes are applied), we cannot use Eq. 2–11c because
$x$ is multiplied by $a$, which is zero. But Eq. 2–11b works:

$$x = v_0 t + 0 = (14 \text{ m/s})(0.50 \text{ s}) = 7.0 \text{ m}.$$

Thus the car travels 7.0 m during the driver's reaction time, until the instant
the brakes are applied. We will use this result as input to part (2).

*Part 2: Braking*

| Known | Wanted |
|---|---|
| $x_0 = 7.0$ m | $x$ |
| $v_0 = 14$ m/s | |
| $v = 0$ | |
| $a = -6.0$ m/s$^2$ | |

Part (2). During the second time interval, the brakes are applied and the car is
brought to rest. The initial position is $x_0 = 7.0$ m (result of part (1)), and other
variables are shown in the second Table in the margin. Equation 2–11a doesn't
contain $x$; Eq. 2–11b contains $x$ but also the unknown $t$. Equation 2–11c,
$v^2 - v_0^2 = 2a(x - x_0)$, is what we want; after setting $x_0 = 7.0$ m, we solve
for $x$, the final position of the car (when it stops):

$$x = x_0 + \frac{v^2 - v_0^2}{2a}$$

$$= 7.0 \text{ m} + \frac{0 - (14 \text{ m/s})^2}{2(-6.0 \text{ m/s}^2)} = 7.0 \text{ m} + \frac{-196 \text{ m}^2/\text{s}^2}{-12 \text{ m/s}^2}$$

$$= 7.0 \text{ m} + 16 \text{ m} = 23 \text{ m}.$$

**FIGURE 2–17** Example 2–9.
Graph of $v$ vs. $t$.

The car traveled 7.0 m while the driver was reacting and another 16 m during
the braking period before coming to a stop, for a total distance traveled of
23 m. Figure 2–17 shows a graph of $v$ vs. $t$: $v$ is constant from $t = 0$ until
$t = 0.50$ s, and after $t = 0.50$ s it decreases linearly to zero.

**NOTE** From the equation above for $x$, we see that the stopping distance after
the driver hit the brakes ($= x - x_0$) increases with the *square* of the initial
speed, not just linearly with speed. If you are traveling twice as fast, it takes
four times the distance to stop.

FIGURE 2–18 Painting of Galileo demonstrating to the Grand Duke of Tuscany his argument for the action of gravity being uniform acceleration. He used an inclined plane to slow down the action. A ball rolling down the plane still accelerates. Tiny bells placed at equal distances along the inclined plane would ring at shorter time intervals as the ball "fell," indicating that the speed was increasing.

FIGURE 2–19 Multiflash photograph of a falling apple, at equal time intervals. The apple falls farther during each successive interval, which means it is accelerating.

# 2–7 Freely Falling Objects

One of the most common examples of uniformly accelerated motion is that of an object allowed to fall freely near the Earth's surface. That a falling object is accelerating may not be obvious at first. And beware of thinking, as was widely believed before the time of Galileo (Fig. 2–18), that heavier objects fall faster than lighter objects and that the speed of fall is proportional to how heavy the object is. *The speed of a falling object is* not *proportional to its mass.*

Galileo made use of his new technique of imagining what would happen in idealized (simplified) cases. For free fall, he postulated that *all objects would fall with the same constant acceleration in the absence of air or other resistance.* He showed that this postulate predicts that for an object falling from rest, the distance traveled will be proportional to the square of the time (Fig. 2–19); that is, $d \propto t^2$. We can see this from Eq. 2–11b for constant acceleration; but Galileo was the first to derive this mathematical relation.

To support his claim that falling objects increase in speed as they fall, Galileo made use of a clever argument: a heavy stone dropped from a height of 2 m will drive a stake into the ground much further than will the same stone dropped from a height of only 0.2 m. Clearly, the stone must be moving faster in the former case.

Galileo claimed that *all* objects, light or heavy, fall with the *same* acceleration, at least in the absence of air. If you hold a piece of paper flat and horizontal in one hand, and a heavier object like a baseball in the other, and release them at the same time as in Fig. 2–20a, the heavier object will reach the ground first. But if you repeat the experiment, this time crumpling the paper into a small wad, you will find (see Fig. 2–20b) that the two objects reach the floor at nearly the same time.

Galileo was sure that air acts as a resistance to very light objects that have a large surface area. But in many ordinary circumstances this air resistance is negligible. In a chamber from which the air has been removed, even light objects like a feather or a horizontally held piece of paper will fall with the same acceleration as any other object (see Fig. 2–21). Such a demonstration in vacuum was not possible in Galileo's time, which makes Galileo's achievement all the greater. Galileo is often called the "father of modern science," not only for the *content* of his science (astronomical discoveries, inertia, free fall) but also for his new methods of *doing* science (idealization and simplification, mathematization of theory, theories that have testable consequences, experiments to test theoretical predictions).

FIGURE 2–20 (a) A ball and a light piece of paper are dropped at the same time. (b) Repeated, with the paper wadded up.

FIGURE 2–21 A rock and a feather are dropped simultaneously (a) in air, (b) in a vacuum.

Air-filled tube
(a)

Evacuated tube
(b)

Galileo's specific contribution to our understanding of the motion of falling objects can be summarized as follows:

**at a given location on the Earth and in the absence of air resistance, all objects fall with the same constant acceleration.**

We call this acceleration the **acceleration due to gravity** at the surface of the Earth, and we give it the symbol $g$. Its magnitude is approximately

$$g = 9.80 \text{ m/s}^2. \qquad \left[\begin{array}{l} \text{acceleration due to gravity} \\ \text{at surface of Earth} \end{array}\right]$$

In British units $g$ is about $32 \text{ ft/s}^2$. Actually, $g$ varies slightly according to latitude and elevation on the Earth's surface, but these variations are so small that we will ignore them for most purposes. (Acceleration of gravity in space beyond the Earth's surface is treated in Chapter 5.) The effects of air resistance are often small, and we will neglect them for the most part. However, air resistance will be noticeable even on a reasonably heavy object if the velocity becomes large.[†] Acceleration due to gravity is a vector, as is any acceleration, and its direction is downward toward the center of the Earth.

When dealing with freely falling objects we can make use of Eqs. 2–11, where for $a$ we use the value of $g$ given above. Also, since the motion is vertical we will substitute $y$ in place of $x$, and $y_0$ in place of $x_0$. We take $y_0 = 0$ unless otherwise specified. *It is arbitrary whether we choose $y$ to be positive in the upward direction or in the downward direction; but we must be consistent about it throughout a problem's solution.*

**EXERCISE E** Return to the Chapter-Opening Question, page 21, and answer it again now, assuming minimal air resistance. Try to explain why you may have answered differently the first time.

**FIGURE 2–22** Example 2–10. (a) An object dropped from a tower falls with progressively greater speed and covers greater distance with each successive second. (See also Fig. 2–19.) (b) Graph of $y$ vs. $t$.

Acceleration due to gravity

$y = 0$

$y_1 = 4.90$ m (After 1.00 s)

$y_2 = 19.6$ m (After 2.00 s)

$+y$

$y_3 = 44.1$ m (After 3.00 s)

$+y$

(a)

(b)

**EXAMPLE 2–10** **Falling from a tower.** Suppose that a ball is dropped $(v_0 = 0)$ from a tower. How far will it have fallen after a time $t_1 = 1.00 \text{ s}$, $t_2 = 2.00 \text{ s}$, and $t_3 = 3.00 \text{ s}$? Ignore air resistance.

**APPROACH** Let us take $y$ as positive downward, so the acceleration is $a = g = +9.80 \text{ m/s}^2$. We set $v_0 = 0$ and $y_0 = 0$. We want to find the position $y$ of the ball after three different time intervals. Equation 2–11b, with $x$ replaced by $y$, relates the given quantities $(t, a, \text{ and } v_0)$ to the unknown $y$.

**SOLUTION** We set $t = t_1 = 1.00 \text{ s}$ in Eq. 2–11b:

$$y_1 = v_0 t_1 + \tfrac{1}{2} a t_1^2$$
$$= 0 + \tfrac{1}{2} a t_1^2 = \tfrac{1}{2}(9.80 \text{ m/s}^2)(1.00 \text{ s})^2 = 4.90 \text{ m}.$$

The ball has fallen a distance of 4.90 m during the time interval $t = 0$ to $t_1 = 1.00 \text{ s}$. Similarly, after 2.00 s $(= t_2)$, the ball's position is

$$y_2 = \tfrac{1}{2} a t_2^2 = \tfrac{1}{2}(9.80 \text{ m/s}^2)(2.00 \text{ s})^2 = 19.6 \text{ m}.$$

Finally, after 3.00 s $(= t_3)$, the ball's position is (see Fig. 2–22)

$$y_3 = \tfrac{1}{2} a t_3^2 = \tfrac{1}{2}(9.80 \text{ m/s}^2)(3.00 \text{ s})^2 = 44.1 \text{ m}.$$

**NOTE** Whenever we say "dropped," it means $v_0 = 0$. Note also the graph of $y$ vs. $t$ (Fig. 2–22b): the curve is not straight but bends upward because $y$ is proportional to $t^2$.

[†]The speed of an object falling in air (or other fluid) does not increase indefinitely. If the object falls far enough, it will reach a maximum velocity called the **terminal velocity** due to air resistance.

**EXAMPLE 2–11** **Thrown down from a tower.** Suppose the ball in Example 2–10 is *thrown* downward with an initial velocity of 3.00 m/s, instead of being dropped. (*a*) What then would be its position after 1.00 s and 2.00 s? (*b*) What would its speed be after 1.00 s and 2.00 s? Compare with the speeds of a dropped ball.

**APPROACH** Again we use Eq. 2–11b, but now $v_0$ is not zero, it is $v_0 = 3.00$ m/s.

**SOLUTION** (*a*) At $t_1 = 1.00$ s, the position of the ball as given by Eq. 2–11b is

$$y = v_0 t + \tfrac{1}{2}at^2 = (3.00 \text{ m/s})(1.00 \text{ s}) + \tfrac{1}{2}(9.80 \text{ m/s}^2)(1.00 \text{ s})^2 = 7.90 \text{ m}.$$

At $t_2 = 2.00$ s (time interval $t = 0$ to $t = 2.00$ s), the position is

$$y = v_0 t + \tfrac{1}{2}at^2 = (3.00 \text{ m/s})(2.00 \text{ s}) + \tfrac{1}{2}(9.80 \text{ m/s}^2)(2.00 \text{ s})^2 = 25.6 \text{ m}.$$

As expected, the ball falls farther each second than if it were dropped with $v_0 = 0$.

(*b*) The velocity is obtained from Eq. 2–11a:

$$
\begin{aligned}
v &= v_0 + at \\
&= 3.00 \text{ m/s} + (9.80 \text{ m/s}^2)(1.00 \text{ s}) = 12.8 \text{ m/s} \quad [\text{at } t_1 = 1.00 \text{ s}] \\
&= 3.00 \text{ m/s} + (9.80 \text{ m/s}^2)(2.00 \text{ s}) = 22.6 \text{ m/s}. \quad [\text{at } t_2 = 2.00 \text{ s}]
\end{aligned}
$$

In Example 2–10, when the ball was dropped $(v_0 = 0)$, the first term $(v_0)$ in these equations was zero, so

$$
\begin{aligned}
v &= 0 + at \\
&= (9.80 \text{ m/s}^2)(1.00 \text{ s}) = 9.80 \text{ m/s} \quad [\text{at } t_1 = 1.00 \text{ s}] \\
&= (9.80 \text{ m/s}^2)(2.00 \text{ s}) = 19.6 \text{ m/s}. \quad [\text{at } t_2 = 2.00 \text{ s}]
\end{aligned}
$$

**NOTE** For both Examples 2–10 and 2–11, the speed increases linearly in time by 9.80 m/s during each second. But the speed of the downwardly thrown ball at any instant is always 3.00 m/s (its initial speed) higher than that of a dropped ball.

**EXAMPLE 2–12** **Ball thrown upward.** A person throws a ball *upward* into the air with an initial velocity of 15.0 m/s. Calculate how high it goes. Ignore air resistance.

**APPROACH** We are not concerned here with the throwing action, but only with the motion of the ball *after* it leaves the thrower's hand (Fig. 2–23) and until it comes back to the hand again. Let us choose $y$ to be positive in the upward direction and negative in the downward direction. (This is a different convention from that used in Examples 2–10 and 2–11, and so illustrates our options.) The acceleration due to gravity is downward and so will have a negative sign, $a = -g = -9.80$ m/s². As the ball rises, its speed decreases until it reaches the highest point (B in Fig. 2–23), where its speed is zero for an instant; then it descends, with increasing speed.

**SOLUTION** We consider the time interval from when the ball leaves the thrower's hand until the ball reaches the highest point. To determine the maximum height, we calculate the position of the ball when its velocity equals zero ($v = 0$ at the highest point). At $t = 0$ (point A in Fig. 2–23) we have $y_0 = 0$, $v_0 = 15.0$ m/s, and $a = -9.80$ m/s². At time $t$ (maximum height), $v = 0$, $a = -9.80$ m/s², and we wish to find $y$. We use Eq. 2–11c, replacing $x$ with $y$: $v^2 = v_0^2 + 2ay$. We solve this equation for $y$:

$$y = \frac{v^2 - v_0^2}{2a} = \frac{0 - (15.0 \text{ m/s})^2}{2(-9.80 \text{ m/s}^2)} = 11.5 \text{ m}.$$

The ball reaches a height of 11.5 m above the hand.

**FIGURE 2–23** An object thrown into the air leaves the thrower's hand at A, reaches its maximum height at B, and returns to the original position at C. Examples 2–12, 2–13, 2–14, and 2–15.

B (v = 0)

$g$ $g$

$v$ $v$

A C

**FIGURE 2–23** (Repeated.)
An object thrown into the air leaves the thrower's hand at A, reaches its maximum height at B, and returns to the original position at C. Examples 2–12, 2–13, 2–14, and 2–15.

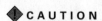

**CAUTION**

*Quadratic equations have two solutions. Sometimes only one corresponds to reality, sometimes both*

**CAUTION**

*(1) Velocity and acceleration are not always in the same direction; the acceleration (of gravity) always points down*
*(2) $a \neq 0$ even at the highest point of a trajectory*

**EXAMPLE 2–13** **Ball thrown upward, II.** In Fig. 2–23, Example 2–12, how long is the ball in the air before it comes back to the hand?

**APPROACH** We need to choose a time interval to calculate how long the ball is in the air before it returns to the hand. We could do this calculation in two parts by first determining the time required for the ball to reach its highest point, and then determining the time it takes to fall back down. However, it is simpler to consider the time interval for the entire motion from A to B to C (Fig. 2–23) in one step and use Eq. 2–11b. We can do this because $y$ is position or displacement, and not the total distance traveled. Thus, at both points A and C, $y = 0$.

**SOLUTION** We use Eq. 2–11b with $a = -9.80 \text{ m/s}^2$ and find

$$y = y_0 + v_0 t + \tfrac{1}{2}at^2$$
$$0 = 0 + (15.0 \text{ m/s})t + \tfrac{1}{2}(-9.80 \text{ m/s}^2)t^2.$$

This equation can be factored (we factor out one $t$):

$$(15.0 \text{ m/s} - 4.90 \text{ m/s}^2 \, t)t = 0.$$

There are two solutions:

$$t = 0 \quad \text{and} \quad t = \frac{15.0 \text{ m/s}}{4.90 \text{ m/s}^2} = 3.06 \text{ s}.$$

The first solution $(t = 0)$ corresponds to the initial point (A) in Fig. 2–23, when the ball was first thrown from $y = 0$. The second solution, $t = 3.06 \text{ s}$, corresponds to point C, when the ball has returned to $y = 0$. Thus the ball is in the air 3.06 s.

**NOTE** We have ignored air resistance in these last two Examples, which could be significant, so our result is only an approximation to a real, practical situation.

We did not consider the throwing action in these Examples. Why? Because during the throw, the thrower's hand is touching the ball and accelerating the ball at a rate unknown to us—the acceleration is *not* g. We consider only the time when the ball is in the air and the acceleration is equal to g.

Every quadratic equation (where the variable is squared) mathematically produces two solutions. In physics, sometimes only one solution corresponds to the real situation, as in Example 2–8, in which case we ignore the "unphysical" solution. But in Example 2–13, both solutions to our equation in $t^2$ are physically meaningful: $t = 0$ and $t = 3.06 \text{ s}$.

**CONCEPTUAL EXAMPLE 2–14** **Two possible misconceptions.** Give examples to show the error in these two common misconceptions: (1) that acceleration and velocity are always in the same direction, and (2) that an object thrown upward has zero acceleration at the highest point (B in Fig. 2–23).

**RESPONSE** Both are wrong. (1) Velocity and acceleration are *not* necessarily in the same direction. When the ball in Fig. 2–23 is moving upward, its velocity is positive (upward), whereas the acceleration is negative (downward). (2) At the highest point (B in Fig. 2–23), the ball has zero velocity for an instant. Is the acceleration also zero at this point? No. The velocity near the top of the arc points upward, then becomes zero for an instant (zero time) at the highest point, and then points downward. Gravity does not stop acting, so $a = -g = -9.80 \text{ m/s}^2$ even there. Thinking that $a = 0$ at point B would lead to the conclusion that upon reaching point B, the ball would stay there: if the acceleration ( = rate of change of velocity) were zero, the velocity would stay zero at the highest point, and the ball would stay up there without falling. Remember: the acceleration of gravity always points down toward the Earth, even when the object is moving up.

**EXAMPLE 2–15** **Ball thrown upward, III.** Let us consider again the ball thrown upward of Examples 2–12 and 2–13, and make more calculations. Calculate (*a*) how much time it takes for the ball to reach the maximum height (point B in Fig. 2–23), and (*b*) the velocity of the ball when it returns to the thrower's hand (point C).

**APPROACH** Again we assume the acceleration is constant, so we can use Eqs. 2–11. We have the maximum height of 11.5 m and initial speed of 15.0 m/s from Example 2–12. Again we take $y$ as positive upward.

**SOLUTION** (*a*) We consider the time interval between the throw $(t = 0, v_0 = 15.0 \text{ m/s})$ and the top of the path $(y = +11.5 \text{ m}, v = 0)$, and we want to find $t$. The acceleration is constant at $a = -g = -9.80 \text{ m/s}^2$. Both Eqs. 2–11a and 2–11b contain the time $t$ with other quantities known. Let us use Eq. 2–11a with $a = -9.80 \text{ m/s}^2$, $v_0 = 15.0 \text{ m/s}$, and $v = 0$:

$$v = v_0 + at;$$

setting $v = 0$ gives $0 = v_0 + at$, which we rearrange to solve for $t$: $at = -v_0$ or

$$t = -\frac{v_0}{a}$$
$$= -\frac{15.0 \text{ m/s}}{-9.80 \text{ m/s}^2} = 1.53 \text{ s}.$$

This is just half the time it takes the ball to go up and fall back to its original position [3.06 s, calculated in Example 2–13]. Thus it takes the same time to reach the maximum height as to fall back to the starting point.

(*b*) Now we consider the time interval from the throw $(t = 0, v_0 = 15.0 \text{ m/s})$ until the ball's return to the hand, which occurs at $t = 3.06 \text{ s}$ (as calculated in Example 2–13), and we want to find $v$ when $t = 3.06 \text{ s}$:

$$v = v_0 + at$$
$$= 15.0 \text{ m/s} - (9.80 \text{ m/s}^2)(3.06 \text{ s}) = -15.0 \text{ m/s}.$$

**NOTE** The ball has the same speed (magnitude of velocity) when it returns to the starting point as it did initially, but in the opposite direction (this is the meaning of the negative sign). And, as we saw in part (*a*), the time is the same up as down. Thus the motion is *symmetrical* about the maximum height.

The acceleration of objects such as rockets and fast airplanes is often given as a multiple of $g = 9.80 \text{ m/s}^2$. For example, a plane pulling out of a dive (see Fig. 2–24) and undergoing 3.00 $g$'s would have an acceleration of $(3.00)(9.80 \text{ m/s}^2) = 29.4 \text{ m/s}^2$.

**PROBLEM SOLVING**
*Acceleration in g's*

**FIGURE 2–24** Several planes, in formation, are just coming out of a downward dive.

**EXERCISE F** Two balls are thrown from a cliff. One is thrown directly up, the other directly down. Both balls have the same initial speed, and both hit the ground below the cliff but at different times. Which ball hits the ground at the greater speed: (a) the ball thrown upward, (b) the ball thrown downward, or (c) both the same? Ignore air resistance.

## Additional Example—Using the Quadratic Formula

**EXAMPLE 2–16** **Ball thrown upward at edge of cliff.** Suppose that the person of Examples 2–12, 2–13, and 2–15 throws the ball upward at 15.0 m/s while standing on the edge of a cliff, so that the ball can fall to the base of the cliff 50.0 m below, as shown in Fig. 2–25a. (a) How long does it take the ball to reach the base of the cliff? (b) What is the total distance traveled by the ball? Ignore air resistance (likely to be significant, so our result is an approximation).

**APPROACH** We again use Eq. 2–11b, with $y$ as $+$ upward, but this time we set $y = -50.0$ m, the bottom of the cliff, which is 50.0 m below the initial position ($y_0 = 0$); hence the minus sign.

**SOLUTION** (a) We use Eq. 2–11b with $a = -9.80$ m/s$^2$, $v_0 = 15.0$ m/s, $y_0 = 0$, and $y = -50.0$ m:

$$y = y_0 + v_0 t + \tfrac{1}{2}at^2$$
$$-50.0\,\text{m} = 0 + (15.0\,\text{m/s})t - \tfrac{1}{2}(9.80\,\text{m/s}^2)t^2.$$

To solve any quadratic equation of the form

$$at^2 + bt + c = 0,$$

where $a$, $b$, and $c$ are constants ($a$ is *not* acceleration here), we use the **quadratic formula** (see Appendix A–4):

$$t = \frac{-b \pm \sqrt{b^2 - 4ac}}{2a}.$$

We rewrite our $y$ equation just above in standard form, $at^2 + bt + c = 0$:

$$(4.90\,\text{m/s}^2)t^2 - (15.0\,\text{m/s})t - (50.0\,\text{m}) = 0.$$

Using the quadratic formula, we find as solutions

$$t = 5.07\,\text{s}$$

and

$$t = -2.01\,\text{s}.$$

The first solution, $t = 5.07$ s, is the answer we are seeking: the time it takes the ball to rise to its highest point and then fall to the base of the cliff. To rise and fall back to the top of the cliff took 3.06 s (Example 2–13); so it took an additional 2.01 s to fall to the base. But what is the meaning of the other solution, $t = -2.01$ s? This is a time before the throw, when our calculation begins, so it isn't relevant here. It is outside our chosen time interval, and so is an *unphysical* solution (also in Example 2–8).

(b) From Example 2–12, the ball moves up 11.5 m, falls 11.5 m back down to the top of the cliff, and then down another 50.0 m to the base of the cliff, for a total distance traveled of 73.0 m. [Note that the *displacement*, however, was $-50.0$ m.] Figure 2–25b shows the $y$ vs. $t$ graph for this situation.

(a)

(b)

**FIGURE 2–25** Example 2–16. (a) A person stands on the edge of a cliff. A ball is thrown upward, then falls back down past the thrower to the base of the cliff, 50.0 m below. (b) The $y$ vs. $t$ graph.

 **CAUTION**

*Sometimes a solution to a quadratic equation does not apply to the actual physical conditions of the Problem*

# 2–8 Graphical Analysis of Linear Motion

## Velocity as Slope

Analysis of motion using graphs can give us additional insight into kinematics. Let us draw a graph of $x$ vs. $t$, making the choice that at $t = 0$, the position of an object is $x = 0$, and the object is moving at a constant velocity, $v = \bar{v} = 11$ m/s (40 km/h). Our graph starts at $x = 0$, $t = 0$ (the origin). The graph of the position increases linearly in time because, by Eq. 2–2, $\Delta x = \bar{v} \Delta t$ and $\bar{v}$ is a constant. So the graph of $x$ vs. $t$ is a straight line, as shown in Fig. 2–26. The small (shaded) triangle on the graph indicates the **slope** of the straight line:

$$\text{slope} = \frac{\Delta x}{\Delta t}.$$

We see, using the definition of average velocity (Eq. 2–2), that the *slope of the x vs. t graph is equal to the velocity*. And, as can be seen from the small triangle on the graph, $\Delta x / \Delta t = (11 \text{ m})/(1.0 \text{ s}) = 11$ m/s, which is the given velocity.

If the object's velocity changes in time, we might have an $x$ vs. $t$ graph like that shown in Fig. 2–27. (Note that this graph is different from showing the "path" of an object on an $x$ vs. $y$ plot.) Suppose the object is at position $x_1$ at time $t_1$, and at position $x_2$ at time $t_2$. $P_1$ and $P_2$ represent these two points on the graph. A straight line drawn from point $P_1(x_1, t_1)$ to point $P_2(x_2, t_2)$ forms the hypotenuse of a right triangle whose sides are $\Delta x$ and $\Delta t$. The ratio $\Delta x / \Delta t$ is the **slope** of the straight line $P_1 P_2$. But $\Delta x / \Delta t$ is also the average velocity of the object during the time interval $\Delta t = t_2 - t_1$. Therefore, we conclude that the *average velocity of an object during any time interval $\Delta t = t_2 - t_1$ is equal to the slope of the straight line* (or **chord**) connecting the two points $(x_1, t_1)$ and $(x_2, t_2)$ on an $x$ vs. $t$ graph.

Consider now a time intermediate between $t_1$ and $t_2$, call it $t_3$, at which moment the object is at $x_3$ (Fig. 2–28). The slope of the straight line $P_1 P_3$ is less than the slope of $P_1 P_2$. Thus the average velocity during the time interval $t_3 - t_1$ is less than during the time interval $t_2 - t_1$.

**FIGURE 2–26** Graph of position vs. time for an object moving at a constant velocity of 11 m/s.

**FIGURE 2–27** Graph of an object's position $x$ vs. time $t$. The slope of the straight line $P_1 P_2$ represents the average velocity of the object during the time interval $\Delta t = t_2 - t_1$.

**FIGURE 2–28** Same position vs. time curve as in Fig. 2–27. Note that the average velocity over the time interval $t_3 - t_1$ (which is the slope of $P_1 P_3$) is less than the average velocity over the time interval $t_2 - t_1$. The slope of the line tangent to the curve at point $P_1$ equals the *instantaneous* velocity at time $t_1$.

Next let us take point $P_3$ in Fig. 2–28 to be closer and closer to point $P_1$. That is, we let the interval $t_3 - t_1$, which we now call $\Delta t$, to become smaller and smaller. The slope of the line connecting the two points becomes closer and closer to the slope of a line **tangent**[†] to the curve at point $P_1$. The average velocity (equal to the slope of the chord) thus approaches the slope of the tangent at point $P_1$. The definition of the instantaneous velocity (Eq. 2–3) is the limiting value of the average velocity as $\Delta t$ approaches zero. Thus the *instantaneous velocity equals the slope of the tangent to the curve of $x$ vs. $t$* at any chosen point (which we can simply call "the slope of the curve" at that point).

**PROBLEM SOLVING**

*Velocity equals slope of x vs. t graph at any instant*

---

[†]The tangent is a straight line that touches the curve only at the one chosen point, without passing across or through the curve at that point.

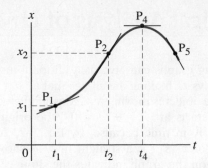

**FIGURE 2–29** Same $x$ vs. $t$ curve as in Figs. 2–27 and 2–28, but here showing the slope at four different points: At $P_4$, the slope is zero, so $v = 0$. At $P_5$ the slope is negative, so $v < 0$.

We can obtain the velocity of an object at any instant from its graph of $x$ vs. $t$. For example, in Fig. 2–29 (which shows the same graph as in Figs. 2–27 and 2–28), as our object moves from $x_1$ to $x_2$, the slope continually increases, so the velocity is increasing. For times after $t_2$, the slope begins to decrease and reaches zero ($v = 0$) where $x$ has its maximum value, at point $P_4$ in Fig. 2–29. Beyond point $P_4$, the slope is negative, as for point $P_5$. The velocity is therefore negative, which makes sense since $x$ is now decreasing—the particle is moving toward decreasing values of $x$, to the left on a standard $xy$ plot.

## Slope and Acceleration

We can also draw a graph of the *velocity*, $v$, vs. time, $t$, as shown in Fig. 2–30. Then the average acceleration over a time interval $\Delta t = t_2 - t_1$ is represented by the slope of the straight line connecting the two points $P_1$ and $P_2$ as shown. [Compare this to the position vs. time graph of Fig. 2–27 for which the slope of the straight line represents the average velocity.] The instantaneous acceleration at any time, say $t_1$, is the slope of the tangent to the $v$ vs. $t$ curve at that time, which is also shown in Fig. 2–30. Using this fact for the situation graphed in Fig. 2–30, as we go from time $t_1$ to time $t_2$ the velocity continually increases, but the acceleration (the rate at which the velocity changes) is decreasing since the slope of the curve is decreasing.

**FIGURE 2–30** A graph of velocity $v$ vs. time $t$. The average acceleration over a time interval $\Delta t = t_2 - t_1$ is the slope of the straight line $P_1 P_2$: $\bar{a} = \Delta v / \Delta t$. The instantaneous acceleration at time $t_1$ is the slope of the $v$ vs. $t$ curve at that instant.

Slope of $P_1 P_2$ is average acceleration during $\Delta t = t_2 - t_1$
Slope of tangent is instantaneous acceleration at $t_1$

$\Delta v = v_2 - v_1$

$\Delta t = t_2 - t_1$

---

**CONCEPTUAL EXAMPLE 2–17** **Analyzing with graphs.** Figure 2–31 shows the velocity as a function of time for two cars accelerating from 0 to 100 km/h in a time of 10.0 s. Compare (*a*) the average acceleration; (*b*) the instantaneous acceleration; and (*c*) the total distance traveled for the two cars.

**RESPONSE** (*a*) Average acceleration is $\Delta v / \Delta t$. Both cars have the same $\Delta v$ (100 km/h) over the same time interval $\Delta t = 10.0$ s, so the average acceleration is the same for both cars. (*b*) Instantaneous acceleration is the slope of the tangent to the $v$ vs. $t$ curve. For the first 4 s or so, the top curve (car A) is steeper than the bottom curve, so car A has a greater acceleration during this interval. The bottom curve is steeper during the last 6 s, so car B has the larger acceleration for this period. (*c*) Except at $t = 0$ and $t = 10.0$ s, car A is always going faster than car B. Since it is going faster, it will go farther in the same time.

**FIGURE 2–31** (below) Example 2–17.

# Summary

[*The Summary that appears at the end of each Chapter in this book gives a brief overview of the main ideas of the Chapter. The Summary cannot serve to give an understanding of the material, which can be accomplished only by a detailed reading of the Chapter.*]

**Kinematics** deals with the description of how objects move. The description of the motion of any object must always be given relative to some particular **reference frame**.

The **displacement** of an object is the change in position of the object.

**Average speed** is the distance traveled divided by the **elapsed time** or **time interval**, $\Delta t$ (the time period over which we choose to make our observations). An object's **average velocity** over a particular time interval is

$$\bar{v} = \frac{\Delta x}{\Delta t}, \tag{2–2}$$

where $\Delta x$ is the displacement during the time interval $\Delta t$.

The **instantaneous velocity**, whose magnitude is the same as the *instantaneous speed*, is defined as the average velocity taken over an infinitesimally short time interval.

Acceleration is the change of velocity per unit time. An object's **average acceleration** over a time interval $\Delta t$ is

$$\bar{a} = \frac{\Delta v}{\Delta t}, \qquad (2\text{--}4)$$

where $\Delta v$ is the change of velocity during the time interval $\Delta t$. **Instantaneous acceleration** is the average acceleration taken over an infinitesimally short time interval.

If an object has position $x_0$ and velocity $v_0$ at time $t = 0$ and moves in a straight line with **constant acceleration**, the velocity $v$ and position $x$ at a later time $t$ are related to the acceleration $a$, the initial position $x_0$, and the initial velocity $v_0$ by Eqs. 2–11:

$$
\begin{aligned}
v &= v_0 + at, \\
x &= x_0 + v_0 t + \tfrac{1}{2} at^2, \\
v^2 &= v_0^2 + 2a(x - x_0), \\
\bar{v} &= \frac{v + v_0}{2}.
\end{aligned}
\qquad (2\text{--}11)
$$

Objects that move vertically near the surface of the Earth, either falling or having been projected vertically up or down, move with the constant downward **acceleration due to gravity**, whose magnitude is $g = 9.80 \text{ m/s}^2$ if air resistance can be ignored. We can apply Eqs. 2–11 for constant acceleration to objects that move up or down freely near the Earth's surface.

The slope of a curve at any point on a graph is the slope of the tangent to the curve at that point. On a graph of position vs. time, the **slope** is equal to the instantaneous velocity. On a graph of velocity vs. time, the slope is the acceleration.

# Questions

1. Does a car speedometer measure speed, velocity, or both? Explain.

2. When an object moves with constant velocity, does its average velocity during any time interval differ from its instantaneous velocity at any instant? Explain.

3. If one object has a greater speed than a second object, does the first necessarily have a greater acceleration? Explain, using examples.

4. Compare the acceleration of a motorcycle that accelerates from 80 km/h to 90 km/h with the acceleration of a bicycle that accelerates from rest to 10 km/h in the same time.

5. Can an object have a northward velocity and a southward acceleration? Explain.

6. Can the velocity of an object be negative when its acceleration is positive? What about vice versa? If yes, give examples in each case.

7. Give an example where both the velocity and acceleration are negative.

8. Can an object be increasing in speed as its acceleration decreases? If so, give an example. If not, explain.

9. Two cars emerge side by side from a tunnel. Car A is traveling with a speed of 60 km/h and has an acceleration of 40 km/h/min. Car B has a speed of 40 km/h and has an acceleration of 60 km/h/min. Which car is passing the other as they come out of the tunnel? Explain your reasoning.

10. A baseball player hits a ball straight up into the air. It leaves the bat with a speed of 120 km/h. In the absence of air resistance, how fast would the ball be traveling when it is caught at the same height above the ground as it left the bat? Explain.

11. As a freely falling object speeds up, what is happening to its acceleration—does it increase, decrease, or stay the same? (a) Ignore air resistance. (b) Consider air resistance.

12. You travel from point A to point B in a car moving at a constant speed of 70 km/h. Then you travel the same distance from point B to another point C, moving at a constant speed of 90 km/h. Is your average speed for the entire trip from A to C equal to 80 km/h? Explain why or why not.

13. Can an object have zero velocity and nonzero acceleration at the same time? Give examples.

14. Can an object have zero acceleration and nonzero velocity at the same time? Give examples.

15. Which of these motions is *not* at constant acceleration: a rock falling from a cliff, an elevator moving from the second floor to the fifth floor making stops along the way, a dish resting on a table? Explain your answers.

16. Describe in words the motion plotted in Fig. 2–32 in terms of velocity, acceleration, etc. [*Hint*: First try to duplicate the motion plotted by walking or moving your hand.]

**FIGURE 2–32** Question 16.

17. Describe in words the motion of the object graphed in Fig. 2–33.

**FIGURE 2–33** Question 17.

# MisConceptual Questions

[List all answers that are valid.]

1. Which of the following should be part of solving any problem in physics? Select all that apply:
   (a) Read the problem carefully.
   (b) Draw a picture of the situation.
   (c) Write down the variables that are given.
   (d) Think about which physics principles to apply.
   (e) Determine which equations can be used to apply the correct physics principles.
   (f) Check the units when you have completed your calculation.
   (g) Consider whether your answer is reasonable.

2. In which of the following cases does a car have a negative velocity and a positive acceleration? A car that is traveling in the
   (a) $-x$ direction at a constant 20 m/s.
   (b) $-x$ direction increasing in speed.
   (c) $+x$ direction increasing in speed.
   (d) $-x$ direction decreasing in speed.
   (e) $+x$ direction decreasing in speed.

3. At time $t = 0$ an object is traveling to the right along the $+x$ axis at a speed of 10.0 m/s with acceleration $-2.0$ m/s$^2$. Which statement is true?
   (a) The object will slow down, eventually coming to a complete stop.
   (b) The object cannot have a negative acceleration and be moving to the right.
   (c) The object will continue to move to the right, slowing down but never coming to a complete stop.
   (d) The object will slow down, momentarily stopping, then pick up speed moving to the left.

4. A ball is thrown straight up. What are the velocity and acceleration of the ball at the highest point in its path?
   (a) $v = 0$, $a = 0$.
   (b) $v = 0$, $a = 9.8$ m/s$^2$ up.
   (c) $v = 0$, $a = 9.8$ m/s$^2$ down.
   (d) $v = 9.8$ m/s up, $a = 0$.
   (e) $v = 9.8$ m/s down, $a = 0$.

5. You drop a rock off a bridge. When the rock has fallen 4 m, you drop a second rock. As the two rocks continue to fall, what happens to their velocities?
   (a) Both increase at the same rate.
   (b) The velocity of the first rock increases faster than the velocity of the second.
   (c) The velocity of the second rock increases faster than the velocity of the first.
   (d) Both velocities stay constant.

6. You drive 4 km at 30 km/h and then another 4 km at 50 km/h. What is your average speed for the whole 8-km trip?
   (a) More than 40 km/h.
   (b) Equal to 40 km/h.
   (c) Less than 40 km/h.
   (d) Not enough information.

7. A ball is dropped from the top of a tall building. At the same instant, a second ball is thrown upward from the ground level. When the two balls pass one another, one on the way up, the other on the way down, compare the magnitudes of their acceleration:
   (a) The acceleration of the dropped ball is greater.
   (b) The acceleration of the ball thrown upward is greater.
   (c) The acceleration of both balls is the same.
   (d) The acceleration changes during the motion, so you cannot predict the exact value when the two balls pass each other.
   (e) The accelerations are in opposite directions.

8. A ball is thrown downward at a speed of 20 m/s. Choosing the $+y$ axis pointing up and neglecting air resistance, which equation(s) could be used to solve for other variables? The acceleration due to gravity is $g = 9.8$ m/s$^2$ downward.
   (a) $v = (20 \text{ m/s}) - gt$.
   (b) $y = y_0 + (-20 \text{ m/s})t - (1/2)gt^2$.
   (c) $v^2 = (20 \text{ m/s})^2 - 2g(y - y_0)$.
   (d) $(20 \text{ m/s}) = (v + v_0)/2$.
   (e) All of the above.

9. A car travels along the $x$ axis with increasing speed. We don't know if to the left or the right. Which of the graphs in Fig. 2–34 most closely represents the motion of the car?

(a)

(b)

(c)

(d)

**FIGURE 2–34**
MisConceptual
Question 9.          (e)

# Problems

[*The Problems at the end of each Chapter are ranked* I, II, *or* III *according to estimated difficulty, with level* I *Problems being easiest. Level* III *are meant as challenges for the best students. The Problems are arranged by Section, meaning that the reader should have read up to and including that Section, but not only that Section—Problems often depend on earlier material. Next is a set of "General Problems" not arranged by Section and not ranked. Finally, there are "Search and Learn" Problems that require rereading parts of the Chapter and sometimes earlier Chapters.*]

(*Note: In Problems, assume a number like* 6.4 *is accurate to* ± 0.1; *and* 950 *is* ± 10 *unless* 950 *is said to be "precisely" or "very nearly"* 950, *in which case assume* 950 ± 1. *See Section 1–4.*)

## 2–1 to 2–3 Speed and Velocity

1. (I) If you are driving 95 km/h along a straight road and you look to the side for 2.0 s, how far do you travel during this inattentive period?

2. (I) What must your car's average speed be in order to travel 235 km in 2.75 h?

3. (I) A particle at $t_1 = -2.0$ s is at $x_1 = 4.8$ cm and at $t_2 = 4.5$ s is at $x_2 = 8.5$ cm. What is its average velocity over this time interval? Can you calculate its average speed from these data? Why or why not?

4. (I) A rolling ball moves from $x_1 = 8.4$ cm to $x_2 = -4.2$ cm during the time from $t_1 = 3.0$ s to $t_2 = 6.1$ s. What is its average velocity over this time interval?

5. (I) A bird can fly 25 km/h. How long does it take to fly 3.5 km?

6. (II) According to a rule-of-thumb, each five seconds between a lightning flash and the following thunder gives the distance to the flash in miles. (*a*) Assuming that the flash of light arrives in essentially no time at all, estimate the speed of sound in m/s from this rule. (*b*) What would be the rule for kilometers?

7. (II) You are driving home from school steadily at 95 km/h for 180 km. It then begins to rain and you slow to 65 km/h. You arrive home after driving 4.5 h. (*a*) How far is your hometown from school? (*b*) What was your average speed?

8. (II) A horse trots away from its trainer in a straight line, moving 38 m away in 9.0 s. It then turns abruptly and gallops halfway back in 1.8 s. Calculate (*a*) its average speed and (*b*) its average velocity for the entire trip, using "away from the trainer" as the positive direction.

9. (II) A person jogs eight complete laps around a 400-m track in a total time of 14.5 min. Calculate (*a*) the average speed and (*b*) the average velocity, in m/s.

10. (II) Every year the Earth travels about $10^9$ km as it orbits the Sun. What is Earth's average speed in km/h?

11. (II) A car traveling 95 km/h is 210 m behind a truck traveling 75 km/h. How long will it take the car to reach the truck?

12. (II) Calculate the average speed and average velocity of a complete round trip in which the outgoing 250 km is covered at 95 km/h, followed by a 1.0-h lunch break, and the return 250 km is covered at 55 km/h.

13. (II) Two locomotives approach each other on parallel tracks. Each has a speed of 155 km/h with respect to the ground. If they are initially 8.5 km apart, how long will it be before they reach each other? (See Fig. 2–35.)

**FIGURE 2–35** Problem 13.

14. (II) Digital bits on a 12.0-cm diameter audio CD are encoded along an outward spiraling path that starts at radius $R_1 = 2.5$ cm and finishes at radius $R_2 = 5.8$ cm. The distance between the centers of neighboring spiral-windings is 1.6 $\mu$m $(= 1.6 \times 10^{-6}$ m$)$. (*a*) Determine the total length of the spiraling path. [*Hint:* Imagine "unwinding" the spiral into a straight path of width 1.6 $\mu$m, and note that the original spiral and the straight path both occupy the same area.] (*b*) To read information, a CD player adjusts the rotation of the CD so that the player's readout laser moves along the spiral path at a constant speed of about 1.2 m/s. Estimate the maximum playing time of such a CD.

15. (III) A bowling ball traveling with constant speed hits the pins at the end of a bowling lane 16.5 m long. The bowler hears the sound of the ball hitting the pins 2.80 s after the ball is released from his hands. What is the speed of the ball, assuming the speed of sound is 340 m/s?

16. (III) An automobile traveling 95 km/h overtakes a 1.30-km-long train traveling in the same direction on a track parallel to the road. If the train's speed is 75 km/h, how long does it take the car to pass it, and how far will the car have traveled in this time? See Fig. 2–36. What are the results if the car and train are traveling in opposite directions?

**FIGURE 2–36** Problem 16.

## 2–4 Acceleration

17. (I) A sports car accelerates from rest to 95 km/h in 4.3 s. What is its average acceleration in m/s²?

18. (I) A sprinter accelerates from rest to 9.00 m/s in 1.38 s. What is her acceleration in (*a*) m/s²; (*b*) km/h²?

19. (II) A sports car moving at constant velocity travels 120 m in 5.0 s. If it then brakes and comes to a stop in 4.0 s, what is the magnitude of its acceleration (assumed constant) in m/s², and in g's $(g = 9.80$ m/s²)?

20. (II) At highway speeds, a particular automobile is capable of an acceleration of about $1.8 \text{ m/s}^2$. At this rate, how long does it take to accelerate from 65 km/h to 120 km/h?

21. (II) A car moving in a straight line starts at $x = 0$ at $t = 0$. It passes the point $x = 25.0 \text{ m}$ with a speed of 11.0 m/s at $t = 3.00 \text{ s}$. It passes the point $x = 385 \text{ m}$ with a speed of 45.0 m/s at $t = 20.0 \text{ s}$. Find (a) the average velocity, and (b) the average acceleration, between $t = 3.00 \text{ s}$ and $t = 20.0 \text{ s}$.

## 2–5 and 2–6 Motion at Constant Acceleration

22. (I) A car slows down from 28 m/s to rest in a distance of 88 m. What was its acceleration, assumed constant?

23. (I) A car accelerates from 14 m/s to 21 m/s in 6.0 s. What was its acceleration? How far did it travel in this time? Assume constant acceleration.

24. (I) A light plane must reach a speed of 35 m/s for takeoff. How long a runway is needed if the (constant) acceleration is $3.0 \text{ m/s}^2$?

25. (II) A baseball pitcher throws a baseball with a speed of 43 m/s. Estimate the average acceleration of the ball during the throwing motion. In throwing the baseball, the pitcher accelerates it through a displacement of about 3.5 m, from behind the body to the point where it is released (Fig. 2–37).

FIGURE 2–37 Problem 25.

26. (II) A world-class sprinter can reach a top speed (of about 11.5 m/s) in the first 18.0 m of a race. What is the average acceleration of this sprinter and how long does it take her to reach that speed?

27. (II) A car slows down uniformly from a speed of 28.0 m/s to rest in 8.00 s. How far did it travel in that time?

28. (II) In coming to a stop, a car leaves skid marks 65 m long on the highway. Assuming a deceleration of $4.00 \text{ m/s}^2$, estimate the speed of the car just before braking.

29. (II) A car traveling at 95 km/h strikes a tree. The front end of the car compresses and the driver comes to rest after traveling 0.80 m. What was the magnitude of the average acceleration of the driver during the collision? Express the answer in terms of "g's," where $1.00 g = 9.80 \text{ m/s}^2$.

30. (II) A car traveling 75 km/h slows down at a constant $0.50 \text{ m/s}^2$ just by "letting up on the gas." Calculate (a) the distance the car coasts before it stops, (b) the time it takes to stop, and (c) the distance it travels during the first and fifth seconds.

31. (II) Determine the stopping distances for an automobile going a constant initial speed of 95 km/h and human reaction time of 0.40 s: (a) for an acceleration $a = -3.0 \text{ m/s}^2$; (b) for $a = -6.0 \text{ m/s}^2$.

32. (II) A driver is traveling 18.0 m/s when she sees a red light ahead. Her car is capable of decelerating at a rate of $3.65 \text{ m/s}^2$. If it takes her 0.350 s to get the brakes on and she is 20.0 m from the intersection when she sees the light, will she be able to stop in time? How far from the beginning of the intersection will she be, and in what direction?

33. (II) A 75-m-long train begins uniform acceleration from rest. The front of the train has a speed of 18 m/s when it passes a railway worker who is standing 180 m from where the front of the train started. What will be the speed of the last car as it passes the worker? (See Fig. 2–38.)

FIGURE 2–38 Problem 33.

34. (II) A space vehicle accelerates uniformly from 85 m/s at $t = 0$ to 162 m/s at $t = 10.0 \text{ s}$. How far did it move between $t = 2.0 \text{ s}$ and $t = 6.0 \text{ s}$?

35. (II) A runner hopes to complete the 10,000-m run in less than 30.0 min. After running at constant speed for exactly 27.0 min, there are still 1200 m to go. The runner must then accelerate at $0.20 \text{ m/s}^2$ for how many seconds in order to achieve the desired time?

36. (III) A fugitive tries to hop on a freight train traveling at a constant speed of 5.0 m/s. Just as an empty box car passes him, the fugitive starts from rest and accelerates at $a = 1.4 \text{ m/s}^2$ to his maximum speed of 6.0 m/s, which he then maintains. (a) How long does it take him to catch up to the empty box car? (b) What is the distance traveled to reach the box car?

37. (III) Mary and Sally are in a foot race (Fig. 2–39). When Mary is 22 m from the finish line, she has a speed of 4.0 m/s and is 5.0 m behind Sally, who has a speed of 5.0 m/s. Sally thinks she has an easy win and so, during the remaining portion of the race, decelerates at a constant rate of $0.40 \text{ m/s}^2$ to the finish line. What constant acceleration does Mary now need during the remaining portion of the race, if she wishes to cross the finish line side-by-side with Sally?

FIGURE 2–39 Problem 37.

38. (III) An unmarked police car traveling a constant 95 km/h is passed by a speeder traveling 135 km/h. Precisely 1.00 s after the speeder passes, the police officer steps on the accelerator; if the police car's acceleration is $2.60 \text{ m/s}^2$, how much time passes before the police car overtakes the speeder (assumed moving at constant speed)?

## 2–7 Freely Falling Objects (neglect air resistance)

39. (I) A stone is dropped from the top of a cliff. It is seen to hit the ground below after 3.55 s. How high is the cliff?

40. (I) Estimate (a) how long it took King Kong to fall straight down from the top of the Empire State Building (380 m high), and (b) his velocity just before "landing."

41. (II) A ball player catches a ball 3.4 s after throwing it vertically upward. With what speed did he throw it, and what height did it reach?

42. (II) A baseball is hit almost straight up into the air with a speed of 25 m/s. Estimate (a) how high it goes, (b) how long it is in the air. (c) What factors make this an estimate?

43. (II) A kangaroo jumps straight up to a vertical height of 1.45 m. How long was it in the air before returning to Earth?

44. (II) The best rebounders in basketball have a vertical leap (that is, the vertical movement of a fixed point on their body) of about 120 cm. (a) What is their initial "launch" speed off the ground? (b) How long are they in the air?

45. (II) An object starts from rest and falls under the influence of gravity. Draw graphs of (a) its speed and (b) the distance it has fallen, as a function of time from $t = 0$ to $t = 5.00$ s. Ignore air resistance.

46. (II) A stone is thrown vertically upward with a speed of 24.0 m/s. (a) How fast is it moving when it is at a height of 13.0 m? (b) How much time is required to reach this height? (c) Why are there two answers to (b)?

47. (II) For an object falling freely from rest, show that the distance traveled *during* each successive second increases in the ratio of successive odd integers (1, 3, 5, etc.). (This was first shown by Galileo.) See Figs. 2–19 and 2–22.

48. (II) A rocket rises vertically, from rest, with an acceleration of 3.2 m/s² until it runs out of fuel at an altitude of 775 m. After this point, its acceleration is that of gravity, downward. (a) What is the velocity of the rocket when it runs out of fuel? (b) How long does it take to reach this point? (c) What maximum altitude does the rocket reach? (d) How much time (total) does it take to reach maximum altitude? (e) With what velocity does it strike the Earth? (f) How long (total) is it in the air?

49. (II) A helicopter is ascending vertically with a speed of 5.40 m/s. At a height of 105 m above the Earth, a package is dropped from the helicopter. How much time does it take for the package to reach the ground? [*Hint*: What is $v_0$ for the package?]

50. (II) Roger sees water balloons fall past his window. He notices that each balloon strikes the sidewalk 0.83 s after passing his window. Roger's room is on the third floor, 15 m above the sidewalk. (a) How fast are the balloons traveling when they pass Roger's window? (b) Assuming the balloons are being released from rest, from what floor are they being released? Each floor of the dorm is 5.0 m high.

51. (II) Suppose you adjust your garden hose nozzle for a fast stream of water. You point the nozzle vertically upward at a height of 1.8 m above the ground (Fig. 2–40). When you quickly turn off the nozzle, you hear the water striking the ground next to you for another 2.5 s. What is the water speed as it leaves the nozzle?

FIGURE 2–40
Problem 51.

52. (III) A baseball is seen to pass upward by a window with a vertical speed of 14 m/s. If the ball was thrown by a person 18 m below on the street, (a) what was its initial speed, (b) what altitude does it reach, (c) when was it thrown, and (d) when does it reach the street again?

53. (III) A falling stone takes 0.31 s to travel past a window 2.2 m tall (Fig. 2–41). From what height above the top of the window did the stone fall?

FIGURE 2–41
Problem 53.

54. (III) A rock is dropped from a sea cliff, and the sound of it striking the ocean is heard 3.4 s later. If the speed of sound is 340 m/s, how high is the cliff?

## 2–8 Graphical Analysis

55. (II) Figure 2–42 shows the velocity of a train as a function of time. (a) At what time was its velocity greatest? (b) During what periods, if any, was the velocity constant? (c) During what periods, if any, was the acceleration constant? (d) When was the magnitude of the acceleration greatest?

FIGURE 2–42 Problem 55.

56. (II) A sports car accelerates approximately as shown in the velocity–time graph of Fig. 2–43. (The short flat spots in the curve represent manual shifting of the gears.) Estimate the car's average acceleration in (a) second gear and (b) fourth gear.

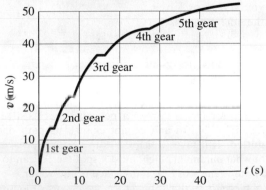

FIGURE 2–43 Problem 56. The velocity of a car as a function of time, starting from a dead stop. The flat spots in the curve represent gear shifts.

**57.** (II) The position of a rabbit along a straight tunnel as a function of time is plotted in Fig. 2–44. What is its instantaneous velocity (a) at $t = 10.0$ s and (b) at $t = 30.0$ s? What is its average velocity (c) between $t = 0$ and $t = 5.0$ s, (d) between $t = 25.0$ s and $t = 30.0$ s, and (e) between $t = 40.0$ s and $t = 50.0$ s?

**FIGURE 2–44** Problems 57, 58, and 59.

**58.** (II) In Fig. 2–44, (a) during what time periods, if any, is the velocity constant? (b) At what time is the velocity greatest? (c) At what time, if any, is the velocity zero? (d) Does the object move in one direction or in both directions during the time shown?

**59.** (III) Sketch the $v$ vs. $t$ graph for the object whose displacement as a function of time is given by Fig. 2–44.

# General Problems

**60.** The acceleration due to gravity on the Moon is about one-sixth what it is on Earth. If an object is thrown vertically upward on the Moon, how many times higher will it go than it would on Earth, assuming the same initial velocity?

**61.** A person who is properly restrained by an over-the-shoulder seat belt has a good chance of surviving a car collision if the deceleration does not exceed 30 "$g$'s" ($1.00\ g = 9.80\ \text{m/s}^2$). Assuming uniform deceleration at 30 $g$'s, calculate the distance over which the front end of the car must be designed to collapse if a crash brings the car to rest from 95 km/h.

**62.** A person jumps out a fourth-story window 18.0 m above a firefighter's safety net. The survivor stretches the net 1.0 m before coming to rest, Fig. 2–45. (a) What was the average deceleration experienced by the survivor when she was slowed to rest by the net? (b) What would you do to make it "safer" (that is, to generate a smaller deceleration): would you stiffen or loosen the net? Explain.

**FIGURE 2–45**
Problem 62.

18.0 m

1.0 m

**63.** Pelicans tuck their wings and free-fall straight down when diving for fish. Suppose a pelican starts its dive from a height of 14.0 m and cannot change its path once committed. If it takes a fish 0.20 s to perform evasive action, at what minimum height must it spot the pelican to escape? Assume the fish is at the surface of the water.

**64.** A bicyclist in the Tour de France crests a mountain pass as he moves at 15 km/h. At the bottom, 4.0 km farther, his speed is 65 km/h. Estimate his average acceleration (in $\text{m/s}^2$) while riding down the mountain.

**65.** Consider the street pattern shown in Fig. 2–46. Each intersection has a traffic signal, and the speed limit is 40 km/h. Suppose you are driving from the west at the speed limit. When you are 10.0 m from the first intersection, all the lights turn green. The lights are green for 13.0 s each. (a) Calculate the time needed to reach the third stoplight. Can you make it through all three lights without stopping? (b) Another car was stopped at the first light when all the lights turned green. It can accelerate at the rate of 2.00 $\text{m/s}^2$ to the speed limit. Can the second car make it through all three lights without stopping? By how many seconds would it make it, or not make it?

**FIGURE 2–46** Problem 65.

**66.** An airplane travels 2100 km at a speed of 720 km/h, and then encounters a tailwind that boosts its speed to 990 km/h for the next 2800 km. What was the total time for the trip? What was the average speed of the plane for this trip? [*Hint*: Does Eq. 2–11d apply?]

**67.** Suppose a car manufacturer tested its cars for front-end collisions by hauling them up on a crane and dropping them from a certain height. (a) Show that the speed just before a car hits the ground, after falling from rest a vertical distance $H$, is given by $\sqrt{2gH}$. What height corresponds to a collision at (b) 35 km/h? (c) 95 km/h?

**68.** A stone is dropped from the roof of a high building. A second stone is dropped 1.30 s later. How far apart are the stones when the second one has reached a speed of 12.0 m/s?

**69.** A person jumps off a diving board 4.0 m above the water's surface into a deep pool. The person's downward motion stops 2.0 m below the surface of the water. Estimate the average deceleration of the person while under the water.

**70.** In putting, the force with which a golfer strikes a ball is planned so that the ball will stop within some small distance of the cup, say 1.0 m long or short, in case the putt is missed. Accomplishing this from an uphill lie (that is, putting the ball downhill, see Fig. 2–47) is more difficult than from a downhill lie. To see why, assume that on a particular green the ball decelerates constantly at $1.8 \text{ m/s}^2$ going downhill, and constantly at $2.6 \text{ m/s}^2$ going uphill. Suppose we have an uphill lie 7.0 m from the cup. Calculate the allowable range of initial velocities we may impart to the ball so that it stops in the range 1.0 m short to 1.0 m long of the cup. Do the same for a downhill lie 7.0 m from the cup. What in your results suggests that the downhill putt is more difficult?

**FIGURE 2–47** Problem 70.

**71.** A stone is thrown vertically upward with a speed of 15.5 m/s from the edge of a cliff 75.0 m high (Fig. 2–48). (a) How much later does it reach the bottom of the cliff? (b) What is its speed just before hitting? (c) What total distance did it travel?

**FIGURE 2–48**
Problem 71.

**72.** In the design of a **rapid transit system**, it is necessary to balance the average speed of a train against the distance between station stops. The more stops there are, the slower the train's average speed. To get an idea of this problem, calculate the time it takes a train to make a 15.0-km trip in two situations: (a) the stations at which the trains must stop are 3.0 km apart (a total of 6 stations, including those at the ends); and (b) the stations are 5.0 km apart (4 stations total). Assume that at each station the train accelerates at a rate of $1.1 \text{ m/s}^2$ until it reaches 95 km/h, then stays at this speed until its brakes are applied for arrival at the next station, at which time it decelerates at $-2.0 \text{ m/s}^2$. Assume it stops at each intermediate station for 22 s.

**73.** A person driving her car at 35 km/h approaches an intersection just as the traffic light turns yellow. She knows that the yellow light lasts only 2.0 s before turning to red, and she is 28 m away from the near side of the intersection (Fig. 2–49). Should she try to stop, or should she speed up to cross the intersection before the light turns red? The intersection is 15 m wide. Her car's maximum deceleration is $-5.8 \text{ m/s}^2$, whereas it can accelerate from 45 km/h to 65 km/h in 6.0 s. Ignore the length of her car and her reaction time.

**FIGURE 2–49** Problem 73.

**74.** A car is behind a truck going 18 m/s on the highway. The car's driver looks for an opportunity to pass, guessing that his car can accelerate at $0.60 \text{ m/s}^2$ and that he has to cover the 20-m length of the truck, plus 10-m extra space at the rear of the truck and 10 m more at the front of it. In the oncoming lane, he sees a car approaching, probably at the speed limit, 25 m/s (55 mph). He estimates that the car is about 500 m away. Should he attempt the pass? Give details.

**75.** Agent Bond is standing on a bridge, 15 m above the road below, and his pursuers are getting too close for comfort. He spots a flatbed truck approaching at 25 m/s, which he measures by knowing that the telephone poles the truck is passing are 25 m apart in this region. The roof of the truck is 3.5 m above the road, and Bond quickly calculates how many poles away the truck should be when he drops down from the bridge onto the truck, making his getaway. How many poles is it?

**76.** A conveyor belt is used to send burgers through a grilling machine. If the grilling machine is 1.2 m long and the burgers require 2.8 min to cook, how fast must the conveyor belt travel? If the burgers are spaced 25 cm apart, what is the rate of burger production (in burgers/min)?

**77.** Two students are asked to find the height of a particular building using a barometer. Instead of using the barometer as an altitude measuring device, they take it to the roof of the building and drop it off, timing its fall. One student reports a fall time of 2.0 s, and the other, 2.3 s. What % difference does the 0.3 s make for the estimates of the building's height?

**78.** Figure 2–50 shows the position vs. time graph for two bicycles, A and B. (a) Identify any instant at which the two bicycles have the same velocity. (b) Which bicycle has the larger acceleration? (c) At which instant(s) are the bicycles passing each other? Which bicycle is passing the other? (d) Which bicycle has the larger instantaneous velocity? (e) Which bicycle has the larger average velocity?

**FIGURE 2–50**
Problem 78.

79. A race car driver must average 200.0 km/h over the course of a time trial lasting ten laps. If the first nine laps were done at an average speed of 196.0 km/h, what average speed must be maintained for the last lap?

80. Two children are playing on two trampolines. The first child bounces up one-and-a-half times higher than the second child. The initial speed up of the second child is 4.0 m/s. (a) Find the maximum height the second child reaches. (b) What is the initial speed of the first child? (c) How long was the first child in the air?

81. If there were no air resistance, how long would it take a free-falling skydiver to fall from a plane at 3200 m to an altitude of 450 m, where she will open her parachute? What would her speed be at 450 m? (In reality, the air resistance will restrict her speed to perhaps 150 km/h.)

82. You stand at the top of a cliff while your friend stands on the ground below you. You drop a ball from rest and see that she catches it 1.4 s later. Your friend then throws the ball up to you, such that it just comes to rest in your hand. What is the speed with which your friend threw the ball?

83. On an audio compact disc (CD), digital bits of information are encoded sequentially along a spiral path. Each bit occupies about $0.28\,\mu m$. A CD player's readout laser scans along the spiral's sequence of bits at a constant speed of about 1.2 m/s as the CD spins. (a) Determine the number $N$ of digital bits that a CD player reads every second. (b) The audio information is sent to each of the two loudspeakers 44,100 times per second. Each of these samplings requires 16 bits, and so you might expect the required bit rate for a CD player to be

$$N_0 = 2\left(44{,}100\,\frac{\text{samplings}}{\text{s}}\right)\left(16\,\frac{\text{bits}}{\text{sampling}}\right) = 1.4 \times 10^6\,\frac{\text{bits}}{\text{s}},$$

where the 2 is for the 2 loudspeakers (the 2 stereo channels). Note that $N_0$ is less than the number $N$ of bits actually read per second by a CD player. The excess number of bits $(= N - N_0)$ is needed for encoding and error-correction. What percentage of the bits on a CD are dedicated to encoding and error-correction?

# Search and Learn

1. Discuss two conditions given in Section 2–7 for being able to use a constant acceleration of magnitude $g = 9.8\ \text{m/s}^2$. Give an example in which one of these conditions would not be met and would not even be a reasonable approximation of motion.

2. In a lecture demonstration, a 3.0-m-long vertical string with ten bolts tied to it at equal intervals is dropped from the ceiling of the lecture hall. The string falls on a tin plate, and the class hears the clink of each bolt as it hits the plate. (a) The sounds will not occur at equal time intervals. Why? (b) Will the time between clinks increase or decrease as the string falls? (c) How could the bolts be tied so that the clinks occur at equal intervals? (Assume the string is vertical with the bottom bolt touching the tin plate when the string is released.)

3. A police car at rest is passed by a speeder traveling at a constant 140 km/h. The police officer takes off in hot pursuit and catches up to the speeder in 850 m, maintaining a constant acceleration. (a) Qualitatively plot the position vs. time graph for both cars from the police car's start to the catch-up point. Calculate (b) how long it took the police officer to overtake the speeder, (c) the required police car acceleration, and (d) the speed of the police car at the overtaking point.

4. Figure 2–51 is a position versus time graph for the motion of an object along the $x$ axis. Consider the time interval from A to B. (a) Is the object moving in the positive or negative $x$ direction? (b) Is the object speeding up or slowing down? (c) Is the acceleration of the object positive or negative? Now consider the time interval from D to E. (d) Is the object moving in the positive or negative $x$ direction? (e) Is the object speeding up or slowing down? (f) Is the acceleration of the object positive or negative? (g) Finally, answer these same three questions for the time interval from C to D.

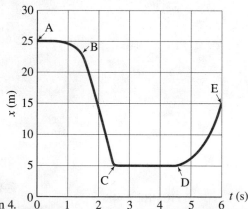

**FIGURE 2–51**
Search and Learn 4.

5. The position of a ball rolling in a straight line is given by $x = 2.0 - 3.6t + 1.7t^2$, where $x$ is in meters and $t$ in seconds. (a) What do the numbers 2.0, 3.6, and 1.7 refer to? (b) What are the units of each of these numbers? (c) Determine the position of the ball at $t = 1.0\,\text{s}$, $2.0\,\text{s}$, and $3.0\,\text{s}$. (d) What is the average velocity over the interval $t = 1.0\,\text{s}$ to $t = 3.0\,\text{s}$?

## ANSWERS TO EXERCISES

**A:** (a) displacement $= -30$ cm; (b) total distance $= 50$ cm.
**B:** (b).
**C:** (a) +; (b) −; (c) −; (d) +.

**D:** (b).
**E:** (e).
**F:** (c).

This snowboarder flying through the air shows an example of motion in two dimensions. In the absence of air resistance, the path would be a perfect parabola. The gold arrow represents the downward acceleration of gravity, $\vec{g}$. Galileo analyzed the motion of objects in 2 dimensions under the action of gravity near the Earth's surface (now called "projectile motion") into its horizontal and vertical components.

We will discuss vectors and how to add them. Besides analyzing projectile motion, we will also see how to work with relative velocity.

# Kinematics in Two Dimensions; Vectors

## CHAPTER-OPENING QUESTION—Guess now!

[*Don't worry about getting the right answer now—you will get another chance later in the Chapter. See also p. 1 of Chapter 1 for more explanation.*]

A small heavy box of emergency supplies is dropped from a moving helicopter at point A as it flies at constant speed in a horizontal direction. Which path in the drawing below best describes the path of the box (neglecting air resistance) as seen by a person standing on the ground?

## CONTENTS

I n Chapter 2 we dealt with motion along a straight line. We now consider the motion of objects that move in paths in two (or three) dimensions. In particular, we discuss an important type of motion known as *projectile motion*: objects projected outward near the Earth's surface, such as struck baseballs and golf balls, kicked footballs, and other projectiles. Before beginning our discussion of motion in two dimensions, we will need a new tool, vectors, and how to add them.

# 3–1 Vectors and Scalars

We mentioned in Chapter 2 that the term *velocity* refers not only to how fast an object is moving but also to its direction. A quantity such as velocity, which has *direction* as well as *magnitude*, is a **vector** quantity. Other quantities that are also vectors are displacement, force, and momentum. However, many quantities have no direction associated with them, such as mass, time, and temperature. They are specified completely by a number and units. Such quantities are called **scalar** quantities.

Drawing a diagram of a particular physical situation is always helpful in physics, and this is especially true when dealing with vectors. On a diagram, each vector is represented by an arrow. The arrow is always drawn so that it points in the direction of the vector quantity it represents. The length of the arrow is drawn proportional to the magnitude of the vector quantity. For example, in Fig. 3–1, green arrows have been drawn representing the velocity of a car at various places as it rounds a curve. The magnitude of the velocity at each point can be read off Fig. 3–1 by measuring the length of the corresponding arrow and using the scale shown (1 cm = 90 km/h).

When we write the symbol for a vector, we will always use boldface type, with a tiny arrow over the symbol. Thus for velocity we write $\vec{v}$. If we are concerned only with the magnitude of the vector, we will write simply $v$, in italics, as we do for other symbols.

Scale for velocity:
1 cm = 90 km/h

**FIGURE 3–1** Car traveling on a road, slowing down to round the curve. The green arrows represent the velocity vector at each position.

# 3–2 Addition of Vectors—Graphical Methods

Because vectors are quantities that have direction as well as magnitude, they must be added in a special way. In this Chapter, we will deal mainly with displacement vectors, for which we now use the symbol $\vec{D}$, and velocity vectors, $\vec{v}$. But the results will apply for other vectors we encounter later.

We use simple arithmetic for adding scalars. Simple arithmetic can also be used for adding vectors if they are in the same direction. For example, if a person walks 8 km east one day, and 6 km east the next day, the person will be 8 km + 6 km = 14 km east of the point of origin. We say that the *net* or *resultant* displacement is 14 km to the east (Fig. 3–2a). If, on the other hand, the person walks 8 km east on the first day, and 6 km west (in the reverse direction) on the second day, then the person will end up 2 km from the origin (Fig. 3–2b), so the resultant displacement is 2 km to the east. In this case, the resultant displacement is obtained by subtraction: 8 km − 6 km = 2 km.

But simple arithmetic cannot be used if the two vectors are not along the same line. For example, suppose a person walks 10.0 km east and then walks 5.0 km north. These displacements can be represented on a graph in which the positive $y$ axis points north and the positive $x$ axis points east, Fig. 3–3. On this graph, we draw an arrow, labeled $\vec{D}_1$, to represent the 10.0-km displacement to the east. Then we draw a second arrow, $\vec{D}_2$, to represent the 5.0-km displacement to the north. Both vectors are drawn to scale, as in Fig. 3–3.

**FIGURE 3–2** Combining vectors in one dimension.

(a)

(b)

**FIGURE 3–3** A person walks 10.0 km east and then 5.0 km north. These two displacements are represented by the vectors $\vec{D}_1$ and $\vec{D}_2$, which are shown as arrows. Also shown is the resultant displacement vector, $\vec{D}_R$, which is the vector sum of $\vec{D}_1$ and $\vec{D}_2$. Measurement on the graph with ruler and protractor shows that $\vec{D}_R$ has a magnitude of 11.2 km and points at an angle $\theta = 27°$ north of east.

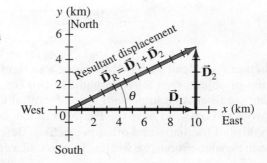

After taking this walk, the person is now 10.0 km east and 5.0 km north of the point of origin. The **resultant displacement** is represented by the arrow labeled $\vec{\mathbf{D}}_R$ in Fig. 3–3. (The subscript R stands for resultant.) Using a ruler and a protractor, you can measure on this diagram that the person is 11.2 km from the origin at an angle $\theta = 27°$ north of east. In other words, the resultant displacement vector has a magnitude of 11.2 km and makes an angle $\theta = 27°$ with the positive $x$ axis. The magnitude (length) of $\vec{\mathbf{D}}_R$ can also be obtained using the theorem of Pythagoras in this case, because $D_1$, $D_2$, and $D_R$ form a right triangle with $D_R$ as the hypotenuse. Thus

$$D_R = \sqrt{D_1^2 + D_2^2} = \sqrt{(10.0 \text{ km})^2 + (5.0 \text{ km})^2}$$
$$= \sqrt{125 \text{ km}^2} = 11.2 \text{ km}.$$

You can use the Pythagorean theorem only when the vectors are *perpendicular* to each other.

The resultant displacement vector, $\vec{\mathbf{D}}_R$, is the sum of the vectors $\vec{\mathbf{D}}_1$ and $\vec{\mathbf{D}}_2$. That is,

$$\vec{\mathbf{D}}_R = \vec{\mathbf{D}}_1 + \vec{\mathbf{D}}_2.$$

This is a *vector* equation. An important feature of adding two vectors that are not along the same line is that the magnitude of the resultant vector is not equal to the sum of the magnitudes of the two separate vectors, but is smaller than their sum. That is,

$$D_R \leq (D_1 + D_2),$$

where the equals sign applies only if the two vectors point in the same direction. In our example (Fig. 3–3), $D_R = 11.2$ km, whereas $D_1 + D_2$ equals 15 km, which is the total distance traveled. Note also that we cannot set $\vec{\mathbf{D}}_R$ equal to 11.2 km, because we have a vector equation and 11.2 km is only a part of the resultant vector, its magnitude. We could write something like this, though: $\vec{\mathbf{D}}_R = \vec{\mathbf{D}}_1 + \vec{\mathbf{D}}_2 = (11.2 \text{ km}, 27° \text{ N of E})$.

Figure 3–3 illustrates the general rules for graphically adding two vectors together, no matter what angles they make, to get their sum. The rules are as follows:

1. On a diagram, draw one of the vectors—call it $\vec{\mathbf{D}}_1$—to scale.
2. Next draw the second vector, $\vec{\mathbf{D}}_2$, to scale, placing its tail at the tip of the first vector and being sure its direction is correct.
3. The arrow drawn from the tail of the first vector to the tip of the second vector represents the *sum*, or **resultant**, of the two vectors.

The length of the resultant vector represents its magnitude. Note that vectors can be moved parallel to themselves on paper (maintaining the same length and angle) to accomplish these manipulations. The length of the resultant can be measured with a ruler and compared to the scale. Angles can be measured with a protractor. This method is known as the **tail-to-tip method of adding vectors**.

The resultant is not affected by the order in which the vectors are added. For example, a displacement of 5.0 km north, to which is added a displacement of 10.0 km east, yields a resultant of 11.2 km and angle $\theta = 27°$ (see Fig. 3–4), the same as when they were added in reverse order (Fig. 3–3). That is, now using $\vec{\mathbf{V}}$ to represent any type of vector,

$$\vec{\mathbf{V}}_1 + \vec{\mathbf{V}}_2 = \vec{\mathbf{V}}_2 + \vec{\mathbf{V}}_1.$$

[Mathematicians call this equation the *commutative* property of vector addition.]

**FIGURE 3–4** If the vectors are added in reverse order, the resultant is the same. (Compare to Fig. 3–3.)

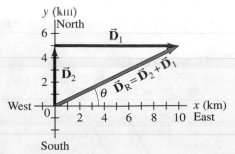

The tail-to-tip method of adding vectors can be extended to three or more vectors. The resultant is drawn from the tail of the first vector to the tip of the last one added. An example is shown in Fig. 3–5; the three vectors could represent displacements (northeast, south, west) or perhaps three forces. Check for yourself that you get the same resultant no matter in which order you add the three vectors.

**FIGURE 3–5** The resultant of three vectors: $\vec{V}_R = \vec{V}_1 + \vec{V}_2 + \vec{V}_3$.

A second way to add two vectors is the **parallelogram method**. It is fully equivalent to the tail-to-tip method. In this method, the two vectors are drawn starting from a common origin, and a parallelogram is constructed using these two vectors as adjacent sides as shown in Fig. 3–6b. The resultant is the diagonal drawn from the common origin. In Fig. 3–6a, the tail-to-tip method is shown, and we can see that both methods yield the same result.

**FIGURE 3–6** Vector addition by two different methods, (a) and (b). Part (c) is incorrect.

(a) Tail-to-tip

(b) Parallelogram

(c) Wrong

◆ **CAUTION**

*Be sure to use the correct diagonal on the parallelogram to get the resultant*

It is a common error to draw the sum vector as the diagonal running between the tips of the two vectors, as in Fig. 3–6c. *This is incorrect*: it does not represent the sum of the two vectors. (In fact, it represents their difference, $\vec{V}_2 - \vec{V}_1$, as we will see in the next Section.)

**CONCEPTUAL EXAMPLE 3–1** **Range of vector lengths.** Suppose two vectors each have length 3.0 units. What is the range of possible lengths for the vector representing the sum of the two?

**RESPONSE** The sum can take on any value from 6.0 (= 3.0 + 3.0) where the vectors point in the same direction, to 0 (= 3.0 − 3.0) when the vectors are antiparallel. Magnitudes between 0 and 6.0 occur when the two vectors are at an angle other than 0° and 180°.

**EXERCISE A** If the two vectors of Example 3–1 are perpendicular to each other, what is the resultant vector length?

# 3–3 Subtraction of Vectors, and Multiplication of a Vector by a Scalar

**FIGURE 3–7** The negative of a vector is a vector having the same length but opposite direction.

Given a vector $\vec{V}$, we define the *negative* of this vector $(-\vec{V})$ to be a vector with the same magnitude as $\vec{V}$ but opposite in direction, Fig. 3–7. Note, however, that no vector is ever negative in the sense of its magnitude: the magnitude of every vector is positive. Rather, a minus sign tells us about its direction.

We can now define the subtraction of one vector from another: the difference between two vectors $\vec{V}_2 - \vec{V}_1$ is defined as

$$\vec{V}_2 - \vec{V}_1 = \vec{V}_2 + (-\vec{V}_1).$$

That is, the difference between two vectors is equal to the sum of the first plus the negative of the second. Thus our rules for addition of vectors can be applied as shown in Fig. 3–8 using the tail-to-tip method.

$$\vec{V}_2 \Big/ - \xrightarrow{\vec{V}_1} \;=\; \vec{V}_2 \Big/ + \xleftarrow{-\vec{V}_1} \;=\; \vec{V}_2 - \vec{V}_1 \; \vec{V}_2$$

**FIGURE 3–8** Subtracting two vectors: $\vec{V}_2 - \vec{V}_1$.

A vector $\vec{V}$ can be multiplied by a scalar $c$. We define their product so that $c\vec{V}$ has the same direction as $\vec{V}$ and has magnitude $cV$. That is, multiplication of a vector by a positive scalar $c$ changes the magnitude of the vector by a factor $c$ but doesn't alter the direction. If $c$ is a negative scalar (such as $-2.0$), the magnitude of the product $c\vec{V}$ is changed by the factor $|c|$ (where $|c|$ means the magnitude of $c$), but the direction is precisely opposite to that of $\vec{V}$. See Fig. 3–9.

**EXERCISE B** What does the "incorrect" vector in Fig. 3–6c represent? (a) $\vec{V}_2 - \vec{V}_1$; (b) $\vec{V}_1 - \vec{V}_2$; (c) something else (specify).

**FIGURE 3–9** Multiplying a vector $\vec{V}$ by a scalar $c$ gives a vector whose magnitude is $c$ times greater and in the same direction as $\vec{V}$ (or opposite direction if $c$ is negative).

$$\vec{V}_2 = 1.5\,\vec{V}$$
$$\vec{V} \qquad \vec{V}_3 = -2.0\,\vec{V}$$

# 3–4 Adding Vectors by Components

Adding vectors graphically using a ruler and protractor is often not sufficiently accurate and is not useful for vectors in three dimensions. We discuss now a more powerful and precise method for adding vectors. But do not forget graphical methods—they are useful for visualizing, for checking your math, and thus for getting the correct result.

## Components

Consider first a vector $\vec{V}$ that lies in a particular plane. It can be expressed as the sum of two other vectors, called the **components** of the original vector. The components are usually chosen to be along two perpendicular directions, such as the $x$ and $y$ axes. The process of finding the components is known as **resolving the vector into its components**. An example is shown in Fig. 3–10; the vector $\vec{V}$ could be a displacement vector that points at an angle $\theta = 30°$ north of east, where we have chosen the positive $x$ axis to be to the east and the positive $y$ axis north. This vector $\vec{V}$ is resolved into its $x$ and $y$ components by drawing dashed lines (AB and AC) out from the tip (A) of the vector, making them perpendicular to the $x$ and $y$ axes. Then the lines 0B and 0C represent the $x$ and $y$ components of $\vec{V}$, respectively, as shown in Fig. 3–10b. These *vector components* are written $\vec{V}_x$ and $\vec{V}_y$. In this book we usually show vector components as arrows, like vectors, but dashed. The *scalar components*, $V_x$ and $V_y$, are the magnitudes of the vector components, with units, accompanied by a positive or negative sign depending on whether they point along the positive or negative $x$ or $y$ axis. As can be seen in Fig. 3–10, $\vec{V}_x + \vec{V}_y = \vec{V}$ by the parallelogram method of adding vectors.

Space is made up of three dimensions, and sometimes it is necessary to resolve a vector into components along three mutually perpendicular directions. In rectangular coordinates the components are $\vec{V}_x$, $\vec{V}_y$, and $\vec{V}_z$.

(a)

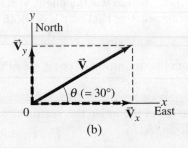

(b)

**FIGURE 3–10** Resolving a vector $\vec{V}$ into its components along a chosen set of $x$ and $y$ axes. The components, once found, themselves represent the vector. That is, the components contain as much information as the vector itself.

To add vectors using the method of components, we need to use the trigonometric functions sine, cosine, and tangent, which we now review.

Given any angle $\theta$, as in Fig. 3–11a, a right triangle can be constructed by drawing a line perpendicular to one of its sides, as in Fig. 3–11b. The longest side of a right triangle, opposite the right angle, is called the hypotenuse, which we label $h$. The side opposite the angle $\theta$ is labeled $o$, and the side adjacent is labeled $a$. We let $h$, $o$, and $a$ represent the lengths of these sides, respectively.

**FIGURE 3–11** Starting with an angle $\theta$ as in (a), we can construct right triangles of different sizes, (b) and (c), but the ratio of the lengths of the sides does not depend on the size of the triangle.

(a)        (b)        (c)

We now define the three trigonometric functions, sine, cosine, and tangent (abbreviated sin, cos, tan), in terms of the right triangle, as follows:

$$\sin \theta = \frac{\text{side opposite}}{\text{hypotenuse}} = \frac{o}{h}$$

$$\cos \theta = \frac{\text{side adjacent}}{\text{hypotenuse}} = \frac{a}{h} \tag{3–1}$$

$$\tan \theta = \frac{\text{side opposite}}{\text{side adjacent}} = \frac{o}{a}.$$

If we make the triangle bigger, but keep the same angles, then the ratio of the length of one side to the other, or of one side to the hypotenuse, remains the same. That is, in Fig. 3–11c we have: $a/h = a'/h'$; $o/h = o'/h'$; and $o/a = o'/a'$. Thus the values of sine, cosine, and tangent do not depend on how big the triangle is. They depend only on the size of the angle. The values of sine, cosine, and tangent for different angles can be found using a scientific calculator, or from the Table in Appendix A.

A useful trigonometric identity is

$$\sin^2 \theta + \cos^2 \theta = 1 \tag{3–2}$$

which follows from the Pythagorean theorem ($o^2 + a^2 = h^2$ in Fig. 3–11). That is:

$$\sin^2 \theta + \cos^2 \theta = \frac{o^2}{h^2} + \frac{a^2}{h^2} = \frac{o^2 + a^2}{h^2} = \frac{h^2}{h^2} = 1.$$

(See Appendix A and inside the rear cover for other details on trigonometric functions and identities.)

The use of trigonometric functions for finding the components of a vector is illustrated in Fig. 3–12, where a vector and its two components are thought of as making up a right triangle. We then see that the sine, cosine, and tangent are as given in Fig. 3–12, where $\theta$ is the angle $\vec{V}$ makes with the $+x$ axis. If we multiply the definition of $\sin \theta = V_y/V$ by $V$ on both sides, we get

$$V_y = V \sin \theta. \tag{3–3a}$$

Similarly, from the definition of $\cos \theta$, we obtain

$$V_x = V \cos \theta. \tag{3–3b}$$

Note that if $\theta$ is not the angle the vector makes with the positive $x$ axis, Eqs. 3–3 are not valid.

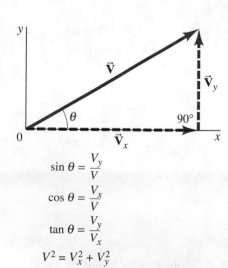

$$\sin \theta = \frac{V_y}{V}$$

$$\cos \theta = \frac{V_x}{V}$$

$$\tan \theta = \frac{V_y}{V_x}$$

$$V^2 = V_x^2 + V_y^2$$

**FIGURE 3–12** Finding the components of a vector using trigonometric functions. The equations are valid only if $\theta$ is the angle $\vec{V}$ makes with the positive $x$ axis.

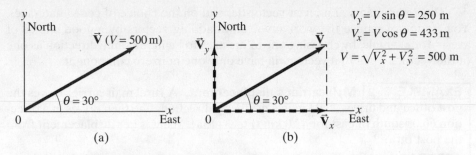

FIGURE 3–13 (a) Vector $\vec{V}$ represents a displacement of 500 m at a 30° angle north of east. (b) The components of $\vec{V}$ are $\vec{V}_x$ and $\vec{V}_y$, whose magnitudes are given on the right in the diagram.

Using Eqs. 3–3, we can calculate $V_x$ and $V_y$ for any vector, such as that illustrated in Fig. 3–10 or Fig. 3–12. Suppose $\vec{V}$ represents a displacement of 500 m in a direction 30° north of east, as shown in Fig. 3–13. Then $V = 500$ m. From a calculator or Tables, $\sin 30° = 0.500$ and $\cos 30° = 0.866$. Then

$$V_x = V \cos \theta = (500\,\text{m})(0.866) = 433\,\text{m (east)},$$
$$V_y = V \sin \theta = (500\,\text{m})(0.500) = 250\,\text{m (north)}.$$

There are two ways to specify a vector in a given coordinate system:

1. We can give its components, $V_x$ and $V_y$.
2. We can give its magnitude $V$ and the angle $\theta$ it makes with the positive $x$ axis.

We can shift from one description to the other using Eqs. 3–3, and, for the reverse, by using the theorem of Pythagoras[†] and the definition of tangent:

$$V = \sqrt{V_x^2 + V_y^2} \tag{3–4a}$$
$$\tan \theta = \frac{V_y}{V_x} \tag{3–4b}$$

as can be seen in Fig. 3–12.

## Adding Vectors

We can now discuss how to add vectors using components. The first step is to resolve each vector into its components. Next we can see, using Fig. 3–14, that the addition of any two vectors $\vec{V}_1$ and $\vec{V}_2$ to give a resultant, $\vec{V}_R = \vec{V}_1 + \vec{V}_2$, implies that

$$\begin{aligned} V_{Rx} &= V_{1x} + V_{2x} \\ V_{Ry} &= V_{1y} + V_{2y}. \end{aligned} \tag{3–5}$$

That is, the sum of the $x$ components equals the $x$ component of the resultant vector, and the sum of the $y$ components equals the $y$ component of the resultant, as can be verified by a careful examination of Fig. 3–14. Note that we do *not* add $x$ components to $y$ components.

If the magnitude and direction of the resultant vector are desired, they can be obtained using Eqs. 3–4.

[†]In three dimensions, the theorem of Pythagoras becomes $V = \sqrt{V_x^2 + V_y^2 + V_z^2}$, where $V_z$ is the component along the third, or $z$, axis.

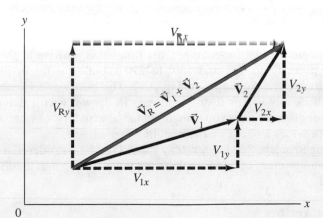

FIGURE 3–14 The components of $\vec{V}_R = \vec{V}_1 + \vec{V}_2$ are $V_{Rx} = V_{1x} + V_{2x}$ and $V_{Ry} = V_{1y} + V_{2y}$.

The components of a given vector depend on the choice of coordinate axes. You can often reduce the work involved in adding vectors by a good choice of axes—for example, by choosing one of the axes to be in the same direction as one of the vectors. Then that vector will have only one nonzero component.

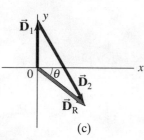

**FIGURE 3–15** Example 3–2. (a) The two displacement vectors, $\vec{D}_1$ and $\vec{D}_2$. (b) $\vec{D}_2$ is resolved into its components. (c) $\vec{D}_1$ and $\vec{D}_2$ are added to obtain the resultant $\vec{D}_R$. The component method of adding the vectors is explained in the Example.

**EXAMPLE 3–2** **Mail carrier's displacement.** A rural mail carrier leaves the post office and drives 22.0 km in a northerly direction. She then drives in a direction 60.0° south of east for 47.0 km (Fig. 3–15a). What is her displacement from the post office?

**APPROACH** We choose the positive $x$ axis to be east and the positive $y$ axis to be north, since those are the compass directions used on most maps. The origin of the $xy$ coordinate system is at the post office. We resolve each vector into its $x$ and $y$ components. We add the $x$ components together, and then the $y$ components together, giving us the $x$ and $y$ components of the resultant.

**SOLUTION** Resolve each displacement vector into its components, as shown in Fig. 3–15b. Since $\vec{D}_1$ has magnitude 22.0 km and points north, it has only a $y$ component:

$$D_{1x} = 0, \qquad D_{1y} = 22.0 \text{ km}.$$

$\vec{D}_2$ has both $x$ and $y$ components:

$$D_{2x} = +(47.0 \text{ km})(\cos 60°) = +(47.0 \text{ km})(0.500) = +23.5 \text{ km}$$
$$D_{2y} = -(47.0 \text{ km})(\sin 60°) = -(47.0 \text{ km})(0.866) = -40.7 \text{ km}.$$

Notice that $D_{2y}$ is negative because this vector component points along the negative $y$ axis. The resultant vector, $\vec{D}_R$, has components:

$$D_{Rx} = D_{1x} + D_{2x} = 0 \text{ km} + 23.5 \text{ km} = +23.5 \text{ km}$$
$$D_{Ry} = D_{1y} + D_{2y} = 22.0 \text{ km} + (-40.7 \text{ km}) = -18.7 \text{ km}.$$

This specifies the resultant vector completely:

$$D_{Rx} = 23.5 \text{ km}, \qquad D_{Ry} = -18.7 \text{ km}.$$

We can also specify the resultant vector by giving its magnitude and angle using Eqs. 3–4:

$$D_R = \sqrt{D_{Rx}^2 + D_{Ry}^2} = \sqrt{(23.5 \text{ km})^2 + (-18.7 \text{ km})^2} = 30.0 \text{ km}$$
$$\tan \theta = \frac{D_{Ry}}{D_{Rx}} = \frac{-18.7 \text{ km}}{23.5 \text{ km}} = -0.796.$$

A calculator with a key labeled INV TAN, or ARC TAN, or TAN$^{-1}$ gives $\theta = \tan^{-1}(-0.796) = -38.5°$. The negative sign means $\theta = 38.5°$ below the $x$ axis, Fig. 3–15c. So, the resultant displacement is 30.0 km directed at 38.5° in a southeasterly direction.

**NOTE** Always be attentive about the quadrant in which the resultant vector lies. An electronic calculator does not fully give this information, but a good diagram does.

**PROBLEM SOLVING**
*Identify the correct quadrant by drawing a careful diagram*

As we saw in Example 3–2, any component that points along the negative $x$ or $y$ axis gets a minus sign. The signs of trigonometric functions depend on which "quadrant" the angle falls in: for example, the tangent is positive in the first and third quadrants (from 0° to 90°, and 180° to 270°), but negative in the second and fourth quadrants; see Appendix A, Fig. A–7. The best way to keep track of angles, and to check any vector result, is always to draw a vector diagram, like Fig. 3–15. A vector diagram gives you something tangible to look at when analyzing a problem, and provides a check on the results.

The following Problem Solving Strategy should not be considered a prescription. Rather it is a summary of things to do to get you thinking and involved in the problem at hand.

## Adding Vectors

Here is a brief summary of how to add two or more vectors using components:

1. **Draw a diagram**, adding the vectors graphically by either the parallelogram or tail-to-tip method.
2. **Choose $x$ and $y$ axes.** Choose them in a way, if possible, that will make your work easier. (For example, choose one axis along the direction of one of the vectors, which then will have only one component.)
3. **Resolve** each vector into its $x$ and $y$ **components**, showing each component along its appropriate ($x$ or $y$) axis as a (dashed) arrow.
4. **Calculate each component** (when not given) using sines and cosines. If $\theta_1$ is the angle that vector $\vec{V}_1$ makes with the positive $x$ axis, then:
$$V_{1x} = V_1 \cos\theta_1, \qquad V_{1y} = V_1 \sin\theta_1.$$

Pay careful attention to **signs**: any component that points along the negative $x$ or $y$ axis gets a minus sign.

5. **Add** the $x$ **components** together to get the $x$ component of the resultant. Similarly for $y$:
$$V_{Rx} = V_{1x} + V_{2x} + \text{any others}$$
$$V_{Ry} = V_{1y} + V_{2y} + \text{any others}.$$

This is the answer: the components of the resultant vector. Check signs to see if they fit the quadrant shown in your diagram (point 1 above).

6. If you want to know the **magnitude and direction** of the resultant vector, use Eqs. 3–4:
$$V_R = \sqrt{V_{Rx}^2 + V_{Ry}^2}, \qquad \tan\theta = \frac{V_{Ry}}{V_{Rx}}.$$

The vector diagram you already drew helps to obtain the correct position (quadrant) of the angle $\theta$.

---

**EXAMPLE 3–3** **Three short trips.** An airplane trip involves three legs, with two stopovers, as shown in Fig. 3–16a. The first leg is due east for 620 km; the second leg is southeast (45°) for 440 km; and the third leg is at 53° south of west, for 550 km, as shown. What is the plane's total displacement?

**APPROACH** We follow the steps in the Problem Solving Strategy above.

**SOLUTION**

1. **Draw a diagram** such as Fig. 3–16a, where $\vec{D}_1$, $\vec{D}_2$, and $\vec{D}_3$ represent the three legs of the trip, and $\vec{D}_R$ is the plane's total displacement.
2. **Choose axes**: Axes are also shown in Fig. 3–16a: $x$ is east, $y$ north.
3. **Resolve components**: It is imperative to draw a good diagram. The components are drawn in Fig. 3–16b. Instead of drawing all the vectors starting from a common origin, as we did in Fig. 3–15b, here we draw them "tail-to-tip" style, which is just as valid and may make it easier to see.
4. **Calculate the components**:
$$\vec{D}_1: \quad D_{1x} = +D_1\cos 0° = D_1 = 620\text{ km}$$
$$D_{1y} = +D_1\sin 0° = 0\text{ km}$$
$$\vec{D}_2: \quad D_{2x} = +D_2\cos 45° = +(440\text{ km})(0.707) = +311\text{ km}$$
$$D_{2y} = -D_2\sin 45° = -(440\text{ km})(0.707) = -311\text{ km}$$
$$\vec{D}_3: \quad D_{3x} = -D_3\cos 53° = -(550\text{ km})(0.602) = -331\text{ km}$$
$$D_{3y} = -D_3\sin 53° = -(550\text{ km})(0.799) = -439\text{ km}.$$

We have given a minus sign to each component that in Fig. 3–16b points in the $-x$ or $-y$ direction. The components are shown in the Table in the margin.

5. **Add the components**: We add the $x$ components together, and we add the $y$ components together to obtain the $x$ and $y$ components of the resultant:
$$D_{Rx} = D_{1x} + D_{2x} + D_{3x} = 620\text{ km} + 311\text{ km} - 331\text{ km} = 600\text{ km}$$
$$D_{Ry} = D_{1y} + D_{2y} + D_{3y} = 0\text{ km} - 311\text{ km} - 439\text{ km} = -750\text{ km}.$$
The $x$ and $y$ components of the resultant are 600 km and $-750$ km, and point respectively to the east and south. This is one way to give the answer.

6. **Magnitude and direction**: We can also give the answer as
$$D_R = \sqrt{D_{Rx}^2 + D_{Ry}^2} = \sqrt{(600)^2 + (-750)^2}\text{ km} = 960\text{ km}$$
$$\tan\theta = \frac{D_{Ry}}{D_{Rx}} = \frac{-750\text{ km}}{600\text{ km}} = -1.25, \qquad \text{so } \theta = -51°.$$

Thus, the total displacement has magnitude 960 km and points 51° below the $x$ axis (south of east), as was shown in our original sketch, Fig. 3–16a.

**FIGURE 3–16** Example 3–3.

| Vector | Components | |
|--------|-----------|------------|
|        | $x$ (km)  | $y$ (km)   |
| $\vec{D}_1$ | 620  | 0     |
| $\vec{D}_2$ | 311  | $-311$ |
| $\vec{D}_3$ | $-331$ | $-439$ |
| $\vec{D}_R$ | 600  | $-750$ |

# 3–5 Projectile Motion

In Chapter 2, we studied the one-dimensional motion of an object in terms of displacement, velocity, and acceleration, including purely vertical motion of a falling object undergoing acceleration due to gravity. Now we examine the more general translational motion of objects moving through the air in two dimensions near the Earth's surface, such as a golf ball, a thrown or batted baseball, kicked footballs, and speeding bullets. These are all examples of **projectile motion** (see Fig. 3–17), which we can describe as taking place in two dimensions if there is no wind.

Although air resistance is often important, in many cases its effect can be ignored, and we will ignore it in the following analysis. We will not be concerned now with the process by which the object is thrown or projected. We consider only its motion *after* it has been projected, and *before* it lands or is caught—that is, we analyze our projected object only when it is moving freely through the air under the action of gravity alone. Then the acceleration of the object is that due to gravity, which acts downward with magnitude $g = 9.80 \text{ m/s}^2$, and we assume it is constant.[†]

Galileo was the first to describe projectile motion accurately. He showed that it could be understood by analyzing the horizontal and vertical components of the motion separately. For convenience, we assume that the motion begins at time $t = 0$ at the origin of an $xy$ coordinate system (so $x_0 = y_0 = 0$).

Let us look at a (tiny) ball rolling off the end of a horizontal table with an initial velocity in the horizontal ($x$) direction, $\vec{\mathbf{v}}_{x0}$. See Fig. 3–18, where an object falling vertically is also shown for comparison. The velocity vector $\vec{\mathbf{v}}$ at each instant points in the direction of the ball's motion at that instant and is thus always tangent to the path. Following Galileo's ideas, we treat the horizontal and vertical components of velocity and acceleration separately, and we can apply the kinematic equations (Eqs. 2–11a through 2–11c) to the $x$ and $y$ components of the motion.

First we examine the vertical ($y$) component of the motion. At the instant the ball leaves the table's top ($t = 0$), it has only an $x$ component of velocity. Once the ball leaves the table (at $t = 0$), it experiences a vertically downward acceleration $g$, the acceleration due to gravity. Thus $v_y$ is initially zero $(v_{y0} = 0)$ but increases continually in the downward direction (until the ball hits the ground). Let us take $y$ to be positive upward. Then the acceleration due to gravity is in the $-y$ direction, so $a_y = -g$. From Eq. 2–11a (using $y$ in place of $x$) we can write $v_y = v_{y0} + a_y t = -gt$ since we set $v_{y0} = 0$. The vertical displacement is given by Eq. 2–11b written in terms of $y$: $y = y_0 + v_{y0} + \frac{1}{2} a_y t^2$. Given $y_0 = 0$, $v_{y0} = 0$, and $a_y = -g$, then $y = -\frac{1}{2}gt^2$.

[†]This restricts us to objects whose distance traveled and maximum height above the Earth are small compared to the Earth's radius (6400 km).

(a)

(b)

**FIGURE 3–17** Photographs of (a) a bouncing ball and (b) a thrown basketball, each showing the characteristic "parabolic" path of projectile motion.

**FIGURE 3–18** Projectile motion of a small ball projected horizontally with initial velocity $\vec{\mathbf{v}} = \vec{\mathbf{v}}_{x0}$. The dashed black line represents the path of the object. The velocity vector $\vec{\mathbf{v}}$ is in the direction of motion at each point, and thus is tangent to the path. The velocity vectors are green arrows, and velocity components are dashed. (A vertically falling object starting from rest at the same place and time is shown at the left for comparison; $v_y$ is the same at each instant for the falling object and the projectile.)

In the horizontal direction, on the other hand, there is no acceleration (we are ignoring air resistance). With $a_x = 0$, the horizontal component of velocity, $v_x$, remains constant, equal to its initial value, $v_{x0}$, and thus has the same magnitude at each point on the path. The horizontal displacement (with $a_x = 0$) is given by $x = v_{x0}t + \frac{1}{2}a_x t^2 = v_{x0}t$. The two vector components, $\vec{v}_x$ and $\vec{v}_y$, can be added vectorially at any instant to obtain the velocity $\vec{v}$ at that time (that is, for each point on the path), as shown in Fig. 3–18.

One result of this analysis, which Galileo himself predicted, is that *an object projected horizontally will reach the ground in the same time as an object dropped vertically*. This is because the vertical motions are the same in both cases, as shown in Fig. 3–18. Figure 3–19 is a multiple-exposure photograph of an experiment that confirms this.

| **EXERCISE C** Two balls having different speeds roll off the edge of a horizontal table at the same time. Which hits the floor sooner, the faster ball or the slower one?

If an object is projected at an upward angle, as in Fig. 3–20, the analysis is similar, except that now there is an initial vertical component of velocity, $v_{y0}$. Because of the downward acceleration of gravity, the upward component of velocity $v_y$ gradually decreases with time until the object reaches the highest point on its path, at which point $v_y = 0$. Subsequently the object moves downward (Fig. 3–20) and $v_y$ increases in the downward direction, as shown (that is, becoming more negative). As before, $v_x$ remains constant.

**FIGURE 3–19** Multiple-exposure photograph showing positions of two balls at equal time intervals. One ball was dropped from rest at the same time the other ball was projected horizontally outward. The vertical position of each ball is seen to be the same at each instant.

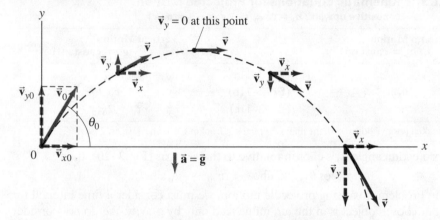

**FIGURE 3–20** Path of a projectile launched with initial velocity $\vec{v}_0$ at angle $\theta_0$ to the horizontal. Path is shown dashed in black, the velocity vectors are green arrows, and velocity components are dashed. The figure does not show where the projectile hits the ground (at that point, projectile motion ceases).

| **EXERCISE D** Where in Fig. 3–20 is (i) $\vec{v} = 0$, (ii) $v_y = 0$, and (iii) $v_x = 0$?

---

**CONCEPTUAL EXAMPLE 3–4** | **Where does the apple land?** A child sits upright in a wagon which is moving to the right at constant speed as shown in Fig. 3–21. The child extends her hand and throws an apple straight upward (from her own point of view, Fig. 3–21a), while the wagon continues to travel forward at constant speed. If air resistance is neglected, will the apple land (a) behind the wagon, (b) in the wagon, or (c) in front of the wagon?

**RESPONSE** The child throws the apple straight up from her own reference frame with initial velocity $\vec{v}_{y0}$ (Fig. 3–21a). But when viewed by someone on the ground, the apple also has an initial horizontal component of velocity equal to the speed of the wagon, $\vec{v}_{x0}$. Thus, to a person on the ground, the apple will follow the path of a projectile as shown in Fig. 3–21b. The apple experiences no horizontal acceleration, so $\vec{v}_{x0}$ will stay constant and equal to the speed of the wagon. As the apple follows its arc, the wagon will be directly under the apple at all times because they have the same horizontal velocity. When the apple comes down, it will drop right into the outstretched hand of the child. The answer is (b).

**FIGURE 3–21** Example 3–4.

(a) Wagon reference frame

(b) Ground reference frame

| **EXERCISE E** Return to the Chapter-Opening Question, page 49, and answer it again now. Try to explain why you may have answered differently the first time. Describe the role of the helicopter in this example of projectile motion.

# 3–6 Solving Projectile Motion Problems

We now work through several Examples of projectile motion quantitatively. We use the kinematic equations (2–11a through 2–11c) separately for the vertical and horizontal components of the motion. These equations are shown separately for the $x$ and $y$ components of the motion in Table 3–1, for the general case of two-dimensional motion at constant acceleration. Note that $x$ and $y$ are the respective displacements, that $v_x$ and $v_y$ are the components of the velocity, and that $a_x$ and $a_y$ are the components of the acceleration, each of which is constant. The subscript 0 means "at $t = 0$."

**TABLE 3–1  General Kinematic Equations for Constant Acceleration in Two Dimensions**

| $x$ component (horizontal) | | $y$ component (vertical) |
|---|---|---|
| $v_x = v_{x0} + a_x t$ | (Eq. 2–11a) | $v_y = v_{y0} + a_y t$ |
| $x = x_0 + v_{x0}t + \frac{1}{2}a_x t^2$ | (Eq. 2–11b) | $y = y_0 + v_{y0}t + \frac{1}{2}a_y t^2$ |
| $v_x^2 = v_{x0}^2 + 2a_x(x - x_0)$ | (Eq. 2–11c) | $v_y^2 = v_{y0}^2 + 2a_y(y - y_0)$ |

We can simplify Eqs. 2–11 to use for projectile motion because we can set $a_x = 0$. See Table 3–2, which assumes $y$ is positive upward, so $a_y = -g = -9.80 \text{ m/s}^2$.

**TABLE 3–2  Kinematic Equations for Projectile Motion**
(*y* positive upward; $a_x = 0$, $a_y = -g = -9.80 \text{ m/s}^2$)

| Horizontal Motion ($a_x = 0$, $v_x = $ constant) | | Vertical Motion† ($a_y = -g = $ constant) |
|---|---|---|
| $v_x = v_{x0}$ | (Eq. 2–11a) | $v_y = v_{y0} - gt$ |
| $x = x_0 + v_{x0}t$ | (Eq. 2–11b) | $y = y_0 + v_{y0}t - \frac{1}{2}gt^2$ |
| | (Eq. 2–11c) | $v_y^2 = v_{y0}^2 - 2g(y - y_0)$ |

† If $y$ is taken positive downward, the minus (−) signs in front of $g$ become + signs.

If the projection angle $\theta_0$ is chosen relative to the $+x$ axis (Fig. 3–20), then
$$v_{x0} = v_0 \cos\theta_0, \quad \text{and} \quad v_{y0} = v_0 \sin\theta_0.$$

**PROBLEM SOLVING**
*Choice of time interval*

In doing Problems involving projectile motion, we must consider a time interval for which our chosen object is in the air, influenced only by gravity. We do not consider the throwing (or projecting) process, nor the time after the object lands or is caught, because then other influences act on the object, and we can no longer set $\vec{a} = \vec{g}$.

---

**PROBLEM SOLVING**

## Projectile Motion

Our approach to solving Problems in Section 2–6 also applies here. Solving Problems involving projectile motion can require creativity, and cannot be done just by following some rules. Certainly you must avoid just plugging numbers into equations that seem to "work."

1. As always, **read** carefully; **choose** the **object** (or objects) you are going to analyze.
2. **Draw** a careful **diagram** showing what is happening to the object.
3. **Choose** an origin and an $xy$ **coordinate system**.
4. Decide on the **time interval**, which for projectile motion can only include motion under the effect of gravity alone, not throwing or landing. The time interval must be the same for the $x$ and $y$ analyses. The $x$ and $y$ motions are connected by the common time, $t$.

5. **Examine** the horizontal ($x$) and vertical ($y$) **motions** separately. If you are given the initial velocity, you may want to resolve it into its $x$ and $y$ components.
6. List the **known** and **unknown** quantities, choosing $a_x = 0$ and $a_y = -g$ or $+g$, where $g = 9.80 \text{ m/s}^2$, and using the + or − sign, depending on whether you choose $y$ positive up or down. Remember that $v_x$ never changes throughout the trajectory, and that $v_y = 0$ at the highest point of any trajectory that returns downward. The velocity just before landing is generally not zero.
7. Think for a minute before jumping into the equations. A little planning goes a long way. **Apply** the **relevant equations** (Table 3–2), combining equations if necessary. You may need to combine components of a vector to get magnitude and direction (Eqs. 3–4).

**EXAMPLE 3–5** **Driving off a cliff.** A movie stunt driver on a motorcycle speeds horizontally off a 50.0-m-high cliff. How fast must the motorcycle leave the cliff top to land on level ground below, 90.0 m from the base of the cliff where the cameras are? Ignore air resistance.

**APPROACH** We explicitly follow the steps of the Problem Solving Strategy on the previous page.

**SOLUTION**

1. and **2. Read, choose the object, and draw a diagram.** Our object is the motor-cycle and driver, taken as a single unit. The diagram is shown in Fig. 3–22.

3. **Choose a coordinate system.** We choose the $y$ direction to be positive upward, with the top of the cliff as $y_0 = 0$. The $x$ direction is horizontal with $x_0 = 0$ at the point where the motorcycle leaves the cliff.

4. **Choose a time interval.** We choose our time interval to begin $(t = 0)$ just as the motorcycle leaves the cliff top at position $x_0 = 0$, $y_0 = 0$. Our time interval ends just before the motorcycle touches the ground below.

5. **Examine $x$ and $y$ motions.** In the horizontal $(x)$ direction, the acceleration $a_x = 0$, so the velocity is constant. The value of $x$ when the motorcycle reaches the ground is $x = +90.0$ m. In the vertical direction, the accelera-tion is the acceleration due to gravity, $a_y = -g = -9.80 \text{ m/s}^2$. The value of $y$ when the motorcycle reaches the ground is $y = -50.0$ m. The initial veloc-ity is horizontal and is our unknown, $v_{x0}$; the initial vertical velocity is zero, $v_{y0} = 0$.

6. **List knowns and unknowns.** See the Table in the margin. Note that in addition to not knowing the initial horizontal velocity $v_{x0}$ (which stays constant until landing), we also do not know the time $t$ when the motorcycle reaches the ground.

7. **Apply relevant equations.** The motorcycle maintains constant $v_x$ as long as it is in the air. The time it stays in the air is determined by the $y$ motion—when it reaches the ground. So we first find the time using the $y$ motion, and then use this time value in the $x$ equations. To find out how long it takes the motorcycle to reach the ground below, we use Eq. 2–11b (Tables 3–1 and 3–2) for the vertical $(y)$ direction with $y_0 = 0$ and $v_{y0} = 0$:

$$y = y_0 + v_{y0}t + \tfrac{1}{2}a_y t^2$$
$$= 0 + 0 + \tfrac{1}{2}(-g)t^2$$

or

$$y = -\tfrac{1}{2}gt^2.$$

We solve for $t$ and set $y = -50.0$ m:

$$t = \sqrt{\frac{2y}{-g}} = \sqrt{\frac{2(-50.0 \text{ m})}{-9.80 \text{ m/s}^2}} = 3.19 \text{ s}.$$

To calculate the initial velocity, $v_{x0}$, we again use Eq. 2–11b, but this time for the horizontal $(x)$ direction, with $a_x = 0$ and $x_0 = 0$:

$$x = x_0 + v_{x0}t + \tfrac{1}{2}a_x t^2$$
$$= 0 + v_{x0}t + 0$$

or

$$x = v_{x0}t.$$

Then

$$v_{x0} = \frac{x}{t} = \frac{90.0 \text{ m}}{3.19 \text{ s}} = 28.2 \text{ m/s},$$

which is about 100 km/h (roughly 60 mi/h).

**NOTE** In the time interval of the projectile motion, the only acceleration is $g$ in the negative $y$ direction. The acceleration in the $x$ direction is zero.

**FIGURE 3–22** Example 3–5.

| Known | Unknown |
|---|---|
| $x_0 = y_0 = 0$ | $v_{x0}$ |
| $x = 90.0$ m | $t$ |
| $y = -50.0$ m | |
| $a_x = 0$ | |
| $a_y = -g = -9.80 \text{ m/s}^2$ | |
| $v_{y0} = 0$ | |

**FIGURE 3–23** Example 3–6.

**EXAMPLE 3–6**   **A kicked football.** A kicked football leaves the ground at an angle $\theta_0 = 37.0°$ with a velocity of 20.0 m/s, as shown in Fig. 3–23. Calculate (*a*) the maximum height, (*b*) the time of travel before the football hits the ground, and (*c*) how far away it hits the ground. Assume the ball leaves the foot at ground level, and ignore air resistance and rotation of the ball.

**APPROACH** This may seem difficult at first because there are so many questions. But we can deal with them one at a time. We take the *y* direction as positive upward, and treat the *x* and *y* motions separately. The total time in the air is again determined by the *y* motion. The *x* motion occurs at constant velocity. The *y* component of velocity varies, being positive (upward) initially, decreasing to zero at the highest point, and then becoming negative as the football falls.

**SOLUTION** We resolve the initial velocity into its components (Fig. 3–23):

$$v_{x0} = v_0 \cos 37.0° = (20.0 \text{ m/s})(0.799) = 16.0 \text{ m/s}$$
$$v_{y0} = v_0 \sin 37.0° = (20.0 \text{ m/s})(0.602) = 12.0 \text{ m/s}.$$

(*a*) To find the maximum height, we consider a time interval that begins just after the football loses contact with the foot until the ball reaches its maximum height. During this time interval, the acceleration is *g* downward. At the maximum height, the velocity is horizontal (Fig. 3–23), so $v_y = 0$. This occurs at a time given by $v_y = v_{y0} - gt$ with $v_y = 0$ (see Eq. 2–11a in Table 3–2), so $v_{y0} = gt$ and

$$t = \frac{v_{y0}}{g} = \frac{(12.0 \text{ m/s})}{(9.80 \text{ m/s}^2)} = 1.224 \text{ s} \approx 1.22 \text{ s}.$$

From Eq. 2–11b, with $y_0 = 0$, we can solve for *y* at this time ($t = v_{y0}/g$):

$$y = v_{y0}t - \tfrac{1}{2}gt^2 = \frac{v_{y0}^2}{g} - \frac{1}{2}\frac{v_{y0}^2}{g} = \frac{v_{y0}^2}{2g} = \frac{(12.0 \text{ m/s})^2}{2(9.80 \text{ m/s}^2)} = 7.35 \text{ m}.$$

The maximum height is 7.35 m. [Solving Eq. 2–11c for *y* gives the same result.]

(*b*) To find the time it takes for the ball to return to the ground, we consider a different time interval, starting at the moment the ball leaves the foot ($t = 0$, $y_0 = 0$) and ending just before the ball touches the ground ($y = 0$ again). We can use Eq. 2–11b with $y_0 = 0$ and also set $y = 0$ (ground level):

$$y = y_0 + v_{y0}t - \tfrac{1}{2}gt^2$$
$$0 = 0 + v_{y0}t - \tfrac{1}{2}gt^2.$$

This equation can be factored:

$$t\left(\tfrac{1}{2}gt - v_{y0}\right) = 0.$$

There are two solutions, $t = 0$ (which corresponds to the initial point, $y_0$), and

$$t = \frac{2v_{y0}}{g} = \frac{2(12.0 \text{ m/s})}{(9.80 \text{ m/s}^2)} = 2.45 \text{ s},$$

which is the total travel time of the football.

(*c*) The total distance traveled in the *x* direction is found by applying Eq. 2–11b with $x_0 = 0$, $a_x = 0$, $v_{x0} = 16.0$ m/s, and $t = 2.45$ s:

$$x = v_{x0}t = (16.0 \text{ m/s})(2.45 \text{ s}) = 39.2 \text{ m}.$$

**NOTE** In (*b*), the time needed for the whole trip, $t = 2v_{y0}/g = 2.45$ s, is double the time to reach the highest point, calculated in (*a*). That is, the time to go up equals the time to come back down to the same level (ignoring air resistance).

**EXERCISE F** In Example 3–6, what is (a) the velocity vector at the maximum height, and (b) the acceleration vector at maximum height?

In Example 3–6, we treated the football as if it were a particle, ignoring its rotation. We also ignored air resistance. Because air resistance is significant on a football, our results are only estimates (mainly overestimates).

**CONCEPTUAL EXAMPLE 3–7** **The wrong strategy.** A boy on a small hill aims his water-balloon slingshot horizontally, straight at a second boy hanging from a tree branch a distance $d$ away, Fig. 3–24. At the instant the water balloon is released, the second boy lets go and falls from the tree, hoping to avoid being hit. Show that he made the wrong move. (He hadn't studied physics yet.) Ignore air resistance.

**RESPONSE** Both the water balloon and the boy in the tree start falling at the same instant, and in a time $t$ they each fall the same vertical distance $y = \frac{1}{2}gt^2$, much like Fig. 3–19. In the time it takes the water balloon to travel the horizontal distance $d$, the balloon will have the same $y$ position as the falling boy. Splat. If the boy had stayed in the tree, he would have avoided the humiliation.

**FIGURE 3–24** Example 3–7.

## Level Horizontal Range

The total distance the football traveled in Example 3–6 is called the horizontal **range** $R$. We now derive a formula for the range, which applies to a projectile that lands at the same level it started $(= y_0)$: that is, $y$ (final) $= y_0$ (see Fig. 3–25a). Looking back at Example 3–6 part (c), we see that $x = R = v_{x0}t$ where (from part b) $t = 2v_{y0}/g$. Thus

$$R = v_{x0}t = v_{x0}\left(\frac{2v_{y0}}{g}\right) = \frac{2v_{x0}v_{y0}}{g} = \frac{2v_0^2 \sin\theta_0 \cos\theta_0}{g}, \qquad [y = y_0]$$

where $v_{x0} = v_0 \cos\theta_0$ and $v_{y0} = v_0 \sin\theta_0$. This can be rewritten, using the trigonometric identity $2\sin\theta \cos\theta = \sin 2\theta$ (Appendix A or inside the rear cover):

$$R = \frac{v_0^2 \sin 2\theta_0}{g}. \qquad [\text{only if } y \text{ (final)} = y_0]$$

Note that the *maximum* range, for a given initial velocity $v_0$, is obtained when $\sin 2\theta$ takes on its maximum value of 1.0, which occurs for $2\theta_0 = 90°$; so

$$\theta_0 = 45° \text{ for maximum range, and } R_{max} = v_0^2/g.$$

The maximum range increases by the square of $v_0$, so doubling the muzzle velocity of a cannon increases its maximum range by a factor of 4.

When air resistance is important, the range is less for a given $v_0$, and the maximum range is obtained at an angle smaller than 45°.

**FIGURE 3–25** (a) The range $R$ of a projectile. (b) There are generally two angles $\theta_0$ that will give the same range. If one angle is $\theta_{01}$, the other is $\theta_{02} = 90° - \theta_{01}$. Example 3–8.

(a)

(b)

**EXAMPLE 3–8** **Range of a cannon ball.** Suppose one of Napoleon's cannons had a muzzle speed, $v_0$, of 60.0 m/s. At what angle should it have been aimed (ignore air resistance) to strike a target 320 m away?

**APPROACH** We use the equation just derived for the range, $R = v_0^2 \sin 2\theta_0/g$, with $R = 320$ m.

**SOLUTION** We solve for $\sin 2\theta_0$ in the range formula:

$$\sin 2\theta_0 = \frac{Rg}{v_0^2} = \frac{(320 \text{ m})(9.80 \text{ m/s}^2)}{(60.0 \text{ m/s})^2} = 0.871.$$

We want to solve for an angle $\theta_0$ that is between 0° and 90°, which means $2\theta_0$ in this equation can be as large as 180°. Thus, $2\theta_0 = 60.6°$ is a solution, so $\theta_0 = 30.3°$. But $2\theta_0 = 180° - 60.6° = 119.4°$ is also a solution (see Appendix A–7), so $\theta_0$ can also be $\theta_0 = 59.7°$. In general we have two solutions (see Fig. 3–25b), which in the present case are given by

$$\theta_0 = 30.3° \quad \text{or} \quad 59.7°.$$

Either angle gives the same range. Only when $\sin 2\theta_0 = 1$ (so $\theta_0 = 45°$) is there a single solution (that is, both solutions are the same).

**FIGURE 3–26** Example 3–9: the football leaves the punter's foot at $y = 0$, and reaches the ground where $y = -1.00$ m.

$y$

$y_0 = 0$

$x$

$y = -1.00$ m

Ground

**EXAMPLE 3–9** **A punt.** Suppose the football in Example 3–6 was punted, and left the punter's foot at a height of 1.00 m above the ground. How far did the football travel before hitting the ground? Set $x_0 = 0$, $y_0 = 0$.

**APPROACH** The only difference here from Example 3–6 is that the football hits the ground *below* its starting point of $y_0 = 0$. That is, the ball hits the ground at $y = -1.00$ m. See Fig. 3–26. Thus we cannot use the range formula which is valid only if $y$ (final) $= y_0$. As in Example 3–6, $v_0 = 20.0$ m/s, $\theta_0 = 37.0°$.

**SOLUTION** With $y = -1.00$ m and $v_{y0} = 12.0$ m/s (see Example 3–6), we use the $y$ version of Eq. 2–11b with $a_y = -g$,

$$y = y_0 + v_{y0}t - \tfrac{1}{2}gt^2,$$

and obtain

$$-1.00 \text{ m} = 0 + (12.0 \text{ m/s})t - (4.90 \text{ m/s}^2)t^2.$$

We rearrange this equation into standard form $(ax^2 + bx + c = 0)$ so we can use the quadratic formula:

$$(4.90 \text{ m/s}^2)t^2 - (12.0 \text{ m/s})t - (1.00 \text{ m}) = 0.$$

The quadratic formula (Appendix A–4) gives

$$t = \frac{12.0 \text{ m/s} \pm \sqrt{(-12.0 \text{ m/s})^2 - 4(4.90 \text{ m/s}^2)(-1.00 \text{ m})}}{2(4.90 \text{ m/s}^2)}$$

$$= 2.53 \text{ s} \quad \text{or} \quad -0.081 \text{ s}.$$

The second solution would correspond to a time prior to the kick, so it doesn't apply. With $t = 2.53$ s for the time at which the ball touches the ground, the horizontal distance the ball traveled is (using $v_{x0} = 16.0$ m/s from Example 3–6):

$$x = v_{x0}t = (16.0 \text{ m/s})(2.53 \text{ s}) = 40.5 \text{ m}.$$

Our assumption in Example 3–6 that the ball leaves the foot at ground level would result in an underestimate of about 1.3 m in the distance our punt traveled.

# *3–7 Projectile Motion Is Parabolic

We now show that the path followed by any projectile is a *parabola*, if we can ignore air resistance and can assume that $\vec{g}$ is constant. To do so, we need to find $y$ as a function of $x$ by eliminating $t$ between the two equations for horizontal and vertical motion (Eq. 2–11b in Table 3–2), and for simplicity we set $x_0 = y_0 = 0$:

$$x = v_{x0}t$$
$$y = v_{y0}t - \tfrac{1}{2}gt^2.$$

From the first equation, we have $t = x/v_{x0}$, and we substitute this into the second one to obtain

$$y = \left(\frac{v_{y0}}{v_{x0}}\right)x - \left(\frac{g}{2v_{x0}^2}\right)x^2. \qquad \textbf{(3–6)}$$

We see that $y$ as a function of $x$ has the form

$$y = Ax - Bx^2,$$

where $A$ and $B$ are constants for any specific projectile motion. This is the standard equation for a parabola. See Figs. 3–17 and 3–27.

The idea that projectile motion is parabolic was, in Galileo's day, at the forefront of physics research. Today we discuss it in Chapter 3 of introductory physics!

**FIGURE 3–27** Examples of projectile motion: a boy jumping, and glowing lava from the volcano Stromboli.

*Some Sections of this book, such as this one, may be considered *optional* at the discretion of the instructor. See the Preface for more details.

# 3–8 Relative Velocity

We now consider how observations made in different frames of reference are related to each other. For example, consider two trains approaching one another, each with a speed of 80 km/h with respect to the Earth. Observers on the Earth beside the train tracks will measure 80 km/h for the speed of each of the trains. Observers on either one of the trains (a different frame of reference) will measure a speed of 160 km/h for the train approaching them.

Similarly, when one car traveling 90 km/h passes a second car traveling in the same direction at 75 km/h, the first car has a speed relative to the second car of 90 km/h − 75 km/h = 15 km/h.

When the velocities are along the same line, simple addition or subtraction is sufficient to obtain the **relative velocity**. But if they are not along the same line, we must make use of vector addition. We emphasize, as mentioned in Section 2–1, that when specifying a velocity, it is important to specify what the reference frame is.

When determining relative velocity, it is easy to make a mistake by adding or subtracting the wrong velocities. It is important, therefore, to draw a diagram and use a careful labeling process. Each velocity is labeled by *two subscripts: the first refers to the object, the second to the reference frame in which it has this velocity.* For example, suppose a boat heads directly across a river, as shown in Fig. 3–28. We let $\vec{v}_{BW}$ be the velocity of the **B**oat with respect to the **W**ater. (This is also what the boat's velocity would be relative to the shore if the water were still.) Similarly, $\vec{v}_{BS}$ is the velocity of the **B**oat with respect to the **S**hore, and $\vec{v}_{WS}$ is the velocity of the **W**ater with respect to the **S**hore (this is the river current). Note that $\vec{v}_{BW}$ is what the boat's motor produces (against the water), whereas $\vec{v}_{BS}$ is equal to $\vec{v}_{BW}$ plus the effect of the current, $\vec{v}_{WS}$. Therefore, the velocity of the boat relative to the shore is (see vector diagram, Fig. 3–28)

$$\vec{v}_{BS} = \vec{v}_{BW} + \vec{v}_{WS}. \tag{3–7}$$

By writing the subscripts using this convention, we see that the inner subscripts (the two W's) on the right-hand side of Eq. 3–7 are the same; also, the outer subscripts on the right of Eq. 3–7 (the B and the S) are the same as the two subscripts for the sum vector on the left, $\vec{v}_{BS}$. By following this convention (first subscript for the object, second for the reference frame), you can write down the correct equation relating velocities in different reference frames.[†]

Equation 3–7 is valid in general and can be extended to three or more velocities. For example, if a fisherman on the boat walks with a velocity $\vec{v}_{FB}$ relative to the boat, his velocity relative to the shore is $\vec{v}_{FS} = \vec{v}_{FB} + \vec{v}_{BW} + \vec{v}_{WS}$. The equations involving relative velocity will be correct when adjacent inner subscripts are identical and when the outermost ones correspond exactly to the two on the velocity on the left of the equation. But this works only with plus signs (on the right), not minus signs.

It is often useful to remember that for any two objects or reference frames, A and B, the velocity of A relative to B has the same magnitude, but opposite direction, as the velocity of B relative to A:

$$\vec{v}_{BA} = -\vec{v}_{AB}. \tag{3–8}$$

For example, if a train is traveling 100 km/h relative to the Earth in a certain direction, objects on the Earth (such as trees) appear to an observer on the train to be traveling 100 km/h in the opposite direction.

**FIGURE 3–28** A boat heads north directly across a river which flows west. Velocity vectors are shown as green arrows:

$\vec{v}_{BS}$ = velocity of **B**oat with respect to the **S**hore,

$\vec{v}_{BW}$ = velocity of **B**oat with respect to the **W**ater,

$\vec{v}_{WS}$ = velocity of **W**ater with respect to the **S**hore (river current).

As it crosses the river, the boat is dragged downstream by the current.

---

[†]We thus can see, for example, that the equation $\vec{v}_{BW} = \vec{v}_{BS} + \vec{v}_{WS}$ is wrong: the inner subscripts are not the same, and the outer ones on the right do not correspond to the subscripts on the left.

**FIGURE 3–29** Example 3–10.

**FIGURE 3–30** Example 3–11. A boat heading directly across a river whose current moves at 1.20 m/s.

**EXAMPLE 3–10** **Heading upstream.** A boat's speed in still water is $v_{BW} = 1.85$ m/s. If the boat is to travel north directly across a river whose westward current has speed $v_{WS} = 1.20$ m/s, at what upstream angle must the boat head? (See Fig. 3–29.)

**APPROACH** If the boat heads straight across the river, the current will drag the boat downstream (westward). To overcome the river's current, the boat must have an upstream (eastward) component of velocity as well as a cross-stream (northward) component. Figure 3–29 has been drawn with $\vec{v}_{BS}$, the velocity of the **B**oat relative to the **S**hore, pointing directly across the river because this is where the boat is supposed to go. (Note that $\vec{v}_{BS} = \vec{v}_{BW} + \vec{v}_{WS}$.)

**SOLUTION** Vector $\vec{v}_{BW}$ points upstream at angle $\theta$ as shown. From the diagram,

$$\sin \theta = \frac{v_{WS}}{v_{BW}} = \frac{1.20 \text{ m/s}}{1.85 \text{ m/s}} = 0.6486.$$

Thus $\theta = 40.4°$, so the boat must head upstream at a 40.4° angle.

**EXAMPLE 3–11** **Heading across the river.** The same boat ($v_{BW} = 1.85$ m/s) now heads directly across the river whose current is still 1.20 m/s. (*a*) What is the velocity (magnitude and direction) of the boat relative to the shore? (*b*) If the river is 110 m wide, how long will it take to cross and how far downstream will the boat be then?

**APPROACH** The boat now heads directly across the river and is pulled downstream by the current, as shown in Fig. 3–30. The boat's velocity with respect to the shore, $\vec{v}_{BS}$, is the sum of its velocity with respect to the water, $\vec{v}_{BW}$, plus the velocity of the water with respect to the shore, $\vec{v}_{WS}$: just as before,

$$\vec{v}_{BS} = \vec{v}_{BW} + \vec{v}_{WS}.$$

**SOLUTION** (*a*) Since $\vec{v}_{BW}$ is perpendicular to $\vec{v}_{WS}$, we can get $v_{BS}$ using the theorem of Pythagoras:

$$v_{BS} = \sqrt{v_{BW}^2 + v_{WS}^2} = \sqrt{(1.85 \text{ m/s})^2 + (1.20 \text{ m/s})^2} = 2.21 \text{ m/s}.$$

We can obtain the angle (note how $\theta$ is defined in Fig. 3–30) from:

$$\tan \theta = v_{WS}/v_{BW} = (1.20 \text{ m/s})/(1.85 \text{ m/s}) = 0.6486.$$

A calculator with a key INV TAN or ARC TAN or TAN$^{-1}$ gives $\theta = \tan^{-1}(0.6486) = 33.0°$. Note that this angle is not equal to the angle calculated in Example 3–10.

(*b*) The travel time for the boat is determined by the time it takes to cross the river. Given the river's width $D = 110$ m, we can use the velocity component in the direction of $D$, $v_{BW} = D/t$. Solving for $t$, we get $t = 110 \text{ m}/1.85 \text{ m/s} = 59.5$ s. The boat will have been carried downstream, in this time, a distance

$$d = v_{WS}t = (1.20 \text{ m/s})(59.5 \text{ s}) = 71.4 \text{ m} \approx 71 \text{ m}.$$

**NOTE** There is no acceleration in this Example, so the motion involves only constant velocities (of the boat or of the river).

# Summary

A quantity such as velocity, that has both a magnitude and a direction, is called a **vector**. A quantity such as mass, that has only a magnitude, is called a **scalar**. On diagrams, vectors are represented by arrows.

Addition of vectors can be done graphically by placing the tail of each successive arrow at the tip of the previous one. The sum, or **resultant vector**, is the arrow drawn from the tail of the first vector to the tip of the last vector. Two vectors can also be added using the parallelogram method.

Vectors can be added more accurately by adding their **components** along chosen axes with the aid of trigonometric functions. A vector of magnitude $V$ making an angle $\theta$ with the $+x$ axis has components

$$V_x = V \cos \theta, \quad V_y = V \sin \theta. \quad \textbf{(3–3)}$$

Given the components, we can find a vector's magnitude and direction from

$$V = \sqrt{V_x^2 + V_y^2}, \quad \tan \theta = \frac{V_y}{V_x}. \quad \textbf{(3–4)}$$

**Projectile motion** is the motion of an object in the air near the Earth's surface under the effect of gravity alone. It can be analyzed as two separate motions if air resistance can be ignored. The horizontal component of motion is at constant velocity, whereas the vertical component is at constant acceleration, $\vec{g}$, just as for an object falling vertically under the action of gravity.

The velocity of an object relative to one frame of reference can be found by vector addition if its velocity relative to a second frame of reference, and the **relative velocity** of the two reference frames, are known.

# Questions

1. One car travels due east at 40 km/h, and a second car travels north at 40 km/h. Are their velocities equal? Explain.

2. Can you conclude that a car is not accelerating if its speedometer indicates a steady 60 km/h? Explain.

3. Give several examples of an object's motion in which a great distance is traveled but the displacement is zero.

4. Can the displacement vector for a particle moving in two dimensions be longer than the length of path traveled by the particle over the same time interval? Can it be less? Discuss.

5. During baseball practice, a player hits a very high fly ball and then runs in a straight line and catches it. Which had the greater displacement, the player or the ball? Explain.

6. If $\vec{V} = \vec{V}_1 + \vec{V}_2$, is $V$ necessarily greater than $V_1$ and/or $V_2$? Discuss.

7. Two vectors have length $V_1 = 3.5$ km and $V_2 = 4.0$ km. What are the maximum and minimum magnitudes of their vector sum?

8. Can two vectors, of unequal magnitude, add up to give the zero vector? Can *three* unequal vectors? Under what conditions?

9. Can the magnitude of a vector ever (*a*) equal, or (*b*) be less than, one of its components?

10. Does the odometer of a car measure a scalar or a vector quantity? What about the speedometer?

11. How could you determine the speed a slingshot imparts to a rock, using only a meter stick, a rock, and the slingshot?

12. In archery, should the arrow be aimed directly at the target? How should your angle of aim depend on the distance to the target?

13. It was reported in World War I that a pilot flying at an altitude of 2 km caught in his bare hands a bullet fired at the plane! Using the fact that a bullet slows down considerably due to air resistance, explain how this incident occurred.

14. You are on the street trying to hit a friend in his dorm window with a water balloon. He has a similar idea and is aiming at you with *his* water balloon. You aim straight at each other and throw at the same instant. Do the water balloons hit each other? Explain why or why not.

15. A projectile is launched at an upward angle of 30° to the horizontal with a speed of 30 m/s. How does the horizontal component of its velocity 1.0 s after launch compare with its horizontal component of velocity 2.0 s after launch, ignoring air resistance? Explain.

16. A projectile has the least speed at what point in its path?

17. Two cannonballs, A and B, are fired from the ground with identical initial speeds, but with $\theta_A$ larger than $\theta_B$. (*a*) Which cannonball reaches a higher elevation? (*b*) Which stays longer in the air? (*c*) Which travels farther? Explain.

18. A person sitting in an enclosed train car, moving at constant velocity, throws a ball straight up into the air in her reference frame. (*a*) Where does the ball land? What is your answer if the car (*b*) accelerates, (*c*) decelerates, (*d*) rounds a curve, (*e*) moves with constant velocity but is open to the air?

19. If you are riding on a train that speeds past another train moving in the same direction on an adjacent track, it appears that the other train is moving backward. Why?

20. Two rowers, who can row at the same speed in still water, set off across a river at the same time. One heads straight across and is pulled downstream somewhat by the current. The other one heads upstream at an angle so as to arrive at a point opposite the starting point. Which rower reaches the opposite side first? Explain.

21. If you stand motionless under an umbrella in a rainstorm where the drops fall vertically, you remain relatively dry. However, if you start running, the rain begins to hit your legs even if they remain under the umbrella. Why?

# MisConceptual Questions

1. You are adding vectors of length 20 and 40 units. Which of the following choices is a possible resultant magnitude?
   (*a*) 0.
   (*b*) 18.
   (*c*) 37.
   (*d*) 64.
   (*e*) 100.

2. The magnitude of a component of a vector must be
   (*a*) less than or equal to the magnitude of the vector.
   (*b*) equal to the magnitude of the vector.
   (*c*) greater than or equal to the magnitude of the vector.
   (*d*) less than, equal to, or greater than the magnitude of the vector.

3. You are in the middle of a large field. You walk in a straight line for 100 m, then turn left and walk 100 m more in a straight line before stopping. When you stop, you are 100 m from your starting point. By how many degrees did you turn?
   (*a*) 90°.
   (*b*) 120°.
   (*c*) 30°.
   (*d*) 180°.
   (*e*) This is impossible. You cannot walk 200 m and be only 100 m away from where you started.

4. A bullet fired from a rifle begins to fall
   (*a*) as soon as it leaves the barrel.
   (*b*) after air friction reduces its speed.
   (*c*) not at all if air resistance is ignored.

5. A baseball player hits a ball that soars high into the air. After the ball has left the bat, and while it is traveling upward (at point P in Fig. 3–31), what is the direction of acceleration? Ignore air resistance.

(*a*)      (*b*)      (*c*)

**FIGURE 3–31**
MisConceptual
Question 5.

6. One ball is dropped vertically from a window. At the same instant, a second ball is thrown horizontally from the same window. Which ball has the greater speed at ground level?
   (*a*) The dropped ball.
   (*b*) The thrown ball.
   (*c*) Neither—they both have the same speed on impact.
   (*d*) It depends on how hard the ball was thrown.

7. You are riding in an enclosed train car moving at 90 km/h. If you throw a baseball straight up, where will the baseball land?
   (a) In front of you.
   (b) Behind you.
   (c) In your hand.
   (d) Can't decide from the given information.

8. Which of the three kicks in Fig. 3–32 is in the air for the longest time? They all reach the same maximum height $h$. Ignore air resistance.
   (a), (b), (c), or (d) all the same time.

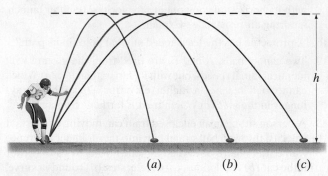

**FIGURE 3–32** MisConceptual Question 8.

9. A baseball is hit high and far. Which of the following statements is true? At the highest point,
   (a) the magnitude of the acceleration is zero.
   (b) the magnitude of the velocity is zero.
   (c) the magnitude of the velocity is the slowest.
   (d) more than one of the above is true.
   (e) none of the above are true.

10. A hunter is aiming horizontally at a monkey who is sitting in a tree. The monkey is so terrified when it sees the gun that it falls off the tree. At that very instant, the hunter pulls the trigger. What will happen?
    (a) The bullet will miss the monkey because the monkey falls down while the bullet speeds straight forward.
    (b) The bullet will hit the monkey because both the monkey and the bullet are falling downward at the same rate due to gravity.
    (c) The bullet will miss the monkey because although both the monkey and the bullet are falling downward due to gravity, the monkey is falling faster.
    (d) It depends on how far the hunter is from the monkey.

11. Which statements are *not* valid for a projectile? Take up as positive.
    (a) The projectile has the same $x$ velocity at any point on its path.
    (b) The acceleration of the projectile is positive and decreasing when the projectile is moving upwards, zero at the top, and increasingly negative as the projectile descends.
    (c) The acceleration of the projectile is a constant negative value.
    (d) The $y$ component of the velocity of the projectile is zero at the highest point of the projectile's path.
    (e) The velocity at the highest point is zero.

12. A car travels 10 m/s east. Another car travels 10 m/s north. The relative speed of the first car with respect to the second is
    (a) less than 20 m/s.
    (b) exactly 20 m/s.
    (c) more than 20 m/s.

For assigned homework and other learning materials, go to the MasteringPhysics website. **MP**

# Problems

## 3–2 to 3–4 Vector Addition

1. (I) A car is driven 225 km west and then 98 km southwest (45°). What is the displacement of the car from the point of origin (magnitude and direction)? Draw a diagram.

2. (I) A delivery truck travels 21 blocks north, 16 blocks east, and 26 blocks south. What is its final displacement from the origin? Assume the blocks are equal length.

3. (I) If $V_x = 9.80$ units and $V_y = -6.40$ units, determine the magnitude and direction of $\vec{V}$.

4. (II) Graphically determine the resultant of the following three vector displacements: (1) 24 m, 36° north of east; (2) 18 m, 37° east of north; and (3) 26 m, 33° west of south.

5. (II) $\vec{V}$ is a vector 24.8 units in magnitude and points at an angle of 23.4° above the negative $x$ axis. (a) Sketch this vector. (b) Calculate $V_x$ and $V_y$. (c) Use $V_x$ and $V_y$ to obtain (again) the magnitude and direction of $\vec{V}$. [*Note:* Part (c) is a good way to check if you've resolved your vector correctly.]

6. (II) Vector $\vec{V}_1$ is 6.6 units long and points along the negative $x$ axis. Vector $\vec{V}_2$ is 8.5 units long and points at +55° to the positive $x$ axis. (a) What are the $x$ and $y$ components of each vector? (b) Determine the sum $\vec{V}_1 + \vec{V}_2$ (magnitude and angle).

7. (II) Figure 3–33 shows two vectors, $\vec{A}$ and $\vec{B}$, whose magnitudes are $A = 6.8$ units and $B = 5.5$ units. Determine $\vec{C}$ if (a) $\vec{C} = \vec{A} + \vec{B}$, (b) $\vec{C} = \vec{A} - \vec{B}$, (c) $\vec{C} = \vec{B} - \vec{A}$. Give the magnitude and direction for each.

**FIGURE 3–33**
Problem 7.

8. (II) An airplane is traveling 835 km/h in a direction 41.5° west of north (Fig. 3–34).
   (a) Find the components of the velocity vector in the northerly and westerly directions. (b) How far north and how far west has the plane traveled after 1.75 h?

**FIGURE 3–34**
Problem 8.

9. (II) Three vectors are shown in Fig. 3–35. Their magnitudes are given in arbitrary units. Determine the sum of the three vectors. Give the resultant in terms of (a) components, (b) magnitude and angle with the +x axis.

**FIGURE 3–35**
Problems 9, 10, 11, 12, and 13.
Vector magnitudes are given in arbitrary units.

10. (II) (a) Given the vectors $\vec{A}$ and $\vec{B}$ shown in Fig. 3–35, determine $\vec{B} - \vec{A}$. (b) Determine $\vec{A} - \vec{B}$ without using your answer in (a). Then compare your results and see if they are opposite.

11. (II) Determine the vector $\vec{A} - \vec{C}$, given the vectors $\vec{A}$ and $\vec{C}$ in Fig. 3–35.

12. (II) For the vectors shown in Fig. 3–35, determine (a) $\vec{B} - 3\vec{A}$, (b) $2\vec{A} - 3\vec{B} + 2\vec{C}$.

13. (II) For the vectors given in Fig. 3–35, determine (a) $\vec{A} - \vec{B} + \vec{C}$, (b) $\vec{A} + \vec{B} - \vec{C}$, and (c) $\vec{C} - \vec{A} - \vec{B}$.

14. (II) Suppose a vector $\vec{V}$ makes an angle $\phi$ with respect to the y axis. What could be the x and y components of the vector $\vec{V}$?

15. (II) The summit of a mountain, 2450 m above base camp, is measured on a map to be 4580 m horizontally from the camp in a direction 38.4° west of north. What are the components of the displacement vector from camp to summit? What is its magnitude? Choose the x axis east, y axis north, and z axis up.

16. (III) You are given a vector in the xy plane that has a magnitude of 90.0 units and a y component of −65.0 units. (a) What are the two possibilities for its x component? (b) Assuming the x component is known to be positive, specify the vector which, if you add it to the original one, would give a resultant vector that is 80.0 units long and points entirely in the −x direction.

### 3–5 and 3–6 Projectile Motion (neglect air resistance)

17. (I) A tiger leaps horizontally from a 7.5-m-high rock with a speed of 3.0 m/s. How far from the base of the rock will she land?

18. (I) A diver running 2.5 m/s dives out horizontally from the edge of a vertical cliff and 3.0 s later reaches the water below. How high was the cliff and how far from its base did the diver hit the water?

19. (II) Estimate by what factor a person can jump farther on the Moon as compared to the Earth if the takeoff speed and angle are the same. The acceleration due to gravity on the Moon is one-sixth what it is on Earth.

20. (II) A ball is thrown horizontally from the roof of a building 7.5 m tall and lands 9.5 m from the base. What was the ball's initial speed?

21. (II) A ball thrown horizontally at 12.2 m/s from the roof of a building lands 21.0 m from the base of the building. How high is the building?

22. (II) A football is kicked at ground level with a speed of 18.0 m/s at an angle of 31.0° to the horizontal. How much later does it hit the ground?

23. (II) A fire hose held near the ground shoots water at a speed of 6.5 m/s. At what angle(s) should the nozzle point in order that the water land 2.5 m away (Fig. 3–36)? Why are there two different angles? Sketch the two trajectories.

**FIGURE 3–36**
Problem 23.

24. (II) You buy a plastic dart gun, and being a clever physics student you decide to do a quick calculation to find its maximum horizontal range. You shoot the gun straight up, and it takes 4.0 s for the dart to land back at the barrel. What is the maximum horizontal range of your gun?

25. (II) A grasshopper hops along a level road. On each hop, the grasshopper launches itself at angle $\theta_0 = 45°$ and achieves a range $R = 0.80$ m. What is the average horizontal speed of the grasshopper as it hops along the road? Assume that the time spent on the ground between hops is negligible.

26. (II) Extreme-sports enthusiasts have been known to jump off the top of El Capitan, a sheer granite cliff of height 910 m in Yosemite National Park. Assume a jumper runs horizontally off the top of El Capitan with speed 4.0 m/s and enjoys a free fall until she is 150 m above the valley floor, at which time she opens her parachute (Fig. 3–37). (a) How long is the jumper in free fall? Ignore air resistance. (b) It is important to be as far away from the cliff as possible before opening the parachute. How far from the cliff is this jumper when she opens her chute?

**FIGURE 3–37**
Problem 26.

27. (II) A projectile is fired with an initial speed of 36.6 m/s at an angle of 42.2° above the horizontal on a long flat firing range. Determine (a) the maximum height reached by the projectile, (b) the total time in the air, (c) the total horizontal distance covered (that is, the range), and (d) the speed of the projectile 1.50 s after firing.

**28.** (II) An athlete performing a long jump leaves the ground at a 27.0° angle and lands 7.80 m away. (*a*) What was the takeoff speed? (*b*) If this speed were increased by just 5.0%, how much longer would the jump be?

**29.** (II) A shot-putter throws the "shot" (mass = 7.3 kg) with an initial speed of 14.4 m/s at a 34.0° angle to the horizontal. Calculate the horizontal distance traveled by the shot if it leaves the athlete's hand at a height of 2.10 m above the ground.

**30.** (II) A baseball is hit with a speed of 27.0 m/s at an angle of 45.0°. It lands on the flat roof of a 13.0-m-tall nearby building. If the ball was hit when it was 1.0 m above the ground, what horizontal distance does it travel before it lands on the building?

**31.** (II) A rescue plane wants to drop supplies to isolated mountain climbers on a rocky ridge 235 m below. If the plane is traveling horizontally with a speed of 250 km/h (69.4 m/s), how far in advance of the recipients (horizontal distance) must the goods be dropped (Fig. 3–38)?

**FIGURE 3–38** Problem 31.

**32.** (III) Suppose the rescue plane of Problem 31 releases the supplies a horizontal distance of 425 m in advance of the mountain climbers. What vertical velocity (up or down) should the supplies be given so that they arrive precisely at the climbers' position (Fig. 3–39)? With what speed do the supplies land?

**FIGURE 3–39** Problem 32.

**33.** (III) A diver leaves the end of a 4.0-m-high diving board and strikes the water 1.3 s later, 3.0 m beyond the end of the board. Considering the diver as a particle, determine: (*a*) her initial velocity, $\vec{v}_0$; (*b*) the maximum height reached; and (*c*) the velocity $\vec{v}_f$ with which she enters the water.

**34.** (III) Show that the time required for a projectile to reach its highest point is equal to the time for it to return to its original height if air resistance is neglible.

**35.** (III) Suppose the kick in Example 3–6 is attempted 36.0 m from the goalposts, whose crossbar is 3.05 m above the ground. If the football is directed perfectly between the goalposts, will it pass over the bar and be a field goal? Show why or why not. If not, from what horizontal distance must this kick be made if it is to score?

**36.** (III) Revisit Example 3–7, and assume that the boy with the slingshot is *below* the boy in the tree (Fig. 3–40) and so aims *upward*, directly at the boy in the tree. Show that again the boy in the tree makes the wrong move by letting go at the moment the water balloon is shot.

**FIGURE 3–40** Problem 36.

**37.** (III) A stunt driver wants to make his car jump over 8 cars parked side by side below a horizontal ramp (Fig. 3–41). (*a*) With what minimum speed must he drive off the horizontal ramp? The vertical height of the ramp is 1.5 m above the cars and the horizontal distance he must clear is 22 m. (*b*) If the ramp is now tilted upward, so that "takeoff angle" is 7.0° above the horizontal, what is the new minimum speed?

**FIGURE 3–41** Problem 37.

## 3–8 Relative Velocity

**38.** (I) A person going for a morning jog on the deck of a cruise ship is running toward the bow (front) of the ship at 2.0 m/s while the ship is moving ahead at 8.5 m/s. What is the velocity of the jogger relative to the water? Later, the jogger is moving toward the stern (rear) of the ship. What is the jogger's velocity relative to the water now?

**39.** (I) Huck Finn walks at a speed of 0.70 m/s across his raft (that is, he walks perpendicular to the raft's motion relative to the shore). The heavy raft is traveling down the Mississippi River at a speed of 1.50 m/s relative to the river bank (Fig. 3–42). What is Huck's velocity (speed and direction) relative to the river bank?

**FIGURE 3–42**
Problem 39.

**40.** (II) Determine the speed of the boat with respect to the shore in Example 3–10.

**41.** (II) Two planes approach each other head-on. Each has a speed of 780 km/h, and they spot each other when they are initially 10.0 km apart. How much time do the pilots have to take evasive action?

**42.** (II) A passenger on a boat moving at 1.70 m/s on a still lake walks up a flight of stairs at a speed of 0.60 m/s, Fig. 3–43. The stairs are angled at 45° pointing in the direction of motion as shown. What is the velocity of the passenger relative to the water?

**FIGURE 3–43** Problem 42.

**43.** (II) A person in the passenger basket of a hot-air balloon throws a ball horizontally outward from the basket with speed 10.0 m/s (Fig. 3–44). What initial velocity (magnitude and direction) does the ball have relative to a person standing on the ground (*a*) if the hot-air balloon is rising at 3.0 m/s relative to the ground during this throw, (*b*) if the hot-air balloon is descending at 3.0 m/s relative to the ground?

**FIGURE 3–44**
Problem 43.

**44.** (II) An airplane is heading due south at a speed of 688 km/h. If a wind begins blowing from the southwest at a speed of 90.0 km/h (average), calculate (*a*) the velocity (magnitude and direction) of the plane, relative to the ground, and (*b*) how far from its intended position it will be after 11.0 min if the pilot takes no corrective action. [*Hint*: First draw a diagram.]

**45.** (II) In what direction should the pilot aim the plane in Problem 44 so that it will fly due south?

**46.** (II) A swimmer is capable of swimming 0.60 m/s in still water. (*a*) If she aims her body directly across a 45-m-wide river whose current is 0.50 m/s, how far downstream (from a point opposite her starting point) will she land? (*b*) How long will it take her to reach the other side?

**47.** (II) (*a*) At what upstream angle must the swimmer in Problem 46 aim, if she is to arrive at a point directly across the stream? (*b*) How long will it take her?

**48.** (II) A boat, whose speed in still water is 2.50 m/s, must cross a 285-m-wide river and arrive at a point 118 m upstream from where it starts (Fig. 3–45). To do so, the pilot must head the boat at a 45.0° upstream angle. What is the speed of the river's current?

**FIGURE 3–45**
Problem 48.

**49.** (II) A child, who is 45 m from the bank of a river, is being carried helplessly downstream by the river's swift current of 1.0 m/s. As the child passes a lifeguard on the river's bank, the lifeguard starts swimming in a straight line (Fig. 3–46) until she reaches the child at a point downstream. If the lifeguard can swim at a speed of 2.0 m/s relative to the water, how long does it take her to reach the child? How far downstream does the lifeguard intercept the child?

**FIGURE 3–46** Problem 49.

**50.** (III) An airplane, whose air speed is 580 km/h, is supposed to fly in a straight path 38.0° N of E. But a steady 82 km/h wind is blowing from the north. In what direction should the plane head? [*Hint*: Use the law of sines, Appendix A–7.]

**51.** (III) Two cars approach a street corner at right angles to each other (Fig. 3–47). Car 1 travels at a speed relative to Earth $v_{1E} = 35$ km/h, and car 2 at $v_{2E} = 55$ km/h. What is the relative velocity of car 1 as seen by car 2? What is the velocity of car 2 relative to car 1?

**FIGURE 3–47**
Problem 51.

# General Problems

**52.** Two vectors, $\vec{V}_1$ and $\vec{V}_2$, add to a resultant $\vec{V}_R = \vec{V}_1 + \vec{V}_2$. Describe $\vec{V}_1$ and $\vec{V}_2$ if (a) $V_R = V_1 + V_2$, (b) $V_R^2 = V_1^2 + V_2^2$, (c) $V_1 + V_2 = V_1 - V_2$.

**53.** On mountainous downhill roads, escape routes are sometimes placed to the side of the road for trucks whose brakes might fail. Assuming a constant upward slope of 26°, calculate the horizontal and vertical components of the acceleration of a truck that slowed from 110 km/h to rest in 7.0 s. See Fig. 3–48.

**FIGURE 3–48** Problem 53.

**54.** A light plane is headed due south with a speed relative to still air of 185 km/h. After 1.00 h, the pilot notices that they have covered only 135 km and their direction is not south but 15.0° east of south. What is the wind velocity?

**55.** An Olympic long jumper is capable of jumping 8.0 m. Assuming his horizontal speed is 9.1 m/s as he leaves the ground, how long is he in the air and how high does he go? Assume that he lands standing upright—that is, the same way he left the ground.

**56.** Romeo is throwing pebbles gently up to Juliet's window, and he wants the pebbles to hit the window with only a horizontal component of velocity. He is standing at the edge of a rose garden 8.0 m below her window and 8.5 m from the base of the wall (Fig. 3–49). How fast are the pebbles going when they hit her window?

**FIGURE 3–49**
Problem 56.

|← 8.5 m →|

**57.** *Apollo* astronauts took a "nine iron" to the Moon and hit a golf ball about 180 m. Assuming that the swing, launch angle, and so on, were the same as on Earth where the same astronaut could hit it only 32 m, estimate the acceleration due to gravity on the surface of the Moon. (We neglect air resistance in both cases, but on the Moon there is none.)

**58.** (a) A long jumper leaves the ground at 45° above the horizontal and lands 8.0 m away. What is her "takeoff" speed $v_0$? (b) Now she is out on a hike and comes to the left bank of a river. There is no bridge and the right bank is 10.0 m away horizontally and 2.5 m vertically below. If she long jumps from the edge of the left bank at 45° with the speed calculated in (a), how long, or short, of the opposite bank will she land (Fig. 3–50)?

**FIGURE 3–50** Problem 58.

**59.** A projectile is shot from the edge of a cliff 115 m above ground level with an initial speed of 65.0 m/s at an angle of 35.0° with the horizontal, as shown in Fig. 3–51. (a) Determine the time taken by the projectile to hit point P at ground level. (b) Determine the distance $X$ of point P from the base of the vertical cliff. At the instant just before the projectile hits point P, find (c) the horizontal and the vertical components of its velocity, (d) the magnitude of the velocity, and (e) the angle made by the velocity vector with the horizontal. (f) Find the maximum height above the cliff top reached by the projectile.

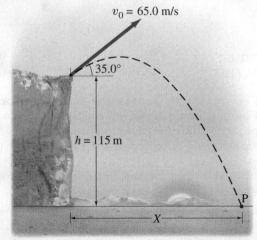

**FIGURE 3–51** Problem 59.

**60.** William Tell must split the apple on top of his son's head from a distance of 27 m. When William aims directly at the apple, the arrow is horizontal. At what angle should he aim the arrow to hit the apple if the arrow travels at a speed of 35 m/s?

**61.** Raindrops make an angle $\theta$ with the vertical when viewed through a moving train window (Fig. 3–52). If the speed of the train is $v_T$, what is the speed of the raindrops in the reference frame of the Earth in which they are assumed to fall vertically?

**FIGURE 3–52**
Problem 61.

**62.** A car moving at 95 km/h passes a 1.00-km-long train traveling in the same direction on a track that is parallel to the road. If the speed of the train is 75 km/h, how long does it take the car to pass the train, and how far will the car have traveled in this time? What are the results if the car and train are instead traveling in opposite directions?

**63.** A hunter aims directly at a target (on the same level) 38.0 m away. (*a*) If the arrow leaves the bow at a speed of 23.1 m/s, by how much will it miss the target? (*b*) At what angle should the bow be aimed so the target will be hit?

**64.** The cliff divers of Acapulco push off horizontally from rock platforms about 35 m above the water, but they must clear rocky outcrops at water level that extend out into the water 5.0 m from the base of the cliff directly under their launch point. See Fig. 3–53. What minimum pushoff speed is necessary to clear the rocks? How long are they in the air?

**FIGURE 3–53**
Problem 64.

**65.** When Babe Ruth hit a homer over the 8.0-m-high right-field fence 98 m from home plate, roughly what was the minimum speed of the ball when it left the bat? Assume the ball was hit 1.0 m above the ground and its path initially made a 36° angle with the ground.

**66.** At serve, a tennis player aims to hit the ball horizontally. What minimum speed is required for the ball to clear the 0.90-m-high net about 15.0 m from the server if the ball is "launched" from a height of 2.50 m? Where will the ball land if it just clears the net (and will it be "good" in the sense that it lands within 7.0 m of the net)? How long will it be in the air? See Fig. 3–54.

**FIGURE 3–54** Problem 66.

**67.** Spymaster Chris, flying a constant 208 km/h horizontally in a low-flying helicopter, wants to drop secret documents into her contact's open car which is traveling 156 km/h on a level highway 78.0 m below. At what angle (with the horizontal) should the car be in her sights when the packet is released (Fig. 3–55)?

**FIGURE 3–55** Problem 67.

**68.** A basketball leaves a player's hands at a height of 2.10 m above the floor. The basket is 3.05 m above the floor. The player likes to shoot the ball at a 38.0° angle. If the shot is made from a horizontal distance of 11.00 m and must be accurate to ±0.22 m (horizontally), what is the range of initial speeds allowed to make the basket?

**69.** A boat can travel 2.20 m/s in still water. (*a*) If the boat points directly across a stream whose current is 1.20 m/s, what is the velocity (magnitude and direction) of the boat relative to the shore? (*b*) What will be the position of the boat, relative to its point of origin, after 3.00 s?

**70.** A projectile is launched from ground level to the top of a cliff which is 195 m away and 135 m high (see Fig. 3–56). If the projectile lands on top of the cliff 6.6 s after it is fired, find the initial velocity of the projectile (magnitude and direction). Neglect air resistance.

**FIGURE 3–56**
Problem 70.

**71.** A basketball is shot from an initial height of 2.40 m (Fig. 3–57) with an initial speed $v_0 = 12$ m/s directed at an angle $\theta_0 = 35°$ above the horizontal. (*a*) How far from the basket was the player if he made a basket? (*b*) At what angle to the horizontal did the ball enter the basket?

**FIGURE 3–57**
Problem 71.

**72.** A rock is kicked horizontally at 15 m/s from a hill with a 45° slope (Fig. 3–58). How long does it take for the rock to hit the ground?

15 m/s

45°

**FIGURE 3–58** Problem 72.

**73.** A batter hits a fly ball which leaves the bat 0.90 m above the ground at an angle of 61° with an initial speed of 28 m/s heading toward centerfield. Ignore air resistance. (*a*) How far from home plate would the ball land if not caught? (*b*) The ball is caught by the centerfielder who, starting at a distance of 105 m from home plate just as the ball was hit, runs straight toward home plate at a constant speed and makes the catch at ground level. Find his speed.

**74.** A ball is shot from the top of a building with an initial velocity of 18 m/s at an angle $\theta = 42°$ above the horizontal. (*a*) What are the horizontal and vertical components of the initial velocity? (*b*) If a nearby building is the same height and 55 m away, how far below the top of the building will the ball strike the nearby building?

**75.** If a baseball pitch leaves the pitcher's hand horizontally at a velocity of 150 km/h, by what % will the pull of gravity change the magnitude of the velocity when the ball reaches the batter, 18 m away? For this estimate, ignore air resistance and spin on the ball.

# Search and Learn

**1.** Here is something to try at a sporting event. Show that the maximum height $h$ attained by an object projected into the air, such as a baseball, football, or soccer ball, is approximately given by

$$h \approx 1.2 t^2 \text{ m},$$

where $t$ is the total time of flight for the object in seconds. Assume that the object returns to the same level as that from which it was launched, as in Fig. 3–59. For example, if you count to find that a baseball was in the air for $t = 5.0$ s, the maximum height attained was $h = 1.2 \times (5.0)^2 = 30$ m. The fun of this relation is that $h$ can be determined without knowledge of the launch speed $v_0$ or launch angle $\theta_0$. Why is that exactly? See Section 3–6.

$v_0$

$\theta_0$

$h$

**FIGURE 3–59** Search and Learn 1.

**2.** Two balls are thrown in the air at different angles, but each reaches the same height. Which ball remains in the air longer? Explain, using equations.

**3.** Show that the speed with which a projectile leaves the ground is equal to its speed just before it strikes the ground at the end of its journey, assuming the firing level equals the landing level.

**4.** The initial angle of projectile A is 30°, while that of projectile B is 60°. Both have the same level horizontal range. How do the initial velocities and flight times (elapsed time from launch until landing) compare for A and B?

**5.** You are driving south on a highway at 12 m/s (approximately 25 mi/h) in a snowstorm. When you last stopped, you noticed that the snow was coming down vertically, but it is passing the windows of the moving car at an angle of 7.0° to the horizontal. Estimate the speed of the vertically falling snowflakes relative to the ground. [*Hint:* Construct a relative velocity diagram similar to Fig. 3–29 or 3–30. Be careful about which angle is the angle given.]

## ANSWERS TO EXERCISES

**A:** $3.0 \sqrt{2} \approx 4.2$ units.
**B:** (*a*).
**C:** They hit at the same time.

**D:** (i) Nowhere; (ii) at the highest point; (iii) nowhere.
**E:** (*d*). It provides the initial velocity of the box.
**F:** (*a*) $v = v_{x0} = 16.0$ m/s, horizontal; (*b*) 9.80 m/s² down.

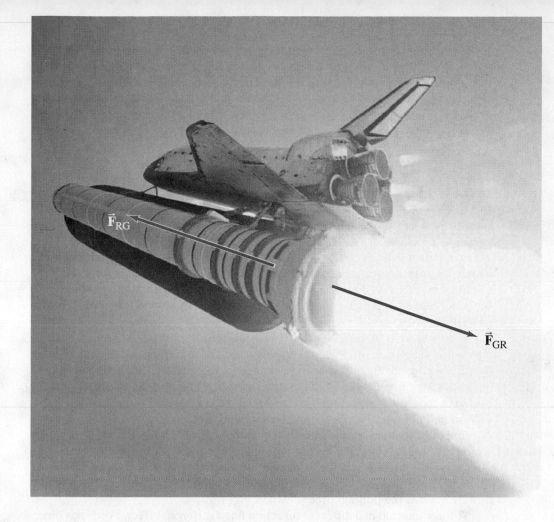

A space shuttle is carried out into space by powerful rockets. They are accelerating, increasing in speed rapidly. To do so, a force must be exerted on them according to Newton's second law, $\Sigma\vec{F} = m\vec{a}$. What exerts this force? The rocket engines exert a force on the gases they push out (expel) from the rear of the rockets (labeled $\vec{F}_{GR}$). According to Newton's third law, these ejected gases exert an equal and opposite force on the rockets in the forward direction. It is this "reaction" force exerted *on* the rockets *by* the gases, labeled $\vec{F}_{RG}$, that accelerates the rockets forward.

# Dynamics: Newton's Laws of Motion

## CHAPTER 4

### CHAPTER-OPENING QUESTIONS—Guess now!

1. A 150-kg football player collides head-on with a 75-kg running back. During the collision, the heavier player exerts a force of magnitude $F_A$ on the smaller player. If the smaller player exerts a force $F_B$ back on the heavier player, which response is most accurate?
   - (a) $F_B = F_A$.
   - (b) $F_B < F_A$.
   - (c) $F_B > F_A$.
   - (d) $F_B = 0$.
   - (e) We need more information.

2. A line by the poet T. S. Eliot (from *Murder in the Cathedral*) has the women of Canterbury say "the earth presses up against our feet." What force is this?
   - (a) Gravity.
   - (b) The normal force.
   - (c) A friction force.
   - (d) Centrifugal force.
   - (e) No force—they are being poetic.

### CONTENTS

e have discussed how motion is described in terms of velocity and acceleration. Now we deal with the question of *why* objects move as they do: What makes an object at rest begin to move? What causes an object to accelerate or decelerate? What is involved when an object moves in a curved path? We can answer in each case that a force is required. In this Chapter[†], we will investigate the connection between force and motion, which is the subject called **dynamics**.

# 4–1 Force

**FIGURE 4–1** A force exerted on a grocery cart—in this case exerted by a person.

Intuitively, we experience **force** as any kind of a push or a pull on an object. When you push a stalled car or a grocery cart (Fig. 4–1), you are exerting a force on it. When a motor lifts an elevator, or a hammer hits a nail, or the wind blows the leaves of a tree, a force is being exerted. We often call these *contact forces* because the force is exerted when one object comes in contact with another object. On the other hand, we say that an object falls because of the *force of gravity* (which is not a contact force).

If an object is at rest, to start it moving requires force—that is, a force is needed to accelerate an object from zero velocity to a nonzero velocity. For an object already moving, if you want to change its velocity—either in direction or in magnitude—a force is required. In other words, to accelerate an object, a force is always required. In Section 4–4 we discuss the precise relation between acceleration and net force, which is Newton's second law.

One way to measure the magnitude (or strength) of a force is to use a spring scale (Fig. 4–2). Normally, such a spring scale is used to find the weight of an object; by weight we mean the force of gravity acting on the object (Section 4–6). The spring scale, once calibrated, can be used to measure other kinds of forces as well, such as the pulling force shown in Fig. 4–2.

A force exerted in a different direction has a different effect. Force has direction as well as magnitude, and is indeed a vector that follows the rules of vector addition discussed in Chapter 3. We can represent any force on a diagram by an arrow, just as we did with velocity. The direction of the arrow is the direction of the push or pull, and its length is drawn proportional to the magnitude of the force.

**FIGURE 4–2** A spring scale used to measure a force.

# 4–2 Newton's First Law of Motion

What is the relationship between force and motion? Aristotle (384–322 B.C.) believed that a force was required to keep an object moving along a horizontal plane. To Aristotle, the natural state of an object was at rest, and a force was believed necessary to keep an object in motion. Furthermore, Aristotle argued, the greater the force on the object, the greater its speed.

Some 2000 years later, Galileo disagreed: he maintained that it is just as natural for an object to be in motion with a constant velocity as it is for it to be at rest.

---

[†]We treat everyday objects in motion here. When velocities are extremely high, close to the speed of light ($3.0 \times 10^8$ m/s), we use the theory of relativity (Chapter 26), and in the submicroscopic world of atoms and molecules we use quantum theory (Chapter 27 *ff*).

To understand Galileo's idea, consider the following observations involving motion along a horizontal plane. To push an object with a rough surface along a tabletop at constant speed requires a certain amount of force. To push an equally heavy object with a very smooth surface across the table at the same speed will require less force. If a layer of oil or other lubricant is placed between the surface of the object and the table, then almost no force is required to keep the object moving. Notice that in each successive step, less force is required. As the next step, we imagine there is no friction at all, that the object does not rub against the table—or there is a perfect lubricant between the object and the table—and theorize that once started, the object would move across the table at constant speed with *no* force applied. A steel ball bearing rolling on a hard horizontal surface approaches this situation. So does a puck on an air table, in which a thin layer of air reduces friction almost to zero.

It was Galileo's genius to imagine such an idealized world—in this case, one where there is no friction—and to see that it could lead to a more accurate and richer understanding of the real world. This idealization led him to his remarkable conclusion that if no force is applied to a moving object, it will continue to move with constant speed in a straight line. An object slows down only if a force is exerted on it. Galileo thus interpreted friction as a force akin to ordinary pushes and pulls.

**FIGURE 4–3** $\vec{\mathbf{F}}$ represents the force applied by the person and $\vec{\mathbf{F}}_{fr}$ represents the force of friction.

To push an object across a table at constant speed requires a force from your hand that can balance the force of friction (Fig. 4–3). When the object moves at constant speed, your pushing force is equal in magnitude to the friction force; but these two forces are in opposite directions, so the *net* force on the object (the vector sum of the two forces) is zero. This is consistent with Galileo's viewpoint, for the object moves with constant velocity when no *net* force is exerted on it.

Upon this foundation laid by Galileo, Isaac Newton (Fig. 4–4) built his great theory of motion. Newton's analysis of motion is summarized in his famous "three laws of motion." In his great work, the *Principia* (published in 1687), Newton readily acknowledged his debt to Galileo. In fact, **Newton's first law of motion** is close to Galileo's conclusions. It states that

> **Every object continues in its state of rest, or of uniform velocity in a straight line, as long as no net force acts on it.**

| NEWTON'S FIRST LAW OF MOTION |
| --- |

The tendency of an object to maintain its state of rest or of uniform velocity in a straight line is called **inertia**. As a result, Newton's first law is often called the **law of inertia**.

**CONCEPTUAL EXAMPLE 4–1** **Newton's first law.** A school bus comes to a sudden stop, and all of the backpacks on the floor start to slide forward. What force causes them to do that?

**RESPONSE** It isn't "force" that does it. By Newton's first law, the backpacks continue their state of motion, maintaining their velocity. The backpacks slow down if a force is applied, such as friction with the floor.

**FIGURE 4–4**
Isaac Newton (1642–1727). Besides developing mechanics, including his three great laws of motion and the law of universal gravitation, he also tried to understand the nature of light.

## Inertial Reference Frames

Newton's first law does not hold in every reference frame. For example, if your reference frame is an accelerating car, an object such as a cup resting on the dashboard may begin to move toward you (it stayed at rest as long as the car's velocity remained constant). The cup accelerated toward you, but neither you nor anything else exerted a force on it in that direction. Similarly, in the reference frame of the decelerating bus in Example 4–1, there was no force pushing the backpacks forward. In accelerating reference frames, Newton's first law does not hold. Physics is easier in reference frames in which Newton's first law *does* hold, and they are called **inertial reference frames** (the law of inertia is valid in them). For most purposes, we usually make the approximation that a reference frame fixed on the Earth is an inertial frame. This is not precisely true, due to the Earth's rotation, but usually it is close enough.

Any reference frame that moves with constant velocity (say, a car or an airplane) relative to an inertial frame is also an inertial reference frame. Reference frames where the law of inertia does *not* hold, such as the accelerating reference frames discussed above, are called **noninertial** reference frames. How can we be sure a reference frame is inertial or not? By checking to see if Newton's first law holds. Thus Newton's first law serves as the definition of inertial reference frames.

# 4–3 Mass

Newton's second law, which we come to in the next Section, makes use of the concept of mass. Newton used the term *mass* as a synonym for "quantity of matter." This intuitive notion of the mass of an object is not very precise because the concept "quantity of matter" is not very well defined. More precisely, we can say that **mass** is a *measure of the inertia* of an object. The more mass an object has, the greater the force needed to give it a particular acceleration. It is harder to start it moving from rest, or to stop it when it is moving, or to change its velocity sideways out of a straight-line path. A truck has much more inertia than a baseball moving at the same speed, and a much greater force is needed to change the truck's velocity at the same rate as the ball's. The truck therefore has much more mass.

To quantify the concept of mass, we must define a standard. In SI units, the unit of mass is the **kilogram** (kg) as we discussed in Chapter 1, Section 1–5.

**CAUTION**

*Distinguish mass from weight*

The terms *mass* and *weight* are often confused with one another, but it is important to distinguish between them. Mass is a property of an object itself (a measure of an object's inertia, or its "quantity of matter"). Weight, on the other hand, is a force, the pull of gravity acting on an object. To see the difference, suppose we take an object to the Moon. The object will weigh only about one-sixth as much as it did on Earth, since the force of gravity is weaker. But its mass will be the same. It will have the same amount of matter as on Earth, and will have just as much inertia—in the absence of friction, it will be just as hard to start it moving on the Moon as on Earth, or to stop it once it is moving. (More on weight in Section 4–6.)

# 4–4 Newton's Second Law of Motion

Newton's first law states that if no net force is acting on an object at rest, the object remains at rest; or if the object is moving, it continues moving with constant speed in a straight line. But what happens if a net force *is* exerted on an object? Newton perceived that the object's velocity will change (Fig. 4–5). A net force exerted on an object may make its velocity increase. Or, if the net force is in a direction opposite to the motion, that force will reduce the object's velocity. If the net force acts sideways on a moving object, the *direction* of the object's velocity changes. That change in the *direction* of the velocity is also an acceleration. So a sideways net force on an object also causes acceleration. In general, we can say that *a net force causes acceleration.*

**FIGURE 4–5** The bobsled accelerates because the team exerts a force.

What precisely is the relationship between acceleration and force? Everyday experience can suggest an answer. Consider the force required to push a cart when friction is small enough to ignore. (If there is friction, consider the *net* force, which is the force you exert minus the force of friction.) If you push the cart horizontally with a gentle but constant force for a certain period of time, you will make the cart accelerate from rest up to some speed, say 3 km/h. If you push with twice the force, the cart will reach 3 km/h in half the time. The acceleration will be twice as great. If you triple the force, the acceleration is tripled, and so on. Thus, the acceleration of an object is directly proportional[†] to the net applied force. But the acceleration depends on the mass of the object as well. If you push an empty grocery cart with the same force as you push one that is filled with groceries, you will find that the full cart accelerates more slowly.

†A review of proportionality is given in Appendix A.

The greater the mass, the less the acceleration for the same net force. The mathematical relation, as Newton argued, is that the acceleration of an object is inversely proportional to its mass. These relationships are found to hold in general and can be summarized as follows:

> **The acceleration of an object is directly proportional to the net force acting on it, and is inversely proportional to the object's mass. The direction of the acceleration is in the direction of the net force acting on the object.**

NEWTON'S SECOND LAW
OF MOTION

This is **Newton's second law of motion**.

Newton's second law can be written as an equation:

$$\vec{a} = \frac{\Sigma \vec{F}}{m},$$

where $\vec{a}$ stands for acceleration, $m$ for the mass, and $\Sigma\vec{F}$ for the *net force* on the object. The symbol $\Sigma$ (Greek "sigma") stands for "sum of"; $\vec{F}$ stands for force, so $\Sigma\vec{F}$ means the *vector sum of all forces* acting on the object, which we define as the **net force**.

We rearrange this equation to obtain the familiar statement of Newton's second law:

$$\Sigma\vec{F} = m\vec{a}. \qquad (4-1)$$

NEWTON'S SECOND LAW
OF MOTION

Newton's second law relates the description of motion to the cause of motion, force. It is one of the most fundamental relationships in physics. From Newton's second law we can make a more precise definition of **force** as *an action capable of accelerating an object*.

Every force $\vec{F}$ is a vector, with magnitude and direction. Equation 4–1 is a vector equation valid in any inertial reference frame. It can be written in component form in rectangular coordinates as

$$\Sigma F_x = ma_x, \qquad \Sigma F_y = ma_y, \qquad \Sigma F_z = ma_z.$$

If the motion is all along a line (one-dimensional), we can leave out the subscripts and simply write $\Sigma F = ma$. Again, $a$ is the acceleration of an object of mass $m$, and $\Sigma F$ includes all the forces acting on that object, and *only* forces acting on that object. (Sometimes the net force $\Sigma F$ is written as $F_{net}$, so $F_{net} = ma$.)

In SI units, with the mass in kilograms, the unit of force is called the **newton** (N). One newton is the force required to impart an acceleration of $1 \, \text{m/s}^2$ to a mass of 1 kg. Thus $1 \, \text{N} = 1 \, \text{kg} \cdot \text{m/s}^2$.

In cgs units, the unit of mass is the gram[†] (g). The unit of force is the *dyne*, which is defined as the net force needed to impart an acceleration of $1 \, \text{cm/s}^2$ to a mass of 1 g. Thus 1 dyne $= 1 \, \text{g} \cdot \text{cm/s}^2$. Because $1 \, \text{g} = 10^{-3} \, \text{kg}$ and $1 \, \text{cm} = 10^{-2} \, \text{m}$, then 1 dyne $= 10^{-5} \, \text{N}$.

In the British system, which we rarely use, the unit of force is the *pound* (abbreviated lb), where $1 \, \text{lb} = 4.44822 \, \text{N} \approx 4.45 \, \text{N}$. The unit of mass is the *slug*, which is defined as that mass which will undergo an acceleration of $1 \, \text{ft/s}^2$ when a force of 1 lb is applied to it. Thus $1 \, \text{lb} = 1 \, \text{slug} \cdot \text{ft/s}^2$. Table 4–1 summarizes the units in the different systems.

It is very important that only one set of units be used in a given calculation or Problem, with the SI being what we almost always use. If the force is given in, say, newtons, and the mass in grams, then before attempting to solve for the acceleration in SI units, we must change the mass to kilograms. For example, if the force is given as 2.0 N along the $x$ axis and the mass is 500 g, we change the latter to 0.50 kg, and the acceleration will then automatically come out in $\text{m/s}^2$ when Newton's second law is used:

$$a_x = \frac{\Sigma F_x}{m} = \frac{2.0 \, \text{N}}{0.50 \, \text{kg}} = \frac{2.0 \, \text{kg} \cdot \text{m/s}^2}{0.50 \, \text{kg}} = 4.0 \, \text{m/s}^2,$$

where we set $1 \, \text{N} = 1 \, \text{kg} \cdot \text{m/s}^2$.

**TABLE 4–1**
**Units for Mass and Force**

| System | Mass | Force |
|---|---|---|
| SI | kilogram (kg) | newton (N) ($= \text{kg} \cdot \text{m/s}^2$) |
| cgs | gram (g) | dyne ($= \text{g} \cdot \text{cm/s}^2$) |
| British | slug | pound (lb) |

Conversion factors: 1 dyne $= 10^{-5} \, \text{N}$;
1 lb $\approx 4.45 \, \text{N}$;
1 slug $\approx 14.6 \, \text{kg}$.

**PROBLEM SOLVING**
*Use a consistent set of units*

---

[†] Be careful not to confuse g for gram with $g$ for the acceleration due to gravity. The latter is always italicized (or boldface when shown as a vector).

**EXAMPLE 4–2** | **ESTIMATE** | **Force to accelerate a fast car.** Estimate the net force needed to accelerate (a) a 1000-kg car at $\frac{1}{2}g$; (b) a 200-gram apple at the same rate.

**APPROACH** We use Newton's second law to find the net force needed for each object; we are given the mass and the acceleration. This is an estimate (the $\frac{1}{2}$ is not said to be precise) so we round off to one significant figure.

**SOLUTION** (a) The car's acceleration is $a = \frac{1}{2}g = \frac{1}{2}(9.8 \text{ m/s}^2) \approx 5 \text{ m/s}^2$. We use Newton's second law to get the net force needed to achieve this acceleration:

$$\Sigma F = ma \approx (1000 \text{ kg})(5 \text{ m/s}^2) = 5000 \text{ N}.$$

(If you are used to British units, to get an idea of what a 5000-N force is, you can divide by 4.45 N/lb and get a force of about 1000 lb.)

(b) For the apple, $m = 200 \text{ g} = 0.2 \text{ kg}$, so

$$\Sigma F = ma \approx (0.2 \text{ kg})(5 \text{ m/s}^2) = 1 \text{ N}.$$

---

**EXAMPLE 4–3** | **Force to stop a car.** What average net force is required to bring a 1500-kg car to rest from a speed of 100 km/h within a distance of 55 m?

**APPROACH** We use Newton's second law, $\Sigma F = ma$, to determine the force, but first we need to calculate the acceleration $a$. We assume the acceleration is constant so that we can use the kinematic equations, Eqs. 2–11, to calculate it.

**FIGURE 4–6** Example 4–3.

**SOLUTION** We assume the motion is along the $+x$ axis (Fig. 4–6). We are given the initial velocity $v_0 = 100 \text{ km/h} = 27.8 \text{ m/s}$ (Section 1–6), the final velocity $v = 0$, and the distance traveled $x - x_0 = 55 \text{ m}$. From Eq. 2–11c, we have

$$v^2 = v_0^2 + 2a(x - x_0),$$

so

$$a = \frac{v^2 - v_0^2}{2(x - x_0)} = \frac{0 - (27.8 \text{ m/s})^2}{2(55 \text{ m})} = -7.0 \text{ m/s}^2.$$

The net force required is then

$$\Sigma F = ma = (1500 \text{ kg})(-7.0 \text{ m/s}^2) = -1.1 \times 10^4 \text{ N},$$

or 11,000 N. The force must be exerted in the direction *opposite* to the initial velocity, which is what the negative sign means.

**NOTE** If the acceleration is not precisely constant, then we are determining an "average" acceleration and we obtain an "average" net force.

---

Newton's second law, like the first law, is valid only in inertial reference frames (Section 4–2). In the noninertial reference frame of a car that begins accelerating, a cup on the dashboard starts sliding—it accelerates—even though the net force on it is zero. Thus $\Sigma \vec{\mathbf{F}} = m\vec{\mathbf{a}}$ does not work in such an accelerating reference frame ($\Sigma \vec{\mathbf{F}} = 0$, but $\vec{\mathbf{a}} \neq 0$ in this noninertial frame).

**EXERCISE A** Suppose you watch a cup slide on the (smooth) dashboard of an accelerating car as we just discussed, but this time from an inertial reference frame outside the car, on the street. From your inertial frame, Newton's laws are valid. What force pushes the cup off the dashboard?

# 4–5 Newton's Third Law of Motion

Newton's second law of motion describes quantitatively how forces affect motion. But where, we may ask, do forces come from? Observations suggest that a force exerted on any object is always exerted *by another object*. A horse pulls a wagon, a person pushes a grocery cart, a hammer pushes on a nail, a magnet attracts a paper clip. In each of these examples, a force is exerted *on* one object, and that force is exerted *by* another object. For example, the force exerted *on* the nail is exerted *by* the hammer.

But Newton realized that things are not so one-sided. True, the hammer exerts a force on the nail (Fig. 4–7). But the nail evidently exerts a force back on the hammer as well, for the hammer's speed is rapidly reduced to zero upon contact. Only a strong force could cause such a rapid deceleration of the hammer. Thus, said Newton, the two objects must be treated on an equal basis. The hammer exerts a force on the nail, and the nail exerts a force back on the hammer. This is the essence of **Newton's third law of motion**:

> **Whenever one object exerts a force on a second object, the second object exerts an equal force in the opposite direction on the first.**

This law is sometimes paraphrased as "to every action there is an equal and opposite reaction." This is perfectly valid. But to avoid confusion, it is very important to remember that the "action" force and the "reaction" force are acting on *different* objects.

As evidence for the validity of Newton's third law, look at your hand when you push against the edge of a desk, Fig. 4–8. Your hand's shape is distorted, clear evidence that a force is being exerted on it. You can *see* the edge of the desk pressing into your hand. You can even *feel* the desk exerting a force on your hand; it hurts! The harder you push against the desk, the harder the desk pushes back on your hand. (You only feel forces exerted *on* you; when you exert a force on another object, what you feel is that object pushing back on you.)

**FIGURE 4–7** A hammer striking a nail. The hammer exerts a force on the nail and the nail exerts a force back on the hammer. The latter force decelerates the hammer and brings it to rest.

NEWTON'S THIRD LAW OF MOTION

⚠ **CAUTION**

*Action and reaction forces act on different objects*

Force exerted on hand by desk

Force exerted on desk by hand

**FIGURE 4–8** If your hand pushes against the edge of a desk (the force vector is shown in red), the desk pushes back against your hand (this force vector is shown in a different color, violet, to remind us that this force acts on a different object).

The force the desk exerts on your hand has the same magnitude as the force your hand exerts on the desk. This is true not only if the desk is at rest but is true even if the desk is accelerating due to the force your hand exerts.

As another demonstration of Newton's third law, consider the ice skater in Fig. 4–9. There is very little friction between her skates and the ice, so she will move freely if a force is exerted on her. She pushes against the wall; and then *she* starts moving backward. The force she exerts on the wall cannot make *her* start moving, because that force acts on the wall. Something had to exert a force *on her* to start her moving, and that force could only have been exerted by the wall. The force with which the wall pushes on her is, by Newton's third law, equal and opposite to the force she exerts on the wall.

When a person throws a package out of a small boat (initially at rest), the boat starts moving in the opposite direction. The person exerts a force on the package. The package exerts an equal and opposite force back on the person, and this force propels the person (and the boat) backward slightly.

**FIGURE 4–9** An example of Newton's third law: when an ice skater pushes against the wall, the wall pushes back and this force causes her to accelerate away.

Force on skater   Force on wall

**FIGURE 4–10** Another example of Newton's third law: the launch of a rocket. The rocket engine pushes the gases downward, and the gases exert an equal and opposite force upward on the rocket, accelerating it upward. (A rocket does *not* accelerate as a result of its expelled gases pushing against the ground.)

**FIGURE 4–11** We can walk forward because, when one foot pushes backward against the ground, the ground pushes forward on that foot (Newton's third law). The two forces shown *act on different objects.*

Horizontal force exerted on the **g**round by **p**erson's foot

$\vec{\mathbf{F}}_{GP}$

Horizontal force exerted on the **p**erson's foot by the **g**round

$\vec{\mathbf{F}}_{PG}$

---

*NEWTON'S THIRD LAW OF MOTION*

---

Rocket propulsion also is explained using Newton's third law (Fig. 4–10). A common misconception is that rockets accelerate because the gases rushing out the back of the engine push against the ground or the atmosphere. Not true. What happens, instead, is that a rocket exerts a strong force on the gases, expelling them; and the gases exert an equal and opposite force *on the rocket*. It is this latter force that propels the rocket forward—the force exerted *on* the rocket *by* the gases (see Chapter-Opening Photo, page 75). Thus, a space vehicle is maneuvered in empty space by firing its rockets in the direction opposite to that in which it needs to accelerate. When the rocket pushes on the gases in one direction, the gases push back on the rocket in the opposite direction. Jet aircraft too accelerate because the gases they thrust out backwards exert a forward force on the engines (Newton's third law).

Consider how we walk. A person begins walking by pushing with the foot backward against the ground. The ground then exerts an equal and opposite force forward on the person (Fig. 4–11), and it is this force, *on* the person, that moves the person forward. (If you doubt this, try walking normally where there is no friction, such as on very smooth slippery ice.) In a similar way, a bird flies forward by exerting a backward force on the air, but it is the air pushing forward (Newton's third law) on the bird's wings that propels the bird forward.

---

**CONCEPTUAL EXAMPLE 4–4** | **What exerts the force to move a car?** What makes a car go forward?

**RESPONSE** A common answer is that the engine makes the car move forward. But it is not so simple. The engine makes the wheels go around. But if the tires are on slick ice or wet mud, they just spin. Friction is needed. On firm ground, the tires push backward against the ground because of friction. By Newton's third law, the ground pushes on the tires in the opposite direction, accelerating the car forward.

---

We tend to associate forces with active objects such as humans, animals, engines, or a moving object like a hammer. It is often difficult to see how an inanimate object at rest, such as a wall or a desk, or the wall of an ice rink (Fig. 4–9), can exert a force. The explanation is that every material, no matter how hard, is elastic (springy) at least to some degree. A stretched rubber band can exert a force on a wad of paper and accelerate it to fly across the room. Other materials may not stretch as readily as rubber, but they do stretch or compress when a force is applied to them. And just as a stretched rubber band exerts a force, so does a stretched (or compressed) wall, desk, or car fender.

From the examples discussed above, we can see how important it is to remember *on* what object a given force is exerted and *by* what object that force is exerted. A force influences the motion of an object only when it is applied *on* that object. A force exerted *by* an object does not influence that same object; it only influences the other object *on* which it is exerted. Thus, to avoid confusion, the two prepositions *on* and *by* must always be used—and used with care.

One way to keep clear which force acts on which object is to use double subscripts. For example, the force exerted *on* the **P**erson *by* the **G**round as the person walks in Fig. 4–11 can be labeled $\vec{\mathbf{F}}_{PG}$. And the force exerted on the ground by the person is $\vec{\mathbf{F}}_{GP}$. By Newton's third law

$$\vec{\mathbf{F}}_{GP} = -\vec{\mathbf{F}}_{PG}. \tag{4–2}$$

$\vec{\mathbf{F}}_{GP}$ and $\vec{\mathbf{F}}_{PG}$ have the same magnitude (Newton's third law), and the minus sign reminds us that these two forces are in opposite directions.

Note carefully that the two forces shown in Fig. 4–11 act on different objects—to emphasize this we used slightly different colors for the vector arrows representing these forces. These two forces would never appear together in a sum of forces in Newton's second law, $\Sigma\vec{\mathbf{F}} = m\vec{\mathbf{a}}$. Why not? Because they act on different objects: $\vec{\mathbf{a}}$ is the acceleration of one particular object, and $\Sigma\vec{\mathbf{F}}$ must include *only* the forces on that *one* object.

Force on sled
exerted by
assistant

Force on
assistant
exerted
by sled

$\vec{\mathbf{F}}_{SA}$
$(= -\vec{\mathbf{F}}_{AS})$

$\vec{\mathbf{F}}_{AS}$

$\vec{\mathbf{F}}_{SG}$     $\vec{\mathbf{F}}_{GS} (= -\vec{\mathbf{F}}_{SG})$

$\vec{\mathbf{F}}_{GA}$     $\vec{\mathbf{F}}_{AG}$

Friction
force on
sled exerted
by ground

Force on
ground
exerted
by sled

Force on
ground
exerted
by assistant

Force on
assistant
exerted
by ground

$(= -\vec{\mathbf{F}}_{AG})$

**FIGURE 4–12** Example 4–5, showing only horizontal forces. Michelangelo has selected a fine block of marble for his next sculpture. Shown here is his assistant pulling it on a sled away from the quarry. Forces on the assistant are shown as red (magenta) arrows. Forces on the sled are purple arrows. Forces acting on the ground are orange arrows. Action–reaction forces that are equal and opposite are labeled by the same subscripts but reversed (such as $\vec{\mathbf{F}}_{GA}$ and $\vec{\mathbf{F}}_{AG}$) and are of different colors because they act on different objects.

---

**CONCEPTUAL EXAMPLE 4–5** **Third law clarification.** Michelangelo's assistant has been assigned the task of moving a block of marble using a sled (Fig. 4–12). He says to his boss, "When I exert a forward force on the sled, the sled exerts an equal and opposite force backward. So how can I ever start it moving? No matter how hard I pull, the backward reaction force always equals my forward force, so the net force must be zero. I'll never be able to move this load." Is he correct?

**RESPONSE** No. Although it is true that the action and reaction forces are equal in magnitude, the assistant has forgotten that they are exerted on different objects. The forward ("action") force is exerted by the assistant on the sled (Fig. 4–12), whereas the backward "reaction" force is exerted by the sled on the assistant. To determine if the *assistant* moves or not, we must consider only the forces *on the assistant* and then apply $\Sigma\vec{\mathbf{F}} = m\vec{\mathbf{a}}$, where $\Sigma\vec{\mathbf{F}}$ is the net force *on the assistant*, $\vec{\mathbf{a}}$ is the acceleration of the assistant, and $m$ is the assistant's mass. There are two forces on the assistant that affect his forward motion; they are shown as bright red (magenta) arrows in Figs. 4–12 and 4–13: they are (1) the horizontal force $\vec{\mathbf{F}}_{AG}$ exerted on the assistant by the ground (the harder he pushes backward against the ground, the harder the ground pushes forward on him— Newton's third law), and (2) the force $\vec{\mathbf{F}}_{AS}$ exerted on the assistant by the sled, pulling backward on him; see Fig. 4–13. If he pushes hard enough on the ground, the force on him exerted by the ground, $\vec{\mathbf{F}}_{AG}$, will be larger than the sled pulling back, $\vec{\mathbf{F}}_{AS}$, and the assistant accelerates forward (Newton's second law). The sled, on the other hand, accelerates forward when the force on *it* exerted by the assistant is greater than the frictional force exerted backward on it by the ground (that is, when $\vec{\mathbf{F}}_{SA}$ has greater magnitude than $\vec{\mathbf{F}}_{SG}$ in Fig. 4–12).

**PROBLEM SOLVING**

*A study of Newton's second and third laws*

Force on
assistant
exerted
by sled

$\vec{\mathbf{F}}_{AS}$

$\vec{\mathbf{F}}_{AG}$

Force on
assistant
exerted
by ground

**FIGURE 4–13** Example 4–5. The horizontal forces on the assistant.

Using double subscripts to clarify Newton's third law can become cumbersome, and we won't usually use them in this way. We will usually use a single subscript referring to what exerts the force on the object being discussed. Nevertheless, if there is any confusion in your mind about a given force, go ahead and use two subscripts to identify *on* what object and *by* what object the force is exerted.

**EXERCISE B** Return to the first Chapter-Opening Question, page 75, and answer it again now. Try to explain why you may have answered differently the first time.

**EXERCISE C** A tennis ball collides head-on with a more massive baseball. (i) Which ball experiences the greater force of impact? (ii) Which experiences the greater acceleration during the impact? (iii) Which of Newton's laws are useful to obtain the correct answers?

**EXERCISE D** If you push on a heavy desk, does it always push back on you? (*a*) No. (*b*) Yes. (*c*) Not unless someone else also pushes on it. (*d*) Yes, if it is out in space. (*e*) A desk never pushes to start with.

# 4–6 Weight—the Force of Gravity; and the Normal Force

As we saw in Chapter 2, Galileo claimed that all objects dropped near the surface of the Earth would fall with the same acceleration, $\vec{g}$, if air resistance was negligible. The force that causes this acceleration is called the *force of gravity* or *gravitational force*. What exerts the gravitational force on an object? It is the Earth, as we will discuss in Chapter 5, and the force acts vertically[†] downward, toward the center of the Earth. Let us apply Newton's second law to an object of mass *m* falling freely due to gravity. For the acceleration, $\vec{a}$, we use the downward acceleration due to gravity, $\vec{g}$. Thus, the **gravitational force** on an object, $\vec{F}_G$, can be written as

$$\vec{F}_G = m\vec{g}. \tag{4–3}$$

The direction of this force is down toward the center of the Earth. The magnitude of the force of gravity on an object, *mg*, is commonly called the object's **weight**.

In SI units, $g = 9.80 \text{ m/s}^2 = 9.80 \text{ N/kg}$,[‡] so the weight of a 1.00-kg mass on Earth is $1.00 \text{ kg} \times 9.80 \text{ m/s}^2 = 9.80 \text{ N}$. We will mainly be concerned with the weight of objects on Earth, but we note that on the Moon, on other planets, or in space, the weight of a given mass will be different than it is on Earth. For example, on the Moon the acceleration due to gravity is about one-sixth what it is on Earth, and a 1.0-kg mass weighs only 1.6 N. Although we will not use British units, we note that for practical purposes on the Earth, a mass of 1.0 kg weighs about 2.2 lb. (On the Moon, 1 kg weighs only about 0.4 lb.)

The force of gravity acts on an object when it is falling. When an object is at rest on the Earth, the gravitational force on it does not disappear, as we know if we weigh it on a spring scale. The same force, given by Eq. 4–3, continues to act. Why, then, doesn't the object move? From Newton's second law, the net force on an object that remains at rest is zero. There must be another force on the object to balance the gravitational force. For an object resting on a table, the table exerts this upward force; see Fig. 4–14a. The table is compressed slightly beneath the object, and due to its elasticity, it pushes up on the object as shown. The force exerted by the table is often called a **contact force**, since it occurs when two objects are in contact. (The force of your hand pushing on a cart is also a contact force.) When a contact force acts *perpendicular* to the common surface of contact, it is referred to as the **normal force** ("normal" means perpendicular); hence it is labeled $\vec{F}_N$ in Fig. 4–14a.

The two forces shown in Fig. 4–14a are both acting on the statue, which remains at rest, so the vector sum of these two forces must be zero (Newton's second law). Hence $\vec{F}_G$ and $\vec{F}_N$ must be of equal magnitude and in opposite directions. But they are *not* the equal and opposite forces spoken of in Newton's third law. The action and reaction forces of Newton's third law act on *different objects*, whereas the two forces shown in Fig. 4–14a act on the *same* object. For each of the forces shown in Fig. 4–14a, we can ask, "What is the reaction force?" The upward force $\vec{F}_N$ on the statue is exerted by the table. The reaction to this force is a force exerted by the statue downward on the table. It is shown in Fig. 4–14b, where it is labeled $\vec{F}'_N$. This force, $\vec{F}'_N$, exerted on the table by the statue, is the reaction force to $\vec{F}_N$ in accord with Newton's third law. What about the other force on the statue, the force of gravity $\vec{F}_G$ exerted by the Earth? Can you guess what the reaction is to this force? We will see in Chapter 5 that the reaction force is also a gravitational force, exerted on the Earth by the statue.

**EXERCISE E** Return to the second Chapter-Opening Question, page 75, and answer it again now. Try to explain why you may have answered differently the first time.

(a)                    (b)

**FIGURE 4–14** (a) The net force on an object at rest is zero according to Newton's second law. Therefore the downward force of gravity ($\vec{F}_G$) on an object at rest must be balanced by an upward force (the normal force $\vec{F}_N$) exerted by the table in this case. (b) $\vec{F}'_N$ is the force exerted on the table by the statue and is the reaction force to $\vec{F}_N$ by Newton's third law. ($\vec{F}'_N$ is shown in a different color to remind us it acts on a different object.) The reaction force to $\vec{F}_G$ is not shown.

**CAUTION**

*Weight and normal force are **not** action–reaction pairs*

---

[†]The concept of "vertical" is tied to gravity. The best definition of *vertical* is that it is the direction in which objects fall. A surface that is "horizontal," on the other hand, is a surface on which a round object won't start rolling: gravity has no effect. Horizontal is perpendicular to vertical.

[‡]Since $1 \text{ N} = 1 \text{ kg·m/s}^2$ (Section 4–4), then $1 \text{ m/s}^2 = 1 \text{ N/kg}$.

**EXAMPLE 4–6** **Weight, normal force, and a box.** A friend has given you a special gift, a box of mass 10.0 kg with a mystery surprise inside. The box is resting on the smooth (frictionless) horizontal surface of a table (Fig. 4–15a). (*a*) Determine the weight of the box and the normal force exerted on it by the table. (*b*) Now your friend pushes down on the box with a force of 40.0 N, as in Fig. 4–15b. Again determine the normal force exerted on the box by the table. (*c*) If your friend pulls upward on the box with a force of 40.0 N (Fig. 4–15c), what now is the normal force exerted on the box by the table?

**APPROACH** The box is at rest on the table, so the net force on the box in each case is zero (Newton's first or second law). The weight of the box has magnitude $mg$ in all three cases.

**SOLUTION** (*a*) The weight of the box is $mg = (10.0 \, \text{kg})(9.80 \, \text{m/s}^2) = 98.0 \, \text{N}$, and this force acts downward. The only other force on the box is the normal force exerted upward on it by the table, as shown in Fig. 4–15a. We chose the upward direction as the positive $y$ direction; then the net force $\Sigma F_y$ on the box is $\Sigma F_y = F_N - mg$; the minus sign means $mg$ acts in the negative $y$ direction ($m$ and $g$ are magnitudes). The box is at rest, so the net force on it must be zero (Newton's second law, $\Sigma F_y = ma_y$, and $a_y = 0$). Thus

$$\Sigma F_y = ma_y$$
$$F_N - mg = 0,$$

so we have

$$F_N = mg.$$

The normal force on the box, exerted by the table, is 98.0 N upward, and has magnitude equal to the box's weight.

(*b*) Your friend is pushing down on the box with a force of 40.0 N. So instead of only two forces acting on the box, now there are three forces acting on the box, as shown in Fig. 4–15b. The weight of the box is still $mg = 98.0 \, \text{N}$. The net force is $\Sigma F_y = F_N - mg - 40.0 \, \text{N}$, and is equal to zero because the box remains at rest ($a = 0$). Newton's second law gives

$$\Sigma F_y = F_N - mg - 40.0 \, \text{N} = 0.$$

We solve this equation for the normal force:

$$F_N = mg + 40.0 \, \text{N} = 98.0 \, \text{N} + 40.0 \, \text{N} = 138.0 \, \text{N},$$

which is greater than in (*a*). The table pushes back with more force when a person pushes down on the box. The normal force is not always equal to the weight!

(*c*) The box's weight is still 98.0 N and acts downward. The force exerted by your friend and the normal force both act upward (positive direction), as shown in Fig. 4–15c. The box doesn't move since your friend's upward force is less than the weight. The net force, again set to zero in Newton's second law because $a = 0$, is

$$\Sigma F_y = F_N - mg + 40.0 \, \text{N} = 0,$$

so

$$F_N = mg - 40.0 \, \text{N} = 98.0 \, \text{N} - 40.0 \, \text{N} = 58.0 \, \text{N}.$$

The table does not push against the full weight of the box because of the upward force exerted by your friend.

**NOTE** The weight of the box ($-mg$) does not change as a result of your friend's push or pull. Only the normal force is affected.

Recall that the normal force is elastic in origin (the table in Fig. 4–15 sags slightly under the weight of the box). The normal force in Example 4–6 is vertical, perpendicular to the horizontal table. The normal force is not always vertical, however. When you push against a wall, for example, the normal force with which the wall pushes back on you is horizontal (Fig. 4–9). For an object on a plane inclined at an angle to the horizontal, such as a skier or car on a hill, the normal force acts perpendicular to the plane and so is not vertical.

(a) $\Sigma F_y = F_N - mg = 0$

(b) $\Sigma F_y = F_N - mg - 40.0 \, \text{N} = 0$

(c) $\Sigma F_y = F_N - mg + 40.0 \, \text{N} = 0$

**FIGURE 4–15** Example 4–6. (a) A 10-kg gift box is at rest on a table. (b) A person pushes down on the box with a force of 40.0 N. (c) A person pulls upward on the box with a force of 40.0 N. The forces are all assumed to act along a line; they are shown slightly displaced in order to be distinguishable. Only forces acting on the box are shown.

⬥ **CAUTION**

*The normal force is not always equal to the weight*

⬥ **CAUTION**

*The normal force, $\vec{F}_N$, is not necessarily vertical*

$\vec{F}_P$  (100.0 N)

$\vec{a}$

$m\vec{g}$  (98.0 N)

**FIGURE 4–16** Example 4–7. The box accelerates upward because $F_P > mg$.

**EXAMPLE 4–7** | **Accelerating the box.** What happens when a person pulls upward on the box in Example 4–6c with a force equal to, or greater than, the box's weight? For example, let $F_P = 100.0 \, \text{N}$ (Fig. 4–16) rather than the 40.0 N shown in Fig. 4–15c.

**APPROACH** We can start just as in Example 4–6, but be ready for a surprise.

**SOLUTION** The net force on the box is

$$\Sigma F_y = F_N - mg + F_P$$
$$= F_N - 98.0 \, \text{N} + 100.0 \, \text{N},$$

and if we set this equal to zero (thinking the acceleration might be zero), we would get $F_N = -2.0 \, \text{N}$. This is nonsense, since the negative sign implies $F_N$ points downward, and the table surely cannot *pull* down on the box (unless there's glue on the table). The least $F_N$ can be is zero, which it will be in this case. What really happens here is that the box accelerates upward ($a \neq 0$) because the net force is not zero. The net force (setting the normal force $F_N = 0$) is

$$\Sigma F_y = F_P - mg = 100.0 \, \text{N} - 98.0 \, \text{N}$$
$$= 2.0 \, \text{N}$$

upward. See Fig. 4–16. We apply Newton's second law and see that the box moves upward with an acceleration

$$a_y = \frac{\Sigma F_y}{m} = \frac{2.0 \, \text{N}}{10.0 \, \text{kg}}$$
$$= 0.20 \, \text{m/s}^2.$$

**EXAMPLE 4–8** | **Apparent weight loss.** A 65-kg woman descends in an elevator that briefly accelerates at 0.20g downward. She stands on a scale that reads in kg. (*a*) During this acceleration, what is her weight and what does the scale read? (*b*) What does the scale read when the elevator descends at a constant speed of 2.0 m/s?

**APPROACH** Figure 4–17 shows all the forces that act on the woman (and *only* those that act on her). The direction of the acceleration is downward, so we choose the positive direction as down (this is the opposite choice from Examples 4–6 and 4–7).

**SOLUTION** (*a*) From Newton's second law,

$$\Sigma F = ma$$
$$mg - F_N = m(0.20g).$$

We solve for $F_N$:

$$F_N = mg - 0.20mg$$
$$= 0.80mg,$$

**FIGURE 4–17** Example 4–8. The acceleration vector is shown in gold to distinguish it from the red force vectors.

$\vec{a}$  $m\vec{g}$

$\vec{F}_N$

and it acts upward. The normal force $\vec{F}_N$ is the force the scale exerts on the person, and is equal and opposite to the force she exerts on the scale: $F'_N = 0.80mg$ downward. Her weight (force of gravity on her) is still $mg = (65 \, \text{kg})(9.8 \, \text{m/s}^2) = 640 \, \text{N}$. But the scale, needing to exert a force of only 0.80mg, will give a reading of $0.80m = 52 \, \text{kg}$.

(*b*) Now there is no acceleration, $a = 0$, so by Newton's second law, $mg - F_N = 0$ and $F_N = mg$. The scale reads her true mass of 65 kg.

**NOTE** The scale in (*a*) gives a reading of 52 kg (as an "apparent mass"), but her mass doesn't change as a result of the acceleration: it stays at 65 kg.

# 4–7 Solving Problems with Newton's Laws: Free-Body Diagrams

Newton's second law tells us that the acceleration of an object is proportional to the *net force* acting on the object. The **net force**, as mentioned earlier, is the *vector sum* of all forces acting on the object. Indeed, extensive experiments have shown that forces do add together as vectors precisely according to the rules we developed in Chapter 3. For example, in Fig. 4–18, two forces of equal magnitude (100 N each) are shown acting on an object at right angles to each other. Intuitively, we can see that the object will start moving at a 45° angle and thus the net force acts at a 45° angle. This is just what the rules of vector addition give. From the theorem of Pythagoras, the magnitude of the resultant force is $F_R = \sqrt{(100\,\text{N})^2 + (100\,\text{N})^2} = 141\,\text{N}$.

(a)   (b)

**FIGURE 4–18** (a) Two horizontal forces, $\vec{\mathbf{F}}_A$ and $\vec{\mathbf{F}}_B$, exerted by workers A and B, act on a crate (we are looking down from above). (b) The sum, or resultant, of $\vec{\mathbf{F}}_A$ and $\vec{\mathbf{F}}_B$ is $\vec{\mathbf{F}}_R$.

**EXAMPLE 4–9 Adding force vectors.** Calculate the sum of the two forces exerted on the boat by workers A and B in Fig. 4–19a.

**APPROACH** We add force vectors like any other vectors as described in Chapter 3. The first step is to choose an *xy* coordinate system (see Fig. 4–19a), and then resolve vectors into their components.

**SOLUTION** The two force vectors are shown resolved into components in Fig. 4–19b. We add the forces using the method of components. The components of $\vec{\mathbf{F}}_A$ are

$$F_{Ax} = F_A \cos 45.0° = (40.0\,\text{N})(0.707) = 28.3\,\text{N},$$
$$F_{Ay} = F_A \sin 45.0° = (40.0\,\text{N})(0.707) = 28.3\,\text{N}.$$

The components of $\vec{\mathbf{F}}_B$ are

$$F_{Bx} = +F_B \cos 37.0° = +(30.0\,\text{N})(0.799) = +24.0\,\text{N},$$
$$F_{By} = -F_B \sin 37.0° = -(30.0\,\text{N})(0.602) = -18.1\,\text{N}.$$

$F_{By}$ is negative because it points along the negative *y* axis. The components of the resultant force are (see Fig. 4–19c)

$$F_{Rx} = F_{Ax} + F_{Bx} = 28.3\,\text{N} + 24.0\,\text{N} = 52.3\,\text{N},$$
$$F_{Ry} = F_{Ay} + F_{By} = 28.3\,\text{N} - 18.1\,\text{N} = 10.2\,\text{N}.$$

To find the magnitude of the resultant force, we use the Pythagorean theorem,

$$F_R = \sqrt{F_{Rx}^2 + F_{Ry}^2} = \sqrt{(52.3)^2 + (10.2)^2}\,\text{N} = 53.3\,\text{N}.$$

The only remaining question is the angle $\theta$ that the net force $\vec{\mathbf{F}}_R$ makes with the *x* axis. We use:

$$\tan \theta = \frac{F_{Ry}}{F_{Rx}} = \frac{10.2\,\text{N}}{52.3\,\text{N}} = 0.195,$$

and $\tan^{-1}(0.195) = 11.0°$. The net force on the boat has magnitude 53.3 N and acts at an 11.0° angle to the *x* axis.

**FIGURE 4–19** Example 4–9: Two force vectors act on a boat.

(a)

(b)

(c)

When solving problems involving Newton's laws and force, it is very important to draw a diagram showing all the forces acting on each object involved. Such a diagram is called a **free-body diagram**, or **force diagram**: choose one object, and draw an arrow to represent each force acting on it. Include *every* force acting on that object. Do not show forces that the chosen object exerts on *other* objects. To help you identify each and every force that is exerted on your chosen object, ask yourself what other objects could exert a force on it. If your problem involves more than one object, a separate free-body diagram is needed for each object. For now, the likely forces that could be acting are *gravity* and *contact forces* (one object pushing or pulling another, normal force, friction). Later we will consider other types of force such as buoyancy, fluid pressure, and electric and magnetic forces.

**PROBLEM SOLVING**
*Free-body diagram*

**FIGURE 4–20** Example 4–10. Which is the correct free-body diagram for a hockey puck sliding across frictionless ice?

**CONCEPTUAL EXAMPLE 4–10** | **The hockey puck.** A hockey puck is sliding at constant velocity across a flat horizontal ice surface that is assumed to be frictionless. Which of the sketches in Fig. 4–20 is the correct free-body diagram for this puck? What would your answer be if the puck slowed down?

**RESPONSE** Did you choose (a)? If so, can you answer the question: what exerts the horizontal force labeled $\vec{\mathbf{F}}$ on the puck? If you say that it is the force needed to maintain the motion, ask yourself: what exerts this force? Remember that another object must exert any force—and there simply isn't any possibility here. Therefore, (a) is wrong. Besides, the force $\vec{\mathbf{F}}$ in Fig. 4–20a would give rise to an acceleration by Newton's second law. It is (b) that is correct. No net force acts on the puck, and the puck slides at constant velocity across the ice.

In the real world, where even smooth ice exerts at least a tiny friction force, then (c) is the correct answer. The tiny friction force is in the direction opposite to the motion, and the puck's velocity decreases, even if very slowly.

## Newton's Laws; Free-Body Diagrams

1. **Draw a sketch** of the situation, after carefully reading the Problem at least twice.

2. Consider only one object (at a time), and draw a **free-body diagram** for that object, showing *all* the forces acting *on* that object. Include any unknown forces that you have to solve for. Do not show any forces that the chosen object exerts on other objects.

    Draw the arrow for each force vector reasonably accurately for direction and magnitude. Label each force acting on the object, including forces you must solve for, according to its source (gravity, person, friction, and so on).

    If several objects are involved, draw a free-body diagram for each object *separately*. For each object, show all the forces acting *on that object* (and *only* forces acting on that object). For each (and every) force, you must be clear about: *on* what object that

force acts, and *by* what object that force is exerted. Only forces acting *on* a given object can be included in $\Sigma \vec{\mathbf{F}} = m\vec{\mathbf{a}}$ for that object.

3. Newton's second law involves vectors, and it is usually important to **resolve vectors** into components. **Choose** $x$ and $y$ **axes** in a way that simplifies the calculation. For example, it often saves work if you choose one coordinate axis to be in the direction of the acceleration (if known).

4. For each object, **apply Newton's second law** to the $x$ and $y$ components separately. That is, the $x$ component of the net force on that object is related to the $x$ component of that object's acceleration: $\Sigma F_x = ma_x$, and similarly for the $y$ direction.

5. **Solve** the equation or equations for the unknown(s). Put in numerical values only at the end, and keep track of units.

This Problem Solving Strategy should not be considered a prescription. Rather it is a summary of things to do that will start you thinking and getting involved in the problem at hand.

When we are concerned only about translational motion, all the forces on a given object can be drawn as acting at the center of the object, thus treating the object as a *point particle*. However, for problems involving rotation or statics, the place *where* each force acts is also important, as we shall see in Chapters 8 and 9.

In the Examples in this Section, we assume that all surfaces are very smooth so that friction can be ignored. (Friction, and Examples using it, are discussed in Section 4–8.)

⚠ **CAUTION**

*Treating an object as a particle*

**EXAMPLE 4–11** **Pulling the mystery box.** Suppose a friend asks to examine the 10.0-kg box you were given (Example 4–6, Fig. 4–15), hoping to guess what is inside; and you respond, "Sure, pull the box over to you." She then pulls the box by the attached cord, as shown in Fig. 4–21a, along the smooth surface of the table. The magnitude of the force exerted by the person is $F_P = 40.0$ N, and it is exerted at a 30.0° angle as shown. Calculate (a) the acceleration of the box, and (b) the magnitude of the upward force $F_N$ exerted by the table on the box. Assume that friction can be neglected.

**APPROACH** We follow the Problem Solving Strategy on the previous page.

**SOLUTION**

1. **Draw a sketch**: The situation is shown in Fig. 4–21a; it shows the box and the force applied by the person, $F_P$.

2. **Free-body diagram**: Figure 4–21b shows the free-body diagram of the box. To draw it correctly, we show *all* the forces acting on the box and *only* the forces acting on the box. They are: the force of gravity $m\vec{\mathbf{g}}$; the normal force exerted by the table $\vec{\mathbf{F}}_N$; and the force exerted by the person $\vec{\mathbf{F}}_P$. We are interested only in translational motion, so we can show the three forces acting at a point, Fig. 4–21c.

3. **Choose axes and resolve vectors**: We expect the motion to be horizontal, so we choose the $x$ axis horizontal and the $y$ axis vertical. The pull of 40.0 N has components

$$F_{Px} = (40.0\,\text{N})(\cos 30.0°) = (40.0\,\text{N})(0.866) = 34.6\,\text{N},$$
$$F_{Py} = (40.0\,\text{N})(\sin 30.0°) = (40.0\,\text{N})(0.500) = 20.0\,\text{N}.$$

In the horizontal ($x$) direction, $\vec{\mathbf{F}}_N$ and $m\vec{\mathbf{g}}$ have zero components. Thus the horizontal component of the net force is $F_{Px}$.

4. (a) **Apply Newton's second law** to get the $x$ component of the acceleration:

$$F_{Px} = ma_x.$$

5. (a) **Solve**:

$$a_x = \frac{F_{Px}}{m} = \frac{(34.6\,\text{N})}{(10.0\,\text{kg})} = 3.46\,\text{m/s}^2.$$

The acceleration of the box is 3.46 m/s² to the right.

(b) Next we want to find $F_N$.

4'. (b) **Apply Newton's second law** to the vertical ($y$) direction, with upward as positive:

$$\Sigma F_y = ma_y$$
$$F_N - mg + F_{Py} = ma_y.$$

5'. (b) **Solve**: We have $mg = (10.0\,\text{kg})(9.80\,\text{m/s}^2) = 98.0$ N and, from point 3 above, $F_{Py} = 20.0$ N. Furthermore, since $F_{Py} < mg$, the box does not move vertically, so $a_y = 0$. Thus

$$F_N - 98.0\,\text{N} + 20.0\,\text{N} = 0,$$

so

$$F_N = 78.0\,\text{N}.$$

**NOTE** $F_N$ is less than $mg$: the table does not push against the full weight of the box because part of the pull exerted by the person is in the upward direction.

**EXERCISE F** A 10.0 kg box is dragged on a horizontal frictionless surface by a horizontal force of 10.0 N. If the applied force is doubled, the normal force on the box will (a) increase; (b) remain the same; (c) decrease.

(a)

(b)

(c)

**FIGURE 4–21** (a) Pulling the box, Example 4–11; (b) is the free-body diagram for the box, and (c) is the free-body diagram considering all the forces to act at a point (translational motion only, which is what we have here).

## Tension in a Flexible Cord

When a flexible cord pulls on an object, the cord is said to be under **tension**, and the force it exerts on the object is the tension $F_T$. If the cord has negligible mass, the force exerted at one end is transmitted undiminished to each adjacent piece of cord along the entire length to the other end. Why? Because $\Sigma \vec{\mathbf{F}} = m\vec{\mathbf{a}} = 0$ for the cord if the cord's mass $m$ is zero (or negligible) no matter what $\vec{\mathbf{a}}$ is. Hence the forces pulling on the cord at its two ends must add up to zero ($F_T$ and $-F_T$). Note that flexible cords and strings can only pull. They can't push because they bend.

**PROBLEM SOLVING**
*Cords can pull but can't push; tension exists throughout a taut cord*

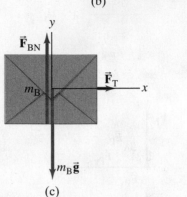

**FIGURE 4–22** Example 4–12. (a) Two boxes, A and B, are connected by a cord. A person pulls horizontally on box A with force $F_P = 40.0$ N. (b) Free-body diagram for box A. (c) Free-body diagram for box B.

Our next Example involves two boxes connected by a cord. We can refer to this group of objects as a system. A **system** is any group of one or more objects we choose to consider and study.

**EXAMPLE 4–12** **Two boxes connected by a cord.** Two boxes, A and B, are connected by a lightweight cord and are resting on a smooth (frictionless) table. The boxes have masses of 12.0 kg and 10.0 kg. A horizontal force $F_P$ of 40.0 N is applied to the 10.0-kg box, as shown in Fig. 4–22a. Find (a) the acceleration of each box, and (b) the tension in the cord connecting the boxes.

**APPROACH** We streamline our approach by not listing each step. We have two boxes so we draw a free-body diagram for each. To draw them correctly, we must consider the forces on *each* box by itself, so that Newton's second law can be applied to each. The person exerts a force $F_P$ on box A. Box A exerts a force $F_T$ on the connecting cord, and the cord exerts an opposite but equal magnitude force $F_T$ back on box A (Newton's third law). The two horizontal forces on box A are shown in Fig. 4–22b, along with the force of gravity $m_A \vec{g}$ downward and the normal force $\vec{F}_{AN}$ exerted upward by the table. The cord is light, so we neglect its mass. The tension at each end of the cord is thus the same. Hence the cord exerts a force $F_T$ on the second box. Figure 4–22c shows the forces on box B, which are $\vec{F}_T$, $m_B \vec{g}$, and the normal force $\vec{F}_{BN}$. There will be only horizontal motion. We take the positive $x$ axis to the right.

**SOLUTION** (a) We apply $\Sigma F_x = ma_x$ to box A:

$$\Sigma F_x = F_P - F_T = m_A a_A. \qquad \text{[box A]}$$

For box B, the only horizontal force is $F_T$, so

$$\Sigma F_x = F_T = m_B a_B. \qquad \text{[box B]}$$

The boxes are connected, and if the cord remains taut and doesn't stretch, then the two boxes will have the same acceleration $a$. Thus $a_A = a_B = a$. We are given $m_A = 10.0$ kg and $m_B = 12.0$ kg. We can add the two equations above to eliminate an unknown $(F_T)$ and obtain

$$(m_A + m_B)a = F_P - F_T + F_T = F_P$$

or

$$a = \frac{F_P}{m_A + m_B} = \frac{40.0 \text{ N}}{22.0 \text{ kg}} = 1.82 \text{ m/s}^2.$$

This is what we sought.

(b) From the equation for box B above $(F_T = m_B a_B)$, the tension in the cord is

$$F_T = m_B a = (12.0 \text{ kg})(1.82 \text{ m/s}^2) = 21.8 \text{ N}.$$

Thus, $F_T < F_P$ (= 40.0 N), as we expect, since $F_T$ acts to accelerate only $m_B$.

*Alternate Solution to (a)* We would have obtained the same result had we considered a single system, of mass $m_A + m_B$, acted on by a net horizontal force equal to $F_P$. (The tension forces $F_T$ would then be considered internal to the system as a whole, and summed together would make zero contribution to the net force on the *whole* system.)

**NOTE** It might be tempting to say that the force the person exerts, $F_P$, acts not only on box A but also on box B. It doesn't. $F_P$ acts only on box A. It affects box B via the tension in the cord, $F_T$, which acts on box B and accelerates it. (You could look at it this way: $F_T < F_P$ because $F_P$ accelerates *both* boxes whereas $F_T$ only accelerates box B.)

**CAUTION**

*For any object, use only the forces on* that *object in calculating* $\Sigma F = ma$

**EXAMPLE 4–13** **Elevator and counterweight (Atwood machine).** A system of two objects suspended over a pulley by a flexible cable, as shown in Fig. 4–23a, is sometimes referred to as an *Atwood machine*. Consider the real-life application of an elevator ($m_E$) and its counterweight ($m_C$). To minimize the work done by the motor to raise and lower the elevator safely, $m_E$ and $m_C$ are made similar in mass. We leave the motor out of the system for this calculation, and assume that the cable's mass is negligible and that the mass of the pulley, as well as any friction, is small and ignorable. These assumptions ensure that the tension $F_T$ in the cable has the same magnitude on both sides of the pulley. Let the mass of the counterweight be $m_C = 1000$ kg. Assume the mass of the empty elevator is 850 kg, and its mass when carrying four passengers is $m_E = 1150$ kg. For the latter case ($m_E = 1150$ kg), calculate (*a*) the acceleration of the elevator and (*b*) the tension in the cable.

**APPROACH** Again we have two objects, and we will need to apply Newton's second law to each of them separately. Each mass has two forces acting on it: gravity downward and the cable tension pulling upward, $\vec{F}_T$. Figures 4–23b and c show the free-body diagrams for the elevator ($m_E$) and for the counterweight ($m_C$). The elevator, being the heavier, will accelerate downward, whereas the counterweight will accelerate upward. The magnitudes of their accelerations will be equal (we assume the cable is massless and doesn't stretch). For the counterweight, $m_C g = (1000 \text{ kg})(9.80 \text{ m/s}^2) = 9800$ N, so $F_T$ must be greater than 9800 N (in order that $m_C$ will accelerate upward). For the elevator, $m_E g = (1150 \text{ kg})(9.80 \text{ m/s}^2) = 11{,}300$ N, which must have greater magnitude than $F_T$ so that $m_E$ accelerates downward. Thus our calculation must give $F_T$ between 9800 N and 11,300 N.

**SOLUTION** (*a*) To find $F_T$ as well as the acceleration $a$, we apply Newton's second law, $\Sigma F = ma$, to each object. We take upward as the positive $y$ direction for both objects. With this choice of axes, $a_C = a$ because $m_C$ accelerates upward, and $a_E = -a$ because $m_E$ accelerates downward. Thus

$$F_T - m_E g = m_E a_E = -m_E a$$
$$F_T - m_C g = m_C a_C = +m_C a.$$

We can subtract the first equation from the second to get

$$(m_E - m_C)g = (m_E + m_C)a,$$

where $a$ is now the only unknown. We solve this for $a$:

$$a = \frac{m_E - m_C}{m_E + m_C} g = \frac{1150 \text{ kg} - 1000 \text{ kg}}{1150 \text{ kg} + 1000 \text{ kg}} g = 0.070g = 0.68 \text{ m/s}^2.$$

The elevator ($m_E$) accelerates downward (and the counterweight $m_C$ upward) at $a = 0.070g = 0.68 \text{ m/s}^2$.

(*b*) The tension in the cable $F_T$ can be obtained from either of the two $\Sigma F = ma$ equations at the start of our solution, setting $a = 0.070g = 0.68 \text{ m/s}^2$:

$$F_T = m_E g - m_E a = m_E(g - a)$$
$$= 1150 \text{ kg} (9.80 \text{ m/s}^2 - 0.68 \text{ m/s}^2) = 10{,}500 \text{ N},$$

or

$$F_T = m_C g + m_C a = m_C(g + a)$$
$$= 1000 \text{ kg} (9.80 \text{ m/s}^2 + 0.68 \text{ m/s}^2) = 10{,}500 \text{ N},$$

which are consistent. As predicted, our result lies between 9800 N and 11,300 N.

**NOTE** We can check our equation for the acceleration $a$ in this Example by noting that if the masses were equal ($m_E = m_C$), then our equation above for $a$ would give $a = 0$, as we should expect. Also, if one of the masses is zero (say, $m_C = 0$), then the other mass ($m_E \neq 0$) would be predicted by our equation to accelerate at $a = g$, again as expected.

(a)

(b)                    (c)

**FIGURE 4–23** Example 4–13. (a) Atwood machine in the form of an elevator–counterweight system. (b) and (c) Free-body diagrams for the two objects.

**PROBLEM SOLVING**
*Check your result by seeing if it works in situations where the answer is easily guessed*

FIGURE 4–24 Example 4–14.

**CONCEPTUAL EXAMPLE 4–14** | **The advantage of a pulley.** A mover is trying to lift a piano (slowly) up to a second-story apartment (Fig. 4–24). He is using a rope looped over two pulleys as shown. What force must he exert on the rope to slowly lift the piano's 1600-N weight?

**RESPONSE** The magnitude of the tension force $F_T$ within the rope is the same at any point along the rope if we assume we can ignore its mass. First notice the forces acting on the lower pulley at the piano. The weight of the piano ($= mg$) pulls down on the pulley. The tension in the rope, looped through this pulley, pulls up *twice*, once on each side of the pulley. Let us apply Newton's second law to the pulley–piano combination (of mass $m$), choosing the upward direction as positive:

$$2F_T - mg = ma.$$

To move the piano with constant speed (set $a = 0$ in this equation) thus requires a tension in the rope, and hence a pull on the rope, of $F_T = mg/2$. The piano mover can exert a force equal to half the piano's weight.

**NOTE** We say the pulley has given a **mechanical advantage** of 2, since without the pulley the mover would have to exert twice the force.

*PHYSICS APPLIED*
*Accelerometer*

FIGURE 4–25 Example 4–15.

(a)

(b)

**EXAMPLE 4–15** | **Accelerometer.** A small mass $m$ hangs from a thin string and can swing like a pendulum. You attach it above the window of your car as shown in Fig. 4–25a. When the car is at rest, the string hangs vertically. What angle $\theta$ does the string make ($a$) when the car accelerates at a constant $a = 1.20 \text{ m/s}^2$, and ($b$) when the car moves at constant velocity, $v = 90 \text{ km/h}$?

**APPROACH** The free-body diagram of Fig. 4–25b shows the pendulum at some angle $\theta$ relative to the vertical, and the forces on it: $m\vec{g}$ downward, and the tension $\vec{F}_T$ in the cord (including its components). These forces do not add up to zero if $\theta \neq 0$; and since we have an acceleration $a$, we expect $\theta \neq 0$.

**SOLUTION** ($a$) The acceleration $a = 1.20 \text{ m/s}^2$ is horizontal ($= a_x$), and the only horizontal force is the $x$ component of $\vec{F}_T$, $F_T \sin \theta$ (Fig. 4–25b). Then from Newton's second law,

$$ma = F_T \sin \theta.$$

The vertical component of Newton's second law gives, since $a_y = 0$,

$$0 = F_T \cos \theta - mg.$$

So

$$mg = F_T \cos \theta.$$

Dividing these two equations, we obtain

$$\tan \theta = \frac{F_T \sin \theta}{F_T \cos \theta} = \frac{ma}{mg} = \frac{a}{g}$$

or

$$\tan \theta = \frac{1.20 \text{ m/s}^2}{9.80 \text{ m/s}^2}$$
$$= 0.122,$$

so

$$\theta = 7.0°.$$

($b$) The velocity is constant, so $a = 0$ and $\tan \theta = 0$. Hence the pendulum hangs vertically ($\theta = 0°$).

**NOTE** This simple device is an **accelerometer**—it can be used to determine acceleration, by measuring the angle $\theta$.

# 4–8 Problems Involving Friction, Inclines

## Friction

Until now we have ignored friction, but it must be taken into account in most practical situations. Friction exists between two solid surfaces because even the smoothest looking surface is quite rough on a microscopic scale, Fig. 4–26. When we try to slide an object across a surface, these microscopic bumps impede the motion. Exactly what is happening at the microscopic level is not yet fully understood. It is thought that the atoms on a bump of one surface may come so close to the atoms of the other surface that attractive electric forces between the atoms could "bond" as a tiny weld between the two surfaces. Sliding an object across a surface is often jerky, perhaps due to the making and breaking of these bonds. Even when a round object rolls across a surface, there is still some friction, called *rolling friction*, although it is generally much less than when an object slides across a surface. We focus now on sliding friction, which is usually called **kinetic friction** (*kinetic* is from the Greek for "moving").

When an object slides along a rough surface, the force of kinetic friction acts opposite to the direction of the object's velocity. The magnitude of the force of kinetic friction depends on the nature of the two sliding surfaces. For given surfaces, experiment shows that the friction force is approximately proportional to the *normal force* between the two surfaces, which is the force that either object exerts on the other and is perpendicular to their common surface of contact (see Fig. 4–27). The force of friction between hard surfaces in many cases depends very little on the total surface area of contact; that is, the friction force on this book is roughly the same whether it is being slid across a table on its wide face or on its spine, assuming the surfaces have the same smoothness. We consider a simple model of friction in which we make this assumption that the friction force is independent of area. Then we write the proportionality between the magnitudes of the friction force $F_{fr}$ and the normal force $F_N$ as an equation by inserting a constant of proportionality, $\mu_k$:

$$F_{fr} = \mu_k F_N. \qquad \text{[kinetic friction]}$$

This relation is not a fundamental law; it is an experimental relation between the magnitude of the friction force $F_{fr}$, which acts parallel to the two surfaces, and the magnitude of the normal force $F_N$, which acts perpendicular to the surfaces. It is *not* a vector equation since the two forces have different directions, perpendicular to one another. The term $\mu_k$ is called the *coefficient of kinetic friction*, and its value depends on the nature of the two surfaces. Measured values for a variety of surfaces are given in Table 4–2. These are only approximate, however, since $\mu$ depends on whether the surfaces are wet or dry, on how much they have been sanded or rubbed, if any burrs remain, and other such factors. But $\mu_k$ (which has no units) is roughly independent of the sliding speed, as well as the area in contact.

**FIGURE 4–26** An object moving to the right on a table. The two surfaces in contact are assumed smooth, but are rough on a microscopic scale.

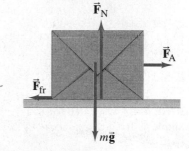

**FIGURE 4–27** When an object is pulled along a surface by an applied force ($\vec{F}_A$), the force of friction $\vec{F}_{fr}$ opposes the motion. The magnitude of $\vec{F}_{fr}$ is proportional to the magnitude of the normal force ($F_N$).

◆ **CAUTION**
$\vec{F}_{fr} \perp \vec{F}_N$

## TABLE 4–2 Coefficients of Friction[†]

| Surfaces | Coefficient of Static Friction, $\mu_s$ | Coefficient of Kinetic Friction, $\mu_k$ |
|---|---|---|
| Wood on wood | 0.4 | 0.2 |
| Ice on ice | 0.1 | 0.03 |
| Metal on metal (lubricated) | 0.15 | 0.07 |
| Steel on steel (unlubricated) | 0.7 | 0.6 |
| Rubber on dry concrete | 1.0 | 0.8 |
| Rubber on wet concrete | 0.7 | 0.5 |
| Rubber on other solid surfaces | 1–4 | 1 |
| Teflon® on Teflon in air | 0.04 | 0.04 |
| Teflon on steel in air | 0.04 | 0.04 |
| Lubricated ball bearings | <0.01 | <0.01 |
| Synovial joints (in human limbs) | 0.01 | 0.01 |

[†] Values are approximate and intended only as a guide.

What we have been discussing up to now is *kinetic friction*, when one object slides over another. There is also **static friction**, which refers to a force parallel to the two surfaces that can arise even when they are not sliding. Suppose an object such as a desk is resting on a horizontal floor. If no horizontal force is exerted on the desk, there also is no friction force. But now suppose you try to push the desk, and it doesn't move. You are exerting a horizontal force, but the desk isn't moving, so there must be another force on the desk keeping it from moving (the net force is zero on an object at rest). This is the force of *static friction* exerted by the floor on the desk. If you push with a greater force without moving the desk, the force of static friction also has increased. If you push hard enough, the desk will eventually start to move, and kinetic friction takes over. At this point, you have exceeded the maximum force of static friction, which is given by $(F_{fr})_{max} = \mu_s F_N$, where $\mu_s$ is the *coefficient of static friction* (Table 4–2). Because the force of static friction can vary from zero to this maximum value, we write

$$F_{fr} \leq \mu_s F_N. \qquad \text{[static friction]}$$

You may have noticed that it is often easier to keep a heavy object sliding than it is to start it sliding in the first place. This is consistent with $\mu_s$ generally being greater than $\mu_k$ (see Table 4–2).

**EXAMPLE 4–16** | **Friction: static and kinetic.** Our 10.0-kg mystery box rests on a horizontal floor. The coefficient of static friction is $\mu_s = 0.40$ and the coefficient of kinetic friction is $\mu_k = 0.30$. Determine the force of friction, $F_{fr}$, acting on the box if a horizontal applied force $F_A$ is exerted on it of magnitude: (a) 0, (b) 10 N, (c) 20 N, (d) 38 N, and (e) 40 N.

**APPROACH** We don't know, right off, if we are dealing with static friction or kinetic friction, nor if the box remains at rest or accelerates. We need to draw a free-body diagram, and then determine in each case whether or not the box will move: the box starts moving if $F_A$ is greater than the maximum static friction force (Newton's second law). The forces on the box are gravity $m\vec{g}$, the normal force exerted by the floor $\vec{F}_N$, the horizontal applied force $\vec{F}_A$, and the friction force $\vec{F}_{fr}$, as shown in Fig. 4–27.

**SOLUTION** The free-body diagram of the box is shown in Fig. 4–27. In the vertical direction there is no motion, so Newton's second law in the vertical direction gives $\Sigma F_y = ma_y = 0$, which tells us $F_N - mg = 0$. Hence the normal force is
$$F_N = mg = (10.0\,\text{kg})(9.80\,\text{m/s}^2) = 98.0\,\text{N}.$$
(a) Because $F_A = 0$ in this first case, the box doesn't move, and $F_{fr} = 0$.

(b) The force of static friction will oppose any applied force up to a maximum of
$$\mu_s F_N = (0.40)(98.0\,\text{N}) = 39\,\text{N}.$$
When the applied force is $F_A = 10\,\text{N}$, the box will not move. Newton's second law gives $\Sigma F_x = F_A - F_{fr} = 0$, so $F_{fr} = 10\,\text{N}$.

(c) An applied force of 20 N is also not sufficient to move the box. Thus $F_{fr} = 20\,\text{N}$ to balance the applied force.

(d) The applied force of 38 N is still not quite large enough to move the box; so the friction force has now increased to 38 N to keep the box at rest.

(e) A force of 40 N will start the box moving since it exceeds the maximum force of static friction, $\mu_s F_N = (0.40)(98\,\text{N}) = 39\,\text{N}$. Instead of static friction, we now have kinetic friction, and its magnitude is
$$F_{fr} = \mu_k F_N = (0.30)(98.0\,\text{N}) = 29\,\text{N}.$$
There is now a net (horizontal) force on the box of magnitude $F = 40\,\text{N} - 29\,\text{N} = 11\,\text{N}$, so the box will accelerate at a rate
$$a_x = \frac{\Sigma F}{m} = \frac{11\,\text{N}}{10.0\,\text{kg}} = 1.1\,\text{m/s}^2$$
as long as the applied force is 40 N. Figure 4–28 shows a graph that summarizes this Example.

**FIGURE 4–27** Repeated for Example 4–16.

**FIGURE 4–28** Example 4–16. Magnitude of the force of friction as a function of the external force applied to an object initially at rest. As the applied force is increased in magnitude, the force of static friction increases in proportion until the applied force equals $\mu_s F_N$. If the applied force increases further, the object will begin to move, and the friction force drops to a roughly constant value characteristic of kinetic friction.

Friction can be a hindrance. It slows down moving objects and causes heating and binding of moving parts in machinery. Friction can be reduced by using lubricants such as oil. More effective in reducing friction between two surfaces is to maintain a layer of air or other gas between them. Devices using this concept, which is not practical for most situations, include air tracks and air tables in which the layer of air is maintained by forcing air through many tiny holes. Another technique to maintain the air layer is to suspend objects in air using magnetic fields ("magnetic levitation").

On the other hand, friction can be helpful. Our ability to walk depends on friction between the soles of our shoes (or feet) and the ground. (Walking involves static friction, not kinetic friction. Why?) The movement of a car, and also its stability, depend on friction. When friction is low, such as on ice, safe walking or driving becomes difficult.

**CONCEPTUAL EXAMPLE 4–17** | **A box against a wall.** You can hold a box against a rough wall (Fig. 4–29) and prevent it from slipping down by pressing hard horizontally. How does the application of a horizontal force keep an object from moving vertically?

**RESPONSE** This won't work well if the wall is slippery. You need friction. Even then, if you don't press hard enough, the box will slip. The horizontal force you apply produces a normal force on the box exerted by the wall (the net force horizontally is zero since the box doesn't move horizontally). The force of gravity $mg$, acting downward on the box, can now be balanced by an upward static friction force whose maximum magnitude is proportional to the normal force. The harder you push, the greater $F_N$ is and the greater $F_{fr}$ can be. If you don't press hard enough, then $mg > \mu_s F_N$ and the box begins to slide down.

FIGURE 4–29 Example 4–17.

**EXERCISE G** If $\mu_s = 0.40$ and $mg = 20\,N$, what minimum force $F$ will keep the box from falling: (a) 100 N; (b) 80 N; (c) 50 N; (d) 20 N; (e) 8 N?

**EXAMPLE 4–18** | **Pulling against friction.** A 10.0-kg box is pulled along a horizontal surface by a force $F_P$ of 40.0 N applied at a 30.0° angle above horizontal. This is like Example 4–11 except now there is friction, and we assume a coefficient of kinetic friction of 0.30. Calculate the acceleration.

**APPROACH** The free-body diagram is shown in Fig. 4–30. It is much like that in Fig. 4–21b, but with one more force, friction.

**SOLUTION** The calculation for the vertical ($y$) direction is just the same as in Example 4–11b, $mg = (10.0\,\text{kg})(9.80\,\text{m/s}^2) = 98.0\,N$ and $F_{Py} = (40.0\,N)(\sin 30.0°) = 20.0\,N$. With $y$ positive upward and $a_y = 0$, we have

$$F_N - mg + F_{Py} = ma_y$$
$$F_N - 98.0\,N + 20.0\,N = 0,$$

so the normal force is $F_N = 78.0\,N$. Now we apply Newton's second law for the horizontal ($x$) direction (positive to the right), and include the friction force:

$$F_{Px} - F_{fr} = ma_x.$$

The friction force is kinetic friction as long as $F_{fr} = \mu_k F_N$ is less than $F_{Px} = (40.0\,N) \cos 30.0° = 34.6\,N$, which it is:

$$F_{fr} = \mu_k F_N = (0.30)(78.0\,N) = 23.4\,N.$$

Hence the box does accelerate:

$$a_x = \frac{F_{Px} - F_{fr}}{m} = \frac{34.6\,N - 23.4\,N}{10.0\,\text{kg}} = 1.1\,\text{m/s}^2.$$

In the absence of friction, as we saw in Example 4–11, the acceleration would be much greater than this.

**NOTE** Our final answer has only two significant figures because our least significant input value ($\mu_k = 0.30$) has two.

FIGURE 4–30 Example 4–18.

(a)

(b)

(c) $m\vec{g}$     (d) $m\vec{g}$

FIGURE 4–31 Example 4–19.

FIGURE 4–32 Example 4–20.

(a)

(b)          (c)

⚠ CAUTION

*Tension in a cord supporting a falling object may not equal object's weight*

**To push or to pull a sled?** Your little sister wants a ride on her sled. If you are on flat ground, will you exert less force if you push her or pull her? See Figs. 4–31a and b. Assume the same angle $\theta$ in each case.

**RESPONSE** Let us draw free-body diagrams for the sled–sister combination, as shown in Figs. 4–31c and d. They show, for the two cases, the forces exerted by you, $\vec{F}$ (an unknown), by the snow, $\vec{F}_N$ and $\vec{F}_{fr}$, and gravity $m\vec{g}$. (a) If you push her, and $\theta > 0$, there is a vertically downward component to your force. Hence the normal force upward exerted by the ground (Fig. 4–31c) will be larger than $mg$ (where $m$ is the mass of sister plus sled). (b) If you pull her, your force has a vertically upward component, so the normal force $F_N$ will be less than $mg$, Fig. 4–31d. Because the friction force is proportional to the normal force, $F_{fr}$ will be less if you pull her. So you exert less force if you pull her.

**EXAMPLE 4–20** **Two boxes and a pulley.** In Fig. 4–32a, two boxes are connected by a cord running over a pulley. The coefficient of kinetic friction between box A and the table is 0.20. We ignore the mass of the cord and pulley and any friction in the pulley, which means we can assume that a force applied to one end of the cord will have the same magnitude at the other end. We wish to find the acceleration, $a$, of the system, which will have the same magnitude for both boxes assuming the cord doesn't stretch. As box B moves down, box A moves to the right.

**APPROACH** The free-body diagrams for each box are shown in Figs. 4–32b and c. The forces on box A are the pulling force of the cord $F_T$, gravity $m_A g$, the normal force exerted by the table $F_N$, and a friction force exerted by the table $F_{fr}$; the forces on box B are gravity $m_B g$, and the cord pulling up, $F_T$.

**SOLUTION** Box A does not move vertically, so Newton's second law tells us the normal force just balances the weight,
$$F_N = m_A g = (5.0\,\text{kg})(9.8\,\text{m/s}^2) = 49\,\text{N}.$$
In the horizontal direction, there are two forces on box A (Fig. 4–32b): $F_T$, the tension in the cord (whose value we don't know), and the force of friction
$$F_{fr} = \mu_k F_N = (0.20)(49\,\text{N}) = 9.8\,\text{N}.$$
The horizontal acceleration (box A) is what we wish to find; we use Newton's second law in the $x$ direction, $\Sigma F_{Ax} = m_A a_x$, which becomes (taking the positive direction to the right and setting $a_{Ax} = a$):
$$\Sigma F_{Ax} = F_T - F_{fr} = m_A a. \qquad \text{[box A]}$$
Next consider box B. The force of gravity $m_B g = (2.0\,\text{kg})(9.8\,\text{m/s}^2) = 19.6\,\text{N}$ pulls downward; and the cord pulls upward with a force $F_T$. So we can write Newton's second law for box B (taking the downward direction as positive):
$$\Sigma F_{By} = m_B g - F_T = m_B a. \qquad \text{[box B]}$$
[Notice that if $a \neq 0$, then $F_T$ is not equal to $m_B g$.]

We have two unknowns, $a$ and $F_T$, and we also have two equations. We solve the box A equation for $F_T$:
$$F_T = F_{fr} + m_A a,$$
and substitute this into the box B equation:
$$m_B g - F_{fr} - m_A a = m_B a.$$
Now we solve for $a$ and put in numerical values:
$$a = \frac{m_B g - F_{fr}}{m_A + m_B} = \frac{19.6\,\text{N} - 9.8\,\text{N}}{5.0\,\text{kg} + 2.0\,\text{kg}} = 1.4\,\text{m/s}^2,$$
which is the acceleration of box A to the right, and of box B down.

If we wish, we can calculate $F_T$ using the third equation up from here:
$$F_T = F_{fr} + m_A a = 9.8\,\text{N} + (5.0\,\text{kg})(1.4\,\text{m/s}^2) = 17\,\text{N}.$$

**NOTE** Box B is not in free fall. It does not fall at $a = g$ because an additional force, $F_T$, is acting upward on it.

## Inclines

Now we consider what happens when an object slides down an incline, such as a hill or ramp. Such problems are interesting because gravity is the accelerating force, yet the acceleration is not vertical. Solving problems is usually easier if we choose the $xy$ coordinate system so the $x$ axis points along the incline (the direction of motion) and the $y$ axis is perpendicular to the incline, as shown in Fig. 4–33. Note also that the normal force is not vertical, but is perpendicular to the sloping surface of the plane, along the $y$ axis in Fig. 4–33.

**EXERCISE H** Is the normal force always perpendicular to an inclined plane? Is it always vertical?

**EXAMPLE 4–21** **The skier.** The skier in Fig. 4–34a has begun descending the 30° slope. If the coefficient of kinetic friction is 0.10, what is her acceleration?

**APPROACH** We choose the $x$ axis along the slope, positive downslope in the direction of the skier's motion. The $y$ axis is perpendicular to the surface. The forces acting on the skier are gravity, $\vec{F}_G = m\vec{g}$, which points vertically downward (*not* perpendicular to the slope), and the two forces exerted on her skis by the snow—the normal force perpendicular to the snowy slope (*not* vertical), and the friction force parallel to the surface. These three forces are shown acting at one point in Fig. 4–34b, which is our free-body diagram for the skier.

**SOLUTION** We have to resolve only one vector into components, the weight $\vec{F}_G$, and its components are shown as dashed lines in Fig. 4–34c. To be general, we use $\theta$ rather than 30° for now. We use the definitions of sine ("side opposite") and cosine ("side adjacent") to obtain the components:

$$F_{Gx} = mg \sin \theta,$$
$$F_{Gy} = -mg \cos \theta$$

where $F_{Gy}$ is in the negative $y$ direction. To calculate the skier's acceleration down the hill, $a_x$, we apply Newton's second law to the $x$ direction:

$$\Sigma F_x = ma_x$$
$$mg \sin \theta - \mu_k F_N = ma_x$$

where the two forces are the $x$ component of the gravity force ($+x$ direction) and the friction force ($-x$ direction). We want to find the value of $a_x$, but we don't yet know $F_N$ in the last equation. Let's see if we can get $F_N$ from the $y$ component of Newton's second law:

$$\Sigma F_y = ma_y$$
$$F_N - mg \cos \theta = ma_y = 0$$

where we set $a_y = 0$ because there is no motion in the $y$ direction (perpendicular to the slope). Thus we can solve for $F_N$:

$$F_N = mg \cos \theta$$

and we can substitute this into our equation above for $ma_x$:

$$mg \sin \theta - \mu_k(mg \cos \theta) = ma_x.$$

There is an $m$ in each term which can be canceled out. Thus (setting $\theta = 30°$ and $\mu_k = 0.10$):

$$a_x = g \sin 30° - \mu_k g \cos 30°$$
$$= 0.50g - (0.10)(0.866)g = 0.41g.$$

The skier's acceleration is 0.41 times the acceleration of gravity, which in numbers[†] is $a = (0.41)(9.8 \text{ m/s}^2) = 4.0 \text{ m/s}^2$.

**NOTE** The mass canceled out, so we have the useful conclusion that *the acceleration doesn't depend on the mass*. That such a cancellation sometimes occurs, and thus may give a useful conclusion as well as saving calculation, is a big advantage of working with the algebraic equations and putting in the numbers only at the end.

---

[†]We used values rounded off to 2 significant figures to obtain $a = 4.0 \text{ m/s}^2$. If we kept all the extra digits in our calculator, we would find $a = 0.4134g \approx 4.1 \text{ m/s}^2$. This difference is within the expected precision (number of significant figures, Section 1–4).

**PROBLEM SOLVING**
*Good choice of coordinate system simplifies the calculation*

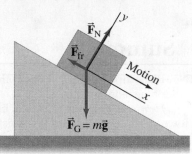

**FIGURE 4–33** Forces on an object sliding down an incline.

**PHYSICS APPLIED**
*Skiing*

**FIGURE 4–34** Example 4–21. Skier descending a slope; $\vec{F}_G = m\vec{g}$ is the force of gravity (weight) on the skier.

(a)

(b)

(c)

**PROBLEM SOLVING**
*It is often helpful to put in numbers only at the end*

In Problems involving a slope or an "inclined plane," avoid making errors in the directions of the normal force and gravity. The normal force on an incline is *not* vertical: it is perpendicular to the slope or plane. And gravity is *not* perpendicular to the slope—gravity acts vertically downward toward the center of the Earth.

# Summary

**Newton's three laws of motion** are the basic classical laws describing motion.

**Newton's first law** (the **law of inertia**) states that if the net force on an object is zero, an object originally at rest remains at rest, and an object in motion remains in motion in a straight line with constant velocity.

**Newton's second law** states that the acceleration of an object is directly proportional to the net force acting on it, and inversely proportional to its mass:

$$\Sigma \vec{F} = m\vec{a}. \tag{4-1}$$

Newton's second law is one of the most important and fundamental laws in classical physics.

**Newton's third law** states that whenever one object exerts a force on a second object, the second object always exerts a force on the first object which is equal in magnitude but opposite in direction:

$$\vec{F}_{AB} = -\vec{F}_{BA} \tag{4-2}$$

where $\vec{F}_{BA}$ is the force on object B exerted by object A.

The tendency of an object to resist a change in its motion is called **inertia**. **Mass** is a measure of the inertia of an object.

**Weight** refers to the **gravitational force** on an object, and is equal to the product of the object's mass $m$ and the acceleration of gravity $\vec{g}$:

$$\vec{F}_G = m\vec{g}. \tag{4-3}$$

**Force**, which is a vector, can be considered as a push or pull; or, from Newton's second law, force can be defined as an action capable of giving rise to acceleration. The **net force** on an object is the vector sum of all forces acting on that object.

When two objects slide over one another, the force of friction that each object exerts on the other can be written approximately as $F_{fr} = \mu_k F_N$, where $F_N$ is the **normal force** (the force each object exerts on the other perpendicular to their contact surfaces), and $\mu_k$ is the coefficient of **kinetic friction**. If the objects are at rest relative to each other, then $F_{fr}$ is just large enough to hold them at rest and satisfies the inequality $F_{fr} < \mu_s F_N$, where $\mu_s$ is the coefficient of **static friction**.

For solving problems involving the forces on one or more objects, it is essential to draw a **free-body diagram** for each object, showing all the forces acting on only that object. Newton's second law can be applied to the vector components for each object.

# Questions

1. Why does a child in a wagon seem to fall backward when you give the wagon a sharp pull forward?

2. A box rests on the (frictionless) bed of a truck. The truck driver starts the truck and accelerates forward. The box immediately starts to slide toward the rear of the truck bed. Discuss the motion of the box, in terms of Newton's laws, as seen (*a*) by Mary standing on the ground beside the truck, and (*b*) by Chris who is riding on the truck (Fig. 4–35).

**FIGURE 4–35**
Question 2.

Box

3. If an object is moving, is it possible for the net force acting on it to be zero? Explain.

4. If the acceleration of an object is zero, are no forces acting on it? Explain.

5. Only one force acts on an object. Can the object have zero acceleration? Can it have zero velocity? Explain.

6. When a golf ball is dropped to the pavement, it bounces back up. (*a*) Is a force needed to make it bounce back up? (*b*) If so, what exerts the force?

7. If you walk along a log floating on a lake, why does the log move in the opposite direction?

8. (*a*) Why do you push down harder on the pedals of a bicycle when first starting out than when moving at constant speed? (*b*) Why do you need to pedal at all when cycling at constant speed?

9. A stone hangs by a fine thread from the ceiling, and a section of the same thread dangles from the bottom of the stone (Fig. 4–36). If a person gives a sharp pull on the dangling thread, where is the thread likely to break: below the stone or above it? What if the person gives a slow and steady pull? Explain your answers.

**FIGURE 4–36**
Question 9.

10. The force of gravity on a 2-kg rock is twice as great as that on a 1-kg rock. Why then doesn't the heavier rock fall faster?

11. (a) You pull a box with a constant force across a frictionless table using an attached rope held horizontally. If you now pull the rope with the same force at an angle to the horizontal (with the box remaining flat on the table), does the acceleration of the box increase, decrease, or remain the same? Explain. (b) What if there is friction?

12. When an object falls freely under the influence of gravity there is a net force $mg$ exerted on it by the Earth. Yet by Newton's third law the object exerts an equal and opposite force on the Earth. Does the Earth move? Explain.

13. Compare the effort (or force) needed to lift a 10-kg object when you are on the Moon with the force needed to lift it on Earth. Compare the force needed to throw a 2-kg object horizontally with a given speed on the Moon and on Earth.

14. According to Newton's third law, each team in a tug of war (Fig. 4–37) pulls with equal force on the other team. What, then, determines which team will win?

**FIGURE 4–37** Question 14. A tug of war. Describe the forces on each of the teams and on the rope.

15. When you stand still on the ground, how large a force does the ground exert on you? Why doesn't this force make you rise up into the air?

16. Whiplash sometimes results from an automobile accident when the victim's car is struck violently from the rear. Explain why the head of the victim seems to be thrown backward in this situation. Is it really?

17. Mary exerts an upward force of 40 N to hold a bag of groceries. Describe the "reaction" force (Newton's third law) by stating (a) its magnitude, (b) its direction, (c) *on* what object it is exerted, and (d) *by* what object it is exerted.

18. A father and his young daughter are ice skating. They face each other at rest and push each other, moving in opposite directions. Which one has the greater final speed? Explain.

19. A heavy crate rests on the bed of a flatbed truck. When the truck accelerates, the crate stays fixed on the truck, so it, too, accelerates. What force causes the crate to accelerate?

20. A block is given a brief push so that it slides up a ramp. After the block reaches its highest point, it slides back down, but the magnitude of its acceleration is less on the descent than on the ascent. Why?

21. Why is the stopping distance of a truck much shorter than for a train going the same speed?

22. What would your bathroom scale read if you weighed yourself on an inclined plane? Assume the mechanism functions properly, even at an angle.

# MisConceptual Questions

1. A truck is traveling horizontally to the right (Fig. 4–38). When the truck starts to slow down, the crate on the (frictionless) truck bed starts to slide. In what direction could the net force be on the crate?
   (a) No direction. The net force is zero.
   (b) Straight down (because of gravity).
   (c) Straight up (the normal force).
   (d) Horizontal and to the right.
   (e) Horizontal and to the left.

**FIGURE 4–38**
MisConceptual Question 1.

2. You are trying to push your stalled car. Although you apply a horizontal force of 400 N to the car, it doesn't budge, and neither do you. Which force(s) must also have a magnitude of 400 N?
   (a) The force exerted by the car on you.
   (b) The friction force exerted by the car on the road.
   (c) The normal force exerted by the road on you.
   (d) The friction force exerted by the road on you.

3. Matt, in the foreground of Fig. 4–39, is able to move the large truck because
   (a) he is stronger than the truck.
   (b) he is heavier in some respects than the truck.
   (c) he exerts a greater force on the truck than the truck exerts back on him.
   (d) the ground exerts a greater friction force on Matt than it does on the truck.
   (e) the truck offers no resistance because its brakes are off.

**FIGURE 4–39** MisConceptual Question 3.

4. A bear sling, Fig. 4–40, is used in some national parks for placing backpackers' food out of the reach of bears. As the backpacker raises the pack by pulling down on the rope, the force $F$ needed:
   (a) decreases as the pack rises until the rope is straight across.
   (b) doesn't change.
   (c) increases until the rope is straight.
   (d) increases but the rope always sags where the pack hangs.

**FIGURE 4–40** MisConceptual Question 4.

5. What causes the boat in Fig. 4–41 to move forward?
   (a) The force the man exerts on the paddle.
   (b) The force the paddle exerts on the water.
   (c) The force the water exerts on the paddle.
   (d) The motion of the water itself.

**FIGURE 4–41** MisConceptual Question 5.

6. A person stands on a scale in an elevator. His apparent weight will be the greatest when the elevator
   (a) is standing still.
   (b) is moving upward at constant velocity.
   (c) is accelerating upward.
   (d) is moving downward at constant velocity.
   (e) is accelerating downward.

7. When a skier skis down a hill, the normal force exerted on the skier by the hill is
   (a) equal to the weight of the skier.
   (b) greater than the weight of the skier.
   (c) less than the weight of the skier.

8. A golf ball is hit with a golf club. While the ball flies through the air, which forces act on the ball? Neglect air resistance.
   (a) The force of the golf club acting on the ball.
   (b) The force of gravity acting on the ball.
   (c) The force of the ball moving forward through the air.
   (d) All of the above.
   (e) Both (a) and (c).

9. Suppose an object is accelerated by a force of 100 N. Suddenly a second force of 100 N in the opposite direction is exerted on the object, so that the forces cancel. The object
   (a) is brought to rest rapidly.
   (b) decelerates gradually to rest.
   (c) continues at the velocity it had before the second force was applied.
   (d) is brought to rest and then accelerates in the direction of the second force.

10. You are pushing a heavy box across a rough floor. When you are initially pushing the box and it is accelerating,
    (a) you exert a force on the box, but the box does not exert a force on you.
    (b) the box is so heavy it exerts a force on you, but you do not exert a force on the box.
    (c) the force you exert on the box is greater than the force of the box pushing back on you.
    (d) the force you exert on the box is equal to the force of the box pushing back on you.
    (e) the force that the box exerts on you is greater than the force you exert on the box.

11. A 50-N crate sits on a horizontal floor where the coefficient of static friction between the crate and the floor is 0.50. A 20-N force is applied to the crate acting to the right. What is the resulting static friction force acting on the crate?
    (a) 20 N to the right.
    (b) 20 N to the left.
    (c) 25 N to the right.
    (d) 25 N to the left.
    (e) None of the above; the crate starts to move.

12. The normal force on an extreme skier descending a very steep slope (Fig. 4–42) can be zero if
    (a) his speed is great enough.
    (b) he leaves the slope (no longer touches the snow).
    (c) the slope is greater than 75°.
    (d) the slope is vertical (90°).

**FIGURE 4–42** MisConceptual Question 12.

13. To pull an old stump out of the ground, you and a friend tie two ropes to the stump. You pull on it with a force of 500 N to the north while your friend pulls with a force of 450 N to the northwest. The total force from the two ropes is
    (a) less than 950 N.
    (b) exactly 950 N.
    (c) more than 950 N.

# Problems

[*It would be wise, before starting the Problems, to reread the Problem Solving Strategies on pages 30, 60, and 88.*]

### 4–4 to 4–6 Newton's Laws, Gravitational Force, Normal Force [Assume no friction.]

1. (I) What force is needed to accelerate a sled (mass = 55 kg) at 1.4 m/s² on horizontal frictionless ice?

2. (I) What is the weight of a 68-kg astronaut (*a*) on Earth, (*b*) on the Moon $\left(g = 1.7 \text{ m/s}^2\right)$, (*c*) on Mars $\left(g = 3.7 \text{ m/s}^2\right)$, (*d*) in outer space traveling with constant velocity?

3. (I) How much tension must a rope withstand if it is used to accelerate a 1210-kg car horizontally along a frictionless surface at 1.20 m/s²?

4. (II) According to a simplified model of a mammalian heart, at each pulse approximately 20 g of blood is accelerated from 0.25 m/s to 0.35 m/s during a period of 0.10 s. What is the magnitude of the force exerted by the heart muscle?

5. (II) Superman must stop a 120-km/h train in 150 m to keep it from hitting a stalled car on the tracks. If the train's mass is $3.6 \times 10^5$ kg, how much force must he exert? Compare to the weight of the train (give as %). How much force does the train exert on Superman?

6. (II) A person has a reasonable chance of surviving an automobile crash if the deceleration is no more than 30 *g*'s. Calculate the force on a 65-kg person accelerating at this rate. What distance is traveled if brought to rest at this rate from 95 km/h?

7. (II) What average force is required to stop a 950-kg car in 8.0 s if the car is traveling at 95 km/h?

8. (II) Estimate the average force exerted by a shot-putter on a 7.0-kg shot if the shot is moved through a distance of 2.8 m and is released with a speed of 13 m/s.

9. (II) A 0.140-kg baseball traveling 35.0 m/s strikes the catcher's mitt, which, in bringing the ball to rest, recoils backward 11.0 cm. What was the average force applied by the ball on the glove?

10. (II) How much tension must a cable withstand if it is used to accelerate a 1200-kg car vertically upward at 0.70 m/s²?

11. (II) A 20.0-kg box rests on a table. (*a*) What is the weight of the box and the normal force acting on it? (*b*) A 10.0-kg box is placed on top of the 20.0-kg box, as shown in Fig. 4–43. Determine the normal force that the table exerts on the 20.0-kg box and the normal force that the 20.0-kg box exerts on the 10.0-kg box.

**FIGURE 4–43**
Problem 11.

12. (II) A 14.0-kg bucket is lowered vertically by a rope in which there is 163 N of tension at a given instant. What is the acceleration of the bucket? Is it up or down?

13. (II) A 75-kg petty thief wants to escape from a third-story jail window. Unfortunately, a makeshift rope made of sheets tied together can support a mass of only 58 kg. How might the thief use this "rope" to escape? Give a quantitative answer.

14. (II) An elevator (mass 4850 kg) is to be designed so that the maximum acceleration is 0.0680*g*. What are the maximum and minimum forces the motor should exert on the supporting cable?

15. (II) Can cars "stop on a dime"? Calculate the acceleration of a 1400-kg car if it can stop from 35 km/h on a dime (diameter = 1.7 cm). How many *g*'s is this? What is the force felt by the 68-kg occupant of the car?

16. (II) A woman stands on a bathroom scale in a motionless elevator. When the elevator begins to move, the scale briefly reads only 0.75 of her regular weight. Calculate the acceleration of the elevator, and find the direction of acceleration.

17. (II) (*a*) What is the acceleration of two falling sky divers (total mass = 132 kg including parachute) when the upward force of air resistance is equal to one-fourth of their weight? (*b*) After opening the parachute, the divers descend leisurely to the ground at constant speed. What now is the force of air resistance on the sky divers and their parachute? See Fig. 4–44.

**FIGURE 4–44**
Problem 17.

18. (II) The cable supporting a 2125-kg elevator has a maximum strength of 21,750 N. What maximum upward acceleration can it give the elevator without breaking?

19. (III) A person jumps from the roof of a house 2.8 m high. When he strikes the ground below, he bends his knees so that his torso decelerates over an approximate distance of 0.70 m. If the mass of his torso (excluding legs) is 42 kg, find (*a*) his velocity just before his feet strike the ground, and (*b*) the average force exerted on his torso by his legs during deceleration.

## 4–7 Newton's Laws and Vectors [Ignore friction.]

**20.** (I) A box weighing 77.0 N rests on a table. A rope tied to the box runs vertically upward over a pulley and a weight is hung from the other end (Fig. 4–45). Determine the force that the table exerts on the box if the weight hanging on the other side of the pulley weighs (a) 30.0 N, (b) 60.0 N, and (c) 90.0 N.

**FIGURE 4–45**
Problem 20.

**21.** (I) Draw the free-body diagram for a basketball player (a) just before leaving the ground on a jump, and (b) while in the air. See Fig. 4–46.

**FIGURE 4–46**
Problem 21.

**22.** (I) Sketch the free-body diagram of a baseball (a) at the moment it is hit by the bat, and again (b) after it has left the bat and is flying toward the outfield. Ignore air resistance.

**23.** (II) Arlene is to walk across a "high wire" strung horizontally between two buildings 10.0 m apart. The sag in the rope when she is at the midpoint is 10.0°, as shown in Fig. 4–47. If her mass is 50.0 kg, what is the tension in the rope at this point?

**FIGURE 4–47** Problem 23.

**24.** (II) A window washer pulls herself upward using the bucket–pulley apparatus shown in Fig. 4–48. (a) How hard must she pull downward to raise herself slowly at constant speed? (b) If she increases this force by 15%, what will her acceleration be? The mass of the person plus the bucket is 72 kg.

**FIGURE 4–48**
Problem 24.

**25.** (II) One 3.2-kg paint bucket is hanging by a massless cord from another 3.2-kg paint bucket, also hanging by a massless cord, as shown in Fig. 4–49. (a) If the buckets are at rest, what is the tension in each cord? (b) If the two buckets are pulled upward with an acceleration of 1.25 m/s² by the upper cord, calculate the tension in each cord.

**FIGURE 4–49**
Problem 25.

**26.** (II) Two snowcats in Antarctica are towing a housing unit north, as shown in Fig. 4–50. The sum of the forces $\vec{F}_A$ and $\vec{F}_B$ exerted on the unit by the horizontal cables is north, parallel to the line L, and $F_A = 4500$ N. Determine $F_B$ and the magnitude of $\vec{F}_A + \vec{F}_B$.

**FIGURE 4–50**
Problem 26.

Top view

**27.** (II) A train locomotive is pulling two cars of the same mass behind it, Fig. 4–51. Determine the ratio of the tension in the coupling (think of it as a cord) between the locomotive and the first car $(F_{T1})$, to that between the first car and the second car $(F_{T2})$, for any nonzero acceleration of the train.

**FIGURE 4–51** Problem 27.

**28.** (II) The two forces $\vec{F}_1$ and $\vec{F}_2$ shown in Fig. 4–52a and b (looking down) act on an 18.5-kg object on a frictionless tabletop. If $F_1 = 10.2$ N and $F_2 = 16.0$ N, find the net force on the object and its acceleration for (a) and (b).

**FIGURE 4–52** Problem 28.

29. (II) At the instant a race began, a 65-kg sprinter exerted a force of 720 N on the starting block at a 22° angle with respect to the ground. (a) What was the horizontal acceleration of the sprinter? (b) If the force was exerted for 0.32 s, with what speed did the sprinter leave the starting block?

30. (II) A 27-kg chandelier hangs from a ceiling on a vertical 4.0-m-long wire. (a) What horizontal force would be necessary to displace its position 0.15 m to one side? (b) What will be the tension in the wire?

31. (II) An object is hanging by a string from your rearview mirror. While you are decelerating at a constant rate from 25 m/s to rest in 6.0 s, (a) what angle does the string make with the vertical, and (b) is it toward the windshield or away from it? [*Hint*: See Example 4–15.]

32. (II) Figure 4–53 shows a block (mass $m_A$) on a smooth horizontal surface, connected by a thin cord that passes over a pulley to a second block ($m_B$), which hangs vertically. (a) Draw a free-body diagram for each block, showing the force of gravity on each, the force (tension) exerted by the cord, and any normal force. (b) Apply Newton's second law to find formulas for the acceleration of the system and for the tension in the cord. Ignore friction and the masses of the pulley and cord.

**FIGURE 4–53**
Problems 32 and 33.
Mass $m_A$ rests on a smooth horizontal surface; $m_B$ hangs vertically.

33. (II) (a) If $m_A = 13.0$ kg and $m_B = 5.0$ kg in Fig. 4–53, determine the acceleration of each block. (b) If initially $m_A$ is at rest 1.250 m from the edge of the table, how long does it take to reach the edge of the table if the system is allowed to move freely? (c) If $m_B = 1.0$ kg, how large must $m_A$ be if the acceleration of the system is to be kept at $\frac{1}{100} g$?

34. (III) Three blocks on a frictionless horizontal surface are in contact with each other as shown in Fig. 4–54. A force $\vec{F}$ is applied to block A (mass $m_A$). (a) Draw a free-body diagram for each block. Determine (b) the acceleration of the system (in terms of $m_A$, $m_B$, and $m_C$), (c) the net force on each block, and (d) the force of contact that each block exerts on its neighbor. (e) If $m_A = m_B = m_C = 10.0$ kg and $F = 96.0$ N, give numerical answers to (b), (c), and (d). Explain how your answers make sense intuitively.

**FIGURE 4–54**
Problem 34.

35. (III) Suppose the pulley in Fig. 4–55 is suspended by a cord C. Determine the tension in this cord after the masses are released and before one hits the ground. Ignore the mass of the pulley and cords.

**FIGURE 4–55**
Problem 35.

## 4–8 Newton's Laws with Friction, Inclines

36. (I) If the coefficient of kinetic friction between a 22-kg crate and the floor is 0.30, what horizontal force is required to move the crate at a steady speed across the floor? What horizontal force is required if $\mu_k$ is zero?

37. (I) A force of 35.0 N is required to start a 6.0-kg box moving across a horizontal concrete floor. (a) What is the coefficient of static friction between the box and the floor? (b) If the 35.0-N force continues, the box accelerates at 0.60 m/s². What is the coefficient of kinetic friction?

38. (I) Suppose you are standing on a train accelerating at $0.20\,g$. What minimum coefficient of static friction must exist between your feet and the floor if you are not to slide?

39. (II) The coefficient of static friction between hard rubber and normal street pavement is about 0.90. On how steep a hill (maximum angle) can you leave a car parked?

40. (II) A flatbed truck is carrying a heavy crate. The coefficient of static friction between the crate and the bed of the truck is 0.75. What is the maximum rate at which the driver can decelerate and still avoid having the crate slide against the cab of the truck?

41. (II) A 2.0-kg silverware drawer does not slide readily. The owner gradually pulls with more and more force, and when the applied force reaches 9.0 N, the drawer suddenly opens, throwing all the utensils to the floor. What is the coefficient of static friction between the drawer and the cabinet?

42. (II) A box is given a push so that it slides across the floor. How far will it go, given that the coefficient of kinetic friction is 0.15 and the push imparts an initial speed of 3.5 m/s?

43. (II) A 1280-kg car pulls a 350-kg trailer. The car exerts a horizontal force of $3.6 \times 10^3$ N against the ground in order to accelerate. What force does the car exert on the trailer? Assume an effective friction coefficient of 0.15 for the trailer.

44. (II) Police investigators, examining the scene of an accident involving two cars, measure 72-m-long skid marks of one of the cars, which nearly came to a stop before colliding. The coefficient of kinetic friction between rubber and the pavement is about 0.80. Estimate the initial speed of that car assuming a level road.

45. (II) Drag-race tires in contact with an asphalt surface have a very high coefficient of static friction. Assuming a constant acceleration and no slipping of tires, estimate the coefficient of static friction needed for a drag racer to cover 1.0 km in 12 s, starting from rest.

46. (II) For the system of Fig. 4–32 (Example 4–20), how large a mass would box A have to have to prevent any motion from occurring? Assume $\mu_s = 0.30$.

47. (II) In Fig. 4–56 the coefficient of static friction between mass $m_A$ and the table is 0.40, whereas the coefficient of kinetic friction is 0.20. (a) What minimum value of $m_A$ will keep the system from starting to move? (b) What value(s) of $m_A$ will keep the system moving at constant speed? [Ignore masses of the cord and the (frictionless) pulley.]

FIGURE 4–56
Problem 47.

48. (II) A small box is held in place against a rough vertical wall by someone pushing on it with a force directed upward at 28° above the horizontal. The coefficients of static and kinetic friction between the box and wall are 0.40 and 0.30, respectively. The box slides down unless the applied force has magnitude 23 N. What is the mass of the box?

49. (II) Two crates, of mass 65 kg and 125 kg, are in contact and at rest on a horizontal surface (Fig. 4–57). A 650-N force is exerted on the 65-kg crate. If the coefficient of kinetic friction is 0.18, calculate (a) the acceleration of the system, and (b) the force that each crate exerts on the other. (c) Repeat with the crates reversed.

FIGURE 4–57
Problem 49.

50. (II) A person pushes a 14.0-kg lawn mower at constant speed with a force of $F = 88.0$ N directed along the handle, which is at an angle of 45.0° to the horizontal (Fig. 4–58). (a) Draw the free-body diagram showing all forces acting on the mower. Calculate (b) the horizontal friction force on the mower, then (c) the normal force exerted vertically upward on the mower by the ground. (d) What force must the person exert on the lawn mower to accelerate it from rest to 1.5 m/s in 2.5 seconds, assuming the same friction force?

FIGURE 4–58
Problem 50.

51. (II) A child on a sled reaches the bottom of a hill with a velocity of 10.0 m/s and travels 25.0 m along a horizontal straightaway to a stop. If the child and sled together have a mass of 60.0 kg, what is the average retarding force on the sled on the horizontal straightaway?

52. (II) (a) A box sits at rest on a rough 33° inclined plane. Draw the free-body diagram, showing all the forces acting on the box. (b) How would the diagram change if the box were sliding down the plane? (c) How would it change if the box were sliding up the plane after an initial shove?

53. (II) A wet bar of soap slides down a ramp 9.0 m long inclined at 8.0°. How long does it take to reach the bottom? Assume $\mu_k = 0.060$.

54. (II) A skateboarder, with an initial speed of 2.0 m/s, rolls virtually friction free down a straight incline of length 18 m in 3.3 s. At what angle $\theta$ is the incline oriented above the horizontal?

55. (II) Uphill escape ramps are sometimes provided to the side of steep downhill highways for trucks with overheated brakes. For a simple 11° upward ramp, what minimum length would be needed for a runaway truck traveling 140 km/h? Note the large size of your calculated length. (If sand is used for the bed of the ramp, its length can be reduced by a factor of about 2.)

56. (II) A 25.0-kg box is released on a 27° incline and accelerates down the incline at 0.30 m/s². Find the friction force impeding its motion. What is the coefficient of kinetic friction?

57. (II) The block shown in Fig. 4–59 has mass $m = 7.0$ kg and lies on a fixed smooth frictionless plane tilted at an angle $\theta = 22.0°$ to the horizontal. (a) Determine the acceleration of the block as it slides down the plane. (b) If the block starts from rest 12.0 m up the plane from its base, what will be the block's speed when it reaches the bottom of the incline?

FIGURE 4–59 Block on inclined plane. Problems 57 and 58.

58. (II) A block is given an initial speed of 4.5 m/s up the 22.0° plane shown in Fig. 4–59. (a) How far up the plane will it go? (b) How much time elapses before it returns to its starting point? Ignore friction.

59. (II) The crate shown in Fig. 4–60 lies on a plane tilted at an angle $\theta = 25.0°$ to the horizontal, with $\mu_k = 0.19$. (a) Determine the acceleration of the crate as it slides down the plane. (b) If the crate starts from rest 8.15 m up along the plane from its base, what will be the crate's speed when it reaches the bottom of the incline?

FIGURE 4–60
Crate on inclined plane. Problems 59 and 60.

60. (II) A crate is given an initial speed of 3.0 m/s up the 25.0° plane shown in Fig. 4–60. (a) How far up the plane will it go? (b) How much time elapses before it returns to its starting point? Assume $\mu_k = 0.12$.

61. (II) A car can decelerate at $-3.80$ m/s² without skidding when coming to rest on a level road. What would its deceleration be if the road is inclined at 9.3° and the car moves uphill? Assume the same static friction coefficient.

62. (II) A skier moves down a 12° slope at constant speed. What can you say about the coefficient of friction, $\mu_k$? Assume the speed is low enough that air resistance can be ignored.

63. (II) The coefficient of kinetic friction for a 22-kg bobsled on a track is 0.10. What force is required to push it down along a 6.0° incline and achieve a speed of 60 km/h at the end of 75 m?

**64. (II)** On an icy day, you worry about parking your car in your driveway, which has an incline of 12°. Your neighbor's driveway has an incline of 9.0°, and the driveway across the street is at 6.0°. The coefficient of static friction between tire rubber and ice is 0.15. Which driveway(s) will be safe to park in?

**65. (III)** Two masses $m_A = 2.0$ kg and $m_B = 5.0$ kg are on inclines and are connected together by a string as shown in Fig. 4–61. The coefficient of kinetic friction between each mass and its incline is $\mu_k = 0.30$. If $m_A$ moves up, and $m_B$ moves down, determine their acceleration. [Ignore masses of the (frictionless) pulley and the cord.]

**FIGURE 4–61** Problem 65.

**66. (III)** A child slides down a slide with a 34° incline, and at the bottom her speed is precisely half what it would have been if the slide had been frictionless. Calculate the coefficient of kinetic friction between the slide and the child.

**67. (III)** (a) Suppose the coefficient of kinetic friction between $m_A$ and the plane in Fig. 4–62 is $\mu_k = 0.15$, and that $m_A = m_B = 2.7$ kg. As $m_B$ moves down, determine the magnitude of the acceleration of $m_A$ and $m_B$, given $\theta = 34°$. (b) What smallest value of $\mu_k$ will keep the system from accelerating? [Ignore masses of the (frictionless) pulley and the cord.]

**FIGURE 4–62**
Problem 67.

# General Problems

**68.** A 2.0-kg purse is dropped from the top of the Leaning Tower of Pisa and falls 55 m before reaching the ground with a speed of 27 m/s. What was the average force of air resistance?

**69.** A crane's trolley at point P in Fig. 4–63 moves for a few seconds to the right with constant acceleration, and the 870-kg load hangs on a light cable at a 5.0° angle to the vertical as shown. What is the acceleration of the trolley and load?

**FIGURE 4–63**
Problem 69.

**70.** A 75.0-kg person stands on a scale in an elevator. What does the scale read (in N and in kg) when (a) the elevator is at rest, (b) the elevator is climbing at a constant speed of 3.0 m/s, (c) the elevator is descending at 3.0 m/s, (d) the elevator is accelerating upward at 3.0 m/s², (e) the elevator is accelerating downward at 3.0 m/s²?

**71.** A city planner is working on the redesign of a hilly portion of a city. An important consideration is how steep the roads can be so that even low-powered cars can get up the hills without slowing down. A particular small car, with a mass of 920 kg, can accelerate on a level road from rest to 21 m/s (75 km/h) in 12.5 s. Using these data, calculate the maximum steepness of a hill.

**72.** If a bicyclist of mass 65 kg (including the bicycle) can coast down a 6.5° hill at a steady speed of 6.0 km/h because of air resistance, how much force must be applied to climb the hill at the same speed (and the same air resistance)?

**73.** Francesca dangles her watch from a thin piece of string while the jetliner she is in accelerates for takeoff, which takes about 16 s. Estimate the takeoff speed of the aircraft if the string makes an angle of 25° with respect to the vertical, Fig. 4–64.

**FIGURE 4–64**
Problem 73.

**74.** Bob traverses a chasm by stringing a rope between a tree on one side of the chasm and a tree on the opposite side, 25 m away, Fig. 4–65. Assume the rope can provide a tension force of up to 29 kN before breaking, and use a "safety factor" of 10 (that is, the rope should only be required to undergo a tension force of 2.9 kN). (a) If Bob's mass is 72.0 kg, determine the distance $x$ that the rope must sag at a point halfway across if it is to be within its recommended safety range. (b) If the rope sags by only one-fourth the distance found in (a), determine the tension force in the rope. Will the rope break?

**FIGURE 4–65** Problem 74.

**75.** Piles of snow on slippery roofs can become dangerous projectiles as they melt. Consider a chunk of snow at the ridge of a roof with a slope of 34°. (*a*) What is the minimum value of the coefficient of static friction that will keep the snow from sliding down? (*b*) As the snow begins to melt, the coefficient of static friction decreases and the snow finally slips. Assuming that the distance from the chunk to the edge of the roof is 4.0 m and the coefficient of kinetic friction is 0.10, calculate the speed of the snow chunk when it slides off the roof. (*c*) If the roof edge is 10.0 m above ground, estimate the speed of the snow when it hits the ground.

**76.** (*a*) What minimum force $F$ is needed to lift the piano (mass $M$) using the pulley apparatus shown in Fig. 4–66? (*b*) Determine the tension in each section of rope: $F_{T1}$, $F_{T2}$, $F_{T3}$, and $F_{T4}$. Assume pulleys are massless and frictionless, and that ropes are massless.

**FIGURE 4–66**
Problem 76.

**77.** In the design of a supermarket, there are to be several ramps connecting different parts of the store. Customers will have to push grocery carts up the ramps and it is desirable that this not be too difficult. The engineer has done a survey and found that almost no one complains if the force required is no more than 18 N. Ignoring friction, at what maximum angle $\theta$ should the ramps be built, assuming a full 25-kg cart?

**78.** A jet aircraft is accelerating at $3.8 \text{ m/s}^2$ as it climbs at an angle of 18° above the horizontal (Fig. 4–67). What is the total force that the cockpit seat exerts on the 75-kg pilot?

**FIGURE 4–67**
Problem 78.

**79.** A 7180-kg helicopter accelerates upward at $0.80 \text{ m/s}^2$ while lifting a 1080-kg frame at a construction site, Fig. 4–68. (*a*) What is the lift force exerted by the air on the helicopter rotors? (*b*) What is the tension in the cable (ignore its mass) which connects the frame to the helicopter? (*c*) What force does the cable exert on the helicopter?

**FIGURE 4–68**
Problem 79.

**80.** An elevator in a tall building is allowed to reach a maximum speed of 3.5 m/s going down. What must the tension be in the cable to stop this elevator over a distance of 2.6 m if the elevator has a mass of 1450 kg including occupants?

**81.** A fisherman in a boat is using a "10-lb test" fishing line. This means that the line can exert a force of 45 N without breaking (1 lb = 4.45 N). (*a*) How heavy a fish can the fisherman land if he pulls the fish up vertically at constant speed? (*b*) If he accelerates the fish upward at $2.0 \text{ m/s}^2$, what maximum weight fish can he land? (*c*) Is it possible to land a 15-lb trout on 10-lb test line? Why or why not?

**82.** A "doomsday" asteroid with a mass of $1.0 \times 10^{10} \text{ kg}$ is hurtling through space. Unless the asteroid's speed is changed by about 0.20 cm/s, it will collide with Earth and cause tremendous damage. Researchers suggest that a small "space tug" sent to the asteroid's surface could exert a gentle constant force of 2.5 N. For how long must this force act?

**83.** Three mountain climbers who are roped together in a line are ascending an icefield inclined at 31.0° to the horizontal (Fig. 4–69). The last climber slips, pulling the second climber off his feet. The first climber is able to hold them both. If each climber has a mass of 75 kg, calculate the tension in each of the two sections of rope between the three climbers. Ignore friction between the ice and the fallen climbers.

**FIGURE 4–69** Problem 83.

**84.** As shown in Fig. 4–70, five balls (masses 2.00, 2.05, 2.10, 2.15, 2.20 kg) hang from a crossbar. Each mass is supported by "5-lb test" fishing line which will break when its tension force exceeds 22.2 N (= 5.00 lb). When this device is placed in an elevator, which accelerates upward, only the lines attached to the 2.05 and 2.00 kg masses do not break. Within what range is the elevator's acceleration?

**FIGURE 4–70**
Problem 84.

**85.** Two rock climbers, Jim and Karen, use safety ropes of similar length. Karen's rope is more elastic, called a *dynamic rope* by climbers. Jim has a *static rope*, not recommended for safety purposes in pro climbing. (*a*) Karen (Fig. 4–71) falls freely about 2.0 m and then the rope stops her over a distance of 1.0 m. Estimate how large a force (assume constant) she will feel from the rope. (Express the result in multiples of her weight.) (*b*) In a similar fall, Jim's rope stretches by only 30 cm. How many times his weight will the rope pull on him? Which climber is more likely to be hurt?

**FIGURE 4–71**
Problem 85.

**86.** A coffee cup on the horizontal dashboard of a car slides forward when the driver decelerates from 45 km/h to rest in 3.5 s or less, but not if she decelerates in a longer time. What is the coefficient of static friction between the cup and the dash? Assume the road and the dashboard are level (horizontal).

**87.** A roller coaster reaches the top of the steepest hill with a speed of 6.0 km/h. It then descends the hill, which is at an average angle of 45° and is 45.0 m long. What will its speed be when it reaches the bottom? Assume $\mu_k = 0.12$.

**88.** A motorcyclist is coasting with the engine off at a steady speed of 20.0 m/s but enters a sandy stretch where the coefficient of kinetic friction is 0.70. Will the cyclist emerge from the sandy stretch without having to start the engine if the sand lasts for 15 m? If so, what will be the speed upon emerging?

**89.** The 70.0-kg climber in Fig. 4–72 is supported in the "chimney" by the friction forces exerted on his shoes and back. The static coefficients of friction between his shoes and the wall, and between his back and the wall, are 0.80 and 0.60, respectively. What is the minimum normal force he must exert? Assume the walls are vertical and that the static friction forces are both at their maximum. Ignore his grip on the rope.

**FIGURE 4–72**
Problem 89.

**90.** A 28.0-kg block is connected to an empty 2.00-kg bucket by a cord running over a frictionless pulley (Fig. 4–73). The coefficient of static friction between the table and the block is 0.45 and the coefficient of kinetic friction between the table and the block is 0.32. Sand is gradually added to the bucket until the system just begins to move. (*a*) Calculate the mass of sand added to the bucket. (*b*) Calculate the acceleration of the system. Ignore mass of cord.

**FIGURE 4–73**
Problem 90.

**91.** A 72-kg water skier is being accelerated by a ski boat on a flat ("glassy") lake. The coefficient of kinetic friction between the skier's skis and the water surface is $\mu_k = 0.25$ (Fig. 4–74). (*a*) What is the skier's acceleration if the rope pulling the skier behind the boat applies a horizontal tension force of magnitude $F_T = 240$ N to the skier ($\theta = 0°$)? (*b*) What is the skier's horizontal acceleration if the rope pulling the skier exerts a force of $F_T = 240$ N on the skier at an upward angle $\theta = 12°$? (*c*) Explain why the skier's acceleration in part (*b*) is greater than that in part (*a*).

**FIGURE 4–74** Problem 91.

**92.** A 75-kg snowboarder has an initial velocity of 5.0 m/s at the top of a 28° incline (Fig. 4–75). After sliding down the 110-m-long incline (on which the coefficient of kinetic friction is $\mu_k = 0.18$), the snowboarder has attained a velocity $v$. The snowboarder then slides along a flat surface (on which $\mu_k = 0.15$) and comes to rest after a distance $x$. Use Newton's second law to find the snowboarder's acceleration while on the incline and while on the flat surface. Then use these accelerations to determine $x$.

**FIGURE 4–75** Problem 92.

**93.** (*a*) If the horizontal acceleration produced briefly by an earthquake is $a$, and if an object is going to "hold its place" on the ground, show that the coefficient of static friction with the ground must be at least $\mu_s = a/g$. (*b*) The famous Loma Prieta earthquake that stopped the 1989 World Series produced ground accelerations of up to $4.0\ \text{m/s}^2$ in the San Francisco Bay Area. Would a chair have started to slide on a floor with coefficient of static friction 0.25?

**94.** Two blocks made of different materials, connected by a thin cord, slide down a plane ramp inclined at an angle $\theta$ to the horizontal, Fig. 4–76 (block B is above block A). The masses of the blocks are $m_A$ and $m_B$, and the coefficients of friction are $\mu_A$ and $\mu_B$. If $m_A = m_B = 5.0\ \text{kg}$, and $\mu_A = 0.20$ and $\mu_B = 0.30$, determine (*a*) the acceleration of the blocks and (*b*) the tension in the cord, for an angle $\theta = 32°$.

**FIGURE 4–76**
Problem 94.

**95.** A car starts rolling down a 1-in-4 hill (1-in-4 means that for each 4 m traveled along the sloping road, the elevation change is 1 m). How fast is it going when it reaches the bottom after traveling 55 m? (*a*) Ignore friction. (*b*) Assume an effective coefficient of friction equal to 0.10.

**96.** A 65-kg ice skater coasts with no effort for 75 m until she stops. If the coefficient of kinetic friction between her skates and the ice is $\mu_k = 0.10$, how fast was she moving at the start of her coast?

**97.** An 18-kg child is riding in a child-restraint chair, securely fastened to the seat of a car (Fig. 4–77). Assume the car has speed 45 km/h when it hits a tree and is brought to rest in 0.20 s. Assuming constant deceleration during the collision, estimate the net horizontal force $F$ that the straps of the restraint chair exert on the child to hold her in the chair.

**FIGURE 4–77**
Problem 97.

# Search and Learn

**1.** (*a*) Finding her car stuck in the mud, a bright graduate of a good physics course ties a strong rope to the back bumper of the car, and the other end to a boulder, as shown in Fig. 4–78a. She pushes at the midpoint of the rope with her maximum effort, which she estimates to be a force $F_P \approx 300\ \text{N}$. The car just begins to budge with the rope at an angle $\theta$, which she estimates to be 5°. With what force is the rope pulling on the car? Neglect the mass of the rope. (*b*) What is the "mechanical advantage" of this technique [Section 4–7]? (*c*) At what angle $\theta$ would this technique become counterproductive? [*Hint*: Consider the forces on a small segment of rope where $\vec{F}_P$ acts, Fig. 4–78b.]

$\vec{F}_P$

B    $\vec{F}_{BR}$    $\vec{F}_{CR}$    C

(a)    $\theta$      $\theta$

$y$

$\vec{F}_P$

$\theta$     $\theta$     $x$

(b)   $\vec{F}_{RB}$     $\vec{F}_{RC}$

**FIGURE 4–78** (a) Getting a car out of the mud, showing the forces on the boulder, on the car, and exerted by the person. (b) The free-body diagram: forces on a small segment of rope.

**2.** (*a*) Show that the minimum stopping distance for an automobile traveling on a level road at speed $v$ is equal to $v^2/(2\,\mu_s\,g)$, where $\mu_s$ is the coefficient of static friction between the tires and the road, and $g$ is the acceleration of gravity. (*b*) What is this distance for a 1200-kg car traveling 95 km/h if $\mu_s = 0.65$? (*c*) What would it be if the car were on the Moon (the acceleration of gravity on the Moon is about $g/6$) but all else stayed the same?

**3.** In the equation for static friction in Section 4–8, what is the significance of the $<$ sign? When should you use the equals sign in the static friction equation?

**4.** Referring to Example 4–21, show that if a skier moves at constant speed straight down a slope of angle $\theta$, then the coefficient of kinetic friction between skis and snow is $\mu_k = \tan\theta$.

## ANSWERS TO EXERCISES

**A:** No force is needed. The car accelerates out from under the cup, which tends to remain at rest. Think of Newton's first law (see Example 4–1).

**B:** (*a*).

**C:** (i) The same; (ii) the tennis ball; (iii) Newton's third law for part (i), second law for part (ii).

**D:** (*b*).

**E:** (*b*).

**F:** (*b*).

**G:** (*c*).

**H:** Yes; no.

The astronauts in the upper left of this photo are working on a space shuttle. As they orbit the Earth—at a rather high speed—they experience apparent weightlessness. The Moon, in the background seen against the blackness of space, also is orbiting the Earth at high speed. Both the Moon and the space shuttle move in nearly circular orbits, and each undergoes a centripetal acceleration. What keeps the Moon and the space shuttle (and its astronauts) from moving off in a straight line away from Earth? It is the force of gravity. Newton's law of universal gravitation states that all objects attract all other objects with a force that depends on their masses and the square of the distance between them.

# Circular Motion; Gravitation

## CHAPTER-OPENING QUESTIONS—Guess now!

**1.** You revolve a ball around you in a horizontal circle at constant speed on a string, as shown here from above. Which path will the ball follow if you let go of the string when the ball is at point P?

**2.** A space station revolves around the Earth as a satellite, 100 km above Earth's surface. What is the net force on an astronaut at rest inside the space station?

(a) Equal to her weight on Earth.
(b) A little less than her weight on Earth.
(c) Less than half her weight on Earth.
(d) Zero (she is weightless).
(e) Somewhat larger than her weight on Earth.

## CONTENTS

A n object moves in a straight line if the net force on it acts along the direction of motion, or the net force is zero. If the net force acts at an angle to the direction of motion at any moment, then the object moves in a curved path. An example of the latter is projectile motion, which we discussed in Chapter 3. Another important case is that of an object moving in a circle, such as a ball at the end of a string being swung in a circle above one's head, or the nearly circular motion of the Moon about the Earth.

**FIGURE 5–1** A small object moving in a circle, showing how the velocity changes. At each point, the instantaneous velocity is in a direction tangent to the circular path.

**FIGURE 5–2** Determining the change in velocity, $\Delta\vec{v}$, for a particle moving in a circle. The length $\Delta\ell$ is the distance along the arc, from A to B.

(a)

(b)

$$\vec{a} = \lim_{\Delta t \to 0} \frac{\Delta\vec{v}}{\Delta t}$$

(c)

Gravitation

In this Chapter, we study the circular motion of objects, and how Newton's laws of motion apply. We also discuss how Newton conceived of another great law by applying the concepts of circular motion to the motion of the Moon and the planets. This is the law of universal gravitation, which was the capstone of Newton's analysis of the physical world.

# 5–1 Kinematics of Uniform Circular Motion

An object that moves in a circle at constant speed $v$ is said to experience **uniform circular motion**. The *magnitude* of the velocity remains constant in this case, but the *direction* of the velocity continuously changes as the object moves around the circle (Fig. 5–1). Because acceleration is defined as the rate of change of velocity, a change in direction of velocity is an acceleration, just as a change in its magnitude is. Thus, an object revolving in a circle is continuously accelerating, even when the speed remains constant ($v_1 = v_2 = v$ in Fig. 5–1). We now investigate this acceleration quantitatively.

Acceleration is defined as

$$\vec{a} = \frac{\vec{v}_2 - \vec{v}_1}{\Delta t} = \frac{\Delta\vec{v}}{\Delta t},$$

where $\Delta\vec{v}$ is the change in velocity during the short time interval $\Delta t$. We will eventually consider the situation in which $\Delta t$ approaches zero and thus obtain the instantaneous acceleration. But for purposes of making a clear drawing, Fig. 5–2, we consider a nonzero time interval. During the time interval $\Delta t$, the particle in Fig. 5–2a moves from point A to point B, covering a distance $\Delta\ell$ *along the arc* which subtends an angle $\Delta\theta$. The change in the velocity vector is $\vec{v}_2 - \vec{v}_1 = \Delta\vec{v}$, and is shown in Fig. 5–2b (note that $\vec{v}_2 = \vec{v}_1 + \Delta\vec{v}$).

Now we let $\Delta t$ be very small, approaching zero. Then $\Delta\ell$ and $\Delta\theta$ are also very small, and $\vec{v}_2$ will be almost parallel to $\vec{v}_1$, Fig. 5–2c; $\Delta\vec{v}$ will be essentially perpendicular to them. Thus $\Delta\vec{v}$ points toward the center of the circle. Since $\vec{a}$, by definition, is in the same direction as $\Delta\vec{v}$ (equation above), it too must point toward the center of the circle. Therefore, this acceleration is called **centripetal acceleration** ("center-pointing" acceleration) or **radial acceleration** (since it is directed along the radius, toward the center of the circle), and we denote it by $\vec{a}_R$.

Now that we have determined the direction, next we find the magnitude of the radial (centripetal) acceleration, $a_R$. Because the line CA in Fig. 5–2a is perpendicular to $\vec{v}_1$, and line CB is perpendicular to $\vec{v}_2$, then the angle $\Delta\theta$ between CA and CB is also the angle between $\vec{v}_1$ and $\vec{v}_2$. Hence the vectors $\vec{v}_1$, $\vec{v}_2$, and $\Delta\vec{v}$ in Fig. 5–2b form a triangle that is geometrically similar[†] to triangle ACB in Fig. 5–2a. If we take $\Delta\theta$ to be very small (letting $\Delta t$ be very small) and set $v = v_1 = v_2$ because the magnitude of the velocity is assumed not to change, we can write

$$\frac{\Delta v}{v} \approx \frac{\Delta\ell}{r}.$$

This is an exact equality when $\Delta t$ approaches zero, for then the arc length $\Delta\ell$ equals the chord length AB. We want to find the instantaneous acceleration, so we let $\Delta t$ approach zero, write the above expression as an equality, and then solve for $\Delta v$:

$$\Delta v = \frac{v}{r}\Delta\ell. \qquad [\Delta t \to 0]$$

To get the centripetal acceleration, $a_R$, we divide $\Delta v$ by $\Delta t$:

$$a_R = \frac{\Delta v}{\Delta t} = \frac{v}{r}\frac{\Delta\ell}{\Delta t}. \qquad [\Delta t \to 0]$$

But $\Delta\ell/\Delta t$ is the linear speed, $v$, of the object, so the radial (centripetal)

[†]Appendix A contains a review of geometry.

acceleration is

$$a_R = \frac{v^2}{r}. \qquad \text{[radial (centripetal) acceleration]} \quad \textbf{(5–1)}$$

[Equation 5–1 is valid at any instant in circular motion, and even when $v$ is not constant.]

To summarize, *an object moving in a circle of radius $r$ at constant speed $v$ has an acceleration whose direction is toward the center of the circle and whose magnitude is* $a_R = v^2/r$. It is not surprising that this acceleration depends on $v$ and $r$. The greater the speed $v$, the faster the velocity changes direction; and the larger the radius, the less rapidly the velocity changes direction.

The acceleration vector points toward the center of the circle when $v$ is constant. But the velocity vector always points in the direction of motion, which is tangential to the circle. Thus the velocity and acceleration vectors are perpendicular to each other at every point in the path for uniform circular motion (Fig. 5–3). This is another example that illustrates the error in thinking that acceleration and velocity are always in the same direction. For an object falling in a vertical path, $\vec{a}$ and $\vec{v}$ are indeed parallel. But in uniform circular motion, $\vec{a}$ and $\vec{v}$ are perpendicular, not parallel (nor were they parallel in projectile motion, Section 3–5).

Circular motion is often described in terms of the **frequency** $f$, the number of revolutions per second. The **period** $T$ of an object revolving in a circle is the time required for one complete revolution. Period and frequency are related by

$$T = \frac{1}{f}. \qquad \textbf{(5–2)}$$

For example, if an object revolves at a frequency of 3 rev/s, then each revolution (= rev) takes $\frac{1}{3}$ s. An object revolving in a circle (of circumference $2\pi r$) at constant speed $v$ travels a distance $2\pi r$ in one revolution which takes a time $T$. Thus

$$v = \frac{\text{distance}}{\text{time}} = \frac{2\pi r}{T}.$$

> ◆ **CAUTION**
> *In uniform circular motion, the speed is constant, but the acceleration is not zero*

> ◆ **CAUTION**
> *The direction of motion ($\vec{v}$) and the acceleration ($\vec{a}$) are not in the same direction; instead, $\vec{a} \perp \vec{v}$*

**FIGURE 5–3** For uniform circular motion, $\vec{a}$ is always perpendicular to $\vec{v}$.

---

**EXAMPLE 5–1** **Acceleration of a revolving ball.** A 150-g ball at the end of a string is revolving uniformly in a horizontal circle of radius 0.600 m, as in Fig. 5–1 or 5–3. The ball makes 2.00 revolutions in a second. What is its centripetal acceleration?

**APPROACH** The centripetal acceleration is $a_R = v^2/r$. We are given $r$, and we can find the speed of the ball, $v$, from the given radius and frequency.

**SOLUTION** If the ball makes 2.00 complete revolutions per second, then the ball travels in a complete circle in a time interval equal to 0.500 s, which is its period $T$. The distance traveled in this time is the circumference of the circle, $2\pi r$, where $r$ is the radius of the circle. Therefore, the ball has speed

$$v = \frac{2\pi r}{T} = \frac{2\pi(0.600\text{ m})}{(0.500\text{ s})} = 7.54\text{ m/s}.$$

The centripetal acceleration[†] is

$$a_R = \frac{v^2}{r} = \frac{(7.54\text{ m/s})^2}{(0.600\text{ m})} = 94.7\text{ m/s}^2.$$

---

**EXERCISE A** In Example 5–1, if the radius is doubled to 1.20 m, but the period stays the same, the centripetal acceleration will change by a factor of:
(a) 2; (b) 4; (c) $\frac{1}{2}$; (d) $\frac{1}{4}$; (e) none of these.

---

[†]Differences in the final digit can depend on whether you keep all digits in your calculator for $v$ (which gives $a_R = 94.7\text{ m/s}^2$), or if you use $v = 7.54$ m/s (which gives $a_R = 94.8\text{ m/s}^2$). Both results are valid since our assumed accuracy is about $\pm 0.1$ m/s (see Section 1–4).

**EXAMPLE 5–2** **Moon's centripetal acceleration.** The Moon's nearly circular orbit around the Earth has a radius of about 384,000 km and a period $T$ of 27.3 days. Determine the acceleration of the Moon toward the Earth.

**APPROACH** Again we need to find the velocity $v$ in order to find $a_R$.

**SOLUTION** In one orbit around the Earth, the Moon travels a distance $2\pi r$, where $r = 3.84 \times 10^8$ m is the radius of its circular path. The time required for one complete orbit is the Moon's period of 27.3 d. The speed of the Moon in its orbit about the Earth is $v = 2\pi r/T$. The period $T$ in seconds is $T = (27.3 \text{ d})(24.0 \text{ h/d})(3600 \text{ s/h}) = 2.36 \times 10^6$ s. Therefore,

$$a_R = \frac{v^2}{r} = \frac{(2\pi r)^2}{T^2 r} = \frac{4\pi^2 r}{T^2} = \frac{4\pi^2 (3.84 \times 10^8 \text{ m})}{(2.36 \times 10^6 \text{ s})^2}$$

$$= 0.00272 \text{ m/s}^2 = 2.72 \times 10^{-3} \text{ m/s}^2.$$

We can write this acceleration in terms of $g = 9.80 \text{ m/s}^2$ (the acceleration of gravity at the Earth's surface) as

$$a_R = 2.72 \times 10^{-3} \text{ m/s}^2 \left( \frac{g}{9.80 \text{ m/s}^2} \right) = 2.78 \times 10^{-4} g$$

$$\approx 0.0003 \, g.$$

**NOTE** The centripetal acceleration of the Moon, $a_R = 2.78 \times 10^{-4} g$, is *not* the acceleration of gravity for objects at the Moon's surface due to the Moon's gravity. Rather, it is the acceleration due to the *Earth's* gravity for any object (such as the Moon) that is 384,000 km from the Earth. Notice how small this acceleration is compared to the acceleration of objects near the Earth's surface.

# 5–2 Dynamics of Uniform Circular Motion

According to Newton's second law $(\Sigma \vec{F} = m\vec{a})$, an object that is accelerating must have a net force acting on it. An object moving in a circle, such as a ball on the end of a string, must therefore have a force applied to it to keep it moving in that circle. That is, a net force is necessary to give it centripetal acceleration. The magnitude of the required force can be calculated using Newton's second law for the radial component, $\Sigma F_R = ma_R$, where $a_R$ is the centripetal acceleration, $a_R = v^2/r$, and $\Sigma F_R$ is the total (or net) force in the radial direction:

$$\Sigma F_R = ma_R = m\frac{v^2}{r}. \qquad \text{[circular motion]} \quad \textbf{(5–3)}$$

**FIGURE 5–4** A force is required to keep an object moving in a circle. If the speed is constant, the force is directed toward the circle's center.

For uniform circular motion ($v$ = constant), the acceleration is $a_R$, which is directed toward the center of the circle at all times. Thus the *net force too must be directed toward the center of the circle* (Fig. 5–4). A net force is necessary because if no net force were exerted on the object, it would not move in a circle but in a straight line, as Newton's first law tells us. The direction of the net force is continually changing so that it is always directed toward the center of the circle. This force is sometimes called a centripetal ("pointing toward the center") force. But be aware that "centripetal force" does not indicate some new kind of force. The term "centripetal force" merely describes the *direction* of the net force needed to provide a circular path: the net force is directed toward the circle's center. The force *must be applied by other objects*. For example, to swing a ball in a circle on the end of a string, you pull on the string and the string exerts the force on the ball. (Try it.) Here, the "centripetal force" that provides the centripetal acceleration is tension in the string. In other cases it can be gravity (on the Moon, for example), a normal force, or even an electric force.

There is a common misconception that an object moving in a circle has an outward force acting on it, a so-called centrifugal ("center-fleeing") force. This is incorrect: *there is no outward force* on the revolving object. Consider, for example, a person swinging a ball on the end of a string around her head (Fig. 5–5). If you have ever done this yourself, you know that you feel a force pulling outward on your hand. The misconception arises when this pull is interpreted as an outward "centrifugal" force pulling on the ball that is transmitted along the string to your hand. This is not what is happening at all. To keep the ball moving in a circle, you pull *inwardly* on the string, and the string exerts this inward force on the ball. The ball exerts an equal and opposite force on the string (Newton's third law), and *this* is the outward force your hand feels (see Fig. 5–5).

The force *on the ball* in Fig. 5–5 is the one exerted *inwardly* on it by you, via the string. To see even more convincing evidence that a "centrifugal force" does not act on the ball, consider what happens when you let go of the string. If a centrifugal force were acting, the ball would fly outward, as shown in Fig. 5–6a. But it doesn't; the ball flies off tangentially (Fig. 5–6b), in the direction of the velocity it had at the moment it was released, because the inward force no longer acts. Try it and see!

**EXERCISE B** Return to Chapter-Opening Question 1, page 109, and answer it again now. Try to explain why you may have answered differently the first time.

---

**EXAMPLE 5–3 | ESTIMATE | Force on revolving ball (horizontal).** Estimate the force a person must exert on a string attached to a 0.150-kg ball to make the ball revolve in a horizontal circle of radius 0.600 m. The ball makes 2.00 revolutions per second ($T = 0.500$ s), as in Example 5–1. Ignore the string's mass.

**APPROACH** First we need to draw the free-body diagram for the ball. The forces acting on the ball are the force of gravity, $m\vec{g}$ downward, and the tension force $\vec{F}_T$ that the string exerts toward the hand at the center (which occurs because the person exerts that same force on the string). The free-body diagram for the ball is shown in Fig. 5–7. The ball's weight complicates matters and makes it impossible to revolve a ball with the cord perfectly horizontal. We estimate the force assuming the weight is small, and letting $\phi \approx 0$ in Fig. 5–7. Then $\vec{F}_T$ will act nearly horizontally and, in any case, provides the force necessary to give the ball its centripetal acceleration.

**SOLUTION** We apply Newton's second law to the radial direction, which we assume is horizontal:

$$(\Sigma F)_R = ma_R,$$

where $a_R = v^2/r$ and $v = 2\pi r/T = 2\pi(0.600\text{ m})/(0.500\text{ s}) = 7.54$ m/s. Thus

$$F_T = m\frac{v^2}{r}$$

$$= (0.150\text{ kg})\frac{(7.54\text{ m/s})^2}{(0.600\text{ m})} \approx 14\text{ N}.$$

**NOTE** We keep only two significant figures in the answer because we ignored the ball's weight; it is $mg = (0.150\text{ kg})(9.80\text{ m/s}^2) = 1.5$ N, about $\frac{1}{10}$ of our result, which is small but maybe not so small as to justify stating a more precise answer for $F_T$.

**FIGURE 5–7** Example 5–3.

**FIGURE 5–5** Swinging a ball on the end of a string (looking down from above).

**FIGURE 5–6** If centrifugal force existed, the revolving ball would fly outward as in (a) when released. In fact, it flies off tangentially as in (b). In (c) sparks fly in straight lines tangentially from the edge of a rotating grinding wheel.

(c)

**FIGURE 5–8** Example 5–4. Free-body diagrams for positions 1 and 2.

⚠️ **CAUTION**

*Circular motion only if cord is under tension*

**FIGURE 5–9** Exercise C.

**EXAMPLE 5–4** **Revolving ball (vertical circle).** A 0.150-kg ball on the end of a 1.10-m-long cord (negligible mass) is swung in a *vertical* circle. (*a*) Determine the minimum speed the ball must have at the top of its arc so that the ball continues moving in a circle. (*b*) Calculate the tension in the cord at the bottom of the arc, assuming the ball is moving at twice the speed of part (*a*).

**APPROACH** The ball moves in a vertical circle and is *not* undergoing uniform circular motion. The radius is assumed constant, but the speed $v$ changes because of gravity. Nonetheless, Eq. 5–1 ($a_R = v^2/r$) is valid at each point along the circle, and we use it at the top and bottom points. The free-body diagram is shown in Fig. 5–8 for both positions.

**SOLUTION** (*a*) At the top (point 1), two forces act on the ball: $m\vec{g}$, the force of gravity, and $\vec{F}_{T1}$, the tension force the cord exerts at point 1. Both act downward, and their vector sum acts to give the ball its centripetal acceleration $a_R$. We apply Newton's second law, for the vertical direction, choosing downward as positive since the acceleration is downward (toward the center):

$$(\Sigma F)_R = ma_R$$

$$F_{T1} + mg = m\frac{v_1^2}{r}. \qquad \text{[at top]}$$

From this equation we can see that the tension force $F_{T1}$ at point 1 will get larger if $v_1$ (ball's speed at top of circle) is made larger, as expected. But we are asked for the *minimum* speed to keep the ball moving in a circle. The cord will remain taut as long as there is tension in it. But if the tension disappears (because $v_1$ is too small) the cord can go limp, and the ball will fall out of its circular path. Thus, the minimum speed will occur if $F_{T1} = 0$ (the ball at the topmost point), for which the equation above becomes

$$mg = m\frac{v_1^2}{r}. \qquad \text{[minimum speed at top]}$$

We solve for $v_1$, keeping an extra digit for use in (*b*):

$$v_1 = \sqrt{gr} = \sqrt{(9.80 \text{ m/s}^2)(1.10 \text{ m})}$$
$$= 3.283 \text{ m/s} \approx 3.28 \text{ m/s}.$$

This is the minimum speed at the top of the circle if the ball is to continue moving in a circular path.

(*b*) When the ball is at the bottom of the circle (point 2 in Fig. 5–8), the cord exerts its tension force $F_{T2}$ upward, whereas the force of gravity, $m\vec{g}$, still acts downward. Choosing *upward* as positive, Newton's second law gives:

$$(\Sigma F)_R = ma_R$$

$$F_{T2} - mg = m\frac{v_2^2}{r}. \qquad \text{[at bottom]}$$

The speed $v_2$ is given as twice that in (*a*), namely 6.566 m/s. We solve for $F_{T2}$:

$$F_{T2} = m\frac{v_2^2}{r} + mg$$
$$= (0.150 \text{ kg})\frac{(6.566 \text{ m/s})^2}{(1.10 \text{ m})} + (0.150 \text{ kg})(9.80 \text{ m/s}^2) = 7.35 \text{ N}.$$

**EXERCISE C** A rider on a Ferris wheel moves in a vertical circle of radius $r$ at constant speed $v$ (Fig. 5–9). Is the normal force that the seat exerts on the rider at the top of the wheel (*a*) less than, (*b*) more than, or (*c*) the same as, the force the seat exerts at the bottom of the wheel?

**CONCEPTUAL EXAMPLE 5–5** | **Tetherball.** The game of tetherball is played with a ball tied to a pole with a cord. After the ball is struck, it revolves around the pole as shown in Fig. 5–10. In what direction is the acceleration of the ball, and what force causes the acceleration, assuming constant speed?

**RESPONSE** If the ball revolves in a horizontal plane as shown, then the acceleration points horizontally toward the center of the ball's circular path (not toward the top of the pole). The force responsible for the acceleration may not be obvious at first, since there seems to be no force pointing directly horizontally. But it is the *net* force (the sum of $m\vec{g}$ and $\vec{F}_T$ here) that must point in the direction of the acceleration. The vertical component of the cord tension, $F_{Ty}$, balances the ball's weight, $m\vec{g}$. The horizontal component of the cord tension, $F_{Tx}$, is the force that produces the centripetal acceleration toward the center.

**FIGURE 5–10** Example 5–5.

**PROBLEM SOLVING**

## Uniform Circular Motion

1. **Draw a free-body diagram**, showing all the forces acting on each object under consideration. Be sure you can identify the source of each force (tension in a cord, Earth's gravity, friction, normal force, and so on). Don't put in something that doesn't belong (like a centrifugal force).

2. **Determine** which of the forces, or which of their components, act to provide the centripetal acceleration—that is, all the **forces or components that act radially**, toward or away from the center of the circular path. The sum of these forces (or components) provides the centripetal acceleration, $a_R = v^2/r$.

3. **Choose a convenient coordinate system**, preferably with one axis along the acceleration direction.

4. **Apply Newton's second law** to the radial component:

$$\Sigma F_R = ma_R = m\frac{v^2}{r}. \quad \text{[radial direction]}$$

# 5–3 Highway Curves: Banked and Unbanked

An example of circular dynamics occurs when an automobile rounds a curve, say to the left. In such a situation, you may feel that you are thrust outward toward the right side door. But there is no mysterious centrifugal force pulling on you. What is happening is that you tend to move in a straight line, whereas the car has begun to follow a curved path. To make you go in the curved path, the seat (friction) or the door of the car (direct contact) exerts a force on you (Fig. 5–11). The car also must have a force exerted on it toward the center of the curve if it is to move in that curve. On a flat road, this force is supplied by friction between the tires and the pavement.

**PHYSICS APPLIED**
*Driving around a curve*

Force on car
(sum of friction forces acting on each tire)

Tendency for passenger to go straight

Force on passenger

**FIGURE 5–11** The road exerts an inward force (friction against the tires) on a car to make it move in a circle. The car exerts an inward force on the passenger.

**FIGURE 5–12** Race car heading into a curve. From the tire marks we see that most cars experienced a sufficient friction force to give them the needed centripetal acceleration for rounding the curve safely. But, we also see tire tracks of cars on which there was not sufficient force—and which unfortunately followed more nearly straight-line paths.

**FIGURE 5–13** Example 5–6. Forces on a car rounding a curve on a flat road. (a) Front view, (b) top view.

(a)

(b)

**PHYSICS APPLIED**

*Antilock brakes*

---

If the wheels and tires of the car are rolling normally without slipping or sliding, the bottom of the tire is at rest against the road at each instant. So the friction force the road exerts on the tires is static friction. But if static friction is not great enough, as under icy conditions or high speed, the static friction force is less than $mv^2/r$ and the car will skid out of a circular path into a more nearly straight path. See Fig. 5–12. Once a car skids or slides, the friction force becomes kinetic friction, which is smaller than static friction.

**EXAMPLE 5–6** **Skidding on a curve.** A 1000-kg car rounds a curve on a flat road of radius 50 m at a speed of 15 m/s (54 km/h). Will the car follow the curve, or will it skid? Assume: (a) the pavement is dry and the coefficient of static friction is $\mu_s = 0.60$; (b) the pavement is icy and $\mu_s = 0.25$.

**APPROACH** The forces on the car are gravity $mg$ downward, the normal force $F_N$ exerted upward by the road, and a horizontal friction force due to the road. They are shown in Fig. 5–13, which is the free-body diagram for the car. The car will follow the curve if the maximum static friction force is greater than the mass times the centripetal acceleration.

**SOLUTION** In the vertical direction ($y$) there is no acceleration. Newton's second law tells us that the normal force $F_N$ on the car is equal to the weight $mg$ since the road is flat:

$$0 = \Sigma F_y = F_N - mg$$

so

$$F_N = mg = (1000 \text{ kg})(9.80 \text{ m/s}^2) = 9800 \text{ N}.$$

In the horizontal direction the only force is friction, and we must compare it to the force needed to produce the centripetal acceleration to see if it is sufficient. The net horizontal force required to keep the car moving in a circle around the curve is

$$(\Sigma F)_R = ma_R = m\frac{v^2}{r} = (1000 \text{ kg})\frac{(15 \text{ m/s})^2}{(50 \text{ m})} = 4500 \text{ N}.$$

Now we compute the maximum total static friction force (the sum of the friction forces acting on each of the four tires) to see if it can be large enough to provide a safe centripetal acceleration. For (a), $\mu_s = 0.60$, and the maximum friction force attainable (recall from Section 4–8 that $F_{fr} \leq \mu_s F_N$) is

$$(F_{fr})_{max} = \mu_s F_N = (0.60)(9800 \text{ N}) = 5880 \text{ N}.$$

Since a force of only 4500 N is needed, and that is, in fact, how much will be exerted by the road as a static friction force, the car can follow the curve. But in (b) the maximum static friction force possible is

$$(F_{fr})_{max} = \mu_s F_N = (0.25)(9800 \text{ N}) = 2450 \text{ N}.$$

The car will skid because the ground cannot exert sufficient force (4500 N is needed) to keep it moving in a curve of radius 50 m at a speed of 54 km/h.

The possibility of skidding is worse if the wheels lock (stop rotating) when the brakes are applied too hard. When the tires are rolling, static friction exists. But if the wheels lock (stop rotating), the tires slide and the friction force, which is now kinetic friction, is less. More importantly, the *direction* of the friction force changes suddenly if the wheels lock. Static friction can point perpendicular to the velocity, as in Fig. 5–13b; but if the car slides, kinetic friction points *opposite* to the velocity. The force no longer points toward the center of the circle, and the car cannot continue in a curved path (see Fig. 5–12). Even worse, if the road is wet or icy, locking of the wheels occurs with less force on the brake pedal since there is less road friction to keep the wheels turning rather than sliding. Antilock brakes (ABS) are designed to limit brake pressure just before the point where sliding would occur, by means of delicate sensors and a fast computer.

**EXERCISE D** To negotiate a flat (unbanked) curve at a *faster* speed, a driver puts a couple of sand bags in his van aiming to increase the force of friction between the tires and the road. Will the sand bags help?

The banking of curves can reduce the chance of skidding. The normal force exerted by a banked road, acting perpendicular to the road, will have a component toward the center of the circle (Fig. 5–14), thus reducing the reliance on friction. For a given banking angle $\theta$, there will be one speed for which no friction at all is required. This will be the case when the horizontal component of the normal force toward the center of the curve, $F_N \sin \theta$ (see Fig. 5–14), is just equal to the force required to give a vehicle its centripetal acceleration—that is, when

$$F_N \sin \theta = m \frac{v^2}{r}. \qquad \text{[no friction required]}$$

The banking angle of a road, $\theta$, is chosen so that this condition holds for a particular speed, called the "design speed."

---

**EXAMPLE 5–7** **Banking angle.** (*a*) For a car traveling with speed $v$ around a curve of radius $r$, determine a formula for the angle at which a road should be banked so that no friction is required. (*b*) What is this angle for a road which has a curve of radius 50 m with a design speed of 50 km/h?

**APPROACH** Even though the road is banked, the car is still moving along a horizontal circle, so the centripetal acceleration needs to be horizontal. We choose our $x$ and $y$ axes as horizontal and vertical so that $a_R$, which is horizontal, is along the $x$ axis. The forces on the car are the Earth's gravity $mg$ downward, and the normal force $F_N$ exerted by the road perpendicular to its surface. See Fig. 5–14, where the components of $F_N$ are also shown. We don't need to consider the friction of the road because we are designing a road to be banked so as to eliminate dependence on friction.

**SOLUTION** (*a*) Since there is no vertical motion, $a_y = 0$ and $\Sigma F_y = ma_y$ gives

$$F_N \cos \theta - mg = 0$$

or

$$F_N = \frac{mg}{\cos \theta}.$$

[Note in this case that $F_N \geq mg$ because $\cos \theta \leq 1$.]
We substitute this relation for $F_N$ into the equation for the horizontal motion,

$$F_N \sin \theta = m \frac{v^2}{r},$$

which becomes

$$\frac{mg}{\cos \theta} \sin \theta = m \frac{v^2}{r}$$

or

$$\tan \theta = \frac{v^2}{rg}.$$

This is the formula for the banking angle $\theta$: no friction needed at this speed $v$.
(*b*) For $r = 50$ m and $v = 50$ km/h ($= 14$ m/s),

$$\tan \theta = \frac{(14 \text{ m/s})^2}{(50 \text{ m})(9.8 \text{ m/s}^2)} = 0.40,$$

so $\theta = \tan^{-1}(0.40) = 22°$.

We have been using the centripetal acceleration $a = v^2/r$ where $r$ is the radius of a circle. For a road, and in many other situations, we don't have a full circle, but only a portion of a circle: $a = v^2/r$ still works and we often call $r$ the **radius of curvature** of that portion of a circle we are dealing with.

**PHYSICS APPLIED**
*Banked curves*

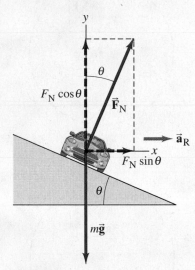

**FIGURE 5–14** Normal force on a car rounding a banked curve, resolved into its horizontal and vertical components. The centripetal acceleration is horizontal (*not* parallel to the sloping road). The friction force on the tires, not shown, could point up or down along the slope, depending on the car's speed. The friction force will be zero for one particular speed.

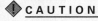
**CAUTION**
*$F_N$ is not always equal to mg*

# *5–4 Nonuniform Circular Motion

(a)

(b)

**FIGURE 5–15** The speed of an object moving in a circle changes if the force on it has a tangential component, $F_{tan}$. Part (a) shows the force $\vec{F}$ and its vector components; part (b) shows the acceleration vector and its vector components.

Circular motion at constant speed occurs when the net force on an object is exerted toward the center of the circle. If the net force is not directed toward the center but is at an angle, as shown in Fig. 5–15a, the force has two components. The component directed toward the center of the circle, $\vec{F}_R$, gives rise to the centripetal acceleration, $\vec{a}_R$, and keeps the object moving in a circle. The component tangent to the circle, $\vec{F}_{tan}$, acts to increase (or decrease) the speed, and thus gives rise to a component of the acceleration tangent to the circle, $\vec{a}_{tan}$. When the speed of the object is changing, a tangential component of force is acting.

When you first start revolving a ball on the end of a string around your head, you must give it tangential acceleration. You do this by pulling on the string with your hand displaced from the center of the circle. In athletics, a hammer thrower accelerates the hammer tangentially in a similar way so that it reaches a high speed before release.

The tangential component of the acceleration, $a_{tan}$, has magnitude equal to the rate of change of the *magnitude* of the object's velocity:

$$a_{tan} = \frac{\Delta v}{\Delta t}.$$

The radial (centripetal) acceleration arises from the change in *direction* of the velocity and, as we have seen (Eq. 5–1), has magnitude

$$a_R = \frac{v^2}{r}.$$

The tangential acceleration always points in a direction tangent to the circle, and is in the direction of motion (parallel to $\vec{v}$, which is always tangent to the circle) if the speed is increasing, as shown in Fig. 5–15b. If the speed is decreasing, $\vec{a}_{tan}$ points antiparallel to $\vec{v}$. In either case, $\vec{a}_{tan}$ and $\vec{a}_R$ are always perpendicular to each other; and *their directions change* continually as the object moves along its circular path. The total vector acceleration $\vec{a}$ is the sum of the two components:

$$\vec{a} = \vec{a}_{tan} + \vec{a}_R.$$

Since $\vec{a}_R$ and $\vec{a}_{tan}$ are always perpendicular to each other, the magnitude of $\vec{a}$ at any moment is

$$a = \sqrt{a_{tan}^2 + a_R^2}.$$

**EXAMPLE 5–8** **Two components of acceleration.** A race car starts from rest in the pit area and accelerates at a uniform rate to a speed of 35 m/s in 11 s, moving on a circular track of radius 500 m. Assuming constant tangential acceleration, find (a) the tangential acceleration, and (b) the radial acceleration, at the instant when the speed is $v = 15$ m/s.

**APPROACH** The tangential acceleration relates to the change in speed of the car, and can be calculated as $a_{tan} = \Delta v / \Delta t$. The centripetal acceleration relates to the change in the *direction* of the velocity vector and is calculated using $a_R = v^2/r$.

**SOLUTION** (a) During the 11-s time interval, we assume the tangential acceleration $a_{tan}$ is constant. Its magnitude is

$$a_{tan} = \frac{\Delta v}{\Delta t} = \frac{(35\ \text{m/s} - 0\ \text{m/s})}{11\ \text{s}} = 3.2\ \text{m/s}^2.$$

(b) When $v = 15$ m/s, the centripetal acceleration is

$$a_R = \frac{v^2}{r} = \frac{(15\ \text{m/s})^2}{(500\ \text{m})} = 0.45\ \text{m/s}^2.$$

**NOTE** The radial (centripetal) acceleration increases continually, whereas the tangential acceleration stays constant.

# 5–5 Newton's Law of Universal Gravitation

Besides developing the three laws of motion, Isaac Newton also examined the motion of the planets and the Moon. In particular, he wondered about the nature of the force that must act to keep the Moon in its nearly circular orbit around the Earth.

Newton was also thinking about the problem of gravity. Since falling objects accelerate, Newton had concluded that they must have a force exerted on them, a force we call the force of gravity. Whenever an object has a force exerted *on* it, that force is exerted *by* some other object. But what *exerts* the force of gravity? Every object on the surface of the Earth feels the force of gravity $F_G$, and no matter where the object is, the force is directed toward the center of the Earth (Fig. 5–16). Newton concluded that it must be the Earth itself that exerts the gravitational force on objects at its surface.

According to legend, Newton noticed an apple drop from a tree. He is said to have been struck with a sudden inspiration: If gravity acts at the tops of trees, and even at the tops of mountains, then perhaps it acts all the way to the Moon! With this idea that it is the Earth's gravity that holds the Moon in its orbit, Newton developed his great theory of gravitation. But there was controversy at the time. Many thinkers had trouble accepting the idea of a force "acting at a distance." Typical forces act through contact—your hand pushes a cart and pulls a wagon, a bat hits a ball, and so on. But gravity acts without contact, said Newton: the Earth exerts a force on a falling apple and on the Moon, even though there is no contact, and the two objects may even be very far apart.[†]

Newton set about determining the magnitude of the gravitational force that the Earth exerts on the Moon as compared to the gravitational force on objects at the Earth's surface. The centripetal acceleration of the Moon, as we calculated in Example 5–2, is $a_R = 0.00272 \text{ m/s}^2$. In terms of the acceleration of gravity at the Earth's surface, $g = 9.80 \text{ m/s}^2$,

$$a_R = \frac{0.00272 \text{ m/s}^2}{9.80 \text{ m/s}^2} g \approx \frac{1}{3600} g.$$

That is, the acceleration of the Moon toward the Earth is about $\frac{1}{3600}$ as great as the acceleration of objects at the Earth's surface. The Moon is 384,000 km from the Earth, which is about 60 times the Earth's radius of 6380 km. That is, the Moon is 60 times farther from the Earth's center than are objects at the Earth's surface. But $60 \times 60 = 60^2 = 3600$. Again that number 3600! Newton concluded that the gravitational force $F_{grav}$ or $F_G$ exerted by the Earth on any object decreases with the square of its distance $r$ from the Earth's center:

$$F_G \propto \frac{1}{r^2}.$$

The Moon is 60 Earth radii away, so it feels a gravitational force only $\frac{1}{60^2} = \frac{1}{3600}$ times as strong as it would if it were at a point on the Earth's surface.

Newton realized that the force of gravity on an object depends not only on distance but also on the object's mass. In fact, it is directly proportional to its mass, as we have seen (Eq. 4–3). According to Newton's third law, when the Earth exerts its gravitational force on any object, such as the Moon, that object exerts an equal and opposite force on the Earth (Fig. 5–17). Because of this *symmetry*, Newton reasoned, the magnitude of the force of gravity must be proportional to *both* masses:

$$F_G \propto \frac{m_E m_{Obj}}{r^2},$$

where $m_E$ and $m_{Obj}$ are the masses of the Earth and the other object, respectively, and $r$ is the distance from the Earth's center to the center of the other object.

[†]To deal with the conceptual difficulty of "action at a distance," the idea of a *gravitational field* was introduced many years later: every object that has mass produces a gravitational field in space. The force one object exerts on a second object is then due to the gravitational field produced by the first object at the position of the second object. We discuss fields in Section 16–7.

**FIGURE 5–16** Anywhere on Earth, whether in Alaska, Peru, or Australia, the force of gravity acts downward toward the Earth's center.

**FIGURE 5–17** The gravitational force one object exerts on a second object is directed toward the first object; and, by Newton's third law, is equal and opposite to the force exerted by the second object on the first. In the case shown, the gravitational force on the Moon due to Earth, $\vec{F}_{ME}$, is equal and opposite to the gravitational force on Earth due to the Moon, $\vec{F}_{EM}$. That is, $\vec{F}_{ME} = -\vec{F}_{EM}$.

Moon
$\vec{F}_{ME}$ Gravitational force exerted on Moon by Earth

$\vec{F}_{EM}$ Gravitational force exerted on Earth by the Moon
Earth

Newton went a step further in his analysis of gravity. In his examination of the orbits of the planets, he concluded that the force required to hold the different planets in their orbits around the Sun seems to diminish as the inverse square of their distance from the Sun. This led him to believe that it is also the gravitational force that acts between the Sun and each of the planets to keep them in their orbits. And if gravity acts between these objects, why not between all objects? Thus he proposed his **law of universal gravitation**, which we can state as follows:

> **Every particle in the universe attracts every other particle with a force that is proportional to the product of their masses and inversely proportional to the square of the distance between them. This force acts along the line joining the two particles.**

The magnitude of the gravitational force can be written as

$$F_G = G \frac{m_1 m_2}{r^2}, \tag{5-4}$$

where $m_1$ and $m_2$ are the masses of the two particles, $r$ is the distance between them, and $G$ is a universal constant which must be measured experimentally.

The value of $G$ must be very small, since we are not aware of any force of attraction between ordinary-sized objects, such as between two baseballs. The force between two ordinary objects was first measured by Henry Cavendish in 1798, over 100 years after Newton published his law. To detect and measure the incredibly small force between ordinary objects, he used an apparatus like that shown in Fig. 5–18. Cavendish confirmed Newton's hypothesis that two objects attract one another and that Eq. 5–4 accurately describes this force. In addition, because Cavendish could measure $F_G$, $m_1$, $m_2$, and $r$ accurately, he was able to determine the value of the constant $G$ as well. The accepted value today is

$$G = 6.67 \times 10^{-11} \, \text{N} \cdot \text{m}^2/\text{kg}^2.$$

(See Table inside front cover for values of all constants to highest known precision.) Equation 5–4 is called an **inverse square law** because the force is inversely proportional to $r^2$.

[Strictly speaking, Eq. 5–4 gives the magnitude of the gravitational force that one particle exerts on a second particle that is a distance $r$ away. For an extended object (that is, not a point), we must consider how to measure the distance $r$. A correct calculation treats each extended body as a collection of particles, and the total force is the sum of the forces due to all the particles. The sum over all these particles is often done using integral calculus, which Newton himself invented. When extended bodies are small compared to the distance between them (as for the Earth–Sun system), little inaccuracy results from considering them as point particles. Newton was able to show that the *gravitational force exerted on a particle outside a uniform sphere is the same as if the entire mass of the sphere was concentrated at its center.*[†] Thus Eq. 5–4 gives the correct force between two uniform spheres where $r$ is the distance between their centers.]

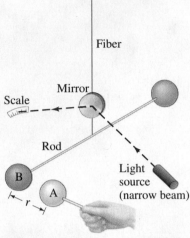

*NEWTON'S*

*LAW*

*OF*

*UNIVERSAL*

*GRAVITATION*

**FIGURE 5–18** Schematic diagram of Cavendish's apparatus. Two spheres are attached to a light horizontal rod, which is suspended at its center by a thin fiber. When a third sphere (labeled A) is brought close to one of the suspended spheres (labeled B), the gravitational force causes the latter to move, and this twists the fiber slightly. The tiny movement is magnified by the use of a narrow light beam directed at a mirror mounted on the fiber. The beam reflects onto a scale. Previous determination of how large a force will twist the fiber a given amount then allows the experimenter to determine the magnitude of the gravitational force between the two objects, A and B.

---

**EXAMPLE 5–9 | ESTIMATE | Can you attract another person gravitationally?** A 50-kg person and a 70-kg person are sitting on a bench close to each other. Estimate the magnitude of the gravitational force each exerts on the other.

**APPROACH** This is an estimate: we let the distance between the centers of the two people be $\frac{1}{2}$ m (about as close as you can get).

**SOLUTION** We use Eq. 5–4, which gives

$$F_G = G \frac{m_1 m_2}{r^2} \approx \frac{(6.67 \times 10^{-11} \, \text{N} \cdot \text{m}^2/\text{kg}^2)(50 \, \text{kg})(70 \, \text{kg})}{(0.5 \, \text{m})^2} \approx 10^{-6} \, \text{N},$$

rounded off to an order of magnitude. Such a force is unnoticeably small unless extremely sensitive instruments are used ($< 1/100,000$ of a pound).

[†]We demonstrate this result in Section 16–12.

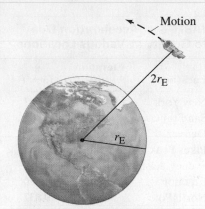

Motion

**EXAMPLE 5–10** **Spacecraft at $2r_E$.** What is the force of gravity acting on a 2000-kg spacecraft when it orbits two Earth radii from the Earth's center (that is, a distance $r_E = 6380$ km above the Earth's surface, Fig. 5–19)? The mass of the Earth is $m_E = 5.98 \times 10^{24}$ kg.

**APPROACH** We could plug all the numbers into Eq. 5–4, but there is a simpler approach. The spacecraft is twice as far from the Earth's center as when it is at the surface of the Earth. Therefore, since the force of gravity $F_G$ decreases as the square of the distance (and $\frac{1}{2^2} = \frac{1}{4}$), the force of gravity on the satellite will be only one-fourth its weight at the Earth's surface.

**SOLUTION** At the surface of the Earth, $F_G = mg$. At a distance from the Earth's center of $2r_E$, $F_G$ is $\frac{1}{4}$ as great:

$$F_G = \tfrac{1}{4}mg = \tfrac{1}{4}(2000 \text{ kg})(9.80 \text{ m/s}^2) = 4900 \text{ N}.$$

**FIGURE 5–19** Example 5–10; a spacecraft in orbit at $r = 2r_E$.

Note carefully that the law of universal gravitation describes a *particular* force (gravity), whereas Newton's second law of motion ($F = ma$) tells how an object accelerates due to *any* type of force.

⬥ **CAUTION**
*Distinguish Newton's second law from the law of universal gravitation*

# 5–6 Gravity Near the Earth's Surface

When Eq. 5–4 is applied to the gravitational force between the Earth and an object at its surface, $m_1$ becomes the mass of the Earth $m_E$, $m_2$ becomes the mass of the object $m$, and $r$ becomes the distance of the object from the Earth's center, which is the radius of the Earth $r_E$. This force of gravity due to the Earth is the weight of the object on Earth, which we have been writing as $mg$. Thus,

$$mg = G\frac{mm_E}{r_E^2}.$$

We can solve this for $g$, the acceleration of gravity at the Earth's surface:

$$g = G\frac{m_E}{r_E^2}. \tag{5–5}$$

Thus, the acceleration of gravity at the surface of the Earth, $g$, is determined by $m_E$ and $r_E$. (Don't confuse $G$ with $g$; they are very different quantities, but are related by Eq. 5–5.)

Until $G$ was measured, the mass of the Earth was not known. But once $G$ was measured, Eq. 5–5 could be used to calculate the Earth's mass, and Cavendish was the first to do so. Since $g = 9.80$ m/s² and the radius of the Earth is $r_E = 6.38 \times 10^6$ m, then, from Eq. 5–5, we obtain the mass of the Earth to be

$$m_E = \frac{gr_E^2}{G} = \frac{(9.80 \text{ m/s}^2)(6.38 \times 10^6 \text{ m})^2}{6.67 \times 10^{-11} \text{ N} \cdot \text{m}^2/\text{kg}^2} = 5.98 \times 10^{24} \text{ kg}.$$

Equation 5–5 can be applied to other planets, where $g$, $m$, and $r$ would refer to that planet.

⬥ **CAUTION**
*Distinguish G from g*

**FIGURE 5–20** Example 5–11. Mount Everest, 8850 m (29,035 ft) above sea level; in the foreground, the author with sherpas at 5500 m (18,000 ft).

**EXAMPLE 5–11** **ESTIMATE** **Gravity on Everest.** Estimate the effective value of $g$ on the top of Mt. Everest, 8850 m (29,035 ft) above sea level (Fig. 5–20). That is, what is the acceleration due to gravity of objects allowed to fall freely at this altitude? Ignore the mass of the mountain itself.

**APPROACH** The force of gravity (and the acceleration due to gravity $g$) depends on the distance from the center of the Earth, so there will be an effective value $g'$ on top of Mt. Everest which will be smaller than $g$ at sea level. We assume the Earth is a uniform sphere (a reasonable "estimate").

**SOLUTION** We use Eq. 5–5, with $r_E$ replaced by $r = 6380 \text{ km} + 8.9 \text{ km} = 6389 \text{ km} = 6.389 \times 10^6$ m:

$$g = G\frac{m_E}{r^2} = \frac{(6.67 \times 10^{-11} \text{ N} \cdot \text{m}^2/\text{kg}^2)(5.98 \times 10^{24} \text{ kg})}{(6.389 \times 10^6 \text{ m})^2} = 9.77 \text{ m/s}^2,$$

which is a reduction of about 3 parts in a thousand (0.3%).

| TABLE 5–1 Acceleration Due to Gravity at Various Locations | | |
|---|---|---|
| Location | Elevation (m) | $g$ (m/s²) |
| New York | 0 | 9.803 |
| San Francisco | 0 | 9.800 |
| Denver | 1650 | 9.796 |
| Pikes Peak | 4300 | 9.789 |
| Sydney, Australia | 0 | 9.798 |
| Equator | 0 | 9.780 |
| North Pole (calculated) | 0 | 9.832 |

**PHYSICS APPLIED**

*Artificial Earth satellites*

Note that Eq. 5–5 does not give precise values for $g$ at different locations because the Earth is not a perfect sphere. The Earth not only has mountains and valleys, and it bulges at the equator, but also its mass is not distributed precisely uniformly. (See Table 5–1.) The Earth's rotation also affects the value of $g$. However, for most practical purposes, when an object is near the Earth's surface, we will simply use $g = 9.80 \text{ m/s}^2$ and write the weight of an object as $mg$.

**EXERCISE E** Suppose you could double the mass of a planet but keep its volume the same. How would the acceleration of gravity, $g$, at the surface change?

# 5–7 Satellites and "Weightlessness"

## Satellite Motion

Artificial satellites circling the Earth are now commonplace (Fig. 5–21). A satellite is put into orbit by accelerating it to a sufficiently high tangential speed with the use of rockets, as shown in Fig. 5–22. If the speed is too high, the spacecraft will not be confined by the Earth's gravity and will escape, never to return. If the speed is too low, it will return to Earth. Satellites are typically put into circular (or nearly circular) orbits, because such orbits require the least takeoff speed.

**FIGURE 5–21** A satellite, the International Space Station, circling the Earth.

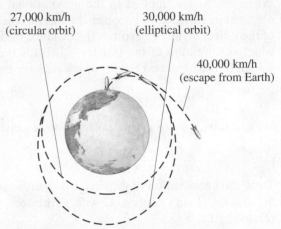

**FIGURE 5–22** Artificial satellites launched at different speeds. Escape velocity from Earth is about 40,000 km/h.

**FIGURE 5–23** A moving satellite "falls" out of a straight-line path toward the Earth.

It is sometimes asked: "What keeps a satellite up?" The answer is: its high speed. If a satellite in orbit stopped moving, it would fall directly to Earth. But at the very high speed a satellite has, it would quickly fly out into space (Fig. 5–23) if it weren't for the gravitational force of the Earth pulling it into orbit. In fact, a satellite in orbit *is* falling (accelerating) toward Earth, but its high tangential speed keeps it from hitting Earth.

For satellites that move in a circle (at least approximately), the needed acceleration is centripetal and equals $v^2/r$. The force that gives a satellite this acceleration is the force of gravity exerted by the Earth, and since a satellite may be at a considerable distance from the Earth, we must use Newton's law of universal gravitation (Eq. 5–4) for the force acting on it. When we apply Newton's second law, $\Sigma F_R = ma_R$ in the radial direction, we find

$$G\frac{mm_E}{r^2} = m\frac{v^2}{r}, \tag{5–6}$$

where $m$ is the mass of the satellite. This equation relates the distance of the satellite from the Earth's center, $r$, to its speed, $v$, in a circular orbit. Note that only one force—gravity—is acting on the satellite, and that $r$ is the sum of the Earth's radius $r_E$ plus the satellite's height $h$ above the Earth: $r = r_E + h$.

If we solve Eq. 5–6 for $v$, we find $v = \sqrt{Gm_E/r}$ and we see that a satellite's speed does not depend on its own mass. Satellites of different mass orbiting at the same distance above Earth have the same speed and period.

PHYSICS APPLIED
*Geosynchronous satellites*

**EXAMPLE 5–12** **Geosynchronous satellite.** A *geosynchronous* satellite is one that stays above the same point on the Earth, which is possible only if it is above a point on the equator. Why? Because the center of a satellite orbit is always at the center of the Earth; so it is not possible to have a satellite orbiting above a fixed point on the Earth at any latitude other than 0°. Geosynchronous satellites are commonly used for TV and radio transmission, for weather forecasting, and as communication relays.[†] Determine (*a*) the height above the Earth's surface such a satellite must orbit, and (*b*) such a satellite's speed. (*c*) Compare to the speed of a satellite orbiting 200 km above Earth's surface.

**APPROACH** To remain above the same point on Earth as the Earth rotates, the satellite must have a period of 24 hours. We can apply Newton's second law, $F = ma$, where $a = v^2/r$ if we assume the orbit is circular.

**SOLUTION** (*a*) The only force on the satellite is the gravitational force due to the Earth. (We can ignore the gravitational force exerted by the Sun. Why?) We apply Eq. 5–6 assuming the satellite moves in a circle:

$$G\frac{m_{Sat}\,m_E}{r^2} = m_{Sat}\frac{v^2}{r}.$$

This equation has two unknowns, $r$ and $v$. So we need a second equation. The satellite revolves around the Earth with the same period that the Earth rotates on its axis, namely once in 24 hours. Thus the speed of the satellite must be

$$v = \frac{2\pi r}{T},$$

where $T = 1\,\text{day} = (24\,\text{h})(3600\,\text{s/h}) = 86{,}400\,\text{s}$. We substitute this into the "satellite equation" above and obtain (after cancelling $m_{Sat}$ on both sides):

$$G\frac{m_E}{r^2} = \frac{(2\pi r)^2}{rT^2}.$$

After cancelling an $r$, we can solve for $r^3$:

$$r^3 = \frac{Gm_E T^2}{4\pi^2} = \frac{(6.67 \times 10^{-11}\,\text{N}\cdot\text{m}^2/\text{kg}^2)(5.98 \times 10^{24}\,\text{kg})(86{,}400\,\text{s})^2}{4\pi^2}$$
$$= 7.54 \times 10^{22}\,\text{m}^3.$$

We take the cube root and find

$$r = 4.22 \times 10^7\,\text{m},$$

or 42,200 km from the Earth's center. We subtract the Earth's radius of 6380 km to find that a geosynchronous satellite must orbit about 36,000 km (about $6r_E$) above the Earth's surface.

(*b*) We solve for $v$ in the satellite equation, Eq. 5–6:

$$v = \sqrt{\frac{Gm_E}{r}} = \sqrt{\frac{(6.67 \times 10^{-11}\,\text{N}\cdot\text{m}^2/\text{kg}^2)(5.98 \times 10^{24}\,\text{kg})}{(4.22 \times 10^7\,\text{m})}} = 3070\,\text{m/s},$$

or about 11,000 km/h ($\approx 7000$ mi/h). We get the same result if we use $v = 2\pi r/T$.

(*c*) The equation in part (*b*) for $v$ shows $v \propto \sqrt{1/r}$. So for $r = r_E + h = 6380\,\text{km} + 200\,\text{km} = 6580\,\text{km}$, we get

$$v' = v\sqrt{\frac{r}{r'}} = (3070\,\text{m/s})\sqrt{\frac{(42{,}200\,\text{km})}{(6580\,\text{km})}} = 7770\,\text{m/s},$$

or about 28,000 km/h ($\approx 17{,}000$ mi/h).

[†]Geosynchronous satellites are useful because receiving and transmitting antennas at a given place on Earth can stay fixed on such a satellite (no tracking and no switching satellites is needed).

(a)    $a = 0$

(b) ⬆ $a = \frac{1}{2}g$ (up)

(c) ⬇ $a = g$ (down)

**FIGURE 5–24** (a) A bag in an elevator at rest exerts a force on a spring scale equal to its weight. (b) In an elevator accelerating upward at $\frac{1}{2}g$, the bag's apparent weight is $1\frac{1}{2}$ times larger than its true weight. (c) In a freely falling elevator, the bag experiences "weightlessness": the scale reads zero.

## Weightlessness

People and other objects in a satellite circling the Earth are said to experience apparent weightlessness. Let us first look at a simpler case: a falling elevator. In Fig. 5–24a, an elevator is at rest with a bag hanging from a spring scale. The scale reading indicates the downward force exerted on it by the bag. This force, exerted *on* the scale, is equal and opposite to the force exerted *by* the scale upward on the bag, and we call its magnitude $w$ (for "weight"). Two forces act on the bag: the downward gravitational force and the upward force exerted by the scale equal to $w$. Because the bag is not accelerating ($a = 0$), when we apply $\Sigma F = ma$ to the bag in Fig. 5–24a we obtain

$$w - mg = 0,$$

where $mg$ is the weight of the bag. Thus, $w = mg$, and since the scale indicates the force $w$ exerted on it by the bag, it registers a force equal to the weight of the bag, as we expect.

Now let the elevator have an acceleration, $a$. Applying Newton's second law, $\Sigma F = ma$, to the bag as seen from an inertial reference frame (the elevator itself is not now an inertial frame) we have

$$w - mg = ma.$$

Solving for $w$, we have

$$w = mg + ma. \qquad\qquad [a \text{ is } + \text{ upward}]$$

We have chosen the positive direction up. Thus, if the acceleration $a$ is up, $a$ is positive; and the scale, which measures $w$, will read more than $mg$. We call $w$ the *apparent weight* of the bag, which in this case would be greater than its actual weight ($mg$). If the elevator accelerates downward, $a$ will be negative and $w$, the apparent weight, will be less than $mg$. The direction of the velocity $\vec{v}$ doesn't matter. Only the direction of the acceleration $\vec{a}$ (and its magnitude) influences the scale reading.

Suppose the elevator's acceleration is $\frac{1}{2}g$ upward; then we find

$$w = mg + m(\tfrac{1}{2}g)$$
$$= \tfrac{3}{2}mg.$$

That is, the scale reads $1\frac{1}{2}$ times the actual weight of the bag (Fig. 5–24b). The apparent weight of the bag is $1\frac{1}{2}$ times its real weight. The same is true of the person: her apparent weight (equal to the normal force exerted on her by the elevator floor) is $1\frac{1}{2}$ times her real weight. We can say that she is experiencing $1\frac{1}{2}$ $g$'s, just as astronauts experience so many $g$'s at a rocket's launch.

If, instead, the elevator's acceleration is $a = -\frac{1}{2}g$ (downward), then $w = mg - \frac{1}{2}mg = \frac{1}{2}mg$. That is, the scale reads half the actual weight. If the elevator is in *free fall* (for example, if the cables break), then $a = -g$ and $w = mg - mg = 0$. The scale reads zero. See Fig. 5–24c. The bag appears weightless. If the person in the elevator accelerating at $-g$ let go of a box, it would not fall to the floor. True, the box would be falling with acceleration $g$. But so would the floor of the elevator and the person. The box would hover right in front of the person. This phenomenon is called **apparent weightlessness** because in the reference frame of the person, objects don't fall or seem to have weight—yet gravity does not disappear. Gravity is still acting on each object, whose weight is still $mg$.

The "weightlessness" experienced by people in a satellite orbit close to the Earth (Fig. 5–25) is the same apparent weightlessness experienced in a freely falling elevator. It may seem strange, at first, to think of a satellite as freely falling. But a satellite is indeed falling toward the Earth, as was shown in Fig. 5–23. The force of gravity causes it to "fall" out of its natural straight-line path. The acceleration of the satellite must be the acceleration due to gravity at that point, because the only force acting on it is gravity. Thus, although the force of gravity acts on objects within the satellite, the objects experience an apparent weightlessness because they, and the satellite, are accelerating together as in free fall.

Figure 5–26 shows some examples of "free fall," or apparent weightlessness, experienced by people on Earth for brief moments.

A completely different situation occurs if a spacecraft is out in space far from the Earth, the Moon, and other attracting bodies. The force of gravity due to the Earth and other celestial bodies will then be quite small because of the distances involved, and persons in such a spacecraft would experience real weightlessness.

**EXERCISE F** Return to Chapter-Opening Question 2, page 109, and answer it again now. Try to explain why you may have answered differently the first time.

(a)      (b)      (c)

FIGURE 5–25 This astronaut is outside the International Space Station. He must feel very free because he is experiencing apparent weightlessness.

FIGURE 5–26 Experiencing "weightlessness" on Earth.

# 5–8 Planets, Kepler's Laws, and Newton's Synthesis

Where did we first get the idea of planets? Have you ever escaped the lights of the city to gaze late at night at the multitude of stars in the night sky? It is a moving experience. Thousands of years ago, the ancients saw this sight every cloudless night, and were fascinated. They noted that the vast majority of stars, bright or dim, seemed to maintain fixed positions relative to each other. The ancients imagined these **fixed stars** as being attached to a huge inverted bowl, or sphere. This **celestial sphere** revolved around the Earth almost exactly once a day (Fig. 5–27), from east to west. Among all the stars that were visible to the naked eye (there were no telescopes until much later, about 1600), the ancients saw five stars that changed position relative to the fixed stars over weeks and months. These five wandering stars were called **planets** (Greek for wandering). Planets were thus visible at night as tiny points of light like other stars.

The ancient idea that the Sun, Moon, and planets revolve around the Earth is called the **geocentric** view (geo = Earth in Greek). It was developed into a fine theoretical system by Ptolemy in the second century B.C. Today we believe in a **heliocentric** system (helios = Sun in Greek), where the Earth is just another planet, between Venus and Mars, orbiting around the Sun. Although a heliocentric view was proposed in ancient times, it was largely ignored until Renaissance Italy of the fifteenth century. The real theory change (see Section 1–1 and Fig. 1–2) began with the heliocentric theory of Nicolaus Copernicus (1473–1543) and then was greatly advanced by the experimental observations of Galileo around 1610 using his newly developed 30× telescope. Galileo observed that the planet Jupiter has moons (like a miniature solar system) and that Venus has phases like our Moon, not explainable by Ptolemy's geocentric system. [Galileo's famous encounter with the Church had little to do with religious faith, but rather with politics, personality conflict, and authority. Today it is generally understood that science and faith are different approaches that are not in conflict.]

FIGURE 5–27 Time exposure showing movement of stars over a period of several hours.

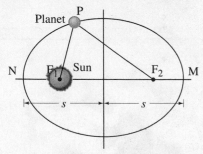

**FIGURE 5–28** *Kepler's first law.* An ellipse is a closed curve such that the sum of the distances from any point P on the curve to two fixed points (called the foci, $F_1$ and $F_2$) remains constant. That is, the sum of the distances, $F_1P + F_2P$, is the same for all points on the curve. A circle is a special case of an ellipse in which the two foci coincide, at the center of the circle.

**FIGURE 5–29** *Kepler's second law.* The two shaded regions have equal areas. The planet moves from point 1 to point 2 in the same time it takes to move from point 3 to point 4. Planets move fastest when closest to the Sun.

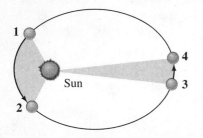

## TABLE 5–2 Planetary Data Applied to Kepler's Third Law

| Planet | Mean Distance to Sun, $s$ ($10^6$ km) | Period, $T$ (Earth yr) | $\dfrac{s^3/T^2}{\left(10^{24}\frac{\text{km}^3}{\text{yr}^2}\right)}$ |
|---|---|---|---|
| Mercury | 57.9 | 0.241 | 3.34 |
| Venus | 108.2 | 0.615 | 3.35 |
| Earth | 149.6 | 1.000 | 3.35 |
| Mars | 227.9 | 1.88 | 3.35 |
| Jupiter | 778.3 | 11.86 | 3.35 |
| Saturn | 1427 | 29.5 | 3.34 |
| Uranus | 2870 | 84.0 | 3.35 |
| Neptune | 4497 | 165 | 3.34 |
| (Pluto)[†] | 5900 | 248 | 3.34 |

[†]Pluto, since its discovery in 1930, was considered a ninth planet. But its small mass and the recent discovery of other objects beyond Neptune with similar masses has led to calling these smaller objects, including Pluto, "dwarf planets." We keep it in the Table to indicate its great distance, and its consistency with Kepler's third law.

## Kepler's Laws

Also about 1600, more than a half century before Newton proposed his three laws of motion and his law of universal gravitation, the German astronomer Johannes Kepler (1571–1630) had worked out a detailed description of the motion of the planets around the Sun. Kepler's work resulted in part from the many years he spent examining data collected (without a telescope) by Tycho Brahe (1546–1601) on the positions of the planets in their motion through the night sky.

Among Kepler's writings were three empirical findings that we now refer to as **Kepler's laws of planetary motion**. These are summarized as follows, with additional explanation in Figs. 5–28 and 5–29.

*Kepler's first law*: The path of each planet around the Sun is an ellipse with the Sun at one focus (Fig. 5–28).

*Kepler's second law*: Each planet moves so that an imaginary line drawn from the Sun to the planet sweeps out equal areas in equal periods of time (Fig. 5–29).

*Kepler's third law*: The ratio of the squares of the periods $T$ of any two planets revolving around the Sun is equal to the ratio of the cubes of their mean distances from the Sun. [The mean distance equals the semimajor axis $s$ (= half the distance from the planet's near point N and far point M from the Sun, Fig. 5–28).] That is, if $T_1$ and $T_2$ represent the periods (the time needed for one revolution about the Sun) for any two planets, and $s_1$ and $s_2$ represent their mean distances from the Sun, then

$$\left(\frac{T_1}{T_2}\right)^2 = \left(\frac{s_1}{s_2}\right)^3.$$

We can rewrite Kepler's third law as

$$\frac{s_1^3}{T_1^2} = \frac{s_2^3}{T_2^2},$$

meaning that $s^3/T^2$ should be the same for each planet. Present-day data are given in Table 5–2; see the last column.

In Examples and Problems we usually will assume the orbits are circles, although it is not quite true in general.

**EXAMPLE 5–13** **Where is Mars?** Mars' period (its "year") was noted by Kepler to be about 687 days (Earth days), which is (687 d/365 d) = 1.88 yr (Earth years). Determine the mean distance of Mars from the Sun using the Earth as a reference.

**APPROACH** We are given the ratio of the periods of Mars and Earth. We can find the distance from Mars to the Sun using Kepler's third law, given the Earth–Sun distance as $1.50 \times 10^{11}$ m (Table 5–2; also Table inside front cover).

**SOLUTION** Let the distance of Mars from the Sun be $s_{MS}$, and the Earth–Sun distance be $s_{ES} = 1.50 \times 10^{11}$ m. From Kepler's third law:

$$\frac{s_{MS}}{s_{ES}} = \left(\frac{T_M}{T_E}\right)^{\frac{2}{3}} = \left(\frac{1.88 \text{ yr}}{1 \text{ yr}}\right)^{\frac{2}{3}} = 1.52.$$

So Mars is 1.52 times the Earth's distance from the Sun, or $2.28 \times 10^{11}$ m.

## Kepler's Third Law Derived, Sun's Mass, Perturbations

We will derive Kepler's third law for the special case of a circular orbit, in which case the mean distance $s$ is the radius $r$ of the circle. (Most planetary orbits are close to a circle.) First, we write Newton's second law of motion, $\Sigma F = ma$. For $F$ we use the law of universal gravitation (Eq. 5–4) for the force between the Sun and a planet of mass $m_1$, and for $a$ the centripetal acceleration, $v^2/r$. We assume the mass of the Sun $M_S$ is much greater than the mass of its planets, so we ignore the effects of the planets on each other. Then

$$\Sigma F = ma$$

$$G\frac{m_1 M_S}{r_1^2} = m_1\frac{v_1^2}{r_1}.$$

Here $m_1$ is the mass of a particular planet, $r_1$ its distance from the Sun, and $v_1$ its speed in orbit; $M_S$ is the mass of the Sun, since it is the gravitational attraction of the Sun that keeps each planet in its orbit. The period $T_1$ of the planet is the time required for one complete orbit, which is a distance equal to $2\pi r_1$, the circumference of a circle. Thus

$$v_1 = \frac{2\pi r_1}{T_1}.$$

We substitute this formula for $v_1$ into the previous equation:

$$G\frac{m_1 M_S}{r_1^2} = m_1\frac{4\pi^2 r_1}{T_1^2}.$$

We rearrange this to get

$$\frac{T_1^2}{r_1^3} = \frac{4\pi^2}{GM_S}. \tag{5–7a}$$

We derived this for planet 1 (say, Mars). The same derivation would apply for a second planet (say, Saturn) orbiting the Sun,

$$\frac{T_2^2}{r_2^3} = \frac{4\pi^2}{GM_S},$$

where $T_2$ and $r_2$ are the period and orbit radius, respectively, for the second planet. Since the right sides of the two previous equations are equal, we have $T_1^2/r_1^3 = T_2^2/r_2^3$ or, rearranging,

$$\left(\frac{T_1}{T_2}\right)^2 = \left(\frac{r_1}{r_2}\right)^3, \tag{5–7b}$$ *Kepler's third law*

which is Kepler's third law. Equations 5–7a and 5–7b are valid also for elliptical orbits if we replace $r$ with the semimajor axis $s$.

---

**EXAMPLE 5–14** **The Sun's mass determined.** Determine the mass of the Sun given the Earth's distance from the Sun as $r_{ES} = 1.5 \times 10^{11}$ m.

**APPROACH** Equation 5–7a relates the mass of the Sun $M_S$ to the period and distance of any planet. We use the Earth.

**SOLUTION** The Earth's period is $T_E = 1\,\text{yr} = \left(365\frac{1}{4}\,\text{d}\right)\left(24\,\text{h/d}\right)\left(3600\,\text{s/h}\right) = 3.16 \times 10^7\,\text{s}$. We solve Eq. 5–7a for $M_S$:

$$M_S = \frac{4\pi^2 r_{ES}^3}{GT_E^2} = \frac{4\pi^2\left(1.5 \times 10^{11}\,\text{m}\right)^3}{\left(6.67 \times 10^{-11}\,\text{N}\cdot\text{m}^2/\text{kg}^2\right)\left(3.16 \times 10^7\,\text{s}\right)^2} = 2.0 \times 10^{30}\,\text{kg}.$$

 **PHYSICS APPLIED**
*Determining the Sun's mass*

Accurate measurements on the orbits of the planets indicated that they did not precisely follow Kepler's laws. For example, slight deviations from perfectly elliptical orbits were observed. Newton was aware that this was to be expected because any planet would be attracted gravitationally not only by the Sun but also (to a much lesser extent) by the other planets. Such deviations, or **perturbations**, in the orbit of Saturn were a hint that helped Newton formulate the law of universal gravitation, that all objects attract each other gravitationally. Observation of other perturbations later led to the discovery of Neptune. Deviations in the orbit of Uranus could not all be accounted for by perturbations due to the other known planets. Careful calculation in the nineteenth century indicated that these deviations could be accounted for if another planet existed farther out in the solar system. The position of this planet was predicted from the deviations in the orbit of Uranus, and telescopes focused on that region of the sky quickly found it; the new planet was called Neptune. Similar but much smaller perturbations of Neptune's orbit led to the discovery of Pluto in 1930.

**PHYSICS APPLIED**
*Perturbations and discovery of planets*

## Other Centers for Kepler's Laws

The derivation of Eq. 5–7b, Kepler's third law, compared two planets revolving around the Sun. But the derivation is general enough to be applied to other systems. For example, we could apply Eq. 5–7b to compare an artificial satellite and our Moon, both revolving around Earth (then $M_S$ would be replaced by $M_E$, the mass of the Earth). Or we could apply Eq. 5–7b to compare two moons revolving around Jupiter. But Kepler's third law, Eq. 5–7b, applies only to objects orbiting the same attracting center. Do not use Eq. 5–7b to compare, say, the Moon's orbit around Earth to the orbit of Mars around the Sun: they depend on different attracting centers.

## Distant Planetary Systems

Starting in the mid-1990s, planets revolving around distant stars (Fig. 5–30) were inferred from the regular "wobble" in position of each star due to the gravitational attraction of the revolving planet(s). Many such "extrasolar" planets are now known.

**FIGURE 5–30** Our solar system (a) is compared to recently discovered planets orbiting (b) the star Upsilon Andromedae with at least three planets. $m_J$ is the mass of Jupiter. (Sizes are not to scale.)

## Newton's Synthesis

Kepler arrived at his laws through careful analysis of experimental data. Fifty years later, Newton was able to show that Kepler's laws could be derived mathematically from the law of universal gravitation and the laws of motion. Newton also showed that for any reasonable form for the gravitational force law, only one that depends on the inverse square of the distance is fully consistent with Kepler's laws. He thus used Kepler's laws as evidence in favor of his law of universal gravitation, Eq. 5–4.

The development by Newton of the law of universal gravitation and the three laws of motion was a major intellectual achievement. With these laws, he was able to describe the motion of objects on Earth and of the far-away planets seen in the night sky. The motions of the planets through the heavens and of objects on Earth were seen to follow the same laws (not recognized previously). For this reason, and also because Newton integrated the results of earlier scientists into his system, we sometimes speak of **Newton's synthesis**.

The laws formulated by Newton are referred to as **causal laws**. By **causality** we mean that one occurrence can cause another. When a rock strikes a window, we infer the rock *causes* the window to break. This idea of "cause and effect" relates to Newton's laws: the acceleration of an object was seen to be *caused* by the net force acting on it.

As a result of Newton's theories, the universe came to be viewed by many as a machine whose parts move in a **deterministic** way. This deterministic view of the universe had to be modified in the twentieth century (Chapter 28).

## Sun/Earth Reference Frames

The geocentric–heliocentric controversy (page 125) may be seen today as a matter of frame of reference. From the reference frame of Earth, we see the Sun and Moon as revolving around us with average periods of 24 h (= definition of 1 day) and almost 25 h, respectively, roughly in circles. The orbits of the planets as seen from Earth are very complicated, however.

In the Sun's reference frame, Earth makes one revolution (= definition of the year) in 365.256 days, in an ellipse that is nearly a circle. The Sun's reference frame has the advantage that the other planets also have simple elliptical orbits. (Or nearly so—each planet's gravity pulls on the others, causing small perturbations.) The Sun's vastly greater mass ($> 10^5 \times$ Earth's) allows it to be an easier reference frame to use.

The Sun itself (and the Earth with it) revolves around the center of our Galaxy (see Fig. 33–2 or 5–49) which itself moves relative to other galaxies. Indeed, there is no one reference frame that we can consider as preferred or central.

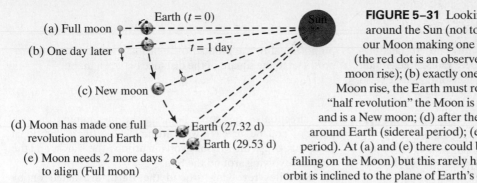

**FIGURE 5–31** Looking down on the plane of Earth's orbit around the Sun (not to scale), above Earth's north pole, showing our Moon making one revolution about Earth: (a) at a Full moon (the red dot is an observer at about 6 PM who can just see the Full moon rise); (b) exactly one day later (for the red dot to see the Moon rise, the Earth must rotate another 50 min); (c) after making a "half revolution" the Moon is in line with the Sun, on the Sun's side, and is a New moon; (d) after the Moon makes one complete revolution around Earth (sidereal period); (e) at the next Full moon (synodic period). At (a) and (e) there could be a **lunar eclipse** (Earth's shadow falling on the Moon) but this rarely happens because the plane of Moon's orbit is inclined to the plane of Earth's orbit, so the Moon is usually above or below the Earth's orbital plane. At (c) there could be a **solar eclipse**, also rare.

# 5–9 Moon Rises an Hour Later Each Day

From the Earth's reference frame, our Moon revolves on average in 24 h, 50 min, which means the Moon rises nearly an hour later each day; and it is at its highest point in the sky about an hour later each day. When the Moon is on the direct opposite side of Earth from the Sun, the Sun's light fully illuminates the Moon and we call it a **Full moon** (Fig. 5–31a). When the Moon is on the same side of the Earth as the Sun, and nearly aligned with both, we see the Moon as a thin sliver—most or all of it is in shadow (= a **New moon**). The phases of the Moon (new, first quarter, full, third quarter) take it from one Full moon to the next Full moon in 29.53 days (= **synodic period**) on average, as seen from the Earth as reference frame (Fig. 5–31e). In the Sun's frame of reference, the Moon revolves around the Earth in 27.32 days (**sidereal period**, Fig. 5–31d). This small difference arises because, when the Moon has made one complete revolution around the Earth, the Earth itself has moved in its orbit relative to the Sun. So the Moon needs more time ($\approx$ 2 days) to be fully aligned with the Sun and Earth and be a Full moon, Fig. 5–31e. The red dot in Figs. 5–31a, b, and e represents an observer at the same location on Earth, which in (a) is when the Full moon is rising and the Sun is just setting.

# 5–10 Types of Forces in Nature

We have already discussed that Newton's law of universal gravitation, Eq. 5–4, describes how a particular type of force—gravity—depends on the masses of the objects involved and the distance between them. Newton's second law, $\Sigma \vec{F} = m\vec{a}$, on the other hand, tells how an object will accelerate due to *any* type of force. But what are the types of forces that occur in nature besides gravity?

In the twentieth century, physicists came to recognize four fundamental forces in nature: (1) the gravitational force; (2) the electromagnetic force (we shall see later that electric and magnetic forces are intimately related); (3) the strong nuclear force (which holds protons and neutrons together to form atomic nuclei); and (4) the weak nuclear force (involved in radioactivity). In this Chapter, we discussed the gravitational force in detail. The nature of the electromagnetic force will be discussed in Chapters 16 to 22. The strong and weak nuclear forces, which are discussed in Chapters 30 to 32, operate at the level of the atomic nucleus and are much less obvious in our daily lives.

Physicists have been working on theories that would unify these four forces—that is, to consider some or all of these forces as different manifestations of the same basic force. So far, the electromagnetic and weak nuclear forces have been theoretically united to form *electroweak* theory, in which the electromagnetic and weak forces are seen as two aspects of a single *electroweak force*. Attempts to further unify the forces, such as in *grand unified theories* (GUT), are hot research topics today.

But where do everyday forces fit? Ordinary forces, other than gravity, such as pushes, pulls, and other contact forces like the normal force and friction, are today considered to be due to the electromagnetic force acting at the atomic level. For example, the force your fingers exert on a pencil is the result of electrical repulsion between the outer electrons of the atoms of your finger and those of the pencil.

# Summary

An object moving in a circle of radius $r$ with constant speed $v$ is said to be in **uniform circular motion**. It has a **radial acceleration** $a_R$ that is directed radially toward the center of the circle (also called **centripetal acceleration**), and has magnitude

$$a_R = \frac{v^2}{r}. \qquad (5\text{-}1)$$

The velocity vector and the acceleration vector $\vec{a}_R$ are continually changing in direction, but are perpendicular to each other at each moment.

A force is needed to keep an object revolving in a circle, and the direction of this force is toward the center of the circle. This force could be due to gravity (as for the Moon), to tension in a cord, to a component of the normal force, or to another type of force or combination of forces.

[*When the speed of circular motion is not constant, the acceleration has two components, tangential as well as centripetal.]

Newton's **law of universal gravitation** states that every particle in the universe attracts every other particle with a force proportional to the product of their masses and inversely proportional to the square of the distance between them:

$$F_G = G \frac{m_1 m_2}{r^2}. \qquad (5\text{-}4)$$

The direction of this force is along the line joining the two particles, and the force is always attractive. It is this gravitational force that keeps the Moon revolving around the Earth, and the planets revolving around the Sun.

Satellites revolving around the Earth are acted on by gravity, but "stay up" because of their high tangential speed.

Newton's three laws of motion, plus his law of universal gravitation, constituted a wide-ranging theory of the universe. With them, motion of objects on Earth and in space could be accurately described. And they provided a theoretical base for **Kepler's laws** of planetary motion.

The four fundamental forces in nature are (1) the gravitational force, (2) the electromagnetic force, (3) the strong nuclear force, and (4) the weak nuclear force. The first two fundamental forces are responsible for nearly all "everyday" forces.

# Questions

1. How many "accelerators" do you have in your car? There are at least three controls in the car which can be used to cause the car to accelerate. What are they? What accelerations do they produce?

2. A car rounds a curve at a steady 50 km/h. If it rounds the same curve at a steady 70 km/h, will its acceleration be any different? Explain.

3. Will the acceleration of a car be the same when a car travels around a sharp curve at a constant 60 km/h as when it travels around a gentle curve at the same speed? Explain.

4. Describe all the forces acting on a child riding a horse on a merry-go-round. Which of these forces provides the centripetal acceleration of the child?

5. A child on a sled comes flying over the crest of a small hill, as shown in Fig. 5–32. His sled does not leave the ground, but he feels the normal force between his chest and the sled decrease as he goes over the hill. Explain this decrease using Newton's second law.

**FIGURE 5–32**
Question 5.

6. Sometimes it is said that water is removed from clothes in the spin dryer by centrifugal force throwing the water outward. Is this correct? Discuss.

7. A girl is whirling a ball on a string around her head in a horizontal plane. She wants to let go at precisely the right time so that the ball will hit a target on the other side of the yard. When should she let go of the string?

8. A bucket of water can be whirled in a vertical circle without the water spilling out, even at the top of the circle when the bucket is upside down. Explain.

9. Astronauts who spend long periods in outer space could be adversely affected by weightlessness. One way to simulate gravity is to shape the spaceship like a cylindrical shell that rotates, with the astronauts walking on the inside surface (Fig. 5–33). Explain how this simulates gravity. Consider (a) how objects fall, (b) the force we feel on our feet, and (c) any other aspects of gravity you can think of.

**FIGURE 5–33**
Question 9.

10. A car maintains a constant speed $v$ as it traverses the hill and valley shown in Fig. 5–34. Both the hill and valley have a radius of curvature $R$. At which point, A, B, or C, is the normal force acting on the car (a) the largest, (b) the smallest? Explain. (c) Where would the driver feel heaviest and (d) lightest? Explain. (e) How fast can the car go without losing contact with the road at A?

**FIGURE 5–34** Question 10.

11. Can a particle with constant speed be accelerating? What if it has constant velocity? Explain.

12. Why do airplanes bank when they turn? How would you compute the banking angle given the airspeed and radius of the turn? [*Hint*: Assume an aerodynamic "lift" force acts perpendicular to the wings. See also Example 5–7.]

13. Does an apple exert a gravitational force on the Earth? If so, how large a force? Consider an apple (a) attached to a tree and (b) falling.

14. Why is more fuel required for a spacecraft to travel from the Earth to the Moon than to return from the Moon to the Earth?

15. Would it require less speed to launch a satellite (a) toward the east or (b) toward the west? Consider the Earth's rotation direction and explain your choice.

16. An antenna loosens and becomes detached from a satellite in a circular orbit around the Earth. Describe the antenna's subsequent motion. If it will land on the Earth, describe where; if not, describe how it could be made to land on the Earth.

17. The Sun is below us at midnight, nearly in line with the Earth's center. Are we then heavier at midnight, due to the Sun's gravitational force on us, than we are at noon? Explain.

18. When will your apparent weight be the greatest, as measured by a scale in a moving elevator: when the elevator (a) accelerates downward, (b) accelerates upward, (c) is in free fall, or (d) moves upward at constant speed? (e) In which case would your apparent weight be the least? (f) When would it be the same as when you are on the ground? Explain.

19. The source of the Mississippi River is closer to the center of the Earth than is its outlet in Louisiana (because the Earth is fatter at the equator than at the poles). Explain how the Mississippi can flow "uphill."

20. People sometimes ask, "What keeps a satellite up in its orbit around the Earth?" How would you respond?

21. Is the centripetal acceleration of Mars in its orbit around the Sun larger or smaller than the centripetal acceleration of the Earth? Explain.

22. The mass of the "planet" Pluto was not known until it was discovered to have a moon. Explain how this enabled an estimate of Pluto's mass.

23. The Earth moves faster in its orbit around the Sun in January than in July. Is the Earth closer to the Sun in January, or in July? Explain. [*Note:* This is not much of a factor in producing the seasons—the main factor is the tilt of the Earth's axis relative to the plane of its orbit.]

# MisConceptual Questions

1. While driving fast around a sharp right turn, you find yourself pressing against the car door. What is happening?
   (a) Centrifugal force is pushing you into the door.
   (b) The door is exerting a rightward force on you.
   (c) Both of the above.
   (d) Neither of the above.

2. Which of the following point towards the center of the circle in uniform circular motion?
   (a) Acceleration.
   (b) Velocity, acceleration, net force.
   (c) Velocity, acceleration.
   (d) Velocity, net force.
   (e) Acceleration, net force.

3. A Ping-Pong ball is shot into a circular tube that is lying flat (horizontal) on a tabletop. When the Ping-Pong ball exits the tube, which path will it follow in Fig. 5–35?

**FIGURE 5–35**
MisConceptual Question 3.

4. A car drives at steady speed around a perfectly circular track.
   (a) The car's acceleration is zero.
   (b) The net force on the car is zero.
   (c) Both the acceleration and net force on the car point outward.
   (d) Both the acceleration and net force on the car point inward.
   (e) If there is no friction, the acceleration is outward.

5. A child whirls a ball in a vertical circle. Assuming the speed of the ball is constant (an approximation), when would the tension in the cord connected to the ball be greatest?
   (a) At the top of the circle.
   (b) At the bottom of the circle.
   (c) A little after the bottom of the circle when the ball is climbing.
   (d) A little before the bottom of the circle when the ball is descending quickly.
   (e) Nowhere; the cord is stretched the same amount at all points.

6. In a rotating vertical cylinder (Rotor-ride) a rider finds herself pressed with her back to the rotating wall. Which is the correct free-body diagram for her (Fig. 5–36)?

(a)　　(b)　　(c)　　(d)　　(e)

**FIGURE 5–36** MisConceptual Question 6.

7. The Moon does not crash into the Earth because:
   (a) the net force on it is zero.
   (b) it is beyond the main pull of the Earth's gravity.
   (c) it is being pulled by the Sun as well as by the Earth.
   (d) it is freely falling but it has a high tangential velocity.

8. Which pulls harder gravitationally, the Earth on the Moon, or the Moon on the Earth? Which accelerates more?
   (a) The Earth on the Moon; the Earth.
   (b) The Earth on the Moon; the Moon.
   (c) The Moon on the Earth; the Earth.
   (d) The Moon on the Earth; the Moon.
   (e) Both the same; the Earth.
   (f) Both the same; the Moon.

9. In the International Space Station which orbits Earth, astronauts experience apparent weightlessness because
   (a) the station is so far away from the center of the Earth.
   (b) the station is kept in orbit by a centrifugal force that counteracts the Earth's gravity.
   (c) the astronauts and the station are in free fall towards the center of the Earth.
   (d) there is no gravity in space.
   (e) the station's high speed nullifies the effects of gravity.

10. Two satellites orbit the Earth in circular orbits of the same radius. One satellite is twice as massive as the other. Which statement is true about the speeds of these satellites?
    (a) The heavier satellite moves twice as fast as the lighter one.
    (b) The two satellites have the same speed.
    (c) The lighter satellite moves twice as fast as the heavier one.
    (d) The ratio of their speeds depends on the orbital radius.

11. A space shuttle in orbit around the Earth carries its payload with its mechanical arm. Suddenly, the arm malfunctions and releases the payload. What will happen to the payload?
    (a) It will fall straight down and hit the Earth.
    (b) It will follow a curved path and eventually hit the Earth.
    (c) It will remain in the same orbit with the shuttle.
    (d) It will drift out into deep space.

*12. A penny is placed on a turntable which is spinning clockwise as shown in Fig. 5–37. If the power to the turntable is turned off, which arrow best represents the direction of the acceleration of the penny at point P while the turntable is still spinning but slowing down?

**FIGURE 5–37** MisConceptual Question 12.

(a)   (b)   (c)   (d)   (e)

For assigned homework and other learning materials, go to the MasteringPhysics website. (MP)

# Problems

## 5–1 to 5–3 Uniform Circular Motion

1. (I) A child sitting 1.20 m from the center of a merry-go-round moves with a speed of 1.10 m/s. Calculate (a) the centripetal acceleration of the child and (b) the net horizontal force exerted on the child (mass = 22.5 kg).

2. (I) A jet plane traveling 1890 km/h (525 m/s) pulls out of a dive by moving in an arc of radius 5.20 km. What is the plane's acceleration in g's?

3. (I) A horizontal force of 310 N is exerted on a 2.0-kg ball as it rotates (at arm's length) uniformly in a horizontal circle of radius 0.90 m. Calculate the speed of the ball.

4. (II) What is the magnitude of the acceleration of a speck of clay on the edge of a potter's wheel turning at 45 rpm (revolutions per minute) if the wheel's diameter is 35 cm?

5. (II) A 0.55-kg ball, attached to the end of a horizontal cord, is revolved in a circle of radius 1.3 m on a frictionless horizontal surface. If the cord will break when the tension in it exceeds 75 N, what is the maximum speed the ball can have?

6. (II) How fast (in rpm) must a centrifuge rotate if a particle 7.00 cm from the axis of rotation is to experience an acceleration of 125,000 g's?

7. (II) A car drives straight down toward the bottom of a valley and up the other side on a road whose bottom has a radius of curvature of 115 m. At the very bottom, the normal force on the driver is twice his weight. At what speed was the car traveling?

8. (II) How large must the coefficient of static friction be between the tires and the road if a car is to round a level curve of radius 125 m at a speed of 95 km/h?

9. (II) What is the maximum speed with which a 1200-kg car can round a turn of radius 90.0 m on a flat road if the coefficient of friction between tires and road is 0.65? Is this result independent of the mass of the car?

10. (II) A bucket of mass 2.00 kg is whirled in a vertical circle of radius 1.20 m. At the lowest point of its motion the tension in the rope supporting the bucket is 25.0 N. (a) Find the speed of the bucket. (b) How fast must the bucket move at the top of the circle so that the rope does not go slack?

11. (II) How many revolutions per minute would a 25-m-diameter Ferris wheel need to make for the passengers to feel "weightless" at the topmost point?

12. (II) A jet pilot takes his aircraft in a vertical loop (Fig. 5–38). (a) If the jet is moving at a speed of 840 km/h at the lowest point of the loop, determine the minimum radius of the circle so that the centripetal acceleration at the lowest point does not exceed 6.0 g's. (b) Calculate the 78-kg pilot's effective weight (the force with which the seat pushes up on him) at the bottom of the circle, and (c) at the top of the circle (assume the same speed).

**FIGURE 5–38** Problem 12.

13. (II) A proposed space station consists of a circular tube that will rotate about its center (like a tubular bicycle tire), Fig. 5–39. The circle formed by the tube has a diameter of 1.1 km. What must be the rotation speed (revolutions per day) if an effect nearly equal to gravity at the surface of the Earth (say, 0.90 g) is to be felt?

**FIGURE 5–39** Problem 13.

**14.** (II) On an ice rink two skaters of equal mass grab hands and spin in a mutual circle once every 2.5 s. If we assume their arms are each 0.80 m long and their individual masses are 55.0 kg, how hard are they pulling on one another?

**15.** (II) A coin is placed 13.0 cm from the axis of a rotating turntable of variable speed. When the speed of the turntable is slowly increased, the coin remains fixed on the turntable until a rate of 38.0 rpm (revolutions per minute) is reached, at which point the coin slides off. What is the coefficient of static friction between the coin and the turntable?

**16.** (II) The design of a new road includes a straight stretch that is horizontal and flat but that suddenly dips down a steep hill at 18°. The transition should be rounded with what minimum radius so that cars traveling 95 km/h will not leave the road (Fig. 5–40)?

**FIGURE 5–40**
Problem 16.

**17.** (II) Two blocks, with masses $m_A$ and $m_B$, are connected to each other and to a central post by thin rods as shown in Fig. 5–41. The blocks revolve about the post at the same frequency $f$ (revolutions per second) on a frictionless horizontal surface at distances $r_A$ and $r_B$ from the post. Derive an algebraic expression for the tension in each rod.

**FIGURE 5–41**
Problem 17.

**18.** (II) Tarzan plans to cross a gorge by swinging in an arc from a hanging vine (Fig. 5–42). If his arms are capable of exerting a force of 1150 N on the vine, what is the maximum speed he can tolerate at the lowest point of his swing? His mass is 78 kg and the vine is 4.7 m long.

**FIGURE 5–42**
Problem 18.

**19.** (II) A 975-kg sports car (including driver) crosses the rounded top of a hill (radius = 88.0 m) at 18.0 m/s. Determine (a) the normal force exerted by the road on the car, (b) the normal force exerted by the car on the 62.0-kg driver, and (c) the car speed at which the normal force on the driver equals zero.

**20.** (II) Highway curves are marked with a suggested speed. If this speed is based on what would be safe in wet weather, estimate the radius of curvature for an unbanked curve marked 50 km/h. Use Table 4–2 (coefficients of friction).

**21.** (III) A pilot performs an evasive maneuver by diving vertically at 270 m/s. If he can withstand an acceleration of 8.0 g's without blacking out, at what altitude must he begin to pull his plane out of the dive to avoid crashing into the sea?

**22.** (III) If a curve with a radius of 95 m is properly banked for a car traveling 65 km/h, what must be the coefficient of static friction for a car not to skid when traveling at 95 km/h?

**23.** (III) A curve of radius 78 m is banked for a design speed of 85 km/h. If the coefficient of static friction is 0.30 (wet pavement), at what range of speeds can a car safely make the curve? [*Hint*: Consider the direction of the friction force when the car goes too slow or too fast.]

## *5–4 Nonuniform Circular Motion

*24. (I) Determine the tangential and centripetal components of the net force exerted on the car (by the ground) in Example 5–8 when its speed is 15 m/s. The car's mass is 950 kg.

*25. (II) A car at the Indianapolis 500 accelerates uniformly from the pit area, going from rest to 270 km/h in a semicircular arc with a radius of 220 m. Determine the tangential and radial acceleration of the car when it is halfway through the arc, assuming constant tangential acceleration. If the curve were flat, what coefficient of static friction would be necessary between the tires and the road to provide this acceleration with no slipping or skidding?

*26. (II) For each of the cases described below, sketch and label the total acceleration vector, the radial acceleration vector, and the tangential acceleration vector. (a) A car is accelerating from 55 km/h to 70 km/h as it rounds a curve of constant radius. (b) A car is going a constant 65 km/h as it rounds a curve of constant radius. (c) A car slows down while rounding a curve of constant radius.

*27. (III) A particle revolves in a horizontal circle of radius 1.95 m. At a particular instant, its acceleration is 1.05 m/s², in a direction that makes an angle of 25.0° to its direction of motion. Determine its speed (a) at this moment, and (b) 2.00 s later, assuming constant tangential acceleration.

## 5–5 and 5–6 Law of Universal Gravitation

**28.** (I) Calculate the force of Earth's gravity on a spacecraft 2.00 Earth radii above the Earth's surface if its mass is 1850 kg.

**29.** (I) At the surface of a certain planet, the gravitational acceleration $g$ has a magnitude of 12.0 m/s². A 24.0-kg brass ball is transported to this planet. What is (a) the mass of the brass ball on the Earth and on the planet, and (b) the weight of the brass ball on the Earth and on the planet?

**30.** (II) At what distance from the Earth will a spacecraft traveling directly from the Earth to the Moon experience zero net force because the Earth and Moon pull in opposite directions with equal force?

**31.** (II) Two objects attract each other gravitationally with a force of $2.5 \times 10^{-10}$ N when they are 0.25 m apart. Their total mass is 4.00 kg. Find their individual masses.

**32.** (II) A hypothetical planet has a radius 2.0 times that of Earth, but has the same mass. What is the acceleration due to gravity near its surface?

33. (II) Calculate the acceleration due to gravity on the Moon, which has radius $1.74 \times 10^6$ m and mass $7.35 \times 10^{22}$ kg.

34. (II) Estimate the acceleration due to gravity at the surface of Europa (one of the moons of Jupiter) given that its mass is $4.9 \times 10^{22}$ kg and making the assumption that its mass per unit volume is the same as Earth's.

35. (II) Given that the acceleration of gravity at the surface of Mars is 0.38 of what it is on Earth, and that Mars' radius is 3400 km, determine the mass of Mars.

36. (II) Find the net force on the Moon ($m_M = 7.35 \times 10^{22}$ kg) due to the gravitational attraction of both the Earth ($m_E = 5.98 \times 10^{24}$ kg) and the Sun ($m_S = 1.99 \times 10^{30}$ kg), assuming they are at right angles to each other, Fig. 5–43.

FIGURE 5–43 Problem 36.
Orientation of Sun (S), Earth (E), and Moon (M) at right angles to each other (not to scale).

37. (II) A hypothetical planet has a mass 2.80 times that of Earth, but has the same radius. What is $g$ near its surface?

38. (II) If you doubled the mass and tripled the radius of a planet, by what factor would $g$ at its surface change?

39. (II) Calculate the effective value of $g$, the acceleration of gravity, at (a) 6400 m, and (b) 6400 km, above the Earth's surface.

40. (II) You are explaining to friends why an astronaut feels weightless orbiting in the space shuttle, and they respond that they thought gravity was just a lot weaker up there. Convince them that it isn't so by calculating how much weaker (in %) gravity is 380 km above the Earth's surface.

41. (II) Every few hundred years most of the planets line up on the same side of the Sun. Calculate the total force on the Earth due to Venus, Jupiter, and Saturn, assuming all four planets are in a line, Fig. 5–44. The masses are $m_V = 0.815\, m_E$, $m_J = 318\, m_E$, $m_{Sat} = 95.1\, m_E$, and the mean distances of the four planets from the Sun are 108, 150, 778, and 1430 million km. What fraction of the Sun's force on the Earth is this?

FIGURE 5–44 Problem 41 (not to scale).

42. (II) Four 7.5-kg spheres are located at the corners of a square of side 0.80 m. Calculate the magnitude and direction of the gravitational force exerted on one sphere by the other three.

43. (II) Determine the distance from the Earth's center to a point outside the Earth where the gravitational acceleration due to the Earth is $\frac{1}{10}$ of its value at the Earth's surface.

44. (II) A certain neutron star has five times the mass of our Sun packed into a sphere about 10 km in radius. Estimate the surface gravity on this monster.

## 5–7 Satellites and Weightlessness

45. (I) A space shuttle releases a satellite into a circular orbit 780 km above the Earth. How fast must the shuttle be moving (relative to Earth's center) when the release occurs?

46. (I) Calculate the speed of a satellite moving in a stable circular orbit about the Earth at a height of 4800 km.

47. (II) You know your mass is 62 kg, but when you stand on a bathroom scale in an elevator, it says your mass is 77 kg. What is the acceleration of the elevator, and in which direction?

48. (II) A 12.0-kg monkey hangs from a cord suspended from the ceiling of an elevator. The cord can withstand a tension of 185 N and breaks as the elevator accelerates. What was the elevator's minimum acceleration (magnitude and direction)?

49. (II) Calculate the period of a satellite orbiting the Moon, 95 km above the Moon's surface. Ignore effects of the Earth. The radius of the Moon is 1740 km.

50. (II) Two satellites orbit Earth at altitudes of 7500 km and 15,000 km above the Earth's surface. Which satellite is faster, and by what factor?

51. (II) What will a spring scale read for the weight of a 58.0-kg woman in an elevator that moves (a) upward with constant speed 5.0 m/s, (b) downward with constant speed 5.0 m/s, (c) with an upward acceleration 0.23 $g$, (d) with a downward acceleration 0.23 $g$, and (e) in free fall?

52. (II) Determine the time it takes for a satellite to orbit the Earth in a circular **near-Earth orbit**. A "near-Earth" orbit is at a height above the surface of the Earth that is very small compared to the radius of the Earth. [*Hint*: You may take the acceleration due to gravity as essentially the same as that on the surface.] Does your result depend on the mass of the satellite?

53. (II) What is the apparent weight of a 75-kg astronaut 2500 km from the center of the Moon in a space vehicle (a) moving at constant velocity and (b) accelerating toward the Moon at 1.8 m/s²? State "direction" in each case.

54. (II) A Ferris wheel 22.0 m in diameter rotates once every 12.5 s (see Fig. 5–9). What is the ratio of a person's apparent weight to her real weight at (a) the top, and (b) the bottom?

55. (II) At what rate must a cylindrical spaceship rotate if occupants are to experience simulated gravity of 0.70 $g$? Assume the spaceship's diameter is 32 m, and give your answer as the time needed for one revolution. (See Question 9, Fig 5–33.)

56. (III) (a) Show that if a satellite orbits very near the surface of a planet with period $T$, the density (= mass per unit volume) of the planet is $\rho = m/V = 3\pi/GT^2$. (b) Estimate the density of the Earth, given that a satellite near the surface orbits with a period of 85 min. Approximate the Earth as a uniform sphere.

## 5–8 Kepler's Laws

**57.** (I) Neptune is an average distance of $4.5 \times 10^9$ km from the Sun. Estimate the length of the Neptunian year using the fact that the Earth is $1.50 \times 10^8$ km from the Sun on average.

**58.** (I) The *asteroid* Icarus, though only a few hundred meters across, orbits the Sun like the planets. Its period is 410 d. What is its mean distance from the Sun?

**59.** (I) Use Kepler's laws and the period of the Moon (27.4 d) to determine the period of an artificial satellite orbiting very near the Earth's surface.

**60.** (II) Determine the mass of the Earth from the known period and distance of the Moon.

**61.** (II) Our Sun revolves about the center of our Galaxy $(m_G \approx 4 \times 10^{41}$ kg$)$ at a distance of about $3 \times 10^4$ light-years $[1\,\text{ly} = (3.00 \times 10^8\,\text{m/s}) \cdot (3.16 \times 10^7\,\text{s/yr}) \cdot (1.00\,\text{yr})]$. What is the period of the Sun's orbital motion about the center of the Galaxy?

**62.** (II) Table 5–3 gives the mean distance, period, and mass for the four largest moons of Jupiter (those discovered by Galileo in 1609). Determine the mass of Jupiter: (*a*) using the data for Io; (*b*) using data for each of the other three moons. Are the results consistent?

**TABLE 5–3  Principal Moons of Jupiter**
(Problems 62 and 63)

| Moon | Mass (kg) | Period (Earth days) | Mean distance from Jupiter (km) |
|------|-----------|---------------------|--------------------------------|
| Io | $8.9 \times 10^{22}$ | 1.77 | $422 \times 10^3$ |
| Europa | $4.9 \times 10^{22}$ | 3.55 | $671 \times 10^3$ |
| Ganymede | $15 \times 10^{22}$ | 7.16 | $1070 \times 10^3$ |
| Callisto | $11 \times 10^{22}$ | 16.7 | $1883 \times 10^3$ |

**63.** (II) Determine the mean distance from Jupiter for each of Jupiter's principal moons, using Kepler's third law. Use the distance of Io and the periods given in Table 5–3. Compare your results to the values in Table 5–3.

**64.** (II) Planet A and planet B are in circular orbits around a distant star. Planet A is 7.0 times farther from the star than is planet B. What is the ratio of their speeds $v_A/v_B$ ?

**65.** (II) *Halley's comet* orbits the Sun roughly once every 76 years. It comes very close to the surface of the Sun on its closest approach (Fig. 5–45). Estimate the greatest distance of the comet from the Sun. Is it still "in" the solar system? What planet's orbit is nearest when it is out there?

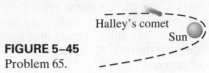

**FIGURE 5–45**
Problem 65.

**66.** (III) The *comet Hale–Bopp* has an orbital period of 2400 years. (*a*) What is its mean distance from the Sun? (*b*) At its closest approach, the comet is about 1.0 AU from the Sun (1 AU = distance from Earth to the Sun). What is the farthest distance? (*c*) What is the ratio of the speed at the closest point to the speed at the farthest point?

# General Problems

**67.** Calculate the centripetal acceleration of the Earth in its orbit around the Sun, and the net force exerted on the Earth. What exerts this force on the Earth? Assume that the Earth's orbit is a circle of radius $1.50 \times 10^{11}$ m.

**68.** A flat puck (mass $M$) is revolved in a circle on a frictionless air hockey table top, and is held in this orbit by a massless cord which is connected to a dangling mass (mass $m$) through a central hole as shown in Fig. 5–46. Show that the speed of the puck is given by $v = \sqrt{mgR/M}$.

**FIGURE 5–46** Problem 68.

**69.** A device for training astronauts and jet fighter pilots is designed to move the trainee in a horizontal circle of radius 11.0 m. If the force felt by the trainee is 7.45 times her own weight, how fast is she revolving? Express your answer in both m/s and rev/s.

**70.** A 1050-kg car rounds a curve of radius 72 m banked at an angle of 14°. If the car is traveling at 85 km/h, will a friction force be required? If so, how much and in what direction?

**71.** In a "Rotor-ride" at a carnival, people rotate in a vertical cylindrically walled "room." (See Fig. 5–47.) If the room radius is 5.5 m, and the rotation frequency 0.50 revolutions per second when the floor drops out, what minimum coefficient of static friction keeps the people from slipping down? People on this ride said they were "pressed against the wall." Is there really an outward force pressing them against the wall? If so, what is its source? If not, what is the proper description of their situation (besides nausea)? [*Hint:* Draw a free-body diagram for a person.]

**FIGURE 5–47**
Problem 71.

**72.** While fishing, you get bored and start to swing a sinker weight around in a circle below you on a 0.25-m piece of fishing line. The weight makes a complete circle every 0.75 s. What is the angle that the fishing line makes with the vertical? [*Hint*: See Fig. 5–10.]

**73.** At what minimum speed must a roller coaster be traveling so that passengers upside down at the top of the circle (Fig. 5–48) do not fall out? Assume a radius of curvature of 8.6 m.

**FIGURE 5–48**
Problem 73.

**74.** Consider a train that rounds a curve with a radius of 570 m at a speed of 160 km/h (approximately 100 mi/h). (a) Calculate the friction force needed on a train passenger of mass 55 kg if the track is not banked and the train does not tilt. (b) Calculate the friction force on the passenger if the train tilts at an angle of 8.0° toward the center of the curve.

**75.** Two equal-mass stars maintain a constant distance apart of $8.0 \times 10^{11}$ m and revolve about a point midway between them at a rate of one revolution every 12.6 yr. (a) Why don't the two stars crash into one another due to the gravitational force between them? (b) What must be the mass of each star?

**76.** How far above the Earth's surface will the acceleration of gravity be half what it is at the surface?

**77.** Is it possible to whirl a bucket of water fast enough in a vertical circle so that the water won't fall out? If so, what is the minimum speed? Define all quantities needed.

**78.** How long would a day be if the Earth were rotating so fast that objects at the equator were apparently weightless?

**79.** The *rings of Saturn* are composed of chunks of ice that orbit the planet. The inner radius of the rings is 73,000 km, and the outer radius is 170,000 km. Find the period of an orbiting chunk of ice at the inner radius and the period of a chunk at the outer radius. Compare your numbers with Saturn's own rotation period of 10 hours and 39 minutes. The mass of Saturn is $5.7 \times 10^{26}$ kg.

**80.** During an *Apollo* lunar landing mission, the command module continued to orbit the Moon at an altitude of about 100 km. How long did it take to go around the Moon once?

**81.** The **Navstar Global Positioning System** (GPS) utilizes a group of 24 satellites orbiting the Earth. Using "triangulation" and signals transmitted by these satellites, the position of a receiver on the Earth can be determined to within an accuracy of a few centimeters. The satellite orbits are distributed around the Earth, allowing continuous navigational "fixes." The satellites orbit at an altitude of approximately 11,000 nautical miles [1 nautical mile = 1.852 km = 6076 ft]. (a) Determine the speed of each satellite. (b) Determine the period of each satellite.

**82.** The *Near Earth Asteroid Rendezvous* (*NEAR*) spacecraft, after traveling 2.1 billion km, is meant to orbit the asteroid Eros with an orbital radius of about 20 km. Eros is roughly 40 km × 6 km × 6 km. Assume Eros has a density (mass/volume) of about $2.3 \times 10^3$ kg/m³. (a) If Eros were a sphere with the same mass and density, what would its radius be? (b) What would $g$ be at the surface of a spherical Eros? (c) Estimate the orbital period of *NEAR* as it orbits Eros, as if Eros were a sphere.

**83.** A train traveling at a constant speed rounds a curve of radius 215 m. A lamp suspended from the ceiling swings out to an angle of 16.5° throughout the curve. What is the speed of the train?

**84.** The Sun revolves around the center of the *Milky Way Galaxy* (Fig. 5–49) at a distance of about 30,000 light-years from the center $(1 \text{ ly} = 9.5 \times 10^{15} \text{ m})$. If it takes about 200 million years to make one revolution, estimate the mass of our Galaxy. Assume that the mass distribution of our Galaxy is concentrated mostly in a central uniform sphere. If all the stars had about the mass of our Sun $(2 \times 10^{30} \text{ kg})$, how many stars would there be in our Galaxy?

**FIGURE 5–49** Edge-on view of our galaxy.
Problem 84.

**85.** A satellite of mass 5500 kg orbits the Earth and has a period of 6600 s. Determine (a) the radius of its circular orbit, (b) the magnitude of the Earth's gravitational force on the satellite, and (c) the altitude of the satellite.

**86.** Astronomers using the Hubble Space Telescope deduced the presence of an extremely massive core in the distant *galaxy* M87, so dense that it could be a black hole (from which no light escapes). They did this by measuring the speed of gas clouds orbiting the core to be 780 km/s at a distance of 60 light-years $(= 5.7 \times 10^{17} \text{ m})$ from the core. Deduce the mass of the core, and compare it to the mass of our Sun.

**87.** Suppose all the mass of the Earth were compacted into a small spherical ball. What radius must the sphere have so that the acceleration due to gravity at the Earth's new surface would equal the acceleration due to gravity at the surface of the Sun?

**88.** A science-fiction tale describes an artificial "planet" in the form of a band completely encircling a sun (Fig. 5–50). The inhabitants live on the inside surface (where it is always noon). Imagine that this sun is exactly like our own, that the distance to the band is the same as the Earth–Sun distance (to make the climate livable), and that the ring rotates quickly enough to produce an apparent gravity of $g$ as on Earth. What will be the period of revolution, this planet's year, in Earth days?

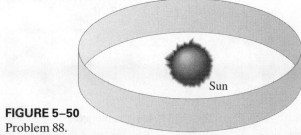

**FIGURE 5–50**
Problem 88.

**89.** An asteroid of mass $m$ is in a circular orbit of radius $r$ around the Sun with a speed $v$. It has an impact with another asteroid of mass $M$ and is kicked into a new circular orbit with a speed of $1.5\,v$. What is the radius of the new orbit in terms of $r$?

**\*90.** Use **dimensional analysis** (Section 1–8) to obtain the form for the centripetal acceleration, $a_R = v^2/r$.

# Search and Learn

1. Reread each Example in this Chapter and identify (i) the object undergoing centripetal acceleration (if any), and (ii) the force, or force component, that causes the circular motion.

2. Redo Example 5–3, precisely this time, by not ignoring the weight of the ball which revolves on a string 0.600 m long. In particular, find the magnitude of $\vec{F}_T$, and the angle it makes with the horizontal. [*Hint*: Set the horizontal component of $\vec{F}_T$ equal to $ma_R$; also, since there is no vertical motion, what can you say about the vertical component of $\vec{F}_T$?]

3. A banked curve of radius $R$ in a new highway is designed so that a car traveling at speed $v_0$ can negotiate the turn safely on glare ice (zero friction). If a car travels too slowly, then it will slip toward the center of the circle. If it travels too fast, it will slip away from the center of the circle. If the coefficient of static friction increases, it becomes possible for a car to stay on the road while traveling at a speed within a range from $v_{min}$ to $v_{max}$. Derive formulas for $v_{min}$ and $v_{max}$ as functions of $\mu_s$, $v_0$, and $R$.

4. *Earth is not quite an inertial frame.* We often make measurements in a reference frame fixed on the Earth, assuming Earth is an inertial reference frame [Section 4–2]. But the Earth rotates, so this assumption is not quite valid. Show that this assumption is off by 3 parts in 1000 by calculating the acceleration of an object at Earth's equator due to Earth's daily rotation, and compare to $g = 9.80 \text{ m/s}^2$, the acceleration due to gravity.

5. A certain white dwarf star was once an average star like our Sun. But now it is in the last stage of its evolution and is the size of our Moon but has the mass of our Sun. (*a*) Estimate the acceleration due to gravity on the surface of this star. (*b*) How much would a 65-kg person weigh on this star? Give as a percentage of the person's weight on Earth. (*c*) What would be the speed of a baseball dropped from a height of 1.0 m when it hit the surface?

6. Jupiter is about 320 times as massive as the Earth. Thus, it has been claimed that a person would be crushed by the force of gravity on a planet the size of Jupiter because people cannot survive more than a few $g$'s. Calculate the number of $g$'s a person would experience at Jupiter's equator, using the following data for Jupiter: mass = $1.9 \times 10^{27}$ kg, equatorial radius = $7.1 \times 10^4$ km, rotation period = 9 hr 55 min. Take the centripetal acceleration into account. [See Sections 5–2, 5–6, and 5–7.]

7. A plumb bob (a mass $m$ hanging on a string) is deflected from the vertical by an angle $\theta$ due to a massive mountain nearby (Fig. 5–51). (*a*) Find an approximate formula for $\theta$ in terms of the mass of the mountain, $m_M$, the distance to its center, $D_M$, and the radius and mass of the Earth. (*b*) Make a rough estimate of the mass of Mt. Everest, assuming it has the shape of a cone 4000 m high and base of diameter 4000 m. Assume its mass per unit volume is 3000 kg per m$^3$. (*c*) Estimate the angle $\theta$ of the plumb bob if it is 5 km from the center of Mt. Everest.

**FIGURE 5–51** Search and Learn 7.

8. (*a*) Explain why a Full moon always rises at sunset. (*b*) Explain how the position of the Moon in Fig. 5–31b cannot be seen yet by the person at the red dot (shown at 6 PM). (*c*) Explain why the red dot is where it is in parts (b) and (e), and show where it should be in part (d). (*d*) PRETTY HARD. Determine the average period of the Moon around the Earth (sidereal period) starting with the synodic period of 29.53 days as observed from Earth. [*Hint*: First determine the angle of the Moon in Fig. 5–31e relative to "horizontal," as in part (a).]

## ANSWERS TO EXERCISES

A: (*a*).
B: (*d*).
C: (*a*).

D: No.
E: $g$ would double.
F: (*b*).

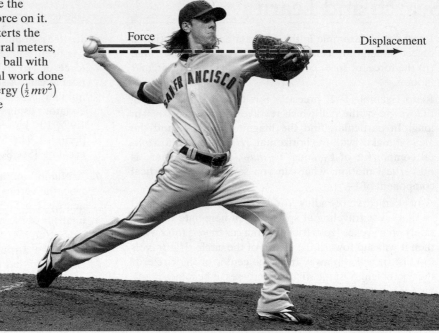

This baseball pitcher is about to accelerate the baseball to a high velocity by exerting a force on it. He will be doing work on the ball as he exerts the force over a displacement of perhaps several meters, from behind his head until he releases the ball with arm outstretched in front of him. The total work done on the ball will be equal to the kinetic energy ($\frac{1}{2}mv^2$) acquired by the ball, a result known as the work-energy principle.

Force          Displacement

# 6
# Work and Energy

**CHAPTER-OPENING QUESTION—Guess now!**

A skier starts at the top of a hill. On which run does her gravitational potential energy change the most: **(a)**, **(b)**, **(c)**, or **(d)**; or are they **(e)** all the same? On which run would her speed at the bottom be the fastest if the runs are icy and we assume no friction or air resistance? Recognizing that there is always some friction, answer the above two questions again. List your four answers now.

● Easy
■ Intermediate
◆ Difficult
◆◆ Very difficult

## CONTENTS

Until now we have been studying the translational motion of an object in terms of Newton's three laws of motion. In that analysis, *force* has played a central role as the quantity determining the motion. In this Chapter and the next, we discuss an alternative analysis of the translational motion of objects in terms of the quantities *energy* and *momentum*. The significance of energy and momentum is that they are *conserved*. That is, in quite general circumstances they remain constant. That conserved quantities exist gives us not only a deeper insight into the nature of the world, but also gives us another way to approach solving practical problems.

The conservation laws of energy and momentum are especially valuable in dealing with systems of many objects, in which a detailed consideration of the forces involved would be difficult or impossible. These laws apply to a wide range of phenomena. They even apply in the atomic and subatomic worlds, where Newton's laws are not sufficient.

This Chapter is devoted to the very important concept of *energy* and the closely related concept of *work*. These two quantities are scalars and so have no direction associated with them, which often makes them easier to work with than vector quantities such as acceleration and force.

**138**

**FIGURE 6–1** A person pulling a crate along the floor. The work done by the force $\vec{F}$ is $W = Fd\cos\theta$, where $\vec{d}$ is the displacement.

# 6–1 Work Done by a Constant Force

The word *work* has a variety of meanings in everyday language. But in physics, work is given a very specific meaning to describe what is accomplished when a force acts on an object, and the object moves through a distance. We consider only translational motion for now and, unless otherwise explained, objects are assumed to be rigid with no complicating internal motion, and can be treated like particles. Then the **work** done on an object by a constant force (constant in both magnitude and direction) is defined to be *the product of the magnitude of the displacement times the component of the force parallel to the displacement*. In equation form, we can write

$$W = F_{\|}d,$$

where $F_{\|}$ is the component of the constant force $\vec{F}$ parallel to the displacement $\vec{d}$. We can also write

$$W = Fd\cos\theta, \tag{6–1}$$

where $F$ is the magnitude of the constant force, $d$ is the magnitude of the displacement of the object, and $\theta$ is the angle between the directions of the force and the displacement (Fig. 6–1). The $\cos\theta$ factor appears in Eq. 6–1 because $F\cos\theta$ $(= F_{\|})$ is the component of $\vec{F}$ that is parallel to $\vec{d}$. Work is a scalar quantity—it has no direction, but only magnitude, which can be positive or negative.

Let us consider the case in which the motion and the force are in the same direction, so $\theta = 0$ and $\cos\theta = 1$; in this case, $W = Fd$. For example, if you push a loaded grocery cart a distance of 50 m by exerting a horizontal force of 30 N on the cart, you do $30\,\text{N} \times 50\,\text{m} = 1500\,\text{N}\cdot\text{m}$ of work on the cart.

As this example shows, in SI units work is measured in newton-meters ($\text{N}\cdot\text{m}$). A special name is given to this unit, the **joule** (J): $1\,\text{J} = 1\,\text{N}\cdot\text{m}$.

[In the cgs system, the unit of work is called the *erg* and is defined as $1\,\text{erg} = 1\,\text{dyne}\cdot\text{cm}$. In British units, work is measured in foot-pounds. Their equivalence is $1\,\text{J} = 10^7\,\text{erg} = 0.7376\,\text{ft}\cdot\text{lb}$.]

A force can be exerted on an object and yet do no work. If you hold a heavy bag of groceries in your hands at rest, you do no work on it. You do exert a force on the bag, but the displacement of the bag is zero, so the work done by you on the bag is $W = 0$. You need both a force and a displacement to do work. You also do no work on the bag of groceries if you carry it as you walk horizontally across the floor at constant velocity, as shown in Fig. 6–2. No horizontal force is required to move the bag at a constant velocity. The person shown in Fig. 6–2 exerts an upward force $\vec{F}_{\text{P}}$ on the bag equal to its weight. But this upward force is perpendicular to the horizontal displacement of the bag and thus is doing no work. This conclusion comes from our definition of work, Eq. 6–1: $W = 0$, because $\theta = 90°$ and $\cos 90° = 0$. Thus, when a particular force is perpendicular to the displacement, no work is done by that force. When you start or stop walking, there is a horizontal acceleration and you do briefly exert a horizontal force, and thus do work on the bag.

**FIGURE 6–2** The person does no work on the bag of groceries because $\vec{F}_{\text{P}}$ is perpendicular to the displacement $\vec{d}$.

**CAUTION**
*Force without work*

When we deal with work, as with force, it is necessary to specify whether you are talking about work done *by* a specific object or done *on* a specific object. It is also important to specify whether the work done is due to one particular force (and which one), or the total (net) work done by the *net force* on the object.

**FIGURE 6–3** Example 6–1. A 50-kg crate is pulled along a floor.

---

**EXAMPLE 6–1** **Work done on a crate.** A person pulls a 50-kg crate 40 m along a horizontal floor by a constant force $F_P = 100$ N, which acts at a 37° angle as shown in Fig. 6–3. The floor is rough and exerts a friction force $\vec{F}_{fr} = 50$ N. Determine (*a*) the work done by each force acting on the crate, and (*b*) the net work done on the crate.

**APPROACH** We choose our coordinate system so that the vector that represents the 40-m displacement is $\vec{x}$ (that is, along the *x* axis). Four forces act on the crate, as shown in the free-body diagram in Fig. 6–3: the force exerted by the person $\vec{F}_P$; the friction force $\vec{F}_{fr}$; the gravitational force exerted by the Earth, $\vec{F}_G = m\vec{g}$; and the normal force $\vec{F}_N$ exerted upward by the floor. The net force on the crate is the vector sum of these four forces.

**SOLUTION** (*a*) The work done by the gravitational force $(\vec{F}_G)$ and by the normal force $(\vec{F}_N)$ is zero, because they are perpendicular to the displacement $\vec{x}$ ($\theta = 90°$ in Eq. 6–1):

$$W_G = mgx \cos 90° = 0$$
$$W_N = F_N x \cos 90° = 0.$$

The work done by $\vec{F}_P$ is

$$W_P = F_P x \cos \theta = (100 \text{ N})(40 \text{ m}) \cos 37° = 3200 \text{ J}.$$

The work done by the friction force is

$$W_{fr} = F_{fr} x \cos 180° = (50 \text{ N})(40 \text{ m})(-1) = -2000 \text{ J}.$$

The angle between the displacement $\vec{x}$ and $\vec{F}_{fr}$ is 180° because they point in opposite directions. Since the force of friction is opposing the motion (and $\cos 180° = -1$), the work done by friction on the crate is *negative*.

(*b*) The net work can be calculated in two equivalent ways.
(1) The net work done on an object is the algebraic sum of the work done by each force, since work is a scalar:

$$W_{net} = W_G + W_N + W_P + W_{fr}$$
$$= 0 + 0 + 3200 \text{ J} - 2000 \text{ J} = 1200 \text{ J}.$$

(2) The net work can also be calculated by first determining the net force on the object and then taking the component of this net force along the displacement: $(F_{net})_x = F_P \cos \theta - F_{fr}$. Then the net work is

$$W_{net} = (F_{net})_x x = (F_P \cos \theta - F_{fr})x$$
$$= (100 \text{ N} \cos 37° - 50 \text{ N})(40 \text{ m}) = 1200 \text{ J}.$$

In the vertical (*y*) direction, there is no displacement and no work done.

---

In Example 6–1 we saw that friction did negative work. In general, the work done by a force is negative whenever the force (or the component of the force, $F_\parallel$) acts in the direction opposite to the direction of motion.

**EXERCISE A** A box is dragged a distance *d* across a floor by a force $\vec{F}_P$ which makes an angle $\theta$ with the horizontal as in Fig. 6–1 or 6–3. If the magnitude of $\vec{F}_P$ is held constant but the angle $\theta$ is increased, the work done by $\vec{F}_P$ (*a*) remains the same; (*b*) increases; (*c*) decreases; (*d*) first increases, then decreases.

## Work

1. **Draw a free-body diagram** showing all the forces acting on the object you choose to study.

2. **Choose** an $xy$ **coordinate system**. If the object is in motion, it may be convenient to choose one of the coordinate directions as the direction of one of the forces, or as the direction of motion. [Thus, for an object on an incline, you might choose one coordinate axis to be parallel to the incline.]

3. **Apply Newton's laws** to determine unknown forces.

4. Find the **work done** *by a specific force* *on* the object by using $W = Fd \cos \theta$ for a constant force. The work done is negative when a force opposes the displacement.

5. To find the **net work** done on the object, either (*a*) find the work done by each force and add the results algebraically; or (*b*) find the net force on the object, $F_{net}$, and then use it to find the net work done, which for constant net force is:

$$W_{net} = F_{net} d \cos \theta.$$

---

**EXAMPLE 6–2** **Work on a backpack.** (*a*) Determine the work a hiker must do on a 15.0-kg backpack to carry it up a hill of height $h = 10.0$ m, as shown in Fig. 6–4a. Determine also (*b*) the work done by gravity on the backpack, and (*c*) the net work done on the backpack. For simplicity, assume the motion is smooth and at constant velocity (i.e., acceleration is zero).

**APPROACH** We explicitly follow the steps of the Problem Solving Strategy above.

**SOLUTION**

1. **Draw a free-body diagram**. The forces on the backpack are shown in Fig. 6–4b: the force of gravity, $m\vec{g}$, acting downward; and $\vec{F}_H$, the force the hiker must exert upward to support the backpack. The acceleration is zero, so horizontal forces on the backpack are negligible.

2. **Choose a coordinate system**. We are interested in the vertical motion of the backpack, so we choose the $y$ coordinate as positive vertically upward.

3. **Apply Newton's laws**. Newton's second law applied in the vertical direction to the backpack gives (with $a_y = 0$)

$$\Sigma F_y = ma_y$$
$$F_H - mg = 0.$$

So,

$$F_H = mg = (15.0\,\text{kg})(9.80\,\text{m/s}^2) = 147\,\text{N}.$$

4. **Work done by a specific force**. (*a*) To calculate the work done by the hiker on the backpack, we use Eq. 6–1, where $\theta$ is shown in Fig. 6–4c,

$$W_H = F_H(d \cos \theta),$$

and we note from Fig. 6–4a that $d \cos \theta = h$. So the work done by the hiker is

$$W_H = F_H(d \cos \theta) = F_H h = mgh = (147\,\text{N})(10.0\,\text{m}) = 1470\,\text{J}.$$

The work done depends only on the elevation change and not on the angle of the hill, $\theta$. The hiker would do the same work to lift the pack vertically by height $h$.

(*b*) The work done by gravity on the backpack is (from Eq. 6–1 and Fig. 6–4c)

$$W_G = mg\,d \cos(180° - \theta).$$

Since $\cos(180° - \theta) = -\cos \theta$ (Appendix A–7), we have

$$W_G = mg(-d \cos \theta)$$
$$= -mgh$$
$$= -(15.0\,\text{kg})(9.80\,\text{m/s}^2)(10.0\,\text{m}) = -1470\,\text{J}.$$

**NOTE** The work done by gravity (which is negative here) does not depend on the angle of the incline, only on the vertical height $h$ of the hill.

5. **Net work done**. (*c*) The *net* work done on the backpack is $W_{net} = 0$, because the net force on the backpack is zero (it is assumed not to accelerate significantly). We can also get the net work done by adding the work done by each force:

$$W_{net} = W_G + W_H = -1470\,\text{J} + 1470\,\text{J} = 0.$$

**NOTE** Even though the *net* work done by all the forces on the backpack is zero, the hiker *does* do work on the backpack equal to 1470 J.

**FIGURE 6–4** Example 6–2.

(a)

(b)

(c)

**PROBLEM SOLVING**

*Work done by gravity depends on height of hill (not on angle)*

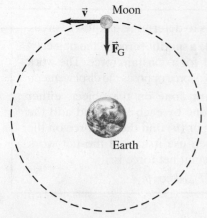

**FIGURE 6–5** Example 6–3.

**FIGURE 6–6** Work done by a force $F$ is (a) approximately equal to the sum of the areas of the rectangles, (b) exactly equal to the area under the curve of $F_\parallel$ vs. $d$.

(a)

(b)

| CONCEPTUAL EXAMPLE 6–3 | **Does the Earth do work on the Moon?** |

The Moon revolves around the Earth in a nearly circular orbit, kept there by the gravitational force exerted by the Earth. Does gravity do (a) positive work, (b) negative work, or (c) no work on the Moon?

**RESPONSE** The gravitational force $\vec{F}_G$ exerted by the Earth on the Moon (Fig. 6–5) acts toward the Earth and provides its centripetal acceleration, inward along the radius of the Moon's orbit. The Moon's displacement at any moment is tangent to the circle, in the direction of its velocity, perpendicular to the radius and perpendicular to the force of gravity. Hence the angle $\theta$ between the force $\vec{F}_G$ and the instantaneous displacement of the Moon is 90°, and the work done by gravity is therefore zero (cos 90° = 0). This is why the Moon, as well as artificial satellites, can stay in orbit without expenditure of fuel: no work needs to be done against the force of gravity.

# *6–2 Work Done by a Varying Force

If the force acting on an object is constant, the work done by that force can be calculated using Eq. 6–1. But in many cases, the force varies in magnitude or direction during a process. For example, as a rocket moves away from Earth, work is done to overcome the force of gravity, which varies as the inverse square of the distance from the Earth's center. Other examples are the force exerted by a spring, which increases with the amount of stretch, or the work done by a varying force that pulls a box or cart up an uneven hill.

The work done by a varying force can be determined graphically. To do so, we plot $F_\parallel$ ($= F \cos\theta$, the component of $\vec{F}$ parallel to the direction of motion at any point) as a function of distance $d$, as in Fig. 6–6a. We divide the distance into small segments $\Delta d$. For each segment, we indicate the average of $F_\parallel$ by a horizontal dashed line. Then the work done for each segment is $\Delta W = F_\parallel \Delta d$, which is the area of a rectangle $\Delta d$ wide and $F_\parallel$ high. The total work done to move the object a total distance $d = d_B - d_A$ is the sum of the areas of the rectangles (five in the case shown in Fig. 6–6a). Usually, the average value of $F_\parallel$ for each segment must be estimated, and a reasonable approximation of the work done can then be made.

If we subdivide the distance into many more segments, $\Delta d$ can be made smaller and our estimate of the work done would be more accurate. In the limit as $\Delta d$ approaches zero, the total area of the many narrow rectangles approaches the area under the curve, Fig. 6–6b. That is, *the work done by a variable force in moving an object between two points is equal to the area under the $F_\parallel$ vs. $d$ curve between those two points.*

# 6–3 Kinetic Energy, and the Work-Energy Principle

*Energy* is one of the most important concepts in science. Yet we cannot give a simple general definition of energy in only a few words. Nonetheless, each specific type of energy can be defined fairly simply. In this Chapter we define translational kinetic energy and some types of potential energy. In later Chapters, we will examine other types of energy, such as that related to heat and electricity. The crucial aspect of energy is that the sum of all types, the *total energy*, is the same after any process as it was before: that is, energy is a conserved quantity.

For the purposes of this Chapter, we can define energy in the traditional way as "the ability to do work." This simple definition is not always applicable,[†] but it is valid for mechanical energy which we discuss in this Chapter. We now define and discuss one of the basic types of energy, kinetic energy.

[†]Energy associated with heat is often not available to do work, as we will discuss in Chapter 15.

A moving object can do work on another object it strikes. A flying cannonball does work on a brick wall it knocks down; a moving hammer does work on a nail it drives into wood. In either case, a moving object exerts a force on a second object which undergoes a displacement. An object in motion has the ability to do work and thus can be said to have energy. The energy of motion is called **kinetic energy**, from the Greek word *kinetikos*, meaning "motion."

FIGURE 6–7 A constant net force $F_{net}$ accelerates a car from speed $v_1$ to speed $v_2$ over a displacement $d$. The net work done is $W_{net} = F_{net} d$.

To obtain a quantitative definition for kinetic energy, let us consider a simple rigid object of mass $m$ (treated as a particle) that is moving in a straight line with an initial speed $v_1$. To accelerate it uniformly to a speed $v_2$, a constant net force $F_{net}$ is exerted on it parallel to its motion over a displacement $d$, Fig. 6–7. Then the net work done on the object is $W_{net} = F_{net} d$. We apply Newton's second law, $F_{net} = ma$, and use Eq. 2–11c $\left(v_2^2 = v_1^2 + 2ad\right)$, which we rewrite as

$$a = \frac{v_2^2 - v_1^2}{2d},$$

where $v_1$ is the initial speed and $v_2$ is the final speed. Substituting this into $F_{net} = ma$, we determine the work done:

$$W_{net} = F_{net} d = mad = m\left(\frac{v_2^2 - v_1^2}{2d}\right)d = m\left(\frac{v_2^2 - v_1^2}{2}\right)$$

or

$$W_{net} = \tfrac{1}{2}mv_2^2 - \tfrac{1}{2}mv_1^2. \tag{6–2}$$

We *define* the quantity $\tfrac{1}{2}mv^2$ to be the **translational kinetic energy (KE)** of the object:

$$\mathrm{KE} = \tfrac{1}{2}mv^2. \tag{6–3}$$

*Kinetic energy (defined)*

(We call this "translational" kinetic energy to distinguish it from rotational kinetic energy, which we will discuss in Chapter 8.) Equation 6–2, derived here for one-dimensional motion with a constant force, is valid in general for translational motion of an object in three dimensions and even if the force varies.

We can rewrite Eq. 6–2 as:

$$W_{net} = \mathrm{KE}_2 - \mathrm{KE}_1$$

or

$$W_{net} = \Delta \mathrm{KE} = \tfrac{1}{2}mv_2^2 - \tfrac{1}{2}mv_1^2. \tag{6–4}$$

*WORK-ENERGY PRINCIPLE*

Equation 6–4 is a useful result known as the **work-energy principle**. It can be stated in words:

> **The net work done on an object is equal to the change in the object's kinetic energy.**

*WORK-ENERGY PRINCIPLE*

Notice that we made use of Newton's second law, $F_{net} = ma$, where $F_{net}$ is the *net* force—the sum of all forces acting on the object. Thus, the work-energy principle is valid only if $W$ is the *net work* done on the object—that is, the work done by all forces acting on the object.

◆ **CAUTION**
*Work-energy valid only for net work*

**FIGURE 6–8** A moving hammer strikes a nail and comes to rest. The hammer exerts a force $F$ on the nail; the nail exerts a force $-F$ on the hammer (Newton's third law). The work done on the nail by the hammer is positive $(W_n = Fd > 0)$. The work done on the hammer by the nail is negative $(W_h = -Fd)$.

The work-energy principle is a very useful reformulation of Newton's laws. It tells us that if (positive) net work $W$ is done on an object, the object's kinetic energy increases by an amount $W$. The principle also holds true for the reverse situation: if the net work $W$ done on an object is negative, the object's kinetic energy decreases by an amount $W$. That is, a net force exerted on an object opposite to the object's direction of motion decreases its speed and its kinetic energy. An example is a moving hammer (Fig. 6–8) striking a nail. The net force on the hammer ($-\vec{F}$ in Fig. 6–8, where $\vec{F}$ is assumed constant for simplicity) acts toward the left, whereas the displacement $\vec{d}$ of the hammer is toward the right. So the net work done on the hammer, $W_h = (F)(d)(\cos 180°) = -Fd$, is negative and the hammer's kinetic energy decreases (usually to zero).

Figure 6–8 also illustrates how energy can be considered the ability to do work. The hammer, as it slows down, does positive work on the nail: $W_n = (+F)(+d) = Fd$ and is positive. The decrease in kinetic energy of the hammer ($= Fd$ by Eq. 6–4) is equal to the work the hammer can do on another object, the nail in this case.

The translational kinetic energy ($= \frac{1}{2}mv^2$) is directly proportional to the mass of the object, and it is also proportional to the *square* of the speed. Thus, if the mass is doubled, the kinetic energy is doubled. But if the speed is doubled, the object has four times as much kinetic energy and is therefore capable of doing four times as much work.

Because of the direct connection between work and kinetic energy, energy is measured in the same units as work: joules in SI units. [The energy unit is ergs in the cgs, and foot-pounds in the British system.] Like work, kinetic energy is a scalar quantity. The kinetic energy of a group of objects is the sum of the kinetic energies of the individual objects.

The work-energy principle can be applied to a particle, and also to an object that can be approximated as a particle, such as an object that is rigid or whose internal motions are insignificant. It is very useful in simple situations, as we will see in the Examples below.

$v_1 = 20$ m/s    $v_2 = 30$ m/s

**FIGURE 6–9** Example 6–4.

---

**EXAMPLE 6–4 | ESTIMATE | Work on a car, to increase its kinetic energy.**
How much net work is required to accelerate a 1000-kg car from 20 m/s to 30 m/s (Fig. 6–9)?

**APPROACH** A car is a complex system. The engine turns the wheels and tires which push against the ground, and the ground pushes back (see Example 4–4). We aren't interested right now in those complications. Instead, we can get a useful result using the work-energy principle, but only if we model the car as a particle or simple rigid object.

**SOLUTION** The net work needed is equal to the increase in kinetic energy:

$$
\begin{aligned}
W &= KE_2 - KE_1 \\
&= \tfrac{1}{2}mv_2^2 - \tfrac{1}{2}mv_1^2 \\
&= \tfrac{1}{2}(1000\text{ kg})(30\text{ m/s})^2 - \tfrac{1}{2}(1000\text{ kg})(20\text{ m/s})^2 \\
&= 2.5 \times 10^5\text{ J}.
\end{aligned}
$$

---

**EXERCISE B** (*a*) Make a guess: will the work needed to accelerate the car in Example 6–4 from rest to 20 m/s be more than, less than, or equal to the work already calculated to accelerate it from 20 m/s to 30 m/s? (*b*) Make the calculation.

(a) $v_1 = 60$ km/h                $v_2 = 0$

$\vec{\mathbf{F}}$

$\vec{\mathbf{d}}$ ($d = 20$ m)

(b) $v_1 = 120$ km/h                $v_2 = 0$

$\vec{\mathbf{F}}$

$\vec{\mathbf{d}}$ ($d =$ ?)

**FIGURE 6–10** Example 6–5. A moving car comes to a stop. Initial velocity is (a) 60 km/h, (b) 120 km/h.

---

**CONCEPTUAL EXAMPLE 6–5** | **Work to stop a car.** A car traveling 60 km/h can brake to a stop in a distance $d$ of 20 m (Fig. 6–10a). If the car is going twice as fast, 120 km/h, what is its stopping distance (Fig. 6–10b)? Assume the maximum braking force is approximately independent of speed.

**RESPONSE** Again we model the car as if it were a particle. Because the net stopping force $F$ is approximately constant, the work needed to stop the car, $Fd$, is proportional to the distance traveled. We apply the work-energy principle, noting that $\vec{\mathbf{F}}$ and $\vec{\mathbf{d}}$ are in opposite directions and that the final speed of the car is zero:

$$W_{\text{net}} = Fd \cos 180° = -Fd.$$

Then

$$-Fd = \Delta KE = \tfrac{1}{2}mv_2^2 - \tfrac{1}{2}mv_1^2$$
$$= 0 - \tfrac{1}{2}mv_1^2.$$

Thus, since the force and mass are constant, we see that the stopping distance, $d$, increases with the square of the speed:

$$d \propto v^2.$$

If the car's initial speed is doubled, the stopping distance is $(2)^2 = 4$ times as great, or 80 m.

| **EXERCISE C** Can kinetic energy ever be negative?

| **EXERCISE D** (*a*) If the kinetic energy of a baseball is doubled, by what factor has its speed increased? (*b*) If its speed is doubled, by what factor does its kinetic energy increase?

# 6–4 Potential Energy

We have just discussed how an object is said to have energy by virtue of its motion, which we call kinetic energy. But it is also possible to have **potential energy**, which is the energy associated with forces that depend on the position or configuration of an object (or objects) relative to the surroundings. Various types of potential energy (PE) can be defined, and each type is associated with a particular force.

The spring of a wind-up toy is an example of an object with potential energy. The spring acquired its potential energy because work was done *on* it by the person winding the toy. As the spring unwinds, it exerts a force and does work to make the toy move.

## Gravitational Potential Energy

Perhaps the most common example of potential energy is *gravitational potential energy*. A heavy brick held high above the ground has potential energy because of its position relative to the Earth. The raised brick has the ability to do work, for if it is released, it will fall to the ground due to the gravitational force, and can do work on, say, a stake, driving it into the ground.

**FIGURE 6–11** A person exerts an upward force $F_{ext} = mg$ to lift a brick from $y_1$ to $y_2$.

Let us seek the form for the gravitational potential energy of an object near the surface of the Earth. For an object of mass $m$ to be lifted vertically, an upward force at least equal to its weight, $mg$, must be exerted on it, say by a person's hand. To lift the object without acceleration, the person exerts an "external force" $F_{ext} = mg$. If it is raised a vertical height $h$, from position $y_1$ to $y_2$ in Fig. 6–11 (upward direction chosen positive), a person does work equal to the product of the "external" force she exerts, $F_{ext} = mg$ upward, multiplied by the vertical displacement $h$. That is,

$$W_{ext} = F_{ext} d \cos 0° = mgh$$
$$= mg(y_2 - y_1). \qquad \textbf{(6–5a)}$$

Gravity is also acting on the object as it moves from $y_1$ to $y_2$, and does work on the object equal to

$$W_G = F_G d \cos \theta = mgh \cos 180°,$$

where $\theta = 180°$ because $\vec{\mathbf{F}}_G$ and $\vec{\mathbf{d}}$ point in opposite directions. So

$$W_G = -mgh$$
$$= -mg(y_2 - y_1). \qquad \textbf{(6–5b)}$$

Next, if we allow the object to start from rest at $y_2$ and fall freely under the action of gravity, it acquires a velocity given by $v^2 = 2gh$ (Eq. 2–11c) after falling a height $h$. It then has kinetic energy $\frac{1}{2}mv^2 = \frac{1}{2}m(2gh) = mgh$, and if it strikes a stake, it can do work on the stake equal to $mgh$ (Section 6–3).

Thus, to raise an object of mass $m$ to a height $h$ *requires* an amount of work equal to $mgh$ (Eq. 6–5a). And once at height $h$, the object has the *ability* to do an amount of work equal to $mgh$. We can say that the work done in lifting the object has been stored as gravitational potential energy.

We therefore define the **gravitational potential energy** of an object, due to Earth's gravity, as the product of the object's weight $mg$ and its height $y$ above some reference level (such as the ground):

$$\text{PE}_G = mgy. \qquad \textbf{(6–6)}$$

The higher an object is above the ground, the more gravitational potential energy it has. We combine Eq. 6–5a with Eq. 6–6:

$$W_{ext} = mg(y_2 - y_1)$$
$$W_{ext} = \text{PE}_2 - \text{PE}_1 = \Delta\text{PE}_G. \qquad \textbf{(6–7a)}$$

That is, the change in gravitational potential energy when an object is moved from height $y_1$ to height $y_2$ is equal to the work done by the net external force that accomplishes this without acceleration.

A more direct way to define the *change in gravitational potential energy*, $\Delta\text{PE}_G$, is that it is equal to the *negative of the work done by gravity* itself (Eq. 6–5b):

$$W_G = -mg(y_2 - y_1)$$
$$W_G = -(\text{PE}_2 - \text{PE}_1) = -\Delta\text{PE}_G$$

or

$$\Delta\text{PE}_G = -W_G. \qquad \textbf{(6–7b)}$$

Gravitational potential energy depends on the *vertical height* of the object *above some reference level* (Eq. 6–6). In some situations, you may wonder from what point to measure the height $y$. The gravitational potential energy of a book held high above a table, for example, depends on whether we measure $y$ from the top of the table, from the floor, or from some other reference point. What is physically important in any situation is the *change* in potential energy, $\Delta\text{PE}$, because that is what is related to the work done, Eqs. 6–7; and it is $\Delta\text{PE}$ that can be measured. We can thus choose to measure $y$ from any reference level that is convenient, but we must choose the reference level at the start and be consistent throughout. The *change* in potential energy between any two points does not depend on this choice.

 **CAUTION**

*Change in PE is what is physically meaningful*

An important result we discussed earlier (see Example 6–2 and Fig. 6–4) concerns the gravity force, which does work only in the vertical direction: the work done by gravity depends only on the vertical height $h$, and not on the path taken, whether it be purely vertical motion or, say, motion along an incline. Thus, from Eqs. 6–7 we see that changes in gravitational potential energy depend only on the change in vertical height and not on the path taken.

Potential energy belongs to a system, and not to a single object alone. Potential energy is associated with a force, and a force on one object is always exerted by some other object. Thus potential energy is a property of the system as a whole. For an object raised to a height $y$ above the Earth's surface, the change in gravitational potential energy is $mgy$. The system here is the object plus the Earth, and properties of both are involved: object ($m$) and Earth ($g$).

◆ **CAUTION**
*Potential energy belongs to a system, not to a single object*

---

**EXAMPLE 6–6** | **Potential energy changes for a roller coaster.** A 1000-kg roller-coaster car moves from point 1, Fig. 6–12, to point 2 and then to point 3. (*a*) What is the gravitational potential energy at points 2 and 3 relative to point 1? That is, take $y = 0$ at point 1. (*b*) What is the change in potential energy when the car goes from point 2 to point 3? (*c*) Repeat parts (*a*) and (*b*), but take the reference point ($y = 0$) to be at point 3.

**APPROACH** We are interested in the potential energy of the car–Earth system. We take upward as the positive $y$ direction, and use the definition of gravitational potential energy to calculate the potential energy.

**SOLUTION** (*a*) We measure heights from point 1 $(y_1 = 0)$, which means initially that the gravitational potential energy is zero. At point 2, where $y_2 = 10\,\text{m}$,

$$\text{PE}_2 = mgy_2 = (1000\,\text{kg})(9.8\,\text{m/s}^2)(10\,\text{m}) = 9.8 \times 10^4\,\text{J}.$$

At point 3, $y_3 = -15\,\text{m}$, since point 3 is below point 1. Therefore,

$$\text{PE}_3 = mgy_3 = (1000\,\text{kg})(9.8\,\text{m/s}^2)(-15\,\text{m}) = -1.5 \times 10^5\,\text{J}.$$

(*b*) In going from point 2 to point 3, the potential energy change $(\text{PE}_{\text{final}} - \text{PE}_{\text{initial}})$ is

$$\text{PE}_3 - \text{PE}_2 = (-1.5 \times 10^5\,\text{J}) - (9.8 \times 10^4\,\text{J}) = -2.5 \times 10^5\,\text{J}.$$

The gravitational potential energy decreases by $2.5 \times 10^5\,\text{J}$.

(*c*) Now we set $y_3 = 0$. Then $y_1 = +15\,\text{m}$ at point 1, so the potential energy initially is

$$\text{PE}_1 = (1000\,\text{kg})(9.8\,\text{m/s}^2)(15\,\text{m}) = 1.5 \times 10^5\,\text{J}.$$

At point 2, $y_2 = 25\,\text{m}$, so the potential energy is

$$\text{PE}_2 = 2.5 \times 10^5\,\text{J}.$$

At point 3, $y_3 = 0$, so the potential energy is zero. The change in potential energy going from point 2 to point 3 is

$$\text{PE}_3 - \text{PE}_2 = 0 - 2.5 \times 10^5\,\text{J} = -2.5 \times 10^5\,\text{J},$$

which is the same as in part (*b*).

**NOTE** Work done by gravity depends only on the vertical height, so changes in gravitational potential energy do not depend on the path taken.

**FIGURE 6–12** Example 6–6.

## Potential Energy Defined in General

There are other kinds of potential energy besides gravitational. Each form of potential energy is associated with a particular force, and can be defined analogously to gravitational potential energy. In general, the *change in potential energy associated with a particular force is equal to the negative of the work done by that force when the object is moved from one point to a second point* (as in Eq. 6–7b for gravity). An alternate and sometimes useful way to define the change in potential energy is: the work required of a net external force to move the object without acceleration between the two points, as in Eq. 6–7a.

*Potential Energy defined*

**FIGURE 6–13** A spring (a) can store energy (elastic PE) when compressed as in (b) and can do work when released (c).

(a)  (b)  (c)

$x = 0$

(a) Unstretched

$\vec{F}_S$  $\vec{F}_{ext}$

(b) Stretched

$\vec{F}_{ext}$  $\vec{F}_S$

(c) Compressed

**FIGURE 6–14** (a) Spring in natural (unstretched) position. (b) Spring is stretched by a person exerting a force $\vec{F}_{ext}$ to the right (positive direction). The spring pulls back with a force $\vec{F}_S$, where $F_S = -kx$. (c) Person compresses the spring $(x < 0)$ by exerting an external force $\vec{F}_{ext}$ to the left; the spring pushes back with a force $F_S = -kx$, where $F_S > 0$ because $x < 0$.

**FIGURE 6–15** As a spring is stretched (or compressed), the magnitude of the force needed increases linearly as $x$ increases: graph of $F = kx$ vs. $x$ from $x = 0$ to $x = x_f$.

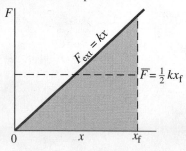

## Potential Energy of Elastic Spring

We now consider potential energy associated with elastic materials, which includes a great variety of practical applications. Consider the simple coil spring shown in Fig. 6–13. The spring has potential energy when compressed (or stretched), because when it is released, it can do work on a ball as shown. To hold a spring either stretched or compressed an amount $x$ from its natural (unstretched) length requires the hand to exert an external force on the spring of magnitude $F_{ext}$ which is directly proportional to $x$. That is,

$$F_{ext} = kx,$$

where $k$ is a constant, called the *spring stiffness constant* (or simply **spring constant**), and is a measure of the stiffness of the particular spring. The stretched or compressed spring itself exerts a force $F_S$ in the opposite direction on the hand, as shown in Fig. 6–14:

$$F_S = -kx. \qquad \text{[spring force]} \quad (6\text{--}8)$$

This force is sometimes called a "restoring force" because the spring exerts its force in the direction opposite the displacement (hence the minus sign), acting to return it to its natural length. Equation 6–8 is known as the **spring equation** and also as **Hooke's law**, and is accurate for springs as long as $x$ is not too great.

To calculate the potential energy of a stretched spring, let us calculate the work required to stretch it (Fig. 6–14b). We might hope to use Eq. 6–1 for the work done on it, $W = Fx$, where $x$ is the amount it is stretched from its natural length. But this would be incorrect since the force $F_{ext}$ $(= kx)$ is not constant but varies over the distance $x$, becoming greater the more the spring is stretched, as shown graphically in Fig. 6–15. So let us use the average force, $\bar{F}$. Since $F_{ext}$ varies linearly, from zero at the unstretched position to $kx$ when stretched to $x$, the average force is $\bar{F} = \frac{1}{2}[0 + kx] = \frac{1}{2}kx$, where $x$ here is the final amount stretched (shown as $x_f$ in Fig. 6–15 for clarity). The work done is then

$$W_{ext} = \bar{F}x = (\tfrac{1}{2}kx)(x) = \tfrac{1}{2}kx^2.$$

Hence the **elastic potential energy**, $PE_{el}$, is proportional to the square of the amount stretched:

$$PE_{el} = \tfrac{1}{2}kx^2. \qquad \text{[elastic spring]} \quad (6\text{--}9)$$

If a spring is *compressed* a distance $x$ from its natural ("equilibrium") length, the average force again has magnitude $\bar{F} = \frac{1}{2}kx$, and again the potential energy is given by Eq. 6–9. Thus $x$ can be either the amount compressed or amount stretched from the spring's natural length.[†] Note that for a spring, we choose the reference point for zero PE at the spring's natural position.

## Potential Energy as Stored Energy

In the above examples of potential energy—from a brick held at a height $y$, to a stretched or compressed spring—an object has the capacity or *potential* to do work even though it is not yet actually doing it. These examples show that energy can be *stored*, for later use, in the form of potential energy (as in Fig. 6–13, for a spring).

Note that there is a single universal formula for the translational kinetic energy of an object, $\frac{1}{2}mv^2$, but there is no single formula for potential energy. Instead, the mathematical form of the potential energy depends on the force involved.

[†]We can also obtain Eq. 6–9 using Section 6–2. The work done, and hence $\Delta PE$, equals the area under the $F$ vs. $x$ graph of Fig. 6–15. This area is a triangle (colored in Fig. 6–15) of altitude $kx$ and base $x$, and hence of area (for a triangle) equal to $\frac{1}{2}(kx)(x) = \frac{1}{2}kx^2$.

# 6–5 Conservative and Nonconservative Forces

The work done against gravity in moving an object from one point to another does not depend on the path taken. For example, it takes the same work (= $mgh$) to lift an object of mass $m$ vertically a height $h$ as to carry it up an incline of the same vertical height, as in Fig. 6–4 (see Example 6–2). Forces such as gravity, for which the work done does not depend on the path taken but only on the initial and final positions, are called **conservative forces**. The elastic force of a spring (or other elastic material), in which $F = -kx$, is also a conservative force. An object that starts at a given point and returns to that same point under the action of a conservative force has no net work done on it because the potential energy is the same at the start and the finish of such a round trip.

Many forces, such as friction and a push or pull exerted by a person, are **nonconservative forces** since any work they do depends on the path. For example, if you push a crate across a floor from one point to another, the work you do depends on whether the path taken is straight or is curved. As shown in Fig. 6–16, if a crate is pushed slowly from point 1 to point 2 along the longer semicircular path, you do more work against friction than if you push it along the straight path.

**FIGURE 6–16** A crate is pushed slowly at constant speed across a rough floor from position 1 to position 2 via two paths, one straight and one curved. The pushing force $\vec{F}_P$ is in the direction of motion at each point. (The friction force opposes the motion.) Hence for a constant magnitude pushing force, the work it does is $W = F_P d$, so if the distance traveled $d$ is greater (as for the curved path), then $W$ is greater. The work done does not depend only on points 1 and 2; it also depends on the path taken.

You do more work on the curved path because the distance is greater and, unlike the gravitational force, the pushing force $\vec{F}_P$ is in the direction of motion at each point. Thus the work done by the person in Fig. 6–16 does not depend *only* on points 1 and 2; it depends also on the path taken. The force of kinetic friction, also shown in Fig. 6–16, always opposes the motion; it too is a nonconservative force, and we discuss how to treat it later in this Chapter (Section 6–9). Table 6–1 lists a few conservative and nonconservative forces.

Because potential energy is energy associated with the position or configuration of objects, potential energy can only make sense if it can be stated uniquely for a given point. This cannot be done with nonconservative forces because the work done depends on the path taken (as in Fig. 6–16). Hence, *potential energy can be defined only for a conservative force*. Thus, although potential energy is always associated with a force, not all forces have a potential energy. For example, there is no potential energy for friction.

| TABLE 6–1 Conservative and Nonconservative Forces | |
|---|---|
| Conservative Forces | Nonconservative Forces |
| Gravitational | Friction |
| Elastic | Air resistance |
| Electric | Tension in cord |
| | Motor or rocket propulsion |
| | Push or pull by a person |

**EXERCISE E** An object acted on by a constant force $F$ moves from point 1 to point 2 and back again. The work done by the force $F$ in this round trip is 60 J. Can you determine from this information if $F$ is a conservative or nonconservative force?

## Work-Energy Extended

We can extend the **work-energy principle** (discussed in Section 6–3) to include potential energy. Suppose several forces act on an object which can undergo translational motion. And suppose only some of these forces are conservative. We write the total (net) work $W_{net}$ as a sum of the work done by conservative forces, $W_C$, and the work done by nonconservative forces, $W_{NC}$:

$$W_{net} = W_C + W_{NC}.$$

Then, from the work-energy principle, Eq. 6–4, we have

$$W_{net} = \Delta KE$$
$$W_C + W_{NC} = \Delta KE$$

where $\Delta KE = KE_2 - KE_1$. Then

$$W_{NC} = \Delta KE - W_C.$$

Work done by a conservative force can be written in terms of potential energy, as we saw in Eq. 6–7b for gravitational potential energy:

$$W_C = -\Delta PE.$$

We combine these last two equations:

$$W_{NC} = \Delta KE + \Delta PE. \tag{6–10}$$

Thus, *the work $W_{NC}$ done by the nonconservative forces acting on an object is equal to the total change in kinetic and potential energies.*

It must be emphasized that *all* the forces acting on an object must be included in Eq. 6–10, either in the potential energy term on the right (if it is a conservative force), or in the work term on the left (but not in both!).

# 6–6 Mechanical Energy and Its Conservation

If we can ignore friction and other nonconservative forces, or if only conservative forces do work on a system, we arrive at a particularly simple and beautiful relation involving energy.

When no nonconservative forces do work, then $W_{NC} = 0$ in the general form of the work-energy principle (Eq. 6–10). Then we have

$$\Delta KE + \Delta PE = 0 \qquad \left[\begin{array}{c}\text{conservative}\\\text{forces only}\end{array}\right] \tag{6–11a}$$

or

$$(KE_2 - KE_1) + (PE_2 - PE_1) = 0. \qquad \left[\begin{array}{c}\text{conservative}\\\text{forces only}\end{array}\right] \tag{6–11b}$$

We now define a quantity $E$, called the **total mechanical energy** of our system, as the sum of the kinetic and potential energies at any moment:

$$E = KE + PE.$$

Now we can rewrite Eq. 6–11b as

$$KE_2 + PE_2 = KE_1 + PE_1 \qquad \left[\begin{array}{c}\text{conservative}\\\text{forces only}\end{array}\right] \tag{6–12a}$$

*CONSERVATION OF MECHANICAL ENERGY*

or

$$E_2 = E_1 = \text{constant.} \qquad \left[\begin{array}{c}\text{conservative}\\\text{forces only}\end{array}\right] \tag{6–12b}$$

Equations 6–12 express a useful and profound principle regarding the total mechanical energy of a system—namely, that it is a **conserved quantity**. The total mechanical energy $E$ remains constant as long as no nonconservative forces do work: KE + PE at some initial time 1 is equal to the KE + PE at any later time 2.

To say it another way, consider Eq. 6–11a which tells us $\Delta PE = -\Delta KE$; that is, if the kinetic energy KE of a system increases, then the potential energy PE must decrease by an equivalent amount to compensate. Thus, the total, KE + PE, remains constant:

**If only conservative forces do work, the total mechanical energy of a system neither increases nor decreases in any process. It stays constant—it is conserved.**

*CONSERVATION OF MECHANICAL ENERGY*

This is the **principle of conservation of mechanical energy** for conservative forces.

In the next Section we shall see the great usefulness of the conservation of mechanical energy principle in a variety of situations, and how it is often easier to use than the kinematic equations or Newton's laws. After that we will discuss how other forms of energy can be included in the general conservation of energy law, such as energy associated with friction.

# 6–7 Problem Solving Using Conservation of Mechanical Energy

A simple example of the conservation of mechanical energy (neglecting air resistance) is a rock allowed to fall due to Earth's gravity from a height $h$ above the ground, as shown in Fig. 6–17. If the rock starts from rest, all of the initial energy is potential energy. As the rock falls, the potential energy $mgy$ decreases (because the rock's height above the ground $y$ decreases), but the rock's kinetic energy increases to compensate, so that the sum of the two remains constant. At any point along the path, the total mechanical energy is given by

$$E = KE + PE = \tfrac{1}{2}mv^2 + mgy$$

where $v$ is its speed at that point. If we let the subscript 1 represent the rock at one point along its path (for example, the initial point), and the subscript 2 represent it at some other point, then we can write

total mechanical energy at point 1 = total mechanical energy at point 2

or (see also Eq. 6–12a)

$$\tfrac{1}{2}mv_1^2 + mgy_1 = \tfrac{1}{2}mv_2^2 + mgy_2. \qquad \text{[gravity only]} \quad \textbf{(6–13)}$$

Just before the rock hits the ground, where we chose $y = 0$, all of the initial potential energy will have been transformed into kinetic energy.

**FIGURE 6–17** The rock's potential energy changes to kinetic energy as it falls. Note bar graphs representing potential energy PE and kinetic energy KE for the three different positions.

---

| **EXAMPLE 6–7** | **Falling rock.** |

If the initial height of the rock in Fig. 6–17 is $y_1 = h = 3.0\,m$, calculate the rock's velocity when it has fallen to 1.0 m above the ground.

**APPROACH** We apply the principle of conservation of mechanical energy, Eq. 6–13, with only gravity acting on the rock. We choose the ground as our reference level ($y = 0$).

**SOLUTION** At the moment of release (point 1) the rock's position is $y_1 = 3.0\,m$ and it is at rest: $v_1 = 0$. We want to find $v_2$ when the rock is at position $y_2 = 1.0\,m$. Equation 6–13 gives

$$\tfrac{1}{2}mv_1^2 + mgy_1 = \tfrac{1}{2}mv_2^2 + mgy_2,$$

The $m$'s cancel out and $v_1 = 0$, so

$$gy_1 = \tfrac{1}{2}v_2^2 + gy_2.$$

Solving for $v_2$ we find

$$v_2 = \sqrt{2g(y_1 - y_2)} = \sqrt{2(9.8\,m/s^2)[(3.0\,m) - (1.0\,m)]} = 6.3\,m/s.$$

The rock's velocity 1.0 m above the ground is 6.3 m/s downward.

**NOTE** The velocity of the rock is independent of the rock's mass.

---

**FIGURE 6–18** A roller-coaster car moving without friction illustrates the conservation of mechanical energy.

Equation 6–13 can be applied to any object moving without friction under the action of gravity. For example, Fig. 6–18 shows a roller-coaster car starting from rest at the top of a hill and coasting without friction to the bottom and up the hill on the other side. True, there is another force besides gravity acting on the car, the normal force exerted by the tracks. But the normal force acts perpendicular to the direction of motion at each point and so does zero work. We ignore rotational motion of the car's wheels and treat the car as a particle undergoing simple translation. Initially, the car has only potential energy. As it coasts down the hill, it loses potential energy and gains in kinetic energy, but the sum of the two remains constant. At the bottom of the hill it has its maximum kinetic energy, and as it climbs up the other side the kinetic energy changes back to potential energy. When the car comes to rest again at the same height from which it started, all of its energy will be potential energy. Given that the gravitational potential energy is proportional to the vertical height, energy conservation tells us that (in the absence of friction) the car comes to rest at a height equal to its original height. If the two hills are the same height, the car will just barely reach the top of the second hill when it stops. If the second hill is lower than the first, not all of the car's kinetic energy will be transformed to potential energy and the car can continue over the top and down the other side. If the second hill is higher, the car will reach a maximum height on it equal to its original height on the first hill. This is true (in the absence of friction) no matter how steep the hill is, since potential energy depends only on the vertical height (Eq. 6–6).

---

**EXAMPLE 6–8** **Roller-coaster car speed using energy conservation.** Assuming the height of the hill in Fig. 6–18 is 40 m, and the roller-coaster car starts from rest at the top, calculate (a) the speed of the roller-coaster car at the bottom of the hill, and (b) at what height it will have half this speed. Take $y = 0$ at the bottom of the hill.

**APPROACH** We use conservation of mechanical energy. We choose point 1 to be where the car starts from rest $(v_1 = 0)$ at the top of the hill $(y_1 = 40\,\text{m})$. In part (a), point 2 is the bottom of the hill, which we choose as our reference level, so $y_2 = 0$. In part (b) we let $y_2$ be the unknown.

**SOLUTION** (a) We use Eq. 6–13 with $v_1 = 0$ and $y_2 = 0$, which gives

$$mgy_1 = \tfrac{1}{2}mv_2^2$$

or

$$v_2 = \sqrt{2gy_1}$$
$$= \sqrt{2(9.8\,\text{m/s}^2)(40\,\text{m})} = 28\,\text{m/s}.$$

(b) Now $y_2$ will be an unknown. We again use conservation of energy,

$$\tfrac{1}{2}mv_1^2 + mgy_1 = \tfrac{1}{2}mv_2^2 + mgy_2,$$

but now $v_2 = \tfrac{1}{2}(28\,\text{m/s}) = 14\,\text{m/s}$ and $v_1 = 0$. Solving for the unknown $y_2$ gives

$$y_2 = y_1 - \frac{v_2^2}{2g} = 40\,\text{m} - \frac{(14\,\text{m/s})^2}{2(9.8\,\text{m/s}^2)} = 30\,\text{m}.$$

That is, the car has a speed of 14 m/s when it is 30 *vertical* meters above the lowest point, both when descending the left-hand hill and when ascending the right-hand hill.

---

The mathematics of the roller-coaster Example 6–8 is almost the same as in Example 6–7. But there is an important difference between them. In Example 6–7 the motion is all vertical and could have been solved using force, acceleration, and the kinematic equations (Eqs. 2–11). For the roller coaster, where the motion is not vertical, we could *not* have used Eqs. 2–11 because $a$ is not constant on the curved track of Example 6–8. But energy conservation readily gives us the answer.

**CONCEPTUAL EXAMPLE 6–9** **Speeds on two water slides.** Two water slides at a pool are shaped differently, but start at the same height h (Fig. 6–19). Two riders start from rest at the same time on different slides. (a) Which rider, Paul or Corinne, is traveling faster at the bottom? (b) Which rider makes it to the bottom first? Ignore friction and assume both slides have the same path length.

**RESPONSE** (a) Each rider's initial potential energy mgh gets transformed to kinetic energy, so the speed v at the bottom is obtained from $\frac{1}{2}mv^2 = mgh$. The mass cancels and so the speed will be the same, regardless of the mass of the rider. Since they descend the same vertical height, they will finish with the same speed. (b) Note that Corinne is consistently at a lower elevation than Paul at any instant, until the end. This means she has converted her potential energy to kinetic energy earlier. Consequently, she is traveling faster than Paul for the whole trip, and because the distance is the same, Corinne gets to the bottom first.

**FIGURE 6–19** Example 6–9.

**FIGURE 6–20** Transformation of energy during a pole vault:
$KE \rightarrow PE_{el} \rightarrow PE_{G}$.

There are many interesting examples of the conservation of energy in sports, such as the pole vault illustrated in Fig. 6–20. We often have to make approximations, but the sequence of events in broad outline for the pole vault is as follows. The initial kinetic energy of the running athlete is transformed into elastic potential energy of the bending pole and, as the athlete leaves the ground, into gravitational potential energy. When the vaulter reaches the top and the pole has straightened out again, the energy has all been transformed into gravitational potential energy (if we ignore the vaulter's low horizontal speed over the bar). The pole does not supply any energy, but it acts as a device to *store* energy and thus aid in the transformation of kinetic energy into gravitational potential energy, which is the net result. The energy required to pass over the bar depends on how high the center of mass (CM) of the vaulter must be raised. By bending their bodies, pole vaulters keep their CM so low that it can actually pass slightly beneath the bar (Fig. 6–21), thus enabling them to cross over a higher bar than would otherwise be possible. (Center of mass is covered in Chapter 7.)

As another example of the conservation of mechanical energy, let us consider an object of mass m connected to a compressed horizontal spring (Fig. 6–13b) whose own mass can be neglected and whose spring stiffness constant is k. When the spring is released, the mass m has speed v at any moment. The potential energy of the system (object plus spring) is $\frac{1}{2}kx^2$, where x is the displacement of the spring from its unstretched length (Eq. 6–9). If neither friction nor any other force is acting, conservation of mechanical energy tells us that

$$\tfrac{1}{2}mv_1^2 + \tfrac{1}{2}kx_1^2 = \tfrac{1}{2}mv_2^2 + \tfrac{1}{2}kx_2^2, \qquad \text{[elastic PE only]} \quad \text{(6–14)}$$

where the subscripts 1 and 2 refer to the velocity and displacement at two different moments.

**FIGURE 6–21** By bending her body, a pole vaulter can keep her center of mass so low that it may even pass below the bar.

(a) $E = \frac{1}{2}kx_1^2$

(b) $E = \frac{1}{2}mv_2^2$

**FIGURE 6–22** Example 6–10.
(a) A dart is pushed against a
spring, compressing it 6.0 cm. The
dart is then released, and in (b) it
leaves the spring at velocity $v_2$.

**FIGURE 6–23** Example 6–11.
A falling ball compresses a spring.

**EXAMPLE 6–10** **Toy dart gun.** A dart of mass 0.100 kg is pressed against the spring of a toy dart gun as shown in Fig. 6–22a. The spring, with spring stiffness constant $k = 250$ N/m and ignorable mass, is compressed 6.0 cm and released. If the dart detaches from the spring when the spring reaches its natural length $(x = 0)$, what speed does the dart acquire?

**APPROACH** The dart is initially at rest (point 1), so $KE_1 = 0$. We ignore friction and use conservation of mechanical energy; the only potential energy is elastic.

**SOLUTION** We use Eq. 6–14 with point 1 being at the maximum compression of the spring, so $v_1 = 0$ (dart not yet released) and $x_1 = -0.060$ m. Point 2 we choose to be the instant the dart flies off the end of the spring (Fig. 6–22b), so $x_2 = 0$ and we want to find $v_2$. Thus Eq. 6–14 can be written

$$0 + \tfrac{1}{2}kx_1^2 = \tfrac{1}{2}mv_2^2 + 0.$$

Then

$$v_2^2 = \frac{kx_1^2}{m} = \frac{(250 \text{ N/m})(-0.060 \text{ m})^2}{(0.100 \text{ kg})} = 9.0 \text{ m}^2/\text{s}^2,$$

and $v_2 = \sqrt{v_2^2} = 3.0$ m/s.

**EXAMPLE 6–11** **Two kinds of potential energy.** A ball of mass $m = 2.60$ kg, starting from rest, falls a vertical distance $h = 55.0$ cm before striking a vertical coiled spring, which it compresses an amount $Y = 15.0$ cm (Fig. 6–23). Determine the spring stiffness constant $k$ of the spring. Assume the spring has negligible mass, and ignore air resistance. Measure all distances from the point where the ball first touches the uncompressed spring $(y = 0$ at this point).

**APPROACH** The forces acting on the ball are the gravitational pull of the Earth and the elastic force exerted by the spring. Both forces are conservative, so we can use conservation of mechanical energy, including both types of potential energy. We must be careful, however: gravity acts throughout the fall (Fig. 6–23), whereas the elastic force does not act until the ball touches the spring (Fig. 6–23b). We choose $y$ positive upward, and $y = 0$ at the end of the spring in its natural (uncompressed) state.

**SOLUTION** We divide this solution into two parts. (An alternate solution follows.)
*Part 1:* Let us first consider the energy changes as the ball falls from a height $y_1 = h = 0.550$ m, Fig. 6–23a, to $y_2 = 0$, just as it touches the spring, Fig. 6–23b. Our system is the ball acted on by gravity plus the spring (which up to this point doesn't do anything). Thus

$$\tfrac{1}{2}mv_1^2 + mgy_1 = \tfrac{1}{2}mv_2^2 + mgy_2$$
$$0 + mgh = \tfrac{1}{2}mv_2^2 + 0.$$

We solve for $v_2 = \sqrt{2gh} = \sqrt{2(9.80 \text{ m/s}^2)(0.550 \text{ m})} = 3.283$ m/s $\approx 3.28$ m/s. This is the speed of the ball just as it touches the top of the spring, Fig. 6–23b.
*Part 2:* As the ball compresses the spring, Figs. 6–23b to c, there are two conservative forces on the ball—gravity and the spring force. So our conservation of energy equation is

$$E_2 \text{ (ball touches spring)} = E_3 \text{ (spring compressed)}$$
$$\tfrac{1}{2}mv_2^2 + mgy_2 + \tfrac{1}{2}ky_2^2 = \tfrac{1}{2}mv_3^2 + mgy_3 + \tfrac{1}{2}ky_3^2.$$

Substituting $y_2 = 0$, $v_2 = 3.283$ m/s, $v_3 = 0$ (the ball comes to rest for an instant), and $y_3 = -Y = -0.150$ m, we have

$$\tfrac{1}{2}mv_2^2 + 0 + 0 = 0 - mgY + \tfrac{1}{2}k(-Y)^2.$$

We know $m$, $v_2$, and $Y$, so we can solve for $k$:

$$k = \frac{2}{Y^2}\left[\tfrac{1}{2}mv_2^2 + mgY\right] = \frac{m}{Y^2}\left[v_2^2 + 2gY\right]$$

$$= \frac{(2.60 \text{ kg})}{(0.150 \text{ m})^2}\left[(3.283 \text{ m/s})^2 + 2(9.80 \text{ m/s}^2)(0.150 \text{ m})\right] = 1590 \text{ N/m}.$$

**Alternate Solution** Instead of dividing the solution into two parts, we can do it all at once. After all, we get to choose what two points are used on the left and right of the energy equation. Let us write the energy equation for points 1 and 3 in Fig. 6–23. Point 1 is the initial point just before the ball starts to fall (Fig. 6–23a), so $v_1 = 0$, and $y_1 = h = 0.550$ m. Point 3 is when the spring is fully compressed (Fig. 6–23c), so $v_3 = 0$, $y_3 = -Y = -0.150$ m. The forces on the ball in this process are gravity and (at least part of the time) the spring. So conservation of energy tells us

$$\tfrac{1}{2}mv_1^2 + mgy_1 + \tfrac{1}{2}k(0)^2 = \tfrac{1}{2}mv_3^2 + mgy_3 + \tfrac{1}{2}ky_3^2$$
$$0 + mgh + 0 = 0 - mgY + \tfrac{1}{2}kY^2$$

where we have set $y = 0$ for the spring at point 1 because it is not acting and is not compressed or stretched. We solve for $k$:

$$k = \frac{2mg(h+Y)}{Y^2} = \frac{2(2.60\text{ kg})(9.80\text{ m/s}^2)(0.550\text{ m} + 0.150\text{ m})}{(0.150\text{ m})^2} = 1590\text{ N/m}$$

just as in our first method of solution.

**PROBLEM SOLVING**
*Quicker Solution*

# 6–8 Other Forms of Energy and Energy Transformations; The Law of Conservation of Energy

Besides the kinetic energy and potential energy of mechanical systems, other forms of energy can be defined as well. These include electric energy, nuclear energy, thermal energy, and the chemical energy stored in food and fuels. These other forms of energy are considered to be kinetic or potential energy at the atomic or molecular level. For example, according to atomic theory, thermal energy is the kinetic energy of rapidly moving molecules—when an object is heated, the molecules that make up the object move faster. On the other hand, the energy stored in food or in a fuel such as gasoline is regarded as potential energy stored by virtue of the relative positions of the atoms within a molecule due to electric forces between the atoms (chemical bonds). The energy in chemical bonds can be released through chemical reactions. This is analogous to a compressed spring which, when released, can do work. Electric, magnetic, and nuclear energies also can be considered examples of kinetic and potential (or stored) energies. We will deal with these other forms of energy in later Chapters.

Energy can be transformed from one form to another. For example, a rock held high in the air has potential energy; as it falls, it loses potential energy and gains in kinetic energy. Potential energy is being transformed into kinetic energy.

Often the transformation of energy involves a transfer of energy from one object to another. The potential energy stored in the spring of Fig. 6–13b is transformed into the kinetic energy of the ball, Fig. 6–13c. Water at the top of a waterfall (Fig. 6–24) or a dam has potential energy, which is transformed into kinetic energy as the water falls. At the base of a dam, the kinetic energy of the water can be transferred to turbine blades and further transformed into electric energy, as discussed later. The potential energy stored in a bent bow can be transformed into kinetic energy of the arrow (Fig. 6–25).

In each of these examples, the transfer of energy is accompanied by the performance of work. The spring of Fig. 6–13 does work on the ball. Water does work on turbine blades. A bow does work on an arrow. This observation gives us a further insight into the relation between work and energy: *work is done when energy is transferred from one object to another.*[†]

[†] If the objects are at different temperatures, heat can flow between them instead, or in addition. See Chapters 14 and 15.

**FIGURE 6–24** Gravitational potential energy of water at the top of Yosemite Falls gets transformed into kinetic energy as the water falls. (Some of the energy is transformed into heat by air resistance, and some into sound.)

**FIGURE 6–25** Potential energy of a bent bow about to be transformed into kinetic energy of an arrow.

One of the great results of physics is that whenever energy is transferred or transformed, it is found that no energy is gained or lost in the process.

This is the **law of conservation of energy**, one of the most important principles in physics; it can be stated as:

**The total energy is neither increased nor decreased in any process. Energy can be transformed from one form to another, and transferred from one object to another, but the total amount remains constant.**

We have already discussed the conservation of energy for mechanical systems involving conservative forces, and we saw how it could be derived from Newton's laws and thus is equivalent to them. But in its full generality, the validity of the law of conservation of energy, encompassing all forms of energy including those associated with nonconservative forces like friction, rests on experimental observation. Even though Newton's laws are found to fail in the submicroscopic world of the atom, the law of conservation of energy has been found to hold in every experimental situation so far tested.

# 6–9 Energy Conservation with Dissipative Forces: Solving Problems

In our applications of energy conservation in Section 6–7, we neglected friction and other nonconservative forces. But in many situations they cannot be ignored. In a real situation, the roller-coaster car in Fig. 6–18, for example, will not in fact reach the same height on the second hill as it had on the first hill because of friction. In this, and in other natural processes, the mechanical energy (sum of the kinetic and potential energies) does not remain constant but decreases. Because frictional forces reduce the mechanical energy (but *not* the total energy), they are called **dissipative forces**. Historically, the presence of dissipative forces hindered the formulation of a comprehensive conservation of energy law until well into the nineteenth century. It was only then that heat, which is always produced when there is friction (try rubbing your hands together), was interpreted in terms of energy. Quantitative studies by nineteenth-century scientists (discussed in Chapters 14 and 15) demonstrated that if heat is considered as a transfer of energy (thermal energy), then the total energy is conserved in any process. For example, if the roller-coaster car in Fig. 6–18 is subject to frictional forces, then the initial total energy of the car will be equal to the kinetic plus potential energy of the car at any subsequent point along its path plus the amount of thermal energy produced in the process (equal to the work done by friction).

Let us recall the general form of the work-energy principle, Eq. 6–10:

$$W_{NC} = \Delta KE + \Delta PE,$$

where $W_{NC}$ is the work done by nonconservative forces such as friction. Consider an object, such as a roller-coaster car, as a particle moving under gravity with nonconservative forces like friction acting on it. When the object moves from some point 1 to another point 2, then

$$W_{NC} = KE_2 - KE_1 + PE_2 - PE_1.$$

We can rewrite this as

$$KE_1 + PE_1 + W_{NC} = KE_2 + PE_2. \tag{6–15}$$

For the case of friction, $W_{NC} = -F_{fr}\,d$, where $d$ is the distance over which the friction (assumed constant) acts as the object moves from point 1 to point 2. ($\vec{F}$ and $\vec{d}$ are in opposite directions, hence the minus sign from $\cos 180° = -1$ in Eq. 6–1.)

With $KE = \frac{1}{2}mv^2$ and $PE = mgy$, Eq. 6–15 with $W_{NC} = -F_{fr}d$ becomes

$$\tfrac{1}{2}mv_1^2 + mgy_1 - F_{fr}d = \tfrac{1}{2}mv_2^2 + mgy_2. \quad \begin{bmatrix} \text{gravity and} \\ \text{friction acting} \end{bmatrix} \quad \textbf{(6–16a)}$$

That is, the initial mechanical energy is reduced by the amount $F_{fr}d$. We could also write this equation as

$$\text{or} \quad \begin{aligned} \tfrac{1}{2}mv_1^2 + mgy_1 &= \tfrac{1}{2}mv_2^2 + mgy_2 + F_{fr}d \\ KE_1 + PE_1 &= KE_2 + PE_2 + F_{fr}d, \end{aligned} \quad \begin{bmatrix} \text{gravity and} \\ \text{friction} \\ \text{acting} \end{bmatrix} \quad \textbf{(6–16b)}$$

and state equally well that the initial mechanical energy of the car (point 1) equals the (reduced) final mechanical energy of the car plus the energy transformed by friction into thermal energy.

Equations 6–16 can be seen to be Eq. 6–13 modified to include nonconservative forces such as friction. As such, they are statements of conservation of energy. When other forms of energy are involved, such as chemical or electrical energy, the total amount of energy is always found to be conserved. Hence the law of conservation of energy is believed to be universally valid.

**EXERCISE F** Return to the Chapter-Opening Question, page 138, and answer it again now. Try to explain why you may have answered differently the first time.

## Work-Energy versus Energy Conservation

The law of conservation of energy is more general and more powerful than the work-energy principle. Indeed, the work-energy principle should not be viewed as a statement of conservation of energy. It is nonetheless useful for mechanical problems; and whether you use it, or use the more powerful conservation of energy, can depend on your *choice of the system* under study. If you choose as your system a particle or rigid object on which external forces do work, then you can use the work-energy principle: the work done by the external forces on your object equals the change in its kinetic energy.

On the other hand, if you choose a system on which no external forces do work, then you need to apply conservation of energy to that system directly.

Consider, for example, a spring connected to a block on a frictionless table (Fig. 6–26). If you choose the block as your system, then the work done on the block by the spring equals the change in kinetic energy of the block: the work-energy principle. (Energy conservation does not apply to this system—the block's energy changes.) If instead you choose the block plus the spring as your system, no external forces do work (since the spring is part of the chosen system). To this system you need to apply conservation of energy: if you compress the spring and then release it, the spring still exerts a force[†] on the block, but the subsequent motion can be discussed in terms of kinetic energy ($\frac{1}{2}mv^2$) plus potential energy ($\frac{1}{2}kx^2$), whose total remains constant.

You may also wonder sometimes whether to approach a problem using work and energy, or instead to use Newton's laws. As a rough guideline, if the force(s) involved are constant, either approach may succeed. If the forces are not constant, and/or the path is not simple, energy may be the better approach because it is a scalar.

Problem solving is not a process that can be done by simply following a set of rules. The Problem Solving Strategy on the next page, like all others, is thus *not* a prescription, but is a summary to help you get started solving problems involving energy.

**PROBLEM SOLVING**
*Choosing the system*

**FIGURE 6–26** A spring connected to a block on a frictionless table. If you choose your system to be the block plus spring, then

$$E = \tfrac{1}{2}mv^2 + \tfrac{1}{2}kx^2$$

is conserved.

**PROBLEM SOLVING**
*Use energy, or Newton's laws?*

[†]The force the spring exerts on the block, and the force the block exerts back on the spring, are not "external" forces—they are within the system.

## Conservation of Energy

1. **Draw a picture** of the physical situation.

2. Determine **the system** for which you will apply energy conservation: the object or objects and the forces acting.

3. Ask yourself what quantity you are looking for, and **choose initial** (point 1) **and final** (point 2) **positions**.

4. If the object under investigation changes its height during the problem, then **choose a reference frame** with a convenient $y = 0$ level for gravitational potential energy; the lowest point in the situation is often a good choice.

   If springs are involved, choose the unstretched spring position to be $x$ (or $y$) $= 0$.

5. **Is mechanical energy conserved?** If no friction or other nonconservative forces act, then conservation of mechanical energy holds:
$$\text{KE}_1 + \text{PE}_1 = \text{KE}_2 + \text{PE}_2. \qquad \textbf{(6–12a)}$$

6. **Apply conservation of energy.** If friction (or other nonconservative forces) are present, then an additional term $(W_{NC})$ will be needed:
$$W_{NC} = \Delta\text{KE} + \Delta\text{PE}. \qquad \textbf{(6–10)}$$
For a constant friction force acting over a distance $d$
$$\text{KE}_1 + \text{PE}_1 = \text{KE}_2 + \text{PE}_2 + F_{fr}d. \qquad \textbf{(6–16b)}$$
For other nonconservative forces use your intuition for the sign of $W_{NC}$: is the total mechanical energy increased or decreased in the process?

7. Use the equation(s) you develop to **solve** for the unknown quantity.

---

**FIGURE 6–27** Example 6–12. Because of friction, a roller-coaster car does not reach the original height on the second hill. (Not to scale.)

**EXAMPLE 6–12 | ESTIMATE | Friction on the roller-coaster car.** The roller-coaster car in Example 6–8 reaches a vertical height of only 25 m on the second hill, where it slows to a momentary stop, Fig. 6–27. It traveled a total distance of 400 m. Determine the thermal energy produced and estimate the average friction force (assume it is roughly constant) on the car, whose mass is 1000 kg.

**APPROACH** We explicitly follow the Problem Solving Strategy above.

**SOLUTION**

1. **Draw a picture.** See Fig. 6–27.

2. **The system.** The system is the roller-coaster car and the Earth (which exerts the gravitational force). The forces acting on the car are gravity and friction. (The normal force also acts on the car, but does no work, so it does not affect the energy.) Gravity is accounted for as potential energy, and friction as a term $F_{fr}d$.

3. **Choose initial and final positions.** We take point 1 to be the instant when the car started coasting (at the top of the first hill), and point 2 to be the instant it stopped at a height of 25 m up the second hill.

4. **Choose a reference frame.** We choose the lowest point in the motion to be $y = 0$ for the gravitational potential energy.

5. **Is mechanical energy conserved?** No. Friction is present.

6. **Apply conservation of energy.** There is friction acting on the car, so we use conservation of energy in the form of Eq. 6–16b, with $v_1 = 0$, $y_1 = 40\,\text{m}$, $v_2 = 0$, $y_2 = 25\,\text{m}$, and $d = 400\,\text{m}$. Thus
$$0 + (1000\,\text{kg})(9.8\,\text{m/s}^2)(40\,\text{m}) = 0 + (1000\,\text{kg})(9.8\,\text{m/s}^2)(25\,\text{m}) + F_{fr}d.$$

7. **Solve.** We solve the above equation for $F_{fr}d$, the energy dissipated to thermal energy:
$$F_{fr}d = mg\,\Delta h = (1000\,\text{kg})(9.8\,\text{m/s}^2)(40\,\text{m} - 25\,\text{m}) = 147{,}000\,\text{J}.$$
The friction force, which acts over a distance of 400 m, averages out to be
$$F_{fr} = (1.47 \times 10^5\,\text{J})/400\,\text{m} = 370\,\text{N}.$$

**NOTE** This result is only a rough average: the friction force at various points depends on the normal force, which varies with slope.

# 6–10 Power

**Power** is defined as the *rate at which work is done*. Average power equals the work done divided by the time to do it. Power can also be defined as the *rate at which energy is transformed*. Thus

$$\overline{P} = \text{average power} = \frac{\text{work}}{\text{time}} = \frac{\text{energy transformed}}{\text{time}}. \quad (6\text{–}17)$$

The power rating of an engine refers to how much chemical or electrical energy can be transformed into mechanical energy per unit time. In SI units, power is measured in joules per second, and this unit is given a special name, the **watt** (W): $1 \text{ W} = 1 \text{ J/s}$. We are most familiar with the watt for electrical devices, such as the rate at which an electric lightbulb or heater changes electric energy into light or thermal energy. But the watt is used for other types of energy transformations as well.

In the British system, the unit of power is the foot-pound per second ($\text{ft} \cdot \text{lb/s}$).[†] For practical purposes, a larger unit is often used, the **horsepower**. One horsepower[†] (hp) is defined as 550 $\text{ft} \cdot \text{lb/s}$, which equals 746 W. An engine's power is usually specified in hp or in kW $\left(1 \text{ kW} \approx 1\frac{1}{3}\text{hp}\right)$.[‡]

To see the distinction between energy and power, consider the following example. A person is limited in the work he or she can do, not only by the total energy required, but also by how fast this energy is transformed: that is, by power. For example, a person may be able to walk a long distance or climb many flights of stairs before having to stop because so much energy has been expended. On the other hand, a person who runs very quickly up stairs may feel exhausted after only a flight or two. He or she is limited in this case by power, the rate at which his or her body can transform chemical energy into mechanical energy.

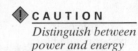

**CAUTION**
*Distinguish between power and energy*

---

**EXAMPLE 6–13** **Stair-climbing power.** A 60-kg jogger runs up a long flight of stairs in 4.0 s (Fig. 6–28). The vertical height of the stairs is 4.5 m. (*a*) Estimate the jogger's power output in watts and horsepower. (*b*) How much energy did this require?

**APPROACH** The work done by the jogger is against gravity, and equals $W = mgy$. To get her average power output, we divide $W$ by the time it took.

**SOLUTION** (*a*) The average power output was

$$\overline{P} = \frac{W}{t} = \frac{mgy}{t} = \frac{(60 \text{ kg})(9.8 \text{ m/s}^2)(4.5 \text{ m})}{4.0 \text{ s}} = 660 \text{ W}.$$

Since there are 746 W in 1 hp, the jogger is doing work at a rate of just under 1 hp. A human cannot do work at this rate for very long.

(*b*) The energy required is $E = \overline{P}t = (660 \text{ J/s})(4.0 \text{ s}) = 2600 \text{ J}$. This result equals $W = mgy$.

**NOTE** The person had to transform more energy than this 2600 J. The total energy transformed by a person or an engine always includes some thermal energy (recall how hot you get running up stairs).

---

Automobiles do work to overcome the force of friction and air resistance, to climb hills, and to accelerate. A car is limited by the rate at which it can do work, which is why automobile engines are rated in horsepower or kilowatts.

**FIGURE 6–28** Example 6–13.

**PHYSICS APPLIED**
*Power needs of a car*

---

[†] The unit was chosen by James Watt (1736–1819), who needed a way to specify the power of his newly developed steam engines. He found by experiment that a good horse can work all day at an average rate of about 360 $\text{ft} \cdot \text{lb/s}$. So as not to be accused of exaggeration in the sale of his steam engines, he multiplied this by $1\frac{1}{2}$ when he defined the hp.

[‡] $1 \text{ kW} = (1000 \text{ W})/(746 \text{ W/hp}) \approx 1\frac{1}{3}\text{hp}$.

A car needs power most when climbing hills and when accelerating. In the next Example, we will calculate how much power is needed in these situations for a car of reasonable size. Even when a car travels on a level road at constant speed, it needs some power just to do work to overcome the retarding forces of internal friction and air resistance. These forces depend on the conditions and speed of the car, but are typically in the range 400 N to 1000 N.

It is often convenient to write power in terms of the net force $F$ applied to an object and its speed $v$. This is readily done because $\overline{P} = W/t$ and $W = Fd$, where $d$ is the distance traveled. Then

$$\overline{P} = \frac{W}{t} = \frac{Fd}{t} = F\overline{v}, \tag{6-18}$$

where $\overline{v} = d/t$ is the average speed of the object.

**FIGURE 6–29** Example 6–14. Calculation of power needed for a car to climb a hill.

**EXAMPLE 6–14** **Power needs of a car.** Calculate the power required of a 1400-kg car under the following circumstances: (*a*) the car climbs a 10° hill (a fairly steep hill) at a steady 80 km/h; and (*b*) the car accelerates along a level road from 90 to 110 km/h in 6.0 s to pass another car. Assume the average retarding force on the car is $F_R = 700$ N throughout. See Fig. 6–29.

**APPROACH** First we must be careful not to confuse $\vec{F}_R$, which is due to air resistance and friction that retards the motion, with the force $\vec{F}$ needed to accelerate the car, which is the frictional force exerted by the road on the tires—the reaction to the motor-driven tires pushing against the road. We must determine the magnitude of the force $F$ before calculating the power.

**SOLUTION** (*a*) To move at a steady speed up the hill, the car must, by Newton's second law, exert a force $F$ equal to the sum of the retarding force, 700 N, and the component of gravity parallel to the hill, $mg \sin 10°$, Fig. 6–29. Thus

$$F = 700\,\text{N} + mg \sin 10°$$
$$= 700\,\text{N} + (1400\,\text{kg})(9.80\,\text{m/s}^2)(0.174) = 3100\,\text{N}.$$

Since $\overline{v} = 80$ km/h $= 22$ m/s[†] and is parallel to $\vec{F}$, then (Eq. 6–18) the power is

$$\overline{P} = F\overline{v} = (3100\,\text{N})(22\,\text{m/s}) = 6.8 \times 10^4\,\text{W} = 68\,\text{kW} = 91\,\text{hp}.$$

(*b*) The car accelerates from 25.0 m/s to 30.6 m/s (90 to 110 km/h) on the flat. The car must exert a force that overcomes the 700-N retarding force plus that required to give it the acceleration

$$\overline{a}_x = \frac{(30.6\,\text{m/s} - 25.0\,\text{m/s})}{6.0\,\text{s}} = 0.93\,\text{m/s}^2.$$

We apply Newton's second law with $x$ being the horizontal direction of motion (no component of gravity):

$$ma_x = \Sigma F_x = F - F_R.$$

We solve for the force required, $F$:

$$F = ma_x + F_R$$
$$= (1400\,\text{kg})(0.93\,\text{m/s}^2) + 700\,\text{N} = 1300\,\text{N} + 700\,\text{N} = 2000\,\text{N}.$$

Since $\overline{P} = F\overline{v}$, the required power increases with speed and the motor must be able to provide a maximum power output in this case of

$$\overline{P} = (2000\,\text{N})(30.6\,\text{m/s}) = 6.1 \times 10^4\,\text{W} = 61\,\text{kW} = 82\,\text{hp}.$$

**NOTE** Even taking into account the fact that only 60 to 80% of the engine's power output reaches the wheels, it is clear from these calculations that an engine of 75 to 100 kW (100 to 130 hp) is adequate from a practical point of view.

[†]Recall 1 km/h = 1000 m/3600 s = 0.278 m/s.

We mentioned in Example 6–14 that only part of the energy output of a car engine reaches the wheels. Not only is some energy wasted in getting from the engine to the wheels, in the engine itself most of the input energy (from the burning of gasoline or other fuel) does not do useful work. An important characteristic of all engines is their overall **efficiency** $e$, defined as the ratio of the useful power output of the engine, $P_{out}$, to the power input, $P_{in}$ (provided by burning of gasoline, for example):

$$e = \frac{P_{out}}{P_{in}}.$$

The efficiency is always less than 1.0 because no engine can create energy, and no engine can even transform energy from one form to another without some energy going to friction, thermal energy, and other nonuseful forms of energy. For example, an automobile engine converts chemical energy released in the burning of gasoline into mechanical energy that moves the pistons and eventually the wheels. But nearly 85% of the input energy is "wasted" as thermal energy that goes into the cooling system or out the exhaust pipe, plus friction in the moving parts. Thus car engines are roughly only about 15% efficient. We will discuss efficiency in more detail in Chapter 15.

# Summary

**Work** is done on an object by a force when the object moves through a distance $d$. If the direction of a constant force $\vec{\mathbf{F}}$ makes an angle $\theta$ with the direction of motion, the work done by this force is

$$W = Fd \cos\theta. \qquad (6\text{--}1)$$

**Energy** can be defined as the ability to do work. In SI units, work and energy are measured in **joules** ($1\,\text{J} = 1\,\text{N}\cdot\text{m}$).

**Kinetic energy** (KE) is energy of motion. An object of mass $m$ and speed $v$ has **translational kinetic energy**

$$\text{KE} = \tfrac{1}{2}mv^2. \qquad (6\text{--}3)$$

The **work-energy principle** states that the *net* work done on an object (by the *net* force) equals the change in kinetic energy of that object:

$$W_{net} = \Delta\text{KE} = \tfrac{1}{2}mv_2^2 - \tfrac{1}{2}mv_1^2. \qquad (6\text{--}4)$$

**Potential energy** (PE) is energy associated with forces that depend on the position or configuration of objects. Gravitational potential energy is

$$\text{PE}_G = mgy, \qquad (6\text{--}6)$$

where $y$ is the height of the object of mass $m$ above an arbitrary reference point. Elastic potential energy is given by

$$\text{PE}_{el} = \tfrac{1}{2}kx^2 \qquad (6\text{--}9)$$

for a stretched or compressed spring, where $x$ is the displacement from the unstretched position and $k$ is the **spring stiffness constant**. Other potential energies include chemical, electrical, and nuclear energy. The *change in potential energy* when an object changes position *is equal to the external work* needed to take the object from one position to the other.

Potential energy is associated only with **conservative forces**, for which the work done by the force in moving an object from one position to another depends only on the two positions and not on the path taken. **Nonconservative forces** like friction are different—work done by them does depend on the path taken and potential energy cannot be defined for them.

The **law of conservation of energy** states that energy can be transformed from one type to another, but the total energy remains constant. It is valid even when friction is present, because the heat generated can be considered a form of energy transfer. When only *conservative forces* act, the **total mechanical energy** is conserved:

$$\text{KE} + \text{PE} = \text{constant}. \qquad (6\text{--}12)$$

When nonconservative forces such as friction act, then

$$W_{NC} = \Delta\text{KE} + \Delta\text{PE}, \qquad (6\text{--}10, 6\text{--}15)$$

where $W_{NC}$ is the work done by nonconservative forces.

**Power** is defined as the rate at which work is done, or the rate at which energy is transformed. The SI unit of power is the **watt** ($1\,\text{W} = 1\,\text{J/s}$).

# Questions

1. In what ways is the word "work" as used in everyday language the same as it is defined in physics? In what ways is it different? Give examples of both.

2. Can a centripetal force ever do work on an object? Explain.

3. Why is it tiring to push hard against a solid wall even though you are doing no work?

4. Can the normal force on an object ever do work? Explain.

5. You have two springs that are identical except that spring 1 is stiffer than spring 2 ($k_1 > k_2$). On which spring is more work done: (*a*) if they are stretched using the same force; (*b*) if they are stretched the same distance?

6. If the speed of a particle triples, by what factor does its kinetic energy increase?

7. List some everyday forces that are not conservative, and explain why they aren't.

**8.** A hand exerts a constant horizontal force on a block that is free to slide on a frictionless surface (Fig. 6–30). The block starts from rest at point A, and by the time it has traveled a distance $d$ to point B it is traveling with speed $v_B$. When the block has traveled another distance $d$ to point C, will its speed be greater than, less than, or equal to $2v_B$? Explain your reasoning.

**FIGURE 6–30** Question 8.

**9.** You lift a heavy book from a table to a high shelf. List the forces on the book during this process, and state whether each is conservative or nonconservative.

**10.** A hill has a height $h$. A child on a sled (total mass $m$) slides down starting from rest at the top. Does the speed at the bottom depend on the angle of the hill if (*a*) it is icy and there is no friction, and (*b*) there is friction (deep snow)? Explain your answers.

**11.** Analyze the motion of a simple swinging pendulum in terms of energy, (*a*) ignoring friction, and (*b*) taking friction into account. Explain why a grandfather clock has to be wound up.

**12.** In Fig. 6–31, water balloons are tossed from the roof of a building, all with the same speed but with different launch angles. Which one has the highest speed when it hits the ground? Ignore air resistance. Explain your answer.

**FIGURE 6–31**
Question 12.

**13.** What happens to the gravitational potential energy when water at the top of a waterfall falls to the pool below?

**14.** Experienced hikers prefer to step over a fallen log in their path rather than stepping on top and stepping down on the other side. Explain.

**15.** The energy transformations in pole vaulting and archery are discussed in this Chapter. In a similar fashion, discuss the energy transformations related to: (*a*) hitting a golf ball; (*b*) serving a tennis ball; and (*c*) shooting a basket in basketball.

**16.** Describe precisely what is "wrong" physically in the famous Escher drawing shown in Fig. 6–32.

**FIGURE 6–32**
Question 16.

**17.** Two identical arrows, one with twice the speed of the other, are fired into a bale of hay. Assuming the hay exerts a constant "frictional" force on the arrows, the faster arrow will penetrate how much farther than the slower arrow? Explain.

**18.** A heavy ball is hung from the ceiling by a steel wire. The instructor pulls the ball back and stands against the wall with the ball against his chin. To avoid injury the instructor is supposed to release the ball without pushing it (Fig. 6–33). Why?

**FIGURE 6–33** Question 18.

**19.** Describe the energy transformations when a child hops around on a pogo stick (there is a spring inside).

**20.** Describe the energy transformations that take place when a skier starts skiing down a hill, but after a time is brought to rest by striking a snowdrift.

**21.** Suppose you lift a suitcase from the floor to a table. The work you do on the suitcase depends on which of the following: (*a*) whether you lift it straight up or along a more complicated path, (*b*) the time the lifting takes, (*c*) the height of the table, and (*d*) the weight of the suitcase?

**22.** Repeat Question 21 for the *power* needed instead of the work.

**23.** Why is it easier to climb a mountain via a zigzag trail rather than to climb straight up?

# MisConceptual Questions

1. You push very hard on a heavy desk, trying to move it. You do work on the desk:
   (a) whether or not it moves, as long as you are exerting a force.
   (b) only if it starts moving.
   (c) only if it doesn't move.
   (d) never—it does work on you.
   (e) None of the above.

2. A satellite in circular orbit around the Earth moves at constant speed. This orbit is maintained by the force of gravity between the Earth and the satellite, yet no work is done on the satellite. How is this possible?
   (a) No work is done if there is no contact between objects.
   (b) No work is done because there is no gravity in space.
   (c) No work is done if the direction of motion is perpendicular to the force.
   (d) No work is done if objects move in a circle.

3. When the speed of your car is doubled, by what factor does its kinetic energy increase?
   (a) $\sqrt{2}$.  (b) 2.  (c) 4.  (d) 8.

4. A car traveling at a velocity $v$ can stop in a minimum distance $d$. What would be the car's minimum stopping distance if it were traveling at a velocity of $2v$?
   (a) $d$.  (b) $\sqrt{2}\,d$.  (c) $2d$.  (d) $4d$.  (e) $8d$.

5. A bowling ball is dropped from a height $h$ onto the center of a trampoline, which launches the ball back up into the air. How high will the ball rise?
   (a) Significantly less than $h$.
   (b) More than $h$. The exact amount depends on the mass of the ball and the springiness of the trampoline.
   (c) No more than $h$—probably a little less.
   (d) Cannot tell without knowing the characteristics of the trampoline.

6. A ball is thrown straight up. At what point does the ball have the most energy? Ignore air resistance.
   (a) At the highest point of its path.
   (b) When it is first thrown.
   (c) Just before it hits the ground.
   (d) When the ball is halfway to the highest point of its path.
   (e) Everywhere; the energy of the ball is the same at all of these points.

7. A car accelerates from rest to 30 km/h. Later, on a highway it accelerates from 30 km/h to 60 km/h. Which takes more energy, going from 0 to 30, or from 30 to 60?
   (a) 0 to 30 km/h.
   (b) 30 to 60 km/h.
   (c) Both are the same.

8. Engines, including car engines, are rated in horsepower. What is horsepower?
   (a) The force needed to start the engine.
   (b) The force needed to keep the engine running at a steady rate.
   (c) The energy the engine needs to obtain from gasoline or some other source.
   (d) The rate at which the engine can do work.
   (e) The amount of work the engine can perform.

9. Two balls are thrown off a building with the same speed, one straight up and one at a 45° angle. Which statement is true if air resistance can be ignored?
   (a) Both hit the ground at the same time.
   (b) Both hit the ground with the same speed.
   (c) The one thrown at an angle hits the ground with a lower speed.
   (d) The one thrown at an angle hits the ground with a higher speed.
   (e) Both (a) and (b).

10. A skier starts from rest at the top of each of the hills shown in Fig. 6–34. On which hill will the skier have the highest speed at the bottom if we ignore friction: (a), (b), (c), (d), or (e) c and d equally?

    (a)        (b)        (c)        (d)

**FIGURE 6–34**
MisConceptual Questions 10 and 11.

11. Answer MisConceptual Question 10 assuming a small amount of friction.

12. A man pushes a block up an incline at a constant speed. As the block moves up the incline,
    (a) its kinetic energy and potential energy both increase.
    (b) its kinetic energy increases and its potential energy remains the same.
    (c) its potential energy increases and its kinetic energy remains the same.
    (d) its potential energy increases and its kinetic energy decreases by the same amount.

13. You push a heavy crate *down* a ramp at a constant velocity. Only four forces act on the crate. Which force does the greatest magnitude of work on the crate?
    (a) The force of friction.
    (b) The force of gravity.
    (c) The normal force.
    (d) The force of you pushing.
    (e) The net force.

14. A ball is thrown straight up. Neglecting air resistance, which statement is *not* true regarding the energy of the ball?
    (a) The potential energy decreases while the ball is going up.
    (b) The kinetic energy decreases while the ball is going up.
    (c) The sum of the kinetic energy and potential energy is constant.
    (d) The potential energy decreases when the ball is coming down.
    (e) The kinetic energy increases when the ball is coming down.

For assigned homework and other learning materials, go to the MasteringPhysics website.

# Problems

## 6–1 Work, Constant Force

**1.** (I) A 75.0-kg firefighter climbs a flight of stairs 28.0 m high. How much work does he do?

**2.** (I) The head of a hammer with a mass of 1.2 kg is allowed to fall onto a nail from a height of 0.50 m. What is the maximum amount of work it could do on the nail? Why do people not just "let it fall" but add their own force to the hammer as it falls?

**3.** (II) How much work did the movers do (horizontally) pushing a 46.0-kg crate 10.3 m across a rough floor without acceleration, if the effective coefficient of friction was 0.50?

**4.** (II) A 1200-N crate rests on the floor. How much work is required to move it at constant speed (a) 5.0 m along the floor against a friction force of 230 N, and (b) 5.0 m vertically?

**5.** (II) What is the minimum work needed to push a 950-kg car 710 m up along a 9.0° incline? Ignore friction.

**6.** (II) Estimate the work you do to mow a lawn 10 m by 20 m with a 50-cm-wide mower. Assume you push with a force of about 15 N.

**7.** (II) In a certain library the first shelf is 15.0 cm off the ground, and the remaining four shelves are each spaced 38.0 cm above the previous one. If the average book has a mass of 1.40 kg with a height of 22.0 cm, and an average shelf holds 28 books (standing vertically), how much work is required to fill all the shelves, assuming the books are all laying flat on the floor to start?

**8.** (II) A **lever** such as that shown in Fig. 6–35 can be used to lift objects we might not otherwise be able to lift. Show that the ratio of output force, $F_O$, to input force, $F_I$, is related to the lengths $\ell_I$ and $\ell_O$ from the pivot by $F_O/F_I = \ell_I/\ell_O$. Ignore friction and the mass of the lever, and assume the work output equals the work input.

**FIGURE 6–35**
A lever. Problem 8.

**9.** (II) A box of mass 4.0 kg is accelerated from rest by a force across a floor at a rate of 2.0 m/s² for 7.0 s. Find the net work done on the box.

**10.** (II) A 380-kg piano slides 2.9 m down a 25° incline and is kept from accelerating by a man who is pushing back on it *parallel to the incline* (Fig. 6–36). Determine: (a) the force exerted by the man, (b) the work done on the piano by the man, (c) the work done on the piano by the force of gravity, and (d) the net work done on the piano. Ignore friction.

**FIGURE 6–36**
Problem 10.

**11.** (II) Recall from Chapter 4, Example 4–14, that you can use a pulley and ropes to decrease the force needed to raise a heavy load (see Fig. 6–37). But for every meter the load is raised, how much rope must be pulled up? Account for this, using energy concepts.

**FIGURE 6–37**
Problem 11.

**12.** (III) A grocery cart with mass of 16 kg is being pushed at constant speed up a 12° ramp by a force $F_P$ which acts at an angle of 17° below the horizontal. Find the work done by each of the forces $(m\vec{g}, \vec{F}_N, \vec{F}_P)$ on the cart if the ramp is 7.5 m long.

## *6–2 Work, Varying Force

**\*13.** (II) The force on a particle, acting along the x axis, varies as shown in Fig. 6–38. Determine the work done by this force to move the particle along the x axis: (a) from $x = 0.0$ to $x = 10.0$ m; (b) from $x = 0.0$ to $x = 15.0$ m.

**FIGURE 6–38** Problem 13.

*14. (III) A 17,000-kg jet takes off from an aircraft carrier via a catapult (Fig. 6–39a). The gases thrust out from the jet's engines exert a constant force of 130 kN on the jet; the force exerted on the jet by the catapult is plotted in Fig. 6–39b. Determine the work done on the jet: (a) by the gases expelled by its engines during launch of the jet; and (b) by the catapult during launch of the jet.

(a)　　　　　(b)

**FIGURE 6–39** Problem 14.

## 6–3 Kinetic Energy; Work-Energy Principle

15. (I) At room temperature, an oxygen molecule, with mass of $5.31 \times 10^{-26}$ kg, typically has a kinetic energy of about $6.21 \times 10^{-21}$ J. How fast is it moving?

16. (I) (a) If the kinetic energy of a particle is tripled, by what factor has its speed increased? (b) If the speed of a particle is halved, by what factor does its kinetic energy change?

17. (I) How much work is required to stop an electron $(m = 9.11 \times 10^{-31}$ kg) which is moving with a speed of $1.10 \times 10^6$ m/s?

18. (I) How much work must be done to stop a 925-kg car traveling at 95 km/h?

19. (II) Two bullets are fired at the same time with the same kinetic energy. If one bullet has twice the mass of the other, which has the greater speed and by what factor? Which can do the most work?

20. (II) A baseball ($m = 145$ g) traveling 32 m/s moves a fielder's glove backward 25 cm when the ball is caught. What was the average force exerted by the ball on the glove?

21. (II) An 85-g arrow is fired from a bow whose string exerts an average force of 105 N on the arrow over a distance of 75 cm. What is the speed of the arrow as it leaves the bow?

22. (II) If the speed of a car is increased by 50%, by what factor will its minimum braking distance be increased, assuming all else is the same? Ignore the driver's reaction time.

23. (II) At an accident scene on a level road, investigators measure a car's skid mark to be 78 m long. It was a rainy day and the coefficient of friction was estimated to be 0.30. Use these data to determine the speed of the car when the driver slammed on (and locked) the brakes. (Why does the car's mass not matter?)

24. (III) One car has twice the mass of a second car, but only half as much kinetic energy. When both cars increase their speed by 8.0 m/s, they then have the same kinetic energy. What were the original speeds of the two cars?

25. (III) A 265-kg load is lifted 18.0 m vertically with an acceleration $a = 0.160\,g$ by a single cable. Determine (a) the tension in the cable; (b) the net work done on the load; (c) the work done by the cable on the load; (d) the work done by gravity on the load; (e) the final speed of the load assuming it started from rest.

## 6–4 and 6–5 Potential Energy

26. (I) By how much does the gravitational potential energy of a 54-kg pole vaulter change if her center of mass rises about 4.0 m during the jump?

27. (I) A spring has a spring constant $k$ of 88.0 N/m. How much must this spring be compressed to store 45.0 J of potential energy?

28. (II) If it requires 6.0 J of work to stretch a particular spring by 2.0 cm from its equilibrium length, how much more work will be required to stretch it an additional 4.0 cm?

29. (II) A 66.5-kg hiker starts at an elevation of 1270 m and climbs to the top of a peak 2660 m high. (a) What is the hiker's change in potential energy? (b) What is the minimum work required of the hiker? (c) Can the actual work done be greater than this? Explain.

30. (II) A 1.60-m-tall person lifts a 1.65-kg book off the ground so it is 2.20 m above the ground. What is the potential energy of the book relative to (a) the ground, and (b) the top of the person's head? (c) How is the work done by the person related to the answers in parts (a) and (b)?

## 6–6 and 6–7 Conservation of Mechanical Energy

31. (I) A novice skier, starting from rest, slides down an icy frictionless 8.0° incline whose vertical height is 105 m. How fast is she going when she reaches the bottom?

32. (I) Jane, looking for Tarzan, is running at top speed (5.0 m/s) and grabs a vine hanging vertically from a tall tree in the jungle. How high can she swing upward? Does the length of the vine affect your answer?

33. (II) A sled is initially given a shove up a frictionless 23.0° incline. It reaches a maximum vertical height 1.22 m higher than where it started at the bottom. What was its initial speed?

34. (II) In the *high jump*, the kinetic energy of an athlete is transformed into gravitational potential energy without the aid of a pole. With what minimum speed must the athlete leave the ground in order to lift his center of mass 2.10 m and cross the bar with a speed of 0.50 m/s?

35. (II) A spring with $k = 83$ N/m hangs vertically next to a ruler. The end of the spring is next to the 15-cm mark on the ruler. If a 2.5-kg mass is now attached to the end of the spring, and the mass is allowed to fall, where will the end of the spring line up with the ruler marks when the mass is at its lowest position?

36. (II) A 0.48-kg ball is thrown with a speed of 8.8 m/s at an upward angle of 36°. (a) What is its speed at its highest point, and (b) how high does it go? (Use conservation of energy.)

37. (II) A 1200-kg car moving on a horizontal surface has speed $v = 85$ km/h when it strikes a horizontal coiled spring and is brought to rest in a distance of 2.2 m. What is the spring stiffness constant of the spring?

38. (II) A 62-kg trampoline artist jumps upward from the top of a platform with a vertical speed of 4.5 m/s. (a) How fast is he going as he lands on the trampoline, 2.0 m below (Fig. 6–40)? (b) If the trampoline behaves like a spring of spring constant $5.8 \times 10^4$ N/m, how far down does he depress it?

**FIGURE 6–40**
Problem 38.

2.0 m

**39.** (II) A vertical spring (ignore its mass), whose spring constant is 875 N/m, is attached to a table and is compressed down by 0.160 m. (a) What upward speed can it give to a 0.380-kg ball when released? (b) How high above its original position (spring compressed) will the ball fly?

**40.** (II) A roller-coaster car shown in Fig. 6–41 is pulled up to point 1 where it is released from rest. Assuming no friction, calculate the speed at points 2, 3, and 4.

**FIGURE 6–41** Problems 40 and 50.

**41.** (II) Chris jumps off a bridge with a bungee cord (a heavy stretchable cord) tied around his ankle, Fig. 6–42. He falls for 15 m before the bungee cord begins to stretch. Chris's mass is 75 kg and we assume the cord obeys Hooke's law, $F = -kx$, with $k = 55$ N/m. If we neglect air resistance, estimate what distance $d$ below the bridge Chris's foot will be before coming to a stop. Ignore the mass of the cord (not realistic, however) and treat Chris as a particle.

**FIGURE 6–42**
Problem 41. (a) Bungee jumper about to jump.
(b) Bungee cord at its unstretched length.
(c) Maximum stretch of cord.

**42.** (II) What should be the spring constant $k$ of a spring designed to bring a 1200-kg car to rest from a speed of 95 km/h so that the occupants undergo a maximum acceleration of 4.0 $g$?

**43.** (III) An engineer is designing a spring to be placed at the bottom of an elevator shaft. If the elevator cable breaks when the elevator is at a height $h$ above the top of the spring, calculate the value that the spring constant $k$ should have so that passengers undergo an acceleration of no more than 5.0 $g$ when brought to rest. Let $M$ be the total mass of the elevator and passengers.

**44.** (III) A block of mass $m$ is attached to the end of a spring (spring stiffness constant $k$), Fig. 6–43. The mass is given an initial displacement $x_0$ from equilibrium, and an initial speed $v_0$. Ignoring friction and the mass of the spring, use energy methods to find (a) its maximum speed, and (b) its maximum stretch from equilibrium, in terms of the given quantities.

**FIGURE 6–43**
Problem 44.

**45.** (III) A cyclist intends to cycle up a 7.50° hill whose vertical height is 125 m. The pedals turn in a circle of diameter 36.0 cm. Assuming the mass of bicycle plus person is 75.0 kg, (a) calculate how much work must be done against gravity. (b) If each complete revolution of the pedals moves the bike 5.10 m along its path, calculate the average force that must be exerted on the pedals tangent to their circular path. Neglect work done by friction and other losses.

## 6–8 and 6–9 Law of Conservation of Energy

**46.** (I) Two railroad cars, each of mass 66,000 kg, are traveling 85 km/h toward each other. They collide head-on and come to rest. How much thermal energy is produced in this collision?

**47.** (I) A 16.0-kg child descends a slide 2.20 m high and, starting from rest, reaches the bottom with a speed of 1.25 m/s. How much thermal energy due to friction was generated in this process?

**48.** (II) A ski starts from rest and slides down a 28° incline 85 m long. (a) If the coefficient of friction is 0.090, what is the ski's speed at the base of the incline? (b) If the snow is level at the foot of the incline and has the same coefficient of friction, how far will the ski travel along the level? Use energy methods.

**49.** (II) A 145-g baseball is dropped from a tree 12.0 m above the ground. (a) With what speed would it hit the ground if air resistance could be ignored? (b) If it actually hits the ground with a speed of 8.00 m/s, what is the average force of air resistance exerted on it?

**50.** (II) Suppose the roller-coaster car in Fig. 6–41 passes point 1 with a speed of 1.30 m/s. If the average force of friction is equal to 0.23 of its weight, with what speed will it reach point 2? The distance traveled is 45.0 m.

**51.** (II) A skier traveling 11.0 m/s reaches the foot of a steady upward 19° incline and glides 15 m up along this slope before coming to rest. What was the average coefficient of friction?

**52.** (II) You drop a ball from a height of 2.0 m, and it bounces back to a height of 1.6 m. (a) What fraction of its initial energy is lost during the bounce? (b) What is the ball's speed just before and just after the bounce? (c) Where did the energy go?

**53.** (II) A 66-kg skier starts from rest at the top of a 1200-m-long trail which drops a total of 230 m from top to bottom. At the bottom, the skier is moving 11.0 m/s. How much energy was dissipated by friction?

**54.** (II) A projectile is fired at an upward angle of 38.0° from the top of a 135-m-high cliff with a speed of 165 m/s. What will be its speed when it strikes the ground below? (Use conservation of energy.)

**55.** (II) The Lunar Module could make a safe landing if its vertical velocity at impact is 3.0 m/s or less. Suppose that you want to determine the greatest height $h$ at which the pilot could shut off the engine if the velocity of the lander relative to the surface at that moment is (a) zero; (b) 2.0 m/s downward; (c) 2.0 m/s upward. Use conservation of energy to determine $h$ in each case. The acceleration due to gravity at the surface of the Moon is 1.62 m/s².

56. (III) Early test flights for the space shuttle used a "glider" (mass of 980 kg including pilot). After a horizontal launch at 480 km/h at a height of 3500 m, the glider eventually landed at a speed of 210 km/h. (a) What would its landing speed have been in the absence of air resistance? (b) What was the average force of air resistance exerted on it if it came in at a constant glide angle of 12° to the Earth's surface?

## 6–10 Power

57. (I) How long will it take a 2750-W motor to lift a 385-kg piano to a sixth-story window 16.0 m above?

58. (I) (a) Show that one British horsepower (550 ft·lb/s) is equal to 746 W. (b) What is the horsepower rating of a 75-W lightbulb?

59. (I) An 85-kg football player traveling 5.0 m/s is stopped in 1.0 s by a tackler. (a) What is the original kinetic energy of the player? (b) What average power is required to stop him?

60. (II) If a car generates 18 hp when traveling at a steady 95 km/h, what must be the average force exerted on the car due to friction and air resistance?

61. (II) An outboard motor for a boat is rated at 35 hp. If it can move a particular boat at a steady speed of 35 km/h, what is the total force resisting the motion of the boat?

62. (II) A shot-putter accelerates a 7.3-kg shot from rest to 14 m/s in 1.5 s. What average power was developed?

63. (II) A driver notices that her 1080-kg car, when in neutral, slows down from 95 km/h to 65 km/h in about 7.0 s on a flat horizontal road. Approximately what power (watts and hp) is needed to keep the car traveling at a constant 80 km/h?

64. (II) How much work can a 2.0-hp motor do in 1.0 h?

65. (II) A 975-kg sports car accelerates from rest to 95 km/h in 6.4 s. What is the average power delivered by the engine?

66. (II) During a workout, football players ran up the stadium stairs in 75 s. The distance along the stairs is 83 m and they are inclined at a 33° angle. If a player has a mass of 82 kg, estimate his average power output on the way up. Ignore friction and air resistance.

67. (II) A pump lifts 27.0 kg of water per minute through a height of 3.50 m. What minimum output rating (watts) must the pump motor have?

68. (II) A ski area claims that its lifts can move 47,000 people per hour. If the average lift carries people about 200 m (vertically) higher, estimate the maximum total power needed.

69. (II) A 65-kg skier grips a moving rope that is powered by an engine and is pulled at constant speed to the top of a 23° hill. The skier is pulled a distance $x = 320$ m along the incline and it takes 2.0 min to reach the top of the hill. If the coefficient of kinetic friction between the snow and skis is $\mu_k = 0.10$, what horsepower engine is required if 30 such skiers (max) are on the rope at one time?

70. (II) What minimum horsepower must a motor have to be able to drag a 370-kg box along a level floor at a speed of 1.20 m/s if the coefficient of friction is 0.45?

71. (III) A bicyclist coasts down a 6.0° hill at a steady speed of 4.0 m/s. Assuming a total mass of 75 kg (bicycle plus rider), what must be the cyclist's power output to climb the same hill at the same speed?

# General Problems

72. Spiderman uses his spider webs to save a runaway train moving about 60 km/h, Fig. 6–44. His web stretches a few city blocks (500 m) before the $10^4$-kg train comes to a stop. Assuming the web acts like a spring, estimate the effective spring constant.

**FIGURE 6–44**
Problem 72.

73. A 36.0-kg crate, starting from rest, is pulled across a floor with a constant horizontal force of 225 N. For the first 11.0 m the floor is frictionless, and for the next 10.0 m the coefficient of friction is 0.20. What is the final speed of the crate after being pulled these 21.0 m?

74. How high will a 1.85-kg rock go from the point of release if thrown straight up by someone who does 80.0 J of work on it? Neglect air resistance.

75. A mass $m$ is attached to a spring which is held stretched a distance $x$ by a force $F$, Fig. 6–45, and then released. The spring pulls the mass to the left, towards its natural equilibrium length. Assuming there is no friction, determine the speed of the mass $m$ when the spring returns: (a) to its normal length $(x = 0)$; (b) to half its original extension $(x/2)$.

**FIGURE 6–45** Problem 75.

76. An elevator cable breaks when a 925-kg elevator is 28.5 m above the top of a huge spring $(k = 8.00 \times 10^4 \,\text{N/m})$ at the bottom of the shaft. Calculate (a) the work done by gravity on the elevator before it hits the spring; (b) the speed of the elevator just before striking the spring; (c) the amount the spring compresses (note that here work is done by both the spring and gravity).

77. (a) A 3.0-g locust reaches a speed of 3.0 m/s during its jump. What is its kinetic energy at this speed? (b) If the locust transforms energy with 35% efficiency, how much energy is required for the jump?

**78.** In a common test for cardiac function (the "stress test"), the patient walks on an inclined treadmill (Fig. 6–46). Estimate the power required from a 75-kg patient when the treadmill is sloping at an angle of 12° and the velocity is 3.1 km/h. (How does this power compare to the power rating of a lightbulb?)

**FIGURE 6–46** Problem 78.

**79.** An airplane pilot fell 370 m after jumping from an aircraft without his parachute opening. He landed in a snowbank, creating a crater 1.1 m deep, but survived with only minor injuries. Assuming the pilot's mass was 88 kg and his speed at impact was 45 m/s, estimate: (a) the work done by the snow in bringing him to rest; (b) the average force exerted on him by the snow to stop him; and (c) the work done on him by air resistance as he fell. Model him as a particle.

**80.** Many cars have "5 mi/h (8 km/h) bumpers" that are designed to compress and rebound elastically without any physical damage at speeds below 8 km/h. If the material of the bumpers permanently deforms after a compression of 1.5 cm, but remains like an elastic spring up to that point, what must be the effective spring constant of the bumper material, assuming the car has a mass of 1050 kg and is tested by ramming into a solid wall?

**81.** In climbing up a rope, a 62-kg athlete climbs a vertical distance of 5.0 m in 9.0 s. What minimum power output was used to accomplish this feat?

**82.** If a 1300-kg car can accelerate from 35 km/h to 65 km/h in 3.8 s, how long will it take to accelerate from 55 km/h to 95 km/h? Assume the power stays the same, and neglect frictional losses.

**83.** A cyclist starts from rest and coasts down a 4.0° hill. The mass of the cyclist plus bicycle is 85 kg. After the cyclist has traveled 180 m, (a) what was the net work done by gravity on the cyclist? (b) How fast is the cyclist going? Ignore air resistance and friction.

**84.** A film of Jesse Owens's famous long jump (Fig. 6–47) in the 1936 Olympics shows that his center of mass rose 1.1 m from launch point to the top of the arc. What minimum speed did he need at launch if he was traveling at 6.5 m/s at the top of the arc?

**FIGURE 6–47**
Problem 84.

**85.** Water flows over a dam at the rate of 680 kg/s and falls vertically 88 m before striking the turbine blades. Calculate (a) the speed of the water just before striking the turbine blades (neglect air resistance), and (b) the rate at which mechanical energy is transferred to the turbine blades, assuming 55% efficiency.

**86.** A 55-kg skier starts from rest at the top of a ski jump, point A in Fig. 6–48, and travels down the ramp. If friction and air resistance can be neglected, (a) determine her speed $v_B$ when she reaches the horizontal end of the ramp at B. (b) Determine the distance $s$ to where she strikes the ground at C.

**FIGURE 6–48**
Problem 86.

**87.** Electric energy units are often expressed in "kilowatt-hours." (a) Show that one kilowatt-hour (kWh) is equal to $3.6 \times 10^6$ J. (b) If a typical family of four uses electric energy at an average rate of 580 W, how many kWh would their electric bill show for one month, and (c) how many joules would this be? (d) At a cost of $0.12 per kWh, what would their monthly bill be in dollars? Does the monthly bill depend on the *rate* at which they use the electric energy?

**88.** If you stand on a bathroom scale, the spring inside the scale compresses 0.60 mm, and it tells you your weight is 760 N. Now if you jump on the scale from a height of 1.0 m, what does the scale read at its peak?

**89.** A 65-kg hiker climbs to the top of a mountain 4200 m high. The climb is made in 4.6 h starting at an elevation of 2800 m. Calculate (a) the work done by the hiker against gravity, (b) the average power output in watts and in horsepower, and (c) assuming the body is 15% efficient, what rate of energy input was required.

**90.** A ball is attached to a horizontal cord of length $\ell$ whose other end is fixed, Fig. 6–49. (a) If the ball is released, what will be its speed at the lowest point of its path? (b) A peg is located a distance $h$ directly below the point of attachment of the cord. If $h = 0.80\ell$, what will be the speed of the ball when it reaches the top of its circular path about the peg?

**FIGURE 6–49**
Problem 90.

**91.** An 18-kg sled starts up a 28° incline with a speed of 2.3 m/s. The coefficient of kinetic friction is $\mu_k = 0.25$. (a) How far up the incline does the sled travel? (b) What condition must you put on the coefficient of static friction if the sled is not to get stuck at the point determined in part (a)? (c) If the sled slides back down, what is its speed when it returns to its starting point?

**92.** A 56-kg student runs at 6.0 m/s, grabs a hanging 10.0-m-long rope, and swings out over a lake (Fig. 6–50). He releases the rope when his velocity is zero. (a) What is the angle θ when he releases the rope? (b) What is the tension in the rope just before he releases it? (c) What is the maximum tension in the rope during the swing?

**FIGURE 6–50**
Problem 92.

θ

10.0 m

**93.** Some electric power companies use water to store energy. Water is pumped from a low reservoir to a high reservoir. To store the energy produced in 1.0 hour by a 180-MW electric power plant, how many cubic meters of water will have to be pumped from the lower to the upper reservoir? Assume the upper reservoir is an average of 380 m above the lower one. Water has a mass of $1.00 \times 10^3$ kg for every $1.0 \, m^3$.

**94.** A softball having a mass of 0.25 kg is pitched horizontally at 120 km/h. By the time it reaches the plate, it may have slowed by 10%. Neglecting gravity, estimate the average force of air resistance during a pitch. The distance between the plate and the pitcher is about 15 m.

# Search and Learn

1. We studied forces earlier and used them to solve Problems. Now we are using energy to solve Problems, even some that could be solved with forces. (a) Give at least three advantages of using energy to solve a Problem. (b) When must you use energy to solve a Problem? (c) When must you use forces to solve a Problem? (d) What information is not available when solving Problems with energy? Look at the Examples in Chapters 6 and 4.

2. The brakes on a truck can overheat and catch on fire if the truck goes down a long steep hill without shifting into a lower gear. (a) Explain why this happens in terms of energy and power. (b) Would it matter if the same elevation change was made going down a steep hill or a gradual hill? Explain your reasoning. [*Hint:* Read Sections 6–4, 6–9, and 6–10 carefully.] (c) Why does shifting into a lower gear help? [*Hint:* Use your own experience, downshifting in a car.] (d) Calculate the thermal energy dissipated from the brakes in an 8000-kg truck that descends a 12° hill. The truck begins braking when its speed is 95 km/h and slows to a speed of 35 km/h in a distance of 0.36 km measured along the road.

3. (a) Only two conservative forces are discussed in this Chapter. What are they, and how are they accounted for when you are dealing with conservation of energy? (b) Not mentioned is the force of water on a swimmer. Is it conservative or nonconservative?

4. Give at least two examples of friction doing positive work. Reread parts of Chapters 4 and 6.

5. Show that on a roller coaster with a circular vertical loop (Fig. 6–51), the difference in your apparent weight at the top of the loop and the bottom of the loop is 6.0 times your weight. Ignore friction. Show also that as long as your speed is above the minimum needed (so the car holds the track), this answer doesn't depend on the size of the loop or how fast you go through it. [Reread Sections 6–6, 5–2, and 4–6.]

**FIGURE 6–51**
Search and Learn 5 and 6.

h

R

6. Suppose that the track in Fig. 6–51 is not frictionless and the values of h and R are given. (See Sections 6–9 and 6–1.) (a) If you measure the velocity of the roller coaster at the top of the hill (of height h) and at the top of the circle (of height 2R), can you determine the work done by friction during the time the roller coaster moves between those two points? Why or why not? (b) Can you determine the average force of friction between those two points? Why or why not? If not, what additional information do you need?

## ANSWERS TO EXERCISES

**A:** (c).
**B:** (a) Less, because $(20)^2 = 400 < (30)^2 - (20)^2 = 500$; (b) $2.0 \times 10^5$ J.
**C:** No, because the speed $v$ would be the square root of a negative number, which is not real.

**D:** (a) $\sqrt{2}$; (b) 4.
**E:** Yes. It is nonconservative, because for a conservative force $W = 0$ in a round trip.
**F:** (e), (e); (e), (c).

Conservation of linear momentum is another great conservation law of physics. Collisions, such as between billiard or pool balls, illustrate this law very nicely: the total vector momentum just before the collision equals the total vector momentum just after the collision. In this photo, the moving cue ball makes a glancing collision with the 11 ball which is initially at rest. After the collision, both balls move at angles, but the sum of their vector momenta equals the initial vector momentum of the incoming cue ball.

We will consider both elastic collisions (where kinetic energy is also conserved) and inelastic collisions. We also examine the concept of center of mass, and how it helps us in the study of complex motion.

$m_1\vec{\mathbf{v}}_1$ (before)

$m_2\vec{\mathbf{v}}_2$ (after)

$m_1\vec{\mathbf{v}}'_1$ (after)

# CHAPTER 7

# Linear Momentum

## CONTENTS

**CHAPTER-OPENING QUESTIONS—Guess now!**

**1.** A railroad car loaded with rocks coasts on a level track without friction. A worker at the back of the car starts throwing the rocks horizontally backward from the car. Then what happens?
 **(a)** The car slows down.
 **(b)** The car speeds up.
 **(c)** First the car speeds up and then it slows down.
 **(d)** The car's speed remains constant.
 **(e)** None of these.

**2.** Which answer would you choose if the rocks fall out through a hole in the floor of the car, one at a time?

The law of conservation of energy, which we discussed in the previous Chapter, is one of several great conservation laws in physics. Among the other quantities found to be conserved are linear momentum, angular momentum, and electric charge. We will eventually discuss all of these because the conservation laws are among the most important ideas in science. In this Chapter we discuss linear momentum and its conservation. The law of conservation of momentum is essentially a reworking of Newton's laws that gives us tremendous physical insight and problem-solving power.

The law of conservation of momentum is particularly useful when dealing with a system of two or more objects that interact with each other, such as in collisions of ordinary objects or nuclear particles.

Our focus up to now has been mainly on the motion of a single object, often thought of as a "particle" in the sense that we have ignored any rotation or internal motion. In this Chapter we will deal with systems of two or more objects, and—toward the end of the Chapter—the concept of center of mass.

# 7–1 Momentum and Its Relation to Force

The **linear momentum** (or "momentum" for short) of an object is defined as the product of its mass and its velocity. Momentum (plural is *momenta*—from Latin) is represented by the symbol $\vec{\mathbf{p}}$. If we let $m$ represent the mass of an object and $\vec{\mathbf{v}}$ represent its velocity, then its momentum $\vec{\mathbf{p}}$ is defined as

$$\vec{\mathbf{p}} = m\vec{\mathbf{v}}. \tag{7-1}$$

Velocity is a vector, so momentum too is a vector. The direction of the momentum is the direction of the velocity, and the magnitude of the momentum is $p = mv$. Because velocity depends on the reference frame, so does momentum; thus the reference frame must be specified. The unit of momentum is that of mass × velocity, which in SI units is kg·m/s. There is no special name for this unit.

Everyday usage of the term *momentum* is in accord with the definition above. According to Eq. 7–1, a fast-moving car has more momentum than a slow-moving car of the same mass; a heavy truck has more momentum than a small car moving with the same speed. The more momentum an object has, the harder it is to stop it, and the greater effect it will have on another object if it is brought to rest by striking that object. A football player is more likely to be stunned if tackled by a heavy opponent running at top speed than by a lighter or slower-moving tackler. A heavy, fast-moving truck can do more damage than a slow-moving motorcycle.

**EXERCISE A** Can a small sports car ever have the same momentum as a large sport-utility vehicle with three times the sports car's mass? Explain.

A force is required to change the momentum of an object, whether to increase the momentum, to decrease it, or to change its direction. Newton originally stated his second law in terms of momentum (although he called the product $mv$ the "quantity of motion"). Newton's statement of the **second law of motion**, translated into modern language, is as follows:

**The rate of change of momentum of an object is equal to the net force applied to it.**

*NEWTON'S SECOND LAW*

We can write this as an equation,

$$\Sigma\vec{\mathbf{F}} = \frac{\Delta\vec{\mathbf{p}}}{\Delta t}, \tag{7-2}$$

*NEWTON'S SECOND LAW*

where $\Sigma\vec{\mathbf{F}}$ is the net force applied to the object (the vector sum of all forces acting on it) and $\Delta\vec{\mathbf{p}}$ is the resulting momentum change that occurs during the time interval[†] $\Delta t$.

**CAUTION**

*The change in the momentum vector is in the direction of the net force*

We can readily derive the familiar form of the second law, $\Sigma\vec{\mathbf{F}} = m\vec{\mathbf{a}}$, from Eq. 7–2 for the case of constant mass. If $\vec{\mathbf{v}}_1$ is the initial velocity of an object and $\vec{\mathbf{v}}_2$ is its velocity after a time interval $\Delta t$ has elapsed, then

$$\Sigma\vec{\mathbf{F}} = \frac{\Delta\vec{\mathbf{p}}}{\Delta t} = \frac{m\vec{\mathbf{v}}_2 - m\vec{\mathbf{v}}_1}{\Delta t} = \frac{m(\vec{\mathbf{v}}_2 - \vec{\mathbf{v}}_1)}{\Delta t} = m\frac{\Delta\vec{\mathbf{v}}}{\Delta t}.$$

By definition, $\vec{\mathbf{a}} = \Delta\vec{\mathbf{v}}/\Delta t$, so

$$\Sigma\vec{\mathbf{F}} = m\vec{\mathbf{a}}. \qquad \text{[constant mass]}$$

Equation 7–2 is a more general statement of Newton's second law than the more familiar version $(\Sigma\vec{\mathbf{F}} = m\vec{\mathbf{a}})$ because it includes the situation in which the mass may change. A change in mass occurs in certain circumstances, such as for rockets which lose mass as they expel burnt fuel.

---

[†]Normally we think of $\Delta t$ as being a small time interval. If it is not small, then Eq. 7–2 is valid if $\Sigma\vec{\mathbf{F}}$ is constant during that time interval, or if $\Sigma\vec{\mathbf{F}}$ is the average net force during that time interval.

FIGURE 7–1 Example 7–1.

**EXAMPLE 7–1** | **ESTIMATE** | **Force of a tennis serve.** For a top player, a tennis ball may leave the racket on the serve with a speed of 55 m/s (about 120 mi/h), Fig. 7–1. If the ball has a mass of 0.060 kg and is in contact with the racket for about 4 ms $(4 \times 10^{-3}\,\text{s})$, estimate the average force on the ball. Would this force be large enough to lift a 60-kg person?

**APPROACH** We write Newton's second law, Eq. 7–2, for the average force as

$$F_{\text{avg}} = \frac{\Delta p}{\Delta t} = \frac{mv_2 - mv_1}{\Delta t},$$

where $mv_1$ and $mv_2$ are the initial and final momenta. The tennis ball is hit when its initial velocity $v_1$ is very nearly zero at the top of the throw, so we set $v_1 = 0$, and we assume $v_2 = 55$ m/s is in the horizontal direction. We ignore all other forces on the ball during this brief time interval, such as gravity, in comparison to the force exerted by the tennis racket.

**SOLUTION** The force exerted on the ball by the racket is

$$F_{\text{avg}} = \frac{\Delta p}{\Delta t} = \frac{mv_2 - mv_1}{\Delta t} = \frac{(0.060\,\text{kg})(55\,\text{m/s}) - 0}{0.004\,\text{s}} \approx 800\,\text{N}.$$

This is a large force, larger than the weight of a 60-kg person, which would require a force $mg = (60\,\text{kg})(9.8\,\text{m/s}^2) \approx 600\,\text{N}$ to lift.

**NOTE** The force of gravity acting on the tennis ball is $mg = (0.060\,\text{kg})(9.8\,\text{m/s}^2) = 0.59\,\text{N}$, which justifies our ignoring it compared to the enormous force the racket exerts.

**NOTE** High-speed photography and radar can give us an estimate of the contact time and the velocity of the ball leaving the racket. But a direct measurement of the force is not practical. Our calculation shows a handy technique for determining an unknown force in the real world.

FIGURE 7–2 Example 7–2.

**EXAMPLE 7–2** | **Washing a car: momentum change and force.** Water leaves a hose at a rate of 1.5 kg/s with a speed of 20 m/s and is aimed at the side of a car, which stops it, Fig. 7–2. (That is, we ignore any splashing back.) What is the force exerted by the water on the car?

**APPROACH** The water leaving the hose has mass and velocity, so it has a momentum $p_{\text{initial}}$ in the horizontal ($x$) direction, and we assume gravity doesn't pull the water down significantly. When the water hits the car, the water loses this momentum $(p_{\text{final}} = 0)$. We use Newton's second law in the momentum form, Eq. 7–2, to find the force that the car exerts on the water to stop it. By Newton's third law, the force exerted by the water on the car is equal and opposite. We have a continuing process: 1.5 kg of water leaves the hose in each 1.0-s time interval. So let us write $F = \Delta p/\Delta t$ where $\Delta t = 1.0$ s, and $mv_{\text{initial}} = (1.5\,\text{kg})(20\,\text{m/s}) = 30\,\text{kg} \cdot \text{m/s}$.

**SOLUTION** The force (assumed constant) that the car must exert to change the momentum of the water is

$$F = \frac{\Delta p}{\Delta t} = \frac{p_{\text{final}} - p_{\text{initial}}}{\Delta t} = \frac{0 - 30\,\text{kg} \cdot \text{m/s}}{1.0\,\text{s}} = -30\,\text{N}.$$

The minus sign indicates that the force exerted by the car on the water is opposite to the water's original velocity. The car exerts a force of 30 N to the left to stop the water, so by Newton's third law, the water exerts a force of 30 N to the right on the car.

**NOTE** Keep track of signs, although common sense helps too. The water is moving to the right, so common sense tells us the force on the car must be to the right.

**EXERCISE B** If the water splashes back from the car in Example 7–2, would the force on the car be larger or smaller?

# 7–2 Conservation of Momentum

The concept of momentum is particularly important because, if no net external force acts on a system, the total momentum of the system is a conserved quantity. This was expressed in Eq. 7–2 for a single object, but it holds also for a system as we shall see.

Consider the head-on collision of two billiard balls, as shown in Fig. 7–3. We assume the net external force on this system of two balls is zero—that is, the only significant forces during the collision are the forces that each ball exerts on the other. Although the momentum of each of the two balls changes as a result of the collision, the *sum* of their momenta is found to be the same before as after the collision. If $m_A \vec{v}_A$ is the momentum of ball A and $m_B \vec{v}_B$ the momentum of ball B, both measured just before the collision, then the total momentum of the two balls before the collision is the vector sum $m_A \vec{v}_A + m_B \vec{v}_B$. Immediately after the collision, the balls each have a different velocity and momentum, which we designate by a "prime" on the velocity: $m_A \vec{v}'_A$ and $m_B \vec{v}'_B$. The total momentum after the collision is the vector sum $m_A \vec{v}'_A + m_B \vec{v}'_B$. No matter what the velocities and masses are, experiments show that the total momentum before the collision is the same as afterward, whether the collision is head-on or not, as long as no net external force acts:

**FIGURE 7–3** Momentum is conserved in a collision of two balls, labeled A and B.

momentum before = momentum after

$$m_A \vec{v}_A + m_B \vec{v}_B = m_A \vec{v}'_A + m_B \vec{v}'_B. \qquad [\Sigma \vec{F}_{ext} = 0] \quad (7\text{–}3)$$

CONSERVATION OF MOMENTUM
(two objects colliding)

That is, the total vector momentum of the system of two colliding balls is conserved: it stays constant. (We saw this result in this Chapter's opening photograph.)

Although the law of conservation of momentum was discovered experimentally, it can be derived from Newton's laws of motion, which we now show.

Let us consider two objects of mass $m_A$ and $m_B$ that have momenta $\vec{p}_A \ (= m_A \vec{v}_A)$ and $\vec{p}_B \ (= m_B \vec{v}_B)$ before they collide and $\vec{p}'_A$ and $\vec{p}'_B$ after they collide, as in Fig. 7–4. During the collision, suppose that the force exerted by object A on object B at any instant is $\vec{F}$. Then, by Newton's third law, the force exerted by object B on object A is $-\vec{F}$. During the brief collision time, we assume no other (external) forces are acting (or that $\vec{F}$ is much greater than any other external forces acting). Over a very short time interval $\Delta t$ we have

$$\vec{F} = \frac{\Delta \vec{p}_B}{\Delta t} = \frac{\vec{p}'_B - \vec{p}_B}{\Delta t}$$

and

$$-\vec{F} = \frac{\Delta \vec{p}_A}{\Delta t} = \frac{\vec{p}'_A - \vec{p}_A}{\Delta t}.$$

We add these two equations together and find

$$0 = \frac{\Delta \vec{p}_B + \Delta \vec{p}_A}{\Delta t} = \frac{(\vec{p}'_B - \vec{p}_B) + (\vec{p}'_A - \vec{p}_A)}{\Delta t}.$$

This means

$$\vec{p}'_B - \vec{p}_B + \vec{p}'_A - \vec{p}_A = 0,$$

or

$$\vec{p}'_A + \vec{p}'_B = \vec{p}_A + \vec{p}_B.$$

This is Eq. 7–3. The total momentum is conserved.

We have put this derivation in the context of a collision. As long as no external forces act, it is valid over any time interval, and conservation of momentum is always valid as long as no external forces act on the chosen system. In the real world, external forces do act: friction on billiard balls, gravity acting on a tennis ball, and so on. So we often want our "observation time" (before and after) to be small. When a racket hits a tennis ball or a bat hits a baseball, both before and after the "collision" the ball moves as a projectile under the action of gravity and air resistance.

**FIGURE 7–4** Collision of two objects. Their momenta before collision are $\vec{p}_A$ and $\vec{p}_B$, and after collision are $\vec{p}'_A$ and $\vec{p}'_B$. At any moment during the collision each exerts a force on the other of equal magnitude but opposite direction.

However, when the bat or racket hits the ball, during the brief time of the collision those external forces are insignificant compared to the collision force the bat or racket exerts on the ball. Momentum is conserved (or very nearly so) as long as we measure $\vec{\mathbf{p}}_A$ and $\vec{\mathbf{p}}_B$ just before the collision and $\vec{\mathbf{p}}'_A$ and $\vec{\mathbf{p}}'_B$ immediately after the collision (Eq. 7–3). We can not wait for external forces to produce their effect before measuring $\vec{\mathbf{p}}'_A$ and $\vec{\mathbf{p}}'_B$.

The above derivation can be extended to include any number of interacting objects. To show this, we let $\vec{\mathbf{p}}$ in Eq. 7–2 ($\Sigma\vec{\mathbf{F}} = \Delta\vec{\mathbf{p}}/\Delta t$) represent the total momentum of a system—that is, the vector sum of the momenta of all objects in the system. (For our two-object system above, $\vec{\mathbf{p}} = m_A\vec{\mathbf{v}}_A + m_B\vec{\mathbf{v}}_B$.) If the net force $\Sigma\vec{\mathbf{F}}$ on the system is zero [as it was above for our two-object system, $\vec{\mathbf{F}} + (-\vec{\mathbf{F}}) = 0$], then from Eq. 7–2, $\Delta\vec{\mathbf{p}} = \Sigma\vec{\mathbf{F}}\,\Delta t = 0$, so the total momentum doesn't change. The general statement of the **law of conservation of momentum** is

**The total momentum of an isolated system of objects remains constant.**

By a **system**, we simply mean a set of objects that we choose, and which may interact with each other. An **isolated system** is one in which the only (significant) forces are those between the objects in the system. The sum of all these "internal" forces within the system will be zero because of Newton's third law. If there are *external forces*—by which we mean forces exerted by objects *outside* the system— and they don't add up to zero, then the total momentum of the system won't be conserved. However, if the system can be redefined so as to include the other objects exerting these forces, then the conservation of momentum principle can apply. For example, if we take as our system a falling rock, it does not conserve momentum because an external force, the force of gravity exerted by the Earth, accelerates the rock and changes its momentum. However, if we include the Earth in the system, the total momentum of rock plus Earth is conserved. (This means that the Earth comes up to meet the rock. But the Earth's mass is so great, its upward velocity is very tiny.)

Although the law of conservation of momentum follows from Newton's second law, as we have seen, it is in fact more general than Newton's laws. In the tiny world of the atom, Newton's laws fail, but the great conservation laws— those of energy, momentum, angular momentum, and electric charge—have been found to hold in every experimental situation tested. It is for this reason that the conservation laws are considered more basic than Newton's laws.

---

**EXAMPLE 7–3** **Railroad cars collide: momentum conserved.** A 10,000-kg railroad car, A, traveling at a speed of 24.0 m/s strikes an identical car, B, at rest. If the cars lock together as a result of the collision, what is their common speed just afterward? See Fig. 7–5.

**APPROACH** We choose our system to be the two railroad cars. We consider a very brief time interval, from just before the collision until just after, so that external forces such as friction can be ignored. Then we apply conservation of momentum.

(a) Before collision

**FIGURE 7–5** Example 7–3.

(b) After collision

**SOLUTION** The initial total momentum is

$$p_{\text{initial}} = m_A v_A + m_B v_B = m_A v_A$$

because car B is at rest initially $(v_B = 0)$. The direction is to the right in the $+x$ direction. After the collision, the two cars become attached, so they will have the same speed, call it $v'$. Then the total momentum after the collision is

$$p_{\text{final}} = (m_A + m_B)v'.$$

We have assumed there are no external forces, so momentum is conserved:

$$p_{\text{initial}} = p_{\text{final}}$$
$$m_A v_A = (m_A + m_B)v'.$$

Solving for $v'$, we obtain

$$v' = \frac{m_A}{m_A + m_B} v_A = \left(\frac{10{,}000 \text{ kg}}{10{,}000 \text{ kg} + 10{,}000 \text{ kg}}\right)(24.0 \text{ m/s}) = 12.0 \text{ m/s},$$

to the right. Their mutual speed after collision is half the initial speed of car A.

**NOTE** We kept symbols until the very end, so we have an equation we can use in other (related) situations.

**NOTE** We haven't included friction here. Why? Because we are examining speeds just before and just after the very brief time interval of the collision, and during that brief time friction can't do much—it is ignorable (but not for long: the cars will slow down because of friction).

**EXERCISE C** In Example 7–3, $m_A = m_B$, so in the last equation, $m_A/(m_A + m_B) = \frac{1}{2}$. Hence $v' = \frac{1}{2}v_A$. What result do you get if (a) $m_B = 3m_A$, (b) $m_B$ is much larger than $m_A$ ($m_B \gg m_A$), and (c) $m_B \ll m_A$?

**EXERCISE D** A 50-kg child runs off a dock at 2.0 m/s (horizontally) and lands in a waiting rowboat of mass 150 kg. At what speed does the rowboat move away from the dock?

The law of conservation of momentum is particularly useful when we are dealing with fairly simple systems such as colliding objects and certain types of "explosions." For example, *rocket propulsion*, which we saw in Chapter 4 can be understood on the basis of action and reaction, can also be explained on the basis of the conservation of momentum. We can consider the rocket plus its fuel as an isolated system if it is far out in space (no external forces). In the reference frame of the rocket before any fuel is ejected, the total momentum of rocket plus fuel is zero. When the fuel burns, the total momentum remains unchanged: the backward momentum of the expelled gases is just balanced by the forward momentum gained by the rocket itself (see Fig. 7–6). Thus, a rocket can accelerate in empty space. There is no need for the expelled gases to push against the Earth or the air (as is sometimes erroneously thought). Similar examples of (nearly) isolated systems where momentum is conserved are the recoil of a gun when a bullet is fired (Example 7–5), and the movement of a rowboat just after a package is thrown from it.

| **CONCEPTUAL EXAMPLE 7–4** | **Falling on or off a sled.** (a) An empty sled is sliding on frictionless ice when Susan drops vertically from a tree down onto the sled. When she lands, does the sled speed up, slow down, or keep the same speed? (b) Later: Susan falls sideways off the sled. When she drops off, does the sled speed up, slow down, or keep the same speed?

**RESPONSE** (a) Because Susan falls vertically onto the sled, she has no initial horizontal momentum. Thus the total horizontal momentum afterward equals the momentum of the sled initially. Since the mass of the system (sled + person) has increased, the speed must decrease.

(b) At the instant Susan falls off, she is moving with the same horizontal speed as she was while on the sled. At the moment she leaves the sled, she has the same momentum she had an instant before. Because her momentum does not change, neither does the sled's (total momentum conserved); the sled keeps the same speed.

**CAUTION**

*A rocket does not push on the Earth; it is propelled by pushing out the gases it burned as fuel*

**FIGURE 7–6** (a) A rocket, containing fuel, at rest in some reference frame. (b) In the same reference frame, the rocket fires and gases are expelled at high speed out the rear. The total vector momentum, $\vec{\mathbf{p}}_{\text{gas}} + \vec{\mathbf{p}}_{\text{rocket}}$, remains zero.

(a)

$$\vec{\mathbf{p}} = 0$$

(b)

$\vec{\mathbf{p}}_{\text{gas}}$     $\vec{\mathbf{p}}_{\text{rocket}}$

(a) Before shooting (at rest)

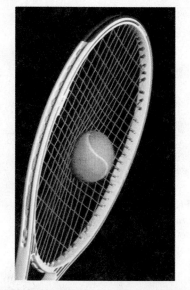

(b) After shooting

**FIGURE 7–7** Example 7–5.

**EXAMPLE 7–5** **Rifle recoil.** Calculate the recoil velocity of a 5.0-kg rifle that shoots a 0.020-kg bullet at a speed of 620 m/s, Fig. 7–7.

**APPROACH** Our system is the rifle and the bullet, both at rest initially, just before the trigger is pulled. The trigger is pulled, an explosion occurs inside the bullet's shell, and we look at the rifle and bullet just as the bullet leaves the barrel (Fig. 7–7b). The bullet moves to the right ($+x$), and the gun recoils to the left. During the very short time interval of the explosion, we can assume the external forces are small compared to the forces exerted by the exploding gunpowder. Thus we can apply conservation of momentum, at least approximately.

**SOLUTION** Let subscript B represent the bullet and R the rifle; the final velocities are indicated by primes. Then momentum conservation in the $x$ direction gives

$$\text{momentum before} = \text{momentum after}$$
$$m_B v_B + m_R v_R = m_B v'_B + m_R v'_R$$
$$0 + 0 = m_B v'_B + m_R v'_R.$$

We solve for the unknown $v'_R$, and find

$$v'_R = -\frac{m_B v'_B}{m_R} = -\frac{(0.020 \text{ kg})(620 \text{ m/s})}{(5.0 \text{ kg})} = -2.5 \text{ m/s}.$$

Since the rifle has a much larger mass, its (recoil) velocity is much less than that of the bullet. The minus sign indicates that the velocity (and momentum) of the rifle is in the negative $x$ direction, opposite to that of the bullet.

**EXERCISE E** Return to the Chapter-Opening Questions, page 170, and answer them again now. Try to explain why you may have answered differently the first time.

# 7–3 Collisions and Impulse

Collisions are a common occurrence in everyday life: a tennis racket or a baseball bat striking a ball, billiard balls colliding, a hammer hitting a nail. When a collision occurs, the interaction between the objects involved is usually far stronger than any external forces. We can then ignore the effects of any other forces during the brief time interval of the collision.

During a collision of two ordinary objects, both objects are deformed, often considerably, because of the large forces involved (Fig. 7–8). When the collision occurs, the force each exerts on the other usually jumps from zero at the moment of contact to a very large force within a very short time, and then rapidly returns to zero again. A graph of the magnitude of the force that one object exerts on the other during a collision, as a function of time, is something like the red curve in Fig. 7–9. The time interval $\Delta t$ is usually very distinct and very small, typically milliseconds for a macroscopic collision.

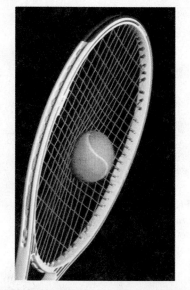

**FIGURE 7–8** Tennis racket striking a ball. Both the ball and the racket strings are deformed due to the large force each exerts on the other.

**FIGURE 7–9** Force as a function of time during a typical collision. $F$ can become very large; $\Delta t$ is typically milliseconds for macroscopic collisions.

From Newton's second law, Eq. 7–2, the *net* force on an object is equal to the rate of change of its momentum:

$$\vec{F} = \frac{\Delta\vec{p}}{\Delta t}.$$

(We have written $\vec{F}$ instead of $\Sigma\vec{F}$ for the net force, which we assume is entirely due to the brief but large average force that acts during the collision.) This equation applies to *each* of the two objects in a collision. We multiply both sides of this equation by the time interval $\Delta t$, and obtain

$$\vec{F}\,\Delta t = \Delta\vec{p}. \tag{7–4}$$

The quantity on the left, the product of the force $\vec{F}$ times the time $\Delta t$ over which the force acts, is called the **impulse**:

$$\text{Impulse} = \vec{F}\,\Delta t. \tag{7–5}$$

We see that the total change in momentum is equal to the impulse. The concept of impulse is useful mainly when dealing with forces that act during a short time interval, as when a bat hits a baseball. The force is generally not constant, and often its variation in time is like that graphed in Figs. 7–9 and 7–10. We can often approximate such a varying force as an average force $\bar{F}$ acting during a time interval $\Delta t$, as indicated by the dashed line in Fig. 7–10. $\bar{F}$ is chosen so that the area shown shaded in Fig. 7–10 (equal to $\bar{F}\times\Delta t$) is equal to the area under the actual curve of $F$ vs. $t$, Fig. 7–9 (which represents the actual impulse).

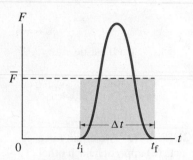

**FIGURE 7–10** The average force $\bar{F}$ acting over a very brief time interval $\Delta t$ gives the same impulse ($\bar{F}\,\Delta t$) as the actual force.

**EXERCISE F** Suppose Fig. 7–9 shows the force on a golf ball vs. time during the time interval when the ball hits a wall. How would the shape of this curve change if a softer rubber ball with the same mass and speed hit the same wall?

**EXAMPLE 7–6 ESTIMATE Karate blow.** Estimate the impulse and the average force delivered by a karate blow that breaks a board (Fig. 7–11). Assume the hand moves at roughly 10 m/s when it hits the board.

**APPROACH** We use the momentum-impulse relation, Eq. 7–4. The hand's speed changes from 10 m/s to zero over a distance of perhaps one cm (roughly how much your hand and the board compress before your hand comes to a stop, and the board begins to give way). The hand's mass should probably include part of the arm, and we take it to be roughly $m \approx 1\,\text{kg}$.

**SOLUTION** The impulse $F\,\Delta t$ equals the change in momentum

$$\bar{F}\,\Delta t = \Delta p = m\,\Delta v \approx (1\,\text{kg})(10\,\text{m/s} - 0) = 10\,\text{kg}\cdot\text{m/s}.$$

We can obtain the force if we know $\Delta t$. The hand is brought to rest over the distance of roughly a centimeter: $\Delta x \approx 1\,\text{cm}$. The average speed during the impact is $\bar{v} = (10\,\text{m/s} + 0)/2 = 5\,\text{m/s}$ and equals $\Delta x/\Delta t$. Thus $\Delta t = \Delta x/\bar{v} \approx (10^{-2}\,\text{m})/(5\,\text{m/s}) = 2\times10^{-3}\,\text{s}$ or 2 ms. The average force is thus (Eq. 7–4) about

$$\bar{F} = \frac{\Delta p}{\Delta t} = \frac{10\,\text{kg}\cdot\text{m/s}}{2\times10^{-3}\,\text{s}} \approx 5000\,\text{N} = 5\,\text{kN}.$$

**FIGURE 7–11** Example 7–6.

# 7–4 Conservation of Energy and Momentum in Collisions

During most collisions, we usually don't know how the collision force varies over time, and so analysis using Newton's second law becomes difficult or impossible. But by making use of the conservation laws for momentum and energy, we can still determine a lot about the motion after a collision, given the motion before the collision. We saw in Section 7–2 that in the collision of two objects such as billiard balls, the total momentum is conserved. If the two objects are very hard and no heat or other energy is produced in the collision, then the total kinetic energy of the two objects is the same after the collision as before. For the brief moment during which the two objects are in contact, some (or all) of the energy is stored momentarily in the form of elastic potential energy.

(a) Approach

(b) Collision

(c) If elastic

(d) If inelastic

**FIGURE 7–12** Two equal-mass objects (a) approach each other with equal speeds, (b) collide, and then (c) bounce off with equal speeds in the opposite directions if the collision is elastic, or (d) bounce back much less or not at all if the collision is inelastic (some of the KE is transformed to other forms of energy such as sound and heat).

**FIGURE 7–13** Two small objects of masses $m_A$ and $m_B$, (a) before the collision and (b) after the collision.

(a)

(b)

But if we compare the total kinetic energy just before the collision with the total kinetic energy just after the collision, and they are found to be the same, then we say that the total kinetic energy is conserved. Such a collision is called an **elastic collision**. If we use the subscripts A and B to represent the two objects, we can write the equation for conservation of total kinetic energy as

$$\text{total KE before} = \text{total KE after}$$
$$\tfrac{1}{2}m_A v_A^2 + \tfrac{1}{2}m_B v_B^2 = \tfrac{1}{2}m_A v_A'^2 + \tfrac{1}{2}m_B v_B'^2. \qquad \text{[elastic collision] (7–6)}$$

Primed quantities (') mean after the collision, and unprimed mean before the collision, just as in Eq. 7–3 for conservation of momentum.

At the atomic level the collisions of atoms and molecules are often elastic. But in the "macroscopic" world of ordinary objects, an elastic collision is an ideal that is never quite reached, since at least a little thermal energy is always produced during a collision (also perhaps sound and other forms of energy). The collision of two hard elastic balls, such as billiard balls, however, is very close to being perfectly elastic, and we often treat it as such.

We do need to remember that even when kinetic energy is not conserved, the *total* energy is always conserved.

Collisions in which kinetic energy is not conserved are said to be **inelastic collisions**. The kinetic energy that is lost is changed into other forms of energy, often thermal energy, so that the total energy (as always) is conserved. In this case,

$$\text{KE}_A + \text{KE}_B = \text{KE}_A' + \text{KE}_B' + \text{thermal and other forms of energy.}$$

See Fig. 7–12, and the details in its caption.

# 7–5 Elastic Collisions in One Dimension

We now apply the conservation laws for momentum and kinetic energy to an elastic collision between two small objects that collide head-on, so all the motion is along a line. To be general, we assume that the two objects are moving, and their velocities are $v_A$ and $v_B$ along the $x$ axis before the collision, Fig. 7–13a. After the collision, their velocities are $v_A'$ and $v_B'$, Fig. 7–13b. For any $v > 0$, the object is moving to the right (increasing $x$), whereas for $v < 0$, the object is moving to the left (toward decreasing values of $x$).

From conservation of momentum, we have

$$m_A v_A + m_B v_B = m_A v_A' + m_B v_B'.$$

Because the collision is assumed to be elastic, kinetic energy is also conserved:

$$\tfrac{1}{2}m_A v_A^2 + \tfrac{1}{2}m_B v_B^2 = \tfrac{1}{2}m_A v_A'^2 + \tfrac{1}{2}m_B v_B'^2.$$

We have two equations, so we can solve for two unknowns. If we know the masses and velocities before the collision, then we can solve these two equations for the velocities after the collision, $v_A'$ and $v_B'$. We derive a helpful result by rewriting the momentum equation as

$$m_A(v_A - v_A') = m_B(v_B' - v_B), \qquad \textbf{(i)}$$

and we rewrite the kinetic energy equation as

$$m_A(v_A^2 - v_A'^2) = m_B(v_B'^2 - v_B^2).$$

Noting that algebraically $(a^2 - b^2) = (a - b)(a + b)$, we write this last equation as

$$m_A(v_A - v_A')(v_A + v_A') = m_B(v_B' - v_B)(v_B' + v_B). \qquad \textbf{(ii)}$$

We divide Eq. (ii) by Eq. (i), and (assuming $v_A \neq v_A'$ and $v_B \neq v_B'$)[†] obtain

$$v_A + v_A' = v_B' + v_B.$$

[†]Note that Eqs. (i) and (ii), which are the conservation laws for momentum and kinetic energy, are both satisfied by the solution $v_A' = v_A$ and $v_B' = v_B$. This is a valid solution, but not very interesting. It corresponds to no collision at all—when the two objects miss each other.

We can rewrite this equation as

$$v_A - v_B = v'_B - v'_A$$

or

$$v_A - v_B = -(v'_A - v'_B). \qquad \text{[head-on (1-D) elastic collision]} \quad (7\text{–}7)$$

◆ CAUTION

*Relative speeds (one dimension only)*

This is an interesting result: it tells us that for any elastic head-on collision, the relative speed of the two objects after the collision $(v'_A - v'_B)$ has the same magnitude (but opposite direction) as before the collision, no matter what the masses are.

Equation 7–7 was derived from conservation of kinetic energy for elastic collisions, and can be used in place of it. Because the $v$'s are not squared in Eq. 7–7, it is simpler to use in calculations than the conservation of kinetic energy equation (Eq. 7–6) directly.

---

**EXAMPLE 7–7** **Equal masses.** Billiard ball A of mass $m$ moving with speed $v_A$ collides head-on with ball B of equal mass. What are the speeds of the two balls after the collision, assuming it is elastic? Assume (*a*) both balls are moving initially ($v_A$ and $v_B$), (*b*) ball B is initially at rest $(v_B = 0)$.

**APPROACH** There are two unknowns, $v'_A$ and $v'_B$, so we need two independent equations. We focus on the time interval from just before the collision until just after. No net external force acts on our system of two balls ($mg$ and the normal force cancel), so momentum is conserved. Conservation of kinetic energy applies as well because we are told the collision is elastic.

**SOLUTION** (*a*) The masses are equal $(m_A = m_B = m)$ so conservation of momentum gives

$$v_A + v_B = v'_A + v'_B.$$

We need a second equation, because there are two unknowns. We could use the conservation of kinetic energy equation, or the simpler Eq. 7–7 derived from it:

$$v_A - v_B = v'_B - v'_A.$$

We add these two equations and obtain

$$v'_B = v_A$$

and then subtract the two equations to obtain

$$v'_A = v_B.$$

That is, the balls exchange velocities as a result of the collision: ball B acquires the velocity that ball A had before the collision, and vice versa.

(*b*) If ball B is at rest initially, so that $v_B = 0$, we have

$$v'_B = v_A$$

and

$$v'_A = 0.$$

That is, ball A is brought to rest by the collision, whereas ball B acquires the original velocity of ball A. See Fig. 7–14.

**NOTE** Our result in part (*b*) is often observed by billiard and pool players, and is valid only if the two balls have equal masses (and no spin is given to the balls).

---

**FIGURE 7–14** In this multiflash photo of a head-on collision between two balls of equal mass, the white cue ball is accelerated from rest by the cue stick and then strikes the red ball, initially at rest. The white ball stops in its tracks, and the (equal-mass) red ball moves off with the same speed as the white ball had before the collision. See Example 7–7, part (*b*).

**EXAMPLE 7–8** **A nuclear collision.** A proton (p) of mass 1.01 u (unified atomic mass units) traveling with a speed of $3.60 \times 10^4$ m/s has an elastic head-on collision with a helium (He) nucleus ($m_{He} = 4.00$ u) initially at rest. What are the velocities of the proton and helium nucleus after the collision? (As mentioned in Chapter 1, 1 u = $1.66 \times 10^{-27}$ kg, but we won't need this fact.) Assume the collision takes place in nearly empty space.

**APPROACH** Like Example 7–7, this is an elastic head-on collision, but now the masses of our two particles are not equal. The only external force could be Earth's gravity, but it is insignificant compared to the powerful forces between the two particles at the moment of collision. So again we use the conservation laws of momentum and of kinetic energy, and apply them to our system of two particles.

**SOLUTION** We use the subscripts p for the proton and He for the helium nucleus. We are given $v_{He} = 0$ and $v_p = 3.60 \times 10^4$ m/s. We want to find the velocities $v'_p$ and $v'_{He}$ after the collision. From conservation of momentum,

$$m_p v_p + 0 = m_p v'_p + m_{He} v'_{He}.$$

Because the collision is elastic, the kinetic energy of our system of two particles is conserved and we can use Eq. 7–7, which becomes

$$v_p - 0 = v'_{He} - v'_p.$$

Thus

$$v'_p = v'_{He} - v_p,$$

and substituting this into our momentum equation displayed above, we get

$$m_p v_p = m_p v'_{He} - m_p v_p + m_{He} v'_{He}.$$

Solving for $v'_{He}$, we obtain

$$v'_{He} = \frac{2m_p v_p}{m_p + m_{He}} = \frac{2(1.01 \text{ u})(3.60 \times 10^4 \text{ m/s})}{(4.00 \text{ u} + 1.01 \text{ u})} = 1.45 \times 10^4 \text{ m/s}.$$

The other unknown is $v'_p$, which we can now obtain from

$$v'_p = v'_{He} - v_p = \left(1.45 \times 10^4 \text{ m/s}\right) - \left(3.60 \times 10^4 \text{ m/s}\right) = -2.15 \times 10^4 \text{ m/s}.$$

The minus sign for $v'_p$ tells us that the proton reverses direction upon collision, and we see that its speed is less than its initial speed (see Fig. 7–15).

**NOTE** This result makes sense: the lighter proton would be expected to "bounce back" from the more massive helium nucleus, but not with its full original velocity as from a rigid wall (which corresponds to extremely large, or infinite, mass).

**FIGURE 7–15** Example 7–8: (a) before collision, (b) after collision.

# 7–6 Inelastic Collisions

Collisions in which kinetic energy is not conserved are called **inelastic collisions**. Some of the initial kinetic energy is transformed into other types of energy, such as thermal or potential energy, so the total kinetic energy after the collision is less than the total kinetic energy before the collision. The inverse can also happen when potential energy (such as chemical or nuclear) is released, in which case the total kinetic energy after the interaction can be greater than the initial kinetic energy. Explosions are examples of this type.

Typical macroscopic collisions are inelastic, at least to some extent, and often to a large extent. If two objects stick together as a result of a collision, the collision is said to be **completely inelastic**. Two colliding balls of putty that stick together or two railroad cars that couple together when they collide are examples of completely inelastic collisions. The kinetic energy in some cases is all transformed to other forms of energy in an inelastic collision, but in other cases only part of it is. In Example 7–3, for instance, we saw that when a traveling railroad car collided with a stationary one, the coupled cars traveled off with some kinetic energy. In a completely inelastic collision, the maximum amount of kinetic energy is transformed to other forms consistent with conservation of momentum. Even though kinetic energy is not conserved in inelastic collisions, the total energy is always conserved, and the total vector momentum is also conserved.

## EXAMPLE 7–9 | Ballistic pendulum.

The *ballistic pendulum* is a device used to measure the speed of a projectile, such as a bullet. The projectile, of mass $m$, is fired into a large block (of wood or other material) of mass $M$, which is suspended like a pendulum. (Usually, $M$ is somewhat greater than $m$.) As a result of the collision, the pendulum and projectile together swing up to a maximum height $h$, Fig. 7–16. Determine the relationship between the initial horizontal speed of the projectile, $v$, and the maximum height $h$.

**APPROACH** We can analyze the process by dividing it into two parts or two time intervals: (1) the time interval from just before to just after the collision itself, and (2) the subsequent time interval in which the pendulum moves from the vertical hanging position to the maximum height $h$.

In part (1), Fig. 7–16a, we assume the collision time is very short, so that the projectile is embedded in the block before the block has moved significantly from its rest position directly below its support. Thus there is effectively no net external force, and we can apply conservation of momentum to this completely inelastic collision.

In part (2), Fig. 7–16b, the pendulum begins to move, subject to a net external force (gravity, tending to pull it back to the vertical position); so for part (2), we cannot use conservation of momentum. But we can use conservation of mechanical energy because gravity is a conservative force (Chapter 6). The kinetic energy immediately after the collision is changed entirely to gravitational potential energy when the pendulum reaches its maximum height, $h$.

**SOLUTION** In part (1) momentum is conserved:

$$\text{total } p \text{ before} = \text{total } p \text{ after}$$
$$mv = (m + M)v', \tag{i}$$

where $v'$ is the speed of the block and embedded projectile just after the collision, before they have moved significantly.

In part (2), mechanical energy is conserved. We choose $y = 0$ when the pendulum hangs vertically, and then $y = h$ when the pendulum–projectile system reaches its maximum height. Thus we write

$$(\text{KE} + \text{PE}) \text{ just after collision} = (\text{KE} + \text{PE}) \text{ at pendulum's maximum height}$$

or

$$\tfrac{1}{2}(m + M)v'^2 + 0 = 0 + (m + M)gh. \tag{ii}$$

We solve for $v'$:

$$v' = \sqrt{2gh}.$$

Inserting this result for $v'$ into Eq. (i) above, and solving for $v$, gives

$$v = \frac{m + M}{m} v' = \frac{m + M}{m}\sqrt{2gh},$$

which is our final result.

**NOTE** The separation of the process into two parts was crucial. Such an analysis is a powerful problem-solving tool. But how do you decide how to make such a division? Think about the conservation laws. They are your *tools*. Start a problem by asking yourself whether the conservation laws apply in the given situation. Here, we determined that momentum is conserved only during the brief collision, which we called part (1). But in part (1), because the collision is inelastic, the conservation of mechanical energy is not valid. Then in part (2), conservation of mechanical energy is valid, but not conservation of momentum.

Note, however, that if there had been significant motion of the pendulum during the deceleration of the projectile in the block, then there *would* have been an external force (gravity) during the collision, so conservation of momentum would not have been valid in part (1).

(a)

(b)

**FIGURE 7–16** Ballistic pendulum. Example 7–9.

PROBLEM SOLVING
*Use the conservation laws to analyze a problem*

**EXAMPLE 7–10** **Railroad cars again.** For the completely inelastic collision of the two railroad cars that we considered in Example 7–3, calculate how much of the initial kinetic energy is transformed to thermal or other forms of energy.

**APPROACH** The railroad cars stick together after the collision, so this is a completely inelastic collision. By subtracting the total kinetic energy after the collision from the total initial kinetic energy, we can find how much energy is transformed to other types of energy.

**SOLUTION** Before the collision, only car A is moving, so the total initial kinetic energy is $\frac{1}{2}m_A v_A^2 = \frac{1}{2}(10,000\,\text{kg})(24.0\,\text{m/s})^2 = 2.88 \times 10^6\,\text{J}$. After the collision, both cars are moving with half the speed, $v' = 12.0\,\text{m/s}$, by conservation of momentum (Example 7–3). So the total kinetic energy afterward is $\text{KE}' = \frac{1}{2}(m_A + m_B)v'^2 = \frac{1}{2}(20,000\,\text{kg})(12.0\,\text{m/s})^2 = 1.44 \times 10^6\,\text{J}$. Hence the energy transformed to other forms is

$$(2.88 \times 10^6\,\text{J}) - (1.44 \times 10^6\,\text{J}) = 1.44 \times 10^6\,\text{J},$$

which is half the original kinetic energy.

**FIGURE 7–17** A recent color-enhanced version of a cloud-chamber photograph made in the early days (1920s) of nuclear physics. Green lines are paths of helium nuclei (He) coming from the left. One He, highlighted in yellow, strikes a proton of the hydrogen gas in the chamber, and both scatter at an angle; the scattered proton's path is shown in red.

# *7–7 Collisions in Two Dimensions

Conservation of momentum and energy can also be applied to collisions in two or three dimensions, where the vector nature of momentum is especially important. One common type of non-head-on collision is that in which a moving object (called the "projectile") strikes a second object initially at rest (the "target"). This is the common situation in games such as billiards and pool, and for experiments in atomic and nuclear physics (the projectiles, from radioactive decay or a high-energy accelerator, strike a stationary target nucleus, Fig. 7–17).

Figure 7–18 shows the incoming projectile, $m_A$, heading along the $x$ axis toward the target object, $m_B$, which is initially at rest. If these are billiard balls, $m_A$ strikes $m_B$ not quite head-on and they go off at the angles $\theta'_A$ and $\theta'_B$, respectively, which are measured relative to $m_A$'s initial direction (the $x$ axis).[†]

**FIGURE 7–18** Object A, the projectile, collides with object B, the target. After the collision, they move off with momenta $\vec{\mathbf{p}}'_A$ and $\vec{\mathbf{p}}'_B$ at angles $\theta'_A$ and $\theta'_B$.

Let us apply the law of conservation of momentum to a collision like that of Fig. 7–18. We choose the $xy$ plane to be the plane in which the initial and final momenta lie. Momentum is a vector, and because the total momentum is conserved, its components in the $x$ and $y$ directions also are conserved. The $x$ component of momentum conservation gives

$$p_{Ax} + p_{Bx} = p'_{Ax} + p'_{Bx}$$

or, with $p_{Bx} = m_B v_{Bx} = 0$,

$$m_A v_A = m_A v'_A \cos \theta'_A + m_B v'_B \cos \theta'_B, \tag{7–8a}$$

where primes (′) refer to quantities *after* the collision. There is no motion in the $y$ direction initially, so the $y$ component of the total momentum is zero before the collision.

[†]The objects may begin to deflect even before they touch if electric, magnetic, or nuclear forces act between them. You might think, for example, of two magnets oriented so that they repel each other: when one moves toward the other, the second moves away before the first one touches it.

The $y$ component equation of momentum conservation is then

$$p_{Ay} + p_{By} = p'_{Ay} + p'_{By}$$

or

$$0 = m_A v'_A \sin \theta'_A + m_B v'_B \sin \theta'_B. \qquad \textbf{(7–8b)}$$

When we have two independent equations, we can solve for two unknowns at most.

---

**EXAMPLE 7–11** **Billiard ball collision in 2-D.** Billiard ball A moving with speed $v_A = 3.0$ m/s in the $+x$ direction (Fig. 7–19) strikes an equal-mass ball B initially at rest. The two balls are observed to move off at 45° to the $x$ axis, ball A above the $x$ axis and ball B below. That is, $\theta'_A = 45°$ and $\theta'_B = -45°$ in Fig. 7–19. What are the speeds of the two balls after the collision?

**APPROACH** There is no net external force on our system of two balls, assuming the table is level (the normal force balances gravity). Thus momentum conservation applies, and we apply it to both the $x$ and $y$ components using the $xy$ coordinate system shown in Fig. 7–19. We get two equations, and we have two unknowns, $v'_A$ and $v'_B$. From symmetry we might guess that the two balls have the same speed. But let us not assume that now. Even though we are not told whether the collision is elastic or inelastic, we can still use conservation of momentum.

**SOLUTION** We apply conservation of momentum for the $x$ and $y$ components, Eqs. 7–8a and b, and we solve for $v'_A$ and $v'_B$. We are given $m_A = m_B (= m)$, so

$$\text{(for } x) \quad mv_A = mv'_A \cos(45°) + mv'_B \cos(-45°)$$

and

$$\text{(for } y) \quad 0 = mv'_A \sin(45°) + mv'_B \sin(-45°).$$

The $m$'s cancel out in both equations (the masses are equal). The second equation yields [recall from trigonometry that $\sin(-\theta) = -\sin\theta$]:

$$v'_B = -v'_A \frac{\sin(45°)}{\sin(-45°)} = -v'_A \left( \frac{\sin 45°}{-\sin 45°} \right) = v'_A.$$

So they do have equal speeds as we guessed at first. The $x$ component equation gives [recall that $\cos(-\theta) = \cos\theta$]:

$$v_A = v'_A \cos(45°) + v'_B \cos(45°) = 2v'_A \cos(45°);$$

solving for $v'_A$ (which also equals $v'_B$) gives

$$v'_A = \frac{v_A}{2\cos(45°)} = \frac{3.0\text{ m/s}}{2(0.707)} = 2.1\text{ m/s}.$$

**FIGURE 7–19** Example 7–11.

---

If we know that a collision is elastic, we can also apply conservation of kinetic energy and obtain a third equation in addition to Eqs. 7–8a and b:

$$KE_A + KE_B = KE'_A + KE'_B$$

or, for the collision shown in Fig. 7–18 or 7–19 (where $KE_B = 0$),

$$\tfrac{1}{2} m_A v_A^2 = \tfrac{1}{2} m_A v'^2_A + \tfrac{1}{2} m_B v'^2_B. \qquad \text{[elastic collision]} \quad \textbf{(7–8c)}$$

If the collision is elastic, we have three independent equations and can solve for three unknowns. If we are given $m_A$, $m_B$, $v_A$ (and $v_B$, if it is not zero), we cannot, for example, predict the final variables, $v'_A$, $v'_B$, $\theta'_A$, and $\theta'_B$, because there are four of them. However, if we measure one of these variables, say $\theta'_A$, then the other three variables ($v'_A$, $v'_B$, and $\theta'_B$) are uniquely determined, and we can determine them using Eqs. 7–8a, b, c.

A note of caution: Eq. 7–7 (page 179) does *not* apply for two-dimensional collisions. It works only when a collision occurs along a line.

**⚠ CAUTION**

*Equation 7–7 applies only in 1-D*

## Momentum Conservation and Collisions

1. Choose your **system**. If the situation is complex, think about how you might break it up into separate parts when one or more conservation laws apply.

2. Consider whether a significant **net external force** acts on your chosen system; if it does, be sure the time interval $\Delta t$ is so short that the effect on momentum is negligible. That is, the forces that act between the interacting objects must be the only significant ones if momentum conservation is to be used. [Note: If this is valid for a portion of the problem, you can use momentum conservation only for that portion.]

3. Draw a **diagram** of the initial situation, just before the interaction (collision, explosion) takes place, and represent the momentum of each object with an arrow and a label. Do the same for the final situation, just after the interaction.

4. Choose a **coordinate system** and "+" and "−" directions. (For a head-on collision, you will need only an $x$ axis.) It is often convenient to choose the $+x$ axis in the direction of one object's initial velocity.

5. Apply the **momentum conservation** equation(s):

   total initial momentum = total final momentum.

   You have one equation for each component ($x$, $y$, $z$): only one equation for a head-on collision. [Don't forget that it is the *total* momentum of the system that is conserved, not the momenta of individual objects.]

6. If the collision is elastic, you can also write down a **conservation of kinetic energy** equation:

   total initial KE = total final KE.

   [Alternatively, you could use Eq. 7–7:

   $$v_A - v_B = v'_B - v'_A,$$

   if the collision is one dimensional (head-on).]

7. Solve for the **unknown(s)**.

8. **Check** your work, check the units, and ask yourself whether the results are reasonable.

# 7–8 Center of Mass (CM)

Momentum is a powerful concept not only for analyzing collisions but also for analyzing the translational motion of real extended objects. Until now, whenever we have dealt with the motion of an extended object (that is, an object that has size), we have assumed that it could be approximated as a point particle or that it undergoes only translational motion. Real extended objects, however, can undergo rotational and other types of motion as well. For example, the diver in Fig. 7–20a undergoes only translational motion (all parts of the object follow the same path), whereas the diver in Fig. 7–20b undergoes both translational and rotational motion. We will refer to motion that is not pure translation as *general motion*.

Observations indicate that even if an object rotates, or several parts of a system of objects move relative to one another, there is one point that moves in the same path that a particle would move if subjected to the same net force. This point is called the **center of mass** (abbreviated CM). The general motion of an extended object (or system of objects) can be considered as *the sum of the translational motion of the CM, plus rotational, vibrational, or other types of motion about the CM*.

As an example, consider the motion of the center of mass of the diver in Fig. 7–20; the CM follows a parabolic path even when the diver rotates, as shown in Fig. 7–20b. This is the same parabolic path that a projected particle follows when acted on only by the force of gravity (projectile motion, Chapter 3). Other points in the rotating diver's body, such as her feet or head, follow more complicated paths.

(a)

(b)

**FIGURE 7–20** The motion of the diver is pure translation in (a), but is translation plus rotation in (b). The black dot represents the diver's CM at each moment.

**FIGURE 7–21** Translation plus rotation: a wrench moving over a smooth horizontal surface. The CM, marked with a red cross, moves in a straight line because no net force acts on the wrench.

Figure 7–21 shows a wrench acted on by zero net force, translating and rotating along a horizontal surface. Note that its CM, marked by a red cross, moves in a straight line, as shown by the dashed white line.

We will show in Section 7–10 that the important properties of the CM follow from Newton's laws if the CM is defined in the following way. We can consider any extended object as being made up of many tiny particles. But first we consider a system made up of only two particles (or small objects), of masses $m_A$ and $m_B$. We choose a coordinate system so that both particles lie on the $x$ axis at positions $x_A$ and $x_B$, Fig. 7–22. The center of mass of this system is defined to be at the position $x_{CM}$, given by

$$x_{CM} = \frac{m_A x_A + m_B x_B}{m_A + m_B} = \frac{m_A x_A + m_B x_B}{M},$$

where $M = m_A + m_B$ is the total mass of the system. The center of mass lies on the line joining $m_A$ and $m_B$. If the two masses are equal $(m_A = m_B = m)$, then $x_{CM}$ is midway between them, because in this case

$$x_{CM} = \frac{m(x_A + x_B)}{2m} = \frac{(x_A + x_B)}{2}.$$

If one mass is greater than the other, then the CM is closer to the larger mass.

If there are more than two particles along a line, there will be additional terms:

$$x_{CM} = \frac{m_A x_A + m_B x_B + m_C x_C + \cdots}{m_A + m_B + m_C + \cdots} = \frac{m_A x_A + m_B x_B + m_C x_C + \cdots}{M}, \quad \textbf{(7–9a)}$$

where $M$ is the total mass of all the particles.

**FIGURE 7–22** The center of mass of a two-particle system lies on the line joining the two masses. Here $m_A > m_B$, so the CM is closer to $m_A$ than to $m_B$.

---

**EXAMPLE 7–12** **CM of three guys on a raft.** On a lightweight (air-filled) "banana boat," three people of roughly equal mass $m$ sit along the $x$ axis at positions $x_A = 1.0\,\text{m}$, $x_B = 5.0\,\text{m}$, and $x_C = 6.0\,\text{m}$, measured from the left-hand end as shown in Fig. 7–23. Find the position of the CM. Ignore the mass of the boat.

**APPROACH** We are given the mass and location of the three people, so we use three terms in Eq. 7–9a. We approximate each person as a point particle. Equivalently, the location of each person is the position of that person's own CM.

**SOLUTION** We use Eq. 7–9a with three terms:

$$x_{CM} = \frac{mx_A + mx_B + mx_C}{m + m + m} = \frac{m(x_A + x_B + x_C)}{3m}$$
$$= \frac{(1.0\,\text{m} + 5.0\,\text{m} + 6.0\,\text{m})}{3} = \frac{12.0\,\text{m}}{3} = 4.0\,\text{m}.$$

The CM is 4.0 m from the left-hand end of the boat.

**FIGURE 7–23** Example 7–12.

**EXERCISE G** Calculate the CM of the three people in Example 7–12, taking the origin at the driver $(x_C = 0)$ on the right. Is the physical location of the CM the same?

Note that the coordinates of the CM depend on the reference frame or coordinate system chosen. But the physical location of the CM is independent of that choice.

If the particles are spread out in two or three dimensions, then we must specify not only the $x$ coordinate of the CM $(x_{CM})$, but also the $y$ and $z$ coordinates, which will be given by formulas like Eq. 7–9a. For example, the $y$ coordinate of the CM will be

$$y_{CM} = \frac{m_A y_A + m_B y_B + \cdots}{m_A + m_B + \cdots} = \frac{m_A y_A + m_B y_B + \cdots}{M} \quad \textbf{(7–9b)}$$

where $M$ is the total mass of all the particles.

**FIGURE 7–24** The force of gravity, considered to act at the CG, causes this object to rotate about the pivot point; if the CG were on a vertical line directly below the pivot, the object would remain at rest.

**FIGURE 7–25** Finding the CG.

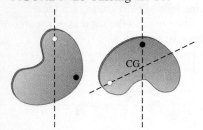

A concept similar to *center of mass* is **center of gravity** (CG). An object's CG is that point at which the force of gravity can be considered to act. The force of gravity actually acts on *all* the different parts or particles of an object, but for purposes of determining the translational motion of an object as a whole, we can assume that the entire weight of the object (which is the sum of the weights of all its parts) acts at the CG. There is a conceptual difference between the center of gravity and the center of mass, but for nearly all practical purposes, they are at the same point.[†]

It is often easier to determine the CM or CG of an extended object experimentally rather than analytically. If an object is suspended from any point, it will swing (Fig. 7–24) due to the force of gravity on it, unless it is placed so its CG lies on a vertical line directly below the point from which it is suspended. If the object is two dimensional, or has a plane of symmetry, it need only be hung from two different pivot points and the respective vertical (plumb) lines drawn. Then the center of gravity will be at the intersection of the two lines, as in Fig. 7–25. If the object doesn't have a plane of symmetry, the CG with respect to the third dimension is found by suspending the object from at least three points whose plumb lines do not lie in the same plane.

For symmetrically shaped objects such as uniform cylinders (wheels), spheres, and rectangular solids, the CM is located at the geometric center of the object.

To locate the center of mass of a group of extended objects, we can use Eqs. 7–9, where the *m*'s are the masses of these objects and the *x*'s, *y*'s, and *z*'s are the coordinates of the CM of each of the objects.

# *7–9 CM for the Human Body

For a group of extended objects, each of whose CM is known, we can find the CM of the group using Eqs. 7–9a and b. As an example, we consider the human body. Table 7–1 indicates the CM and hinge points (joints) for the different components of a "representative" person. Of course, there are wide variations among people, so these data represent only a very rough average. The numbers represent a *percentage* of the total height, which is regarded as 100 units; similarly, the total mass is 100 units. For example, if a person is 1.70 m tall, his or her shoulder joint would be $(1.70\text{ m})(81.2/100) = 1.38\text{ m}$ above the floor.

**TABLE 7–1 Center of Mass of Parts of Typical Human Body, given as %**
(full height and mass = 100 units)

| Distance of Hinge Points from Floor (%) | Hinge Points (•) (Joints) | Center of Mass (×) (% Height Above Floor) | | Percent Mass |
|---|---|---|---|---|
| | | Head | 93.5% | 6.9% |
| 91.2% | Base of skull on spine | Trunk and neck | 71.1% | 46.1% |
| 81.2% | Shoulder joint | Upper arms | 71.7% | 6.6% |
| | elbow 62.2%[‡] | Lower arms | 55.3% | 4.2% |
| | wrist 46.2%[‡] | Hands | 43.1% | 1.7% |
| 52.1% | Hip joint | Upper legs (thighs) | 42.5% | 21.5% |
| 28.5% | Knee joint | | | |
| | | Lower legs | 18.2% | 9.6% |
| 4.0% | Ankle joint | Feet | 1.8% | 3.4% |
| | | Body CM = | 58.0% | 100.0% |

[‡] For arm hanging vertically.

[†]There would be a difference between the CM and CG only in the unusual case of an object so large that the acceleration due to gravity, *g*, was different at different parts of the object.

**EXAMPLE 7–13** **A leg's CM.** Determine the position of the CM of a whole leg (*a*) when stretched out, and (*b*) when bent at 90°. See Fig. 7–26. Assume the person is 1.70 m tall.

(a)

**APPROACH** Our system consists of three objects: upper leg, lower leg, and foot. The location of the CM of each object, as well as the mass of each, is given in Table 7–1, where they are expressed in percentage units. To express the results in meters, these percentage values need to be multiplied by (1.70 m/100). When the leg is stretched out, the problem is one dimensional and we can solve for the *x* coordinate of the CM. When the leg is bent, the problem is two dimensional and we need to find both the *x* and *y* coordinates.

**SOLUTION** (*a*) We determine the distances from the hip joint using Table 7–1 and obtain the numbers (%) shown in Fig. 7–26a. Using Eq. 7–9a, we obtain ($ul$ = upper leg, etc.)

(b)

$$x_{CM} = \frac{m_{ul}x_{ul} + m_{\ell\ell}x_{\ell\ell} + m_f x_f}{m_{ul} + m_{\ell\ell} + m_f}$$

$$= \frac{(21.5)(9.6) + (9.6)(33.9) + (3.4)(50.3)}{21.5 + 9.6 + 3.4} = 20.4 \text{ units.}$$

Thus, the center of mass of the leg and foot is 20.4 units from the hip joint, or $52.1 - 20.4 = 31.7$ units from the base of the foot. Since the person is 1.70 m tall, this is $(1.70 \text{ m})(31.7/100) = 0.54 \text{ m}$ above the bottom of the foot.

**FIGURE 7–26** Example 7–13: finding the CM of a leg in two different positions using percentages from Table 7–1. ($\otimes$ represents the calculated CM.)

(*b*) We use an *xy* coordinate system, as shown in Fig. 7–26b. First, we calculate how far to the right of the hip joint the CM lies, accounting for all three parts:

$$x_{CM} = \frac{(21.5)(9.6) + (9.6)(23.6) + (3.4)(23.6)}{21.5 + 9.6 + 3.4} = 14.9 \text{ units.}$$

For our 1.70-m-tall person, this is $(1.70 \text{ m})(14.9/100) = 0.25 \text{ m}$ from the hip joint. Next, we calculate the distance, $y_{CM}$, of the CM above the floor:

$$y_{CM} = \frac{(3.4)(1.8) + (9.6)(18.2) + (21.5)(28.5)}{3.4 + 9.6 + 21.5} = 23.0 \text{ units,}$$

or $(1.70 \text{ m})(23.0/100) = 0.39 \text{ m}$. Thus, the CM is located 39 cm above the floor and 25 cm to the right of the hip joint.

**NOTE** The CM lies outside the body in (*b*).

**FIGURE 7–27** A high jumper's CM may actually pass beneath the bar.

Knowing the CM of the body when it is in various positions is of great use in studying body mechanics. One simple example from athletics is shown in Fig. 7–27. If high jumpers can get into the position shown, their CM can pass below the bar which their bodies go over, meaning that for a particular takeoff speed, they can clear a higher bar. This is indeed what they try to do.

**PHYSICS APPLIED**
*The high jump*

# *7–10 CM and Translational Motion

As mentioned in Section 7–8, a major reason for the importance of the concept of center of mass is that the motion of the CM for a system of particles (or an extended object) is directly related to the net force acting on the system as a whole. We now show this, taking the simple case of one-dimensional motion (*x* direction) and only three particles, but the extension to more objects and to three dimensions follows the same reasoning.

Suppose the three particles lie on the $x$ axis and have masses $m_A$, $m_B$, $m_C$, and positions $x_A$, $x_B$, $x_C$. From Eq. 7–9a for the center of mass, we can write

$$Mx_{CM} = m_A x_A + m_B x_B + m_C x_C,$$

where $M = m_A + m_B + m_C$ is the total mass of the system. If these particles are in motion (say, along the $x$ axis with velocities $v_A$, $v_B$, and $v_C$, respectively), then in a short time interval $\Delta t$ each particle and the CM will have traveled a distance $\Delta x = v\Delta t$, so that

$$Mv_{CM} \Delta t = m_A v_A \Delta t + m_B v_B \Delta t + m_C v_C \Delta t.$$

We cancel $\Delta t$ and get

$$Mv_{CM} = m_A v_A + m_B v_B + m_C v_C. \tag{7–10}$$

Since $m_A v_A + m_B v_B + m_C v_C$ is the sum of the momenta of the particles of the system, it represents the *total momentum* of the system. Thus we see from Eq. 7–10 that *the total (linear) momentum of a system of particles is equal to the product of the total mass M and the velocity of the center of mass of the system*. Or, *the linear momentum of an extended object is the product of the object's mass and the velocity of its CM.*

If forces are acting on the particles, then the particles may be accelerating. In a short time interval $\Delta t$, each particle's velocity will change by an amount $\Delta v = a \Delta t$. If we use the same reasoning as we did to obtain Eq. 7–10, we find

$$Ma_{CM} = m_A a_A + m_B a_B + m_C a_C.$$

According to Newton's second law, $m_A a_A = F_A$, $m_B a_B = F_B$, and $m_C a_C = F_C$, where $F_A$, $F_B$, and $F_C$ are the net forces on the three particles, respectively. Thus we get for the system as a whole $Ma_{CM} = F_A + F_B + F_C$, or

| NEWTON'S SECOND LAW (for a system) |

$$Ma_{CM} = F_{net}. \tag{7–11}$$

That is, *the sum of all the forces acting on the system is equal to the total mass of the system times the acceleration of its center of mass*. This is **Newton's second law** for a system of particles. It also applies to an extended object (which can be thought of as a collection of particles). Thus the center of mass of a system of particles (or of an object) with total mass $M$ moves as if all its mass were concentrated at the center of mass and all the external forces acted at that point. We can thus treat the *translational motion* of any object or system of objects as the motion of a particle (see Figs. 7–20 and 7–21). This result simplifies our analysis of the motion of complex systems and extended objects. Although the motion of various parts of the system may be complicated, we may often be satisfied with knowing the motion of the center of mass. This result also allows us to solve certain types of problems very easily, as illustrated by the following Example.

**CONCEPTUAL EXAMPLE 7–14** **A two-stage rocket.** A rocket is shot into the air as shown in Fig. 7–28. At the moment the rocket reaches its highest point, a horizontal distance $d$ from its starting point, a prearranged explosion separates it into two parts of equal mass. Part I is stopped in midair by the explosion, and it falls vertically to Earth. Where does part II land? Assume $\vec{g}$ = constant.

**FIGURE 7–28** Example 7–14.

**RESPONSE** After the rocket is fired, the path of the CM of the system continues to follow the parabolic trajectory of a projectile acted on by only a constant gravitational force. The CM will thus land at a point $2d$ from the starting point. Since the masses of I and II are equal, the CM must be midway between them at any time. Therefore, part II lands a distance $3d$ from the starting point.

**NOTE** If part I had been given a kick up or down, instead of merely falling, the solution would have been more complicated.

**EXERCISE H** A woman stands up in a rowboat and walks from one end of the boat to the other. How does the boat move, as seen from the shore?

An interesting application is the discovery of nearby stars (see Section 5–8) that seem to "wobble." What could cause such a wobble? It could be that a planet orbits the star, and each exerts a gravitational force on the other. The planets are too small and too far away to be observed directly by telescopes. But the slight wobble in the motion of the star suggests that both the planet and the star (its sun) orbit about their mutual center of mass, and hence the star appears to have a wobble. Irregularities in the star's motion can be measured to high accuracy, yielding information on the size of the planets' orbits and their masses. See Fig. 5–30 in Chapter 5.

**PHYSICS APPLIED**
*Distant planets discovered*

# Summary

The **linear momentum**, $\vec{p}$, of an object is defined as the product of its mass times its velocity,

$$\vec{p} = m\vec{v}. \tag{7–1}$$

In terms of momentum, **Newton's second law** can be written as

$$\Sigma\vec{F} = \frac{\Delta\vec{p}}{\Delta t}. \tag{7–2}$$

That is, the rate of change of momentum of an object equals the net force exerted on it.

When the net external force on a system of objects is zero, the total momentum remains constant. This is the **law of conservation of momentum**. Stated another way, the total momentum of an isolated system of objects remains constant.

The law of conservation of momentum is very useful in dealing with **collisions**. In a collision, two (or more) objects interact with each other over a very short time interval, and the force each exerts on the other during this time interval is very large compared to any other forces acting.

The **impulse** delivered by a force on an object is defined as

$$\text{Impulse} = \vec{F}\,\Delta t, \tag{7–5}$$

where $\vec{F}$ is the average force acting during the (usually very short) time interval $\Delta t$. The impulse is equal to the change in momentum of the object:

$$\text{Impulse} = \vec{F}\,\Delta t = \Delta\vec{p}. \tag{7–4}$$

Total momentum is conserved in *any* collision as long as any net external force is zero or negligible. If $m_A\vec{v}_A$ and $m_B\vec{v}_B$ are the momenta of two objects before the collision and $m_A\vec{v}'_A$

and $m_B\vec{v}'_B$ are their momenta after, then momentum conservation tells us that

$$m_A\vec{v}_A + m_B\vec{v}_B = m_A\vec{v}'_A + m_B\vec{v}'_B \tag{7–3}$$

for this two-object system.

Total energy is also conserved. But this may not be helpful unless kinetic energy is conserved, in which case the collision is called an **elastic collision** and we can write

$$\tfrac{1}{2}m_A v_A^2 + \tfrac{1}{2}m_B v_B^2 = \tfrac{1}{2}m_A v_A'^2 + \tfrac{1}{2}m_B v_B'^2. \tag{7–6}$$

If kinetic energy is not conserved, the collision is called **inelastic**. Macroscopic collisions are generally inelastic. A **completely inelastic** collision is one in which the colliding objects stick together after the collision.

The **center of mass** (CM) of an extended object (or group of objects) is that point at which the net force can be considered to act, for purposes of determining the translational motion of the object as a whole. The $x$ component of the CM for objects with mass $m_A$, $m_B$, ..., is given by

$$x_{CM} = \frac{m_A x_A + m_B x_B + \cdots}{m_A + m_B + \cdots}. \tag{7–9a}$$

[*The center of mass of a system of total mass $M$ moves in the same path that a particle of mass $M$ would move if subjected to the same net external force. In equation form, this is Newton's second law for a system of particles (or extended objects):

$$Ma_{CM} = F_{net} \tag{7–11}$$

where $M$ is the total mass of the system, $a_{CM}$ is the acceleration of the CM of the system, and $F_{net}$ is the total (net) external force acting on all parts of the system.]

# Questions

1. We claim that momentum is conserved. Yet most moving objects eventually slow down and stop. Explain.

2. A light object and a heavy object have the same kinetic energy. Which has the greater momentum? Explain.

3. When a person jumps from a tree to the ground, what happens to the momentum of the person upon striking the ground?

4. When you release an inflated but untied balloon, why does it fly across the room?

5. Explain, on the basis of conservation of momentum, how a fish propels itself forward by swishing its tail back and forth.

6. Two children float motionlessly in a space station. The 20-kg girl pushes on the 40-kg boy and he sails away at 1.0 m/s. The girl (a) remains motionless; (b) moves in the same direction at 1.0 m/s; (c) moves in the opposite direction at 1.0 m/s; (d) moves in the opposite direction at 2.0 m/s; (e) none of these.

7. According to Eq. 7–4, the longer the impact time of an impulse, the smaller the force can be for the same momentum change, and hence the smaller the deformation of the object on which the force acts. On this basis, explain the value of air bags, which are intended to inflate during an automobile collision and reduce the possibility of fracture or death.

8. If a falling ball were to make a perfectly elastic collision with the floor, would it rebound to its original height? Explain.

9. A boy stands on the back of a rowboat and dives into the water. What happens to the boat as he leaves it? Explain.

10. It is said that in ancient times a rich man with a bag of gold coins was stranded on the surface of a frozen lake. Because the ice was frictionless, he could not push himself to shore and froze to death. What could he have done to save himself had he not been so miserly?

11. The speed of a tennis ball on the return of a serve can be just as fast as the serve, even though the racket isn't swung very fast. How can this be?

12. Is it possible for an object to receive a larger impulse from a small force than from a large force? Explain.

13. In a collision between two cars, which would you expect to be more damaging to the occupants: if the cars collide and remain together, or if the two cars collide and rebound backward? Explain.

14. A very elastic "superball" is dropped from a height $h$ onto a hard steel plate (fixed to the Earth), from which it rebounds at very nearly its original speed. (a) Is the momentum of the ball conserved during any part of this process? (b) If we consider the ball and the Earth as our system, during what parts of the process is momentum conserved? (c) Answer part (b) for a piece of putty that falls and sticks to the steel plate.

15. Cars used to be built as rigid as possible to withstand collisions. Today, though, cars are designed to have "crumple zones" that collapse upon impact. What is the advantage of this new design?

16. At a hydroelectric power plant, water is directed at high speed against turbine blades on an axle that turns an electric generator. For maximum power generation, should the turbine blades be designed so that the water is brought to a dead stop, or so that the water rebounds?

17. A squash ball hits a wall at a 45° angle as shown in Fig. 7–29. What is the direction (a) of the change in momentum of the ball, (b) of the force on the wall?

**FIGURE 7–29** Question 17.

18. Why can a batter hit a pitched baseball farther than a ball he himself has tossed up in the air?

19. Describe a collision in which all kinetic energy is lost.

20. If a 20-passenger plane is not full, sometimes passengers are told they must sit in certain seats and may not move to empty seats. Why might this be?

21. Why do you tend to lean backward when carrying a heavy load in your arms?

22. Why is the CM of a 1-m length of pipe at its midpoint, whereas this is not true for your arm or leg?

23. How can a rocket change direction when it is far out in space and essentially in a vacuum?

24. Bob and Jim decide to play tug-of-war on a frictionless (icy) surface. Jim is considerably stronger than Bob, but Bob weighs 160 lb whereas Jim weighs 145 lb. Who loses by crossing over the midline first? Explain.

*25. In one type of nuclear radioactive decay, an electron and a recoil nucleus are emitted but often do not separate along the same line. Use conservation of momentum in two dimensions to explain why this implies the emission of at least one other particle (it came to be called a "neutrino").

*26. Show on a diagram how your CM shifts when you move from a lying position to a sitting position.

*27. If only an external force can change the momentum of the center of mass of an object, how can the internal force of the engine accelerate a car?

*28. A rocket following a parabolic path through the air suddenly explodes into many pieces. What can you say about the motion of this system of pieces?

# MisConceptual Questions

1. A truck going 15 km/h has a head-on collision with a small car going 30 km/h. Which statement best describes the situation?
   (a) The truck has the greater change of momentum because it has the greater mass.
   (b) The car has the greater change of momentum because it has the greater speed.
   (c) Neither the car nor the truck changes its momentum in the collision because momentum is conserved.
   (d) They both have the same change in magnitude of momentum because momentum is conserved.
   (e) None of the above is necessarily true.

2. A small boat coasts at constant speed under a bridge. A heavy sack of sand is dropped from the bridge onto the boat. The speed of the boat
   (a) increases.
   (b) decreases.
   (c) does not change.
   (d) Without knowing the mass of the boat and the sand, we can't tell.

3. Two identical billiard balls traveling at the same speed have a head-on collision and rebound. If the balls had twice the mass, but maintained the same size and speed, how would the rebound be different?
   (a) At a higher speed.
   (b) At slower speed.
   (c) No difference.

4. An astronaut is a short distance away from her space station without a tether rope. She has a large wrench. What should she do with the wrench to move toward the space station?
   (a) Throw it directly away from the space station.
   (b) Throw it directly toward the space station.
   (c) Throw it toward the station without letting go of it.
   (d) Throw it parallel to the direction of the station's orbit.
   (e) Throw it opposite to the direction of the station's orbit.

5. A space vehicle, in circular orbit around the Earth, collides with a small asteroid which ends up in the vehicle's storage bay. For this collision,
   (a) only momentum is conserved.
   (b) only kinetic energy is conserved.
   (c) both momentum and kinetic energy are conserved.
   (d) neither momentum nor kinetic energy is conserved.

6. A golf ball and an equal-mass bean bag are dropped from the same height and hit the ground. The bean bag stays on the ground while the golf ball rebounds. Which experiences the greater impulse from the ground?
   (a) The golf ball.
   (b) The bean bag.
   (c) Both the same.
   (d) Not enough information.

7. You are lying in bed and want to shut your bedroom door. You have a bouncy ball and a blob of clay, both with the same mass. Which one would be more effective to throw at your door to close it?
   (a) The bouncy ball.
   (b) The blob of clay.
   (c) Both the same.
   (d) Neither will work.

8. A baseball is pitched horizontally toward home plate with a velocity of 110 km/h. In which of the following scenarios does the change in momentum of the baseball have the largest magnitude?
   (a) The catcher catches the ball.
   (b) The ball is popped straight up at a speed of 110 km/h.
   (c) The baseball is hit straight back to the pitcher at a speed of 110 km/h.
   (d) Scenarios (a) and (b) have the same change in momentum.
   (e) Scenarios (a), (b), and (c) have the same change in momentum.

9. A small car and a heavy pickup truck are both out of gas. The truck has twice the mass of the car. After you push first the car and then the truck for the *same amount of time* with the same force, what can you say about the momentum and kinetic energy (KE) of the car and the truck? Ignore friction.
   (a) They have the same momentum and the same KE.
   (b) The car has more momentum and more KE than the truck.
   (c) The truck has more momentum and more KE than the car.
   (d) They have the same momentum, but the car has more kinetic energy than the truck.
   (e) They have the same kinetic energy, but the truck has more momentum than the car.

10. Answer the previous Question (# 9) but now assume that you push both the car and the truck for the *same distance* with the same force. [*Hint*: See also Chapter 6.]

11. A railroad tank car contains milk and rolls at a constant speed along a level track. The milk begins to leak out the bottom. The car then
    (a) slows down.
    (b) speeds up.
    (c) maintains a constant speed.
    (d) Need more information about the rate of the leak.

12. A bowling ball hangs from a 1.0-m-long cord, Fig. 7–30: (i) A 200-gram putty ball moving 5.0 m/s hits the bowling ball and sticks to it, causing the bowling ball to swing up; (ii) a 200-gram rubber ball moving 5.0 m/s hits the bowling ball and bounces straight back at nearly 5.0 m/s, causing the bowling ball to swing up. Describe what happens.
    (a) The bowling ball swings up by the same amount in both (i) and (ii).
    (b) The ball swings up farther in (i) than in (ii).
    (c) The ball swings up farther in (ii) than in (i).
    (d) Not enough information is given; we need the contact time between the rubber ball and the bowling ball.

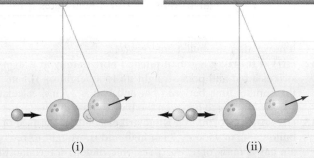

FIGURE 7–30 MisConceptual Question 12.

# Problems

## 7–1 and 7–2 Momentum and Its Conservation

1. (I) What is the magnitude of the momentum of a 28-g sparrow flying with a speed of 8.4 m/s?

2. (I) A constant friction force of 25 N acts on a 65-kg skier for 15 s on level snow. What is the skier's change in velocity?

3. (I) A 7150-kg railroad car travels alone on a level friction-less track with a constant speed of 15.0 m/s. A 3350-kg load, initially at rest, is dropped onto the car. What will be the car's new speed?

4. (I) A 110-kg tackler moving at 2.5 m/s meets head-on (and holds on to) an 82-kg halfback moving at 5.0 m/s. What will be their mutual speed immediately after the collision?

5. (II) Calculate the force exerted on a rocket when the propel-ling gases are being expelled at a rate of 1300 kg/s with a speed of $4.5 \times 10^4$ m/s.

6. (II) A 7700-kg boxcar traveling 14 m/s strikes a second car at rest. The two stick together and move off with a speed of 5.0 m/s. What is the mass of the second car?

7. (II) A child in a boat throws a 5.30-kg package out horizon-tally with a speed of 10.0 m/s, Fig. 7–31. Calculate the velocity of the boat immediately after, assuming it was initially at rest. The mass of the child is 24.0 kg and the mass of the boat is 35.0 kg.

$v = 10.0$ m/s

**FIGURE 7–31**
Problem 7.

8. (II) An atomic nucleus at rest decays radioactively into an alpha particle and a different nucleus. What will be the speed of this recoiling nucleus if the speed of the alpha particle is $2.8 \times 10^5$ m/s? Assume the recoiling nucleus has a mass 57 times greater than that of the alpha particle.

9. (II) An atomic nucleus initially moving at 320 m/s emits an alpha particle in the direction of its velocity, and the remain-ing nucleus slows to 280 m/s. If the alpha particle has a mass of 4.0 u and the original nucleus has a mass of 222 u, what speed does the alpha particle have when it is emitted?

10. (II) An object at rest is suddenly broken apart into two fragments by an explosion. One fragment acquires twice the kinetic energy of the other. What is the ratio of their masses?

11. (II) A 22-g bullet traveling 240 m/s penetrates a 2.0-kg block of wood and emerges going 150 m/s. If the block is stationary on a frictionless surface when hit, how fast does it move after the bullet emerges?

12. (III) A 0.145-kg baseball pitched horizontally at 27.0 m/s strikes a bat and pops straight up to a height of 31.5 m. If the contact time between ball and bat is 2.5 ms, calculate the average force between the ball and bat during contact.

13. (III) Air in a 120-km/h wind strikes head-on the face of a building 45 m wide by 75 m high and is brought to rest. If air has a mass of 1.3 kg per cubic meter, determine the average force of the wind on the building.

14. (III) A 725-kg two-stage rocket is traveling at a speed of $6.60 \times 10^3$ m/s away from Earth when a predesigned explosion separates the rocket into two sections of equal mass that then move with a speed of $2.80 \times 10^3$ m/s relative to each other along the original line of motion. (a) What is the speed and direction of each section (relative to Earth) after the explosion? (b) How much energy was supplied by the explosion? [Hint: What is the change in kinetic energy as a result of the explosion?]

## 7–3 Collisions and Impulse

15. (I) A 0.145-kg baseball pitched at 31.0 m/s is hit on a hori-zontal line drive straight back at the pitcher at 46.0 m/s. If the contact time between bat and ball is $5.00 \times 10^{-3}$ s, calculate the force (assumed to be constant) between the ball and bat.

16. (II) A golf ball of mass 0.045 kg is hit off the tee at a speed of 38 m/s. The golf club was in contact with the ball for $3.5 \times 10^{-3}$ s. Find (a) the impulse imparted to the golf ball, and (b) the average force exerted on the ball by the golf club.

17. (II) A 12-kg hammer strikes a nail at a velocity of 7.5 m/s and comes to rest in a time interval of 8.0 ms. (a) What is the impulse given to the nail? (b) What is the average force acting on the nail?

18. (II) A tennis ball of mass $m = 0.060$ kg and speed $v = 28$ m/s strikes a wall at a 45° angle and rebounds with the same speed at 45° (Fig. 7–32). What is the impulse (magnitude and direction) given to the ball?

**FIGURE 7–32**
Problem 18.

19. (II) A 125-kg astronaut (including space suit) acquires a speed of 2.50 m/s by pushing off with her legs from a 1900-kg space capsule. (a) What is the change in speed of the space capsule? (b) If the push lasts 0.600 s, what is the average force exerted by each on the other? As the refer-ence frame, use the position of the capsule before the push. (c) What is the kinetic energy of each after the push?

20. (II) Rain is falling at the rate of 2.5 cm/h and accumu-lates in a pan. If the raindrops hit at 8.0 m/s, estimate the force on the bottom of a 1.0-m² pan due to the impacting rain which we assume does not rebound. Water has a mass of $1.00 \times 10^3$ kg per m³.

21. (II) A 95-kg fullback is running at 3.0 m/s to the east and is stopped in 0.85 s by a head-on tackle by a tackler running due west. Calculate (a) the original momentum of the fullback, (b) the impulse exerted on the fullback, (c) the impulse exerted on the tackler, and (d) the average force exerted on the tackler.

22. (II) With what impulse does a 0.50-kg newspaper have to be thrown to give it a velocity of 3.0 m/s?

*23. (III) Suppose the force acting on a tennis ball (mass 0.060 kg) points in the $+x$ direction and is given by the graph of Fig. 7–33 as a function of time. (a) Use graphical methods (count squares) to estimate the total impulse given the ball. (b) Estimate the velocity of the ball after being struck, assuming the ball is being served so it is nearly at rest initially. [*Hint*: See Section 6–2.]

**FIGURE 7–33**
Problem 23.

24. (III) (a) Calculate the impulse experienced when a 55-kg person lands on firm ground after jumping from a height of 2.8 m. (b) Estimate the average force exerted on the person's feet by the ground if the landing is stiff-legged, and again (c) with bent legs. With stiff legs, assume the body moves 1.0 cm during impact, and when the legs are bent, about 50 cm. [*Hint*: The average net force on him, which is related to impulse, is the vector sum of gravity and the force exerted by the ground. See Fig. 7–34.] We will see in Chapter 9 that the force in (b) exceeds the ultimate strength of bone (Table 9–2).

**FIGURE 7–34**
Problem 24.

## 7–4 and 7–5 Elastic Collisions

25. (II) A ball of mass 0.440 kg moving east ($+x$ direction) with a speed of 3.80 m/s collides head-on with a 0.220-kg ball at rest. If the collision is perfectly elastic, what will be the speed and direction of each ball after the collision?

26. (II) A 0.450-kg hockey puck, moving east with a speed of 5.80 m/s, has a head-on collision with a 0.900-kg puck initially at rest. Assuming a perfectly elastic collision, what will be the speed and direction of each puck after the collision?

27. (II) A 0.060-kg tennis ball, moving with a speed of 5.50 m/s, has a head-on collision with a 0.090-kg ball initially moving in the same direction at a speed of 3.00 m/s. Assuming a perfectly elastic collision, determine the speed and direction of each ball after the collision.

28. (II) Two billiard balls of equal mass undergo a perfectly elastic head-on collision. If one ball's initial speed was 2.00 m/s, and the other's was 3.60 m/s in the opposite direction, what will be their speeds and directions after the collision?

29. (II) A 0.280-kg croquet ball makes an elastic head-on collision with a second ball initially at rest. The second ball moves off with half the original speed of the first ball. (a) What is the mass of the second ball? (b) What fraction of the original kinetic energy ($\Delta KE/KE$) gets transferred to the second ball?

30. (II) A ball of mass $m$ makes a head-on elastic collision with a second ball (at rest) and rebounds with a speed equal to 0.450 its original speed. What is the mass of the second ball?

31. (II) A ball of mass 0.220 kg that is moving with a speed of 5.5 m/s collides head-on and elastically with another ball initially at rest. Immediately after the collision, the incoming ball bounces backward with a speed of 3.8 m/s. Calculate (a) the velocity of the target ball after the collision, and (b) the mass of the target ball.

32. (II) Determine the fraction of kinetic energy lost by a neutron ($m_1 = 1.01\,\text{u}$) when it collides head-on and elastically with a target particle at rest which is (a) $^1_1\text{H}$ ($m = 1.01\,\text{u}$); (b) $^2_1\text{H}$ (heavy hydrogen, $m = 2.01\,\text{u}$); (c) $^{12}_6\text{C}$ ($m = 12.00\,\text{u}$); (d) $^{208}_{82}\text{Pb}$ (lead, $m = 208\,\text{u}$).

## 7–6 Inelastic Collisions

33. (I) In a ballistic pendulum experiment, projectile 1 results in a maximum height $h$ of the pendulum equal to 2.6 cm. A second projectile (of the same mass) causes the pendulum to swing twice as high, $h_2 = 5.2$ cm. The second projectile was how many times faster than the first?

34. (II) (a) Derive a formula for the fraction of kinetic energy lost, $\Delta KE/KE$, in terms of $m$ and $M$ for the ballistic pendulum collision of Example 7–9. (b) Evaluate for $m = 18.0$ g and $M = 380$ g.

35. (II) A 28-g rifle bullet traveling 190 m/s embeds itself in a 3.1-kg pendulum hanging on a 2.8-m-long string, which makes the pendulum swing upward in an arc. Determine the vertical and horizontal components of the pendulum's maximum displacement.

36. (II) An internal explosion breaks an object, initially at rest, into two pieces, one of which has 1.5 times the mass of the other. If 5500 J is released in the explosion, how much kinetic energy does each piece acquire?

37. (II) A 980-kg sports car collides into the rear end of a 2300-kg SUV stopped at a red light. The bumpers lock, the brakes are locked, and the two cars skid forward 2.6 m before stopping. The police officer, estimating the coefficient of kinetic friction between tires and road to be 0.80, calculates the speed of the sports car at impact. What was that speed?

38. (II) You drop a 14-g ball from a height of 1.5 m and it only bounces back to a height of 0.85 m. What was the total impulse on the ball when it hit the floor? (Ignore air resistance.)

39. Croquet ball A moving at 4.3 m/s makes a head on collision with ball B of equal mass and initially at rest. Immediately after the collision ball B moves forward at 3.0 m/s. What fraction of the initial kinetic energy is lost in the collision?

40. (II) A wooden block is cut into two pieces, one with three times the mass of the other. A depression is made in both faces of the cut, so that a firecracker can be placed in it with the block reassembled. The reassembled block is set on a rough-surfaced table, and the fuse is lit. When the firecracker explodes inside, the two blocks separate and slide apart. What is the ratio of distances each block travels?

41. (II) A 144-g baseball moving 28.0 m/s strikes a stationary 5.25-kg brick resting on small rollers so it moves without significant friction. After hitting the brick, the baseball bounces straight back, and the brick moves forward at 1.10 m/s. (a) What is the baseball's speed after the collision? (b) Find the total kinetic energy before and after the collision.

**42.** (III) A pendulum consists of a mass $M$ hanging at the bottom end of a massless rod of length $\ell$, which has a frictionless pivot at its top end. A mass $m$, moving as shown in Fig. 7–35 with velocity $v$, impacts $M$ and becomes embedded. What is the smallest value of $v$ sufficient to cause the pendulum (with embedded mass $m$) to swing clear over the top of its arc?

**FIGURE 7–35**
Problem 42.

**43.** (III) A bullet of mass $m = 0.0010 \, \text{kg}$ embeds itself in a wooden block with mass $M = 0.999 \, \text{kg}$, which then compresses a spring ($k = 140 \, \text{N/m}$) by a distance $x = 0.050 \, \text{m}$ before coming to rest. The coefficient of kinetic friction between the block and table is $\mu = 0.50$. (a) What is the initial velocity (assumed horizontal) of the bullet? (b) What fraction of the bullet's initial kinetic energy is dissipated (in damage to the wooden block, rising temperature, etc.) in the collision between the bullet and the block?

## *7–7 Collisions in Two Dimensions

**\*44.** (II) Billiard ball A of mass $m_A = 0.120 \, \text{kg}$ moving with speed $v_A = 2.80 \, \text{m/s}$ strikes ball B, initially at rest, of mass $m_B = 0.140 \, \text{kg}$. As a result of the collision, ball A is deflected off at an angle of $30.0°$ with a speed $v'_A = 2.10 \, \text{m/s}$. (a) Taking the $x$ axis to be the original direction of motion of ball A, write down the equations expressing the conservation of momentum for the components in the $x$ and $y$ directions separately. (b) Solve these equations for the speed, $v'_B$, and angle, $\theta'_B$, of ball B after the collision. Do not assume the collision is elastic.

**\*45.** (II) A radioactive nucleus at rest decays into a second nucleus, an electron, and a neutrino. The electron and neutrino are emitted at right angles and have momenta of $9.6 \times 10^{-23} \, \text{kg·m/s}$ and $6.2 \times 10^{-23} \, \text{kg·m/s}$, respectively. Determine the magnitude and the direction of the momentum of the second (recoiling) nucleus.

**\*46.** (III) Billiard balls A and B, of equal mass, move at right angles and meet at the origin of an $xy$ coordinate system as shown in Fig. 7–36. Initially ball A is moving along the $y$ axis at $+2.0 \, \text{m/s}$, and ball B is moving to the right along the $x$ axis with speed $+3.7 \, \text{m/s}$. After the collision (assumed elastic), ball B is moving along the positive $y$ axis (Fig. 7–36) with velocity $v'_B$. What is the final direction of ball A, and what are the speeds of the two balls?

**FIGURE 7–36**
Problem 46.
(Ball A after the collision is not shown.)

**\*47.** (III) An atomic nucleus of mass $m$ traveling with speed $v$ collides elastically with a target particle of mass $2m$ (initially at rest) and is scattered at $90°$. (a) At what angle does the target particle move after the collision? (b) What are the final speeds of the two particles? (c) What fraction of the initial kinetic energy is transferred to the target particle?

**\*48.** (III) A neon atom ($m = 20.0 \, \text{u}$) makes a perfectly elastic collision with another atom at rest. After the impact, the neon atom travels away at a $55.6°$ angle from its original direction and the unknown atom travels away at a $-50.0°$ angle. What is the mass (in u) of the unknown atom? [*Hint*: You could use the law of sines.]

## 7–8 Center of Mass (CM)

**49.** (I) The distance between a carbon atom ($m = 12 \, \text{u}$) and an oxygen atom ($m = 16 \, \text{u}$) in the CO molecule is $1.13 \times 10^{-10} \, \text{m}$. How far from the carbon atom is the center of mass of the molecule?

**50.** (I) Find the center of mass of the three-mass system shown in Fig. 7–37 relative to the 1.00-kg mass.

**FIGURE 7–37** Problem 50.

**51.** (II) The CM of an empty 1250-kg car is 2.40 m behind the front of the car. How far from the front of the car will the CM be when two people sit in the front seat 2.80 m from the front of the car, and three people sit in the back seat 3.90 m from the front? Assume that each person has a mass of 65.0 kg.

**52.** (II) Three cubes, of side $\ell_0$, $2\ell_0$, and $3\ell_0$, are placed next to one another (in contact) with their centers along a straight line as shown in Fig. 7–38. What is the position, along this line, of the CM of this system? Assume the cubes are made of the same uniform material.

**FIGURE 7–38**
Problem 52.

**53.** (II) A (lightweight) pallet has a load of ten identical cases of tomato paste (see Fig. 7–39), each of which is a cube of length $\ell$. Find the center of gravity in the horizontal plane, so that the crane operator can pick up the load without tipping it.

**FIGURE 7–39**
Problem 53.

**54.** (III) Determine the CM of the uniform thin L-shaped construction brace shown in Fig. 7–40.

**FIGURE 7–40** Problem 54.
This L-shaped object has uniform thickness $d$ (not shown).

**55.** (III) A uniform circular plate of radius $2R$ has a circular hole of radius $R$ cut out of it. The center $C'$ of the smaller circle is a distance $0.80R$ from the center $C$ of the larger circle, Fig. 7–41. What is the position of the center of mass of the plate? [*Hint*: Try subtraction.]

**FIGURE 7–41**
Problem 55.

### *7–9  CM for the Human Body

*56. (I) Assume that your proportions are the same as those in Table 7–1, and calculate the mass of one of your legs.

*57. (I) Determine the CM of an outstretched arm using Table 7–1.

*58. (II) Use Table 7–1 to calculate the position of the CM of an arm bent at a right angle. Assume that the person is 155 cm tall.

*59. (II) When a high jumper is in a position such that his arms and lower legs are hanging vertically, and his thighs, trunk, and head are horizontal just above the bar, estimate how far below the torso's median line the CM will be. Will this CM be outside the body? Use Table 7–1.

*60. (III) Repeat Problem 59 assuming the body bends at the hip joint by about 15°. Estimate, using Fig. 7–27 as a model.

### *7–10  CM and Translational Motion

*61. (II) The masses of the Earth and Moon are $5.98 \times 10^{24}$ kg and $7.35 \times 10^{22}$ kg, respectively, and their centers are separated by $3.84 \times 10^8$ m. (*a*) Where is the CM of the Earth–Moon system located? (*b*) What can you say about the motion of the Earth–Moon system about the Sun, and of the Earth and Moon separately about the Sun?

*62. (II) A mallet consists of a uniform cylindrical head of mass 2.30 kg and a diameter 0.0800 m mounted on a uniform cylindrical handle of mass 0.500 kg and length 0.240 m, as shown in Fig. 7–42. If this mallet is tossed, spinning, into the air, how far above the bottom of the handle is the point that will follow a parabolic trajectory?

**FIGURE 7–42**
Problem 62.

24.0 cm

8.00 cm

*63. (II) A 52-kg woman and a 72-kg man stand 10.0 m apart on nearly frictionless ice. (*a*) How far from the woman is their CM? (*b*) If each holds one end of a rope, and the man pulls on the rope so that he moves 2.5 m, how far from the woman will he be now? (*c*) How far will the man have moved when he collides with the woman?

*64. (II) Suppose that in Example 7–14 (Fig. 7–28), $m_{II} = 3m_I$. (*a*) Where then would $m_{II}$ land? (*b*) What if $m_I = 3m_{II}$?

*65. (II) Two people, one of mass 85 kg and the other of mass 55 kg, sit in a rowboat of mass 58 kg. With the boat initially at rest, the two people, who have been sitting at opposite ends of the boat, 3.0 m apart from each other, now exchange seats. How far and in what direction will the boat move?

*66. (III) A huge balloon and its gondola, of mass $M$, are in the air and stationary with respect to the ground. A passenger, of mass $m$, then climbs out and slides down a rope with speed $v$, measured with respect to the balloon. With what speed and direction (relative to Earth) does the balloon then move? What happens if the passenger stops?

## General Problems

67. Two astronauts, one of mass 55 kg and the other 85 kg, are initially at rest together in outer space. They then push each other apart. How far apart are they when the lighter astronaut has moved 12 m?

68. Two asteroids strike head-on: before the collision, asteroid A $(m_A = 7.5 \times 10^{12}$ kg) has velocity 3.3 km/s and asteroid B $(m_B = 1.45 \times 10^{13}$ kg) has velocity 1.4 km/s in the opposite direction. If the asteroids stick together, what is the velocity (magnitude and direction) of the new asteroid after the collision?

69. A ball is dropped from a height of 1.60 m and rebounds to a height of 1.20 m. Approximately how many rebounds will the ball make before losing 90% of its energy?

70. A 4800-kg open railroad car coasts at a constant speed of 7.60 m/s on a level track. Snow begins to fall vertically and fills the car at a rate of 3.80 kg/min. Ignoring friction with the tracks, what is the car's speed after 60.0 min? (See Section 7–2.)

71. Two bumper cars in an amusement park ride collide elastically as one approaches the other directly from the rear (Fig. 7–43). Car A has a mass of 435 kg and car B 495 kg, owing to differences in passenger mass. If car A approaches at 4.50 m/s and car B is moving at 3.70 m/s, calculate (*a*) their velocities after the collision, and (*b*) the change in momentum of each.

**FIGURE 7–43** Problem 71:
(a) before collision, (b) after collision.

**72.** A gun fires a bullet vertically into a 1.40-kg block of wood at rest on a thin horizontal sheet, Fig. 7–44. If the bullet has a mass of 25.0 g and a speed of 230 m/s, how high will the block rise into the air after the bullet becomes embedded in it?

**FIGURE 7–44**
Problem 72.

**73.** You have been hired as an expert witness in a court case involving an automobile accident. The accident involved car A of mass 1500 kg which crashed into stationary car B of mass 1100 kg. The driver of car A applied his brakes 15 m before he skidded and crashed into car B. After the collision, car A slid 18 m while car B slid 30 m. The coefficient of kinetic friction between the locked wheels and the road was measured to be 0.60. Show that the driver of car A was exceeding the 55-mi/h (90-km/h) speed limit before applying the brakes.

**74.** A meteor whose mass was about $1.5 \times 10^8$ kg struck the Earth $(m_E = 6.0 \times 10^{24}$ kg) with a speed of about 25 km/s and came to rest in the Earth. (*a*) What was the Earth's recoil speed (relative to Earth at rest before the collision)? (*b*) What fraction of the meteor's kinetic energy was transformed to kinetic energy of the Earth? (*c*) By how much did the Earth's kinetic energy change as a result of this collision?

**75.** A 28-g bullet strikes and becomes embedded in a 1.35-kg block of wood placed on a horizontal surface just in front of the gun. If the coefficient of kinetic friction between the block and the surface is 0.28, and the impact drives the block a distance of 8.5 m before it comes to rest, what was the muzzle speed of the bullet?

**76.** You are the design engineer in charge of the crashworthiness of new automobile models. Cars are tested by smashing them into fixed, massive barriers at 45 km/h. A new model of mass 1500 kg takes 0.15 s from the time of impact until it is brought to rest. (*a*) Calculate the average force exerted on the car by the barrier. (*b*) Calculate the average deceleration of the car in *g*'s.

**77.** A 0.25-kg skeet (clay target) is fired at an angle of 28° to the horizontal with a speed of 25 m/s (Fig. 7–45). When it reaches the maximum height, *h*, it is hit from below by a 15-g pellet traveling vertically upward at a speed of 230 m/s. The pellet is embedded in the skeet. (*a*) How much higher, *h'*, does the skeet go up? (*b*) How much extra distance, Δ*x*, does the skeet travel because of the collision?

**78.** Two balls, of masses $m_A = 45$ g and $m_B = 65$ g, are suspended as shown in Fig. 7–46. The lighter ball is pulled away to a 66° angle with the vertical and released. (*a*) What is the velocity of the lighter ball before impact? (*b*) What is the velocity of each ball after the elastic collision? (*c*) What will be the maximum height of each ball after the elastic collision?

**FIGURE 7–46**
Problem 78.

**79.** A block of mass $m = 2.50$ kg slides down a 30.0° incline which is 3.60 m high. At the bottom, it strikes a block of mass $M = 7.00$ kg which is at rest on a horizontal surface, Fig. 7–47. (Assume a smooth transition at the bottom of the incline.) If the collision is elastic, and friction can be ignored, determine (*a*) the speeds of the two blocks after the collision, and (*b*) how far back up the incline the smaller mass will go.

**FIGURE 7–47**
Problem 79.

**80.** The space shuttle launches an 850-kg satellite by ejecting it from the cargo bay. The ejection mechanism is activated and is in contact with the satellite for 4.8 s to give it a velocity of 0.30 m/s in the *x* direction relative to the shuttle. The mass of the shuttle is 92,000 kg. (*a*) Determine the component of velocity $v_f$ of the shuttle in the minus *x* direction resulting from the ejection. (*b*) Find the average force that the shuttle exerts on the satellite during the ejection.

**81.** Astronomers estimate that a 2.0-km-diameter asteroid collides with the Earth once every million years. The collision could pose a threat to life on Earth. (*a*) Assume a spherical asteroid has a mass of 3200 kg for each cubic meter of volume and moves toward the Earth at 15 km/s. How much destructive energy could be released when it embeds itself in the Earth? (*b*) For comparison, a nuclear bomb could release about $4.0 \times 10^{16}$ J. How many such bombs would have to explode simultaneously to release the destructive energy of the asteroid collision with the Earth?

**82.** An astronaut of mass 210 kg including his suit and jet pack wants to acquire a velocity of 2.0 m/s to move back toward his space shuttle. Assuming the jet pack can eject gas with a velocity of 35 m/s, what mass of gas will need to be ejected?

**FIGURE 7–45** Problem 77.

83. Two blocks of mass $m_A$ and $m_B$, resting on a frictionless table, are connected by a stretched spring and then released (Fig. 7–48). (a) Is there a net external force on the system before release? (b) Determine the ratio of their speeds, $v_A/v_B$. (c) What is the ratio of their kinetic energies? (d) Describe the motion of the CM of this system. Ignore mass of spring.

**FIGURE 7–48** Problem 83.

84. A golf ball rolls off the top of a flight of concrete steps of total vertical height 4.00 m. The ball hits four times on the way down, each time striking the horizontal part of a different step 1.00 m lower. If all collisions are perfectly elastic, what is the bounce height on the fourth bounce when the ball reaches the bottom of the stairs?

85. A massless spring with spring constant $k$ is placed between a block of mass $m$ and a block of mass $3m$. Initially the blocks are at rest on a frictionless surface and they are held together so that the spring between them is compressed by an amount $D$ from its equilibrium length. The blocks are then released and the spring pushes them off in opposite directions. Find the speeds of the two blocks when they detach from the spring.

*86. A novice pool player is faced with the corner pocket shot shown in Fig. 7–49. Relative dimensions are also shown. Should the player worry that this might be a "scratch shot," in which the cue ball will also fall into a pocket? Give details. Assume equal-mass balls and an elastic collision. Ignore spin.

**FIGURE 7–49**
Problem 86.

# Search and Learn

1. Consider the Examples in this Chapter involving $\Sigma \vec{F}_{ext} = \Delta \vec{p}/\Delta t$. Provide some general guidelines as to when it is best to solve the problem using $\Sigma \vec{F}_{ext} = 0$ so $\Sigma \vec{p}_i = \Sigma \vec{p}_f$, and when to use the principle of impulse instead so that $\Sigma \vec{F}_{ext} \Delta t = \Delta \vec{p}$.

2. A 6.0-kg object moving in the $+x$ direction at 6.5 m/s collides head-on with an 8.0-kg object moving in the $-x$ direction at 4.0 m/s. Determine the final velocity of each object if: (a) the objects stick together; (b) the collision is elastic; (c) the 6.0-kg object is at rest after the collision; (d) the 8.0-kg object is at rest after the collision; (e) the 6.0-kg object has a velocity of 4.0 m/s in the $-x$ direction after the collision. Finally, (f) are the results in (c), (d), and (e) "reasonable"? Explain.

3. In a physics lab, a cube slides down a frictionless incline as shown in Fig. 7–50 and elastically strikes another cube at the bottom that is only one-half its mass. If the incline is 35 cm high and the table is 95 cm off the floor, where does each cube land? [*Hint*: Both leave the incline moving horizontally.]

**FIGURE 7–50**
Search and Learn 3.

4. **The gravitational slingshot effect.** Figure 7–51 shows the planet Saturn moving in the negative $x$ direction at its orbital speed (with respect to the Sun) of 9.6 km/s. The mass of Saturn is $5.69 \times 10^{26}$ kg. A spacecraft with mass 825 kg approaches Saturn. When far from Saturn, it moves in the $+x$ direction at 10.4 km/s. The gravitational attraction of Saturn (a conservative force) acting on the spacecraft causes it to swing around the planet (orbit shown as dashed line) and head off in the opposite direction. Estimate the final speed of the spacecraft after it is far enough away to be considered free of Saturn's gravitational pull.

**FIGURE 7–51**
Search and Learn 4.

5. Take the general case of an object of mass $m_A$ and velocity $v_A$ elastically striking a stationary $(v_B = 0)$ object of mass $m_B$ head-on. (a) Show that the final velocities $v'_A$ and $v'_B$ are given by

$$v'_A = \left( \frac{m_A - m_B}{m_A + m_B} \right) v_A, \qquad v'_B = \left( \frac{2m_A}{m_A + m_B} \right) v_A.$$

(b) What happens in the extreme case when $m_A$ is much smaller than $m_B$? Cite a common example of this. (c) What happens in the extreme case when $m_A$ is much larger than $m_B$? Cite a common example of this. (d) What happens in the case when $m_A = m_B$? Cite a common example.

# ANSWERS TO EXERCISES

A: Yes, if the sports car's speed is three times greater.
B: Larger ($\Delta p$ is greater).
C: (a) 6.0 m/s; (b) almost zero; (c) almost 24.0 m/s.
D: 0.50 m/s.

E: (b); (d).
F: The curve would be wider and less high.
G: $x_{CM} = -2.0$ m; yes.
H: The boat moves in the opposite direction.

You too can experience rapid rotation—if your stomach can take the high angular velocity and centripetal acceleration of some of the faster amusement park rides. If not, try the slower merry-go-round or Ferris wheel. Rotating carnival rides have rotational kinetic energy as well as angular momentum. Angular acceleration is produced by a net torque, and rotating objects have rotational kinetic energy.

# CHAPTER 8

# Rotational Motion

## CONTENTS

**CHAPTER-OPENING QUESTION—Guess now!**

A solid ball and a solid cylinder roll down a ramp. They both start from rest at the same time and place. Which gets to the bottom first?

  (a) They get there at the same time.

  (b) They get there at almost exactly the same time except for frictional differences.

  (c) The ball gets there first.

  (d) The cylinder gets there first.

  (e) Can't tell without knowing the mass and radius of each.

Until now, we have been concerned mainly with translational motion. We discussed the kinematics and dynamics of translational motion (the role of force). We also discussed the energy and momentum for translational motion. In this Chapter we will deal with rotational motion. We will discuss the kinematics of rotational motion and then its dynamics (involving torque), as well as rotational kinetic energy and angular momentum (the rotational analog of linear momentum). Our understanding of the world around us will be increased significantly—from rotating bicycle wheels and compact discs to amusement park rides, a spinning skater, the rotating Earth, and a centrifuge—and there may be a few surprises.

We will consider mainly the rotation of rigid objects about a fixed axis. A **rigid object** is an object with a definite shape that doesn't change, so that the particles composing it stay in fixed positions relative to one another. Any real object is capable of vibrating or deforming when a force is exerted on it. But these effects are often very small, so the concept of an ideal rigid object is very useful as a good approximation.

# 8–1 Angular Quantities

The motion of a rigid object can be analyzed as the translational motion of the object's center of mass, plus rotational motion *about* its center of mass (Section 7–8). We have already discussed translational motion in detail, so now we focus on purely rotational motion. By *purely rotational motion* we mean that all points in the object move in circles, such as the point P in the rotating wheel of Fig. 8–1, and that the centers of these circles all lie on one line called the **axis of rotation**. In Fig. 8–1 the axis of rotation is perpendicular to the page and passes through point O.

Every point in an object rotating about a fixed axis moves in a circle (shown dashed in Fig. 8–1 for point P) whose center is on the axis of rotation and whose radius is $r$, the distance of that point from the axis of rotation. A straight line drawn from the axis to any point in the object sweeps out the same angle $\theta$ in the same time interval.

To indicate the **angular position** of a rotating object, or how far it has rotated, we specify the angle $\theta$ of some particular line in the object (red in Fig. 8–1) with respect to a reference line, such as the $x$ axis in Fig. 8–1. A point in the object, such as P in Fig. 8–1, moves through an angle $\theta$ when it travels the distance $\ell$ measured along the circumference of its circular path. Angles are commonly measured in degrees, but the mathematics of circular motion is much simpler if we use the *radian* for angular measure. One **radian** (abbreviated rad) is defined as the angle subtended by an arc whose length is equal to the radius. For example, in Fig. 8–1b, point P is a distance $r$ from the axis of rotation, and it has moved a distance $\ell$ along the arc of a circle. The arc length $\ell$ is said to "subtend" the angle $\theta$. In radians, any angle $\theta$ is given by

$$\theta = \frac{\ell}{r}, \qquad [\theta \text{ in radians}] \quad \textbf{(8–1a)}$$

where $r$ is the radius of the circle, and $\ell$ is the arc length subtended by the angle $\theta$ specified in radians. If $\ell = r$, then $\theta = 1$ rad.

The radian is dimensionless since it is the ratio of two lengths. Nonetheless when giving an angle in radians, we always mention rad to remind us it is not degrees. It is often useful to rewrite Eq. 8–1a in terms of arc length $\ell$:

$$\ell = r\theta. \qquad \textbf{(8–1b)}$$

Radians can be related to degrees in the following way. In a complete circle there are 360°, which must correspond to an arc length equal to the circumference of the circle, $\ell = 2\pi r$. For a full circle, $\theta = \ell/r = 2\pi r/r = 2\pi$ rad. Thus

$$360° = 2\pi \text{ rad}.$$

One radian is then $360°/2\pi \approx 360°/6.28 \approx 57.3°$. An object that makes one complete revolution (rev) has rotated through 360°, or $2\pi$ radians:

$$1 \text{ rev} = 360° = 2\pi \text{ rad}.$$

---

**EXAMPLE 8–1** **Bike wheel.** A bike wheel rotates 4.50 revolutions. How many radians has it rotated?

**APPROACH** All we need is a conversion of units using

$$1 \text{ revolution} = 360° = 2\pi \text{ rad} = 6.28 \text{ rad}.$$

**SOLUTION**

$$4.50 \text{ revolutions} = (4.50 \text{ rev})\left(2\pi \frac{\text{rad}}{\text{rev}}\right) = 9.00\pi \text{ rad} = 28.3 \text{ rad}.$$

(a)

(b)

**FIGURE 8–1** Looking at a wheel that is rotating counterclockwise about an axis through the wheel's center at O (axis perpendicular to the page). Each point, such as point P, moves in a circular path; $\ell$ is the distance P travels as the wheel rotates through the angle $\theta$.

⚠ **CAUTION**

*Use radians in calculating, not degrees*

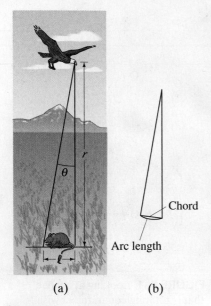

(a)                    (b)

**FIGURE 8–2** (a) Example 8–2.
(b) For small angles, arc length and
the chord length (straight line) are
nearly equal.

**FIGURE 8–3** A wheel rotates
about its axle from (a) initial
position $\theta_1$ to (b) final position $\theta_2$.
The angular displacement is
$\Delta\theta = \theta_2 - \theta_1$.

(a)

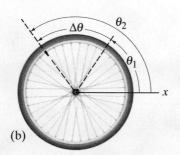

(b)

---

**EXAMPLE 8–2** | **Birds of prey—in radians.** A particular bird's eye can just distinguish objects that subtend an angle no smaller than about $3 \times 10^{-4}$ rad. (a) How many degrees is this? (b) How small an object can the bird just distinguish when flying at a height of 100 m (Fig. 8–2a)?

**APPROACH** For (a) we use the relation $360° = 2\pi$ rad. For (b) we use Eq. 8–1b, $\ell = r\theta$, to find the arc length.

**SOLUTION** (a) We convert $3 \times 10^{-4}$ rad to degrees:

$$(3 \times 10^{-4}\,\text{rad})\left(\frac{360°}{2\pi\,\text{rad}}\right) = 0.017°.$$

(b) We use Eq. 8–1b, $\ell = r\theta$. For small angles, the arc length $\ell$ and the chord length are approximately† the same (Fig. 8–2b). Since $r = 100$ m and $\theta = 3 \times 10^{-4}$ rad, we find

$$\ell = r\theta = (100\,\text{m})(3 \times 10^{-4}\,\text{rad}) = 3 \times 10^{-2}\,\text{m} = 3\,\text{cm}.$$

A bird can distinguish a small mouse (about 3 cm long) from a height of 100 m. That is good eyesight.

**NOTE** Had the angle been given in degrees, we would first have had to convert it to radians to make this calculation. Equations 8–1 are valid *only* if the angle is specified in radians. Degrees (or revolutions) won't work.

To describe rotational motion, we make use of angular quantities, such as angular velocity and angular acceleration. These are defined in analogy to the corresponding quantities in linear motion, and are chosen to describe the rotating object as a whole, so they are the same for each point in the rotating object. Each point in a rotating object may also have translational velocity and acceleration, but they have different values for different points in the object.

When an object such as the bicycle wheel in Fig. 8–3 rotates from some initial position, specified by $\theta_1$, to some final position, $\theta_2$, its *angular displacement* is

$$\Delta\theta = \theta_2 - \theta_1.$$

The *angular velocity* (denoted by $\omega$, the Greek lowercase letter omega) is defined in analogy with linear (translational) velocity that was discussed in Chapter 2. Instead of linear displacement, we use the angular displacement. Thus the **average angular velocity** of an object rotating about a fixed axis is defined as

$$\bar{\omega} = \frac{\Delta\theta}{\Delta t}, \tag{8–2a}$$

where $\Delta\theta$ is the angle through which the object has rotated in the time interval $\Delta t$. The **instantaneous angular velocity** is the limit of this ratio as $\Delta t$ approaches zero:

$$\omega = \lim_{\Delta t \to 0} \frac{\Delta\theta}{\Delta t}. \tag{8–2b}$$

Angular velocity is generally specified in radians per second (rad/s). Note that *all points in a rigid object rotate with the same angular velocity*, since every position in the object moves through the same angle in the same time interval.

An object such as the wheel in Fig. 8–3 can rotate about a fixed axis either clockwise or counterclockwise. The direction can be specified with a + or − sign. The usual convention is to choose the angular displacement $\Delta\theta$ and angular velocity $\omega$ as positive when the wheel rotates counterclockwise. If the rotation is clockwise, then $\theta$ would decrease, so $\Delta\theta$ and $\omega$ would be negative.

---

†Even for an angle as large as 15°, the error in making this estimate is only 1%, but for larger angles the error increases rapidly. (The **chord** is the straight-line distance between the ends of the arc.)

*Angular acceleration* (denoted by $\alpha$, the Greek lowercase letter alpha), in analogy to linear acceleration, is defined as the change in angular velocity divided by the time required to make this change. The **average angular acceleration** is defined as

$$\bar{\alpha} = \frac{\omega_2 - \omega_1}{\Delta t} = \frac{\Delta\omega}{\Delta t}, \tag{8-3a}$$

where $\omega_1$ is the angular velocity initially, and $\omega_2$ is the angular velocity after a time interval $\Delta t$. **Instantaneous angular acceleration** is defined as the limit of this ratio as $\Delta t$ approaches zero:

$$\alpha = \lim_{\Delta t \to 0} \frac{\Delta\omega}{\Delta t}. \tag{8-3b}$$

Since $\omega$ is the same for all points of a rotating object, Eq. 8–3 tells us that $\alpha$ also will be the same for all points. Thus, $\omega$ and $\alpha$ are properties of the rotating object as a whole. With $\omega$ measured in radians per second and $t$ in seconds, $\alpha$ has units of radians per second squared $(\text{rad/s}^2)$.

Each point or particle of a rotating object has, at any moment, a linear velocity $v$ and a linear acceleration $a$. We can now relate the linear quantities at each point, $v$ and $a$, to the angular quantities, $\omega$ and $\alpha$, for a rigid object rotating about a fixed axis. Consider a point P located a distance $r$ from the axis of rotation, as in Fig. 8–4. If the object rotates with angular velocity $\omega$, any point will have a linear velocity whose direction is tangent to its circular path. The magnitude of that point's linear velocity is $v = \Delta\ell/\Delta t$. From Eq. 8–1b, a change in rotation angle $\Delta\theta$ (in radians) is related to the linear distance traveled by $\Delta\ell = r\,\Delta\theta$. Hence

$$v = \frac{\Delta\ell}{\Delta t} = r\frac{\Delta\theta}{\Delta t}$$

or (since $\Delta\theta/\Delta t = \omega$)

$$v = r\omega. \tag{8-4}$$

In this very useful Eq. 8–4, $r$ is the distance of a point from the rotation axis and $\omega$ is given in rad/s. Thus, although $\omega$ is the same for every point in the rotating object at any instant, the linear velocity $v$ is greater for points farther from the axis (Fig. 8–5). Note that Eq. 8–4 is valid both instantaneously and on average.

| **CONCEPTUAL EXAMPLE 8–3** | **Is the lion faster than the horse?** On a rotating carousel or merry-go-round, one child sits on a horse near the outer edge and another child sits on a lion halfway out from the center. (*a*) Which child has the greater linear velocity? (*b*) Which child has the greater angular velocity? |

**RESPONSE** (*a*) The *linear* velocity is the distance traveled divided by the time interval. In one rotation the child on the outer edge travels a longer distance than the child near the center, but the time interval is the same for both. Thus the child at the outer edge, on the horse, has the greater linear velocity.

(*b*) The *angular* velocity is the angle of rotation of the carousel as a whole divided by the time interval. For example, in one rotation both children rotate through the same angle (360° or $2\pi$ radians). The two children have the same angular velocity.

If the angular velocity of a rotating object changes, the object as a whole—and each point in it—has an angular acceleration. Each point also has a linear acceleration whose direction is tangent to that point's circular path. We use Eq. 8–4 ($v = r\omega$) to see that the angular acceleration $\alpha$ is related to the tangential linear acceleration $a_{\text{tan}}$ of a point in the rotating object by

$$a_{\text{tan}} = \frac{\Delta v}{\Delta t} = r\frac{\Delta\omega}{\Delta t}$$

or (using Eq. 8–3)

$$a_{\text{tan}} = r\alpha. \tag{8-5}$$

In this equation, $r$ is the radius of the circle in which the particle is moving, and the subscript "tan" in $a_{\text{tan}}$ stands for "tangential."

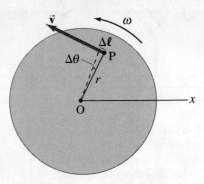

**FIGURE 8–4** A point P on a rotating wheel has a linear velocity $\vec{v}$ at any moment.

**FIGURE 8–5** A wheel rotating uniformly counterclockwise. Two points on the wheel, at distances $r_A$ and $r_B$ from the center, have the same angular velocity $\omega$ because they travel through the same angle $\theta$ in the same time interval. But the two points have different linear velocities because they travel different distances in the same time interval. Since $v = r\omega$ and $r_B > r_A$, then $v_B > v_A$.

FIGURE 8–6 On a rotating wheel whose angular speed is increasing, a point P has both tangential and radial (centripetal) components of linear acceleration. (See also Chapter 5.)

The total linear acceleration of a point in the rotating object is the vector sum of two components:

$$\vec{a} = \vec{a}_{tan} + \vec{a}_R,$$

where the radial component, $\vec{a}_R$, is the radial or "centripetal" acceleration and its direction is toward the center of the point's circular path; see Fig. 8–6. We saw in Chapter 5 (Eq. 5–1) that a particle moving in a circle of radius $r$ with linear speed $v$ has a radial acceleration $a_R = v^2/r$. We can rewrite this in terms of $\omega$ using Eq. 8–4:

$$a_R = \frac{v^2}{r} = \frac{(r\omega)^2}{r} = \omega^2 r. \qquad (8-6)$$

Thus the centripetal acceleration is greater the farther you are from the axis of rotation: the children farthest out on a carousel feel the greatest acceleration.

Equations 8–1, 8–4, 8–5, and 8–6 relate the angular quantities describing the rotation of an object to the linear quantities for each point of a rotating object. Table 8–1 summarizes these relationships.

### TABLE 8–1  Linear and Rotational Quantities

| Linear | Type | Rotational | Relation‡ |
|--------|------|------------|-----------|
| $x$ | displacement | $\theta$ | $x = r\theta$ |
| $v$ | velocity | $\omega$ | $v = r\omega$ |
| $a_{tan}$ | acceleration | $\alpha$ | $a_{tan} = r\alpha$ |

‡ You must use radians.

FIGURE 8–7 Examples 8–4 and 8–5. The total acceleration vector is $\vec{a} = \vec{a}_{tan} + \vec{a}_R$, at $t = 8.0$ s.

(a)

(b)

---

**EXAMPLE 8–4  Angular and linear velocities.** A carousel is initially at rest. At $t = 0$ it is given a constant angular acceleration $\alpha = 0.060$ rad/s$^2$, which increases its angular velocity for 8.0 s. At $t = 8.0$ s, determine (a) the angular velocity of the carousel, and (b) the linear velocity of a child (Fig. 8–7a) located 2.5 m from the center, point P in Fig. 8–7b.

**APPROACH** The angular acceleration $\alpha$ is constant, so we can use $\alpha = \Delta\omega/\Delta t$ (Eq. 8–3a) to solve for $\omega$ after a time $t = 8.0$ s. With this $\omega$, we determine the linear velocity using Eq. 8–4, $v = r\omega$.

**SOLUTION** (a) In Eq. 8–3a, $\bar{\alpha} = (\omega_2 - \omega_1)/\Delta t$, we put $\Delta t = 8.0$ s, $\bar{\alpha} = 0.060$ rad/s$^2$, and $\omega_1 = 0$. Solving for $\omega_2$, we get

$$\omega_2 = \omega_1 + \bar{\alpha}\,\Delta t = 0 + (0.060\,\text{rad/s}^2)(8.0\,\text{s}) = 0.48\,\text{rad/s}.$$

During the 8.0-s time interval, the carousel accelerates from $\omega_1 = 0$ to $\omega_2 = 0.48$ rad/s.

(b) The linear velocity of the child with $r = 2.5$ m at time $t = 8.0$ s is found using Eq. 8–4:

$$v = r\omega = (2.5\,\text{m})(0.48\,\text{rad/s}) = 1.2\,\text{m/s}.$$

Note that the "rad" has been omitted in the final result because it is dimensionless (and only a reminder)—it is a ratio of two distances, Eq. 8–1a.

---

**EXAMPLE 8–5  Angular and linear accelerations.** For the child on the rotating carousel of Example 8–4, determine that child's (a) tangential (linear) acceleration, (b) centripetal acceleration, (c) total acceleration.

**APPROACH** We use the relations discussed above, Eqs. 8–5 and 8–6.

**SOLUTION** (a) The child's tangential acceleration is given by Eq. 8–5:

$$a_{tan} = r\alpha = (2.5\,\text{m})(0.060\,\text{rad/s}^2) = 0.15\,\text{m/s}^2,$$

and it is the same throughout the 8.0-s acceleration period.

(b) The child's centripetal acceleration at $t = 8.0$ s is given by Eq. 8–6:

$$a_R = \frac{v^2}{r} = \frac{(1.2\,\text{m/s})^2}{(2.5\,\text{m})} = 0.58\,\text{m/s}^2.$$

(c) The two components of linear acceleration calculated in parts (a) and (b) are perpendicular to each other. Thus the total linear acceleration at $t = 8.0$ s has magnitude

$$a = \sqrt{a_{\text{tan}}^2 + a_R^2} = \sqrt{(0.15 \text{ m/s}^2)^2 + (0.58 \text{ m/s}^2)^2} = 0.60 \text{ m/s}^2.$$

**NOTE** The linear acceleration at this chosen instant is mostly centripetal, and keeps the child moving in a circle with the carousel. The tangential component that speeds up the circular motion is smaller.

**NOTE** The direction of the linear acceleration (magnitude calculated above as $0.60 \text{ m/s}^2$) is at the angle $\theta$ shown in Fig. 8–7b:

$$\theta = \tan^{-1}\left(\frac{a_{\text{tan}}}{a_R}\right) = \tan^{-1}\left(\frac{0.15 \text{ m/s}^2}{0.58 \text{ m/s}^2}\right) = 0.25 \text{ rad},$$

so $\theta \approx 15°$.

We can relate the angular velocity $\omega$ to the frequency of rotation, $f$. The **frequency** is the number of complete revolutions (rev) per second, as we saw in Chapter 5. One revolution (of a wheel, say) corresponds to an angle of $2\pi$ radians, and thus $1 \text{ rev/s} = 2\pi \text{ rad/s}$. Hence, in general, the frequency $f$ is related to the angular velocity $\omega$ by

$$f = \frac{\omega}{2\pi}$$

or

$$\omega = 2\pi f. \tag{8–7}$$

The unit for frequency, revolutions per second (rev/s), is given the special name the hertz (Hz). That is,

$$1 \text{ Hz} = 1 \text{ rev/s}.$$

Note that "revolution" is not really a unit, so we can also write $1 \text{ Hz} = 1 \text{ s}^{-1}$.

The time required for one complete revolution is called the **period** $T$, and it is related to the frequency by

$$T = \frac{1}{f}. \tag{8–8}$$

If a particle rotates at a frequency of three revolutions per second, then the period of each revolution is $\frac{1}{3}$ s.

**EXERCISE A** In Example 8–4 we found that the carousel, after 8.0 s, rotates at an angular velocity $\omega = 0.48 \text{ rad/s}$, and continues to do so after $t = 8.0$ s because the acceleration ceased. What are the frequency and period of the carousel when rotating at this constant angular velocity $\omega = 0.48 \text{ rad/s}$?

# 8–2 Constant Angular Acceleration

In Chapter 2, we derived the useful kinematic equations (Eqs. 2–11) that relate acceleration, velocity, distance, and time for the special case of uniform linear acceleration. Those equations were derived from the definitions of linear velocity and acceleration, assuming constant acceleration. The definitions of angular velocity and angular acceleration (Eqs. 8–2 and 8–3) are just like those for their linear counterparts, except that $\theta$ replaces the linear displacement $x$, $\omega$ replaces $v$, and $\alpha$ replaces $a$. Therefore, the angular equations for **constant angular acceleration** will be analogous to Eqs. 2–11 with $x$ replaced by $\theta$, $v$ by $\omega$, and $a$ by $\alpha$, and they can be derived in exactly the same way.

We summarize these angular equations here, opposite their linear equivalents, Eqs. 2–11 (for simplicity we choose $\theta_0 = 0$ and $x_0 = 0$ at the initial time $t_0 = 0$):

| | Angular | Linear | |
|---|---|---|---|
| *Kinematic equations* | $\omega = \omega_0 + \alpha t$ | $v = v_0 + at$ | [constant $\alpha$, $a$] **(8–9a)** |
| *for constant* | $\theta = \omega_0 t + \frac{1}{2}\alpha t^2$ | $x = v_0 t + \frac{1}{2}at^2$ | [constant $\alpha$, $a$] **(8–9b)** |
| *angular acceleration* | $\omega^2 = \omega_0^2 + 2\alpha\theta$ | $v^2 = v_0^2 + 2ax$ | [constant $\alpha$, $a$] **(8–9c)** |
| [$x_0 = 0$, $\theta_0 = 0$] | $\overline{\omega} = \dfrac{\omega + \omega_0}{2}$ | $\overline{v} = \dfrac{v + v_0}{2}$ | [constant $\alpha$, $a$] **(8–9d)** |

Note that $\omega_0$ represents the angular velocity at $t_0 = 0$, whereas $\theta$ and $\omega$ represent the angular position and velocity, respectively, at time $t$. Since the angular acceleration is constant, $\alpha = \overline{\alpha}$.

**PHYSICS APPLIED**
*Centrifuge*

---

**EXAMPLE 8–6** **Centrifuge acceleration.** A centrifuge rotor is accelerated for 30 s from rest to 20,000 rpm (revolutions per minute). (*a*) What is its average angular acceleration? (*b*) Through how many revolutions has the centrifuge rotor turned during its acceleration period, assuming constant angular acceleration?

**APPROACH** To determine $\overline{\alpha} = \Delta\omega/\Delta t$, we need the initial and final angular velocities. For (*b*), we use Eqs. 8–9 (recall that one revolution corresponds to $\theta = 2\pi$ rad).

**SOLUTION** (*a*) The initial angular velocity is $\omega_0 = 0$. The final angular velocity is

$$\omega = 2\pi f = (2\pi \text{ rad/rev})\frac{(20,000 \text{ rev/min})}{(60 \text{ s/min})} = 2100 \text{ rad/s}.$$

Then, since $\overline{\alpha} = \Delta\omega/\Delta t$ and $\Delta t = 30$ s, we have

$$\overline{\alpha} = \frac{\omega - \omega_0}{\Delta t} = \frac{2100 \text{ rad/s} - 0}{30 \text{ s}} = 70 \text{ rad/s}^2.$$

That is, every second the rotor's angular velocity increases by 70 rad/s, or by $(70 \text{ rad/s})(1 \text{ rev}/2\pi \text{ rad}) = 11$ revolutions per second.

(*b*) To find $\theta$ we could use either Eq. 8–9b or 8–9c (or both to check our answer). The former gives

$$\theta = \omega_0 t + \tfrac{1}{2}\alpha t^2 = 0 + \tfrac{1}{2}(70 \text{ rad/s}^2)(30 \text{ s})^2 = 3.15 \times 10^4 \text{ rad},$$

where we have kept an extra digit because this is an intermediate result. To find the total number of revolutions, we divide by $2\pi$ rad/rev and obtain

$$\frac{3.15 \times 10^4 \text{ rad}}{2\pi \text{ rad/rev}} = 5.0 \times 10^3 \text{ rev}.$$

**NOTE** Let us calculate $\theta$ using Eq. 8–9c:

$$\theta = \frac{\omega^2 - \omega_0^2}{2\alpha} = \frac{(2100 \text{ rad/s})^2 - 0}{2(70 \text{ rad/s}^2)} = 3.15 \times 10^4 \text{ rad}$$

which checks our answer above from Eq. 8–9b perfectly.

# 8–3 Rolling Motion (Without Slipping)

The rolling motion of a ball or wheel is familiar in everyday life: a ball rolling across the floor, or the wheels and tires of a car or bicycle rolling along the pavement. Rolling *without slipping* depends on static friction between the rolling object and the ground. The friction is static because the rolling object's point of contact with the ground is at rest at each moment.

Rolling without slipping involves both rotation and translation. There is a simple relation between the linear speed $v$ of the axle and the angular velocity $\omega$ of the rotating wheel or sphere: namely, $v = r\omega$ (where $r$ is the radius) as we now show.

**204** CHAPTER 8 Rotational Motion

Figure 8–8a shows a wheel rolling to the right without slipping. At the instant shown, point P on the wheel is in contact with the ground and is momentarily at rest. (If P was not at rest, the wheel would be slipping.) The velocity of the axle at the wheel's center C is $\vec{v}$. In Fig. 8–8b we have put ourselves in the reference frame of the wheel—that is, we are moving to the right with velocity $\vec{v}$ relative to the ground. In this reference frame the axle C is at rest, whereas the ground and point P are moving to the left with velocity $-\vec{v}$ as shown. In Fig. 8–8b we are seeing pure rotation. So we can use Eq. 8–4 to obtain $v = r\omega$, where $r$ is the radius of the wheel. This is the same $v$ as in Fig. 8–8a, so we see that the linear speed $v$ of the axle relative to the ground is related to the angular velocity $\omega$ of the wheel by

$$v = r\omega. \qquad \text{[rolling without slipping]}$$

This relationship is valid only if there is no slipping.

**FIGURE 8–8** (a) A wheel rolling to the right. Its center C moves with velocity $\vec{v}$. Point P is at rest at the instant shown. (b) The same wheel as seen from a reference frame in which the axle of the wheel C is at rest—that is, we are moving to the right with velocity $\vec{v}$ relative to the ground. Point P, which was at rest in (a), here in (b) is moving to the left with velocity $-\vec{v}$ as shown. (See also Section 3–8 on relative velocity.) Thus $v = r\omega$.

**EXAMPLE 8–7** **Bicycle.** A bicycle slows down uniformly from $v_0 = 8.40$ m/s to rest over a distance of 115 m, Fig. 8–9. Each wheel and tire has an overall diameter of 68.0 cm. Determine (a) the angular velocity of the wheels at the initial instant $(t = 0)$; (b) the total number of revolutions each wheel rotates before coming to rest; (c) the angular acceleration of the wheel; and (d) the time it took to come to a stop.

**APPROACH** We assume the bicycle wheels roll without slipping and the tire is in firm contact with the ground. The speed of the bike $v$ and the angular velocity of the wheels $\omega$ are related by $v = r\omega$. The bike slows down uniformly, so the angular acceleration is constant and we can use Eqs. 8–9.

**SOLUTION** (a) The initial angular velocity of the wheel, whose radius is 34.0 cm, is

$$\omega_0 = \frac{v_0}{r} = \frac{8.40 \text{ m/s}}{0.340 \text{ m}} = 24.7 \text{ rad/s}.$$

(b) In coming to a stop, the bike passes over 115 m of ground. The circumference of the wheel is $2\pi r$, so each revolution of the wheel corresponds to a distance traveled of $2\pi r = (2\pi)(0.340 \text{ m})$. Thus the number of revolutions the wheel makes in coming to a stop is

$$\frac{115 \text{ m}}{2\pi r} = \frac{115 \text{ m}}{(2\pi)(0.340 \text{ m})} = 53.8 \text{ rev}.$$

(c) The angular acceleration of the wheel can be obtained from Eq. 8–9c, for which we set $\omega = 0$ and $\omega_0 = 24.7$ rad/s. Because each revolution corresponds to $2\pi$ radians of angle, then $\theta = 2\pi$ rad/rev $\times$ 53.8 rev $(= 338$ rad$)$ and

$$\alpha = \frac{\omega^2 - \omega_0^2}{2\theta} = \frac{0 - (24.7 \text{ rad/s})^2}{2(2\pi \text{ rad/rev})(53.8 \text{ rev})} = -0.902 \text{ rad/s}^2.$$

(d) Equation 8–9a or b allows us to solve for the time. The first is easier:

$$t = \frac{\omega - \omega_0}{\alpha} = \frac{0 - 24.7 \text{ rad/s}}{-0.902 \text{ rad/s}^2} = 27.4 \text{ s}.$$

**NOTE** When the bike tire completes one revolution, the bike advances linearly a distance equal to the outer circumference $(2\pi r)$ of the tire, as long as there is no slipping or sliding.

$v_0 = 8.40$ m/s

115 m

Bike as seen from the ground at $t = 0$

**FIGURE 8–9** Example 8–7.

# 8–4 Torque

We have so far discussed rotational kinematics—the description of rotational motion in terms of angular position, angular velocity, and angular acceleration. Now we discuss the dynamics, or causes, of rotational motion. Just as we found analogies between linear and rotational motion for the description of motion, so rotational equivalents for dynamics exist as well.

To make an object start rotating about an axis clearly requires a force. But the direction of this force, and where it is applied, are also important. Take, for example, an ordinary situation such as the overhead view of the door in Fig. 8–10. If you apply a force $\vec{F}_A$ perpendicular to the door as shown, you will find that the greater the magnitude, $F_A$, the more quickly the door opens. But now if you apply the same force at a point closer to the hinge—say, $\vec{F}_B$ in Fig. 8–10—the door will not open so quickly. The effect of the force is less: *where* the force acts, as well as its magnitude and direction, affects how quickly the door opens. Indeed, if only this one force acts, the angular acceleration of the door is proportional not only to the magnitude of the force, but is also directly proportional to *the perpendicular distance from the axis of rotation to the line along which the force acts*. This distance is called the **lever arm**, or **moment arm**, of the force, and is labeled $r_A$ and $r_B$ for the two forces in Fig. 8–10. Thus, if $r_A$ in Fig. 8–10 is three times larger than $r_B$, then the angular acceleration of the door will be three times as great, assuming that the magnitudes of the forces are the same. To say it another way, if $r_A = 3r_B$, then $F_B$ must be three times as large as $F_A$ to give the same angular acceleration. (Figure 8–11 shows two examples of tools whose long lever arms are very effective.)

**FIGURE 8–10** Top view of a door. Applying the same force with different lever arms, $r_A$ and $r_B$. If $r_A = 3r_B$, then to create the same effect (angular acceleration), $F_B$ needs to be three times $F_A$.

The angular acceleration, then, is proportional to the product of the *force times the lever arm*. This product is called the *moment of the force* about the axis, or, more commonly, it is called the **torque**, and is represented by $\tau$ (Greek lowercase letter tau). Thus, the angular acceleration $\alpha$ of an object is directly proportional to the net applied torque $\tau$:

$$\alpha \propto \tau,$$

and we see that it is torque that gives rise to angular acceleration. This is the rotational analog of Newton's second law for linear motion, $a \propto F$.

We defined the lever arm as the *perpendicular* distance from the axis of rotation to the line of action of the force—that is, the distance which is perpendicular both to the axis of rotation and to an imaginary line drawn along the direction of the force. We do this to take into account the effect of forces acting at an angle. It is clear that a force applied at an angle, such as $\vec{F}_C$ in Fig. 8–12, will be less effective than the same magnitude force applied perpendicular to the door, such as $\vec{F}_A$ (Fig. 8–12a). And if you push on the end of the door so that the force is directed at the hinge (the axis of rotation), as indicated by $\vec{F}_D$, the door will not rotate at all.

The lever arm for a force such as $\vec{F}_C$ is found by drawing a line along the direction of $\vec{F}_C$ (this is the "line of action" of $\vec{F}_C$). Then we draw another line, perpendicular to this line of action, that goes to the axis of rotation and is perpendicular also to it. The length of this second line is the lever arm for $\vec{F}_C$ and is labeled $r_C$ in Fig. 8–12b. The lever arm for $\vec{F}_A$ is the full distance from the hinge to the doorknob, $r_A$ (just as in Fig. 8–10). Thus $r_C$ is much smaller than $r_A$.

**FIGURE 8–11** (a) A plumber can exert greater torque using a wrench with a long lever arm. (b) A tire iron too can have a long lever arm.

(a)                    (b)

**FIGURE 8–12** (a) Forces acting at different angles at the doorknob. (b) The lever arm is defined as the perpendicular distance from the axis of rotation (the hinge) to the line of action of the force ($r_C$ for the force $\vec{F}_C$).

The magnitude of the torque associated with $\vec{F}_C$ is then $r_C F_C$. This short lever arm $r_C$ and the corresponding smaller torque associated with $\vec{F}_C$ are consistent with the observation that $\vec{F}_C$ is less effective in accelerating the door than is $\vec{F}_A$ with its larger lever arm. When the lever arm is defined in this way, experiment shows that the relation $\alpha \propto \tau$ is valid in general. Notice in Fig. 8–12 that the line of action of the force $\vec{F}_D$ passes through the hinge, and hence its lever arm is zero. Consequently, zero torque is associated with $\vec{F}_D$ and it gives rise to no angular acceleration, in accord with everyday experience (you can't get a door to start moving by pushing directly at the hinge).

In general, then, we can write the magnitude of the torque about a given axis as

$$\tau = r_\perp F, \tag{8–10a}$$

where $r_\perp$ is the lever arm, and the perpendicular symbol ($\perp$) reminds us that we must use the distance from the axis of rotation that is perpendicular to the line of action of the force (Fig. 8–13a).

An equivalent way of determining the torque associated with a force is to resolve the force into components parallel and perpendicular to the line that connects the axis to the point of application of the force, as shown in Fig. 8–13b. The component $F_\parallel$ exerts no torque since it is directed at the rotation axis (its lever arm is zero). Hence the torque will be equal to $F_\perp$ times the distance $r$ from the axis to the point of application of the force:

$$\tau = r F_\perp. \tag{8–10b}$$

This gives the same result as Eq. 8–10a because $F_\perp = F \sin\theta$ and $r_\perp = r \sin\theta$. Thus

$$\tau = r F \sin\theta \tag{8–10c}$$

in either case. [Note that $\theta$ is the angle between the directions of $\vec{F}$ and $r$ (radial line from the axis to the point where $\vec{F}$ acts).] We can use any of Eqs. 8–10 to calculate the torque, whichever is easiest.

Because torque is a distance times a force, it is measured in units of $m \cdot N$ in SI units,[†] $cm \cdot dyne$ in the cgs system, and $ft \cdot lb$ in the English system.

---

**EXAMPLE 8–8** **Biceps torque.** The biceps muscle exerts a vertical force on the lower arm, bent as shown in Figs. 8–14a and b. For each case, calculate the torque about the axis of rotation through the elbow joint, assuming the muscle is attached 5.0 cm from the elbow as shown.

**APPROACH** The force is given, and the lever arm in (a) is given. In (b) we have to take into account the angle to get the lever arm.

**SOLUTION** (a) $F = 700\,N$ and $r_\perp = 0.050\,m$, so

$$\tau = r_\perp F = (0.050\,m)(700\,N) = 35\,m \cdot N.$$

(b) Because the arm is at an angle below the horizontal, the lever arm is shorter (Fig. 8–14c) than in part (a): $r_\perp = (0.050\,m)(\sin 60°)$, where $\theta = 60°$ is the angle between $\vec{F}$ and $r$. $F$ is still 700 N, so

$$\tau = (0.050\,m)(0.866)(700\,N) = 30\,m \cdot N.$$

The arm can exert less torque at this angle than when it is at 90°. Weight machines at gyms are often designed to take this variation with angle into account.

**NOTE** In (b), we could instead have used $\tau = r F_\perp$. As shown in Fig. 8–14d, $F_\perp = F \sin 60°$. Then $\tau = r F_\perp = r F \sin\theta = (0.050\,m)(700\,N)(0.866)$ gives the same result.

---

(a)

(b)

**FIGURE 8–13** Torque = $r_\perp F = r F_\perp$.

**FIGURE 8–14** Example 8–8.

(a)

700 N

Axis at elbow

5.0 cm

(b)

700 N

Axis

$r_\perp$

30°

60°

(c)

(d)

**FIGURE 8–15** Exercise B.

**FIGURE 8–16** Only the component of $\vec{F}$ that acts in the plane perpendicular to the rotation axis, $\vec{F}_\perp$, acts to accelerate the wheel about the axis. The component parallel to the axis, $\vec{F}_\parallel$, would tend to move the axis itself, which we assume is held fixed.

**EXERCISE B** Two forces ($F_B = 20\,\text{N}$ and $F_A = 30\,\text{N}$) are applied to a meter stick which can rotate about its left end, Fig. 8–15. Force $\vec{F}_B$ is applied perpendicularly at the midpoint. Which force exerts the greater torque: $F_A$, $F_B$, or both the same?

When more than one torque acts on an object, the angular acceleration $\alpha$ is found to be proportional to the *net* torque. If all the torques acting on an object tend to rotate it in the same direction about a fixed axis of rotation, the net torque is the sum of the torques. But if, say, one torque acts to rotate an object in one direction, and a second torque acts to rotate the object in the opposite direction, the net torque is the difference of the two torques. We normally assign a positive sign to torques that act to rotate the object counterclockwise (just as $\theta$ is usually positive counterclockwise), and a negative sign to torques that act to rotate the object clockwise.

## *Forces that Act to Tilt the Axis

We have been considering only rotation about a fixed axis, and so we considered only forces that act in a plane perpendicular to the axis of rotation. If there is a force (or component of a force) acting parallel to the axis of rotation, it will tend to tilt the axis of rotation—the component $\vec{F}_\parallel$ in Fig. 8–16 is an example. Since we are assuming the axis remains fixed in direction, either there can be no such forces or else the axis must be mounted in bearings or hinges that hold the axis fixed. Thus, only a force, or component of a force ($\vec{F}_\perp$ in Fig. 8–16), in a plane perpendicular to the axis will give rise to rotational acceleration about the axis.

# 8–5 Rotational Dynamics; Torque and Rotational Inertia

We discussed in Section 8–4 that the angular acceleration $\alpha$ of a rotating object is proportional to the net torque $\tau$ applied to it:

$$\alpha \propto \Sigma\tau.$$

We write $\Sigma\tau$ to remind us that it is the *net* torque (sum of all torques acting on the object) that is proportional to $\alpha$. This corresponds to Newton's second law for translational motion, $a \propto \Sigma F$. In the translational case, the acceleration is not only proportional to the net force, but it is also inversely proportional to the inertia of the object, which we call its mass, $m$. Thus we wrote $a = \Sigma F / m$. But what plays the role of mass for the rotational case? That is what we now set out to determine. At the same time, we will see that the relation $\alpha \propto \Sigma\tau$ follows directly from Newton's second law, $\Sigma F = ma$.

We first examine a very simple case: a particle of mass $m$ revolving in a circle of radius $r$ at the end of a string or rod whose mass we can ignore compared to $m$ (Fig. 8–17). Consider a force $F$ that acts on the mass $m$ tangent to the circle as shown. The torque that gives rise to an angular acceleration is $\tau = rF$. If we use Newton's second law for linear quantities, $\Sigma F = ma$, and Eq. 8–5 relating the angular acceleration to the tangential linear acceleration, $a_{\text{tan}} = r\alpha$, then we have

$$F = ma$$
$$= mr\alpha.$$

When we multiply both sides of this equation by $r$, we find that the torque

$$\tau = rF = r(mr\alpha),$$

or

$$\tau = mr^2\alpha. \qquad \text{[single particle]} \quad \textbf{(8–11)}$$

Here at last we have a direct relation between the angular acceleration and the applied torque $\tau$. The quantity $mr^2$ represents the *rotational inertia* of the particle and is called its **moment of inertia**.

**FIGURE 8–17** A mass $m$ revolving in a circle of radius $r$ about a fixed point C.

Now let us consider a rotating rigid object, such as a wheel rotating about a fixed axis (an axle) through its center. We can think of the wheel as consisting of many particles located at various distances from the axis of rotation. We can apply Eq. 8–11 to each particle of the object, and then sum over all the particles. The sum of the various torques is the net torque, $\Sigma\tau$, so we obtain:

$$\Sigma\tau = (\Sigma mr^2)\alpha \qquad (8\text{--}12)$$

where we factored out $\alpha$ because it is the same for all the particles of a rigid object. The sum $\Sigma mr^2$ represents the sum of the masses of each particle in the object multiplied by the square of the distance of that particle from the axis of rotation. If we assign each particle a number (1, 2, 3, … ), then $\Sigma mr^2 = m_1 r_1^2 + m_2 r_2^2 + m_3 r_3^2 + \cdots$. This sum is called the **moment of inertia** (or *rotational inertia*) $I$ of the object:

$$I = \Sigma mr^2 = m_1 r_1^2 + m_2 r_2^2 + \cdots. \qquad (8\text{--}13)$$

Combining Eqs. 8–12 and 8–13, we can write

$$\Sigma\tau = I\alpha. \qquad (8\text{--}14)$$

NEWTON'S SECOND LAW FOR ROTATION

This is the rotational equivalent of Newton's second law. It is valid for the rotation of a rigid object about a fixed axis. [It is also valid when the object is rotating while translating with acceleration, as long as $I$ and $\alpha$ are calculated about the center of mass of the object, and the rotation axis through the CM doesn't change direction. A ball rolling down a ramp is an example.]

We see that the moment of inertia, $I$, which is a measure of the rotational inertia of an object, plays the same role for rotational motion that mass does for translational motion. As can be seen from Eq. 8–13, the rotational inertia of a rigid object depends not only on its mass, but also on how that mass is distributed with respect to the axis. For example, a large-diameter cylinder will have greater rotational inertia than one of equal mass but smaller diameter, Fig. 8–18. The former will be harder to start rotating, and harder to stop. When the mass is concentrated farther from the axis of rotation, the rotational inertia is greater. For rotational motion, the mass of an object can *not* be considered as concentrated at its center of mass.

**FIGURE 8–18** A large-diameter cylinder has greater rotational inertia than one of smaller diameter but equal mass.

---

**EXAMPLE 8–9** **Two weights on a bar: different axis, different $I$.** Two small "weights," of mass 5.0 kg and 7.0 kg, are mounted 4.0 m apart on a light rod (whose mass can be ignored), as shown in Fig. 8–19. Calculate the moment of inertia of the system (*a*) when rotated about an axis halfway between the weights, Fig. 8–19a, and (*b*) when rotated about an axis 0.50 m to the left of the 5.0-kg mass (Fig. 8–19b).

**APPROACH** In each case, the moment of inertia of the system is found by summing over the two parts using Eq. 8–13.

**SOLUTION** (*a*) Both weights are the same distance, 2.0 m, from the axis of rotation. Thus

$$\begin{aligned} I &= \Sigma mr^2 = (5.0\,\text{kg})(2.0\,\text{m})^2 + (7.0\,\text{kg})(2.0\,\text{m})^2 \\ &= 20\,\text{kg}\cdot\text{m}^2 + 28\,\text{kg}\cdot\text{m}^2 = 48\,\text{kg}\cdot\text{m}^2. \end{aligned}$$

(*b*) The 5.0-kg mass is now 0.50 m from the axis, and the 7.0-kg mass is 4.50 m from the axis. Then

$$\begin{aligned} I &= \Sigma mr^2 = (5.0\,\text{kg})(0.50\,\text{m})^2 + (7.0\,\text{kg})(4.5\,\text{m})^2 \\ &= 1.3\,\text{kg}\cdot\text{m}^2 + 142\,\text{kg}\cdot\text{m}^2 = 143\,\text{kg}\cdot\text{m}^2. \end{aligned}$$

**NOTE** This Example illustrates two important points. First, the moment of inertia of a given system is different for different axes of rotation. Second, we see in part (*b*) that mass close to the axis of rotation contributes little to the total moment of inertia; here, the 5.0-kg object contributed less than 1% to the total.

**FIGURE 8–19** Example 8–9: calculating the moment of inertia.

**CAUTION**

*I depends on axis of rotation and on distribution of mass*

| Object | Location of axis | | Moment of inertia |
|--------|-----------------|---|---------------------|
| (a) **Thin hoop,** radius $R$ | Through center | | $MR^2$ |
| (b) **Thin hoop,** radius $R$ width $w$ | Through central diameter | | $\frac{1}{2}MR^2 + \frac{1}{12}Mw^2$ |
| (c) **Solid cylinder,** radius $R$ | Through center | | $\frac{1}{2}MR^2$ |
| (d) **Hollow cylinder,** inner radius $R_1$ outer radius $R_2$ | Through center | | $\frac{1}{2}M(R_1^2 + R_2^2)$ |
| (e) **Uniform sphere,** radius $R$ | Through center | | $\frac{2}{5}MR^2$ |
| (f) **Long uniform rod,** length $\ell$ | Through center | | $\frac{1}{12}M\ell^2$ |
| (g) **Long uniform rod,** length $\ell$ | Through end | | $\frac{1}{3}M\ell^2$ |
| (h) **Rectangular thin plate,** length $\ell$, width $w$ | Through center | | $\frac{1}{12}M(\ell^2 + w^2)$ |

**FIGURE 8–20** Moments of inertia for various objects of uniform composition, each with mass $M$.

For most ordinary objects, the mass is distributed continuously, and the calculation of the moment of inertia, $\Sigma mr^2$, can be difficult. Expressions can, however, be worked out (using calculus) for the moments of inertia of regularly shaped objects in terms of the dimensions of the objects. Figure 8–20 gives these expressions for a number of solids rotated about the axes specified. The only one for which the result is obvious is that for the thin hoop or ring rotated about an axis passing through its center perpendicular to the plane of the hoop (Fig. 8–20a). For a hoop, all the mass is concentrated at the same distance from the axis, $R$. Thus $\Sigma mr^2 = (\Sigma m)R^2 = MR^2$, where $M$ is the total mass of the hoop. In Fig. 8–20, we use capital $R$ to refer to the outer radius of an object (in (d) also the inner radius).

When calculation is difficult, $I$ can be determined experimentally by measuring the angular acceleration $\alpha$ about a fixed axis due to a known net torque, $\Sigma\tau$, and applying Newton's second law, $I = \Sigma\tau/\alpha$, Eq. 8–14.

# 8–6 Solving Problems in Rotational Dynamics

When working with torque and angular acceleration (Eq. 8–14), it is important to use a consistent set of units, which in SI is: $\alpha$ in rad/s$^2$; $\tau$ in m·N; and the moment of inertia, $I$, in kg·m$^2$.

## Rotational Motion

1. As always, draw a clear and complete **diagram**.
2. Choose the object or objects that will be the **system** to be studied.
3. Draw a **free-body diagram** for the object under consideration (or for each object, if more than one), showing all (and only) the forces acting on that object and exactly where they act, so you can determine the torque due to each. Gravity acts at the CM of the object (Section 7–8).
4. Identify the axis of rotation and determine the **torques** about it. Choose positive and negative

directions of rotation (counterclockwise and clockwise), and assign the correct sign to each torque.

5. Apply **Newton's second law for rotation**, $\Sigma\tau = I\alpha$. If the moment of inertia is not given, and it is not the unknown sought, you need to determine it first. Use consistent units, which in SI are: $\alpha$ in $rad/s^2$; $\tau$ in $m \cdot N$; and $I$ in $kg \cdot m^2$.
6. Also apply **Newton's second law for translation**, $\Sigma\vec{F} = m\vec{a}$, and **other** laws or principles as needed.
7. **Solve** the resulting equation(s) for the unknown(s).
8. Do a rough **estimate** to determine if your answer is reasonable.

**EXAMPLE 8–10** **A heavy pulley.** A 15.0-N force (represented by $\vec{F}_T$) is applied to a cord wrapped around a pulley of mass $M = 4.00\,kg$ and radius $R = 33.0\,cm$, Fig. 8–21. The pulley accelerates uniformly from rest to an angular speed of 30.0 rad/s in 3.00 s. If there is a frictional torque $\tau_{fr} = 1.10\,m \cdot N$ at the axle, determine the moment of inertia of the pulley. The pulley rotates about its center.

**APPROACH** We follow the steps of the Problem Solving Strategy above.

**SOLUTION**
1. **Draw a diagram.** The pulley and the attached cord are shown in Fig. 8–21.
2. **Choose the system:** the pulley.
3. **Draw a free-body diagram.** The force that the cord exerts on the pulley is shown as $\vec{F}_T$ in Fig. 8–21. The friction force acts all around the axle, retarding the motion, as suggested by $\vec{F}_{fr}$ in Fig. 8–21. We are given only its torque, which is what we need. Two other forces could be included in the diagram: the force of gravity $mg$ down and whatever force keeps the axle in place (they balance each other). They do not contribute to the torque (their lever arms are zero) and so we omit them to keep our diagram simple.
4. **Determine the torques.** The cord exerts a force $\vec{F}_T$ that acts at the edge of the pulley, so its lever arm is $R$. The torque exerted by the cord equals $RF_T$ and is counterclockwise, which we choose to be positive. The frictional torque is given as $\tau_{fr} = 1.10\,m \cdot N$; it opposes the motion and is negative.
5. **Apply Newton's second law for rotation.** The net torque is
$$\Sigma\tau = RF_T - \tau_{fr} = (0.330\,m)(15.0\,N) - 1.10\,m \cdot N = 3.85\,m \cdot N.$$
The angular acceleration $\alpha$ is found from the given data that it takes 3.00 s to accelerate the pulley from rest to $\omega = 30.0\,rad/s$:
$$\alpha = \frac{\Delta\omega}{\Delta t} = \frac{30.0\,rad/s - 0}{3.00\,s} = 10.0\,rad/s^2.$$
Newton's second law, $\Sigma\tau = I\alpha$, can be solved for $I$ which is the unknown: $I = \Sigma\tau/\alpha$.
6. **Other calculations:** None needed.
7. **Solve for unknowns.** From Newton's second law,
$$I = \frac{\Sigma\tau}{\alpha} = \frac{3.85\,m \cdot N}{10.0\,rad/s^2} = 0.385\,kg \cdot m^2.$$
8. **Do a rough estimate.** We can do a rough estimate of the moment of inertia by assuming the pulley is a uniform cylinder and using Fig. 8–20c:
$$I \approx \tfrac{1}{2}MR^2 = \tfrac{1}{2}(4.00\,kg)(0.330\,m)^2 = 0.218\,kg \cdot m^2.$$
This is the same order of magnitude as our result, but numerically somewhat less. This makes sense, though, because a pulley is not usually a uniform cylinder but instead has more of its mass concentrated toward the outside edge. Such a pulley would be expected to have a greater moment of inertia than a solid cylinder of equal mass. A thin hoop, Fig. 8–20a, ought to have a greater $I$ than our pulley, and indeed it does: $I = MR^2 = 0.436\,kg \cdot m^2$.

**FIGURE 8–21** Example 8–10.

**PROBLEM SOLVING**
*Usefulness and power of rough estimates*

(a)

(b)

**FIGURE 8–22** Example 8–11. (a) Pulley and falling bucket of mass $m$. This is also the free-body diagram for the pulley. (b) Free-body diagram for the bucket.

## Additional Example—a bit more challenging

**EXAMPLE 8–11** **Pulley and bucket.** Consider again the pulley in Example 8–10. But instead of a constant 15.0-N force being exerted on the cord, we now have a bucket of weight $w = 15.0$ N (mass $m = w/g = 1.53$ kg) hanging from the cord. See Fig. 8–22a. We assume the cord has negligible mass and does not stretch or slip on the pulley. Calculate the angular acceleration $\alpha$ of the pulley and the linear acceleration $a$ of the bucket. Assume the same frictional torque $\tau_{fr} = 1.10$ m·N acts.

**APPROACH** This situation looks a lot like Example 8–10, Fig. 8–21. But there is a big difference: the tension in the cord is now an unknown, and it is no longer equal to the weight of the bucket if the bucket accelerates. Our system has two parts: the bucket, which can undergo translational motion (Fig. 8–22b is its free-body diagram); and the pulley. The pulley does not translate, but it can rotate. We apply the rotational version of Newton's second law to the pulley, $\Sigma\tau = I\alpha$, and the linear version to the bucket, $\Sigma F = ma$.

**SOLUTION** Let $F_T$ be the tension in the cord. Then a force $F_T$ acts at the edge of the pulley, and we apply Newton's second law, Eq. 8–14, for the rotation of the pulley:

$$I\alpha = \Sigma\tau = RF_T - \tau_{fr}. \qquad \text{[pulley]}$$

Next we look at the (linear) motion of the bucket of mass $m$. Figure 8–22b, the free-body diagram for the bucket, shows that two forces act on the bucket: the force of gravity $mg$ acts downward, and the tension of the cord $F_T$ pulls upward. Applying Newton's second law, $\Sigma F = ma$, for the bucket, we have (taking downward as positive):

$$mg - F_T = ma. \qquad \text{[bucket]}$$

Note that the tension $F_T$, which is the force exerted on the edge of the pulley, is *not* equal to the weight of the bucket ($= mg = 15.0$ N). There must be a net force on the bucket if it is accelerating, so $F_T < mg$. We can also see this from the last equation above, $F_T = mg - ma$.

To obtain $\alpha$, we note that the tangential acceleration of a point on the edge of the pulley is the same as the acceleration of the bucket if the cord doesn't stretch or slip. Hence we can use Eq. 8–5, $a_{tan} = a = R\alpha$. Substituting $F_T = mg - ma = mg - mR\alpha$ into the first equation above (Newton's second law for rotation of the pulley), we obtain

$$I\alpha = \Sigma\tau = RF_T - \tau_{fr} = R(mg - mR\alpha) - \tau_{fr} = mgR - mR^2\alpha - \tau_{fr}.$$

The unknown $\alpha$ appears on the left and in the second term on the far right, so we bring that term to the left side and solve for $\alpha$:

$$\alpha = \frac{mgR - \tau_{fr}}{I + mR^2}.$$

The numerator $(mgR - \tau_{fr})$ is the net torque, and the denominator $(I + mR^2)$ is the total rotational inertia of the system. With $mg = 15.0$ N ($m = 1.53$ kg) and, from Example 8–10, $I = 0.385$ kg·m² and $\tau_{fr} = 1.10$ m·N, then

$$\alpha = \frac{(15.0 \text{ N})(0.330 \text{ m}) - 1.10 \text{ m·N}}{0.385 \text{ kg·m}^2 + (1.53 \text{ kg})(0.330 \text{ m})^2} = 6.98 \text{ rad/s}^2.$$

The angular acceleration is somewhat less in this case than the 10.0 rad/s² of Example 8–10. Why? Because $F_T$ ($= mg - ma = 15.0$ N $- ma$) is less than the 15.0-N force in Example 8–10. The linear acceleration of the bucket is

$$a = R\alpha = (0.330 \text{ m})(6.98 \text{ rad/s}^2) = 2.30 \text{ m/s}^2.$$

**NOTE** The tension in the cord $F_T$ is less than $mg$ because the bucket accelerates.

# 8–7 Rotational Kinetic Energy

The quantity $\frac{1}{2}mv^2$ is the kinetic energy of an object undergoing translational motion. An object rotating about an axis is said to have **rotational kinetic energy**. By analogy with translational kinetic energy, we might expect this to be given by the expression $\frac{1}{2}I\omega^2$, where $I$ is the moment of inertia of the object and $\omega$ is its angular velocity. We can indeed show that this is true.

Consider any rigid rotating object as made up of many tiny particles, each of mass $m$. If we let $r$ represent the distance of any one particle from the axis of rotation, then its linear velocity is $v = r\omega$. The total kinetic energy of the whole object will be the sum of the kinetic energies of all its particles:

$$\text{KE} = \Sigma(\tfrac{1}{2}mv^2) = \Sigma(\tfrac{1}{2}mr^2\omega^2)$$
$$= \tfrac{1}{2}(\Sigma mr^2)\omega^2.$$

We have factored out the $\tfrac{1}{2}$ and the $\omega^2$ since they are the same for every particle of a rigid object. Since $\Sigma mr^2 = I$, the moment of inertia, we see that the kinetic energy of a rigid rotating object is

$$\text{rotational KE} = \tfrac{1}{2}I\omega^2. \qquad (8\text{–}15)$$

The units are joules, as with all other forms of energy.

An object that rotates while its center of mass (CM) undergoes translational motion will have both translational and rotational kinetic energy. Equation 8–15 gives the rotational kinetic energy if the rotation axis is fixed. If the object is moving, such as a wheel rolling down a hill, this equation is still valid as long as the rotation axis is fixed in direction. Then the total kinetic energy is

$$\text{KE} = \tfrac{1}{2}Mv_{\text{CM}}^2 + \tfrac{1}{2}I_{\text{CM}}\omega^2, \qquad (8\text{–}16)$$

where $v_{\text{CM}}$ is the linear velocity of the center of mass, $I_{\text{CM}}$ is the moment of inertia about an axis through the center of mass, $\omega$ is the angular velocity about this axis, and $M$ is the total mass of the object.

---

**EXAMPLE 8–12** **Sphere rolling down an incline.** What will be the speed of a solid sphere of mass $M$ and radius $R$ when it reaches the bottom of an incline if it starts from rest at a vertical height $H$ and rolls without slipping? See Fig. 8–23. (Assume sufficient static friction so no slipping occurs: we will see shortly that static friction does no work.) Compare your result to that for an object *sliding* down a frictionless incline.

**APPROACH** We use the law of conservation of energy with gravitational potential energy, now including rotational kinetic energy as well as translational KE.

**SOLUTION** The total energy at any point a vertical distance $y$ above the base of the incline is

$$E = \tfrac{1}{2}Mv^2 + \tfrac{1}{2}I_{\text{CM}}\omega^2 + Mgy,$$

where $v$ is the speed of the center of mass, and $Mgy$ is the gravitational potential energy. Applying conservation of energy, we equate the total energy at the top $(y = H, v = 0, \omega = 0)$ to the total energy at the bottom $(y = 0)$:

$$E_{\text{top}} = E_{\text{bottom}}$$
$$0 + 0 + MgH = \tfrac{1}{2}Mv^2 + \tfrac{1}{2}I_{\text{CM}}\omega^2 + 0. \qquad \text{[energy conservation]}$$

The moment of inertia of a solid sphere about an axis through its center of mass is $I_{\text{CM}} = \tfrac{2}{5}MR^2$, Fig. 8–20e. Since the sphere rolls without slipping, we have $\omega = v/R$ (recall Fig. 8–8). Hence

$$MgH = \tfrac{1}{2}Mv^2 + \tfrac{1}{2}\left(\tfrac{2}{5}MR^2\right)\left(\frac{v^2}{R^2}\right).$$

Canceling the $M$'s and $R$'s, we obtain

$$\left(\tfrac{1}{2} + \tfrac{1}{5}\right)v^2 = gH$$

or

$$v = \sqrt{\tfrac{10}{7}gH}. \qquad \text{[rolling sphere]}$$

We can compare this result for the speed of a rolling sphere to that for an object sliding down a plane without rotating and without friction, $\tfrac{1}{2}mv^2 = mgH$ (see our energy conservation equation above, removing the rotational term). For the sliding object, $v = \sqrt{2gH}$, which is greater than our result for a rolling sphere $(2 > 10/7)$. An object sliding without friction or rotation transforms its initial potential energy entirely into translational kinetic energy (none into rotational kinetic energy), so the speed of its center of mass is greater.

**NOTE** Our result for the rolling sphere shows (perhaps surprisingly) that $v$ is independent of both the mass $M$ and the radius $R$ of the sphere.

**FIGURE 8–23** A sphere rolling down a hill has both translational and rotational kinetic energy. Example 8–12.

**PROBLEM SOLVING**

*Rotational energy adds to other forms of energy to get the total energy which is conserved*

**Which is fastest?** Several objects roll without slipping down an incline of vertical height $H$, all starting from rest at the same moment. The objects are a thin hoop (or a plain wedding band), a spherical marble, a solid cylinder (a D-cell battery), and an empty soup can. In addition, a greased box slides down without friction. In what order do they reach the bottom of the incline?

**RESPONSE** We use conservation of energy with gravitational potential energy plus rotational and translational kinetic energy. The sliding box would be fastest because the potential energy loss ($MgH$) is transformed completely into translational kinetic energy of the box, whereas for rolling objects the initial potential energy is shared between translational and rotational kinetic energies, and so the speed of the CM is less. For each of the rolling objects we can state that the decrease in potential energy equals the increase in translational plus rotational kinetic energy:

$$MgH = \tfrac{1}{2}Mv^2 + \tfrac{1}{2}I_{CM}\omega^2.$$

For all our rolling objects, the moment of inertia $I_{CM}$ is a numerical factor times the mass $M$ and the radius $R^2$ (Fig. 8–20). The mass $M$ is in each term, so the translational speed $v$ doesn't depend on $M$; nor does it depend on the radius $R$ since $\omega = v/R$, so $R^2$ cancels out for all the rolling objects. Thus the speed $v$ at the bottom of the incline depends only on that numerical factor in $I_{CM}$ which expresses how the mass is distributed. The hoop, with all its mass concentrated at radius $R$ $\left(I_{CM} = MR^2\right)$, has the largest moment of inertia; hence it will have the lowest speed and will arrive at the bottom behind the D-cell $\left(I_{CM} = \tfrac{1}{2}MR^2\right)$, which in turn will be behind the marble $\left(I_{CM} = \tfrac{2}{5}MR^2\right)$. The empty can, which is mainly a hoop plus a thin disk, has most of its mass concentrated at $R$; so it will be a bit faster than the pure hoop but slower than the D-cell. See Fig. 8–24.

**NOTE** The rolling objects do not even have to have the same radius: the speed at the bottom does not depend on the object's mass $M$ or radius $R$, but only on the shape (and the height of the incline $H$).

**FIGURE 8–24** Example 8–13.

**FIGURE 8–25** A sphere rolling to the right on a plane surface. The point in contact with the ground at any moment, point P, is momentarily at rest. Point A to the left of P is moving nearly vertically upward at the instant shown, and point B to the right is moving nearly vertically downward. An instant later, point B will touch the plane and be at rest momentarily. Thus no work is done by the force of static friction.

If there had been little or no static friction between the rolling objects and the plane in these Examples, the round objects would have slid rather than rolled, or a combination of both. Static friction must be present to make a round object roll. We did not need to take friction into account in the energy equation for the rolling objects because it is *static* friction and does no work—the point of contact of a sphere at each instant does not slide, but moves perpendicular to the plane (first down and then up as shown in Fig. 8–25) as it rolls. Thus, no work is done by the static friction force because the force and the motion (displacement) are perpendicular. The reason the rolling objects in Examples 8–12 and 8–13 move down the slope more slowly than if they were sliding is *not* because friction slows them down. Rather, it is because some of the gravitational potential energy is converted to rotational kinetic energy, leaving less for the translational kinetic energy.

**EXERCISE C** Return to the Chapter-Opening Question, page 198, and answer it again now. Try to explain why you may have answered differently the first time.

## Work Done by Torque

The work done on an object rotating about a fixed axis, such as the pulleys in Figs. 8–21 and 8–22, can be written using angular quantities. As shown in Fig. 8–26, a force $F$ exerting a torque $\tau = rF$ on a wheel does work $W = F\,\Delta\ell$ in rotating the wheel a small distance $\Delta\ell$ at the point of application of $\vec{F}$. The wheel has rotated through a small angle $\Delta\theta = \Delta\ell/r$ (Eq. 8–1). Hence

$$W = F\,\Delta\ell = Fr\,\Delta\theta.$$

Because $\tau = rF$, then

$$W = \tau\,\Delta\theta \qquad\qquad (8\text{–}17)$$

is the work done by the torque $\tau$ when rotating the wheel through an angle $\Delta\theta$. Finally, power $P$ is the rate work is done:

$$P = W/\Delta t = \tau\,\Delta\theta/\Delta t = \tau\omega,$$

which is analogous to the translational version, $P = Fv$ (see Eq. 6–18).

**FIGURE 8–26** Torque $\tau = rF$ does work when rotating a wheel equal to $W = F\,\Delta\ell = Fr\,\Delta\theta = \tau\,\Delta\theta$.

# 8–8 Angular Momentum and Its Conservation

Throughout this Chapter we have seen that if we use the appropriate angular variables, the kinematic and dynamic equations for rotational motion are analogous to those for ordinary linear motion. We saw in the previous Section, for example, that rotational kinetic energy can be written as $\frac{1}{2}I\omega^2$, which is analogous to the translational kinetic energy, $\frac{1}{2}mv^2$. In like manner, the linear momentum, $p = mv$, has a rotational analog. It is called **angular momentum**, $L$. For a symmetrical object rotating about a fixed axis through the CM, the angular momentum is

$$L = I\omega, \qquad\qquad (8\text{–}18)$$

where $I$ is the moment of inertia and $\omega$ is the angular velocity about the axis of rotation. The SI units for $L$ are $kg \cdot m^2/s$, which has no special name.

We saw in Chapter 7 (Section 7–1) that Newton's second law can be written not only as $\Sigma F = ma$ but also more generally in terms of momentum (Eq. 7–2), $\Sigma F = \Delta p/\Delta t$. In a similar way, the rotational equivalent of Newton's second law, which we saw in Eq. 8–14 can be written as $\Sigma\tau = I\alpha$, can also be written in terms of angular momentum:

$$\Sigma\tau = \frac{\Delta L}{\Delta t}, \qquad\qquad (8\text{–}19)$$

where $\Sigma\tau$ is the net torque acting to rotate the object, and $\Delta L$ is the change in angular momentum in a time interval $\Delta t$. Equation 8–14, $\Sigma\tau = I\alpha$, is a special case of Eq. 8–19 when the moment of inertia is constant. This can be seen as follows. If an object has angular velocity $\omega_0$ at time $t = 0$, and angular velocity $\omega$ after a time interval $\Delta t$, then its angular acceleration (Eq. 8–3) is

$$\alpha = \frac{\Delta\omega}{\Delta t} = \frac{\omega - \omega_0}{\Delta t}.$$

Then from Eq. 8–19, we have

$$\Sigma\tau = \frac{\Delta L}{\Delta t} = \frac{I\omega - I\omega_0}{\Delta t} = \frac{I(\omega - \omega_0)}{\Delta t} = I\frac{\Delta\omega}{\Delta t} = I\alpha,$$

which is Eq. 8–14.

Angular momentum is an important concept in physics because, under certain conditions, it is a conserved quantity. We can see from Eq. 8–19 that if the net torque $\Sigma\tau$ on an object is zero, then $\Delta L/\Delta t$ equals zero. That is, $\Delta L = 0$, so $L$ does not change. This is the **law of conservation of angular momentum** for a rotating object:

> **The total angular momentum of a rotating object remains constant if the net torque acting on it is zero.**

The law of conservation of angular momentum is one of the great conservation laws of physics, along with those for energy and linear momentum.

When there is zero net torque acting on an object, and the object is rotating about a fixed axis or about an axis through its center of mass whose direction doesn't change, we can write

$$I\omega = I_0\omega_0 = \text{constant}. \qquad\qquad (8\text{–}20)$$

$I_0$ and $\omega_0$ are the moment of inertia and angular velocity, respectively, about that axis at some initial time ($t = 0$), and $I$ and $\omega$ are their values at some other time. The parts of the object may alter their positions relative to one another, so that $I$ changes. But then $\omega$ changes as well, so that the product $I\omega$ remains constant.

*I* large, ω small      *I* small, ω large

**FIGURE 8–27** A skater spinning on ice, illustrating conservation of angular momentum: (a) *I* is large and ω is small; (b) *I* is smaller so ω is larger.

**FIGURE 8–28** A diver rotates faster when arms and legs are tucked in than when they are outstretched. Angular momentum is conserved.

**FIGURE 8–29** Example 8–14.

Many interesting phenomena can be understood on the basis of conservation of angular momentum. Consider a skater doing a spin on the tips of her skates, Fig. 8–27. She rotates at a relatively low speed when her arms are outstretched; when she brings her arms in close to her body, she suddenly spins much faster. From the definition of moment of inertia, $I = \Sigma mr^2$, it is clear that when she pulls her arms in closer to the axis of rotation, $r$ is reduced for the arms so her moment of inertia is reduced. Since the angular momentum $I\omega$ remains constant (we ignore the small torque due to friction), if $I$ decreases, then the angular velocity ω must increase. If the skater reduces her moment of inertia by a factor of 2, she will then rotate with twice the angular velocity.

**EXERCISE D** When a spinning figure skater pulls in her arms, her moment of inertia decreases; to conserve angular momentum, her angular velocity increases. Does her rotational kinetic energy also increase? If so, where does the energy come from?

A similar example is the diver shown in Fig. 8–28. The push as she leaves the board gives her an initial angular momentum about her center of mass. When she curls herself into the tuck position, she rotates quickly one or more times. She then stretches out again, increasing her moment of inertia which reduces the angular velocity to a small value, and then she enters the water. The change in moment of inertia from the straight position to the tuck position can be a factor of as much as $3\frac{1}{2}$.

Note that for angular momentum to be conserved, the net torque must be zero; but the net force does not necessarily have to be zero. The net force on the diver in Fig. 8–28, for example, is not zero (gravity is acting), but the net torque about her CM is zero because the force of gravity acts at her center of mass.

---

**EXAMPLE 8–14** **Clutch.** A simple clutch consists of two cylindrical plates that can be pressed together to connect two sections of an axle, as needed, in a piece of machinery. The two plates have masses $M_A = 6.0$ kg and $M_B = 9.0$ kg, with equal radii $R = 0.60$ m. They are initially separated (Fig. 8–29). Plate $M_A$ is accelerated from rest to an angular velocity $\omega_1 = 7.2$ rad/s in time $\Delta t = 2.0$ s. Calculate (a) the angular momentum of $M_A$, and (b) the torque required to accelerate $M_A$ from rest to $\omega_1$. (c) Next, plate $M_B$, initially at rest but free to rotate without friction, is placed in firm contact with freely rotating plate $M_A$, and the two plates then both rotate at a constant angular velocity $\omega_2$, which is considerably less than $\omega_1$. Why does this happen, and what is $\omega_2$?

**APPROACH** We use angular momentum, $L = I\omega$ (Eq. 8–18), plus Newton's second law for rotation, Eq. 8–19.

**SOLUTION** (a) The angular momentum of $M_A$, a cylinder, is

$$L_A = I_A\omega_1 = \tfrac{1}{2}M_A R^2 \omega_1 = \tfrac{1}{2}(6.0\text{ kg})(0.60\text{ m})^2(7.2\text{ rad/s}) = 7.8\text{ kg}\cdot\text{m}^2/\text{s}.$$

(b) The plate started from rest so the torque, assumed constant, was

$$\tau = \frac{\Delta L}{\Delta t} = \frac{7.8\text{ kg}\cdot\text{m}^2/\text{s} - 0}{2.0\text{ s}} = 3.9\text{ m}\cdot\text{N}.$$

(c) Initially, before contact, $M_A$ is rotating at constant $\omega_1$ (we ignore friction). When plate B comes in contact, why is their joint rotation speed less? You might think in terms of the torque each exerts on the other upon contact. But quantitatively, it's easier to use conservation of angular momentum, Eq. 8–20, since no external torques are assumed to act. Thus

angular momentum before = angular momentum after
$$I_A\omega_1 = (I_A + I_B)\omega_2.$$

Solving for $\omega_2$ we find (after cancelling factors of $R^2$)

$$\omega_2 = \left(\frac{I_A}{I_A + I_B}\right)\omega_1 = \left(\frac{M_A}{M_A + M_B}\right)\omega_1 = \left(\frac{6.0\text{ kg}}{15.0\text{ kg}}\right)(7.2\text{ rad/s}) = 2.9\text{ rad/s}.$$

**EXAMPLE 8–15** **ESTIMATE** **Neutron star.** Astronomers detect stars that are rotating extremely rapidly, known as neutron stars. A neutron star is believed to form from the inner core of a larger star that collapsed, under its own gravitation, to a star of very small radius and very high density. Before collapse, suppose the core of such a star is the size of our Sun $(R \approx 7 \times 10^5 \text{ km})$ with mass 2.0 times as great as the Sun, and is rotating at a frequency of 1.0 revolution every 100 days. If it were to undergo gravitational collapse to a neutron star of radius 10 km, what would its rotation frequency be? Assume the star is a uniform sphere at all times, and loses no mass.

**APPROACH** We assume the star is isolated (no external forces), so we can use conservation of angular momentum for this process.

**SOLUTION** From conservation of angular momentum, Eq. 8–20,

$$I_1 \omega_1 = I_2 \omega_2,$$

where the subscripts 1 and 2 refer to initial (normal star) and final (neutron star), respectively. Then, assuming no mass is lost in the process $(M_1 = M_2)$,

$$\omega_2 = \left(\frac{I_1}{I_2}\right)\omega_1 = \left(\frac{\frac{2}{5}M_1 R_1^2}{\frac{2}{5}M_2 R_2^2}\right)\omega_1 = \frac{R_1^2}{R_2^2}\omega_1.$$

The frequency $f = \omega/2\pi$, so

$$f_2 = \frac{\omega_2}{2\pi} = \frac{R_1^2}{R_2^2}f_1$$

$$= \left(\frac{7 \times 10^5 \text{ km}}{10 \text{ km}}\right)^2 \left(\frac{1.0 \text{ rev}}{100 \text{ d} (24 \text{ h/d})(3600 \text{ s/h})}\right) \approx 6 \times 10^2 \text{ rev/s},$$

which is 600 Hz or $(600 \text{ rev/s})(60 \text{ s/min}) = 36,000 \text{ rpm}$.

# *8–9 Vector Nature of Angular Quantities

Up to now we have considered only the magnitudes of angular quantities such as $\omega$, $\alpha$, and $L$. But they have a vector aspect too, and now we consider the directions. In fact, we have to *define* the directions for rotational quantities. We consider first the angular velocity, $\vec{\omega}$.

Consider the rotating wheel shown in Fig. 8–30a. The linear velocities of different particles of the wheel point in all different directions. The only unique direction in space associated with the rotation is along the axis of rotation, perpendicular to the actual motion. We therefore choose the axis of rotation to be the direction of the angular velocity vector, $\vec{\omega}$. Actually, there is still an ambiguity since $\vec{\omega}$ could point in either direction along the axis of rotation (up or down in Fig. 8–30a). The convention we use, called **the right-hand rule**, is this: when the fingers of the right hand are curled around the rotation axis and point in the direction of the rotation, then the thumb points in the direction of $\vec{\omega}$. This is shown in Fig. 8–30b. Note that $\vec{\omega}$ points in the direction a right-handed screw would move when turned in the direction of rotation. Thus, if the rotation of the wheel in Fig. 8–30a is counterclockwise, the direction of $\vec{\omega}$ is upward as shown in Fig. 8–30b. If the wheel rotates clockwise, then $\vec{\omega}$ points in the opposite direction, downward. Note that no part of the rotating object moves in the direction of $\vec{\omega}$.

If the axis of rotation is fixed, then $\vec{\omega}$ can change only in magnitude. Thus $\vec{\alpha} = \Delta\vec{\omega}/\Delta t$ must also point along the axis of rotation. If the rotation is counterclockwise as in Fig. 8–30a and the magnitude of $\omega$ is increasing, then $\vec{\alpha}$ points upward; but if $\omega$ is decreasing (the wheel is slowing down), $\vec{\alpha}$ points downward. If the rotation is clockwise, $\vec{\alpha}$ points downward if $\omega$ is increasing, and $\vec{\alpha}$ points upward if $\omega$ is decreasing.

**FIGURE 8–30** (a) Rotating wheel. (b) Right-hand rule for obtaining the direction of $\vec{\omega}$.

FIGURE 8–30 (Repeated.)
(a) Rotating wheel. (b) Right-hand rule for obtaining the direction of $\vec{\omega}$.

FIGURE 8–31 (a) A person standing on a circular platform, initially at rest, begins walking along the edge at speed $v$. The platform, mounted on nearly friction-free bearings, begins rotating in the opposite direction, so that (b) the total angular momentum remains zero ($\vec{L}_{platform} = -\vec{L}_{person}$).

(a)

(b)

FIGURE 8–32 Example 8–16.

Angular momentum, like linear momentum, is a vector quantity. For a symmetrical object rotating about a symmetry axis (such as a wheel, cylinder, hoop, or sphere), we can write the vector angular momentum as

$$\vec{L} = I\vec{\omega}. \tag{8–21}$$

The angular velocity vector $\vec{\omega}$ (and therefore also $\vec{L}$) points along the axis of rotation in the direction given by the right-hand rule (Fig. 8–30b).

The vector nature of angular momentum can be used to explain a number of interesting (and sometimes surprising) phenomena. For example, consider a person standing at rest on a circular platform capable of rotating without friction about an axis through its center (that is, a simplified merry-go-round). If the person now starts to walk along the edge of the platform, Fig. 8–31a, the platform starts rotating in the opposite direction. Why? One explanation is that the person's foot exerts a force on the platform. Another explanation (and this is the most useful analysis here) is that this is an example of the conservation of angular momentum. If the person starts walking counterclockwise, the person's angular momentum will point upward along the axis of rotation (remember how we defined the direction of $\vec{\omega}$ using the right-hand rule). The magnitude of the person's angular momentum will be $L = I\omega = (mr^2)(v/r)$, where $v$ is the person's speed (relative to the Earth, not to the platform), $r$ is his distance from the rotation axis, $m$ is his mass, and $mr^2$ is his moment of inertia if we consider him a particle (mass concentrated at one point, Eq. 8–11). The platform rotates in the opposite direction, so its angular momentum points downward. If the total angular momentum of the system is initially zero (person and platform at rest), it will remain zero after the person starts walking. That is, the upward angular momentum of the person just balances the oppositely directed downward angular momentum of the platform (Fig. 8–31b), so the total vector angular momentum remains zero. Even though the person exerts a force (and torque) on the platform, the platform exerts an equal and opposite torque on the person. So the net torque on the *system* of person plus platform is zero (ignoring friction), and the total angular momentum remains constant.

---

**CONCEPTUAL EXAMPLE 8–16** **Spinning bicycle wheel.** Your physics teacher is holding a spinning bicycle wheel while he stands with feet fixed on a stationary frictionless turntable (Fig. 8–32). What will happen if the teacher suddenly flips the bicycle wheel over so that it is spinning in the opposite direction?

**RESPONSE** We consider the system of turntable, teacher, and bicycle wheel. The total angular momentum initially is $\vec{L}$ vertically upward. That is also what the system's angular momentum must be afterward, since $\vec{L}$ is conserved when there is no net torque. Thus, if the wheel's angular momentum after being flipped over is $-\vec{L}$ downward, then the angular momentum of teacher plus turntable will have to be $+2\vec{L}$ upward. We can safely predict that the turntable (and teacher) will begin revolving in the same direction that the bicycle wheel was spinning originally.

---

**EXERCISE E** In Example 8–16, what if he moves the axis only 90° so it is horizontal? (*a*) The same direction and speed as above; (*b*) the same as above, but slower; (*c*) the opposite result.

**EXERCISE F** Suppose you are standing on the edge of a large freely rotating turntable. If you walk toward the center, (*a*) the turntable slows down; (*b*) the turntable speeds up; (*c*) its rotation speed is unchanged; (*d*) the turntable stops; (*e*) you need to know the walking speed to answer.

One final note: the motion of particles and objects in **rotating frames of reference** is extremely interesting, though a bit advanced and so is treated at the end of the book in Appendix C.

# Summary

When a rigid object rotates about a fixed axis, each point of the object moves in a circular path. Lines drawn perpendicularly from the rotation axis to various points in the object all sweep out the same angle $\theta$ in any given time interval.

Angles are conventionally measured in **radians**, where one radian is the angle subtended by an arc whose length is equal to the radius, or

$$2\pi \text{ rad} = 360°$$
$$1 \text{ rad} \approx 57.3°.$$

**Angular velocity**, $\omega$, is defined as the rate of change of angular position:

$$\omega = \frac{\Delta\theta}{\Delta t}. \tag{8–2}$$

All parts of a rigid object rotating about a fixed axis have the same angular velocity at any instant.

**Angular acceleration**, $\alpha$, is defined as the rate of change of angular velocity:

$$\alpha = \frac{\Delta\omega}{\Delta t}. \tag{8–3}$$

The linear velocity $v$ and acceleration $a$ of a point located a distance $r$ from the axis of rotation are related to $\omega$ and $\alpha$ by

$$v = r\omega, \tag{8–4}$$
$$a_{\text{tan}} = r\alpha, \tag{8–5}$$
$$a_{\text{R}} = \omega^2 r, \tag{8–6}$$

where $a_{\text{tan}}$ and $a_{\text{R}}$ are the tangential and radial (centripetal) components of the linear acceleration, respectively.

The frequency $f$ is related to $\omega$ by

$$\omega = 2\pi f, \tag{8–7}$$

and to the period $T$ by

$$T = 1/f. \tag{8–8}$$

If a rigid object undergoes uniformly accelerated rotational motion ($\alpha$ = constant), equations analogous to those for linear motion are valid:

$$\omega = \omega_0 + \alpha t, \qquad \theta = \omega_0 t + \tfrac{1}{2}\alpha t^2,$$
$$\omega^2 = \omega_0^2 + 2\alpha\theta, \qquad \overline{\omega} = \frac{\omega + \omega_0}{2}. \tag{8–9}$$

The **torque** due to a force $\vec{F}$ exerted on a rigid object is equal to

$$\tau = r_\perp F = rF_\perp = rF \sin\theta, \tag{8–10}$$

where $r_\perp$, called the **lever arm**, is the perpendicular distance from the axis of rotation to the line along which the force acts, and $\theta$ is the angle between $\vec{F}$ and $r$.

The rotational equivalent of Newton's second law is

$$\Sigma\tau = I\alpha, \tag{8–14}$$

where $I = \Sigma mr^2$ is the **moment of inertia** of the object about the axis of rotation. $I$ depends not only on the mass of the object but also on how the mass is distributed relative to the axis of rotation. For a uniform solid cylinder or sphere of radius $R$ and mass $M$, $I$ has the form $I = \tfrac{1}{2}MR^2$ or $\tfrac{2}{5}MR^2$, respectively (see Fig. 8–20).

The **rotational kinetic energy** of an object rotating about a fixed axis with angular velocity $\omega$ is

$$\text{KE} = \tfrac{1}{2}I\omega^2. \tag{8–15}$$

For an object both translating and rotating, the total kinetic energy is the sum of the translational kinetic energy of the object's center of mass plus the rotational kinetic energy of the object about its center of mass:

$$\text{KE} = \tfrac{1}{2}Mv_{\text{CM}}^2 + \tfrac{1}{2}I_{\text{CM}}\omega^2 \tag{8–16}$$

as long as the rotation axis is fixed in direction.

The **angular momentum** $L$ of an object rotating about a fixed rotation axis is given by

$$L = I\omega. \tag{8–18}$$

Newton's second law, in terms of angular momentum, is

$$\Sigma\tau = \frac{\Delta L}{\Delta t}. \tag{8–19}$$

If the net torque on an object is zero, $\Delta L/\Delta t = 0$, so $L$ = constant. This is the **law of conservation of angular momentum** for a rotating object.

The following Table summarizes angular (or rotational) quantities, comparing them to their translational analogs.

| Translation | Rotation | Connection |
|---|---|---|
| $x$ | $\theta$ | $x = r\theta$ |
| $v$ | $\omega$ | $v = r\omega$ |
| $a$ | $\alpha$ | $a_{\text{tan}} = r\alpha$ |
| $m$ | $I$ | $I = \Sigma mr^2$ |
| $F$ | $\tau$ | $\tau = rF\sin\theta$ |
| $\text{KE} = \tfrac{1}{2}mv^2$ | $\tfrac{1}{2}I\omega^2$ | |
| $p = mv$ | $L = I\omega$ | |
| $W = Fd$ | $W = \tau\theta$ | |
| $\Sigma F = ma$ | $\Sigma\tau = I\alpha$ | |
| $\Sigma F = \dfrac{\Delta p}{\Delta t}$ | $\Sigma\tau = \dfrac{\Delta L}{\Delta t}$ | |

[*Angular velocity, angular acceleration, and angular momentum are vectors. For a rigid object rotating about a fixed axis, the vectors $\vec{\omega}$, $\vec{\alpha}$, and $\vec{L}$ point along the rotation axis. The direction of $\vec{\omega}$ or $\vec{L}$ is given by the **right-hand rule**.]

# Questions

1. A bicycle odometer (which counts revolutions and is calibrated to report distance traveled) is attached near the wheel axle and is calibrated for 27-inch wheels. What happens if you use it on a bicycle with 24-inch wheels?

2. Suppose a disk rotates at constant angular velocity. (*a*) Does a point on the rim have radial and/or tangential acceleration? (*b*) If the disk's angular velocity increases uniformly, does the point have radial and/or tangential acceleration? (*c*) For which cases would the magnitude of either component of linear acceleration change?

3. Can a small force ever exert a greater torque than a larger force? Explain.

4. Why is it more difficult to do a sit-up with your hands behind your head than when your arms are stretched out in front of you? A diagram may help you to answer this.

5. If the net force on a system is zero, is the net torque also zero? If the net torque on a system is zero, is the net force zero? Explain and give examples.

6. Mammals that depend on being able to run fast have slender lower legs with flesh and muscle concentrated high, close to the body (Fig. 8–33). On the basis of rotational dynamics, explain why this distribution of mass is advantageous.

**FIGURE 8–33** Question 6. A gazelle.

7. This book has three symmetry axes through its center, all mutually perpendicular. The book's moment of inertia would be smallest about which of the three? Explain.

8. Can the mass of a rigid object be considered concentrated at its CM for rotational motion? Explain.

9. The moment of inertia of a rotating solid disk about an axis through its CM is $\frac{1}{2}MR^2$ (Fig. 8–20c). Suppose instead that a parallel axis of rotation passes through a point on the edge of the disk. Will the moment of inertia be the same, larger, or smaller? Explain why.

10. Two inclines have the same height but make different angles with the horizontal. The same steel ball rolls without slipping down each incline. On which incline will the speed of the ball at the bottom be greater? Explain.

11. Two spheres look identical and have the same mass. However, one is hollow and the other is solid. Describe an experiment to determine which is which.

12. A sphere and a cylinder have the same radius and the same mass. They start from rest at the top of an incline. (*a*) Which reaches the bottom first? (*b*) Which has the greater speed at the bottom? (*c*) Which has the greater total kinetic energy at the bottom? (*d*) Which has the greater rotational kinetic energy? Explain your answers.

13. Why do tightrope walkers (Fig. 8–34) carry a long, narrow rod?

**FIGURE 8–34** Question 13.

14. We claim that momentum and angular momentum are conserved. Yet most moving or rotating objects eventually slow down and stop. Explain.

15. Can the diver of Fig. 8–28 do a somersault without having any initial rotation when she leaves the board? Explain.

16. When a motorcyclist leaves the ground on a jump and leaves the throttle on (so the rear wheel spins), why does the front of the cycle rise up?

17. A shortstop may leap into the air to catch a ball and throw it quickly. As he throws the ball, the upper part of his body rotates. If you look quickly you will notice that his hips and legs rotate in the opposite direction (Fig. 8–35). Explain.

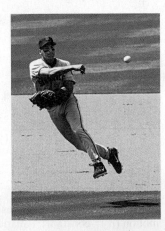

**FIGURE 8–35** Question 17. A shortstop in the air, throwing the ball.

*18. The angular velocity of a wheel rotating on a horizontal axle points west. In what direction is the linear velocity of a point on the top of the wheel? If the angular acceleration points east, describe the tangential linear acceleration of this point at the top of the wheel. Is the angular speed increasing or decreasing?

*19. In what direction is the Earth's angular velocity vector as it rotates daily about its axis, north or south?

*20. On the basis of the law of conservation of angular momentum, discuss why a helicopter must have more than one rotor (or propeller). Discuss one or more ways the second propeller can operate in order to keep the helicopter stable.

# MisConceptual Questions

1. Bonnie sits on the outer rim of a merry-go-round, and Jill sits midway between the center and the rim. The merry-go-round makes one complete revolution every 2 seconds. Jill's linear velocity is:
   (a) the same as Bonnie's.
   (b) twice Bonnie's.
   (c) half of Bonnie's.
   (d) one-quarter of Bonnie's.
   (e) four times Bonnie's.

2. An object at rest begins to rotate with a constant angular acceleration. If this object rotates through an angle $\theta$ in time $t$, through what angle did it rotate in the time $\frac{1}{2}t$?
   (a) $\frac{1}{2}\theta$.  (b) $\frac{1}{4}\theta$.  (c) $\theta$.  (d) $2\theta$.  (e) $4\theta$.

3. A car speedometer that is supposed to read the linear speed of the car uses a device that actually measures the angular speed of the tires. If larger-diameter tires are mounted on the car instead, how will that affect the speedometer reading? The speedometer
   (a) will still read the speed accurately.
   (b) will read low.
   (c) will read high.

4. The solid dot shown in Fig. 8–36 is a pivot point. The board can rotate about the pivot. Which force shown exerts the largest magnitude torque on the board?

(e) 500 N
(d) 800 N
(b) 500 N   (c) 500 N
(a) 1000 N

**FIGURE 8–36** MisConceptual Question 4.

5. Consider a force $F = 80\,\text{N}$ applied to a beam as shown in Fig. 8–37. The length of the beam is $\ell = 5.0\,\text{m}$, and $\theta = 37°$, so that $x = 3.0\,\text{m}$ and $y = 4.0\,\text{m}$. Of the following expressions, which ones give the correct torque produced by the force $\vec{F}$ around point P?
   (a) 80 N.
   (b) (80 N)(5.0 m).
   (c) (80 N)(5.0 m)(sin 37°).
   (d) (80 N)(4.0 m).
   (e) (80 N)(3.0 m).
   (f) (48 N)(5.0 m).
   (g) (48 N)(4.0 m)(sin 37°).

**FIGURE 8–37**
MisConceptual Question 5.

6. Two spheres have the same radius and equal mass. One sphere is solid, and the other is hollow and made of a denser material. Which one has the bigger moment of inertia about an axis through its center?
   (a) The solid one.
   (b) The hollow one.
   (c) Both the same.

7. Two wheels having the same radius and mass rotate at the same angular velocity (Fig. 8–38). One wheel is made with spokes so nearly all the mass is at the rim. The other is a solid disk. How do their rotational kinetic energies compare?
   (a) They are nearly the same.
   (b) The wheel with spokes has about twice the KE.
   (c) The wheel with spokes has higher KE, but not twice as high.
   (d) The solid wheel has about twice the KE.
   (e) The solid wheel has higher KE, but not twice as high.

**FIGURE 8–38**
MisConceptual
Question 7.

8. If you used 1000 J of energy to throw a ball, would it travel faster if you threw the ball (ignoring air resistance)
   (a) so that it was also rotating?
   (b) so that it wasn't rotating?
   (c) It makes no difference.

9. A small solid sphere and a small thin hoop are rolling along a horizontal surface with the same translational speed when they encounter a 20° rising slope. If these two objects roll up the slope without slipping, which will rise farther up the slope?
   (a) The sphere.
   (b) The hoop.
   (c) Both the same.
   (d) More information about the objects' mass and diameter is needed.

10. A small mass $m$ on a string is rotating without friction in a circle. The string is shortened by pulling it through the axis of rotation without any external torque, Fig. 8–39. What happens to the angular velocity of the object?
   (a) It increases.
   (b) It decreases.
   (c) It remains the same.

**FIGURE 8–39**
MisConceptual
Questions 10 and 11.

11. A small mass $m$ on a string is rotating without friction in a circle. The string is shortened by pulling it through the axis of rotation without any external torque, Fig. 8–39. What happens to the tangential velocity of the object?
   (a) It increases.
   (b) It decreases.
   (c) It remains the same.

12. If there were a great migration of people toward the Earth's equator, the length of the day would
   (a) increase because of conservation of angular momentum.
   (b) decrease because of conservation of angular momentum.
   (c) decrease because of conservation of energy.
   (d) increase because of conservation of energy.
   (e) remain unaffected.

13. Suppose you are sitting on a rotating stool holding a 2-kg mass in each outstretched hand. If you suddenly drop the masses, your angular velocity will
   (a) increase.    (b) decrease.    (c) stay the same.

# Problems

## 8–1 Angular Quantities

1. (I) Express the following angles in radians: (a) 45.0°, (b) 60.0°, (c) 90.0°, (d) 360.0°, and (e) 445°. Give as numerical values and as fractions of $\pi$.

2. (I) The Sun subtends an angle of about 0.5° to us on Earth, 150 million km away. Estimate the radius of the Sun.

3. (I) A laser beam is directed at the Moon, 380,000 km from Earth. The beam diverges at an angle $\theta$ (Fig. 8–40) of $1.4 \times 10^{-5}$ rad. What diameter spot will it make on the Moon?

**FIGURE 8–40** Problem 3.

4. (I) The blades in a blender rotate at a rate of 6500 rpm. When the motor is turned off during operation, the blades slow to rest in 4.0 s. What is the angular acceleration as the blades slow down?

5. (II) The platter of the **hard drive** of a computer rotates at 7200 rpm (rpm = revolutions per minute = rev/min). (a) What is the angular velocity (rad/s) of the platter? (b) If the reading head of the drive is located 3.00 cm from the rotation axis, what is the linear speed of the point on the platter just below it? (c) If a single bit requires 0.50 $\mu$m of length along the direction of motion, how many bits per second can the writing head write when it is 3.00 cm from the axis?

6. (II) A child rolls a ball on a level floor 3.5 m to another child. If the ball makes 12.0 revolutions, what is its diameter?

7. (II) (a) A grinding wheel 0.35 m in diameter rotates at 2200 rpm. Calculate its angular velocity in rad/s. (b) What are the linear speed and acceleration of a point on the edge of the grinding wheel?

8. (II) A bicycle with tires 68 cm in diameter travels 9.2 km. How many revolutions do the wheels make?

9. (II) Calculate the angular velocity (a) of a clock's second hand, (b) its minute hand, and (c) its hour hand. State in rad/s. (d) What is the angular acceleration in each case?

10. (II) A rotating merry-go-round makes one complete revolution in 4.0 s (Fig. 8–41). (a) What is the linear speed of a child seated 1.2 m from the center? (b) What is her acceleration (give components)?

**FIGURE 8–41**
Problem 10.

11. (II) What is the linear speed, due to the Earth's rotation, of a point (a) on the equator, (b) on the Arctic Circle (latitude 66.5° N), and (c) at a latitude of 42.0° N?

12. (II) Calculate the angular velocity of the Earth (a) in its orbit around the Sun, and (b) about its axis.

13. (II) How fast (in rpm) must a centrifuge rotate if a particle 8.0 cm from the axis of rotation is to experience an acceleration of 100,000 $g$'s?

14. (II) A 61-cm-diameter wheel accelerates uniformly about its center from 120 rpm to 280 rpm in 4.0 s. Determine (a) its angular acceleration, and (b) the radial and tangential components of the linear acceleration of a point on the edge of the wheel 2.0 s after it has started accelerating.

15. (II) In traveling to the Moon, astronauts aboard the *Apollo* spacecraft put the spacecraft into a slow rotation to distribute the Sun's energy evenly (so one side would not become too hot). At the start of their trip, they accelerated from no rotation to 1.0 revolution every minute during a 12-min time interval. Think of the spacecraft as a cylinder with a diameter of 8.5 m rotating about its cylindrical axis. Determine (a) the angular acceleration, and (b) the radial and tangential components of the linear acceleration of a point on the skin of the ship 6.0 min after it started this acceleration.

16. (II) A turntable of radius $R_1$ is turned by a circular rubber roller of radius $R_2$ in contact with it at their outer edges. What is the ratio of their angular velocities, $\omega_1/\omega_2$?

## 8–2 and 8–3 Constant Angular Acceleration; Rolling

17. (I) An automobile engine slows down from 3500 rpm to 1200 rpm in 2.5 s. Calculate (a) its angular acceleration, assumed constant, and (b) the total number of revolutions the engine makes in this time.

18. (I) A centrifuge accelerates uniformly from rest to 15,000 rpm in 240 s. Through how many revolutions did it turn in this time?

19. (I) Pilots can be tested for the stresses of flying high-speed jets in a whirling "human centrifuge," which takes 1.0 min to turn through 23 complete revolutions before reaching its final speed. (a) What was its angular acceleration (assumed constant), and (b) what was its final angular speed in rpm?

20. (II) A cooling fan is turned off when it is running at 850 rev/min. It turns 1250 revolutions before it comes to a stop. (a) What was the fan's angular acceleration, assumed constant? (b) How long did it take the fan to come to a complete stop?

21. (II) A wheel 31 cm in diameter accelerates uniformly from 240 rpm to 360 rpm in 6.8 s. How far will a point on the edge of the wheel have traveled in this time?

22. (II) The tires of a car make 75 revolutions as the car reduces its speed uniformly from 95 km/h to 55 km/h. The tires have a diameter of 0.80 m. (a) What was the angular acceleration of the tires? If the car continues to decelerate at this rate, (b) how much more time is required for it to stop, and (c) how far does it go?

**23.** (II) A small rubber wheel is used to drive a large pottery wheel. The two wheels are mounted so that their circular edges touch. The small wheel has a radius of 2.0 cm and accelerates at the rate of 7.2 rad/s², and it is in contact with the pottery wheel (radius 27.0 cm) without slipping. Calculate (a) the angular acceleration of the pottery wheel, and (b) the time it takes the pottery wheel to reach its required speed of 65 rpm.

## 8–4 Torque

**24.** (I) A 52-kg person riding a bike puts all her weight on each pedal when climbing a hill. The pedals rotate in a circle of radius 17 cm. (a) What is the maximum torque she exerts? (b) How could she exert more torque?

**25.** (II) Calculate the net torque about the axle of the wheel shown in Fig. 8–42. Assume that a friction torque of 0.60 m·N opposes the motion.

FIGURE 8–42 Problem 25.

**26.** (II) A person exerts a horizontal force of 42 N on the end of a door 96 cm wide. What is the magnitude of the torque if the force is exerted (a) perpendicular to the door and (b) at a 60.0° angle to the face of the door?

**27.** (II) Two blocks, each of mass $m$, are attached to the ends of a massless rod which pivots as shown in Fig. 8–43. Initially the rod is held in the horizontal position and then released. Calculate the magnitude and direction of the net torque on this system when it is first released.

FIGURE 8–43 Problem 27.

**28.** (II) The bolts on the cylinder head of an engine require tightening to a torque of 95 m·N. If a wrench is 28 cm long, what force perpendicular to the wrench must the mechanic exert at its end? If the six-sided bolt head is 15 mm across (Fig. 8–44), estimate the force applied near each of the six points by a wrench.

FIGURE 8–44 Problem 28.

**29.** (II) Determine the net torque on the 2.0-m-long uniform beam shown in Fig. 8–45. All forces are shown. Calculate about (a) point C, the CM, and (b) point P at one end.

FIGURE 8–45
Problem 29.

## 8–5 and 8–6 Rotational Dynamics

**30.** (I) Determine the moment of inertia of a 10.8-kg sphere of radius 0.648 m when the axis of rotation is through its center.

**31.** (I) Estimate the moment of inertia of a bicycle wheel 67 cm in diameter. The rim and tire have a combined mass of 1.1 kg. The mass of the hub (at the center) can be ignored (why?).

**32.** (II) A merry-go-round accelerates from rest to 0.68 rad/s in 34 s. Assuming the merry-go-round is a uniform disk of radius 7.0 m and mass 31,000 kg, calculate the net torque required to accelerate it.

**33.** (II) An oxygen molecule consists of two oxygen atoms whose total mass is $5.3 \times 10^{-26}$ kg and whose moment of inertia about an axis perpendicular to the line joining the two atoms, midway between them, is $1.9 \times 10^{-46}$ kg·m². From these data, estimate the effective distance between the atoms.

**34.** (II) A grinding wheel is a uniform cylinder with a radius of 8.50 cm and a mass of 0.380 kg. Calculate (a) its moment of inertia about its center, and (b) the applied torque needed to accelerate it from rest to 1750 rpm in 5.00 s. Take into account a frictional torque that has been measured to slow down the wheel from 1500 rpm to rest in 55.0 s.

**35.** (II) The forearm in Fig. 8–46 accelerates a 3.6-kg ball at 7.0 m/s² by means of the triceps muscle, as shown. Calculate (a) the torque needed, and (b) the force that must be exerted by the triceps muscle. Ignore the mass of the arm.

**36.** (II) Assume that a 1.00-kg ball is thrown solely by the action of the forearm, which rotates about the elbow joint under the action of the triceps muscle, Fig. 8–46. The ball is accelerated uniformly from rest to 8.5 m/s in 0.38 s, at which point it is released. Calculate (a) the angular acceleration of the arm, and (b) the force required of the triceps muscle. Assume that the forearm has a mass of 3.7 kg and rotates like a uniform rod about an axis at its end.

FIGURE 8–46
Problems 35 and 36.

**37.** (II) A softball player swings a bat, accelerating it from rest to 2.6 rev/s in a time of 0.20 s. Approximate the bat as a 0.90-kg uniform rod of length 0.95 m, and compute the torque the player applies to one end of it.

**38.** (II) A small 350-gram ball on the end of a thin, light rod is rotated in a horizontal circle of radius 1.2 m. Calculate (a) the moment of inertia of the ball about the center of the circle, and (b) the torque needed to keep the ball rotating at constant angular velocity if air resistance exerts a force of 0.020 N on the ball. Ignore air resistance on the rod and its moment of inertia.

**39.** (II) Calculate the moment of inertia of the array of point objects shown in Fig. 8–47 about (a) the y axis, and (b) the x axis. Assume $m = 2.2$ kg, $M = 3.4$ kg, and the objects are wired together by very light, rigid pieces of wire. The array is rectangular and is split through the middle by the x axis. (c) About which axis would it be harder to accelerate this array?

**FIGURE 8–47** Problem 39.

**40.** (II) A potter is shaping a bowl on a potter's wheel rotating at constant angular velocity of 1.6 rev/s (Fig. 8–48). The friction force between her hands and the clay is 1.5 N total. (a) How large is her torque on the wheel, if the diameter of the bowl is 9.0 cm? (b) How long would it take for the potter's wheel to stop if the only torque acting on it is due to the potter's hands? The moment of inertia of the wheel and the bowl is 0.11 kg·m².

**FIGURE 8–48**
Problem 40.

**41.** (II) A dad pushes tangentially on a small hand-driven merry-go-round and is able to accelerate it from rest to a frequency of 15 rpm in 10.0 s. Assume the merry-go-round is a uniform disk of radius 2.5 m and has a mass of 560 kg, and two children (each with a mass of 25 kg) sit opposite each other on the edge. Calculate the torque required to produce the acceleration, neglecting frictional torque. What force is required at the edge?

**42.** (II) A 0.72-m-diameter solid sphere can be rotated about an axis through its center by a torque of 10.8 m·N which accelerates it uniformly from rest through a total of 160 revolutions in 15.0 s. What is the mass of the sphere?

**43.** (II) Let us treat a helicopter rotor blade as a long thin rod, as shown in Fig. 8–49. (a) If each of the three rotor helicopter blades is 3.75 m long and has a mass of 135 kg, calculate the moment of inertia of the three rotor blades about the axis of rotation. (b) How much torque must the motor apply to bring the blades from rest up to a speed of 6.0 rev/s in 8.0 s?

**FIGURE 8–49**
Problem 43.

**44.** (II) A centrifuge rotor rotating at 9200 rpm is shut off and is eventually brought uniformly to rest by a frictional torque of 1.20 m·N. If the mass of the rotor is 3.10 kg and it can be approximated as a solid cylinder of radius 0.0710 m, through how many revolutions will the rotor turn before coming to rest, and how long will it take?

**45.** (II) To get a flat, uniform cylindrical satellite spinning at the correct rate, engineers fire four tangential rockets as shown in Fig. 8–50. Suppose that the satellite has a mass of 3600 kg and a radius of 4.0 m, and that the rockets each add a mass of 250 kg. What is the steady force required of each rocket if the satellite is to reach 32 rpm in 5.0 min, starting from rest?

**FIGURE 8–50**
Problem 45.

**46.** (III) Two blocks are connected by a light string passing over a pulley of radius 0.15 m and moment of inertia $I$. The blocks move (towards the right) with an acceleration of 1.00 m/s² along their frictionless inclines (see Fig. 8–51). (a) Draw free-body diagrams for each of the two blocks and the pulley. (b) Determine $F_{TA}$ and $F_{TB}$, the tensions in the two parts of the string. (c) Find the net torque acting on the pulley, and determine its moment of inertia, $I$.

**FIGURE 8–51** Problem 46.

**47. (III)** An *Atwood machine* consists of two masses, $m_A = 65$ kg and $m_B = 75$ kg, connected by a massless inelastic cord that passes over a pulley free to rotate, Fig. 8–52. The pulley is a solid cylinder of radius $R = 0.45$ m and mass 6.0 kg. (a) Determine the acceleration of each mass. (b) What % error would be made if the moment of inertia of the pulley is ignored? [*Hint*: The tensions $F_{TA}$ and $F_{TB}$ are not equal. We discussed the Atwood machine in Example 4–13, assuming $I = 0$ for the pulley.]

**FIGURE 8–52** Problem 47. Atwood machine.

**48. (III)** A hammer thrower accelerates the hammer (mass = 7.30 kg) from rest within four full turns (revolutions) and releases it at a speed of 26.5 m/s. Assuming a uniform rate of increase in angular velocity and a horizontal circular path of radius 1.20 m, calculate (a) the angular acceleration, (b) the (linear) tangential acceleration, (c) the centripetal acceleration just before release, (d) the net force being exerted on the hammer by the athlete just before release, and (e) the angle of this force with respect to the radius of the circular motion. Ignore gravity.

## 8–7 Rotational Kinetic Energy

**49. (I)** An automobile engine develops a torque of 265 m·N at 3350 rpm. What is the horsepower of the engine?

**50. (I)** A centrifuge rotor has a moment of inertia of $3.25 \times 10^{-2}$ kg·m². How much energy is required to bring it from rest to 8750 rpm?

**51. (I)** Calculate the translational speed of a cylinder when it reaches the foot of an incline 7.20 m high. Assume it starts from rest and rolls without slipping.

**52. (II)** A bowling ball of mass 7.25 kg and radius 10.8 cm rolls without slipping down a lane at 3.10 m/s. Calculate its total kinetic energy.

**53. (II)** Estimate the kinetic energy of the Earth with respect to the Sun as the sum of two terms, (a) that due to its daily rotation about its axis, and (b) that due to its yearly revolution about the Sun. [Assume the Earth is a uniform sphere with mass $= 6.0 \times 10^{24}$ kg, radius $= 6.4 \times 10^6$ m, and is $1.5 \times 10^8$ km from the Sun.]

**54. (II)** A rotating uniform cylindrical platform of mass 220 kg and radius 5.5 m slows down from 3.8 rev/s to rest in 16 s when the driving motor is disconnected. Estimate the power output of the motor (hp) required to maintain a steady speed of 3.8 rev/s.

**55. (II)** A merry-go-round has a mass of 1440 kg and a radius of 7.50 m. How much net work is required to accelerate it from rest to a rotation rate of 1.00 revolution per 7.00 s? Assume it is a solid cylinder.

**56. (II)** A sphere of radius $r = 34.5$ cm and mass $m = 1.80$ kg starts from rest and rolls without slipping down a 30.0° incline that is 10.0 m long. (a) Calculate its translational and rotational speeds when it reaches the bottom. (b) What is the ratio of translational to rotational kinetic energy at the bottom? Avoid putting in numbers until the end so you can answer: (c) do your answers in (a) and (b) depend on the radius of the sphere or its mass?

**57. (II)** A ball of radius $r$ rolls on the inside of a track of radius $R$ (see Fig. 8–53). If the ball starts from rest at the vertical edge of the track, what will be its speed when it reaches the lowest point of the track, rolling without slipping?

**FIGURE 8–53** Problem 57.

**58. (II)** Two masses, $m_A = 32.0$ kg and $m_B = 38.0$ kg, are connected by a rope that hangs over a pulley (as in Fig. 8–54). The pulley is a uniform cylinder of radius $R = 0.311$ m and mass 3.1 kg. Initially $m_A$ is on the ground and $m_B$ rests 2.5 m above the ground. If the system is released, use conservation of energy to determine the speed of $m_B$ just before it strikes the ground. Assume the pulley bearing is frictionless.

**FIGURE 8–54** Problem 58.

**59. (III)** A 1.80-m-long pole is balanced vertically with its tip on the ground. It starts to fall and its lower end does not slip. What will be the speed of the upper end of the pole just before it hits the ground? [*Hint*: Use conservation of energy.]

## 8–8 Angular Momentum

**60. (I)** What is the angular momentum of a 0.270-kg ball revolving on the end of a thin string in a circle of radius 1.35 m at an angular speed of 10.4 rad/s?

**61. (I)** (a) What is the angular momentum of a 2.8-kg uniform cylindrical grinding wheel of radius 28 cm when rotating at 1300 rpm? (b) How much torque is required to stop it in 6.0 s?

**62. (II)** A person stands, hands at his side, on a platform that is rotating at a rate of 0.90 rev/s. If he raises his arms to a horizontal position, Fig. 8–55, the speed of rotation decreases to 0.60 rev/s. (a) Why? (b) By what factor has his moment of inertia changed?

**FIGURE 8–55** Problem 62.

**63.** (II) A nonrotating cylindrical disk of moment of inertia $I$ is dropped onto an identical disk rotating at angular speed $\omega$. Assuming no external torques, what is the final common angular speed of the two disks?

**64.** (II) A diver (such as the one shown in Fig. 8–28) can reduce her moment of inertia by a factor of about 3.5 when changing from the straight position to the tuck position. If she makes 2.0 rotations in 1.5 s when in the tuck position, what is her angular speed (rev/s) when in the straight position?

**65.** (II) A figure skater can increase her spin rotation rate from an initial rate of 1.0 rev every 1.5 s to a final rate of 2.5 rev/s. If her initial moment of inertia was $4.6 \text{ kg} \cdot \text{m}^2$, what is her final moment of inertia? How does she physically accomplish this change?

**66.** (II) (a) What is the angular momentum of a figure skater spinning at 3.0 rev/s with arms in close to her body, assuming her to be a uniform cylinder with a height of 1.5 m, a radius of 15 cm, and a mass of 48 kg? (b) How much torque is required to slow her to a stop in 4.0 s, assuming she does *not* move her arms?

**67.** (II) A person of mass 75 kg stands at the center of a rotating merry-go-round platform of radius 3.0 m and moment of inertia $820 \text{ kg} \cdot \text{m}^2$. The platform rotates without friction with angular velocity 0.95 rad/s. The person walks radially to the edge of the platform. (a) Calculate the angular velocity when the person reaches the edge. (b) Calculate the rotational kinetic energy of the system of platform plus person before and after the person's walk.

**68.** (II) A potter's wheel is rotating around a vertical axis through its center at a frequency of 1.5 rev/s. The wheel can be considered a uniform disk of mass 5.0 kg and diameter 0.40 m. The potter then throws a 2.6-kg chunk of clay, approximately shaped as a flat disk of radius 7.0 cm, onto the center of the rotating wheel. What is the frequency of the wheel after the clay sticks to it? Ignore friction.

**69.** (II) A 4.2-m-diameter merry-go-round is rotating freely with an angular velocity of 0.80 rad/s. Its total moment of inertia is $1360 \text{ kg} \cdot \text{m}^2$. Four people standing on the ground, each of mass 65 kg, suddenly step onto the edge of the merry-go-round. (a) What is the angular velocity of the merry-go-round now? (b) What if the people were on it initially and then jumped off in a radial direction (relative to the merry-go-round)?

**70.** (II) A uniform horizontal rod of mass $M$ and length $\ell$ rotates with angular velocity $\omega$ about a vertical axis through its center. Attached to each end of the rod is a small mass $m$. Determine the angular momentum of the system about the axis.

**71.** (II) Suppose our Sun eventually collapses into a white dwarf, losing about half its mass in the process, and winding up with a radius 1.0% of its existing radius. Assuming the lost mass carries away no angular momentum, (a) what would the Sun's new rotation rate be? Take the Sun's current period to be about 30 days. (b) What would be its final kinetic energy in terms of its initial kinetic energy of today?

**72.** (II) A uniform disk turns at 3.3 rev/s around a frictionless central axis. A nonrotating rod, of the same mass as the disk and length equal to the disk's diameter, is dropped onto the freely spinning disk, Fig. 8–56. They then turn together around the axis with their centers superposed. What is the angular frequency in rev/s of the combination?

**FIGURE 8–56**
Problem 72.

**73.** (III) An asteroid of mass $1.0 \times 10^5$ kg, traveling at a speed of 35 km/s relative to the Earth, hits the Earth at the equator tangentially, in the direction of Earth's rotation, and is embedded there. Use angular momentum to estimate the percent change in the angular speed of the Earth as a result of the collision.

## *8–9 Angular Quantities as Vectors

**74.** (III) Suppose a 65-kg person stands at the edge of a 5.5-m diameter merry-go-round turntable that is mounted on frictionless bearings and has a moment of inertia of $1850 \text{ kg} \cdot \text{m}^2$. The turntable is at rest initially, but when the person begins running at a speed of 4.0 m/s (with respect to the turntable) around its edge, the turntable begins to rotate in the opposite direction. Calculate the angular velocity of the turntable.

# ❙General Problems

**75.** A merry-go-round with a moment of inertia equal to $1260 \text{ kg} \cdot \text{m}^2$ and a radius of 2.5 m rotates with negligible friction at 1.70 rad/s. A child initially standing still next to the merry-go-round jumps onto the edge of the platform straight toward the axis of rotation, causing the platform to slow to 1.35 rad/s. What is her mass?

**76.** A 1.6-kg grindstone in the shape of a uniform cylinder of radius 0.20 m acquires a rotational rate of 24 rev/s from rest over a 6.0-s interval at constant angular acceleration. Calculate the torque delivered by the motor.

**77.** On a 12.0-cm-diameter audio compact disc (CD), digital bits of information are encoded sequentially along an outward spiraling path. The spiral starts at radius $R_1 = 2.5$ cm and winds its way out to radius $R_2 = 5.8$ cm. To read the digital information, a CD player rotates the CD so that the player's readout laser scans along the spiral's sequence of bits at a constant linear speed of 1.25 m/s. Thus the player must accurately adjust the rotational frequency $f$ of the CD as the laser moves outward. Determine the values for $f$ (in units of rpm) when the laser is located at $R_1$ and when it is at $R_2$.

**78.** (*a*) A yo-yo is made of two solid cylindrical disks, each of mass 0.050 kg and diameter 0.075 m, joined by a (concentric) thin solid cylindrical hub of mass 0.0050 kg and diameter 0.013 m. Use conservation of energy to calculate the linear speed of the yo-yo just before it reaches the end of its 1.0-m-long string, if it is released from rest. (*b*) What fraction of its kinetic energy is rotational?

**79.** A cyclist accelerates from rest at a rate of $1.00 \text{ m/s}^2$. How fast will a point at the top of the rim of the tire (diameter = 68.0 cm) be moving after 2.25 s? [*Hint*: At any moment, the lowest point on the tire is in contact with the ground and is at rest—see Fig. 8–57.]

**FIGURE 8–57**
Problem 79.

This point on tire
at rest momentarily

**80.** Suppose David puts a 0.60-kg rock into a sling of length 1.5 m and begins whirling the rock in a nearly horizontal circle, accelerating it from rest to a rate of 75 rpm after 5.0 s. What is the torque required to achieve this feat, and where does the torque come from?

**81. Bicycle gears**: (*a*) How is the angular velocity $\omega_R$ of the rear wheel of a bicycle related to the angular velocity $\omega_F$ of the front sprocket and pedals? Let $N_F$ and $N_R$ be the number of teeth on the front and rear sprockets, respectively, Fig. 8–58. The teeth are spaced the same on both sprockets and the rear sprocket is firmly attached to the rear wheel. (*b*) Evaluate the ratio $\omega_R/\omega_F$ when the front and rear sprockets have 52 and 13 teeth, respectively, and (*c*) when they have 42 and 28 teeth.

**FIGURE 8–58**
Problem 81.

**82.** Figure 8–59 illustrates an $H_2O$ molecule. The O—H bond length is 0.096 nm and the H—O—H bonds make an angle of 104°. Calculate the moment of inertia of the $H_2O$ molecule (assume the atoms are points) about an axis passing through the center of the oxygen atom (*a*) perpendicular to the plane of the molecule, and (*b*) in the plane of the molecule, bisecting the H—O—H bonds.

**FIGURE 8–59**
Problem 82.

**83.** A hollow cylinder (hoop) is rolling on a horizontal surface at speed $v = 3.0 \text{ m/s}$ when it reaches a 15° incline. (*a*) How far up the incline will it go? (*b*) How long will it be on the incline before it arrives back at the bottom?

**84.** Determine the angular momentum of the Earth (*a*) about its rotation axis (assume the Earth is a uniform sphere), and (*b*) in its orbit around the Sun (treat the Earth as a particle orbiting the Sun).

**85.** A wheel of mass $M$ has radius $R$. It is standing vertically on the floor, and we want to exert a horizontal force $F$ at its axle so that it will climb a step against which it rests (Fig. 8–60). The step has height $h$, where $h < R$. What minimum force $F$ is needed?

**FIGURE 8–60** Problem 85.

**86.** If the coefficient of static friction between a car's tires and the pavement is 0.65, calculate the minimum torque that must be applied to the 66-cm-diameter tire of a 1080-kg automobile in order to "lay rubber" (make the wheels spin, slipping as the car accelerates). Assume each wheel supports an equal share of the weight.

**87.** A 4.00-kg mass and a 3.00-kg mass are attached to opposite ends of a very light 42.0-cm-long horizontal rod (Fig. 8–61). The system is rotating at angular speed $\omega = 5.60 \text{ rad/s}$ about a vertical axle at the center of the rod. Determine (*a*) the kinetic energy KE of the system, and (*b*) the net force on each mass.

**FIGURE 8–61** Problem 87.

**88.** A small mass $m$ attached to the end of a string revolves in a circle on a frictionless tabletop. The other end of the string passes through a hole in the table (Fig. 8–62). Initially, the mass revolves with a speed $v_1 = 2.4 \text{ m/s}$ in a circle of radius $r_1 = 0.80 \text{ m}$. The string is then pulled slowly through the hole so that the radius is reduced to $r_2 = 0.48 \text{ m}$. What is the speed, $v_2$, of the mass now?

**FIGURE 8–62**
Problem 88.

**89.** A uniform rod of mass $M$ and length $\ell$ can pivot freely (i.e., we ignore friction) about a hinge attached to a wall, as in Fig. 8–63. The rod is held horizontally and then released. At the moment of release, determine (a) the angular acceleration of the rod, and (b) the linear acceleration of the tip of the rod. Assume that the force of gravity acts at the center of mass of the rod, as shown. [Hint: See Fig. 8–20g.]

**FIGURE 8–63**
Problem 89.

**90.** Suppose a star the size of our Sun, but with mass 8.0 times as great, were rotating at a speed of 1.0 revolution every 9.0 days. If it were to undergo gravitational collapse to a neutron star of radius 12 km, losing $\frac{3}{4}$ of its mass in the process, what would its rotation speed be? Assume the star is a uniform sphere at all times. Assume also that the thrown-off mass carries off either (a) no angular momentum, or (b) its proportional share $\left(\frac{3}{4}\right)$ of the initial angular momentum.

**91.** A large spool of rope rolls on the ground with the end of the rope lying on the top edge of the spool. A person grabs the end of the rope and walks a distance $\ell$, holding onto it, Fig. 8–64. The spool rolls behind the person without slipping. What length of rope unwinds from the spool? How far does the spool's center of mass move?

**FIGURE 8–64**
Problem 91.

**92.** The Moon orbits the Earth such that the same side always faces the Earth. Determine the ratio of the Moon's spin angular momentum (about its own axis) to its orbital angular momentum. (In the latter case, treat the Moon as a particle orbiting the Earth.)

**93.** A spherical asteroid with radius $r = 123\,\text{m}$ and mass $M = 2.25 \times 10^{10}\,\text{kg}$ rotates about an axis at four revolutions per day. A "tug" spaceship attaches itself to the asteroid's south pole (as defined by the axis of rotation) and fires its engine, applying a force $F$ tangentially to the asteroid's surface as shown in Fig. 8–65. If $F = 285\,\text{N}$, how long will it take the tug to rotate the asteroid's axis of rotation through an angle of 5.0° by this method?

**FIGURE 8–65**
Problem 93.

$F = 285\,\text{N}$

**94.** Most of our Solar System's mass is contained in the Sun, and the planets possess almost all of the Solar System's angular momentum. This observation plays a key role in theories attempting to explain the formation of our Solar System. Estimate the fraction of the Solar System's total angular momentum that is possessed by planets using a simplified model which includes only the large outer planets with the most angular momentum. The central Sun (mass $1.99 \times 10^{30}\,\text{kg}$, radius $6.96 \times 10^8\,\text{m}$) spins about its axis once every 25 days and the planets Jupiter, Saturn, Uranus, and Neptune move in nearly circular orbits around the Sun with orbital data given in the Table below. Ignore each planet's spin about its own axis.

| Planet | Mean Distance from Sun ($\times 10^6\,\text{km}$) | Orbital Period (Earth Years) | Mass ($\times 10^{25}\,\text{kg}$) |
|---|---|---|---|
| Jupiter | 778 | 11.9 | 190 |
| Saturn | 1427 | 29.5 | 56.8 |
| Uranus | 2870 | 84.0 | 8.68 |
| Neptune | 4500 | 165 | 10.2 |

**95.** Water drives a waterwheel (or turbine) of radius $R = 3.0\,\text{m}$ as shown in Fig. 8–66. The water enters at a speed $v_1 = 7.0\,\text{m/s}$ and exits from the waterwheel at a speed $v_2 = 3.8\,\text{m/s}$. (a) If 85 kg of water passes through per second, what is the rate at which the water delivers angular momentum to the waterwheel? (b) What is the torque the water applies to the waterwheel? (c) If the water causes the waterwheel to make one revolution every 5.5 s, how much power is delivered to the wheel?

**FIGURE 8–66**
Problem 95.

**96.** The radius of the roll of paper shown in Fig. 8–67 is 7.6 cm and its moment of inertia is $I = 3.3 \times 10^{-3}\,\text{kg}\cdot\text{m}^2$. A force of 3.5 N is exerted on the end of the roll for 1.3 s, but the paper does not tear so it begins to unroll. A constant friction torque of 0.11 m·N is exerted on the roll which gradually brings it to a stop. Assuming that the paper's thickness is negligible, calculate (a) the length of paper that unrolls during the time that the force is applied (1.3 s) and (b) the length of paper that unrolls from the time the force ends to the time when the roll has stopped moving.

**FIGURE 8–67**
Problem 96.

# Search and Learn

1. Why are Eqs. 8–4 and 8–5 valid for radians but not for revolutions or degrees? Read Section 8–1 and follow the derivations carefully to find the answer.

2. **Total solar eclipses** can happen on Earth because of amazing coincidences: for one, the sometimes near-perfect alignment of Earth, Moon, and Sun. Secondly, using the information inside the front cover, calculate the angular diameters (in radians) of the Sun and the Moon, as seen from Earth, and then comment.

3. Two uniform spheres simultaneously start rolling (from rest) down an incline. One sphere has twice the radius and twice the mass of the other. (a) Which reaches the bottom of the incline first? (b) Which has the greater speed there? (c) Which has the greater total kinetic energy at the bottom? Explain your answers.

4. A bicyclist traveling with speed $v = 8.2$ m/s on a flat road is making a turn with a radius $r = 13$ m. There are three forces acting on the cyclist and cycle: the normal force ($\vec{F}_N$) and friction force ($\vec{F}_{fr}$) exerted by the road on the tires; and $m\vec{g}$, the total weight of the cyclist and cycle. Ignore the small mass of the wheels. (a) Explain carefully why the angle $\theta$ the bicycle makes with the vertical (Fig. 8–68) must be given by $\tan\theta = F_{fr}/F_N$ if the cyclist is to maintain balance. (b) Calculate $\theta$ for the values given. [Hint: Consider the "circular" translational motion of the bicycle and rider.] (c) If the coefficient of static friction between tires and road is $\mu_s = 0.65$, what is the minimum turning radius?

5. Model a figure skater's body as a solid cylinder and her arms as thin rods, making reasonable estimates for the dimensions. Then calculate the ratio of the angular speeds for a spinning skater with outstretched arms, and with arms held tightly against her body. Check Sections 8–5 and 8–8.

6. One possibility for a low-pollution automobile is for it to use energy stored in a heavy rotating **flywheel**. Suppose such a car has a total mass of 1100 kg, uses a uniform cylindrical flywheel of diameter 1.50 m and mass 270 kg, and should be able to travel 350 km without needing a flywheel "spinup." (a) Make reasonable assumptions (average frictional retarding force on car = 450 N, thirty acceleration periods from rest to 95 km/h, equal uphill and downhill, and that energy can be put back into the flywheel as the car goes downhill), and estimate what total energy needs to be stored in the flywheel. (b) What is the angular velocity of the flywheel when it has a full "energy charge"? (c) About how long would it take a 150-hp motor to give the flywheel a full energy charge before a trip?

*7. A person stands on a platform, initially at rest, that can rotate freely without friction. The moment of inertia of the person plus the platform is $I_P$. The person holds a spinning bicycle wheel with its axis horizontal. The wheel has moment of inertia $I_W$ and angular velocity $\omega_W$. What will be the angular velocity $\omega_P$ of the platform if the person moves the axis of the wheel so that it points (a) vertically upward, (b) at a 60° angle to the vertical, (c) vertically downward? (d) What will $\omega_P$ be if the person reaches up and stops the wheel in part (a)? See Sections 8–8 and 8–9.

(a)  (b)

**FIGURE 8–68** Search and Learn 4.

Our whole built environment, from modern bridges to skyscrapers, has required architects and engineers to determine the forces and stresses within these structures. The object is to keep these structures standing, or "static"—that is, not in motion, especially not falling down.

The study of statics applies equally well to the human body, including balance, the forces in muscles, joints, and bones, and ultimately the possibility of fracture.

CHAPTER

# 9

# Static Equilibrium; Elasticity and Fracture

## CONTENTS

**CHAPTER-OPENING QUESTION—Guess now!**

The diving board shown here is held by two supports at A and B. Which statement is true about the forces exerted *on* the diving board at A and B?

**(a)** $\vec{\mathbf{F}}_A$ is down, $\vec{\mathbf{F}}_B$ is up, and $F_B$ is larger than $F_A$.

**(b)** Both forces are up and $F_B$ is larger than $F_A$.

**(c)** $\vec{\mathbf{F}}_A$ is down, $\vec{\mathbf{F}}_B$ is up, and $F_A$ is larger than $F_B$.

**(d)** Both forces are down and approximately equal.

**(e)** $\vec{\mathbf{F}}_B$ is down, $\vec{\mathbf{F}}_A$ is up, and they are equal.

In this Chapter, we will study a special case in mechanics—when the net force and the net torque on an object, or system of objects, are both zero. In this case both the linear acceleration and the angular acceleration of the object or system are zero. The object is either at rest, or its center of mass is moving at constant velocity. We will be concerned mainly with the first situation, in which the object or objects are all at rest, or *static* (= not moving).

The net force and the net torque can be zero, but this does not imply that no forces at all act on the objects. In fact it is virtually impossible to find an object on which no forces act. Just how and where these forces act can be very important, both for buildings and other structures, and in the human body.

Sometimes, as we shall see in this Chapter, the forces may be so great that the object is seriously *deformed*, or it may even *fracture* (break)—and avoiding such problems gives this field of *statics* even greater importance.

**Statics** is concerned with the calculation of the forces acting on and within structures that are in *equilibrium*. Determination of these forces, which occupies us in the first part of this Chapter, then allows a determination of whether the structures can sustain the forces without significant deformation or fracture, subjects we discuss later in this Chapter. These techniques can be applied in a wide range of fields. Architects and engineers must be able to calculate the forces on the structural components of buildings, bridges, machines, vehicles, and other structures, since any material will buckle or break if too much force is applied (Fig. 9–1). In the human body a knowledge of the forces in muscles and joints is of great value for doctors, physical therapists, and athletes.

# 9–1 The Conditions for Equilibrium

Objects in daily life have at least one force acting on them (gravity). If they are at rest, then there must be other forces acting on them as well so that the net force is zero. A book at rest on a table, for example, has two forces acting on it, the downward force of gravity and the normal force the table exerts upward on it (Fig. 9–2). Because the book is at rest, Newton's second law tells us that the net force on it is zero. Thus the upward force exerted by the table on the book must be equal in magnitude to the force of gravity acting downward on the book. Such an object is said to be in **equilibrium** (Latin for "equal forces" or "balance") under the action of these two forces.

Do not confuse the two forces in Fig. 9–2 with the equal and opposite forces of Newton's third law, which act on different objects. In Fig. 9–2, both forces act on the same object; and they happen to add up to zero.

**FIGURE 9–1** Elevated walkway collapse in a Kansas City hotel in 1981. How a simple physics calculation could have prevented the tragic loss of over 100 lives is considered in Example 9–12.

**FIGURE 9–2** The book is in equilibrium; the net force on it is zero.

---

**EXAMPLE 9–1** **Straightening teeth.** The wire band shown in Fig. 9–3a has a tension $F_T$ of 2.0 N along it. It therefore exerts forces of 2.0 N on the high-lighted tooth (to which it is attached) in the two directions shown. Calculate the resultant force on the tooth due to the wire, $F_R$.

(a)

(b)

**FIGURE 9–3** Forces on a tooth. Example 9–1.

**APPROACH** Since the two forces $F_T$ are equal, their sum will be directed along the line that bisects the angle between them, which we have chosen to be the $y$ axis. The $x$ components of the two forces add up to zero.

**SOLUTION** The $y$ component of each force is $(2.0\ \text{N})(\cos 70°) = 0.68\ \text{N}$; adding the two together, we get a resultant force $F_R = 1.4\ \text{N}$ as shown in Fig. 9–3b. We assume that the tooth is in equilibrium because the gums exert a nearly equal magnitude force in the opposite direction. Actually that is not quite so since the objective is to move the tooth ever so slowly.

**NOTE** If the wire is firmly attached to the tooth, the tension to the right, say, can be made larger than that to the left, and the resultant force would correspondingly be directed more toward the right.

**PHYSICS APPLIED**
*Braces for teeth*

## The First Condition for Equilibrium

For an object to be at rest, Newton's second law tells us that the sum of the forces acting on it must add up to zero. Since force is a vector, the components of the net force must each be zero. Hence, a condition for equilibrium is that

$$\Sigma F_x = 0, \qquad \Sigma F_y = 0, \qquad \Sigma F_z = 0. \qquad (9\text{--}1)$$

We will mainly be dealing with forces that act in a plane, so we usually need only the $x$ and $y$ components. We must remember that if a particular force component points along the negative $x$ or $y$ axis, it must have a negative sign. Equations 9–1 represent the **first condition for equilibrium.**

We saw in Chapter 4 that to solve Problems involving forces, we need to draw a *free-body diagram*, indicating *all* the forces on a given object (see Section 4–7).

**EXAMPLE 9–2** **Chandelier cord tension.** Calculate the tensions $\vec{F}_A$ and $\vec{F}_B$ in the two cords that are connected to the vertical cord supporting the 200-kg chandelier in Fig. 9–4. Ignore the mass of the cords.

**APPROACH** We need a free-body diagram, but for which object? If we choose the chandelier, the cord supporting it must exert a force equal to the chandelier's weight $mg = (200 \text{ kg})(9.80 \text{ m/s}^2) = 1960 \text{ N}$. But the forces $\vec{F}_A$ and $\vec{F}_B$ don't get involved. Instead, let us choose as our object the point where the three cords join (it could be a knot). The free-body diagram is then as shown in Fig. 9–4a. The three forces—$\vec{F}_A$, $\vec{F}_B$, and the tension in the vertical cord equal to the weight of the 200-kg chandelier—act at this point where the three cords join. For this junction point we write $\Sigma F_x = 0$ and $\Sigma F_y = 0$, since the problem is laid out in two dimensions. The directions of $\vec{F}_A$ and $\vec{F}_B$ are known, since tension in a cord can only be along the cord—any other direction would cause the cord to bend, as already pointed out in Chapter 4. Thus, our unknowns are the magnitudes $F_A$ and $F_B$.

**SOLUTION** We first resolve $\vec{F}_A$ into its horizontal ($x$) and vertical ($y$) components. Although we don't know the value of $F_A$, we can write (see Fig. 9–4b) $F_{Ax} = -F_A \cos 60°$ and $F_{Ay} = F_A \sin 60°$. $\vec{F}_B$ has only an $x$ component. In the vertical direction, we have the downward force exerted by the vertical cord equal to the weight of the chandelier $mg = (200 \text{ kg})(g)$, and the vertical component of $\vec{F}_A$ upward:

$$\Sigma F_y = 0$$
$$F_A \sin 60° - (200 \text{ kg})(g) = 0$$

so

$$F_A = \frac{(200 \text{ kg})g}{\sin 60°} = (231 \text{ kg})g = (231 \text{ kg})(9.80 \text{ m/s}^2) = 2260 \text{ N}.$$

In the horizontal direction, with $\Sigma F_x = 0$,

$$\Sigma F_x = F_B - F_A \cos 60° = 0.$$

Thus

$$F_B = F_A \cos 60° = (231 \text{ kg})(g)(0.500) = (115 \text{ kg})g = 1130 \text{ N}.$$

The magnitudes of $\vec{F}_A$ and $\vec{F}_B$ determine the strength of cord or wire that must be used. In this case, the cord must be able to support a mass of more than 230 kg.

**NOTE** We didn't insert the value of $g$, the acceleration due to gravity, until the end. In this way we found the magnitude of the force in terms of $g$ times the number of kilograms (which may be a more familiar quantity than newtons).

**EXERCISE A** In Example 9–2, $F_A$ has to be greater than the chandelier's weight, $mg$. Why?

## The Second Condition for Equilibrium

Although Eqs. 9–1 are a necessary condition for an object to be in equilibrium, they are not always a sufficient condition. Figure 9–5 shows an object on which the net force is zero. Although the two forces labeled $\vec{F}$ add up to give zero net force on the object, they do give rise to a net torque that will rotate the object.

(a)

(b)

**FIGURE 9–4** Example 9–2.

**FIGURE 9–5** Although the net force on it is zero, the ruler will move (rotate). A pair of equal forces acting in opposite directions but at different points on an object (as shown here) is referred to as a *couple*.

Referring to Eq. 8–14, $\Sigma\tau = I\alpha$, we see that if an object is to remain at rest, the net torque applied to it (calculated about *any* axis) must be zero. Thus we have the **second condition for equilibrium**: that the sum of the torques acting on an object, as calculated about any axis, must be zero:

$$\Sigma\tau = 0. \tag{9-2}$$

This condition will ensure that the angular acceleration, $\alpha$, about any axis will be zero. If the object is not rotating initially ($\omega = 0$), it will not start rotating. Equations 9–1 and 9–2 are the only requirements for an object to be in equilibrium.

We will mainly consider cases in which the forces all act in a plane (we call it the $xy$ plane). In such cases the torque is calculated about an axis that is perpendicular to the $xy$ plane. *The choice of this axis is arbitrary.* If the object is at rest, then $\Sigma\tau = 0$ is valid about any axis. Therefore we can choose any axis that makes our calculation easier. Once the axis is chosen, all torques must be calculated about that axis.

**CAUTION**

*Axis choice for $\Sigma\tau = 0$ is arbitrary. All torques must be calculated about the same axis.*

| CONCEPTUAL EXAMPLE 9–3 | **A lever.** The bar in Fig. 9–6 is being used as a lever to pry up a large rock. The small rock acts as a fulcrum (pivot point). The force $F_P$ required at the long end of the bar can be quite a bit smaller than the rock's weight $mg$, since it is the *torques* that balance in the rotation about the fulcrum. If, however, the leverage isn't sufficient, and the large rock isn't budged, what are two ways to increase the lever arm?

**RESPONSE** One way is to increase the lever arm of the force $F_P$ by slipping a pipe over the end of the bar and thereby pushing with a longer lever arm. A second way is to move the fulcrum closer to the large rock. This may change the long lever arm $R$ only a little, but it changes the short lever arm $r$ by a substantial fraction and therefore changes the ratio of $R/r$ dramatically. In order to pry the rock, the torque due to $F_P$ must at least balance the torque due to $mg$; that is, $mgr = F_P R$ and

$$\frac{r}{R} = \frac{F_P}{mg}.$$

With $r$ smaller, the weight $mg$ can be balanced with less force $F_P$. The ratio $R/r$ is the **mechanical advantage** of the system. A lever is a "simple machine." We discussed another simple machine, the pulley, in Chapter 4, Example 4–14.

**PHYSICS APPLIED**
*The lever*

**FIGURE 9–6** Example 9–3. A lever can "multiply" your force.

**EXERCISE B** For simplicity, we wrote the equation in Example 9–3 as if the lever were perpendicular to the forces. Would the equation be valid even for a lever at an angle as shown in Fig. 9–6?

# 9–2 Solving Statics Problems

The subject of statics is important because it allows us to calculate certain forces on (or within) a structure when some of the forces on it are already known. We will mainly consider situations in which all the forces act in a plane, so we can have two force equations ($x$ and $y$ components) and one torque equation, for a total of three equations. Of course, you do not have to use all three equations if they are not needed. When using a torque equation, a torque that tends to rotate the object counterclockwise is usually considered positive, whereas a torque that tends to rotate it clockwise is considered negative. (But the opposite convention would be OK too.)

One of the forces that acts on objects is the force of gravity. As we discussed in Section 7–8, we can consider the force of gravity on an object as acting at its center of gravity (CG) or center of mass (CM), which for practical purposes are the same point. For uniform symmetrically shaped objects, the CG is at the geometric center. For more complicated objects, the CG can be determined as discussed in Section 7–8.

**PROBLEM SOLVING**
$\tau > 0$ *counterclockwise*
$\tau < 0$ *clockwise*

There is no single technique for attacking statics problems, but the following procedure may be helpful.

## Statics

1. Choose one object at a time for consideration. Make a careful **free-body diagram** by showing all the forces acting on that object, including gravity, and the points at which these forces act. If you aren't sure of the direction of a force, choose a direction; if the actual direction of the force (or component of a force) is opposite, your eventual calculation will give a result with a minus sign.

2. Choose a convenient **coordinate system**, and resolve the forces into their components.

3. Using letters to represent unknowns, write down the **equilibrium equations** for the **forces**:

$$\Sigma F_x = 0$$

and

$$\Sigma F_y = 0,$$

assuming all the forces act in a plane.

4. For the **torque equation**,

$$\Sigma \tau = 0,$$

choose any axis perpendicular to the $xy$ plane that might make the calculation easier. (For example, you can reduce the number of unknowns in the resulting equation by choosing the axis so that one of the unknown forces acts through that axis; then this force will have zero lever arm and produce zero torque, and so won't appear in the torque equation.) Pay careful attention to determining the lever arm for each force correctly. Give each torque a + or − sign to indicate torque direction. For example, if torques tending to rotate the object counterclockwise are positive, then those tending to rotate it clockwise are negative.

5. **Solve** these equations for the unknowns. Three equations allow a maximum of three unknowns to be solved for. They can be forces, distances, or even angles.

---

**PHYSICS APPLIED**
*Balancing a seesaw*

**EXAMPLE 9–4** **Balancing a seesaw.** A board of mass $M = 4.0\,\text{kg}$ serves as a seesaw for two children, as shown in Fig. 9–7a. Child A has a mass of 30 kg and sits 2.5 m from the pivot point, P (his center of gravity is 2.5 m from the pivot). At what distance $x$ from the pivot must child B, of mass 25 kg, place herself to balance the seesaw? Assume the board is uniform and centered over the pivot.

**APPROACH** We follow the steps of the Problem Solving Strategy above.

**SOLUTION**

1. **Free-body diagram.** We choose the board as our object, and assume it is horizontal. Its free-body diagram is shown in Fig. 9–7b. The forces acting on the board are the forces exerted downward on it by each child, $\vec{\mathbf{F}}_A$ and $\vec{\mathbf{F}}_B$, the upward force exerted by the pivot $\vec{\mathbf{F}}_N$, and the force of gravity on the board $(= M\vec{\mathbf{g}})$ which acts at the center of the uniform board.

2. **Coordinate system.** We choose $y$ to be vertical, with positive upward, and $x$ horizontal to the right, with origin at the pivot.

3. **Force equation.** All the forces are in the $y$ (vertical) direction, so

$$\Sigma F_y = 0$$
$$F_N - m_A g - m_B g - Mg = 0,$$

where $F_A = m_A g$ and $F_B = m_B g$.

**FIGURE 9–7** (a) Two children on a seesaw, Example 9–4. (b) Free-body diagram of the board.

**4. Torque equation.** Let us calculate the torque about an axis through the board at the pivot point, P. Then the lever arms for $F_N$ and for the weight of the board are zero, and they will contribute zero torque about point P. Thus the torque equation will involve only the forces $\vec{F}_A$ and $\vec{F}_B$, which are equal to the weights of the children. The torque exerted by each child will be $mg$ times the appropriate lever arm, which here is the distance of each child from the pivot point. $\vec{F}_A$ tends to rotate the board counterclockwise $(+)$ and $\vec{F}_B$ clockwise $(-)$, so the torque equation is

$$\Sigma\tau = 0$$
$$m_A\, g(2.5\text{ m}) - m_B\, gx + Mg(0\text{ m}) + F_N(0\text{ m}) = 0$$

or

$$m_A\, g(2.5\text{ m}) - m_B\, gx = 0,$$

where two terms were dropped because their lever arms were zero.

**5. Solve.** We solve the torque equation for $x$ and find

$$x = \frac{m_A}{m_B}(2.5\text{ m}) = \frac{30\text{ kg}}{25\text{ kg}}(2.5\text{ m}) = 3.0\text{ m}.$$

To balance the seesaw, child B must sit so that her CG is 3.0 m from the pivot point. This makes sense: since she is lighter, she must sit farther from the pivot than the heavier child in order to provide torques of equal magnitude.

**EXERCISE C** We did not need to use the force equation to solve Example 9–4 because of our choice of the axis. Use the force equation to find the force exerted by the pivot.

---

**EXAMPLE 9–5** **Forces on a beam and supports.** A uniform 1500-kg beam, 20.0 m long, supports a 15,000-kg printing press 5.0 m from the right support column (Fig. 9–8). Calculate the force on each of the vertical support columns.

**APPROACH** We analyze the forces on the beam (the force the beam exerts on each column is equal and opposite to the force exerted by the column on the beam). We label these forces $\vec{F}_A$ and $\vec{F}_B$ in Fig. 9–8. The weight of the beam itself acts at its center of gravity, 10.0 m from either end. We choose a convenient axis for writing the torque equation: the point of application of $\vec{F}_A$ (labeled P), so $\vec{F}_A$ will not enter the equation (its lever arm will be zero) and we will have an equation in only one unknown, $F_B$.

**SOLUTION** The torque equation, $\Sigma\tau = 0$, with the counterclockwise direction as positive, gives

$$\Sigma\tau = -(10.0\text{ m})(1500\text{ kg})g - (15.0\text{ m})(15,000\text{ kg})g + (20.0\text{ m})F_B = 0.$$

Solving for $F_B$, we find $F_B = (12,000\text{ kg})g = 118,000\text{ N}$. To find $F_A$, we use $\Sigma F_y = 0$, with $+y$ upward:

$$\Sigma F_y = F_A - (1500\text{ kg})g - (15,000\text{ kg})g + F_B = 0.$$

Putting in $F_B = (12,000\text{ kg})g$, we find that $F_A = (4500\text{ kg})g = 44,100\text{ N}$.

**FIGURE 9–8** A 1500-kg beam supports a 15,000-kg machine. Example 9–5.

---

Figure 9–9 shows a uniform beam that extends beyond its support like a diving board. Such a beam is called a **cantilever**. The forces acting on the beam in Fig. 9–9 are those due to the supports, $\vec{F}_A$ and $\vec{F}_B$, and the force of gravity which acts at the CG, 5.0 m to the right of the right hand support. If you follow the procedure of the last Example and calculate $F_A$ and $F_B$, assuming they point upward as shown in Fig. 9–9, you will find that $F_A$ comes out negative. If the beam has a mass of 1200 kg and a weight $mg = 12,000\text{ N}$, then $F_B = 15,000\text{ N}$ and $F_A = -3000\text{ N}$ (see Problem 10). Whenever an unknown force comes out negative, it merely means that the force actually points in the opposite direction from what you assumed. Thus in Fig. 9–9, $\vec{F}_A$ actually must pull downward (by means of bolts, screws, fasteners, and/or glue). To see why $\vec{F}_A$ has to act downward, note that the board's weight acting at the CG would otherwise rotate the board clockwise about support B.

**PHYSICS APPLIED**

*Cantilever*

**PROBLEM SOLVING**

*If a force comes out negative*

**FIGURE 9–9** A cantilever.

$y$

$x$

$F_{Hy}$   $\vec{F}_H$   $\vec{F}_T$   $F_{Ty}$

$F_{Hx}$   $\theta$

Hinge   $F_{Tx}$

$m\vec{g}$

$M\vec{g}$

Arlene's Book Store

**FIGURE 9–10** Example 9–6.

**EXERCISE D** Return to the Chapter-Opening Question, page 230, and answer it again now. Try to explain why you may have answered differently the first time.

Our next Example involves a beam that is attached to a wall by a hinge and is supported by a cable or cord (Fig. 9–10). It is important to remember that a flexible cable can support a force only along its length. (If there were a component of force perpendicular to the cable, it would bend because it is flexible.) But for a rigid device, such as the hinge in Fig. 9–10, the force can be in any direction and we can know the direction only after solving the equations. (The hinge is assumed small and smooth, so it can exert no internal torque on the beam.)

**EXAMPLE 9–6** **Hinged beam and cable.** A uniform beam, 2.20 m long with mass $m = 25.0$ kg, is mounted by a small hinge on a wall as shown in Fig. 9–10. The beam is held in a horizontal position by a cable that makes an angle $\theta = 30.0°$. The beam supports a sign of mass $M = 28.0$ kg suspended from its end. Determine the components of the force $\vec{F}_H$ that the (smooth) hinge exerts on the beam, and the tension $F_T$ in the supporting cable.

**APPROACH** Figure 9–10 is the free-body diagram for the beam, showing all the forces acting on the beam. It also shows the components of $\vec{F}_T$ and a guess for the direction of $\vec{F}_H$. We have three unknowns, $F_{Hx}$, $F_{Hy}$, and $F_T$ (we are given $\theta$), so we will need all three equations, $\Sigma F_x = 0$, $\Sigma F_y = 0$, $\Sigma \tau = 0$.

**SOLUTION** The sum of the forces in the vertical ($y$) direction is

$$\Sigma F_y = 0$$
$$F_{Hy} + F_{Ty} - mg - Mg = 0. \qquad \textbf{(i)}$$

In the horizontal ($x$) direction, the sum of the forces is

$$\Sigma F_x = 0$$
$$F_{Hx} - F_{Tx} = 0. \qquad \textbf{(ii)}$$

For the torque equation, we choose the axis at the point where $\vec{F}_T$ and $M\vec{g}$ act. Then our torque equation will contain only one unknown, $F_{Hy}$, because the lever arms for $\vec{F}_T$, $M\vec{g}$, and $F_{Hx}$ are zero. We choose torques that tend to rotate the beam counterclockwise as positive. The weight $mg$ of the (uniform) beam acts at its center, so we have

$$\Sigma \tau = 0$$
$$-(F_{Hy})(2.20 \text{ m}) + mg(1.10 \text{ m}) = 0.$$

We solve for $F_{Hy}$:

$$F_{Hy} = \left(\frac{1.10 \text{ m}}{2.20 \text{ m}}\right)mg = (0.500)(25.0 \text{ kg})(9.80 \text{ m/s}^2) = 123 \text{ N}. \qquad \textbf{(iii)}$$

Next, since the tension $\vec{F}_T$ in the cable acts along the cable ($\theta = 30.0°$), we see from Fig. 9–10 that $\tan\theta = F_{Ty}/F_{Tx}$, or

$$F_{Ty} = F_{Tx}\tan\theta = F_{Tx}(\tan 30.0°). \qquad \textbf{(iv)}$$

Equation (i) above gives

$$F_{Ty} = (m + M)g - F_{Hy} = (53.0 \text{ kg})(9.80 \text{ m/s}^2) - 123 \text{ N} = 396 \text{ N}.$$

Equations (iv) and (ii) give

$$F_{Tx} = F_{Ty}/\tan 30.0° = 396 \text{ N}/\tan 30.0° = 686 \text{ N};$$
$$F_{Hx} = F_{Tx} = 686 \text{ N}.$$

The components of $\vec{F}_H$ are $F_{Hy} = 123$ N and $F_{Hx} = 686$ N. The tension in the wire is $F_T = \sqrt{F_{Tx}^2 + F_{Ty}^2} = \sqrt{(686 \text{ N})^2 + (396 \text{ N})^2} = 792 \text{ N}.$[†]

[†]Our calculation used numbers rounded off to 3 significant figures. If you keep an extra digit, or leave the numbers in your calculator, you get $F_{Ty} = 396.5$ N, $F_{Tx} = 686.8$ N, and $F_T = 793$ N, all within the expected precision of 3 significant figures (Section 1–4).

***Alternate Solution*** Let us see the effect of choosing a different axis for calculating torques, such as an axis through the hinge. Then the lever arm for $F_H$ is zero, and the torque equation ($\Sigma\tau = 0$) becomes

$$-mg(1.10\text{ m}) - Mg(2.20\text{ m}) + F_{Ty}(2.20\text{ m}) = 0.$$

We solve this for $F_{Ty}$ and find

$$F_{Ty} = \frac{m}{2}g + Mg = (12.5\text{ kg} + 28.0\text{ kg})(9.80\text{ m/s}^2) = 397\text{ N}.$$

We get the same result, within the precision of our significant figures.

**NOTE** It doesn't matter which axis we choose for $\Sigma\tau = 0$. Using a second axis can serve as a check.

## *A More Difficult Example—The Ladder

**EXAMPLE 9–7** **Ladder.** A 5.0-m-long ladder leans against a wall at a point 4.0 m above a cement floor as shown in Fig. 9–11. The ladder is uniform and has mass $m = 12.0$ kg. Assuming the wall is frictionless, but the floor is not, determine the forces exerted on the ladder by the floor and by the wall.

**APPROACH** Figure 9–11 is the free-body diagram for the ladder, showing all the forces acting on the ladder. The wall, since it is frictionless, can exert a force only perpendicular to the wall, and we label that force $\vec{F}_W$. The cement floor exerts a force $\vec{F}_C$ which has both horizontal and vertical force components: $F_{Cx}$ is frictional and $F_{Cy}$ is the normal force. Finally, gravity exerts a force $mg = (12.0\text{ kg})(9.80\text{ m/s}^2) = 118\text{ N}$ on the ladder at its midpoint, since the ladder is uniform.

**SOLUTION** Again we use the equilibrium conditions, $\Sigma F_x = 0$, $\Sigma F_y = 0$, $\Sigma\tau = 0$. We will need all three since there are three unknowns: $F_W$, $F_{Cx}$, and $F_{Cy}$. The $y$ component of the force equation is

$$\Sigma F_y = F_{Cy} - mg = 0,$$

so immediately we have

$$F_{Cy} = mg = 118\text{ N}.$$

The $x$ component of the force equation is

$$\Sigma F_x = F_{Cx} - F_W = 0.$$

To determine both $F_{Cx}$ and $F_W$, we need a torque equation. If we choose to calculate torques about an axis through the point where the ladder touches the cement floor, then $\vec{F}_C$, which acts at this point, will have a lever arm of zero and so won't enter the equation. The ladder touches the floor a distance $x_0 = \sqrt{(5.0\text{ m})^2 - (4.0\text{ m})^2} = 3.0\text{ m}$ from the wall (right triangle, $c^2 = a^2 + b^2$). The lever arm for $mg$ is half this, or 1.5 m, and the lever arm for $F_W$ is 4.0 m, Fig. 9–11. The torque equation about the ladder's contact point on the cement is

$$\Sigma\tau = (4.0\text{ m})F_W - (1.5\text{ m})mg = 0.$$

Thus

$$F_W = \frac{(1.5\text{ m})(12.0\text{ kg})(9.8\text{ m/s}^2)}{4.0\text{ m}} = 44\text{ N}.$$

Then, from the $x$ component of the force equation,

$$F_{Cx} = F_W = 44\text{ N}.$$

Since the components of $\vec{F}_C$ are $F_{Cx} = 44$ N and $F_{Cy} = 118$ N, then

$$F_C = \sqrt{(44\text{ N})^2 + (118\text{ N})^2} = 126\text{ N} \approx 130\text{ N}$$

(rounded off to two significant figures), and it acts at an angle to the floor of

$$\theta = \tan^{-1}(118\text{ N}/44\text{ N}) = 70°.$$

**NOTE** The force $\vec{F}_C$ does *not* have to act along the ladder's direction because the ladder is rigid and not flexible like a cord or cable (see page 89, Chapter 4).

**FIGURE 9–11** A ladder leaning against a wall. Example 9–7. The force $\vec{F}_C$ that the cement floor exerts on the ladder need not be along the ladder which (unlike a cord) is rigid.

**FIGURE 9–12** The biceps (flexor) and triceps (extensor) muscles in the human arm.

Tendon

Triceps muscle (extensor)

Biceps muscle (flexor)

Insertion

🏃 **PHYSICS APPLIED**
*Forces in muscles and joints*

**FIGURE 9–13** Example 9–8, forces on forearm.

$\vec{\mathbf{F}}_M$

$\vec{\mathbf{F}}_J$

5.0 cm

CG

(2.0 kg)$\vec{\mathbf{g}}$

(5.0 kg)$\vec{\mathbf{g}}$

|←15 cm→|

(a)

|←—— 35 cm ——→|

45°

(b)

🏃 **PHYSICS APPLIED**
*Muscle insertion and lever arm*

🏃 **PHYSICS APPLIED**
*Forces on the spine, and back pain*

# 9–3 Applications to Muscles and Joints

The techniques we have been discussing for calculating forces on objects in equilibrium can be applied to the human body to study forces on muscles, bones, joints. Generally a muscle is attached, via tendons, to two different bones, as in Fig. 9–12. The points of attachment are called *insertions*. Two bones are flexibly connected at a *joint*, such as at the elbow, knee, hip, and shoulder. A muscle exerts a pull when its fibers contract under stimulation by a nerve, but a muscle can *not* exert a push. Muscles that tend to bring two limbs closer together, such as the biceps muscle in the upper arm (Fig. 9–12), are called *flexors*; those that act to extend a limb outward, such as the triceps muscle in Fig. 9–12, are called *extensors*. You use the flexor muscle in the upper arm when lifting an object in your hand; you use the extensor muscle when throwing a ball.

---

**EXAMPLE 9–8** **Force exerted by biceps muscle.** How much force must the biceps muscle exert when a 5.0-kg ball is held in the hand (*a*) with the arm horizontal as in Fig. 9–13a, and (*b*) when the arm is at a 45° angle as in Fig. 9–13b? The biceps muscle is connected to the forearm by a tendon attached 5.0 cm from the elbow joint. Assume that the mass of forearm and hand together is 2.0 kg and their CG is as shown.

**APPROACH** The free-body diagram for the forearm is shown in Fig. 9–13; the forces are the weights of the arm and ball, the upward force $\vec{\mathbf{F}}_M$ exerted by the muscle, and a force $\vec{\mathbf{F}}_J$ exerted at the joint by the bone in the upper arm (all assumed to act vertically). We wish to find the magnitude of $\vec{\mathbf{F}}_M$, which can be done using the torque equation and by choosing our axis through the joint so that $\vec{\mathbf{F}}_J$ contributes zero torque.

**SOLUTION** (*a*) We calculate torques about the point where $\vec{\mathbf{F}}_J$ acts in Fig. 9–13a. The $\Sigma \tau = 0$ equation gives

$$(0.050\,\text{m})F_M - (0.15\,\text{m})(2.0\,\text{kg})g - (0.35\,\text{m})(5.0\,\text{kg})g = 0.$$

We solve for $F_M$:

$$F_M = \frac{(0.15\,\text{m})(2.0\,\text{kg})g + (0.35\,\text{m})(5.0\,\text{kg})g}{0.050\,\text{m}} = (41\,\text{kg})g = 400\,\text{N}.$$

(*b*) The lever arm, as calculated about the joint, is reduced by the factor cos 45° for all three forces. Our torque equation will look like the one just above, except that each term will have its lever arm reduced by the same factor, which will cancel out. The same result is obtained, $F_M = 400\,\text{N}$.

**NOTE** The force required of the muscle (400 N) is quite large compared to the weight of the object lifted ($= mg = 49\,\text{N}$). Indeed, the muscles and joints of the body are generally subjected to quite large forces.

**NOTE** Forces exerted on joints can be large and even painful or injurious. Using $\Sigma F_y = 0$ we calculate for this case $F_J = F_M - (2.0\,\text{kg})g - (5.0\,\text{kg})g = 330\,\text{N}$.

---

The point of insertion of a muscle varies from person to person. A slight increase in the distance of the joint to the point of insertion of the biceps muscle from 5.0 cm to 5.5 cm can be a considerable advantage for lifting and throwing. Champion athletes are often found to have muscle insertions farther from the joint than the average person, and if this applies to one muscle, it usually applies to all.

As another example of the large forces acting within the human body, we consider the muscles used to support the body (trunk) when you bend forward (Fig. 9–14a). The lowest vertebra on the spinal column (fifth lumbar vertebra) acts as a fulcrum for this bending position. The "erector spinae" muscles in the back that support the trunk act at an effective angle of about 12° to the axis of the spine. Let us assume the trunk makes an angle of 30° with the horizontal.

Figure 9–14b is a simplified schematic drawing showing the forces on the upper body. The force exerted by the back muscles is represented by $\vec{\mathbf{F}}_M$, the force exerted on the base of the spine at the lowest vertebra is $\vec{\mathbf{F}}_V$, and $\vec{\mathbf{w}}_H$, $\vec{\mathbf{w}}_A$, and $\vec{\mathbf{w}}_T$ represent the weights of the **h**ead, freely hanging **a**rms, and **t**runk, respectively. The values shown are approximations. The distances (in cm) refer to a person 180 cm tall, but are approximately in the same ratio of $1:2:3$ for an average person of any height, and the result in the following Example is then independent of the height of the person.

(a)

---

**EXAMPLE 9–9** **Forces on your back.** Calculate the magnitude and direction of the force $\vec{\mathbf{F}}_V$ acting on the fifth lumbar vertebra as represented in Fig. 9–14b.

**APPROACH** We use the model of the upper body described above and shown in Fig. 9–14b. We can calculate $F_M$ using the torque equation if we take the axis at the base of the spine (point S); with this choice, the other unknown, $F_V$, doesn't appear in the equation because its lever arm is zero. To figure the lever arms, we need to use trigonometric functions.

**SOLUTION** For $\vec{\mathbf{F}}_M$, the lever arm (perpendicular distance from axis to line of action of the force) will be the real distance to where the force acts (48 cm) multiplied by $\sin 12°$, as shown in Fig. 9–14c. The lever arms for $\vec{\mathbf{w}}_H$, $\vec{\mathbf{w}}_A$, and $\vec{\mathbf{w}}_T$ can be seen from Fig. 9–14b to be their respective distances from S times $\sin 60°$. $F_M$ tends to rotate the trunk counterclockwise, which we take to be positive. Then $\vec{\mathbf{w}}_H$, $\vec{\mathbf{w}}_A$, $\vec{\mathbf{w}}_T$ will contribute negative torques. Thus $\Sigma\tau = 0$ gives

$$(0.48\text{ m})(\sin 12°)(F_M) - (0.72\text{ m})(\sin 60°)(w_H)$$
$$- (0.48\text{ m})(\sin 60°)(w_A) - (0.36\text{ m})(\sin 60°)(w_T) = 0.$$

Solving for $F_M$ and putting in the values for $w_H$, $w_A$, $w_T$ given in Fig. 9–14b, we find

$$F_M = \frac{(0.72\text{ m})(0.07w) + (0.48\text{ m})(0.12w) + (0.36\text{ m})(0.46w)}{(0.48\text{ m})(\sin 12°)}(\sin 60°)$$
$$= 2.37w \approx 2.4w,$$

where $w$ is the total weight of the body. To get the components of $\vec{\mathbf{F}}_V$ we use the $x$ and $y$ components of the force equation (noting that $30° - 12° = 18°$):

$$\Sigma F_y = F_{Vy} - F_M \sin 18° - w_H - w_A - w_T = 0$$

so

$$F_{Vy} = 1.38w \approx 1.4w,$$

and

$$\Sigma F_x = F_{Vx} - F_M \cos 18° = 0$$

so

$$F_{Vx} = 2.25w \approx 2.3w,$$

where we keep 3 significant figures for calculating, but round off to 2 for giving the answer. Then

$$F_V = \sqrt{F_{Vx}^2 + F_{Vy}^2} = 2.6w.$$

The angle $\theta$ that $F_V$ makes with the horizontal is given by $\tan\theta = F_{Vy}/F_{Vx} = 0.61$, so $\theta = 32°$.

**NOTE** The force on the lowest vertebra is over $2\frac{1}{2}$ times the total body weight! This force is exerted by the "sacral" bone at the base of the spine, through the somewhat flexible *intervertebral disk*. The disks at the base of the spine are clearly being compressed under very large forces. [If the body was less bent over (say, the 30° angle in Fig. 9–14b becomes 40° or 50°), then the stress on the lower back will be less (see Problem 33).]

$w_H = 0.07w$
(head)

$w_A = 0.12w$
(2 arms)

$w_T = 0.46w$
(trunk)

$w$ = Total weight of person

(b)

Lever arm for $\vec{\mathbf{F}}_M$

Axis for $\Sigma\tau$ calculation

(c)

**FIGURE 9–14** (a) A person bending over. (b) Forces on the back exerted by the back muscles ($\vec{\mathbf{F}}_M$) and by the vertebrae ($\vec{\mathbf{F}}_V$) when a person bends over. (c) Finding the lever arm for $\vec{\mathbf{F}}_M$.

If the person in Fig. 9–14 has a mass of 90 kg and is holding 20 kg in his hands (this increases $w_A$ to $0.34w$), then $F_V$ is increased to almost four times the person's weight ($3.7w$). For this 200-lb person, the force on the disk would be over 700 lb! With such strong forces acting, it is little wonder that so many people suffer from low back pain at one time or another.

# 9–4 Stability and Balance

An object in static equilibrium, if left undisturbed, will undergo no translational or rotational acceleration since the sum of all the forces and the sum of all the torques acting on it are zero. However, if the object is displaced slightly, three outcomes are possible: (1) the object returns to its original position, in which case it is said to be in **stable equilibrium**; (2) the object moves even farther from its original position, and it is said to be in **unstable equilibrium**; or (3) the object remains in its new position, and it is said to be in **neutral equilibrium**.

Consider the following examples. A ball suspended freely from a string is in stable equilibrium, for if it is displaced to one side, it will return to its original position (Fig. 9–15a) due to the net force and torque exerted on it. On the other hand, a pencil standing on its point is in unstable equilibrium. If its center of gravity is directly over its tip (Fig. 9–15b), the net force and net torque on it will be zero. But if it is displaced ever so slightly as shown—say, by a slight vibration or tiny air current—there will be a torque on it, and this torque acts to make the pencil continue to fall in the direction of the original displacement. Finally, an example of an object in neutral equilibrium is a sphere resting on a horizontal tabletop. If it is moved slightly to one side, it will remain in its new position—no net torque acts on it.

**FIGURE 9–15** (a) Stable equilibrium, and (b) unstable equilibrium.

Net force

(a)

(b)

In most situations, such as in the design of structures and in working with the human body, we are interested in maintaining stable equilibrium, or *balance*, as we sometimes say. In general, an object whose center of gravity (CG) is below its point of support, such as a ball on a string, will be in stable equilibrium. If the CG is above the base of support, we have a more complicated situation. Consider a standing refrigerator (Fig. 9–16a). If it is tipped slightly, it will return to its original position due to the torque on it as shown in Fig. 9–16b. But if it is tipped too far, Fig. 9–16c, it will fall over. The critical point is reached when the CG shifts from one side of the pivot point to the other. When the CG is on one side, the torque pulls the object back onto its original base of support, Fig. 9–16b. If the object is tipped further, the CG goes past the pivot point and the torque causes the object to topple, Fig. 9–16c. In general, *an object whose center of gravity is above its base of support will be stable if a vertical line projected downward from the CG falls within the base of support.* This is because the normal force upward on the object (which balances out gravity) can be exerted only within the area of contact, so if the force of gravity acts beyond this area, a net torque will act to topple the object.

CG   CG   CG

(a)       (b)       (c)

**FIGURE 9–16** Equilibrium of a refrigerator resting on a flat floor.

Stability, then, can be relative. A brick lying on its widest face is more stable than a brick standing on its end, for it will take more of an effort to tip it over. In the extreme case of the pencil in Fig. 9–15b, the base is practically a point and the slightest disturbance will topple it. In general, the larger the base and the lower the CG, the more stable the object.

In this sense, humans are less stable than four-legged mammals, which have a larger base of support because of their four legs, and most also have a lower center of gravity. When walking and performing other kinds of movement, a person continually shifts the body so that its CG is over the feet, although in the normal adult this requires no conscious thought. Even as simple a movement as bending over requires moving the hips backward so that the CG remains over the feet, and you do this repositioning without thinking about it. To see this, position yourself with your heels and back to a wall and try to touch your toes. You won't be able to do it without falling. Persons carrying heavy loads automatically adjust their posture so that the CG of the total mass is over their feet, Fig. 9–17.

**FIGURE 9–17** Humans adjust their posture to achieve stability when carrying loads.

Total CG

**PHYSICS APPLIED**
*Humans and balance*

# 9–5 Elasticity; Stress and Strain

In the first part of this Chapter we studied how to calculate the forces on objects in equilibrium. In this Section we study the effects of these forces: any object changes shape under the action of applied forces. If the forces are great enough, the object will break, or *fracture*, as we will discuss in Section 9–6.

## Elasticity and Hooke's Law

If a force is exerted on an object, such as the vertically suspended metal rod shown in Fig. 9–18, the length of the object changes. If the amount of elongation, $\Delta\ell$, is small compared to the length of the object, experiment shows that $\Delta\ell$ is proportional to the force exerted on the object. This proportionality can be written as an equation:

$$F = k\,\Delta\ell. \tag{9–3}$$

Here $F$ represents the force pulling on the object, $\Delta\ell$ is the change in length, and $k$ is a proportionality constant. Equation 9–3, which is sometimes called **Hooke's law**[†] after Robert Hooke (1635–1703), who first noted it, is found to be valid for almost any solid material from iron to bone—but it is valid only up to a point. For if the force is too great, the object stretches excessively and eventually breaks.

Figure 9–19 shows a typical graph of applied force versus elongation. Up to a point called the **proportional limit**, Eq. 9–3 is a good approximation for many common materials, and the curve is a straight line. Beyond this point, the graph deviates from a straight line, and no simple relationship exists between $F$ and $\Delta\ell$. Nonetheless, up to a point farther along the curve called the **elastic limit**, the object will return to its original length if the applied force is removed. The region from the origin to the elastic limit is called the *elastic region*. If the object is stretched beyond the elastic limit, it enters the *plastic region*: it does not return to the original length upon removal of the external force, but remains permanently deformed (such as a bent paper clip). The maximum elongation is reached at the *breaking point*. The maximum force that can be applied without breaking is called the **ultimate strength** of the material (actually, force per unit area, as we discuss in Section 9–6).

**FIGURE 9–18** Hooke's law: $\Delta\ell \propto$ applied force.

**FIGURE 9–19** Applied force vs. elongation for a typical metal under tension.

## Young's Modulus

The amount of elongation of an object, such as the rod shown in Fig. 9–18, depends not only on the force applied to it, but also on the material of which it is made and on its dimensions. That is, the constant $k$ in Eq. 9–3 can be written in terms of these factors.

If we compare rods made of the same material but of different lengths and cross-sectional areas, it is found that for the same applied force, the amount of stretch (again assumed small compared to the total length) is proportional to the original length and inversely proportional to the cross-sectional area. That is, the longer the object, the more it elongates for a given force; and the thicker it is, the less it elongates. These findings can be combined with Eq. 9–3 to yield

$$\Delta\ell = \frac{1}{E}\frac{F}{A}\,\ell_0, \tag{9–4}$$

where $\ell_0$ is the original length of the object, $A$ is the cross-sectional area, and $\Delta\ell$ is the change in length due to the applied force $F$. $E$ is a constant of proportionality[‡] known as the **elastic modulus**, or **Young's modulus**; its value depends only on the material.

[†] The term "law" applied to this relation is historical, but today it is not really appropriate. First of all, it is only an approximation, and second, it refers only to a limited set of phenomena. Most physicists today prefer to reserve the word "law" for those relations that are deeper and more encompassing and precise, such as Newton's laws of motion or the law of conservation of energy.

[‡] The fact that $E$ is in the denominator, so $1/E$ is the actual proportionality constant, is merely a convention. When we rewrite Eq. 9–4 to get Eq. 9–5, $E$ is found in the numerator.

The value of Young's modulus for various materials is given in Table 9–1 (the shear modulus and bulk modulus in this Table are discussed later in this Section). Because $E$ is a property only of the material and is independent of the object's size or shape, Eq. 9–4 is far more useful for practical calculation than Eq. 9–3.

**TABLE 9–1 Elastic Moduli**

| Material | Young's Modulus, $E$ (N/m$^2$) | Shear Modulus, $G$ (N/m$^2$) | Bulk Modulus, $B$ (N/m$^2$) |
|---|---|---|---|
| *Solids* | | | |
| Iron, cast | $100 \times 10^9$ | $40 \times 10^9$ | $90 \times 10^9$ |
| Steel | $200 \times 10^9$ | $80 \times 10^9$ | $140 \times 10^9$ |
| Brass | $100 \times 10^9$ | $35 \times 10^9$ | $80 \times 10^9$ |
| Aluminum | $70 \times 10^9$ | $25 \times 10^9$ | $70 \times 10^9$ |
| Concrete | $20 \times 10^9$ | | |
| Brick | $14 \times 10^9$ | | |
| Marble | $50 \times 10^9$ | | $70 \times 10^9$ |
| Granite | $45 \times 10^9$ | | $45 \times 10^9$ |
| Wood (pine) (parallel to grain) | $10 \times 10^9$ | | |
| (perpendicular to grain) | $1 \times 10^9$ | | |
| Nylon | $\approx 3 \times 10^9$ | | |
| Bone (limb) | $15 \times 10^9$ | $80 \times 10^9$ | |
| *Liquids* | | | |
| Water | | | $2.0 \times 10^9$ |
| Alcohol (ethyl) | | | $1.0 \times 10^9$ |
| Mercury | | | $2.5 \times 10^9$ |
| *Gases*[†] | | | |
| Air, H$_2$, He, CO$_2$ | | | $1.01 \times 10^5$ |

[†]At normal atmospheric pressure; no variation in temperature during process.

---

**EXAMPLE 9–10** **Tension in piano wire.** A 1.60-m-long steel piano wire has a diameter of 0.20 cm. How great is the tension in the wire if it stretches 0.25 cm when tightened?

**APPROACH** We assume Hooke's law holds, and use it in the form of Eq. 9–4, finding $E$ for steel in Table 9–1.

**SOLUTION** We solve for $F$ in Eq. 9–4 and note that the area of the wire is $A = \pi r^2 = (3.14)(0.0010\,\text{m})^2 = 3.14 \times 10^{-6}\,\text{m}^2$. Then

$$F = E \frac{\Delta \ell}{\ell_0} A$$

$$= (2.0 \times 10^{11}\,\text{N/m}^2)\left(\frac{0.0025\,\text{m}}{1.60\,\text{m}}\right)(3.14 \times 10^{-6}\,\text{m}^2)$$

$$= 980\,\text{N}.$$

**NOTE** The large tension in all the wires in a piano must be supported by a strong frame.

**EXERCISE E** Two steel wires have the same length and are under the same tension. But wire A has twice the diameter of wire B. Which of the following is true? (*a*) Wire B stretches twice as much as wire A. (*b*) Wire B stretches four times as much as wire A. (*c*) Wire A stretches twice as much as wire B. (*d*) Wire A stretches four times as much as wire B. (*e*) Both wires stretch the same amount.

## Stress and Strain

From Eq. 9–4, we see that the change in length of an object is directly proportional to the product of the object's length $\ell_0$ and the force per unit area $F/A$ applied to it. It is general practice to define the force per unit area as the **stress**:

$$\text{stress} = \frac{\text{force}}{\text{area}} = \frac{F}{A},$$

which has SI units of $N/m^2$. Also, the **strain** is defined to be the ratio of the change in length to the original length:

$$\text{strain} = \frac{\text{change in length}}{\text{original length}} = \frac{\Delta\ell}{\ell_0},$$

and is dimensionless (no units). Strain is thus the fractional change in length of the object, and is a measure of how much the object has been deformed. Stress is applied to the material by external agents, whereas strain is the material's response to the stress. Equation 9–4 can be rewritten as

$$\frac{F}{A} = E\frac{\Delta\ell}{\ell_0} \qquad\qquad (9\text{–}5)$$

or

$$E = \frac{F/A}{\Delta\ell/\ell_0} = \frac{\text{stress}}{\text{strain}}.$$

Thus we see that the strain is directly proportional to the stress, in the linear (elastic) region of Fig. 9–19.

## Tension, Compression, and Shear Stress

The rod shown in Fig. 9–20a is said to be under *tension* or **tensile stress**. Not only is there a force pulling down on the rod at its lower end, but since the rod is in equilibrium we know that the support at the top is exerting an equal[†] upward force on the rod at its upper end, Fig. 9–20a. In fact, this tensile stress exists throughout the material. Consider, for example, the lower half of a suspended rod as shown in Fig. 9–20b. This lower half is in equilibrium, so there must be an upward force on it to balance the downward force at its lower end. What exerts this upward force? It must be the upper part of the rod. Thus we see that external forces applied to an object give rise to internal forces, or stress, within the material itself.

Strain or deformation due to tensile stress is but one type of stress to which materials can be subjected. There are two other common types of stress: compressive and shear. **Compressive stress** is the exact opposite of tensile stress. Instead of being stretched, the material is compressed: the forces act inwardly on the object. Columns that support a weight, such as the columns of a Greek temple (Fig. 9–21), are subjected to compressive stress. Equations 9–4 and 9–5 apply equally well to compression and tension, and the values for the modulus $E$ are usually the same.

**FIGURE 9–20** Stress exists *within* the material.

---

[†]Or a greater force if the weight of the rod cannot be ignored compared to $F$.

**FIGURE 9–21** This Greek temple, in Agrigento, Sicily, built 2500 years ago, shows the post-and-beam construction. The columns are under compression.

FIGURE 9–22 The three types of
stress for rigid objects.

Tension
(a)

Compression
(b)

Shear
(c)

Figure 9–22 compares tensile and compressive stresses as well as the third type, shear stress. An object under **shear stress** has equal and opposite forces applied *across* its opposite faces. A simple example is a book or brick firmly attached to a tabletop, on which a force is exerted parallel to the top surface. The table exerts an equal and opposite force along the bottom surface. Although the dimensions of the object do not change significantly, the shape of the object does change, Fig. 9–22c. An equation similar to Eq. 9–4 can be applied to calculate shear strain:

$$\Delta \ell = \frac{1}{G}\frac{F}{A}\ell_0 \tag{9–6}$$

but $\Delta \ell$, $\ell_0$, and $A$ must be reinterpreted as indicated in Fig. 9–22c. Note that $A$ is the area of the surface *parallel* to the applied force (and not perpendicular as for tension and compression), and $\Delta \ell$ is *perpendicular* to $\ell_0$. The constant of proportionality $G$ is called the **shear modulus** and is generally one-half to one-third the value of Young's modulus $E$ (see Table 9–1). Figure 9–23 suggests why $\Delta \ell \propto \ell_0$: the fatter book shifts more for the same shearing force.

FIGURE 9–23 The fatter book (a) shifts more than the thinner book (b) with the same applied shear force.

(a)

(b)

## Volume Change—Bulk Modulus

If an object is subjected to inward forces from all sides, its volume will decrease. A common situation is an object submerged in a fluid. In this case, the fluid exerts a pressure on the object in all directions, as we shall see in Chapter 10. *Pressure* is defined as force per unit area, and thus is the equivalent of stress. For this situation the change in volume, $\Delta V$, is proportional to the original volume, $V_0$, and to the change in the pressure, $\Delta P$. We thus obtain a relation of the same form as Eq. 9–4 but with a proportionality constant called the **bulk modulus** $B$:

$$\frac{\Delta V}{V_0} = -\frac{1}{B}\Delta P \tag{9–7}$$

or

$$B = -\frac{\Delta P}{\Delta V / V_0}.$$

The minus sign means the volume *decreases* with an increase in pressure.

Values for the bulk modulus are given in Table 9–1. Since liquids and gases do not have a fixed shape, only the bulk modulus (not the Young's or shear moduli) applies to them.

# 9–6 Fracture

If the stress on a solid object is too great, the object fractures, or breaks (Fig. 9–24). Table 9–2 lists the ultimate strengths for tension, compression, and shear for a variety of materials. These values give the maximum force per unit area, or stress, that an object can withstand under each of these three types of stress for various types of material. They are, however, representative values only, and the actual value for a given specimen can differ considerably. It is therefore necessary to maintain a **safety factor** of from 3 to perhaps 10 or more—that is, the actual stresses on a structure should not exceed one-tenth to one-third of the values given in the Table. You may encounter tables of "allowable stresses" in which appropriate safety factors have already been included.

**FIGURE 9–24** Fracture as a result of the three types of stress.

**TABLE 9–2 Ultimate Strengths of Materials** (force/area)

| Material | Tensile Strength (N/m$^2$) | Compressive Strength (N/m$^2$) | Shear Strength (N/m$^2$) |
|---|---|---|---|
| Iron, cast | $170 \times 10^6$ | $550 \times 10^6$ | $170 \times 10^6$ |
| Steel | $500$–$2500 \times 10^6$ | $500 \times 10^6$ | $250 \times 10^6$ |
| Brass | $250 \times 10^6$ | $250 \times 10^6$ | $200 \times 10^6$ |
| Aluminum | $200 \times 10^6$ | $200 \times 10^6$ | $200 \times 10^6$ |
| Concrete | $2 \times 10^6$ | $20 \times 10^6$ | $2 \times 10^6$ |
| Brick | | $35 \times 10^6$ | |
| Marble | | $80 \times 10^6$ | |
| Granite | | $170 \times 10^6$ | |
| Wood (pine) (parallel to grain) | $40 \times 10^6$ | $35 \times 10^6$ | $5 \times 10^6$ |
| (perpendicular to grain) | | $10 \times 10^6$ | |
| Nylon | $500 \times 10^6$ | | |
| Bone (limb) | $130 \times 10^6$ | $170 \times 10^6$ | |

**EXAMPLE 9–11** **ESTIMATE** **Breaking the piano wire.** The steel piano wire we discussed in Example 9–10 was 1.60 m long with a diameter of 0.20 cm. Approximately what tension force would break it?

**APPROACH** We set the tensile stress $F/A$ equal to the ultimate tensile strength of steel given in Table 9–2, and we choose the highest value which represents high-carbon steel.

**SOLUTION** The wire's area is $A = \pi r^2$, where $r = 0.10\,\text{cm} = 1.0 \times 10^{-3}\,\text{m}$. Table 9–2 tells us

$$\frac{F}{A} = 2500 \times 10^6\,\text{N/m}^2,$$

so the wire would likely break if the force exceeded

$$F = (2500 \times 10^6\,\text{N/m}^2)(\pi)(1.0 \times 10^{-3}\,\text{m})^2 = 8000\,\text{N}.$$

As can be seen in Table 9–2, concrete (like stone and brick) is reasonably strong under compression but extremely weak under tension. Thus concrete can be used as vertical columns placed under compression, but is of little value as a beam because it cannot withstand the tensile forces that result from the inevitable sagging of the lower edge of a beam (see Fig. 9–25).

**FIGURE 9–25** A beam sags, at least a little (but is exaggerated here), even under its own weight. The beam thus changes shape: the upper edge is compressed, and the lower edge is under tension (elongated). Shearing stress also occurs within the beam.

**FIGURE 9–26** Steel rods around which concrete is poured for strength.

**PHYSICS APPLIED**
*Reinforced concrete
and
prestressed concrete*

*Reinforced concrete*, in which iron rods are embedded in the concrete (Fig. 9–26), is much stronger. But the concrete on the lower edge of a loaded beam still tends to crack because it is weak under tension. This problem is solved with *prestressed concrete*, which also contains iron rods or a wire mesh, but during the pouring of the concrete, the rods or wire are held under tension. After the concrete dries, the tension on the iron is released, putting the concrete under compression. The amount of compressive stress is carefully predetermined so that when loads are applied to the beam, the compression on the lower edge is never allowed to be reduced so far as to put the concrete into tension.

**PHYSICS APPLIED**
*A tragic collapse*

**FIGURE 9–27** Example 9–12.

(a)    (b)

(c) Force on pin A exerted by vertical rod

(d) Forces on pins at A exerted by vertical rods

**CONCEPTUAL EXAMPLE 9–12** | **A tragic substitution.** Two walkways, one above the other, are suspended from vertical rods attached to the ceiling of a high hotel lobby, Fig. 9–27a. The original design called for single rods 14 m long, but when such long rods proved to be unwieldy to install, it was decided to replace each long rod with two shorter ones as shown schematically in Fig. 9–27b. Determine the net force exerted by the rods on the supporting pin A (assumed to be the same size) for each design. Assume each vertical rod supports a mass $m$ of each bridge.

**RESPONSE** The single long vertical rod in Fig. 9–27a exerts an upward force equal to $mg$ on pin A to support the mass $m$ of the upper bridge. Why? Because the pin is in equilibrium, and the other force that balances this is the downward force $mg$ exerted on it by the upper bridge (Fig. 9–27c). There is thus a shear stress on the pin because the rod pulls up on one side of the pin, and the bridge pulls down on the other side. The situation when two shorter rods support the bridges (Fig. 9–27b) is shown in Fig. 9–27d, in which only the connections at the upper bridge are shown. The lower rod exerts a force of $mg$ downward on the lower of the two pins because it supports the lower bridge. The upper rod exerts a force of $2mg$ on the upper pin (labelled A) because the upper rod supports both bridges. Thus we see that when the builders substituted two shorter rods for each single long one, the stress in the supporting pin A was *doubled*. What perhaps seemed like a simple substitution did, in fact, lead to a tragic collapse in 1981 with a loss of life of over 100 people (see Fig. 9–1). Having a feel for physics, and being able to make simple calculations based on physics, can have a great effect, literally, on people's lives.

# *9–7 Spanning a Space: Arches and Domes

There are a great many areas where the arts and humanities overlap the sciences, and this is especially clear in architecture, where the forces in the materials that make up a structure need to be understood to avoid excessive deformation and collapse. Many of the features we admire in the architecture of the past were introduced not simply for their decorative effect, but for technical reasons. One example is the development of methods to span a space, from the simple beam supported by columns, to arches and domes.

The first important architectural invention was the post-and-beam (or post-and-lintel) construction, in which two upright posts or columns support a horizontal beam. Before steel was introduced in the nineteenth century, the length of a beam was quite limited because the strongest building materials were then stone and brick. Hence the width of a span was limited by the size of available stones. Equally important, stone and brick, though strong under compression—are very weak under tension and shear; all three types of stress occur in a beam (see Fig. 9–25). The minimal space that could be spanned using stone is shown by the closely spaced columns of the great Greek temples (Fig. 9–21).

The semicircular **arch** (Figs. 9–28a and b) was introduced by the ancient Romans 2000 years ago. Aside from its aesthetic appeal, it was a tremendous technological innovation. The advantage of the "true" or semicircular arch is that, if well designed, its wedge-shaped stones experience stress which is mainly compressive even when supporting a large load such as the wall and roof of a cathedral.

(a)

(b)

**FIGURE 9–28** (a) Round arches in the Roman Forum, 2000 years old. The one in the background is the Arch of Titus. (b) An arch is used here to good effect for a bridge over a chasm on the California coast.

Because the stones are forced to squeeze against one another, they are mainly under compression (see Fig. 9–29). A round arch consisting of many well-shaped stones could span a very wide space. However, because the arch transfers horizontal as well as vertical forces to the supports, considerable buttressing on the sides is needed, as we discuss shortly.

**FIGURE 9–29** Stones in a round arch (see Fig. 9–28a) are mainly under compression.

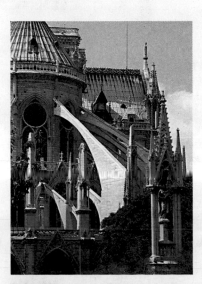

**FIGURE 9–30** Flying buttresses (on the cathedral of Notre Dame, in Paris).

The pointed arch came into use about A.D. 1100 and became the hallmark of the great Gothic cathedrals. It too was an important technical innovation, and was first used to support heavy loads such as the tower and central arch of a cathedral. Apparently the builders realized that, because of the steepness of the pointed arch, the forces due to the weight above could be brought down more nearly vertically, so less horizontal buttressing would be needed. The pointed arch reduced the load on the walls, so there could be more window openings and light. The smaller buttressing needed was provided on the outside by graceful flying buttresses (Fig. 9–30).

The technical innovation of the pointed arch was achieved not through calculation but through experience and intuition; it was not until much later that detailed calculations, such as those presented earlier in this Chapter, came into use.

2.0 m

$6.0 \times 10^4$ N          $6.0 \times 10^4$ N

2.0 m

$6.0 \times 10^4$ N          $6.0 \times 10^4$ N

**FIGURE 9–31** (a) Forces in a round arch, compared (b) with those in a pointed arch.

4.0 m

8.0 m

$F_H$          $F_H$

$F_V = 6.0 \times 10^4$ N          $F_V = 6.0 \times 10^4$ N

(a) ⊢—— 8.0 m ——⊣          (b) ⊢—— 8.0 m ——⊣

**FIGURE 9–32** Interior of the Pantheon in Rome, built almost 2000 years ago. This view, showing the great dome and its central opening for light, was painted about 1740 by Panini. Photographs do not capture its grandeur as well as this painting does.

**FIGURE 9–33** The skyline of Florence, showing Brunelleschi's dome on the cathedral.

To make an accurate analysis of a stone arch is quite difficult in practice. But if we make some simplifying assumptions, we can show why the horizontal component of the force at the base is less for a pointed arch than for a round one. Figure 9–31 shows a round arch and a pointed arch, each with an 8.0-m span. The height of the round arch is thus 4.0 m, whereas that of the pointed arch is larger and has been chosen to be 8.0 m. Each arch supports a weight of $12.0 \times 10^4$ N (= 12,000 kg × $g$) which, for simplicity, we have divided into two parts (each $6.0 \times 10^4$ N) acting on the two halves of each arch as shown. To be in equilibrium, each of the supports must exert an upward force of $6.0 \times 10^4$ N. For rotational equilibrium, each support also exerts a horizontal force, $F_H$, at the base of the arch, and it is this we want to calculate. We focus only on the right half of each arch. We set equal to zero the total torque calculated about the apex of the arch due to the three forces exerted on that half arch. The torque equation ($\Sigma \tau = 0$) contains three terms: the weight above, the support $F_V$ below, and the horizontal force $F_H$, which for the round arch (see Fig. 9–31a) is

$$-(2.0 \text{ m})(6.0 \times 10^4 \text{ N}) + (4.0 \text{ m})(6.0 \times 10^4 \text{ N}) - (4.0 \text{ m})(F_H) = 0.$$

Thus $F_H = 3.0 \times 10^4$ N for the round arch. For the pointed arch, the torque equation is (see Fig. 9–31b)

$$-(2.0 \text{ m})(6.0 \times 10^4 \text{ N}) + (4.0 \text{ m})(6.0 \times 10^4 \text{ N}) - (8.0 \text{ m})(F_H) = 0.$$

Solving, we find that $F_H = 1.5 \times 10^4$ N—only half as much as for the round arch! From this calculation we can see that the horizontal buttressing force required for a pointed arch is less because the arch is higher, and there is therefore a longer lever arm for this force. Indeed, the steeper the arch, the less the horizontal component of the force needs to be, and hence the more nearly vertical is the force exerted at the base of the arch.

Whereas an arch spans a two-dimensional space, a **dome**—which is basically an arch rotated about a vertical axis—spans a three-dimensional space. The Romans built the first large domes. Their shape was hemispherical and some still stand, such as that of the Pantheon in Rome (Fig. 9–32), built nearly 2000 years ago.

Fourteen centuries later, a new cathedral was being built in Florence.[†] It was to have a dome 43 m in diameter to rival that of the Pantheon, whose construction has remained a mystery. The new dome was to rest on a "drum" with no external abutments. Filippo Brunelleschi (1377–1446) designed a pointed dome (Fig. 9–33), since a pointed dome, like a pointed arch, exerts a smaller side thrust against its base. A dome, like an arch, is not stable until all the stones are in place. To support smaller domes during construction, wooden frameworks were used. But no trees big enough or strong enough could be found to span the 43-m space required. Brunelleschi built the dome in horizontal layers, each bonded to the previous one, holding it in place until the last stone of the circle was placed. Each closed ring was then strong enough to support the next layer. An amazing feat. Only in the twentieth century were larger domes built.

[†]Supervised by Arnolfo di Cambio, who wanted a dome but was unable to design it himself. But he was confident that Florentine genius would find a way, given that the ancient Romans had done it with the Pantheon, which still stands.

**EXAMPLE 9–13** **A modern dome.** The $1.2 \times 10^6$ kg dome of the Small Sports Palace in Rome (Fig. 9–34a) is supported by 36 buttresses positioned at a 38° angle so that they connect smoothly with the dome. Calculate the components of the force, $F_H$ and $F_V$, that each buttress exerts on the dome so that the force acts purely in compression—that is, at a 38° angle (Fig. 9–34b).

(a)

(b)

**FIGURE 9–34** Example 9–13. (a) The dome of the Small Sports Palace in Rome, built by Pier Luigi Nervi for the 1960 Olympics. (b) The force components each buttress exerts on the dome.

**APPROACH** We can find the vertical component $F_V$ exerted upward by each buttress because each supports $\frac{1}{36}$ of the dome's weight. We find $F_H$ knowing that the buttress needs to be under compression so $\vec{F} = \vec{F}_V + \vec{F}_H$ acts at a 38° angle.

**SOLUTION** The vertical load on *each* buttress is $\frac{1}{36}$ of the total weight. Thus

$$F_V = \frac{mg}{36} = \frac{(1.2 \times 10^6 \text{ kg})(9.8 \text{ m/s}^2)}{36} = 330{,}000 \text{ N}.$$

The force must act at a 38° angle at the base of the dome in order to be purely compressive. Thus

$$\tan 38° = \frac{F_V}{F_H};$$

$$F_H = \frac{F_V}{\tan 38°} = \frac{330{,}000 \text{ N}}{\tan 38°} = 420{,}000 \text{ N}.$$

**NOTE** For each buttress to exert this 420,000-N horizontal force, a prestressed-concrete tension ring surrounds the base of the buttresses beneath the ground (see Problem 58 and Fig. 9–77).

# Summary

An object at rest is said to be in **equilibrium**. The subject concerned with the determination of the forces within a structure at rest is called **statics**.

The two necessary conditions for an object to be in equilibrium are (1) the vector sum of all the forces on it must be zero, and (2) the sum of all the torques (calculated about any arbitrary axis) must also be zero. For a two-dimensional problem we can write

$$\Sigma F_x = 0, \qquad \Sigma F_y = 0, \qquad \Sigma \tau = 0. \qquad \text{(9–1, 9–2)}$$

It is important when doing statics problems to apply the equilibrium conditions to only one object at a time.

An object in static equilibrium is said to be in (*a*) **stable**, (*b*) **unstable**, or (*c*) **neutral equilibrium**, depending on whether a slight displacement leads to (*a*) a return to the original position, (*b*) further movement away from the original position, or (*c*) rest in the new position. An object in stable equilibrium is also said to be in *balance*.

**Hooke's law** applies to many elastic solids, and states that the change in length of an object is proportional to the applied force:

$$F = k \, \Delta \ell. \qquad \text{(9–3)}$$

If the force is too great, the object will exceed its **elastic limit**, which means it will no longer return to its original shape when the distorting force is removed. If the force is even greater, the **ultimate strength** of the material can be exceeded, and the object will **fracture**. The force per unit area acting on an object is the **stress**, and the resulting fractional change in length is the **strain**.

The stress on an object is present within the object and can be of three types: **compression**, **tension**, or **shear**. The ratio of stress to strain is called the **elastic modulus** of the material. **Young's modulus** applies for compression and tension, and the **shear modulus** for shear. **Bulk modulus** applies to an object whose volume changes as a result of pressure on all sides. All three moduli are constants for a given material when distorted within the elastic region.

[*Arches and domes are special ways to span a space that allow the stresses to be managed well.]

# Questions

1. Describe several situations in which an object is not in equilibrium, even though the net force on it is zero.

2. A bungee jumper momentarily comes to rest at the bottom of the dive before he springs back upward. At that moment, is the bungee jumper in equilibrium? Explain.

3. You can find the center of gravity of a meter stick by resting it horizontally on your two index fingers, and then slowly drawing your fingers together. First the meter stick will slip on one finger, and then on the other, but eventually the fingers meet at the CG. Why does this work?

4. Your doctor's scale has arms on which weights slide to counter your weight, Fig. 9–35. These weights are much lighter than you are. How does this work?

**FIGURE 9–35** Question 4.

5. A ground retaining wall is shown in Fig. 9–36a. The ground, particularly when wet, can exert a significant force $F$ on the wall. (a) What force produces the torque to keep the wall upright? (b) Explain why the retaining wall in Fig. 9–36b would be much less likely to overturn than that in Fig. 9–36a.

(a)              (b)

**FIGURE 9–36** Question 5.

6. Can the sum of the torques on an object be zero while the net force on the object is nonzero? Explain.

7. A ladder, leaning against a wall, makes a 60° angle with the ground. When is it more likely to slip: when a person stands on the ladder near the top or near the bottom? Explain.

8. A uniform meter stick supported at the 25-cm mark is in equilibrium when a 1-kg rock is suspended at the 0-cm end (as shown in Fig. 9–37). Is the mass of the meter stick greater than, equal to, or less than the mass of the rock? Explain your reasoning.

**FIGURE 9–37** Question 8.

9. Why do you tend to lean backward when carrying a heavy load in your arms?

10. Figure 9–38 shows a cone. Explain how to lay it on a flat table so that it is in (a) stable equilibrium, (b) unstable equilibrium, (c) neutral equilibrium.

**FIGURE 9–38** Question 10.

11. Place yourself facing the edge of an open door. Position your feet astride the door with your nose and abdomen touching the door's edge. Try to rise on your tiptoes. Why can't this be done?

12. Why is it not possible to sit upright in a chair and rise to your feet without first leaning forward?

13. Why is it more difficult to do sit-ups when your knees are bent than when your legs are stretched out?

14. Explain why touching your toes while you are seated on the floor with outstretched legs produces less stress on the lower spinal column than when touching your toes from a standing position. Use a diagram.

15. Which configuration of bricks, Fig. 9–39a or Fig. 9–39b, is the more likely to be stable? Why?

(a)              (b)

**FIGURE 9–39** Question 15. The dots indicate the CG of each brick (assumed uniform). The fractions $\frac{1}{4}$ and $\frac{1}{2}$ indicate what portion of each brick is hanging beyond its support.

16. Name the type of equilibrium for each position of the ball in Fig. 9–40.

**FIGURE 9–40** Question 16.

17. Is the Young's modulus for a bungee cord smaller or larger than that for an ordinary rope?

18. Examine how a pair of scissors or shears cuts through a piece of cardboard. Is the name "shears" justified? Explain.

19. Materials such as ordinary concrete and stone are very weak under tension or shear. Would it be wise to use such a material for either of the supports of the cantilever shown in Fig. 9–9? If so, which one(s)? Explain.

# MisConceptual Questions

1. A 60-kg woman stands on the very end of a uniform board, of length $\ell$, which is supported one-quarter of the way from one end and is balanced (Fig. 9–41). What is the mass of the board?
   (a) 15 kg.  (b) 20 kg.  (c) 30 kg.  (d) 60 kg.  (e) 120 kg.

**FIGURE 9–41**
MisConceptual
Question 1.

2. When you apply the torque equation $\Sigma \tau = 0$ to an object in equilibrium, the axis about which torques are calculated
   (a) must be located at a pivot.
   (b) must be located at the object's center of gravity.
   (c) should be located at the edge of the object.
   (d) can be located anywhere.

3. A uniform beam is hinged at one end and held in a horizontal position by a cable, as shown in Fig. 9–42. The tension in the cable
   (a) must be at least half the weight of the beam, no matter what the angle of the cable.
   (b) could be less than half the beam's weight for some angles.
   (c) will be half the beam's weight for all angles.
   (d) will equal the beam's weight for all angles.

**FIGURE 9–42**
MisConceptual Question 3:
beam and cable.

4. A heavy ball suspended by a cable is pulled to the side by a horizontal force $\vec{\mathbf{F}}$ as shown in Fig. 9–43. If angle $\theta$ is small, the magnitude of the force $F$ can be less than the weight of the ball because:
   (a) the force holds up only part of the ball's weight.
   (b) even though the ball is stationary, it is not really in equilibrium.
   (c) $\vec{\mathbf{F}}$ is equal to only the $x$ component of the tension in the cable.
   (d) the original statement is not true. To move the ball, $\vec{\mathbf{F}}$ must be at least equal to the ball's weight.

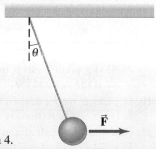

**FIGURE 9–43**
MisConceptual Question 4.

5. Two children are balanced on opposite sides of a seesaw. If one child leans inward toward the pivot point, her side will
   (a) rise.   (b) fall.   (c) neither rise nor fall.

6. A 10.0-N weight is suspended by two cords as shown in Fig. 9–44. What can you say about the tension in the two cords?
   (a) The tension in both cords is 5.0 N.
   (b) The tension in both cords is equal but not 5.0 N.
   (c) The tension in cord A is greater than that in cord B.
   (d) The tension in cord B is greater than that in cord A.

**FIGURE 9–44**
MisConceptual Question 6.

7. As you increase the force that you apply while pulling on a rope, which of the following is NOT affected?
   (a) The stress on the rope.
   (b) The strain on the rope.
   (c) The Young's modulus of the rope.
   (d) All of the above.
   (e) None of the above.

8. A woman is balancing on a high wire which is tightly strung, as shown in Fig. 9–45. The tension in the wire is
   (a) about half the woman's weight.
   (b) about twice the woman's weight.
   (c) about equal to the woman's weight.
   (d) much less than the woman's weight.
   (e) much more than the woman's weight.

**FIGURE 9–45**
MisConceptual
Question 8.

9. A parking garage is designed for two levels of cars. To make more money, the owner decides to double the size of the garage in each dimension (length, width, and number of levels). For the support columns to hold up four floors instead of two, how should he change the columns' diameter?
   (a) Double the area of the columns by increasing their diameter by a factor of 2.
   (b) Double the area of the columns by increasing their diameter by a factor of $\sqrt{2}$.
   (c) Quadruple the area of the columns by increasing their diameter by a factor of 2.
   (d) Increase the area of the columns by a factor of 8 by increasing their diameter by a factor of $2\sqrt{2}$.
   (e) He doesn't need to increase the diameter of the columns.

10. A rubber band is stretched by 1.0 cm when a force of 0.35 N is applied to each end. If instead a force of 0.70 N is applied to each end, estimate how far the rubber band will stretch from its unstretched length:
    (a) 0.25 cm.  (b) 0.5 cm.  (c) 1.0 cm.  (d) 2.0 cm.  (e) 4.0 cm.

# Problems

## 9–1 and 9–2 Equilibrium

1. (I) Three forces are applied to a tree sapling, as shown in Fig. 9–46, to stabilize it. If $\vec{F}_A = 385$ N and $\vec{F}_B = 475$ N, find $\vec{F}_C$ in magnitude and direction.

**FIGURE 9–46** Problem 1.

2. (I) Calculate the mass $m$ needed in order to suspend the leg shown in Fig. 9–47. Assume the leg (with cast) has a mass of 15.0 kg, and its CG is 35.0 cm from the hip joint; the cord holding the sling is 78.0 cm from the hip joint.

**FIGURE 9–47** Problem 2.

3. (I) A tower crane (Fig. 9–48a) must always be carefully balanced so that there is no net torque tending to tip it. A particular crane at a building site is about to lift a 2800-kg air-conditioning unit. The crane's dimensions are shown in Fig. 9–48b. (a) Where must the crane's 9500-kg counterweight be placed when the load is lifted from the ground? (The counterweight is usually moved automatically via sensors and motors to precisely compensate for the load.) (b) Determine the maximum load that can be lifted with this counterweight when it is placed at its full extent. Ignore the mass of the beam.

(a)

Counterweight
$M = 9500$ kg

|←3.4 m→|———7.7 m———→|

$m = 2800$ kg

**FIGURE 9–48**
(b)    Problem 3.

4. (I) What is the mass of the diver in Fig. 9–49 if she exerts a torque of 1800 m·N on the board, relative to the left (A) support post?

**FIGURE 9–49**
Problems 4 and 5.

5. (II) Calculate the forces $F_A$ and $F_B$ that the supports exert on the diving board of Fig. 9–49 when a 52-kg person stands at its tip. (a) Ignore the weight of the board. (b) Take into account the board's mass of 28 kg. Assume the board's CG is at its center.

6. (II) Figure 9–50 shows a pair of forceps used to hold a thin plastic rod firmly. If the thumb and finger each squeeze with a force $F_T = F_F = 11.0$ N, what force do the forceps jaws exert on the plastic rod?

**FIGURE 9–50**
Problem 6.

7. (II) Two cords support a chandelier in the manner shown in Fig. 9–4 except that the upper cord makes an angle of 45° with the ceiling. If the cords can sustain a force of 1660 N without breaking, what is the maximum chandelier weight that can be supported?

8. (II) The two trees in Fig. 9–51 are 6.6 m apart. A backpacker is trying to lift his pack out of the reach of bears. Calculate the magnitude of the force $\vec{F}$ that he must exert downward to hold a 19-kg backpack so that the rope sags at its midpoint by (a) 1.5 m, (b) 0.15 m.

**FIGURE 9–51**
Problems 8 and 70.

9. (II) A 110-kg horizontal beam is supported at each end. A 320-kg piano rests a quarter of the way from one end. What is the vertical force on each of the supports?

10. (II) Calculate $F_A$ and $F_B$ for the uniform cantilever shown in Fig. 9–9 whose mass is 1200 kg.

11. (II) A 75-kg adult sits at one end of a 9.0-m-long board. His 25-kg child sits on the other end. (a) Where should the pivot be placed so that the board is balanced, ignoring the board's mass? (b) Find the pivot point if the board is uniform and has a mass of 15 kg.

**12. (II)** Find the tension in the two cords shown in Fig. 9–52. Neglect the mass of the cords, and assume that the angle $\theta$ is 33° and the mass $m$ is 190 kg.

**FIGURE 9–52**
Problem 12.

**13. (II)** Find the tension in the two wires supporting the traffic light shown in Fig. 9–53.

**FIGURE 9–53**
Problem 13.

**14. (II)** How close to the edge of the 24.0-kg table shown in Fig. 9–54 can a 66.0-kg person sit without tipping it over?

**FIGURE 9–54**
Problem 14.

**15. (II)** The force required to pull the cork out of the top of a wine bottle is in the range of 200 to 400 N. What range of forces $F$ is required to open a wine bottle with the bottle opener shown in Fig. 9–55?

**FIGURE 9–55**
Problem 15.

**16. (II)** Calculate $F_A$ and $F_B$ for the beam shown in Fig. 9–56. The downward forces represent the weights of machinery on the beam. Assume the beam is uniform and has a mass of 280 kg.

**FIGURE 9–56**
Problem 16.

**17. (II)** Three children are trying to balance on a seesaw, which includes a fulcrum rock acting as a pivot at the center, and a very light board 3.2 m long (Fig. 9–57). Two playmates are already on either end. Boy A has a mass of 45 kg, and boy B a mass of 35 kg. Where should girl C, whose mass is 25 kg, place herself so as to balance the seesaw?

**FIGURE 9–57** Problem 17.

**18. (II)** A shop sign weighing 215 N hangs from the end of a uniform 155-N beam as shown in Fig. 9–58. Find the tension in the supporting wire (at 35.0°), and the horizontal and vertical forces exerted by the hinge on the beam at the wall. [*Hint*: First draw a free-body diagram.]

**FIGURE 9–58**
Problem 18.

**19. (II)** A traffic light hangs from a pole as shown in Fig. 9–59. The uniform aluminum pole AB is 7.20 m long and has a mass of 12.0 kg. The mass of the traffic light is 21.5 kg. Determine (*a*) the tension in the horizontal massless cable CD, and (*b*) the vertical and horizontal components of the force exerted by the pivot A on the aluminum pole.

**FIGURE 9–59**
Problem 19.

**20. (II)** A uniform steel beam has a mass of 940 kg. On it is resting half of an identical beam, as shown in Fig. 9–60. What is the vertical support force at each end?

**FIGURE 9–60** Problem 20.

**21.** (II) A 2500-kg trailer is attached to a stationary truck at point B, Fig. 9–61. Determine the normal force exerted by the road on the rear tires at A, and the vertical force exerted on the trailer by the support B.

**FIGURE 9–61** Problem 21.

**22.** (II) A 20.0-m-long uniform beam weighing 650 N rests on walls A and B, as shown in Fig. 9–62. (a) Find the maximum weight of a person who can walk to the extreme end D without tipping the beam. Find the forces that the walls A and B exert on the beam when the person is standing: (b) at D; (c) 2.0 m to the right of A.

**FIGURE 9–62** Problem 22.

**23.** (II) A 0.75-kg sheet is centered on a clothesline as shown in Fig. 9–63. The clothesline on either side of the hanging sheet makes an angle of 3.5° with the horizontal. Calculate the tension in the clothesline (ignore its mass) on either side of the sheet. Why is the tension so much greater than the weight of the sheet?

**FIGURE 9–63**
Problem 23.

**24.** (II) A 172-cm-tall person lies on a light (massless) board which is supported by two scales, one under the top of her head and one beneath the bottom of her feet (Fig. 9–64). The two scales read, respectively, 35.1 and 31.6 kg. What distance is the center of gravity of this person from the bottom of her feet?

**FIGURE 9–64** Problem 24.

**25.** (II) A man doing push-ups pauses in the position shown in Fig. 9–65. His mass $m = 68$ kg. Determine the normal force exerted by the floor (a) on each hand; (b) on each foot.

**FIGURE 9–65** Problem 25.

**26.** (III) Two wires run from the top of a pole 2.6 m tall that supports a volleyball net. The two wires are anchored to the ground 2.0 m apart, and each is 2.0 m from the pole (Fig. 9–66). The tension in each wire is 115 N. What is the tension in the net, assumed horizontal and attached at the top of the pole?

**FIGURE 9–66**
Problem 26.

**27.** (III) A uniform rod AB of length 5.0 m and mass $M = 3.8$ kg is hinged at A and held in equilibrium by a light cord, as shown in Fig. 9–67. A load $W = 22$ N hangs from the rod at a distance $d$ so that the tension in the cord is 85 N. (a) Draw a free-body diagram for the rod. (b) Determine the vertical and horizontal forces on the rod exerted by the hinge. (c) Determine $d$ from the appropriate torque equation.

**FIGURE 9–67**
Problem 27.

**28.** (III) You are on a pirate ship and being forced to walk the plank (Fig. 9–68). You are standing at the point marked C. The plank is nailed onto the deck at point A, and rests on the support 0.75 m away from A. The center of mass of the uniform plank is located at point B. Your mass is 65 kg and the mass of the plank is 45 kg. What is the minimum downward force the nails must exert on the plank to hold it in place?

**FIGURE 9–68**
Problem 28.

**29.** (III) A door 2.30 m high and 1.30 m wide has a mass of 13.0 kg. A hinge 0.40 m from the top and another hinge 0.40 m from the bottom each support half the door's weight (Fig. 9–69). Assume that the center of gravity is at the geometrical center of the door, and determine the horizontal and vertical force components exerted by each hinge on the door.

**FIGURE 9–69**
Problem 29.

*30. (III) A uniform ladder of mass $m$ and length $\ell$ leans at an angle $\theta$ against a frictionless wall, Fig. 9–70. If the coefficient of static friction between the ladder and the ground is $\mu_s$, determine a formula for the minimum angle at which the ladder will not slip.

**FIGURE 9–70**
Problem 30.

## 9–3 Muscles and Joints

**31.** (I) Suppose the point of insertion of the biceps muscle into the lower arm shown in Fig. 9–13a (Example 9–8) is 6.0 cm instead of 5.0 cm; how much mass could the person hold with a muscle exertion of 450 N?

**32.** (I) Approximately what magnitude force, $F_M$, must the extensor muscle in the upper arm exert on the lower arm to hold a 7.3-kg shot put (Fig. 9–71)? Assume the lower arm has a mass of 2.3 kg and its CG is 12.0 cm from the elbow-joint pivot.

**FIGURE 9–71**
Problem 32.

**33.** (II) Redo Example 9–9, assuming now that the person is less bent over so that the 30° in Fig. 9–14b is instead 45°. What will be the magnitude of $F_V$ on the vertebra?

**34.** (II) (a) Calculate the magnitude of the force, $F_M$, required of the "deltoid" muscle to hold up the outstretched arm shown in Fig. 9–72. The total mass of the arm is 3.3 kg. (b) Calculate the magnitude of the force $F_J$ exerted by the shoulder joint on the upper arm and the angle (to the horizontal) at which it acts.

**FIGURE 9–72** Problems 34 and 35.

**35.** (II) Suppose the hand in Problem 34 holds an 8.5-kg mass. What force, $F_M$, is required of the deltoid muscle, assuming the mass is 52 cm from the shoulder joint?

**36.** (II) The Achilles tendon is attached to the rear of the foot as shown in Fig. 9–73. When a person elevates himself just barely off the floor on the "ball of one foot," estimate the tension $F_T$ in the Achilles tendon (pulling upward), and the (downward) force $F_B$ exerted by the lower leg bone on the foot. Assume the person has a mass of 72 kg and $D$ is twice as long as $d$.

**FIGURE 9–73**
Problem 36.

**37.** (II) If 25 kg is the maximum mass $m$ that a person can hold in a hand when the arm is positioned with a 105° angle at the elbow as shown in Fig. 9–74, what is the maximum force $F_{max}$ that the biceps muscle exerts on the forearm? Assume the forearm and hand have a total mass of 2.0 kg with a CG that is 15 cm from the elbow, and that the biceps muscle attaches 5.0 cm from the elbow.

**FIGURE 9–74**
Problem 37.

## 9–4 Stability and Balance

**38.** (II) The Leaning Tower of Pisa is 55 m tall and about 7.7 m in radius. The top is 4.5 m off center. Is the tower in stable equilibrium? If so, how much farther can it lean before it becomes unstable? Assume the tower is of uniform composition.

**39.** (III) Four bricks are to be stacked at the edge of a table, each brick overhanging the one below it, so that the top brick extends as far as possible beyond the edge of the table. (a) To achieve this, show that successive bricks must extend no more than (starting at the top) $\frac{1}{2}$, $\frac{1}{4}$, $\frac{1}{6}$, and $\frac{1}{8}$ of their length beyond the one below (Fig. 9–75a). (b) Is the top brick completely beyond the base? (c) Determine a general formula for the maximum total distance spanned by $n$ bricks if they are to remain stable. (d) A builder wants to construct a corbeled arch (Fig. 9–75b) based on the principle of stability discussed in (a) and (c) above. What minimum number of bricks, each 0.30 m long and uniform, is needed if the arch is to span 1.0 m?

(a)                                                    (b)

**FIGURE 9–75** Problem 39.

## 9–5 Elasticity; Stress and Strain

**40.** (I) A nylon string on a tennis racket is under a tension of 275 N. If its diameter is 1.00 mm, by how much is it lengthened from its untensioned length of 30.0 cm?

**41.** (I) A marble column of cross-sectional area 1.4 m² supports a mass of 25,000 kg. (a) What is the stress within the column? (b) What is the strain?

**42.** (I) By how much is the column in Problem 41 shortened if it is 8.6 m high?

**43.** (I) A sign (mass 1700 kg) hangs from the bottom end of a vertical steel girder with a cross-sectional area of 0.012 m². (a) What is the stress within the girder? (b) What is the strain on the girder? (c) If the girder is 9.50 m long, how much is it lengthened? (Ignore the mass of the girder itself.)

**44.** (II) One liter of alcohol $(1000 \text{ cm}^3)$ in a flexible container is carried to the bottom of the sea, where the pressure is $2.6 \times 10^6 \text{ N/m}^2$. What will be its volume there?

**45.** (II) How much pressure is needed to compress the volume of an iron block by 0.10%? Express your answer in N/m², and compare it to atmospheric pressure $(1.0 \times 10^5 \text{ N/m}^2)$.

**46.** (II) A 15-cm-long tendon was found to stretch 3.7 mm by a force of 13.4 N. The tendon was approximately round with an average diameter of 8.5 mm. Calculate Young's modulus of this tendon.

**47.** (II) A steel wire 2.3 mm in diameter stretches by 0.030% when a mass is suspended from it. How large is the mass?

**48.** (II) At depths of 2000 m in the sea, the pressure is about 200 times atmospheric pressure $(1 \text{ atm} = 1.0 \times 10^5 \text{ N/m}^2)$. By what percentage does the interior space of an iron bathysphere's volume change at this depth?

**49.** (III) A scallop forces open its shell with an elastic material called abductin, whose Young's modulus is about $2.0 \times 10^6 \text{ N/m}^2$. If this piece of abductin is 3.0 mm thick and has a cross-sectional area of 0.50 cm², how much potential energy does it store when compressed 1.0 mm?

## 9–6 Fracture

**50.** (I) The femur bone in the human leg has a minimum effective cross section of about 3.0 cm² $(= 3.0 \times 10^{-4} \text{ m}^2)$. How much compressive force can it withstand before breaking?

**51.** (II) (a) What is the maximum tension possible in a 1.00-mm-diameter nylon tennis racket string? (b) If you want tighter strings, what do you do to prevent breakage: use thinner or thicker strings? Why? What causes strings to break when they are hit by the ball?

**52.** (II) If a compressive force of $3.3 \times 10^4$ N is exerted on the end of a 22-cm-long bone of cross-sectional area 3.6 cm², (a) will the bone break, and (b) if not, by how much does it shorten?

**53.** (II) (a) What is the minimum cross-sectional area required of a vertical steel cable from which is suspended a 270-kg chandelier? Assume a safety factor of 7.0. (b) If the cable is 7.5 m long, how much does it elongate?

**54.** (II) Assume the supports of the uniform cantilever shown in Fig. 9–76 ($m = 2900$ kg) are made of wood. Calculate the minimum cross-sectional area required of each, assuming a safety factor of 9.0.

**FIGURE 9–76**
Problem 54.

**55.** (II) An iron bolt is used to connect two iron plates together. The bolt must withstand shear forces up to about 3300 N. Calculate the minimum diameter for the bolt, based on a safety factor of 7.0.

**56.** (III) A steel cable is to support an elevator whose total (loaded) mass is not to exceed 3100 kg. If the maximum acceleration of the elevator is 1.8 m/s², calculate the diameter of cable required. Assume a safety factor of 8.0.

## *9–7 Arches and Domes

***57.** (II) How high must a pointed arch be if it is to span a space 8.0 m wide and exert one-third the horizontal force at its base that a round arch would?

***58.** (II) The subterranean tension ring that exerts the balancing horizontal force on the abutments for the dome in Fig. 9–34 is 36-sided, so each segment makes a 10° angle with the adjacent one (Fig. 9–77). Calculate the tension $F$ that must exist in each segment so that the required force of $4.2 \times 10^5$ N can be exerted at each corner (Example 9–13).

**FIGURE 9–77**
Problem 58.

# General Problems

**59.** A woman holds a 2.0-m-long uniform 10.0-kg pole as shown in Fig. 9–78. (a) Determine the forces she must exert with each hand (magnitude and direction). To what position should she move her left hand so that neither hand has to exert a force greater than (b) 150 N? (c) 85 N?

**FIGURE 9–78**
Problem 59.

**60.** A cube of side $\ell$ rests on a rough floor. It is subjected to a steady horizontal pull $F$, exerted a distance $h$ above the floor as shown in Fig. 9–79. As $F$ is increased, the block will either begin to slide, or begin to tip over. Determine the coefficient of static friction $\mu_s$ so that (a) the block begins to slide rather than tip; (b) the block begins to tip. [*Hint*: Where will the normal force on the block act if it tips?]

**FIGURE 9–79**
Problem 60.

**61.** A 50-story building is being planned. It is to be 180.0 m high with a base 46.0 m by 76.0 m. Its total mass will be about $1.8 \times 10^7$ kg, and its weight therefore about $1.8 \times 10^8$ N. Suppose a 200-km/h wind exerts a force of 950 N/m² over the 76.0-m-wide face (Fig. 9–80). Calculate the torque about the potential pivot point, the rear edge of the building (where $\vec{\mathbf{F}}_E$ acts in Fig. 9–80), and determine whether the building will topple. Assume the total force of the wind acts at the midpoint of the building's face, and that the building is not anchored in bedrock. [*Hint:* $\vec{\mathbf{F}}_E$ in Fig. 9–80 represents the force that the Earth would exert on the building in the case where the building would just begin to tip.]

**FIGURE 9–80** Forces on a building subjected to wind ($\vec{\mathbf{F}}_A$), gravity ($m\vec{\mathbf{g}}$), and the force $\vec{\mathbf{F}}_E$ on the building due to the Earth if the building were just about to tip. Problem 61.

**62.** The center of gravity of a loaded truck depends on how the truck is packed. If it is 4.0 m high and 2.4 m wide, and its CG is 2.2 m above the ground, how steep a slope can the truck be parked on without tipping over (Fig. 9–81)?

**FIGURE 9–81** Problem 62.

**63.** A uniform meter stick with a mass of 180 g is supported horizontally by two vertical strings, one at the 0-cm mark and the other at the 90-cm mark (Fig. 9–82). What is the tension in the string (*a*) at 0 cm? (*b*) at 90 cm?

**FIGURE 9–82** Problem 63.

**64.** There is a maximum height of a uniform vertical column made of any material that can support itself without buckling, and it is independent of the cross-sectional area (why?). Calculate this height for (*a*) steel (density $7.8 \times 10^3$ kg/m³), and (*b*) granite (density $2.7 \times 10^3$ kg/m³).

**65.** When a mass of 25 kg is hung from the middle of a fixed straight aluminum wire, the wire sags to make an angle of 12° with the horizontal as shown in Fig. 9–83. Determine the radius of the wire.

**FIGURE 9–83** Problem 65.

**66.** A 65.0-kg painter is on a uniform 25-kg scaffold supported from above by ropes (Fig. 9–84). There is a 4.0-kg pail of paint to one side, as shown. Can the painter walk safely to both ends of the scaffold? If not, which end(s) is dangerous, and how close to the end can he approach safely?

**FIGURE 9–84** Problem 66.

**67.** A 15.0-kg ball is supported from the ceiling by rope A. Rope B pulls downward and to the side on the ball. If the angle of A to the vertical is 22° and if B makes an angle of 53° to the vertical (Fig. 9–85), find the tensions in ropes A and B.

**FIGURE 9–85** Problem 67.

**68.** The roof over a 9.0-m × 10.0-m room in a school has a total mass of 13,600 kg. The roof is to be supported by vertical wooden "2 × 4s" (actually about 4.0 cm × 9.0 cm) equally spaced along the 10.0-m sides. How many supports are required on each side, and how far apart must they be? Consider only compression, and assume a safety factor of 12.

**69.** A 25-kg object is being lifted by two people pulling on the ends of a 1.15-mm-diameter nylon cord that goes over two 3.00-m-high poles 4.0 m apart, as shown in Fig. 9–86. How high above the floor will the object be when the cord breaks?

**FIGURE 9–86** Problem 69.

**70.** A 23.0-kg backpack is suspended midway between two trees by a light cord as in Fig. 9–51. A bear grabs the backpack and pulls vertically downward with a constant force, so that each section of cord makes an angle of 27° below the horizontal. Initially, without the bear pulling, the angle was 15°; the tension in the cord with the bear pulling is double what it was when he was not. Calculate the force the bear is exerting on the backpack.

**71.** Two identical, uniform beams are symmetrically set up against each other (Fig. 9–87) on a floor with which they have a coefficient of friction $\mu_s = 0.50$. What is the minimum angle the beams can make with the floor and still not fall?

**FIGURE 9–87**
Problem 71.

**72.** A steel rod of radius $R = 15$ cm and length $\ell_0$ stands upright on a firm surface. A 65-kg man climbs atop the rod. (*a*) Determine the percent decrease in the rod's length. (*b*) When a metal is compressed, each atom moves closer to its neighboring atom by exactly the same fractional amount. If iron atoms in steel are normally $2.0 \times 10^{-10}$ m apart, by what distance did this interatomic spacing have to change in order to produce the normal force required to support the man? [*Note*: Neighboring atoms repel each other, and this repulsion accounts for the observed normal force.]

**73.** A home mechanic wants to raise the 280-kg engine out of a car. The plan is to stretch a rope vertically from the engine to a branch of a tree 6.0 m above, and back to the bumper (Fig. 9–88). When the mechanic climbs up a stepladder and pulls horizontally on the rope at its midpoint, the engine rises out of the car. (*a*) How much force must the mechanic exert to hold the engine 0.50 m above its normal position? (*b*) What is the system's mechanical advantage?

6.0 m

**FIGURE 9–88**
Problem 73.

**74.** A 2.0-m-high box with a 1.0-m-square base is moved across a rough floor as in Fig. 9–89. The uniform box weighs 250 N and has a coefficient of static friction with the floor of 0.60. What minimum force must be exerted on the box to make it slide? What is the maximum height $h$ above the floor that this force can be applied without tipping the box over? Note that as the box tips, the normal force and the friction force will act at the lowest corner.

1.0 m

$\vec{\mathbf{F}}$

CG

2.0 m

$h$

$\vec{\mathbf{F}}_N$

$\vec{\mathbf{F}}_{fr}$

**FIGURE 9–89**
Problem 74.

**75.** A tightly stretched horizontal "high wire" is 36 m long. It sags vertically 2.1 m when a 60.0-kg tightrope walker stands at its center. What is the tension in the wire? Is it possible to increase the tension in the wire so that there is no sag?

**76.** Parachutists whose chutes have failed to open have been known to survive if they land in deep snow. Assume that a 75-kg parachutist hits the ground with an area of impact of 0.30 m$^2$ at a velocity of 55 m/s, and that the ultimate strength of body tissue is $5 \times 10^5$ N/m$^2$. Assume that the person is brought to rest in 1.0 m of snow. Show that the person may escape serious injury.

**77.** If the left vertical support column in Example 9–5 is made of steel, what is its cross-sectional area? Assume that a safety factor of 3 was used in its design to avoid fracture.

**78.** The mobile in Fig. 9–90 is in equilibrium. Object B has mass of 0.748 kg. Determine the masses of objects A, C, and D. (Neglect the weights of the crossbars.)

30.00 cm | 7.50 cm

15.00 cm | 5.00 cm

A

17.50 cm | 5.00 cm

B

D

C

**FIGURE 9–90**
Problem 78.

**79.** In a mountain-climbing technique called the "Tyrolean traverse," a rope is anchored on both ends (to rocks or strong trees) across a deep chasm, and then a climber traverses the rope while attached by a sling as in Fig. 9–91. This technique generates tremendous forces in the rope and anchors, so a basic understanding of physics is crucial for safety. A typical climbing rope can undergo a tension force of perhaps 29 kN before breaking, and a "safety factor" of 10 is usually recommended. The length of rope used in the Tyrolean traverse must allow for some "sag" to remain in the recommended safety range. Consider a 75-kg climber at the center of a Tyrolean traverse, spanning a 25-m chasm. (*a*) To be within its recommended safety range, what minimum distance $x$ must the rope sag? (*b*) If the Tyrolean traverse is set up incorrectly so that the rope sags by only one-fourth the distance found in (*a*), determine the tension in the rope. Ignore stretching of the rope. Will the rope break?

25 m

$x$

75 kg

**FIGURE 9–91**
Problem 79.

**80.** When a wood shelf of mass 6.6 kg is fastened inside a slot in a vertical support as shown in Fig. 9–92, the support exerts a torque on the shelf. (*a*) Draw a free-body diagram for the shelf, assuming three vertical forces (two exerted by the support slot—explain why). Then calculate (*b*) the magnitudes of the three forces and (*c*) the torque exerted by the support (about the left end of the shelf).

**FIGURE 9–92**
Problem 80.

32.0 cm
3.0 cm
2.0 cm

**81.** A cubic crate of side $s = 2.0$ m is top-heavy: its CG is 18 cm above its true center. How steep an incline can the crate rest on without tipping over? [*Hint*: The normal force would act at the lowest corner.]

# Search and Learn

1. Stand facing a wall with your toes tight against the wall, and go up on your tiptoes. Then turn around and push your back against the wall with your heels tight against the wall. Using the ideas of Section 9–4, explain why you can or cannot perform these motions.

2. From what minimum height must a 1.2-kg rectangular brick 15.0 cm × 6.0 cm × 4.0 cm be dropped above a rigid steel floor in order to break the brick? Assume the brick strikes the floor directly on its largest face, and that the compression of the brick is much greater than that of the steel (that is, ignore compression of the steel). State other simplifying assumptions that may be necessary.

3. Suppose a 65-kg person jumps from a height of 3.0 m down to the ground. (*a*) What is the speed of the person just before landing (Chapter 2)? (*b*) Estimate the average force on the person's feet exerted by the ground to bring the person to rest, if the knees are bent so the person's CG moves a distance $d = 50$ cm during the deceleration period (Fig. 9–93). [*Hint*: This force exerted by the ground ≠ net force. You may want to consult Chapters 2, 4, and 7, and be sure to draw a careful free-body diagram of the person.] (*c*) Estimate the decelerating force if the person lands stiff-legged so $d \approx 1.0$ cm. (*d*) Estimate the stress in the tibia (a lower leg bone of area = $3.0 \times 10^{-4}$ m²), and determine whether or not the bone will break if the landing is made with bent legs ($d = 50$ cm). (*e*) Estimate the stress and determine if the tibia will break in a stiff-legged landing ($d = 1.0$ cm).

$v = ?$

$v = 0$

**FIGURE 9–93**
Search and Learn 3.  $y = 0$ —

4. In Example 9–6, the torque is calculated around the axis where the cable is attached to the beam. (*a*) By using the $\Sigma \tau = 0$ equation with this axis of rotation, how do you know that the vertical force on the hinge points up and not down? (*b*) What advantage would be gained in solving part (*a*) if the axis of rotation were chosen around the hinge instead of around the point where the cable is attached? (*c*) Show that you get the same answer as in Example 9–6 if you solve the problem as in part (*b*). (*d*) In general, do you see any patterns in Sections 9–2 and 9–3 for choosing the axis of rotation to solve Problems in this Chapter?

5. Consider a ladder with a painter climbing up it (Fig. 9–94). The mass of the uniform ladder is 12.0 kg, and the mass of the painter is 55.0 kg. If the ladder begins to slip at its base when the painter's feet are 70% of the way up the length of the ladder, what is the coefficient of static friction between the ladder and the floor? Assume the wall is frictionless.

$\vec{F}_W$

$\vec{F}_{Cy}$  $\vec{F}_C$

$m\vec{g}$

$M\vec{g}$

$\vec{F}_{Cx}$

4.0 m

2.8 m

3.0 m

2.1 m

**FIGURE 9–94**
Search and Learn 5.

## ANSWERS TO EXERCISES

**A:** $F_A$ also has a component to balance the sideways force $F_B$.

**B:** Yes: $\cos\theta$ (angle of bar with ground) appears on both sides and cancels out.

**C:** $F_N = m_A g + m_B g + M g$
= $(30 \text{ kg} + 25 \text{ kg} + 4.0 \text{ kg}) g = 560$ N.

**D:** (*a*).

**E:** (*b*).

1929

2009

We start our study with fluids at rest, such as water in a glass or a lake. Pressure in a fluid increases with depth, a fact that allows less dense objects to float—the pressure underneath is higher than on top. When fluids flow, such as water or air, interesting effects occur because the pressure in the fluid is lower where the fluid velocity is higher (Bernoulli's principle).

The great mass of a glacier's ice (photos here) moves slowly, like a viscous liquid. The dark lines are "moraines," made up of rock broken off mountain walls by the moving ice, and represent streamlines. The two photos, taken in 1929 and 2009 by Italian expeditions to the mountain K2 (on the right in the distance), show the same glacier has become less thick, presumably due to global warming.

# CHAPTER 10

# Fluids

## CONTENTS

**CHAPTER-OPENING QUESTIONS—Guess now!**

**1.** Which container has the largest pressure at the bottom? Assume each container holds the same volume of water.

**(a)**   **(b)**   **(c)**   **(d)**   **(e)**

The pressures are equal.

**2.** Two balloons are tied and hang with their nearest edges about 3 cm apart. If you blow between the balloons (not *at* the balloons, but at the opening between them), what will happen?

**(a)** Nothing.

**(b)** The balloons will move closer together.

**(c)** The balloons will move farther apart.

In previous Chapters we considered objects that were solid and assumed to maintain their shape except for a small amount of elastic deformation. We sometimes treated objects as point particles. Now we are going to shift our attention to materials that are very deformable and can flow. Such "fluids" include liquids and gases. We will examine fluids both at rest (fluid statics) and in motion (fluid dynamics).

# 10–1 Phases of Matter

The three common **phases**, or **states**, of matter are solid, liquid, and gas. A simple way to distinguish these three phases is as follows. A **solid** maintains a generally fixed size and shape; usually it requires a large force to change the volume or shape of a solid[†] (although a thin object might bend). A **liquid** does not maintain a fixed shape—it takes on the shape of its container, and it can flow; but like a solid it is not readily compressible, and its volume can be changed significantly only by a very large force. A **gas** has neither a fixed shape nor a fixed volume—it will expand to fill its container. For example, when air is pumped into an automobile tire, the air does not all run to the bottom of the tire as a liquid would; it spreads out to fill the whole volume of the tire.

Because liquids and gases do not maintain a fixed shape, they both have the ability to flow. They are thus referred to collectively as **fluids**.

The division of matter into three phases is not always simple. How, for example, should butter be classified? Furthermore, a fourth phase of matter can be distinguished, the **plasma** phase, which occurs only at very high temperatures and consists of ionized atoms (electrons separated from the nuclei). Some scientists believe that *colloids* (suspensions of tiny particles in a liquid) should also be considered a separate phase of matter. **Liquid crystals**, used in TV, cell phone, and computer screens, can be considered a phase of matter in between solids and liquids. For now, we will be interested in the three ordinary phases of matter.

# 10–2 Density and Specific Gravity

It is sometimes said that iron is "heavier" than wood. This cannot really be true since a large log clearly weighs more than an iron nail. What we should say is that iron is more *dense* than wood.

The **density**, $\rho$, of a substance ($\rho$ is the lowercase Greek letter rho) is defined as its mass per unit volume:

$$\rho = \frac{m}{V}, \tag{10–1}$$

where $m$ is the mass of a sample of the substance and $V$ its volume. Density is a characteristic property of any pure substance. Objects made of a particular pure substance, such as pure gold, can have any size or mass, but the density will be the same for each.

We can use the concept of density, Eq. 10–1, to write the mass of an object as

$$m = \rho V,$$

and the weight of an object as

$$mg = \rho V g.$$

The SI unit for density is $kg/m^3$. Sometimes densities are given in $g/cm^3$. Note that a density given in $g/cm^3$ must be multiplied by 1000 to give the result in $kg/m^3$ $[1\,kg/m^3 = 1000\,g/(100\,cm)^3 = 10^3\,g/10^6\,cm^3 = 10^{-3}\,g/cm^3]$. For example, the density of aluminum is $\rho = 2.70\,g/cm^3$, which equals $2700\,kg/m^3$. The densities of various substances are given in Table 10–1. The Table specifies temperature and atmospheric pressure because they affect density (the effect is slight for liquids and solids). Note that air is about 1000 times less dense than water.

---

**EXAMPLE 10–1 Mass, given volume and density.** What is the mass of a solid iron wrecking ball of radius 18 cm?

**APPROACH** First we use the standard formula $V = \frac{4}{3}\pi r^3$ (see inside rear cover) to obtain the sphere's volume. Then Eq. 10–1 and Table 10–1 give us the mass $m$.

**SOLUTION** The volume of the sphere is

$$V = \frac{4}{3}\pi r^3 = \frac{4}{3}(3.14)(0.18\,m)^3 = 0.024\,m^3.$$

From Table 10–1, the density of iron is $\rho = 7800\,kg/m^3$, so Eq. 10–1 gives

$$m = \rho V = (7800\,kg/m^3)(0.024\,m^3) = 190\,kg.$$

---

**TABLE 10–1
Densities of Substances[‡]**

| Substance | Density, $\rho$ (kg/m³) |
|---|---|
| *Solids* | |
| Aluminum | $2.70 \times 10^3$ |
| Iron and steel | $7.8\ \times 10^3$ |
| Copper | $8.9\ \times 10^3$ |
| Lead | $11.3\ \times 10^3$ |
| Gold | $19.3\ \times 10^3$ |
| Concrete | $2.3\ \times 10^3$ |
| Granite | $2.7\ \times 10^3$ |
| Wood (typical) | $0.3 – 0.9 \times 10^3$ |
| Glass, common | $2.4 – 2.8 \times 10^3$ |
| Ice ($H_2O$) | $0.917 \times 10^3$ |
| Bone | $1.7 – 2.0 \times 10^3$ |
| *Liquids* | |
| Water (4°C) | $1.000 \times 10^3$ |
| Sea water | $1.025 \times 10^3$ |
| Blood, plasma | $1.03\ \times 10^3$ |
| Blood, whole | $1.05\ \times 10^3$ |
| Mercury | $13.6\ \times 10^3$ |
| Alcohol, ethyl | $0.79\ \times 10^3$ |
| Gasoline | $0.7 – 0.8 \times 10^3$ |
| *Gases* | |
| Air | 1.29 |
| Helium | 0.179 |
| Carbon dioxide | 1.98 |
| Water (steam) (100°C) | 0.598 |

[‡]Densities are given at 0°C and 1 atm pressure unless otherwise specified.

---

[†]Section 9–5.

The **specific gravity** of a substance is defined as the ratio of the density of that substance to the density of water at 4.0°C. Because specific gravity (abbreviated SG) is a ratio, it is a simple number without dimensions or units. For example (see Table 10–1), the specific gravity of lead is 11.3 [$(11.3 \times 10^3 \text{ kg/m}^3)/(1.00 \times 10^3 \text{ kg/m}^3)$]. The SG of alcohol is 0.79.

The concepts of density and specific gravity are especially helpful in the study of fluids because we are not always dealing with a fixed volume or mass.

# 10–3 Pressure in Fluids

Pressure and force are related, but they are not the same thing. **Pressure** is defined as force per unit area, where the force $F$ is understood to be the magnitude of the force acting perpendicular to the surface area $A$:

$$\text{pressure} = P = \frac{F}{A}. \qquad \textbf{(10–2)}$$

Although force is a vector, pressure is a scalar. Pressure has magnitude only. The SI unit of pressure is $\text{N/m}^2$. This unit has the official name **pascal** (Pa), in honor of Blaise Pascal (see Section 10–5); that is, $1 \text{ Pa} = 1 \text{ N/m}^2$. However, for simplicity, we will often use $\text{N/m}^2$. Other units sometimes used are $\text{dynes/cm}^2$, and $\text{lb/in.}^2$ (pounds per square inch, abbreviated "psi"). Several other units for pressure are discussed in Sections 10–4 and 10–6, along with conversions between them (see also the Table inside the front cover).

---

**EXAMPLE 10–2** | **Calculating pressure.** A 60-kg person's two feet cover an area of 500 cm². (*a*) Determine the pressure exerted by the two feet on the ground. (*b*) If the person stands on one foot, what will be the pressure under that foot?

**APPROACH** Assume the person is at rest. Then the ground pushes up on her with a force equal to her weight $mg$, and she exerts a force $mg$ on the ground where her feet (or foot) contact it. Because $1 \text{ cm}^2 = (10^{-2} \text{ m})^2 = 10^{-4} \text{ m}^2$, then $500 \text{ cm}^2 = 0.050 \text{ m}^2$.

**SOLUTION** (*a*) The pressure on the ground exerted by the two feet is

$$P = \frac{F}{A} = \frac{mg}{A} = \frac{(60 \text{ kg})(9.8 \text{ m/s}^2)}{(0.050 \text{ m}^2)} = 12 \times 10^3 \text{ N/m}^2.$$

(*b*) If the person stands on one foot, the force is still equal to the person's weight, but the area will be half as much, so the pressure will be twice as much: $24 \times 10^3 \text{ N/m}^2$.

---

**FIGURE 10–1** Pressure is the same in every direction in a nonmoving fluid at a given depth. If this weren't true, the fluid would be in motion.

**FIGURE 10–2** If there were a component of force parallel to the solid surface of the container, the liquid would move in response to it. For a liquid at rest, $F_\| = 0$.

Pressure is particularly useful for dealing with fluids. It is an experimental observation that *a fluid exerts pressure in every direction*. This is well known to swimmers and divers who feel the water pressure on all parts of their bodies. At any depth in a fluid at rest, the pressure is the same in all directions at that given depth. To see why, consider a tiny cube of the fluid (Fig. 10–1) which is so small that we can consider it a point and can ignore the force of gravity on it. The pressure on one side of it must equal the pressure on the opposite side. If this weren't true, there would be a net force on the cube and it would start moving. If the fluid is not flowing, then the pressures must be equal.

For a fluid at rest, the force due to fluid pressure always acts *perpendicular* to any solid surface it touches. If there were a component of the force parallel to the surface, as shown in Fig. 10–2, then according to Newton's third law the solid surface would exert a force back on the fluid, which would cause the fluid to flow—in contradiction to our assumption that the fluid is at rest. Thus the force due to the pressure in a fluid at rest is always perpendicular to the surface.

We now calculate quantitatively how the pressure in a liquid of uniform density varies with depth. Let us look at a depth $h$ below the surface of the liquid as shown in Fig. 10–3 (that is, the liquid's top surface is a height $h$ above this level). The pressure due to the liquid at this depth $h$ is due to the weight of the column of liquid above it. Thus the force due to the weight of liquid acting on the area $A$ is $F = mg = (\rho V)g = \rho A h g$, where $Ah$ is the volume of the column of liquid, $\rho$ is the density of the liquid (assumed to be constant), and $g$ is the acceleration of gravity. The pressure $P$ due to the weight of liquid is then

$$P = \frac{F}{A} = \frac{\rho A h g}{A}$$

$$P = \rho g h. \qquad\qquad \text{[liquid]} \quad \textbf{(10–3a)}$$

**FIGURE 10–3** Calculating the pressure at a depth $h$ in a liquid, due to the weight of the liquid above.

Note that the area $A$ doesn't affect the pressure at a given depth. The fluid pressure is directly proportional to the density of the liquid and to the depth within the liquid. In general, *the pressure at equal depths within a uniform liquid is the same.*

**EXERCISE A** Return to Chapter-Opening Question 1, page 260, and answer it again now. Try to explain why you may have answered differently the first time.

Equation 10–3a is extremely useful. It is valid for fluids whose density is constant and does not change with depth—that is, if the fluid is *incompressible*. This is usually a good approximation for liquids (although at great depths in the ocean, the density of water is increased some by compression due to the great weight of water above).

If the density of a fluid does vary, a useful relation can be found by considering a thin horizontal slab of the fluid of thickness $\Delta h = h_2 - h_1$. The pressure on the top of the slab, at depth $h_1$, is $P_1 = \rho g h_1$. The pressure on the bottom of the slab (pushing upward), at depth $h_2$, is $P_2 = \rho g h_2$. The difference in pressure is

$$\Delta P = P_2 - P_1 = \rho g (h_2 - h_1)$$

or

$$\Delta P = \rho g \, \Delta h. \qquad [\rho \approx \text{constant over } \Delta h] \quad \textbf{(10–3b)}$$

Equation 10–3b tells us how the pressure changes over a small change in depth ($\Delta h$) within a fluid, even if compressible.

Gases are very compressible, and density can vary significantly with depth. For this more general case, in which $\rho$ may vary, we need to use Eq. 10–3b where $\Delta h$ should be small if $\rho$ varies significantly with depth (or height).

**EXAMPLE 10–3** **Pressure at a faucet.** The surface of the water in a storage tank is 30 m above a water faucet in the kitchen of a house, Fig. 10–4. Calculate the difference in water pressure between the faucet and the surface of the water in the tank.

**APPROACH** Water is practically incompressible, so $\rho$ is constant even for a $\Delta h = 30 \text{ m}$ when used in Eq. 10–3b. Only $\Delta h$ matters; we can ignore the "route" of the pipe and its bends.

**SOLUTION** We assume the atmospheric pressure at the surface of the water in the storage tank is the same as at the faucet. So, the water pressure difference between the faucet and the surface of the water in the tank is

$$\Delta P = \rho g \, \Delta h = (1.0 \times 10^3 \text{ kg/m}^3)(9.8 \text{ m/s}^2)(30 \text{ m}) - 2.9 \times 10^5 \text{ N/m}^2.$$

**NOTE** The height $\Delta h$ is sometimes called the **pressure head**. In this Example, the head of water is 30 m at the faucet. The very different diameters of the tank and faucet don't affect the result—only height does.

**EXERCISE B** A dam holds back a lake that is 85 m deep at the dam. If the lake is 20 km long, how much thicker should the dam be than if the lake were smaller, only 1.0 km long?

**PHYSICS APPLIED**
*Water supply*

**FIGURE 10–4** Example 10–3.

$\Delta h = 30 \text{ m}$

# 10–4 Atmospheric Pressure and Gauge Pressure

## Atmospheric Pressure

The pressure of the Earth's atmosphere, as in any fluid, changes with depth. But the Earth's atmosphere is somewhat complicated: not only does the density of air vary greatly with altitude but there is no distinct top surface to the atmosphere from which $h$ (in Eq. 10–3a) could be measured. We can, however, calculate the approximate difference in pressure between two altitudes above Earth's surface using Eq. 10–3b.

The pressure of the air at a given place varies slightly according to the weather. At sea level, the pressure of the atmosphere on average is $1.013 \times 10^5 \, \text{N/m}^2$ (or $14.7 \, \text{lb/in.}^2$). This value lets us define a commonly used unit of pressure, the **atmosphere** (abbreviated atm):

$$1 \, \text{atm} = 1.013 \times 10^5 \, \text{N/m}^2 = 101.3 \, \text{kPa}.$$

Another unit of pressure sometimes used (in meteorology and on weather maps) is the **bar**, which is defined as

$$1 \, \text{bar} = 1.000 \times 10^5 \, \text{N/m}^2.$$

Thus standard atmospheric pressure is slightly more than 1 bar.

The pressure due to the weight of the atmosphere is exerted on all objects immersed in this great sea of air, including our bodies. How does a human body withstand the enormous pressure on its surface? The answer is that living cells maintain an internal pressure that closely equals the external pressure, just as the pressure inside a balloon closely matches the outside pressure of the atmosphere. An automobile tire, because of its rigidity, can maintain internal pressures much greater than the external pressure.

(a)          (b)

**FIGURE 10–5** Example 10–4.

> **CONCEPTUAL EXAMPLE 10–4**   **Finger holds water in a straw.** You insert a straw of length $\ell$ into a tall glass of water. You place your finger over the top of the straw, capturing some air above the water but preventing any additional air from getting in or out, and then you lift the straw from the water. You find that the straw retains most of the water (Fig. 10–5a). Does the air in the space between your finger and the top of the water have a pressure $P$ that is greater than, equal to, or less than, the atmospheric pressure $P_0$ outside the straw?
>
> **RESPONSE** Consider the forces on the column of water (Fig. 10–5b). Atmospheric pressure outside the straw pushes upward on the water at the bottom of the straw, gravity pulls the water downward, and the air pressure inside the top of the straw pushes downward on the water. Since the water is in equilibrium, the upward force due to atmospheric pressure $P_0$ must balance the two downward forces. The only way this is possible is for the air pressure $P$ inside the straw at the top to be *less than* the atmosphere pressure outside the straw. (When you initially remove the straw from the water glass, a little water may leave the bottom of the straw, thus increasing the volume of trapped air and reducing its density and pressure.)

## Gauge Pressure

It is important to note that tire gauges, and most other pressure gauges, register the pressure above and beyond atmospheric pressure. This is called **gauge pressure**. Thus, to get the **absolute pressure**, $P$, we must add the atmospheric pressure, $P_0$, to the gauge pressure, $P_G$:

$$P = P_G + P_0.$$

If a tire gauge registers 220 kPa, the absolute pressure within the tire is 220 kPa + 101 kPa = 321 kPa, equivalent to about 3.2 atm (2.2 atm gauge pressure).

# 10–5 Pascal's Principle

The Earth's atmosphere exerts a pressure on all objects with which it is in contact, including other fluids. External pressure acting on a fluid is transmitted throughout that fluid. For instance, according to Eq. 10–3a, the pressure due to the water at a depth of 100 m below the surface of a lake is $P = \rho g \, \Delta h = (1000 \, \text{kg/m}^3)(9.8 \, \text{m/s}^2)(100 \, \text{m}) = 9.8 \times 10^5 \, \text{N/m}^2$, or 9.7 atm. However, the total pressure at this point is due to the pressure of water plus the pressure of the air above it. Hence the total pressure (if the lake is near sea level) is 9.7 atm + 1.0 atm = 10.7 atm. This is just one example of a general principle attributed to the French philosopher and scientist Blaise Pascal (1623–1662). **Pascal's principle** states that *if an external pressure is applied to a confined fluid, the pressure at every point within the fluid increases by that amount.*

A number of practical devices make use of Pascal's principle. One example is the hydraulic lift, illustrated in Fig. 10–6a, in which a small input force is used to exert a large output force by making the area of the output piston larger than the area of the input piston. To see how this works, we assume the input and output pistons are at the same height (at least approximately). Then the external input force $F_{in}$, by Pascal's principle, increases the pressure equally throughout. Therefore, at the same level (see Fig. 10–6a),

$$P_{out} = P_{in}$$

where the input quantities are represented by the subscript "in" and the output by "out." Since $P = F/A$, we write the above equality as

$$\frac{F_{out}}{A_{out}} = \frac{F_{in}}{A_{in}},$$

or

$$\frac{F_{out}}{F_{in}} = \frac{A_{out}}{A_{in}}.$$

The quantity $F_{out}/F_{in}$ is called the **mechanical advantage** of the hydraulic lift, and it is equal to the ratio of the areas. For example, if the area of the output piston is 20 times that of the input cylinder, the force is multiplied by a factor of 20. Thus a force of 200 lb could lift a 4000-lb car.

**FIGURE 10–6** Applications of Pascal's principle: (a) hydraulic lift; (b) hydraulic brakes in a car.

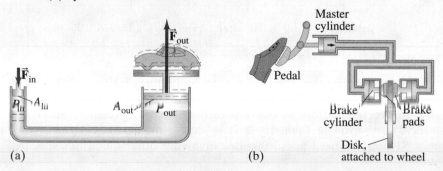

(a)        (b)

Master cylinder

Pedal

Brake cylinder

Brake pads

Disk, attached to wheel

Figure 10–6b illustrates the brake system of a car. When the driver presses the brake pedal, the pressure in the master cylinder increases. This pressure increase occurs throughout the brake fluid, thus pushing the brake pads against the disk attached to the car's wheel.

**(a)** Open-tube manometer

$P_0$

$\Delta h$

$P$
(Pressure being measured)

Air pressure

Flexible chamber

**(b)** Aneroid gauge (used mainly for air pressure, and then called an aneroid barometer)

Scale reading, gauge pressure

Atmospheric pressure

Spring

Pressure of air in tire

**(c)** Tire gauge

**FIGURE 10–7** Pressure gauges: (a) open-tube manometer, (b) aneroid gauge, and (c) common tire pressure gauge.

# 10–6 Measurement of Pressure; Gauges and the Barometer

Many devices have been invented to measure pressure, some of which are shown in Fig. 10–7. The simplest is the *open-tube* **manometer** (Fig. 10–7a) which is a U-shaped tube partially filled with a liquid, usually mercury or water. The pressure $P$ being measured is related (by Eq. 10–3b) to the difference in height $\Delta h$ of the two levels of the liquid by the relation

$$P = P_0 + \rho g \, \Delta h, \qquad \text{[manometer]} \quad \textbf{(10–3c)}$$

where $P_0$ is atmospheric pressure (acting on the top of the liquid in the left-hand tube), and $\rho$ is the density of the liquid. Note that the quantity $\rho g \, \Delta h$ is the gauge pressure—the amount by which $P$ exceeds atmospheric pressure $P_0$. If the liquid in the left-hand column were lower than that in the right-hand column, $P$ would have to be less than atmospheric pressure (and $\Delta h$ would be negative).

Instead of calculating the product $\rho g \, \Delta h$, sometimes only the change in height $\Delta h$ is specified. In fact, pressures are sometimes specified as so many "millimeters of mercury" (mm-Hg) or "mm of water" (mm-H$_2$O). The unit mm-Hg is equivalent to a pressure of 133 N/m$^2$, because $\rho g \, \Delta h$ for 1 mm ($= 1.0 \times 10^{-3}$ m) of mercury gives

$$\rho g \, \Delta h = (13.6 \times 10^3 \, \text{kg/m}^3)(9.80 \, \text{m/s}^2)(1.00 \times 10^{-3} \, \text{m})$$
$$= 1.33 \times 10^2 \, \text{N/m}^2.$$

The unit mm-Hg is also called the **torr** in honor of Evangelista Torricelli (1608–1647), a student of Galileo's who invented the barometer (see top of next page). Conversion factors among the various units of pressure (an incredible nuisance!) are given in Table 10–2. It is important that only N/m$^2$ = Pa, the proper SI unit, be used in calculations involving other quantities specified in SI units.

Another type of pressure gauge is the **aneroid gauge** (Fig. 10–7b) in which the pointer is linked to the flexible ends of an evacuated thin metal chamber. In electronic gauges, the pressure may be applied to a thin metal diaphragm whose resulting deformation is translated into an electrical signal by a transducer. A common tire gauge uses a spring, as shown in Fig. 10–7c.

**PROBLEM SOLVING**

*Use SI unit in calculations:*
1 Pa = 1 N/m$^2$

## TABLE 10–2 Conversion Factors Between Different Units of Pressure

| In Terms of 1 Pa = 1 N/m² | 1 atm in Different Units |
|---|---|
| 1 atm = $1.013 \times 10^5$ N/m² | 1 atm = $1.013 \times 10^5$ N/m² |
| = $1.013 \times 10^5$ Pa = 101.3 kPa | |
| 1 bar = $1.000 \times 10^5$ N/m² | 1 atm = 1.013 bar |
| 1 dyne/cm² = 0.1 N/m² | 1 atm = $1.013 \times 10^6$ dyne/cm² |
| 1 lb/in.² = $6.90 \times 10^3$ N/m² | 1 atm = 14.7 lb/in.² |
| 1 lb/ft² = 47.9 N/m² | 1 atm = $2.12 \times 10^3$ lb/ft² |
| 1 cm-Hg = $1.33 \times 10^3$ N/m² | 1 atm = 76.0 cm-Hg |
| 1 mm-Hg = 133 N/m² | 1 atm = 760 mm-Hg |
| 1 torr = 133 N/m² | 1 atm = 760 torr |
| 1 mm-H₂O (4°C) = 9.80 N/m² | 1 atm = $1.03 \times 10^4$ mm-H₂O (4°C) |
| | ≈ 10 m of water |

Atmospheric pressure can be measured by a modified kind of mercury manometer with one end closed, called a mercury **barometer** (Fig. 10–8). The glass tube is completely filled with mercury and then inverted into the bowl of mercury. If the tube is long enough, the level of the mercury will drop, leaving a vacuum at the top of the tube, since atmospheric pressure can support a column of mercury only about 76 cm high (exactly 76.0 cm at standard atmospheric pressure). That is, a column of mercury 76 cm high exerts the same pressure as the atmosphere[†]:

$$P = \rho g \, \Delta h$$
$$= (13.6 \times 10^3 \, \text{kg/m}^3)(9.80 \, \text{m/s}^2)(0.760 \, \text{m}) = 1.013 \times 10^5 \, \text{N/m}^2 = 1.00 \, \text{atm}.$$

**FIGURE 10–8** A mercury barometer, invented by Torricelli, is shown here when the air pressure is standard atmospheric, 76.0 cm-Hg.

Household barometers are usually of the aneroid type (Fig. 10–7b), either mechanical (with dial) or electronic.

A calculation similar to that just done will show that atmospheric pressure can maintain a column of water 10.3 m high in a tube whose top is under vacuum (Fig. 10–9). No matter how good a vacuum pump is, water cannot be made to rise more than about 10 m under normal atmospheric pressure. To pump water out of deep mine shafts with a vacuum pump requires multiple stages for depths greater than 10 m. Galileo studied this problem, and his student Torricelli was the first to explain it. The point is that a pump does not really suck water up a tube—it merely reduces the pressure at the top of the tube. Atmospheric air pressure *pushes* the water up the tube if the top end is at low pressure (under a vacuum), just as it is air pressure that pushes (or maintains) the mercury 76 cm high in a barometer. [Force pumps, Section 10–14, can push higher.]

**FIGURE 10–9** A water barometer: a full tube of water (longer than 10 m), closed at the top, is inserted into a tub of water. When the submerged bottom end of the tube is unplugged, some water flows out of the tube into the tub, leaving a vacuum at the top of the tube above the water's upper surface. Why? Because air pressure can support a column of water only 10 m high.

| CONCEPTUAL EXAMPLE 10–5 | **Suction.** A novice engineer proposes suction cup shoes for space shuttle astronauts working on the exterior of a spacecraft. Having just studied this Chapter, you gently remind him of the fallacy of this plan. What is it?

**RESPONSE** Suction cups work by pushing out the air underneath the cup. What holds the suction cup in place is the air pressure outside it. (This can be a substantial force when on Earth. For example, a 10-cm-diameter suction cup has an area of $7.9 \times 10^{-3}$ m². The force of the atmosphere on it is $(7.9 \times 10^{-3} \, \text{m}^2)(1.0 \times 10^5 \, \text{N/m}^2) \approx 800 \, \text{N}$, about 180 lbs!) But in outer space, there is no air pressure to push the suction cup onto the spacecraft.

We sometimes mistakenly think of suction as something we actively do. For example, we intuitively think that we pull the soda up through a straw. Instead, what we do is lower the pressure at the top of the straw, and the atmosphere *pushes* the soda up the straw.

[†]This calculation confirms the entry in Table 10–2, 1 atm = 76.0 cm-Hg.

# 10–7 Buoyancy and Archimedes' Principle

Objects submerged in a fluid appear to weigh less than they do when outside the fluid. For example, a large rock that you would have difficulty lifting off the ground can often be easily lifted from the bottom of a stream. When you lift the rock through the surface of the water, it suddenly seems to be much heavier. Many objects, such as wood, float on the surface of water. These are two examples of **buoyancy**. In each example, the force of gravity is acting downward. But in addition, an upward *buoyant force* is exerted by the liquid. The buoyant force on fish and underwater divers almost exactly balances the force of gravity downward, and allows them to "hover" in equilibrium.

The buoyant force occurs because the pressure in a fluid increases with depth. Thus the upward pressure on the bottom surface of a submerged object is greater than the downward pressure on its top surface. To see this effect, consider a cylinder of height $\Delta h$ whose top and bottom ends have an area $A$ and which is completely submerged in a fluid of density $\rho_F$, as shown in Fig. 10–10. The fluid exerts a pressure $P_1 = \rho_F g h_1$ at the top surface of the cylinder (Eq. 10–3a).

**FIGURE 10–10** Determination of the buoyant force.

The force due to this pressure on top of the cylinder is $F_1 = P_1 A = \rho_F g h_1 A$, and it is directed downward. Similarly, the fluid exerts an upward force on the bottom of the cylinder equal to $F_2 = P_2 A = \rho_F g h_2 A$. The net force on the cylinder exerted by the fluid pressure, which is the **buoyant force**, $\vec{F}_B$, acts upward and has the magnitude

$$
\begin{aligned}
F_B = F_2 - F_1 &= \rho_F g A (h_2 - h_1) \\
&= \rho_F g A \, \Delta h \\
&= \rho_F V g \\
&= m_F g,
\end{aligned}
$$

where $V = A \, \Delta h$ is the volume of the cylinder; the product $\rho_F V$ is the mass of the fluid displaced, and $\rho_F V g = m_F g$ is the weight of fluid which takes up a volume equal to the volume of the cylinder. Thus the buoyant force on the cylinder is equal to the weight of fluid displaced by the cylinder.

This result is valid no matter what the shape of the object. Its discovery is credited to Archimedes (287?–212 B.C.), and it is called **Archimedes' principle**:

**the buoyant force on an object immersed in a fluid is equal to the weight of the fluid displaced by that object.**

By "fluid displaced," we mean a volume of fluid equal to the submerged volume of the object (or that part of the object that is submerged). If the object is placed in a glass or tub initially filled to the brim with water, the water that flows over the top represents the water displaced by the object.

FIGURE 10–11
Archimedes' principle.

We can derive Archimedes' principle in general by the following simple but elegant argument. The irregularly shaped object D shown in Fig. 10–11a is acted on by the force of gravity (its weight, $m\vec{\mathbf{g}}$, downward) and the buoyant force, $\vec{\mathbf{F}}_B$, upward. We wish to determine $F_B$. To do so, we next consider a body (D' in Fig. 10–11b), this time made of the fluid itself, with the same shape and size as the original object, and located at the same depth. You might think of this body of fluid as being separated from the rest of the fluid by an imaginary membrane. The buoyant force $F_B$ on this body of fluid will be exactly the same as that on the original object since the surrounding fluid, which exerts $F_B$, is in exactly the same configuration. This body of fluid D' is in equilibrium (the fluid as a whole is at rest). Therefore, $F_B = m'g$, where $m'g$ is the weight of the body of fluid D'. Hence the buoyant force $F_B$ is equal to the weight of the body of fluid whose volume equals the volume of the original submerged object, which is Archimedes' principle.

Archimedes' discovery was made by experiment. What we have done is show that Archimedes' principle can be derived from Newton's laws.

---

**CONCEPTUAL EXAMPLE 10–6** | **Two pails of water.** Consider two identical pails of water filled to the brim. One pail contains only water, the other has a piece of wood floating in it. Which pail has the greater weight?

**RESPONSE** Both pails weigh the same. Recall Archimedes' principle: the wood displaces a volume of water with weight equal to the weight of the wood. Some water will overflow the pail, but Archimedes' principle tells us the spilled water has weight equal to the weight of the wood; so the pails have the same weight.

---

**EXAMPLE 10–7** | **Recovering a submerged statue.** A 70-kg ancient statue lies at the bottom of the sea. Its volume is $3.0 \times 10^4 \, cm^3$. How much force is needed to lift it (without acceleration)?

**APPROACH** The force $F$ needed to lift the statue is equal to the statue's weight $mg$ minus the buoyant force $F_B$. Figure 10–12 is the free-body diagram.

**SOLUTION** We apply Newton's second law, $\Sigma F = ma = 0$, which gives $F + F_B - mg = 0$ or

$$F = mg - F_B.$$

The buoyant force on the statue due to the water is equal to the weight of $3.0 \times 10^4 \, cm^3 = 3.0 \times 10^{-2} \, m^3$ of water (for seawater, $\rho = 1.025 \times 10^3 \, kg/m^3$):

$$F_B = m_{H_2O} g = \rho_{H_2O} Vg = (1.025 \times 10^3 \, kg/m^3)(3.0 \times 10^{-2} \, m^3)(9.8 \, m/s^2)$$
$$= 3.0 \times 10^2 \, N,$$

where we use the chemical symbol for water, $H_2O$, as a subscript. The weight of the statue is $mg = (70 \, kg)(9.8 \, m/s^2) = 6.9 \times 10^2 \, N$. Hence the force $F$ needed to lift it is $690 \, N - 300 \, N = 390 \, N$. It is as if the statue had a mass of only $(390 \, N)/(9.8 \, m/s^2) = 40 \, kg$.

**NOTE** Here $F = 390 \, N$ is the force needed to lift the statue without acceleration when it is under water. As the statue comes *out* of the water, the force $F$ increases, reaching 690 N when the statue is fully out of the water.

**FIGURE 10–12** Example 10–7. The force needed to lift the statue is $\vec{\mathbf{F}}$.

Archimedes is said to have discovered his principle in his bath while thinking how he might determine whether the king's new crown was pure gold or a fake. Gold has a specific gravity of 19.3, somewhat higher than that of most metals, but a determination of specific gravity or density is not readily done directly because, even if the mass is known, the volume of an irregularly shaped object is not easily calculated. However, if the object is weighed in air ($= w$) and also "weighed" while it is under water ($= w'$), the density can be determined using Archimedes' principle, as the following Example shows. The quantity $w'$ is called the **apparent weight** in water, and is what a scale reads when the object is submerged in water (see Fig. 10–13); $w'$ equals the true weight ($w = mg$) minus the buoyant force.

**EXAMPLE 10–8** **Archimedes: Is the crown gold?** When a crown of mass 14.7 kg is submerged in water, an accurate scale reads only 13.4 kg. Is the crown made of gold?

**APPROACH** If the crown is gold, its density and specific gravity must be very high, SG = 19.3 (see Section 10–2 and Table 10–1). We determine the specific gravity using Archimedes' principle and the two free-body diagrams shown in Fig. 10–13.

**SOLUTION** The *apparent weight* of the submerged object (the crown) is $w'$ (what the scale reads), and is the force pulling down on the scale hook. By Newton's third law, $w'$ equals the force $F'_T$ that the scale exerts on the crown in Fig. 10–13b. The sum of the forces on the crown is zero, so $w'$ equals the actual weight $w$ ($= mg$) minus the buoyant force $F_B$:

$$w' = F'_T = w - F_B$$

so

$$w - w' = F_B.$$

Let $V$ be the volume of the completely submerged object and $\rho_O$ the object's density (so $\rho_O V$ is its mass), and let $\rho_F$ be the density of the fluid (water). Then $(\rho_F V)g$ is the weight of fluid displaced ($= F_B$). Now we can write

$$w = mg = \rho_O Vg$$
$$w - w' = F_B = \rho_F Vg.$$

We divide these two equations and obtain

$$\frac{w}{w - w'} = \frac{\rho_O Vg}{\rho_F Vg} = \frac{\rho_O}{\rho_F}.$$

We see that $w/(w - w')$ is equal to the specific gravity of the object (the crown) if the fluid in which it is submerged is water ($\rho_F = 1.00 \times 10^3 \text{ kg/m}^3$). Thus

$$\frac{\rho_O}{\rho_{H_2O}} = \frac{w}{w - w'} = \frac{(14.7 \text{ kg})g}{(14.7 \text{ kg} - 13.4 \text{ kg})g} = \frac{14.7 \text{ kg}}{1.3 \text{ kg}} = 11.3.$$

This corresponds to a density of 11,300 kg/m³. The crown is not gold, but seems to be made of lead (see Table 10–1).

**FIGURE 10–13** (a) A scale reads the mass of an object in air—in this case the crown of Example 10–8. All objects are at rest, so the tension $F_T$ in the connecting cord equals the weight $w$ of the object: $F_T = mg$. We show the free-body diagram of the crown, and $F_T$ is what causes the scale reading (it is equal to the net downward force on the scale, by Newton's third law). (b) Submerged, the crown has an additional force on it, the buoyant force $F_B$. The net force is zero, so $F'_T + F_B = mg$ ($= w$). The scale now reads $m' = 13.4$ kg, where $m'$ is related to the effective weight by $w' = m'g$. Thus $F'_T = w' = w - F_B$.

$w = (14.7 \text{ kg})g$

$(\vec{F}_T = -m\vec{g})$

$m\vec{g}$

(a)

$w' = (13.4 \text{ kg})g$

$\vec{F}'_T$  $\vec{F}_B$

$\vec{w} = m\vec{g}$

(b)

Archimedes' principle applies equally well to objects that float, such as wood. In general, *an object floats on a fluid if its density* $(\rho_O)$ *is less than that of the fluid* $(\rho_F)$. This is readily seen from Fig. 10–14a, where a submerged log of mass $m_O$ will experience a net upward force and float to the surface if $F_B > m_O g$; that is, if $\rho_F V g > \rho_O V g$ or $\rho_F > \rho_O$. At equilibrium—that is, when floating—the buoyant force on an object has magnitude equal to the weight of the object. For example, a log whose specific gravity is 0.60 and whose volume is 2.0 m³ has a mass

$$m_O = \rho_O V = (0.60 \times 10^3 \text{ kg/m}^3)(2.0 \text{ m}^3) = 1200 \text{ kg}.$$

If the log is fully submerged, it will displace a mass of water

$$m_F = \rho_F V = (1000 \text{ kg/m}^3)(2.0 \text{ m}^3) = 2000 \text{ kg}.$$

Hence the buoyant force on the log will be greater than its weight, and it will float upward to the surface (Fig. 10–14). The log will come to equilibrium when it displaces 1200 kg of water, which means that 1.2 m³ of its volume will be submerged. This 1.2 m³ corresponds to 60% of the volume of the log ($= 1.2/2.0 = 0.60$), so 60% of the log is submerged.

In general when an object floats, we have $F_B = m_O g$, which we can write as (see Fig. 10–15)

$$F_B = m_O g$$
$$\rho_F V_{\text{displ}} g = \rho_O V_O g,$$

where $V_O$ is the full volume of the object and $V_{\text{displ}}$ is the volume of fluid it displaces (= volume submerged). Thus

$$\frac{V_{\text{displ}}}{V_O} = \frac{\rho_O}{\rho_F}.$$

That is, the fraction of the object submerged is given by the ratio of the object's density to that of the fluid. If the fluid is water, this fraction equals the specific gravity of the object.

---

**EXAMPLE 10–9** **Hydrometer calibration.** A **hydrometer** is a simple instrument used to measure the specific gravity of a liquid by indicating how deeply the instrument sinks in the liquid. A particular hydrometer (Fig. 10–16) consists of a glass tube, weighted at the bottom, which is 25.0 cm long and 2.00 cm² in cross-sectional area, and has a mass of 45.0 g. How far from the weighted end should the 1.000 mark be placed?

**APPROACH** The hydrometer will float in water if its density $\rho$ is less than $\rho_{H_2O} = 1.000$ g/cm³, the density of water. The fraction of the hydrometer submerged ($V_{\text{displaced}}/V_{\text{total}}$) is equal to the density ratio $\rho/\rho_{H_2O}$.

**SOLUTION** The hydrometer has an overall density

$$\rho = \frac{m}{V} = \frac{45.0 \text{ g}}{(2.00 \text{ cm}^2)(25.0 \text{ cm})} = 0.900 \text{ g/cm}^3.$$

Thus, when placed in water, it will come to equilibrium when 0.900 of its volume is submerged. Since it is of uniform cross section, $(0.900)(25.0 \text{ cm}) = 22.5$ cm of its length will be submerged. The specific gravity of water is defined to be 1.000, so the mark should be placed 22.5 cm from the weighted end.

**NOTE** Hydrometers can be used to measure the density of liquids like car antifreeze coolant, car battery acid (a measure of its charge), wine fermenting in casks, and many others.

---

**EXERCISE C** Which of the following objects, submerged in water, experiences the largest magnitude of the buoyant force? (*a*) A 1-kg helium balloon; (*b*) 1 kg of wood; (*c*) 1 kg of ice; (*d*) 1 kg of iron; (*e*) all the same.

**EXERCISE D** Which of the following objects, submerged in water, experiences the largest magnitude of the buoyant force? (*a*) A 1-m³ helium balloon; (*b*) 1 m³ of wood; (*c*) 1 m³ of ice; (*d*) 1 m³ of iron; (*e*) all the same.

**FIGURE 10–14** (a) The fully submerged log accelerates upward because $F_B > m_O g$. It comes to equilibrium (b) when $\Sigma F = 0$, so $F_B = m_O g = (1200 \text{ kg})g$. Then 1200 kg, or 1.2 m³, of water is displaced.

**FIGURE 10–15** An object floating in equilibrium: $F_B = m_O g$.

**FIGURE 10–16** A hydrometer. Example 10–9.

Archimedes' principle is also useful in geology. According to the theories of plate tectonics and continental drift, the continents float on a fluid "sea" of slightly deformable rock (mantle rock). Some interesting calculations can be done using very simple models, which we consider in the Problems at the end of the Chapter.

Air is a fluid, and it too exerts a buoyant force. Ordinary objects weigh less in air than they do in a vacuum. Because the density of air is so small, the effect for ordinary solids is slight. There are objects, however, that *float* in air—helium-filled balloons, for example, because the density of helium is less than the density of air.

**EXAMPLE 10–10** **Helium balloon.** What volume $V$ of helium is needed if a balloon is to lift a load of 180 kg (including the weight of the empty balloon)?

**APPROACH** The buoyant force on the helium balloon, $F_B$, which is equal to the weight of displaced air, must be at least equal to the weight of the helium plus the weight of the balloon and load (Fig. 10–17). Table 10–1 gives the density of helium as $0.179 \text{ kg/m}^3$.

**SOLUTION** The buoyant force must have a minimum value of

$$F_B = (m_{He} + 180 \text{ kg})g.$$

This equation can be written in terms of density using Archimedes' principle:

$$\rho_{air} V g = (\rho_{He} V + 180 \text{ kg})g.$$

Solving now for $V$, we find

$$V = \frac{180 \text{ kg}}{\rho_{air} - \rho_{He}} = \frac{180 \text{ kg}}{(1.29 \text{ kg/m}^3 - 0.179 \text{ kg/m}^3)} = 160 \text{ m}^3.$$

**NOTE** This is the minimum volume needed near the Earth's surface, where $\rho_{air} = 1.29 \text{ kg/m}^3$. To reach a high altitude, a greater volume would be needed since the density of air decreases with altitude.

**FIGURE 10–17** Example 10–10.

**CONCEPTUAL EXAMPLE 10–11** **Throwing a rock overboard.** A rowboat carrying a large granite rock floats in a small lake. If the rock (SG $\approx$ 3, Table 10–1) is thrown overboard and sinks, does the lake level drop, rise, or stay the same?

**RESPONSE** Together the boat and rock float, so the buoyant force on them equals their total weight. The boat and rock displace a mass of water whose weight is equal to the weight of boat plus rock. When the rock is thrown into the lake, it displaces only its own volume, which is smaller than the volume of water the rock displaced when in the boat ($\approx \frac{1}{3}$ as much because the rock's density is $\approx$ 3 times greater than water). So less lake water is displaced and the water level of the lake *drops* when the rock is in the lake.

Maybe numbers can help. Suppose the boat and the rock each has a mass of 60 kg. Then the boat carrying the rock displaces 120 kg of water, which is a volume of $0.12 \text{ m}^3$ ($\rho = 1000 \text{ kg/m}^3$ for water, Table 10–1). When the rock is thrown into the lake, the boat alone now displaces $0.06 \text{ m}^3$. The rock displaces only its own volume of $0.02 \text{ m}^3$ ($\rho = m/V \approx 3$ so $V \approx 0.06 \text{ m}^3/3$). Thus a total of $0.08 \text{ m}^3$ of water is displaced. Less water is displaced so the water level of the lake drops.

**EXERCISE E** If you throw a flat 60-kg aluminum plate into water, the plate sinks. But if that aluminum is shaped into a rowboat, it floats. Explain.

# 10–8 Fluids in Motion; Flow Rate and the Equation of Continuity

We now turn to the subject of fluids in motion, which is called **fluid dynamics**, or (especially if the fluid is water) **hydrodynamics**.

We can distinguish two main types of fluid flow. If the flow is smooth, such that neighboring layers of the fluid slide by each other smoothly, the flow is said to be **streamline** or **laminar flow**.[†] In streamline flow, each particle of the fluid follows a smooth path, called a **streamline**, and these paths do not cross one another (Fig. 10–18a).

[†]The word laminar means "in layers."

(a)

(b)

**FIGURE 10–18** (a) Streamline, or laminar, flow; (b) turbulent flow. The photos show airflow around an airfoil or airplane wing (more in Section 10–10).

Above a certain speed, the flow becomes turbulent. **Turbulent flow** is characterized by erratic, small, whirlpool-like circles called *eddy currents* or *eddies* (Fig. 10–18b). Eddies absorb a great deal of energy, and although a certain amount of internal friction called **viscosity** is present even during streamline flow, it is much greater when the flow is turbulent. A few tiny drops of ink or food coloring dropped into a moving liquid can quickly reveal whether the flow is streamline or turbulent.

Let us consider the steady laminar flow of a fluid through an enclosed tube or pipe as shown in Fig. 10–19. First we determine how the speed of the fluid changes when the diameter of the tube changes. The mass **flow rate** is defined as the mass $\Delta m$ of fluid that passes a given point per unit time $\Delta t$:

$$\text{mass flow rate} = \frac{\Delta m}{\Delta t}.$$

**FIGURE 10–19** Fluid flow through a pipe of varying diameter.

In Fig. 10–19, the volume of fluid passing point 1 (through area $A_1$) in a time $\Delta t$ is $A_1 \Delta \ell_1$, where $\Delta \ell_1$ is the distance the fluid moves in time $\Delta t$. The velocity[†] of fluid (density $\rho_1$) passing point 1 is $v_1 = \Delta \ell_1 / \Delta t$. Then the mass flow rate $\Delta m_1 / \Delta t$ through area $A_1$ is

$$\frac{\Delta m_1}{\Delta t} = \frac{\rho_1 \Delta V_1}{\Delta t} = \frac{\rho_1 A_1 \Delta \ell_1}{\Delta t} = \rho_1 A_1 v_1,$$

where $\Delta V_1 = A_1 \Delta \ell_1$ is the volume of mass $\Delta m_1$. Similarly, at point 2 (through area $A_2$), the flow rate is $\rho_2 A_2 v_2$. Since no fluid flows in or out the sides of the tube, the flow rates through $A_1$ and $A_2$ must be equal. Thus

$$\frac{\Delta m_1}{\Delta t} = \frac{\Delta m_2}{\Delta t},$$

and

$$\rho_1 A_1 v_1 = \rho_2 A_2 v_2. \tag{10–4a}$$

This is called the **equation of continuity**.

If the fluid is incompressible ($\rho$ doesn't change with pressure), which is an excellent approximation for liquids under most circumstances (and sometimes for gases as well), then $\rho_1 = \rho_2$, and the equation of continuity becomes

$$A_1 v_1 = A_2 v_2. \qquad [\rho = \text{constant}] \tag{10–4b}$$

The product $Av$ represents the *volume rate of flow* (volume of fluid passing a given point per second), since $\Delta V / \Delta t = A \Delta \ell / \Delta t = Av$, which in SI units is m³/s. Equation 10–4b tells us that where the cross-sectional area is large, the velocity is small, and where the area is small, the velocity is large. That this is reasonable can be seen by looking at a river. A river flows slowly through a meadow where it is broad, but speeds up to torrential speed when passing through a narrow gorge.

[†]If there were no viscosity, the velocity would be the same across a cross section of the tube. Real fluids have viscosity, and this internal friction causes different layers of the fluid to flow at different speeds. In this case $v_1$ and $v_2$ represent the average speeds at each cross section.

$v$ = valves
$c$ = capillaries

FIGURE 10–20 Human circulatory system.

FIGURE 10–21 Example 10–13.

FIGURE 10–19 (Repeated.)
Fluid flow through a pipe of varying diameter.

**EXAMPLE 10–12** **ESTIMATE** **Blood flow.** In humans, blood flows from the heart into the aorta, from which it passes into the major arteries, Fig. 10–20. These branch into the small arteries (arterioles), which in turn branch into myriads of tiny capillaries. The blood returns to the heart via the veins. The radius of the aorta is about 1.2 cm, and the blood passing through it has a speed of about 40 cm/s. A typical capillary has a radius of about $4 \times 10^{-4}$ cm, and blood flows through it at a speed of about $5 \times 10^{-4}$ m/s. Estimate the number of capillaries that are in the body.

**APPROACH** We assume the density of blood doesn't vary significantly from the aorta to the capillaries. By the equation of continuity, the volume flow rate in the aorta must equal the volume flow rate through *all* the capillaries. The total area of all the capillaries is given by the area of a typical capillary multiplied by the total number $N$ of capillaries.

**SOLUTION** Let $A_1$ be the area of the aorta and $A_2$ be the area of *all* the capillaries through which blood flows. Then $A_2 = N \pi r_{cap}^2$, where $r_{cap} \approx 4 \times 10^{-4}$ cm is the estimated average radius of one capillary. From the equation of continuity (Eq. 10–4b), we have

$$v_2 A_2 = v_1 A_1$$
$$v_2 N \pi r_{cap}^2 = v_1 \pi r_{aorta}^2$$

so

$$N = \frac{v_1}{v_2} \frac{r_{aorta}^2}{r_{cap}^2} = \left( \frac{0.40\,\text{m/s}}{5 \times 10^{-4}\,\text{m/s}} \right) \left( \frac{1.2 \times 10^{-2}\,\text{m}}{4 \times 10^{-6}\,\text{m}} \right)^2 \approx 7 \times 10^9,$$

or on the order of 10 billion capillaries.

**EXAMPLE 10–13** **Heating duct to a room.** What area must a heating duct have if air moving 3.0 m/s along it can replenish the air every 15 minutes in a room of volume 300 $m^3$? Assume the air's density remains constant.

**APPROACH** We apply the equation of continuity at constant density, Eq. 10–4b, to the air that flows through the duct (point 1 in Fig. 10–21) and then into the room (point 2). The volume flow rate in the room equals the volume of the room divided by the 15-min replenishing time.

**SOLUTION** Consider the room as a large section of the duct, Fig. 10–21, and think of air equal to the volume of the room as passing by point 2 in $t = 15$ min $= 900$ s. Reasoning in the same way we did to obtain Eq. 10–4a (changing $\Delta t$ to $t$), we write $v_2 = \ell_2/t$ so $A_2 v_2 = A_2 \ell_2/t = V_2/t$, where $V_2$ is the volume of the room. Then the equation of continuity becomes $A_1 v_1 = A_2 v_2 = V_2/t$ and

$$A_1 = \frac{V_2}{v_1 t} = \frac{300\,\text{m}^3}{(3.0\,\text{m/s})(900\,\text{s})} = 0.11\,\text{m}^2.$$

**NOTE** If the duct is square, then each side has length $\ell = \sqrt{A} = 0.33$ m, or 33 cm. A rectangular duct 20 cm $\times$ 55 cm will also do.

# 10–9 Bernoulli's Equation

Have you ever wondered why an airplane can fly, or how a sailboat can move against the wind? These are examples of a principle worked out by Daniel Bernoulli (1700–1782) concerning fluids in motion. In essence, **Bernoulli's principle** states that *where the velocity of a fluid is high, the pressure is low, and where the velocity is low, the pressure is high.* For example, if the pressure in the fluid is measured at points 1 and 2 of Fig. 10–19, it will be found that the pressure is lower at point 2, where the velocity is greater, than it is at point 1, where the velocity is smaller. At first glance, this might seem strange; you might expect that the greater speed at point 2 would imply a higher pressure. But this cannot be the case:

if the pressure in the fluid at point 2 were higher than at point 1, this higher pressure would slow the fluid down, whereas in fact it has sped up in going from point 1 to point 2. Thus the pressure at point 2 must be less than at point 1, to be consistent with the fact that the fluid accelerates.

To help clarify any misconceptions, a faster fluid might indeed exert a greater force bouncing off an obstacle placed in its path. But that is not what we mean by the pressure in a fluid. We are examining smooth streamline flow, with no obstacles that interrupt the flow. The fluid pressure is exerted on the walls of a tube or pipe, or on the surface of a material the fluid passes over.

Bernoulli developed an equation that expresses this principle quantitatively. To derive Bernoulli's equation, we assume the flow is steady and laminar, the fluid is incompressible, and the viscosity is small enough to be ignored. To be general, we assume the fluid is flowing in a tube of nonuniform cross section that varies in height above some reference level, Fig. 10–22. We will consider the volume of fluid shown in color and calculate the work done to move it from the position shown in Fig. 10–22a to that shown in Fig. 10–22b. In this process, fluid entering area $A_1$ flows a distance $\Delta\ell_1$ and forces the fluid at area $A_2$ to move a distance $\Delta\ell_2$. The fluid to the left of area $A_1$ exerts a pressure $P_1$ on our section of fluid and does an amount of work

$$W_1 = F_1\,\Delta\ell_1 = P_1 A_1\,\Delta\ell_1,$$

(since $P = F/A$). At point 2, the work done on our section of fluid is

$$W_2 = -P_2 A_2\,\Delta\ell_2.$$

The negative sign is present because the force exerted on the fluid is opposite to the displacement. Work is also done on the fluid by the force of gravity. The net effect of the process shown in Fig. 10–22 is to move a mass $m$ of volume $A_1\,\Delta\ell_1$ ($= A_2\,\Delta\ell_2$, since the fluid is incompressible) from point 1 to point 2, so the work done by gravity is

$$W_3 = -mg(y_2 - y_1),$$

where $y_1$ and $y_2$ are heights of the center of the tube above some (arbitrary) reference level. In the case shown in Fig. 10–22, this term is negative since the motion is uphill against the force of gravity. The net work $W$ done on the fluid is thus

$$W = W_1 + W_2 + W_3$$
$$W = P_1 A_1\,\Delta\ell_1 - P_2 A_2\,\Delta\ell_2 - mgy_2 + mgy_1.$$

According to the work-energy principle (Section 6–3), the net work done on a system is equal to its change in kinetic energy. Hence

$$\tfrac{1}{2}mv_2^2 - \tfrac{1}{2}mv_1^2 = P_1 A_1\,\Delta\ell_1 - P_2 A_2\,\Delta\ell_2 - mgy_2 + mgy_1.$$

The mass $m$ has volume $A_1\,\Delta\ell_1 = A_2\,\Delta\ell_2$ for an incompressible fluid. Thus we can substitute $m = \rho A_1\,\Delta\ell_1 = \rho A_2\,\Delta\ell_2$, and then divide through by $A_1\,\Delta\ell_1 = A_2\,\Delta\ell_2$, to obtain

$$\tfrac{1}{2}\rho v_2^2 - \tfrac{1}{2}\rho v_1^2 = P_1 - P_2 - \rho g y_2 + \rho g y_1,$$

which we rearrange to get

$$P_2 + \tfrac{1}{2}\rho v_2^2 + \rho g y_2 = P_1 + \tfrac{1}{2}\rho v_1^2 + \rho g y_1. \qquad \textbf{(10–5)} \quad \textit{Bernoulli's equation}$$

This is **Bernoulli's equation**. Since points 1 and 2 can be any two points along a tube of flow, Bernoulli's equation can be written as

$$P + \tfrac{1}{2}\rho v^2 + \rho g y = \text{constant}$$

at every point in the fluid, where $y$ is the height of the center of the tube above a fixed reference level. [Note that if there is no flow $(v_1 = v_2 = 0)$, then Eq. 10–5 reduces to the hydrostatic equation, Eq. 10–3b or c: $P_1 - P_2 = \rho g(y_2 - y_1)$.]

**FIGURE 10–22** Fluid flow: for derivation of Bernoulli's equation.

Bernoulli's equation is an expression of the law of energy conservation, since we derived it from the work-energy principle.

**EXERCISE F** As water in a level pipe passes from a narrow cross section of pipe to a wider cross section, how does the pressure against the walls change?

**PHYSICS APPLIED**
*Hot-water heating system*

**EXAMPLE 10–14** | **Flow and pressure in a hot-water heating system.** Water circulates throughout a house in a hot-water heating system. If the water is pumped at a speed of 0.50 m/s through a 4.0-cm-diameter pipe in the basement under a pressure of 3.0 atm, what will be the flow speed and pressure in a 2.6-cm-diameter pipe on the second floor 5.0 m above? Assume the pipes do not divide into branches.

**APPROACH** We use the equation of continuity at constant density to determine the flow speed on the second floor, and then Bernoulli's equation to find the pressure.

**SOLUTION** We take $v_2$ in the equation of continuity, Eq. 10–4, as the flow speed on the second floor, and $v_1$ as the flow speed in the basement. Noting that the areas are proportional to the radii squared $(A = \pi r^2)$, we obtain

$$v_2 = \frac{v_1 A_1}{A_2} = \frac{v_1 \pi r_1^2}{\pi r_2^2} = (0.50 \text{ m/s}) \frac{(0.020 \text{ m})^2}{(0.013 \text{ m})^2} = 1.2 \text{ m/s}.$$

To find the pressure on the second floor, we use Bernoulli's equation (Eq. 10–5):

$$\begin{aligned}
P_2 &= P_1 + \rho g(y_1 - y_2) + \tfrac{1}{2}\rho(v_1^2 - v_2^2) \\
&= (3.0 \times 10^5 \text{ N/m}^2) + (1.0 \times 10^3 \text{ kg/m}^3)(9.8 \text{ m/s}^2)(-5.0 \text{ m}) \\
&\quad + \tfrac{1}{2}(1.0 \times 10^3 \text{ kg/m}^3)[(0.50 \text{ m/s})^2 - (1.2 \text{ m/s})^2] \\
&= (3.0 \times 10^5 \text{ N/m}^2) - (4.9 \times 10^4 \text{ N/m}^2) - (6.0 \times 10^2 \text{ N/m}^2) \\
&= 2.5 \times 10^5 \text{ N/m}^2 = 2.5 \text{ atm}.
\end{aligned}$$

**NOTE** The velocity term contributes very little in this case.

# 10–10 Applications of Bernoulli's Principle: Torricelli, Airplanes, Baseballs, Blood Flow

Bernoulli's equation can be applied to many situations. One example is to calculate the velocity, $v_1$, of a liquid flowing out of a spigot at the bottom of a reservoir, Fig. 10–23. We choose point 2 in Eq. 10–5 to be the top surface of the liquid. Assuming the diameter of the reservoir is large compared to that of the spigot, $v_2$ will be almost zero. Points 1 (the spigot) and 2 (top surface) are open to the atmosphere, so the pressure at both points is equal to atmospheric pressure: $P_1 = P_2$. Then Bernoulli's equation becomes

$$\tfrac{1}{2}\rho v_1^2 + \rho g y_1 = \rho g y_2$$

or

$$v_1 = \sqrt{2g(y_2 - y_1)}. \tag{10–6}$$

**FIGURE 10–23** Torricelli's theorem: $v_1 = \sqrt{2g(y_2 - y_1)}$.

This result is called **Torricelli's theorem.** Although it is seen to be a special case of Bernoulli's equation, it was discovered a century earlier by Evangelista Torricelli. Equation 10–6 tells us that the liquid leaves the spigot with the same speed that a freely falling object would attain if falling from the same height. This should not be too surprising since the derivation of Bernoulli's equation relies on the conservation of energy.

Another special case of Bernoulli's equation arises when a fluid is flowing horizontally with no appreciable change in height; that is, $y_1 = y_2$. Then Eq. 10–5 becomes

$$P_1 + \tfrac{1}{2}\rho v_1^2 = P_2 + \tfrac{1}{2}\rho v_2^2, \tag{10-7}$$

which tells us quantitatively that the speed is high where the pressure is low, and vice versa. It explains many common phenomena, some of which are illustrated in Figs. 10–24 to 10–30. The pressure in the air blown at high speed across the top of the vertical tube of a perfume atomizer (Fig. 10–24a) is less than the normal air pressure acting on the surface of the liquid in the bowl. Thus atmospheric pressure in the bowl pushes the perfume up the tube because of the lower pressure at the top. A Ping-Pong ball can be made to float above a blowing jet of air (a hair dryer or a vacuum cleaner that can also blow air), Fig. 10–24b; if the ball begins to leave the jet of air, the higher pressure in the still air outside the jet pushes the ball back in.

**EXERCISE G** Return to Chapter-Opening Question 2, page 260, and answer it again now. Try to explain why you may have answered differently the first time. Try it and see.

**FIGURE 10–24** Examples of Bernoulli's principle: (a) atomizer, (b) Ping-Pong ball in jet of air.

## Airplane Wings and Dynamic Lift

Airplanes experience a "lift" force on their wings, keeping them up in the air, if they are moving at a sufficiently high speed relative to the air and the wing is tilted upward at a small angle (the "attack angle"). See Fig. 10–25, where streamlines of air are shown rushing by the wing (we are in the reference frame of the wing, as if sitting on the wing). The upward tilt, as well as the rounded upper surface of the wing, causes the streamlines to be forced upward and to be crowded together above the wing. The area of air flowing between any two streamlines is smaller as the streamlines get closer together, so from the equation of continuity $(A_1 v_1 = A_2 v_2)$, the air speed increases above the wing where the streamlines are squished together. (Recall also how the crowded streamlines in a pipe constriction, Fig. 10–19, indicate the velocity is higher in the constriction.) Thus the air speed is greater above the wing than below it, so the pressure above the wing is less than the pressure below the wing (Bernoulli's principle). Hence there is a net upward force on the wing called **dynamic lift**. Experiments show that the speed of air above the wing can even be double the speed of the air below it. (Friction between the air and wing exerts a *drag force*, toward the rear, which must be overcome by the plane's engines.)

A flat wing, or one with symmetric cross section, will experience lift as long as the front of the wing is tilted upward (attack angle). The wing shown in Fig. 10–25 can experience lift even if the attack angle is zero, because the rounded upper surface deflects air up, squeezing the streamlines together. Airplanes can fly upside down, experiencing lift, if the attack angle is sufficient to deflect streamlines up and closer together.

Our picture considers streamlines; but if the attack angle is larger than about 15°, turbulence sets in (Fig. 10–18b) leading to greater drag and less lift, causing the plane to "stall" and then to drop.

From another point of view, the upward tilt of a wing means the air moving horizontally in front of the wing is deflected downward; the change in momentum of the rebounding air molecules results in an upward force on the wing (Newton's third law).

**FIGURE 10–25** Lift on an airplane wing. We are in the reference frame of the wing, seeing the air flow by.

Lower pressure

Higher pressure

**PHYSICS APPLIED**
*Airplanes and dynamic lift*

## Sailboats

A sailboat can move "against" the wind, with the aid of the Bernoulli effect, by setting the sails at an angle, as shown in Fig. 10–26. The air traveling rapidly over the bulging front surface of the mainsail exerts a smaller pressure than the relatively still air behind the sail. The result is a net force on the sail, $\vec{\mathbf{F}}_{\text{wind}}$, as shown in Fig. 10–26b. This force would tend to make the boat move sideways if it weren't for the keel that extends vertically downward beneath the water: the water exerts a force ($\vec{\mathbf{F}}_{\text{water}}$) on the keel nearly perpendicular to the keel. The resultant of these two forces ($\vec{\mathbf{F}}_R$) is almost directly forward as shown.

**FIGURE 10–26** Sailboat (a) sailing against the wind with (b) analysis.

(a)

(b)

FIGURE 10–27 Looking down on a pitched baseball heading toward home plate. We are in the reference frame of the baseball, with the air flowing by.

FIGURE 10–28 Rear of the head and shoulders showing arteries leading to the brain and to the arms. High blood velocity past the constriction in the left subclavian artery causes low pressure in the left vertebral artery, in which a reverse (downward) blood flow can then occur, resulting in a TIA, a loss of blood to the brain.

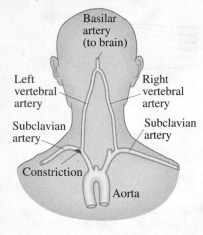

*Smoke up a chimney*
*Underground air circulation*

FIGURE 10–30 Bernoulli's principle explains air flow in underground burrows.

## Baseball Curve

Why a spinning pitched baseball (or tennis ball) curves can also be explained using Bernoulli's principle. It is simplest if we put ourselves in the reference frame of the ball, with the air rushing by, just as we did for the airplane wing. Suppose the ball is rotating counterclockwise as seen from above, Fig. 10–27. A thin layer of air ("boundary layer") is being dragged around by the ball. We are looking down on the ball, and at point A in Fig. 10–27, this boundary layer tends to slow down the oncoming air. At point B, the air rotating with the ball adds its speed to that of the oncoming air, so the air speed is higher at B than at A. The higher speed at B means the pressure is lower at B than at A, resulting in a net force toward B. The ball's path curves toward the left (as seen by the pitcher).

## Lack of Blood to the Brain—TIA

In medicine, one of many applications of Bernoulli's principle is to explain a TIA, a *transient ischemic attack* (meaning a temporary lack of blood supply to the brain). A person suffering a TIA may experience symptoms such as dizziness, double vision, headache, and weakness of the limbs. A TIA can occur as follows. Blood normally flows up to the brain at the back of the head via the two vertebral arteries—one going up each side of the neck—which meet to form the basilar artery just below the brain, as shown in Fig. 10–28. Each vertebral artery connects to the subclavian artery, as shown, before the blood passes to the arms. When an arm is exercised vigorously, blood flow increases to meet the needs of the arm's muscles. If the subclavian artery on one side of the body is partially blocked, however, as in arteriosclerosis (hardening of the arteries), the blood velocity will have to be higher on that side to supply the needed blood. (Recall the equation of continuity: smaller area means larger velocity for the same flow rate, Eqs. 10–4.) The increased blood velocity past the opening to the vertebral artery results in lower pressure (Bernoulli's principle). Thus, blood rising in the vertebral artery on the "good" side at normal pressure can be *diverted down* into the other vertebral artery because of the low pressure on that side, instead of passing upward to the brain. Hence the blood supply to the brain is reduced.

## Other Applications

A **venturi tube** is essentially a pipe with a narrow constriction (the throat). The flowing fluid speeds up as it passes through this constriction, so the pressure is lower in the throat. A **venturi meter**, Fig. 10–29, is used to measure the flow speed of gases and liquids, including blood velocity in arteries. The velocity $v_1$ can be determined by measuring the pressure $P_1$ and $P_2$, the areas $A_1$ and $A_2$, as well as the density of the fluid. (The formula is given in Problem 56.)

FIGURE 10–29 Venturi meter.

Why does smoke go up a chimney? It's partly because hot air rises (it's less dense and therefore buoyant). But Bernoulli's principle also plays a role. When wind blows across the top of a chimney, the pressure is less there than inside the house. Hence, air and smoke are pushed up the chimney by the higher indoor pressure. Even on an apparently still night there is usually enough ambient air flow at the top of a chimney to assist upward flow of smoke.

If gophers, prairie dogs, rabbits, and other animals that live underground are to avoid suffocation, the air must circulate in their burrows. The burrows always have at least two entrances (Fig. 10–30). The speed of air flow across different holes will usually be slightly different. This results in a slight pressure difference, which forces a flow of air through the burrow via Bernoulli's principle. The flow of air is enhanced if one hole is higher than the other (animals often build mounds) since wind speed tends to increase with height.

Bernoulli's equation ignores the effects of friction (viscosity) and the compressibility of the fluid. The energy that is transformed to internal (or potential) energy due to compression and to thermal energy by friction can be taken into account by adding terms to Eq. 10–5. These terms are difficult to calculate theoretically and are normally determined empirically for given situations. They do not significantly alter the explanations for the phenomena described above.

# * 10–11 Viscosity

Real fluids have a certain amount of internal friction called **viscosity**, as mentioned in Section 10–8. Viscosity exists in both liquids and gases, and is essentially a frictional force between adjacent layers of fluid as the layers move past one another. In liquids, viscosity is due to the electrical cohesive forces between the molecules. In gases, it arises from collisions between the molecules.

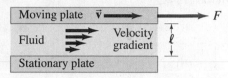

**FIGURE 10–31** Determination of viscosity.

The viscosity of different fluids can be expressed quantitatively by a *coefficient of viscosity*, $\eta$ (the Greek lowercase letter eta), which is defined in the following way. A thin layer of fluid is placed between two flat plates. One plate is stationary and the other is made to move, Fig. 10–31. The fluid directly in contact with each plate is held to the surface by the adhesive force between the molecules of the liquid and those of the plate. Thus the upper surface of the fluid moves with the same speed $v$ as the upper plate, whereas the fluid in contact with the stationary plate remains stationary. The stationary layer of fluid retards the flow of the layer just above it, which in turn retards the flow of the next layer, and so on. Thus the velocity varies continuously from 0 to $v$, as shown. The increase in velocity divided by the distance over which this change is made—equal to $v/\ell$—is called the *velocity gradient*. To move the upper plate requires a force, which you can verify by moving a flat plate across a puddle of syrup on a table. For a given fluid, it is found that the force required, $F$, is proportional to the area of fluid in contact with each plate, $A$, and to the speed, $v$, and is inversely proportional to the separation, $\ell$, of the plates: $F \propto vA/\ell$. For different fluids, the more viscous the fluid, the greater is the required force. The proportionality constant for this equation is defined as the coefficient of viscosity, $\eta$:

$$F = \eta A \frac{v}{\ell}. \qquad (10\text{–}8)$$

Solving for $\eta$, we find $\eta = F\ell/vA$. The SI unit for $\eta$ is $\mathrm{N\cdot s/m^2} = \mathrm{Pa\cdot s}$ (pascal·second). In the cgs system, the unit is $\mathrm{dyne\cdot s/cm^2}$, which is called a *poise* (P). Viscosities are often given in centipoise ($1\,\mathrm{cP} = 10^{-2}\,\mathrm{P} = 10^{-3}\,\mathrm{Pa\cdot s}$). Table 10–3 lists the coefficient of viscosity for various fluids. The temperature is also specified, since it has a strong effect; the viscosity of liquids such as motor oil, for example, decreases rapidly as temperature increases.[‡]

**TABLE 10–3 Coefficients of Viscosity**

| Fluid (temperature in °C) | Coefficient of Viscosity, $\eta$ (Pa·s)[†] |
|---|---|
| Water (0°) | $1.8 \times 10^{-3}$ |
| (20°) | $1.0 \times 10^{-3}$ |
| (100°) | $0.3 \times 10^{-3}$ |
| Whole blood (37°) | $\approx 4 \times 10^{-3}$ |
| Blood plasma (37°) | $\approx 1.5 \times 10^{-3}$ |
| Ethyl alcohol (20°) | $1.2 \times 10^{-3}$ |
| Engine oil (30°) (SAE 10) | $200 \times 10^{-3}$ |
| Glycerine (20°) | $1500 \times 10^{-3}$ |
| Air (20°) | $0.018 \times 10^{-3}$ |
| Hydrogen (0°) | $0.009 \times 10^{-3}$ |
| Water vapor (100°) | $0.013 \times 10^{-3}$ |

[†]$1\,\mathrm{Pa\cdot s} = 10\,\mathrm{poise\ (P)} = 1000\,\mathrm{cP}$.

# * 10–12 Flow in Tubes: Poiseuille's Equation, Blood Flow

If a fluid had no viscosity, it could flow through a level tube or pipe without a force being applied. Viscosity acts like a sort of friction (between fluid layers moving at slightly different speeds), so a pressure difference between the ends of a level tube is necessary for the steady flow of any real fluid, be it water or oil in a pipe, or blood in the circulatory system of a human.

[‡]The Society of Automotive Engineers assigns numbers to represent the viscosity of oils: 30-weight (SAE 30) is more viscous than 10-weight. Multigrade oils, such as 20–50, are designed to maintain viscosity as temperature increases; 20–50 means the oil acts like 20-weight when cool and is like 50-weight when it is hot (engine running temperature). In other words, the viscosity does not drop precipitously as the oil warms up, as a simple 20-weight oil would.

Wall of artery

(a)

Artery wall thickening

Blockage

(b)

**FIGURE 10–32** A cross section of a human artery that (a) is healthy, (b) is partly blocked as a result of arteriosclerosis.

**PHYSICS APPLIED**

*Medicine— blood flow and heart disease*

**FIGURE 10–33** Spherical water droplets, dew on a blade of grass.

**FIGURE 10–34** U-shaped wire apparatus holding a film of liquid to measure surface tension ($\gamma = F/2\ell$).

(a) Top view

(b) Edge view (magnified)

The French scientist J. L. Poiseuille (1799–1869), who was interested in the physics of blood circulation (and after whom the "poise" is named), determined how the variables affect the flow rate of an incompressible fluid undergoing laminar flow in a cylindrical tube. His result, known as **Poiseuille's equation**, is:

$$Q = \frac{\pi R^4 (P_1 - P_2)}{8\eta\ell},$$   (10–9)

where $R$ is the inside radius of the tube, $\ell$ is the tube length, $P_1 - P_2$ is the pressure difference between the ends, $\eta$ is the coefficient of viscosity, and $Q$ is the volume rate of flow (volume of fluid flowing past a given point per unit time which in SI has units of m$^3$/s). Equation 10–9 applies only to laminar (streamline) flow.

Poiseuille's equation tells us that the flow rate $Q$ is directly proportional to the "pressure gradient," $(P_1 - P_2)/\ell$, and it is inversely proportional to the viscosity of the fluid. This is just what we might expect. It may be surprising, however, that $Q$ also depends on the *fourth* power of the tube's radius. This means that for the same pressure gradient, if the tube radius is halved, the flow rate is decreased by a factor of 16! Thus the rate of flow, or alternately the pressure required to maintain a given flow rate, is greatly affected by only a small change in tube radius.

An interesting example of this $R^4$ dependence is *blood flow* in the human body. Poiseuille's equation is valid only for the streamline flow of an incompressible fluid. So it cannot be precisely accurate for blood whose flow is not without turbulence and that contains blood cells (whose diameter is almost equal to that of a capillary). Nonetheless, Poiseuille's equation does give a reasonable first approximation. Because the radius of arteries is reduced as a result of arteriosclerosis (thickening and hardening of artery walls, Fig. 10–32) and by cholesterol buildup, the pressure gradient must be increased to maintain the same flow rate. If the radius is reduced by half, the heart would have to increase the pressure by a factor of about $2^4 = 16$ in order to maintain the same blood-flow rate. The heart must work much harder under these conditions, but usually cannot maintain the original flow rate. Thus, high blood pressure is an indication both that the heart is working harder and that the blood-flow rate is reduced.

# *10–13 Surface Tension and Capillarity

The *surface* of a liquid at rest behaves in an interesting way, almost as if it were a stretched membrane under tension. For example, a drop of water on the end of a dripping faucet, or hanging from a thin branch in the early morning dew (Fig. 10–33), forms into a nearly spherical shape as if it were a tiny balloon filled with water. A steel needle can be made to float on the surface of water even though it is denser than the water. The surface of a liquid acts like it is under tension, and this tension, acting along the surface, arises from the attractive forces between the molecules. This effect is called **surface tension**. More specifically, a quantity called the *surface tension*, $\gamma$ (the Greek letter gamma), is defined as the force $F$ per unit length $\ell$ that acts perpendicular to any line or cut in a liquid surface, tending to pull the surface closed:

$$\gamma = \frac{F}{\ell}.$$   (10–10)

To understand this, consider the U-shaped apparatus shown in Fig. 10–34 which encloses a thin film of liquid (such as a liquid soap film). Because of surface tension, a force $F$ is required to pull the movable wire and thus increase the surface area of the liquid. The liquid contained by the wire apparatus is a thin film having both a top and a bottom surface. Hence the total length of the surface being increased is $2\ell$, and the surface tension is $\gamma = F/2\ell$. A delicate apparatus of this type can be used to measure the surface tension of various liquids. The surface tension of water is 0.072 N/m at 20°C. Table 10–4 (next page) gives the values for several substances. Note that temperature has a considerable effect on the surface tension.

(a)

(b)

**FIGURE 10–35** (a) Water strider. (b) Paper clip (light coming through window blinds).

Because of surface tension, some insects (Fig. 10–35a) can walk on water, and objects more dense than water, such as a paper clip (Fig. 10–35b), can float on the surface. Figure 10–36a shows how the surface tension can support the weight $w$ of an object. Actually, the object sinks slightly into the fluid, so $w$ is the "effective weight" of that object—its true weight less the buoyant force.

**EXAMPLE 10–15 | ESTIMATE | Insect walks on water.** The base of an insect's leg is approximately spherical in shape, with a radius of about $2.0 \times 10^{-5}$ m. The 0.0030-g mass of the insect is supported equally by its six legs. Estimate the angle $\theta$ at which the surface tension force acts (see Fig. 10–36) for an insect on the surface of water. Assume the water temperature is 20°C.

**APPROACH** Since the insect is in equilibrium, the upward surface tension force is equal to the pull of gravity downward on each leg. We ignore buoyant forces for this estimate.

**SOLUTION** For each leg, we assume the surface tension force acts all around a circle of radius $r$, at an angle $\theta$, as shown in Fig. 10–36a. Only the vertical component, $\gamma \cos \theta$, acts to balance the weight $mg$. We set the length $\ell$ in Eq. 10–10 equal to the circumference of the circle, $\ell \approx 2\pi r$. Then the net upward force due to surface tension is $F_y \approx (\gamma \cos \theta)\ell \approx 2\pi r \gamma \cos \theta$. We set this surface tension force equal to one-sixth the weight of the insect since it has six legs:

$$2\pi r \gamma \cos \theta \approx \tfrac{1}{6} mg$$
$$(6.28)(2.0 \times 10^{-5}\,\text{m})(0.072\,\text{N/m}) \cos \theta \approx \tfrac{1}{6}(3.0 \times 10^{-6}\,\text{kg})(9.8\,\text{m/s}^2)$$
$$\cos \theta \approx 0.54.$$

So $\theta \approx 57°$.

**NOTE** If $\cos \theta$ had come out greater than 1, the surface tension would not have been great enough to support the insect's weight. If the insect is very light, it will sink less into the water and $\theta$ (Fig. 10–36a) will be larger than calculated above.

**NOTE** Our estimate ignored the buoyant force and ignored any difference between the radius of the insect's "foot" and the radius of the surface depression.

Soaps and detergents lower the surface tension of water. This is desirable for washing and cleaning since the high surface tension of pure water prevents it from penetrating easily between the fibers of material and into tiny crevices. Substances that reduce the surface tension of a liquid are called *surfactants*.

## *Capillarity

Surface tension plays a role in another interesting phenomenon, *capillarity*. It is a common observation that water in a glass container rises up slightly where it touches the glass, Fig. 10–37a. The water is said to "wet" the glass. Mercury, on the other hand, is depressed when it touches the glass, Fig. 10–37b; the mercury does not wet the glass. Whether a liquid wets a solid surface is determined by the relative strength of the cohesive forces between the molecules of the liquid compared to the adhesive forces between the molecules of the liquid and those of the container. **Cohesion** refers to the force between molecules of the same type, whereas **adhesion** refers to the force between molecules of different types. Water wets glass because the water molecules are more strongly attracted to the glass molecules than they are to other water molecules. The opposite is true for mercury: the cohesive forces are stronger than the adhesive forces.

**TABLE 10–4 Surface Tension of Some Substances**

| Substance | Surface Tension (N/m) |
|---|---|
| Mercury (20°C) | 0.44 |
| Blood, whole (37°C) | 0.058 |
| Blood, plasma (37°C) | 0.073 |
| Alcohol, ethyl (20°C) | 0.023 |
| Water   (0°C) | 0.076 |
| (20°C) | 0.072 |
| (100°C) | 0.059 |
| Benzene (20°C) | 0.029 |
| Soap solution (20°C) | $\approx 0.025$ |
| Oxygen ($-193$°C) | 0.016 |

$F = \gamma\ell$  $F = \gamma\ell$

(a)

(b)

**FIGURE 10–36** Surface tension acting on (a) a sphere, and (b) an insect leg. Example 10–15.

⊕ **PHYSICS APPLIED**
*Soaps and detergents*

**FIGURE 10–37** (a) Water "wets" the surface of glass, whereas (b) mercury does not "wet" the glass.

| Water | Mercury |
|---|---|
| (a) | (b) |

(a)
Glass tube
in water

(b)
Glass tube
in mercury

**FIGURE 10–38** Capillarity.

**FIGURE 10–39** One kind of pump (reciprocating type): the intake valve opens and air (or fluid that is being pumped) fills the empty space when the piston moves to the left. When the piston moves to the right (not shown), the outlet valve opens and fluid is forced out.

Piston

Outlet

Intake

PHYSICS APPLIED

*Heart as a pump*

In tubes having very small diameters, liquids are observed to rise or fall relative to the level of the surrounding liquid. This phenomenon is called **capillarity**, and such thin tubes are called **capillaries**. Whether the liquid rises or falls (Fig. 10–38) depends on the relative strengths of the adhesive and cohesive forces. Thus water rises in a glass tube, whereas mercury falls. The actual amount of rise (or fall) depends on the surface tension—which is what keeps the liquid surface from breaking apart.

# *10–14 Pumps, and the Heart

We conclude this Chapter with a brief discussion of pumps, including the heart. Pumps can be classified into categories according to their function. A *vacuum pump* is designed to reduce the pressure (usually of air) in a given vessel. A *force pump*, on the other hand, is a pump that is intended to increase the pressure—for example, to lift a liquid (such as water from a well) or to push a fluid through a pipe. Figure 10–39 illustrates the principle behind a simple reciprocating pump. It could be a vacuum pump, in which case the intake is connected to the vessel to be evacuated. A similar mechanism is used in some force pumps, and in this case the fluid is forced under increased pressure through the outlet.

Another type of pump is the centrifugal pump, shown in Fig. 10–40. It, or any force pump, can be used as a *circulating pump*—that is, to circulate a fluid around a closed path, such as the cooling water or lubricating oil in an automobile.

**FIGURE 10–40** Centrifugal pump: the rotating blades force fluid through the outlet pipe; this kind of pump is used in vacuum cleaners and as a water pump in automobiles.

Intake

Outlet

Rotor

The heart of a human (and of other animals as well) is essentially a circulating pump. The action of a human heart is shown in Fig. 10–41. There are actually two separate paths for blood flow. The longer path takes blood to the parts of the body, via the arteries, bringing oxygen to body tissues and picking up carbon dioxide, which it carries back to the heart via veins. This blood is then pumped to the lungs (the second path), where the carbon dioxide is released and oxygen is taken up. The oxygen-laden blood is returned to the heart, where it is again pumped to the tissues of the body.

**FIGURE 10–41** Pumping human heart. (a) In the diastole phase, the heart relaxes between beats. Blood moves into the heart; both atria fill rapidly. (b) When the atria contract, the systole or pumping phase begins. The contraction pushes the blood through the mitral and tricuspid valves into the ventricles. (c) The contraction of the ventricles forces the blood through the semilunar valves into the pulmonary artery, which leads to the lungs, and to the aorta (the body's largest artery), which leads to the arteries serving all the body. (d) When the heart relaxes, the semilunar valves close; blood fills the atria, beginning the cycle again.

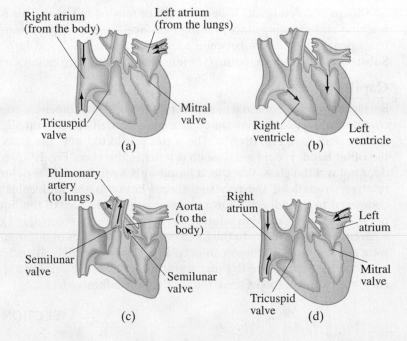

Blood pressure is measured using one of the types of gauge mentioned earlier (Section 10–6), and it is usually calibrated in mm-Hg. The gauge is attached to a closed, air-filled cuff that is wrapped around the upper arm at the level of the heart, Fig. 10–42. Two values of blood pressure are measured: the maximum pressure when the heart is pumping, called *systolic pressure*; and the pressure when the heart is in the resting part of the cycle, called *diastolic pressure*. Initially, the air pressure in the cuff is increased high above the systolic pressure by a pump, compressing the main (brachial) artery in the arm and briefly cutting off the flow of blood. The air pressure is then reduced slowly until blood again begins to flow into the arm; it can be detected by listening with a stethoscope to the characteristic tapping sound[†] of the blood returning to the forearm. At this point, systolic pressure is just equal to the air pressure in the arm cuff which can be read off the gauge. The air pressure is subsequently reduced further, and the tapping sound disappears when blood at low pressure can enter the artery. At this point, the gauge indicates the diastolic pressure. Normal systolic pressure is around 120 mm-Hg, whereas normal diastolic pressure is around 70 or 80 mm-Hg. Blood pressure is reported in the form 120/70.

**PHYSICS APPLIED**
*Blood pressure*

**FIGURE 10–42** Device for measuring blood pressure.

[†]When the blood starts flowing through the constriction caused by the tight cuff, its velocity is high and the flow is turbulent. It is the turbulence that causes the tapping sound.

# Summary

The three common phases of matter are **solid**, **liquid**, and **gas**. Liquids and gases are collectively called **fluids**, meaning they have the ability to flow. The **density** of a material is defined as its mass per unit volume:

$$\rho = \frac{m}{V}. \tag{10–1}$$

**Specific gravity (SG)** is the ratio of the density of the material to the density of water (at 4°C).

**Pressure** is defined as force per unit area:

$$P = \frac{F}{A}. \tag{10–2}$$

The pressure $P$ at a depth $h$ in a liquid of constant density $\rho$, due to the weight of the liquid, is given by

$$P = \rho g h, \tag{10–3a}$$

where $g$ is the acceleration due to gravity.

**Pascal's principle** says that an external pressure applied to a confined fluid is transmitted throughout the fluid.

Pressure is measured using a **manometer** or other type of gauge. A **barometer** is used to measure atmospheric pressure. Standard **atmospheric pressure** (average at sea level) is $1.013 \times 10^5 \text{ N/m}^2$. **Gauge pressure** is the total (absolute) pressure minus atmospheric pressure.

**Archimedes' principle** states that an object submerged wholly or partially in a fluid is buoyed up by a force equal to the weight of fluid it displaces ($F_B = m_F g = \rho_F V_{displ} g$).

Fluid flow can be characterized either as **streamline** (also called **laminar**), in which the layers of fluid move smoothly and regularly along paths called **streamlines**, or as **turbulent**, in which case the flow is not smooth and regular but is characterized by irregularly shaped whirlpools.

Fluid flow rate is the mass or volume of fluid that passes a given point per unit time. The **equation of continuity** states that for an incompressible fluid flowing in an enclosed tube, the product of the velocity of flow and the cross-sectional area of the tube remains constant:

$$Av = \text{constant}. \tag{10–4}$$

**Bernoulli's principle** tells us that where the velocity of a fluid is high, the pressure in it is low, and where the velocity is low, the pressure is high. For steady laminar flow of an incompressible and nonviscous fluid, **Bernoulli's equation**, which is based on the law of conservation of energy, is

$$P_2 + \tfrac{1}{2}\rho v_2^2 + \rho g y_2 = P_1 + \tfrac{1}{2}\rho v_1^2 + \rho g y_1, \tag{10–5}$$

for two points along the flow.

[***Viscosity** refers to friction within a fluid and is essentially a frictional force between adjacent layers of fluid as they move past one another.]

[*Liquid surfaces hold together as if under tension (**surface tension**), allowing drops to form and objects like needles and insects to stay on the surface.]

# Questions

1. If one material has a higher density than another, must the molecules of the first be heavier than those of the second? Explain.

2. Consider what happens when you push both a pin and the blunt end of a pen against your skin with the same force. Decide what determines whether your skin is cut—the net force applied to it or the pressure.

3. A small amount of water is boiled in a 1-gallon metal can. The can is removed from the heat and the lid put on. As the can cools, it collapses and looks crushed. Explain.

4. An ice cube floats in a glass of water filled to the brim. What can you say about the density of ice? As the ice melts, will the water overflow? Explain.

5. Will an ice cube float in a glass of alcohol? Why or why not?

6. A submerged can of Coke® will sink, but a can of Diet Coke® will float. (Try it!) Explain.

7. Why don't ships made of iron sink?

8. A barge filled high with sand approaches a low bridge over the river and cannot quite pass under it. Should sand be added to, or removed from, the barge? [*Hint:* Consider Archimedes' principle.]

9. Explain why helium weather balloons, which are used to measure atmospheric conditions at high altitudes, are normally released while filled to only 10–20% of their maximum volume.

10. Will an empty balloon have precisely the same apparent weight on a scale as a balloon filled with air? Explain.

11. Why do you float higher in salt water than in fresh water?

12. Why does the stream of water from a faucet become narrower as it falls (Fig. 10–43)?

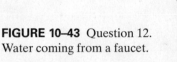

**FIGURE 10–43** Question 12. Water coming from a faucet.

13. Children are told to avoid standing too close to a rapidly moving train because they might get sucked under it. Is this possible? Explain.

14. A tall Styrofoam cup is filled with water. Two holes are punched in the cup near the bottom, and water begins rushing out. If the cup is dropped so it falls freely, will the water continue to flow from the holes? Explain.

15. Why do airplanes normally take off into the wind?

16. Two ships moving in parallel paths close to one another risk colliding. Why?

17. If you dangle two pieces of paper vertically, a few inches apart (Fig. 10–44), and blow between them, how do you think the papers will move? Try it and see. Explain.

**FIGURE 10–44** Question 17.

18. Why does the canvas top of a convertible bulge out when the car is traveling at high speed? [*Hint:* The windshield deflects air upward, pushing streamlines closer together.]

19. Roofs of houses are sometimes "blown" off (or are they pushed off?) during a tornado or hurricane. Explain using Bernoulli's principle.

20. Explain how the tube in Fig. 10–45, known as a **siphon**, can transfer liquid from one container to a lower one even though the liquid must flow uphill for part of its journey. (Note that the tube must be filled with liquid to start with.)

**FIGURE 10–45** Question 20. A siphon.

*21. When blood pressure is measured, why must the arm cuff be held at the level of the heart?

# MisConceptual Questions

1. You hold a piece of wood in one hand and a piece of iron in the other. Both pieces have the same volume, and you hold them fully under water at the same depth. At the moment you let go of them, which one experiences the greater buoyancy force?
   (*a*) The piece of wood.
   (*b*) The piece of iron.
   (*c*) They experience the same buoyancy force.
   (*d*) More information is needed.

2. Three containers are filled with water to the same height and have the same surface area at the base, but the total weight of water is different for each (Fig. 10–46). In which container does the water exert the greatest force on the bottom of the container?
   (*a*) Container A.
   (*b*) Container B.
   (*c*) Container C.
   (*d*) All three are equal.

**FIGURE 10–46**
MisConceptual Question 2.

3. Beaker A is filled to the brim with water. Beaker B is the same size and contains a small block of wood which floats when the beaker is filled with water to the brim. Which beaker weighs more?
   (*a*) Beaker A.
   (*b*) Beaker B.
   (*c*) The same for both.

4. Why does an ocean liner float?
   (*a*) It is made of steel, which floats.
   (*b*) Its very big size changes the way water supports it.
   (*c*) It is held up in the water by large Styrofoam compartments.
   (*d*) The average density of the ocean liner is less than that of seawater.
   (*e*) Remember the *Titanic*—ocean liners do not float.

5. A rowboat floats in a swimming pool, and the level of the water at the edge of the pool is marked. Consider the following situations. (i) The boat is removed from the water. (ii) The boat in the water holds an iron anchor which is removed from the boat and placed on the shore. For each situation, the level of the water will
   (*a*) rise.    (*b*) fall.    (*c*) stay the same.

**6.** You put two ice cubes in a glass and fill the glass to the rim with water. As the ice melts, the water level
  (a) drops below the rim.
  (b) rises and water spills out of the glass.
  (c) remains the same.
  (d) drops at first, then rises until a little water spills out.

**7.** Hot air is less dense than cold air. Could a hot-air balloon be flown on the Moon, where there is no atmosphere?
  (a) No, there is no cold air to displace, so no buoyancy force would exist.
  (b) Yes, warm air always rises, especially in a weak gravitational field like that of the Moon.
  (c) Yes, but the balloon would have to be filled with helium instead of hot air.

**8.** An object that can float in both water and in oil (whose density is less than that of water) experiences a buoyant force that is
  (a) greater when it is floating in oil than when floating in water.
  (b) greater when it is floating in water than when floating in oil.
  (c) the same when it is floating in water or in oil.

**9.** As water flows from a low elevation to a higher elevation through a pipe that changes in diameter,
  (a) the water pressure will increase.
  (b) the water pressure will decrease.
  (c) the water pressure will stay the same.
  (d) Need more information to determine how the water pressure changes.

**10.** Water flows in a horizontal pipe that is narrow but then widens and the speed of the water becomes less. The pressure in the water moving in the pipe is
  (a) greater in the wide part.
  (b) greater in the narrow part.
  (c) the same in both parts.
  (d) greater where the speed is higher.
  (e) greater where the speed is lower.

**11.** When a baseball curves to the right (a curveball), air is flowing
  (a) faster over the left side than over the right side.
  (b) faster over the right side than over the left side.
  (c) faster over the top than underneath.
  (d) at the same speed all around the baseball, but the ball curves as a result of the way the wind is blowing on the field.

**12.** How is the smoke drawn up a chimney affected when a wind is blowing outside?
  (a) Smoke rises more rapidly in the chimney.
  (b) Smoke rises more slowly in the chimney.
  (c) Smoke is forced back down the chimney.
  (d) Smoke is unaffected.

For assigned homework and other learning materials, go to the MasteringPhysics website.

# Problems

### 10–2 Density and Specific Gravity

**1.** (I) The approximate volume of the granite monolith known as El Capitan in Yosemite National Park (Fig. 10–47) is about $10^8 \, m^3$. What is its approximate mass?

**FIGURE 10–47** Problem 1.

**2.** (I) What is the approximate mass of air in a living room 5.6 m × 3.6 m × 2.4 m?

**3.** (I) If you tried to smuggle gold bricks by filling your backpack, whose dimensions are 54 cm × 31 cm × 22 cm, what would its mass be?

**4.** (I) State your mass and then estimate your volume. [*Hint*: Because you can swim on or just under the surface of the water in a swimming pool, you have a pretty good idea of your density.]

**5.** (II) A bottle has a mass of 35.00 g when empty and 98.44 g when filled with water. When filled with another fluid, the mass is 89.22 g. What is the specific gravity of this other fluid?

**6.** (II) If 4.0 L of antifreeze solution (specific gravity = 0.80) is added to 5.0 L of water to make a 9.0-L mixture, what is the specific gravity of the mixture?

**7.** (III) The Earth is not a uniform sphere, but has regions of varying density. Consider a simple model of the Earth divided into three regions—inner core, outer core, and mantle. Each region is taken to have a unique constant density (the average density of that region in the real Earth):

| Region | Radius (km) | Density (kg/m³) |
|---|---|---|
| Inner Core | 0–1220 | 13,000 |
| Outer Core | 1220–3480 | 11,100 |
| Mantle | 3480–6380 | 4400 |

(a) Use this model to predict the average density of the entire Earth. (b) If the radius of the Earth is 6380 km and its mass is $5.98 \times 10^{24}$ kg, determine the actual average density of the Earth and compare it (as a percent difference) with the one you determined in (a).

## 10–3 to 10–6 Pressure; Pascal's Principle

**8.** (I) Estimate the pressure needed to raise a column of water to the same height as a 46-m-tall pine tree.

**9.** (I) Estimate the pressure exerted on a floor by (a) one pointed heel of area = 0.45 cm², and (b) one wide heel of area 16 cm², Fig. 10–48. The person wearing the shoes has a mass of 56 kg.

**FIGURE 10–48** Problem 9.

**10.** (I) What is the difference in blood pressure (mm-Hg) between the top of the head and bottom of the feet of a 1.75-m-tall person standing vertically?

**11.** (I) (a) Calculate the total force of the atmosphere acting on the top of a table that measures 1.7 m × 2.6 m. (b) What is the total force acting upward on the underside of the table?

**12.** (II) How high would the level be in an alcohol barometer at normal atmospheric pressure?

**13.** (II) In a movie, Tarzan evades his captors by hiding under water for many minutes while breathing through a long, thin reed. Assuming the maximum pressure difference his lungs can manage and still breathe is −85 mm-Hg, calculate the deepest he could have been.

**14.** (II) The maximum gauge pressure in a hydraulic lift is 17.0 atm. What is the largest-size vehicle (kg) it can lift if the diameter of the output line is 25.5 cm?

**15.** (II) The gauge pressure in each of the four tires of an automobile is 240 kPa. If each tire has a "footprint" of 190 cm² (area touching the ground), estimate the mass of the car.

**16.** (II) (a) Determine the total force and the absolute pressure on the bottom of a swimming pool 28.0 m by 8.5 m whose uniform depth is 1.8 m. (b) What will be the pressure against the *side* of the pool near the bottom?

**17.** (II) A house at the bottom of a hill is fed by a full tank of water 6.0 m deep and connected to the house by a pipe that is 75 m long at an angle of 61° from the horizontal (Fig. 10–49). (a) Determine the water gauge pressure at the house. (b) How high could the water shoot if it came vertically out of a broken pipe in front of the house?

**FIGURE 10–49** Problem 17.

**18.** (II) Water and then oil (which don't mix) are poured into a U-shaped tube, open at both ends. They come to equilibrium as shown in Fig. 10–50. What is the density of the oil? [*Hint*: Pressures at points a and b are equal. Why?]

**FIGURE 10–50**
Problem 18.

**19.** (II) How high would the atmosphere extend if it were of uniform density throughout, equal to half the present density at sea level?

**20.** (II) Determine the minimum gauge pressure needed in the water pipe leading into a building if water is to come out of a faucet on the fourteenth floor, 44 m above that pipe.

**21.** (II) A **hydraulic press** for compacting powdered samples has a large cylinder which is 10.0 cm in diameter, and a small cylinder with a diameter of 2.0 cm (Fig. 10–51). A lever is attached to the small cylinder as shown. The sample, which is placed on the large cylinder, has an area of 4.0 cm². What is the pressure on the sample if 320 N is applied to the lever?

**FIGURE 10–51** Problem 21.

**22.** (II) An open-tube mercury manometer is used to measure the pressure in an oxygen tank. When the atmospheric pressure is 1040 mbar, what is the absolute pressure (in Pa) in the tank if the height of the mercury in the open tube is (a) 18.5 cm higher, (b) 5.6 cm lower, than the mercury in the tube connected to the tank? See Fig. 10–7a.

## 10–7 Buoyancy and Archimedes' Principle

**23.** (II) What fraction of a piece of iron will be submerged when it floats in mercury?

**24.** (II) A geologist finds that a Moon rock whose mass is 9.28 kg has an apparent mass of 6.18 kg when submerged in water. What is the density of the rock?

**25.** (II) A crane lifts the 18,000-kg steel hull of a sunken ship out of the water. Determine (a) the tension in the crane's cable when the hull is fully submerged in the water, and (b) the tension when the hull is completely out of the water.

**26.** (II) A spherical balloon has a radius of 7.15 m and is filled with helium. How large a cargo can it lift, assuming that the skin and structure of the balloon have a mass of 930 kg? Neglect the buoyant force on the cargo volume itself.

**27.** (II) What is the likely identity of a metal (see Table 10–1) if a sample has a mass of 63.5 g when measured in air and an apparent mass of 55.4 g when submerged in water?

28. (II) Calculate the true mass (in vacuum) of a piece of aluminum whose apparent mass is 4.0000 kg when weighed in air.

29. (II) Because gasoline is less dense than water, drums containing gasoline will float in water. Suppose a 210-L steel drum is completely full of gasoline. What total volume of steel can be used in making the drum if the gasoline-filled drum is to float in fresh water?

30. (II) A scuba diver and her gear displace a volume of 69.6 L and have a total mass of 72.8 kg. (a) What is the buoyant force on the diver in seawater? (b) Will the diver sink or float?

31. (II) The specific gravity of ice is 0.917, whereas that of seawater is 1.025. What percent of an iceberg is above the surface of the water?

32. (II) Archimedes' principle can be used to determine the specific gravity of a solid using a known liquid (Example 10–8). The reverse can be done as well. (a) As an example, a 3.80-kg aluminum ball has an apparent mass of 2.10 kg when submerged in a particular liquid: calculate the density of the liquid. (b) Determine a formula for finding the density of a liquid using this procedure.

33. (II) A 32-kg child decides to make a raft out of empty 1.0-L soda bottles and duct tape. Neglecting the mass of the duct tape and plastic in the bottles, what minimum number of soda bottles will the child need to be able stay dry on the raft?

34. (II) An undersea research chamber is spherical with an external diameter of 5.20 m. The mass of the chamber, when occupied, is 74,400 kg. It is anchored to the sea bottom by a cable. What is (a) the buoyant force on the chamber, and (b) the tension in the cable?

35. (II) A 0.48-kg piece of wood floats in water but is found to sink in alcohol (SG = 0.79), in which it has an apparent mass of 0.047 kg. What is the SG of the wood?

36. (II) A two-component model used to determine percent body fat in a human body assumes that a fraction $f$ ($< 1$) of the body's total mass $m$ is composed of fat with a density of $0.90 \text{ g/cm}^3$, and that the remaining mass of the body is composed of fat-free tissue with a density of $1.10 \text{ g/cm}^3$. If the specific gravity of the entire body's density is $X$, show that the percent body fat ($= f \times 100$) is given by

$$\% \text{ Body fat} = \frac{495}{X} - 450.$$

37. (II) On dry land, an athlete weighs 70.2 kg. The same athlete, when submerged in a swimming pool and hanging from a scale, has an "apparent weight" of 3.4 kg. Using Example 10–8 as a guide, (a) find the total volume $V$ of the submerged athlete. (b) Assume that when submerged, the athlete's body contains a residual volume $V_R = 1.3 \times 10^{-3} \text{ m}^3$ of air (mainly in the lungs). Taking $V - V_R$ to be the actual volume of the athlete's body, find the body's specific gravity, SG. (c) What is the athlete's percent body fat assuming it is given by the formula $(495/\text{SG}) - 450$?

38. (II) How many helium-filled balloons would it take to lift a person? Assume the person has a mass of 72 kg and that each helium-filled balloon is spherical with a diameter of 33 cm.

39. (III) A scuba tank, when fully submerged, displaces 15.7 L of seawater. The tank itself has a mass of 14.0 kg and, when "full," contains 3.00 kg of air. Assuming only its weight and the buoyant force act on the tank, determine the net force (magnitude and direction) on the fully submerged tank at the beginning of a dive (when it is full of air) and at the end of a dive (when it no longer contains any air).

40. (III) A 3.65-kg block of wood (SG = 0.50) floats on water. What minimum mass of lead, hung from the wood by a string, will cause the block to sink?

## 10–8 to 10–10 Fluid Flow, Bernoulli's Equation

41. (I) A 12-cm-radius air duct is used to replenish the air of a room 8.2 m × 5.0 m × 3.5 m every 12 min. How fast does the air flow in the duct?

42. (I) Calculate the average speed of blood flow in the major arteries of the body, which have a total cross-sectional area of about 2.0 cm². Use the data of Example 10–12.

43. (I) How fast does water flow from a hole at the bottom of a very wide, 4.7-m-deep storage tank filled with water? Ignore viscosity.

44. (I) Show that Bernoulli's equation reduces to the hydrostatic variation of pressure with depth (Eq. 10–3b) when there is no flow $(v_1 = v_2 = 0)$.

45. (II) What is the volume rate of flow of water from a 1.85-cm-diameter faucet if the pressure head is 12.0 m?

46. (II) A fish tank has dimensions 36 cm wide by 1.0 m long by 0.60 m high. If the filter should process all the water in the tank once every 3.0 h, what should the flow speed be in the 3.0-cm-diameter input tube for the filter?

47. (II) What gauge pressure in the water pipes is necessary if a fire hose is to spray water to a height of 16 m?

48. (II) A $\frac{5}{8}$-in. (inside) diameter garden hose is used to fill a round swimming pool 6.1 m in diameter. How long will it take to fill the pool to a depth of 1.4 m if water flows from the hose at a speed of 0.40 m/s?

49. (II) A 180-km/h wind blowing over the flat roof of a house causes the roof to lift off the house. If the house is 6.2 m × 12.4 m in size, estimate the weight of the roof. Assume the roof is not nailed down.

50. (II) A 6.0-cm-diameter horizontal pipe gradually narrows to 4.5 cm. When water flows through this pipe at a certain rate, the gauge pressure in these two sections is 33.5 kPa and 22.6 kPa, respectively. What is the volume rate of flow?

51. (II) Estimate the air pressure inside a category 5 hurricane, where the wind speed is 300 km/h (Fig. 10–52).

**FIGURE 10–52** Problem 51.

52. (II) What is the lift (in newtons) due to Bernoulli's principle on a wing of area 88 m² if the air passes over the top and bottom surfaces at speeds of 280 m/s and 150 m/s, respectively?

53. (II) Water at a gauge pressure of 3.8 atm at street level flows into an office building at a speed of 0.78 m/s through a pipe 5.0 cm in diameter. The pipe tapers down to 2.8 cm in diameter by the top floor, 16 m above (Fig. 10–53), where the faucet has been left open. Calculate the flow velocity and the gauge pressure in the pipe on the top floor. Assume no branch pipes and ignore viscosity.

**FIGURE 10–53**
Problem 53.

54. (II) Show that the power needed to drive a fluid through a pipe with uniform cross-section is equal to the volume rate of flow, $Q$, times the pressure difference, $P_1 - P_2$. Ignore viscosity.

55. (III) In Fig. 10–54, take into account the speed of the top surface of the tank and show that the speed of fluid leaving an opening near the bottom is

$$v_1 = \sqrt{\frac{2gh}{\left(1 - A_1^2/A_2^2\right)}},$$

where $h = y_2 - y_1$, and $A_1$ and $A_2$ are the areas of the opening and of the top surface, respectively. Assume $A_1 \ll A_2$ so that the flow remains nearly steady and laminar.

**FIGURE 10–54**
Problem 55.

56. (III) (a) Show that the flow speed measured by a venturi meter (see Fig. 10–29) is given by the relation

$$v_1 = A_2 \sqrt{\frac{2(P_1 - P_2)}{\rho(A_1^2 - A_2^2)}}.$$

(b) A venturi meter is measuring the flow of water; it has a main diameter of 3.5 cm tapering down to a throat diameter of 1.0 cm. If the pressure difference is measured to be 18 mm-Hg, what is the speed of the water entering the venturi throat?

57. (III) A fire hose exerts a force on the person holding it. This is because the water accelerates as it goes from the hose through the nozzle. How much force is required to hold a 7.0-cm-diameter hose delivering 420 L/min through a 0.75-cm-diameter nozzle?

## *10–11 Viscosity

*58. (II) A viscometer consists of two concentric cylinders, 10.20 cm and 10.60 cm in diameter. A liquid fills the space between them to a depth of 12.0 cm. The outer cylinder is fixed, and a torque of 0.024 m · N keeps the inner cylinder turning at a steady rotational speed of 57 rev/min. What is the viscosity of the liquid?

## *10–12 Flow in Tubes; Poiseuille's Equation

*59. (I) Engine oil (assume SAE 10, Table 10–3) passes through a fine 1.80-mm-diameter tube that is 10.2 cm long. What pressure difference is needed to maintain a flow rate of 6.2 mL/min?

*60. (I) A gardener feels it is taking too long to water a garden with a $\frac{3}{8}$-in.-diameter hose. By what factor will the time be cut using a $\frac{5}{8}$-in.-diameter hose instead? Assume nothing else is changed.

*61. (II) What diameter must a 15.5-m-long air duct have if the ventilation and heating system is to replenish the air in a room 8.0 m × 14.0 m × 4.0 m every 15.0 min? Assume the pump can exert a gauge pressure of $0.710 \times 10^{-3}$ atm.

*62. (II) What must be the pressure difference between the two ends of a 1.6-km section of pipe, 29 cm in diameter, if it is to transport oil $(\rho = 950 \text{ kg/m}^3, \ \eta = 0.20 \text{ Pa} \cdot \text{s})$ at a rate of 650 cm³/s?

*63. (II) Poiseuille's equation does not hold if the flow velocity is high enough that turbulence sets in. The onset of turbulence occurs when the **Reynolds number**, $Re$, exceeds approximately 2000. $Re$ is defined as

$$Re = \frac{2\bar{v}r\rho}{\eta},$$

where $\bar{v}$ is the average speed of the fluid, $\rho$ is its density, $\eta$ is its viscosity, and $r$ is the radius of the tube in which the fluid is flowing. (a) Determine if blood flow through the aorta is laminar or turbulent when the average speed of blood in the aorta $(r = 0.80 \text{ cm})$ during the resting part of the heart's cycle is about 35 cm/s. (b) During exercise, the blood-flow speed approximately doubles. Calculate the Reynolds number in this case, and determine if the flow is laminar or turbulent.

*64. (II) Assuming a constant pressure gradient, if blood flow is reduced by 65%, by what factor is the radius of a blood vessel decreased?

*65. (II) Calculate the pressure drop per cm along the aorta using the data of Example 10–12 and Table 10–3.

*66. (III) A patient is to be given a blood transfusion. The blood is to flow through a tube from a raised bottle to a needle inserted in the vein (Fig. 10–55). The inside diameter of the 25-mm-long needle is 0.80 mm, and the required flow rate is 2.0 cm³ of blood per minute. How high $h$ should the bottle be placed above the needle? Obtain $\rho$ and $\eta$ from the Tables. Assume the blood pressure is 78 torr above atmospheric pressure.

**FIGURE 10–55**
Problems 66 and 74.

*67. (I) If the force $F$ needed to move the wire in Fig. 10–34 is $3.4 \times 10^{-3}$ N, calculate the surface tension $\gamma$ of the enclosed fluid. Assume $\ell = 0.070$ m.

*68. (I) Calculate the force needed to move the wire in Fig. 10–34 if it holds a soapy solution (Table 10–4) and the wire is 21.5 cm long.

*69. (II) The surface tension of a liquid can be determined by measuring the force $F$ needed to just lift a circular platinum ring of radius $r$ from the surface of the liquid. (a) Find a formula for $\gamma$ in terms of $F$ and $r$. (b) At 30°C, if $F = 6.20 \times 10^{-3}$ N and $r = 2.9$ cm, calculate $\gamma$ for the tested liquid.

*70. (II) If the base of an insect's leg has a radius of about $3.0 \times 10^{-5}$ m and the insect's mass is 0.016 g, would you expect the six-legged insect to remain on top of the water? Why or why not?

*71. (III) Estimate the diameter of a steel needle that can just barely remain on top of water due to surface tension.

*72. (II) A physician judges the health of a heart by measuring the pressure with which it pumps blood. If the physician mistakenly attaches the pressurized cuff around a standing patient's calf (about 1 m below the heart) instead of the arm (Fig. 10–42), what error (in Pa) would be introduced in the heart's blood pressure measurement?

# General Problems

73. A 3.2-N force is applied to the plunger of a hypodermic needle. If the diameter of the plunger is 1.3 cm and that of the needle is 0.20 mm, (a) with what force does the fluid leave the needle? (b) What force on the plunger would be needed to push fluid into a vein where the gauge pressure is 75 mm-Hg? Answer for the instant just before the fluid starts to move.

74. Intravenous transfusions are often made under gravity, as shown in Fig. 10–55. Assuming the fluid has a density of $1.00$ g/cm$^3$, at what height $h$ should the bottle be placed so the liquid pressure is (a) 52 mm-Hg, and (b) 680 mm-H$_2$O? (c) If the blood pressure is 75 mm-Hg above atmospheric pressure, how high should the bottle be placed so that the fluid just barely enters the vein?

75. A beaker of water rests on an electronic balance that reads 975.0 g. A 2.6-cm-diameter solid copper ball attached to a string is submerged in the water, but does not touch the bottom. What are the tension in the string and the new balance reading?

76. Estimate the difference in air pressure between the top and the bottom of the Empire State Building in New York City. It is 380 m tall and is located at sea level. Express as a fraction of atmospheric pressure at sea level.

77. A hydraulic lift is used to jack a 960-kg car 42 cm off the floor. The diameter of the output piston is 18 cm, and the input force is 380 N. (a) What is the area of the input piston? (b) What is the work done in lifting the car 42 cm? (c) If the input piston moves 13 cm in each stroke, how high does the car move up for each stroke? (d) How many strokes are required to jack the car up 42 cm? (e) Show that energy is conserved.

78. When you ascend or descend a great deal when driving in a car, your ears "pop," which means that the pressure behind the eardrum is being equalized to that outside. If this did not happen, what would be the approximate force on an eardrum of area 0.20 cm$^2$ if a change in altitude of 1250 m takes place?

79. Giraffes are a wonder of cardiovascular engineering. Calculate the difference in pressure (in atmospheres) that the blood vessels in a giraffe's head must accommodate as the head is lowered from a full upright position to ground level for a drink. The height of an average giraffe is about 6 m.

80. How high should the pressure head be if water is to come from a faucet at a speed of 9.2 m/s? Ignore viscosity.

81. Suppose a person can reduce the pressure in his lungs to $-75$ mm-Hg gauge pressure. How high can water then be "sucked" up a straw?

82. A bicycle pump is used to inflate a tire. The initial tire (gauge) pressure is 210 kPa (30 psi). At the end of the pumping process, the final pressure is 310 kPa (45 psi). If the diameter of the plunger in the cylinder of the pump is 2.5 cm, what is the range of the force that needs to be applied to the pump handle from beginning to end?

83. Estimate the pressure on the mountains underneath the Antarctic ice sheet, which is typically 2 km thick.

84. A simple model (Fig. 10–56) considers a continent as a block (density $\approx 2800$ kg/m$^3$) floating in the mantle rock around it (density $\approx 3300$ kg/m$^3$). Assuming the continent is 35 km thick (the average thickness of the Earth's continental crust), estimate the height of the continent above the surrounding mantle rock.

FIGURE 10–56 Problem 84.

85. A ship, carrying fresh water to a desert island in the Caribbean, has a horizontal cross-sectional area of 2240 m$^2$ at the waterline. When unloaded, the ship rises 8.25 m higher in the sea. How much water (m$^3$) was delivered?

86. A raft is made of 12 logs lashed together. Each is 45 cm in diameter and has a length of 6.5 m. How many people can the raft hold before they start getting their feet wet, assuming the average person has a mass of 68 kg? Do not neglect the weight of the logs. Assume the specific gravity of wood is 0.60.

87. Estimate the total mass of the Earth's atmosphere, using the known value of atmospheric pressure at sea level.

**88.** During each heartbeat, approximately 70 cm$^3$ of blood is pushed from the heart at an average pressure of 105 mm-Hg. Calculate the power output of the heart, in watts, assuming 70 beats per minute.

**89.** Four lawn sprinkler heads are fed by a 1.9-cm-diameter pipe. The water comes out of the heads at an angle of 35° above the horizontal and covers a radius of 6.0 m. (*a*) What is the velocity of the water coming out of each sprinkler head? (Assume zero air resistance.) (*b*) If the output diameter of each head is 3.0 mm, how many liters of water do the four heads deliver per second? (*c*) How fast is the water flowing inside the 1.9-cm-diameter pipe?

**90.** One arm of a U-shaped tube (open at both ends) contains water, and the other alcohol. If the two fluids meet at exactly the bottom of the U, and the alcohol is at a height of 16.0 cm, at what height will the water be?

**91.** The contraction of the left ventricle (chamber) of the heart pumps blood to the body. Assuming that the inner surface of the left ventricle has an area of 82 cm$^2$ and the maximum pressure in the blood is 120 mm-Hg, estimate the force exerted by that ventricle at maximum pressure.

**92.** An airplane has a mass of $1.7 \times 10^6$ kg, and the air flows past the lower surface of the wings at 95 m/s. If the wings have a surface area of 1200 m$^2$, how fast must the air flow over the upper surface of the wing if the plane is to stay in the air?

**93.** A drinking fountain shoots water about 12 cm up in the air from a nozzle of diameter 0.60 cm (Fig. 10–57). The pump at the base of the unit (1.1 m below the nozzle) pushes water into a 1.2-cm-diameter supply pipe that goes up to the nozzle. What gauge pressure does the pump have to provide? Ignore the viscosity; your answer will therefore be an underestimate.

**FIGURE 10–57**
Problem 93.

**94.** A hurricane-force wind of 180 km/h blows across the face of a storefront window. Estimate the force on the 2.0 m × 3.0 m window due to the difference in air pressure inside and outside the window. Assume the store is airtight so the inside pressure remains at 1.0 atm. (This is why you should not tightly seal a building in preparation for a hurricane.)

**\*95.** Blood is placed in a bottle 1.40 m above a 3.8-cm-long needle, of inside diameter 0.40 mm, from which it flows at a rate of 4.1 cm$^3$/min. What is the viscosity of this blood?

**96.** You are watering your lawn with a hose when you put your finger over the hose opening to increase the distance the water reaches. If you are holding the hose horizontally, and the distance the water reaches increases by a factor of 4, what fraction of the hose opening did you block?

**97.** A copper (Cu) weight is placed on top of a 0.40-kg block of wood (density = $0.60 \times 10^3$ kg/m$^3$) floating in water, as shown in Fig. 10–58. What is the mass of the copper if the top of the wood block is exactly at the water's surface?

**FIGURE 10–58** Problem 97.

**98.** You need to siphon water from a clogged sink. The sink has an area of 0.38 m$^2$ and is filled to a height of 4.0 cm. Your siphon tube rises 45 cm above the bottom of the sink and then descends 85 cm to a pail as shown in Fig. 10–59. The siphon tube has a diameter of 2.3 cm. (*a*) Assuming that the water level in the sink has almost zero velocity, use Bernoulli's equation to estimate the water velocity when it enters the pail. (*b*) Estimate how long it will take to empty the sink. Ignore viscosity.

**FIGURE 10–59**
Problem 98.

**\*99.** If cholesterol buildup reduces the diameter of an artery by 25%, by what % will the blood flow rate be reduced, assuming the same pressure difference?

# Search and Learn

1. A 5.0-kg block and 4.0 kg of water in a 0.50-kg container are placed symmetrically on a board that can balance at the center (Fig. 10–60). A solid aluminum cube of sides 10.0 cm is lowered into the water. How much of the aluminum must be under water to make this system balance? How would your answer change for a lead cube of the same size? Explain. (See Sections 10–7 and 9–1.)

5.0 kg    10.0 cm    4.5 kg

**FIGURE 10–60** Search and Learn 1.

2. (a) Show that the buoyant force $F_B$ on a partially submerged object such as a ship acts at the center of gravity of the fluid before it is displaced, Fig. 10–61. This point is called the **center of buoyancy**. (b) To ensure that a ship is in stable equilibrium, would it be better if its center of buoyancy was above, below, or at the same point as its center of gravity? Explain. (See Section 10–7 and Chapter 9.)

$\vec{F}_B$

$m\vec{g}$

**FIGURE 10–61**
Search and Learn 2.

3. In working out his principle, Pascal showed dramatically how force can be multiplied with fluid pressure. He placed a long, thin tube of radius $r = 0.30$ cm vertically into a wine barrel of radius $R = 21$ cm, Fig. 10–62. He found that when the barrel was filled with water and the tube filled to a height of 12 m, the barrel burst. Calculate (a) the mass of water in the tube, and (b) the net force exerted by the water in the barrel on the lid just before rupture.

$r = 0.30$ cm

12 m

$R = 21$ cm

**FIGURE 10–62** Search and Learn 3 (not to scale).

4. (a) When submerged in water, two objects with different volumes have the same *apparent* weight. When taken out of water, compare their weights in air. (b) Which object has the greater density?

5. A tub of water rests on a scale as shown in Fig. 10–63. The weight of the tub plus water is 100 N. A 50-N concrete brick is then lowered down from a fixed arm into the water but does not touch the tub. What does the scale read now? [*Hint*: Draw two free-body diagrams, one for the brick and a second one for the tub + water + brick.]

50 N

**FIGURE 10–63** Search and Learn 5.

6. What approximations are made in the derivation of Bernoulli's equation? Qualitatively, how do you think Bernoulli's equation would change if each of these approximations was not made? (See Sections 10–8, 10–9, 10–11, and 10–12.)

*7. Estimate the density of the water 5.4 km deep in the sea. (See Table 9–1 and Section 9–5 regarding bulk modulus.) By what fraction does it differ from the density at the surface?

## ANSWERS TO EXERCISES

A: (d).
B: The same. Pressure depends on depth, not on length.
C: (a).
D: (e).

E: The rowboat is shaped to have a lot of empty, air-filled space, so its "average" density is much lower than that of water (unless the boat becomes full of water, in which case it sinks). Steel ships float for the same reason.
F: Increases.
G: (b).

An object attached to a coil spring can exhibit oscillatory motion. Many kinds of oscillatory motion are sinusoidal in time, or nearly so, and are referred to as simple harmonic motion. Real systems generally have at least some friction, causing the motion to be damped. The automobile spring shown here has a shock absorber (yellow) that purposefully dampens the oscillation to make for a smooth ride. When an external sinusoidal force is exerted on a system able to oscillate, resonance occurs if the driving force is at or near the natural frequency of oscillation.

Vibrations can give rise to waves—such as water waves or waves traveling along a cord—which travel outward from their source.

# Oscillations and Waves

## CONTENTS

## CHAPTER-OPENING QUESTIONS—Guess now!

**1.** A simple pendulum consists of a mass $m$ (the "bob") hanging on the end of a thin string of length $\ell$ and negligible mass. The bob is pulled sideways so the string makes a 5.0° angle to the vertical; when released, it oscillates back and forth at a frequency $f$. If the pendulum is started at a 10.0° angle instead, its frequency would be

(a) twice as great.
(b) half as great.
(c) the same, or very close to it.
(d) not quite twice as great.
(e) a bit more than half as great.

**2.** You drop a rock into a pond, and water waves spread out in circles.
(a) The waves carry water outward, away from where the rock hit. That moving water carries energy outward.
(b) The waves only make the water move up and down. No energy is carried outward from where the rock hit.
(c) The waves only make the water move up and down, but the waves do carry energy outward, away from where the rock hit.

Many objects vibrate or oscillate—an object on the end of a spring, a tuning fork, the balance wheel of an old watch, a pendulum, a plastic ruler held firmly over the edge of a table and gently struck, the strings of a guitar or piano. Spiders detect prey by the vibrations of their webs; cars oscillate up and down when they hit a bump; buildings and bridges vibrate when heavy trucks pass or the wind is fierce. Indeed, because most solids are elastic (see Section 9–5), they vibrate (at least briefly) when given an impulse. Electrical oscillations occur in radio and television sets. At the atomic level, atoms oscillate within a molecule, and the atoms of a solid oscillate about their relatively fixed positions.

Because it is so common in everyday life and occurs in so many areas of physics, oscillatory (or vibrational) motion is of great importance. Mechanical oscillations or vibrations are fully described on the basis of Newtonian mechanics.

Vibrations and wave motion are intimately related. Waves—whether ocean waves, waves on a string, earthquake waves, or sound waves in air—have as their source a vibration. In the case of sound, not only is the source a vibrating object, but so is the detector—the eardrum or the membrane of a microphone. Indeed, when a wave travels through a medium, the medium oscillates (such as air for sound waves). In the second half of this Chapter, after we discuss oscillations, we will discuss simple waves such as those on water or on a string. In Chapter 12 we will study sound waves, and in later Chapters we will encounter other forms of wave motion, including electromagnetic waves and light.

# 11–1 Simple Harmonic Motion— Spring Oscillations

When an object **vibrates** or **oscillates** back and forth, over the same path, each oscillation taking the same amount of time, the motion is **periodic**. The simplest form of periodic motion is represented by an object oscillating on the end of a uniform coil spring. Because many other types of oscillatory motion closely resemble this system, we will look at it in detail. We assume that the mass of the spring can be ignored, and that the spring is mounted horizontally, as shown in Fig. 11–1a, so that the object of mass $m$ slides without friction on the horizontal surface. Any spring has a natural length at which it exerts no force on the mass $m$. The position of the mass at this point is called the **equilibrium position**. If the mass is moved either to the left, which compresses the spring, or to the right, which stretches it, the spring exerts a force on the mass that acts in the direction of returning the mass to the equilibrium position; hence it is called a *restoring force*. We consider the common situation where we can assume the restoring force $F$ is directly proportional to the displacement $x$ the spring has been stretched (Fig. 11–1b) or compressed (Fig. 11–1c) from the equilibrium position:

$$F = -kx. \qquad \text{[force exerted by spring]} \quad \textbf{(11–1)}$$

Note that the equilibrium position has been chosen at $x = 0$ and the minus sign in Eq. 11–1 indicates that the restoring force is always in the direction opposite to the displacement $x$. For example, if we choose the positive direction to the right in Fig. 11–1, $x$ is positive when the spring is stretched (Fig. 11–1b), but the direction of the restoring force is to the left (negative direction). If the spring is compressed, $x$ is negative (to the left) but the force $F$ acts toward the right (Fig. 11–1c).

Equation 11–1 is often referred to as Hooke's law (Sections 6–4 and 9–5), and is accurate only if the spring is not compressed to where the coils are close to touching, or stretched beyond the elastic region (see Fig. 9–19). Hooke's law works not only for springs but for other oscillating solids as well; it thus has wide applicability, even though it is valid only over a certain range of $F$ and $x$ values.

The proportionality constant $k$ in Eq. 11–1 is called the *spring constant* for that particular spring, or its *spring stiffness constant* (units = N/m). To stretch the spring a distance $x$, an (external) force must be exerted on the free end of the spring with a magnitude at least equal to

$$F_{\text{ext}} = +kx. \qquad \text{[external force on spring]}$$

The greater the value of $k$, the greater the force needed to stretch a spring a given distance. That is, the stiffer the spring, the greater the spring constant $k$.

Note that the force $F$ in Eq. 11–1 is *not* a constant, but varies with position. Therefore the acceleration of the mass $m$ is not constant, so we *cannot* use the equations for constant acceleration developed in Chapter 2.

(a)

(b)

(c)

**FIGURE 11–1** An object of mass $m$ oscillating at the end of a uniform spring. The force $\vec{F}$ on the object at the different positions is shown *above* the object.

 **CAUTION**

*Eqs. 2–11 for constant acceleration do not apply to a spring*

(a) $x = -A$    $x = 0$

(b) $x = 0$

(c) $x = 0$    $x = A$

(d) $x = 0$

(e) $x = -A$    $x = 0$

**FIGURE 11–2** An object oscillating on a frictionless surface, indicating the force on the object and its velocity at different positions of its oscillation cycle.

⚠ **CAUTION**

*For vertical spring, measure displacement (x or y) from the vertical equilibrium position*

Let us examine what happens when our uniform spring is initially compressed a distance $x = -A$, as shown in Fig. 11–2a, and then our object of mass $m$ is released on the frictionless surface. The spring exerts a force on the mass that accelerates it toward the equilibrium position. Because the mass has inertia, it passes the equilibrium position with considerable speed. Indeed, as the mass reaches the equilibrium position, the force on it decreases to zero, but its speed at this point is a maximum, $v_{max}$ (Fig. 11–2b). As the mass moves farther to the right, the force on it acts to slow it down, and it stops for an instant at $x = A$ (Fig. 11–2c). It then begins moving back in the opposite direction, accelerating until it passes the equilibrium point (Fig. 11–2d), and then slows down until it reaches zero speed at the original starting point, $x = -A$ (Fig. 11–2e). It then repeats the motion, moving back and forth symmetrically between $x = A$ and $x = -A$.

**EXERCISE A** A mass is oscillating on a frictionless surface at the end of a horizontal spring. Where, if anywhere, is the acceleration of the mass zero (see Fig. 11–2)? (a) At $x = -A$; (b) at $x = 0$; (c) at $x = +A$; (d) at both $x = -A$ and $x = +A$; (e) nowhere.

To discuss oscillatory motion, we need to define a few terms. The distance $x$ of the mass from the equilibrium point at any moment is the **displacement** (with a + or − sign). The maximum displacement—the greatest distance from the equilibrium point—is called the **amplitude**, $A$. One **cycle** refers to the complete to-and-fro motion from some initial point back to that same point—say, from $x = -A$ to $x = +A$ and back to $x = -A$. The **period**, $T$, is defined as the time required to complete one cycle. Finally, the **frequency**, $f$, is the number of complete cycles per second. Frequency is generally specified in hertz (Hz), where $1\,\text{Hz} = 1$ cycle per second $(\text{s}^{-1})$. Given their definitions, frequency and period are inversely related, as we saw earlier (Eqs. 5–2 and 8–8):

$$ f = \frac{1}{T} \quad \text{and} \quad T = \frac{1}{f}. \tag{11–2} $$

For example, if the frequency is 2 cycles per second, then each cycle takes $\frac{1}{2}$s.

**EXERCISE B** If an oscillating mass has a frequency of 1.25 Hz, it makes 100 oscillations in (a) 12.5 s, (b) 125 s, (c) 80 s, (d) 8.0 s.

The oscillation of a spring hung vertically is similar to that of a horizontal spring; but because of gravity, the length of a vertical spring with a mass $m$ on the end will be longer at equilibrium than when that same spring is horizontal. See Fig. 11–3. The spring is in equilibrium when $\Sigma F = 0 = mg - kx_0$, so the spring stretches an extra amount $x_0 = mg/k$ to be in equilibrium. If $x$ is measured from this new equilibrium position, Eq. 11–1 can be used directly with the same value of $k$.

**FIGURE 11–3**
(a) Free spring, hung vertically.
(b) Mass $m$ attached to spring in new equilibrium position, which occurs when $\Sigma F = 0 = mg - kx_0$.

(a)          (b)

**EXAMPLE 11–1** **Car springs.** When a family of four with a total mass of 200 kg step into their 1200-kg car, the car's springs compress 3.0 cm. (a) What is the spring constant of the car's springs (Fig. 11–4), assuming they act as a single spring? (b) How far will the car lower if loaded with 300 kg rather than 200 kg?

**APPROACH** We use Hooke's law: the weight of the people, $mg$, causes a 3.0-cm displacement.

**SOLUTION** (a) The added force of $(200\,\text{kg})(9.8\,\text{m/s}^2) = 1960\,\text{N}$ causes the springs to compress $3.0 \times 10^{-2}\,\text{m}$. Therefore (Eq. 11–1), the spring constant is

$$k = \frac{F}{x} = \frac{1960\,\text{N}}{3.0 \times 10^{-2}\,\text{m}} = 6.5 \times 10^4\,\text{N/m}.$$

(b) If the car is loaded with 300 kg, Hooke's law gives

$$x = \frac{F}{k} = \frac{(300\,\text{kg})(9.8\,\text{m/s}^2)}{(6.5 \times 10^4\,\text{N/m})} = 4.5 \times 10^{-2}\,\text{m},$$

or 4.5 cm.

**NOTE** In (b), we could have obtained $x$ without solving for $k$: since $x$ is proportional to $F$, if 200 kg compresses the spring 3.0 cm, then 1.5 times the force will compress the spring 1.5 times as much, or 4.5 cm.

**FIGURE 11–4** Photo of a car's spring. (Also visible is the shock absorber, in blue—see Section 11–5.)

Any oscillating system for which the net restoring force is directly proportional to the negative of the displacement (as in Eq. 11–1, $F = -kx$) is said to exhibit **simple harmonic motion** (SHM).[†] Such a system is often called a **simple harmonic oscillator** (SHO). We saw in Section 9–5 that most solid materials stretch or compress according to Eq. 11–1 as long as the displacement is not too great. Because of this, many natural oscillations are simple harmonic, or sufficiently close to it that they can be treated using this SHM model.

**CONCEPTUAL EXAMPLE 11–2** **Is the motion simple harmonic?** Which of the following forces would cause an object to move in simple harmonic motion? (a) $F = -0.5x^2$, (b) $F = -2.3y$, (c) $F = 8.6x$, (d) $F = -4\theta$?

**RESPONSE** Both (b) and (d) will give simple harmonic motion because they give the force as minus a constant times a displacement. The displacement need not be $x$, but the minus sign is required to restore the system to equilibrium, which is why (c) does not produce SHM.

# 11–2 Energy in Simple Harmonic Motion

With forces that are not constant, such as here with simple harmonic motion, it is often convenient and useful to use the energy approach, as we saw in Chapter 6.

To stretch or compress a spring, work has to be done. Hence potential energy is stored in a stretched or compressed spring. We have already seen in Section 6–4 that elastic potential energy is given by

$$PE = \tfrac{1}{2}kx^2.$$

The total mechanical energy $E$ is the sum of the kinetic and potential energies,

$$E = \tfrac{1}{2}mv^2 + \tfrac{1}{2}kx^2, \tag{11–3}$$

where $v$ is the speed of the mass $m$ at a distance $x$ from the equilibrium position.

[†]The word "harmonic" refers to the motion being sinusoidal, which we discuss in Section 11–3. It is "simple" when the motion is sinusoidal of a single frequency. This can happen only if friction or other forces are not acting.

$E = \frac{1}{2}kA^2$

(a) $\quad x = -A \quad x = 0 \quad x = A$
$\qquad\quad (v = 0)$

$E = \frac{1}{2}mv_{max}^2$

(b) $\quad x = -A \quad x = 0 \quad x = A$

$E = \frac{1}{2}kA^2$

(c) $\quad x = -A \quad x = 0 \quad x = A$
$\qquad\qquad\qquad (v = 0)$

$E = \frac{1}{2}mv^2 + \frac{1}{2}kx^2$

(d) $\quad x = -A \quad x = 0 \;|\; x = A$
$\qquad\qquad\qquad x$

**FIGURE 11–5** Energy changes from potential energy to kinetic energy and back again as the spring oscillates. Energy bar graphs (on the right) were used in Section 6–7.

SHM can occur only if friction is negligible so that the total mechanical energy $E$ remains constant. As the mass oscillates back and forth, the energy continuously changes from potential energy to kinetic energy, and back again (Fig. 11–5). At the extreme points, $x = -A$ and $x = A$ (Fig. 11–5a, c), all the energy is stored in the spring as potential energy (and is the same whether the spring is compressed or stretched to the full amplitude). At these extreme points, the mass stops for an instant as it changes direction, so $v = 0$ and

$$E = \frac{1}{2}m(0)^2 + \frac{1}{2}kA^2 = \frac{1}{2}kA^2. \qquad (11\text{–}4a)$$

Thus, the **total mechanical energy of a simple harmonic oscillator is proportional to the square of the amplitude**. At the equilibrium point, $x = 0$ (Fig. 11–5b), all the energy is kinetic:

$$E = \frac{1}{2}mv_{max}^2 + \frac{1}{2}k(0)^2 = \frac{1}{2}mv_{max}^2, \qquad (11\text{–}4b)$$

where $v_{max}$ is the maximum speed during the motion (which occurs at $x = 0$). At intermediate points (Fig. 11–5d), the energy is part kinetic and part potential; because energy is conserved (we use Eqs. 11–3 and 11–4a),

$$\frac{1}{2}mv^2 + \frac{1}{2}kx^2 = \frac{1}{2}kA^2. \qquad (11\text{–}4c)$$

From this conservation of energy equation, we can obtain the velocity as a function of position. Solving for $v^2$, we have

$$v^2 = \frac{k}{m}(A^2 - x^2) = \frac{k}{m}A^2\left(1 - \frac{x^2}{A^2}\right).$$

From Eqs. 11–4a and 11–4b, we have $\frac{1}{2}mv_{max}^2 = \frac{1}{2}kA^2$, so $v_{max}^2 = (k/m)A^2$ or

$$v_{max} = \sqrt{\frac{k}{m}}\, A. \qquad (11\text{–}5a)$$

Inserting this equation into the equation just above it and taking the square root, we have

$$v = \pm v_{max}\sqrt{1 - \frac{x^2}{A^2}}. \qquad (11\text{–}5b)$$

This gives the velocity of the object at any position $x$. The object moves back and forth, so its velocity can be either in the $+$ or $-$ direction, but its magnitude depends only on its position $x$.

**CONCEPTUAL EXAMPLE 11–3** | **Doubling the amplitude.** Suppose the spring in Fig. 11–5 is stretched twice as far (to $x = 2A$). What happens to (a) the energy of the system, (b) the maximum velocity of the oscillating mass, (c) the maximum acceleration of the mass?

**RESPONSE** (a) From Eq. 11–4a, the total energy is proportional to the square of the amplitude $A$, so stretching it twice as far quadruples the energy ($2^2 = 4$). You may protest, "I did work stretching the spring from $x = 0$ to $x = A$. Don't I do the same work stretching it from $A$ to $2A$?" No. The force you exert is proportional to the displacement $x$, so for the second displacement, from $x = A$ to $2A$, you do more work than for the first displacement ($x = 0$ to $A$).

(b) From Eq. 11–5a, we can see that when the amplitude is doubled, the maximum velocity must be doubled.

(c) Since the force is twice as great when we stretch the spring twice as far ($F = kx$), the acceleration is also twice as great: $a \propto F \propto x$.

**EXERCISE C** Suppose the spring in Fig. 11–5 is compressed to $x = -A$, but is given a push to the right so that the initial speed of the mass $m$ is $v_0$. What effect does this push have on (a) the energy of the system, (b) the maximum velocity, (c) the maximum acceleration?

**EXAMPLE 11–4** **Spring calculations.** A spring stretches 0.150 m when a 0.300-kg mass is gently suspended from it as in Fig. 11–3b. The spring is then set up horizontally with the 0.300-kg mass resting on a frictionless table as in Fig. 11–5. The mass is pulled so that the spring is stretched 0.100 m from the equilibrium point, and released from rest. Determine: (a) the spring stiffness constant $k$; (b) the amplitude of the horizontal oscillation $A$; (c) the magnitude of the maximum velocity $v_{max}$; (d) the magnitude of the velocity $v$ when the mass is 0.050 m from equilibrium; and (e) the magnitude of the maximum acceleration $a_{max}$ of the mass.

**APPROACH** Wow, a lot of questions, but we can take them one by one. When the 0.300-kg mass hangs at rest from the spring as in Fig. 11–3b, we apply Newton's second law for the vertical forces: $\Sigma F = 0 = mg - kx_0$, so $k = mg/x_0$. For the horizontal oscillations, the amplitude is given, the velocities are found using conservation of energy, and the acceleration is found from $F = ma$.

**SOLUTION** (a) The spring stretches 0.150 m due to the 0.300-kg load, so

$$k = \frac{F}{x_0} = \frac{mg}{x_0} = \frac{(0.300\text{ kg})(9.80\text{ m/s}^2)}{0.150\text{ m}} = 19.6\text{ N/m}.$$

(b) The spring is now horizontal (on a table). It is stretched 0.100 m from equilibrium and is given no initial speed, so $A = 0.100$ m.

(c) The maximum velocity $v_{max}$ is attained as the mass passes through the equilibrium point where all the energy is kinetic. By comparing the total energy (see Eq. 11–3) at equilibrium with that at full extension, conservation of energy tells us that

$$\tfrac{1}{2}mv_{max}^2 + 0 = 0 + \tfrac{1}{2}kA^2,$$

where $A = 0.100$ m. Solving for $v_{max}$ (or using Eq. 11–5a), we have

$$v_{max} = A\sqrt{\frac{k}{m}} = (0.100\text{ m})\sqrt{\frac{19.6\text{ N/m}}{0.300\text{ kg}}} = 0.808\text{ m/s}.$$

(d) We use conservation of energy, or Eq. 11–5b derived from it, and find that

$$v = v_{max}\sqrt{1 - \frac{x^2}{A^2}} = (0.808\text{ m/s})\sqrt{1 - \frac{(0.050\text{ m})^2}{(0.100\text{ m})^2}} = 0.700\text{ m/s}.$$

(e) By Newton's second law, $F = ma$. So the maximum acceleration occurs where the force is greatest—that is, when $x = A = 0.100$ m. Thus

$$a_{max} = \frac{F_{max}}{m} = \frac{kA}{m} = \frac{(19.6\text{ N/m})(0.100\text{ m})}{0.300\text{ kg}} = 6.53\text{ m/s}^2.$$

**NOTE** We cannot use the kinematic equations, Eqs. 2–11, because the acceleration is not constant in SHM.

**EXAMPLE 11–5** **Energy calculations.** For the simple harmonic oscillator of Example 11–4, determine (a) the total energy, and (b) the kinetic and potential energies at half amplitude ($x = \pm A/2$).

**APPROACH** We use conservation of energy for a mass–spring system, Eqs. 11–3 and 11–4.

**SOLUTION** (a) With $k = 19.6$ N/m and $A = 0.100$ m, the total energy $E$ from Eq. 11 4a is

$$E = \tfrac{1}{2}kA^2 = \tfrac{1}{2}(19.6\text{ N/m})(0.100\text{ m})^2 = 9.80 \times 10^{-2}\text{ J}.$$

(b) At $x = A/2 = 0.050$ m, we have

$$\text{PE} = \tfrac{1}{2}kx^2 = \tfrac{1}{2}(19.6\text{ N/m})(0.050\text{ m})^2 = 2.45 \times 10^{-2}\text{ J}.$$

By conservation of energy, the kinetic energy must be

$$\text{KE} = E - \text{PE} = 7.35 \times 10^{-2}\text{ J}.$$

# 11–3 The Period and Sinusoidal Nature of SHM

The period of a simple harmonic oscillator is found to depend on the stiffness of the spring and also on the mass $m$ that is oscillating. But—strange as it may seem—the *period does not depend on the amplitude*. You can find this out for yourself by using a watch and timing 10 or 20 cycles of an oscillating spring for a small amplitude and then for a large amplitude.

The period $T$ is given by (see derivation on next page):

$$T = 2\pi\sqrt{\frac{m}{k}}. \qquad \textbf{(11–6a)}$$

We see that the larger the mass, the longer the period; and the stiffer the spring (larger $k$), the shorter the period. This makes sense since a larger mass means more inertia and therefore slower response (smaller acceleration). And larger $k$ means greater force and therefore quicker response (larger acceleration). Notice that Eq. 11–6a is not a direct proportion: the period varies as the *square root* of $m/k$. For example, the mass must be quadrupled to double the period. Equation 11–6a is fully in accord with experiment and is valid not only for a spring, but for all kinds of simple harmonic motion—that is, for motion subject to a restoring force proportional to displacement, Eq. 11–1.

We can write the frequency using $f = 1/T$ (Eq. 11–2):

$$f = \frac{1}{T} = \frac{1}{2\pi}\sqrt{\frac{k}{m}}. \qquad \textbf{(11–6b)}$$

**EXERCISE D** By how much should the mass on the end of a spring be changed to halve the frequency of its oscillations? (*a*) No change; (*b*) doubled; (*c*) quadrupled; (*d*) halved; (*e*) quartered.

---

**EXAMPLE 11–6  ESTIMATE  Spider web.** A spider of mass 0.30 g waits in its web of negligible mass (Fig. 11–6). A slight movement causes the web to vibrate with a frequency of about 15 Hz. (*a*) Estimate the value of the spring stiffness constant $k$ for the web. (*b*) At what frequency would you expect the web to vibrate if an insect of mass 0.10 g were trapped in addition to the spider?

**APPROACH** We can only make a rough estimate because a spider's web is fairly complicated and may vibrate with a mixture of frequencies. We use SHM as an approximate model.

**SOLUTION** (*a*) The frequency of SHM is given by Eq. 11–6b,

$$f = \frac{1}{2\pi}\sqrt{\frac{k}{m}}.$$

We solve for $k$:

$$k = (2\pi f)^2 m$$
$$= (2\pi)^2 (15\,\text{s}^{-1})^2 (3.0 \times 10^{-4}\,\text{kg}) = 2.7\,\text{N/m}.$$

(*b*) The total mass is now $0.10\,\text{g} + 0.30\,\text{g} = 4.0 \times 10^{-4}\,\text{kg}$. We could substitute $m = 4.0 \times 10^{-4}\,\text{kg}$ into Eq. 11–6b. Instead, we notice that the frequency decreases with the square root of the mass. Since the new mass is 4/3 times the first mass, the frequency changes by a factor of $1/\sqrt{4/3} = \sqrt{3/4}$. Thus $f = (15\,\text{Hz})(\sqrt{3/4}) = 13\,\text{Hz}$.

**NOTE** Check this result by direct substitution of $k$, found in part (*a*), and the new mass $m$ into Eq. 11–6b.

**FIGURE 11–6** Example 11–6. A spider waits for its prey (on the left).

> **EXAMPLE 11–7 | ESTIMATE | A vibrating floor.** A large motor in a factory causes the floor to vibrate up and down at a frequency of 10 Hz. The amplitude of the floor's motion near the motor is about 3.0 mm. Estimate the maximum acceleration of the floor near the motor.
>
> **APPROACH** Assuming the motion of the floor is roughly SHM, we can make an estimate for the maximum acceleration using $F = ma$ and Eq. 11–6b.
>
> **SOLUTION** The maximum acceleration occurs when the force $(F = kx)$ is largest, which is when $x = A$. Thus, $a_{max} = F_{max}/m = kA/m = (k/m)A$. From Eq. 11–6b, $(k/m) = (2\pi f)^2$, so
>
> $$a_{max} = \frac{F_{max}}{m} = \left(\frac{k}{m}\right)A = (2\pi f)^2 A = (2\pi)^2 (10\,\mathrm{s}^{-1})^2 (3.0 \times 10^{-3}\,\mathrm{m}) = 12\,\mathrm{m/s^2}.$$
>
> **NOTE** The maximum acceleration is a little over $g$, so when the floor accelerates down, objects sitting on the floor will actually lose contact with the floor momentarily, which will cause noise and serious wear.

**PHYSICS APPLIED**
*Unwanted floor vibrations*

## Period and Frequency—Derivation

We can derive a formula for the period of simple harmonic motion (SHM) by comparing SHM to an object rotating uniformly in a circle. From this same "reference circle" we can obtain a second useful result—a formula for the position of an oscillating mass as a function of time. There is nothing actually rotating in a circle when a spring oscillates linearly, but it is the mathematical similarity that we find useful.

Consider a small object of mass $m$ revolving counterclockwise in a circle of radius $A$, with constant speed $v_{max}$, on top of a table as shown in Fig. 11–7. As viewed from above, the motion is a circle in the $xy$ plane. But a person who looks at the motion from the edge of the table sees an oscillatory motion back and forth, and this one-dimensional motion corresponds precisely to simple harmonic motion, as we shall now see.

What the person sees, and what we are interested in, is the projection of the circular motion onto the $x$ axis (Fig. 11–7b). To see that this $x$ motion is analogous to SHM, let us calculate the magnitude of the $x$ component of the velocity $v_{max}$, which is labeled $v$ in Fig. 11–7. The two triangles involving $\theta$ in Fig. 11–7a are similar, so

$$\frac{v}{v_{max}} = \frac{\sqrt{A^2 - x^2}}{A}$$

or

$$v = v_{max} \sqrt{1 - \frac{x^2}{A^2}}.$$

This is exactly the equation for the speed of a mass oscillating with SHM, as we saw in Eq. 11–5b. Thus the projection on the $x$ axis of an object revolving in a circle has the same motion as a mass undergoing SHM.

We can now determine the period of SHM because it is equal to the time for our object revolving in a circle to make one complete revolution. First we note that the velocity $v_{max}$ is equal to the circumference of the circle (distance) divided by the period $T$:

$$v_{max} = \frac{2\pi A}{T} = 2\pi A f. \tag{11–7}$$

We solve for the period $T$ in terms of $A$:

$$T = \frac{2\pi A}{v_{max}}.$$

From Eq. 11–5a, $A/v_{max} = \sqrt{m/k}$. Thus

$$T = 2\pi \sqrt{\frac{m}{k}},$$

which is Eq. 11–6a, the formula we were looking for. The period depends on the mass $m$ and the spring stiffness constant $k$, but not on the amplitude $A$.

(a)

(b)

**FIGURE 11–7** (a) Circular motion of a small (red) object. (b) Side view of circular motion ($x$ component) is simple harmonic motion.

## Position as a Function of Time

We now use the reference circle to find the position of a mass undergoing simple harmonic motion as a function of time. From Fig. 11–7, we see that $\cos \theta = x/A$, so the projection of the object's position on the $x$ axis is

$$x = A \cos \theta.$$

The mass in the reference circle (Fig. 11–7) is rotating with uniform angular velocity $\omega$. We then can write $\theta = \omega t$, where $\theta$ is in radians (Section 8–1). Thus

$$x = A \cos \omega t. \tag{11–8a}$$

Furthermore, since the angular velocity $\omega$ (specified in radians per second) can be written as $\omega = 2\pi f$, where $f$ is the frequency (Eq. 8–7), we then write

$$x = A \cos(2\pi f t), \tag{11–8b}$$

or in terms of the period $T$,

$$x = A \cos(2\pi t/T). \tag{11–8c}$$

⚠ **CAUTION**

*t is a variable (time);*
*T is a constant for a given situation*

Notice in Eq. 11–8c that when $t = T$ (that is, after a time equal to one period), we have the cosine of $2\pi$ (or 360°), which is the same as the cosine of zero. This makes sense since the motion repeats itself after a time $t = T$.

Because the cosine function varies between 1 and −1, Eqs. 11–8 tell us that $x$ varies between $A$ and $-A$, as it must. If a pen is attached to a vibrating mass as a sheet of paper is moved at a steady rate beneath it (Fig. 11–8), a sinusoidal curve will be drawn that accurately follows Eqs. 11–8.

**FIGURE 11–8** Position as a function of time for a simple harmonic oscillator, $x = A \cos(2\pi t/T)$.

---

**EXAMPLE 11–8** **Starting with $x = A \cos \omega t$.** The displacement of an object is described by the following equation, where $x$ is in meters and $t$ is in seconds:

$$x = (0.30 \text{ m}) \cos(8.0 \, t).$$

Determine the oscillating object's (a) amplitude, (b) frequency, (c) period, (d) maximum speed, and (e) maximum acceleration.

**APPROACH** We start by comparing the given equation for $x$ with Eq. 11–8b, $x = A \cos(2\pi f t)$.

**SOLUTION** From $x = A \cos(2\pi f t)$, we see by inspection that (a) the amplitude $A = 0.30$ m, and (b) $2\pi f = 8.0 \text{ s}^{-1}$; so $f = (8.0 \text{ s}^{-1}/2\pi) = 1.27$ Hz. (c) Then $T = 1/f = 0.79$ s. (d) The maximum speed (see Eq. 11–7) is

$$v_{max} = 2\pi A f$$
$$= (2\pi)(0.30 \text{ m})(1.27 \text{ s}^{-1}) = 2.4 \text{ m/s}.$$

(e) The maximum acceleration, by Newton's second law, is $a_{max} = F_{max}/m = kA/m$, because $F (= kx)$ is greatest when $x$ is greatest. From Eq. 11–6b we see that $k/m = (2\pi f)^2$. Hence

$$a_{max} = \frac{k}{m} A = (2\pi f)^2 A$$
$$= (2\pi)^2 (1.27 \text{ s}^{-1})^2 (0.30 \text{ m}) = 19 \text{ m/s}^2.$$

---

## Sinusoidal Motion

Equation 11–8a, $x = A \cos \omega t$, assumes that the oscillating object starts from rest $(v = 0)$ at its maximum displacement $(x = A)$ at $t = 0$. Other equations for SHM are also possible, depending on the initial conditions (when you choose $t$ to be zero).

$x(t)$

**FIGURE 11–9** Sinusoidal nature of SHM, position as a function of time. In this case, $x = A \sin(2\pi t/T)$ because at $t = 0$ the mass is at the equilibrium position $x = 0$ and has (or is given) an initial speed at $t = 0$ that carries it to $x = A$ at $t = \frac{1}{4}T$.

For example, if at $t = 0$ the object is at the equilibrium position and the oscillations are begun by giving the object a push to the right $(+x)$, the equation would be

$$x = A \sin \omega t = A \sin(2\pi t/T).$$

This curve, shown in Fig. 11–9, has the same shape as the cosine curve shown in Fig. 11–8, except it is shifted to the right by a quarter cycle. Hence at $t = 0$ it starts out at $x = 0$ instead of at $x = A$.

Both sine and cosine curves are referred to as being **sinusoidal** (having the shape of a sine function). Thus simple harmonic motion[†] is said to be sinusoidal because the position varies as a sinusoidal function of time.

## *Velocity and Acceleration as Functions of Time

Figure 11–10a, like Fig. 11–8, shows a graph of displacement $x$ vs. time $t$, as given by Eqs. 11–8. We can also find the velocity $v$ as a function of time from Fig. 11–7a. For the position shown (red dot in Fig. 11–7a), the magnitude of $v$ is $v_{max} \sin \theta$, but $\vec{v}$ points to the left, so $v = -v_{max} \sin \theta$. Again setting $\theta = \omega t = 2\pi ft = 2\pi t/T$, we have

$$v = -v_{max} \sin \omega t = -v_{max} \sin(2\pi ft) = -v_{max} \sin(2\pi t/T). \quad \text{(11–9)}$$

Just after $t = 0$, the velocity is negative (points to the left) and remains so until $t = \frac{1}{2}T$ (corresponding to $\theta = 180° = \pi$ radians). After $t = \frac{1}{2}T$ until $t = T$ the velocity is positive. The velocity as a function of time (Eq. 11–9) is plotted in Fig. 11–10b. From Eqs. 11–6b and 11–7,

$$v_{max} = 2\pi Af = A\sqrt{\frac{k}{m}}.$$

For a given spring–mass system, the maximum speed $v_{max}$ is higher if the amplitude is larger, and always occurs as the mass passes the equilibrium point.

Newton's second law and Eqs. 11–8 give us the acceleration as a function of time:

$$a = \frac{F}{m} = \frac{-kx}{m} = -\left(\frac{kA}{m}\right)\cos \omega t = -a_{max}\cos(2\pi t/T) \quad \text{(11–10)}$$

where the maximum acceleration is

$$a_{max} = kA/m.$$

Equation 11–10 is plotted in Fig. 11–10c. Because the acceleration of a SHO is *not* constant, the equations for uniformly accelerated motion do *not* apply to SHM.

(a)

(b)

(c)

**FIGURE 11–10** Graphs showing (a) displacement $x$ as a function of time $t$: $x = A\cos(2\pi t/T)$; (b) velocity as a function of time: $v = -v_{max}\sin(2\pi t/T)$, where $v_{max} = A\sqrt{k/m}$; (c) acceleration as a function of time: $a = -a_{max}\cos(2\pi t/T)$, where $a_{max} = Ak/m$.

**FIGURE 11–11** Strobe-light photo of an oscillating pendulum, at equal time intervals.

# 11–4 The Simple Pendulum

A **simple pendulum** consists of a small object (the pendulum bob) suspended from the end of a lightweight cord, Fig. 11–11. We assume that the cord does not stretch and that its mass can be ignored relative to that of the bob. The motion of a simple pendulum moving back and forth with negligible friction resembles simple harmonic motion: the pendulum bob oscillates along the arc of a circle with equal amplitude on either side of its equilibrium point, and as it passes through the equilibrium point (where it would hang vertically) it has its maximum speed. But is it really undergoing SHM? That is, is the restoring force proportional to its displacement? Let us find out.

[†]Simple harmonic motion can be *defined* as motion that is sinusoidal. This definition is fully consistent with our earlier definition in Section 11–1.

**FIGURE 11–12** Simple pendulum, and a free-body diagram.

**TABLE 11–1**
**Sin $\theta$ at Small Angles**

| $\theta$ (degrees) | $\theta$ (radians) | $\sin \theta$ | % Difference |
|---|---|---|---|
| 0 | 0 | 0 | 0 |
| 1° | 0.01745 | 0.01745 | 0.005% |
| 5° | 0.08727 | 0.08716 | 0.1% |
| 10° | 0.17453 | 0.17365 | 0.5% |
| 15° | 0.26180 | 0.25882 | 1.1% |
| 20° | 0.34907 | 0.34202 | 2.0% |
| 30° | 0.52360 | 0.50000 | 4.5% |

**FIGURE 11–13** The swinging motion of this elaborate lamp, hanging by a very long cord from the ceiling of the cathedral at Pisa, is said to have been observed by Galileo and to have inspired him to the conclusion that the period of a pendulum does not depend on amplitude.

PHYSICS APPLIED
*Pendulum clock*

The displacement $s$ of the pendulum along the arc is given by $s = \ell\theta$, where $\theta$ is the angle (in radians) that the cord makes with the vertical and $\ell$ is the length of the cord (Fig. 11–12). If the restoring force is proportional to $s$ or to $\theta$, the motion will be simple harmonic. The restoring force is the net force on the bob, which equals the component of the weight ($mg$) tangent to the arc:

$$F = -mg \sin\theta,$$

where $g$ is the acceleration due to gravity. The minus sign here, as in Eq. 11–1, means the force is in the direction opposite to the angular displacement $\theta$. Since $F$ is proportional to the sine of $\theta$ and not to $\theta$ itself, the motion is *not* SHM. However, if $\theta$ is small, then $\sin\theta$ is very nearly equal to $\theta$ when the angle is specified in radians. This can be seen by noting in Fig. 11–12 that the arc length $s$ ($= \ell\theta$) is nearly the same length as the chord ($= \ell\sin\theta$) indicated by the horizontal straight dashed line, *if $\theta$ is small*. For angles less than 15°, the difference between $\theta$ (in radians) and $\sin\theta$ is less than 1%—see Table 11–1. Thus, to a very good approximation for small angles,

$$F = -mg\sin\theta \approx -mg\theta.$$

Substituting $s = \ell\theta$, or $\theta = s/\ell$, we have

$$F \approx -\frac{mg}{\ell}s.$$

Thus, for small displacements, the motion can be modeled as being approximately simple harmonic, because this approximate equation fits Hooke's law, $F = -kx$, where in place of $x$ we have arc length $s$. The effective force constant is $k = mg/\ell$. If we substitute $k = mg/\ell$ into Eq. 11–6a, we obtain the period of a simple pendulum:

$$T = 2\pi\sqrt{\frac{m}{k}} = 2\pi\sqrt{\frac{m}{mg/\ell}}$$

or

$$T = 2\pi\sqrt{\frac{\ell}{g}}. \qquad \text{[$\theta$ small]} \quad \textbf{(11–11a)}$$

The frequency is $f = 1/T$, so

$$f = \frac{1}{2\pi}\sqrt{\frac{g}{\ell}}. \qquad \text{[$\theta$ small]} \quad \textbf{(11–11b)}$$

The mass $m$ of the pendulum bob does not appear in these formulas for $T$ and $f$. Thus we have the surprising result that the period and frequency of a simple pendulum do not depend on the mass of the pendulum bob. You may have noticed this if you pushed a small child and then a large one on the same swing.

We also see from Eq. 11–11a that the period of a pendulum does not depend on the amplitude (like any SHM, Section 11–3), as long as the amplitude $\theta$ is small. Galileo is said to have first noted this fact while watching a swinging lamp in the cathedral at Pisa (Fig. 11–13). This discovery led to the invention of the pendulum clock, the first really precise timepiece, which became the standard for centuries.

**EXERCISE E** Return to Chapter-Opening Question 1, page 292, and answer it again now. Try to explain why you may have answered differently the first time.

**EXERCISE F** If a simple pendulum is taken from sea level to the top of a high mountain and started at the same angle of 5°, it would oscillate at the top of the mountain (*a*) slightly slower; (*b*) slightly faster; (*c*) at exactly the same frequency; (*d*) not at all—it would stop; (*e*) none of these.

Because a pendulum does not undergo *precisely* SHM, the period does depend slightly on the amplitude—the more so for large amplitudes. The accuracy of a pendulum clock would be affected, after many swings, by the decrease in amplitude due to friction. But the mainspring in a pendulum clock (or the falling weight in a grandfather clock) supplies energy to compensate for the friction and to maintain the amplitude constant, so that the timing remains precise.

**EXAMPLE 11–9** | **Measuring g.** A geologist uses a simple pendulum that has a length of 37.10 cm and a frequency of 0.8190 Hz at a particular location on the Earth. What is the acceleration due to gravity at this location?

**APPROACH** We can use the length $\ell$ and frequency $f$ of the pendulum in Eq. 11–11b, which contains our unknown, $g$.

**SOLUTION** We solve Eq. 11–11b for $g$ and obtain

$$g = (2\pi f)^2 \ell = (2\pi)^2 (0.8190 \text{ s}^{-1})^2 (0.3710 \text{ m}) = 9.824 \text{ m/s}^2.$$

# 11–5 Damped Harmonic Motion

The amplitude of any real oscillating spring or swinging pendulum slowly decreases in time until the oscillations stop altogether. Figure 11–14 shows a typical graph of the displacement as a function of time. This is called **damped harmonic motion**. The damping[†] is generally due to the resistance of air and to internal friction within the oscillating system. The energy that is dissipated to thermal energy results in a decreased amplitude of oscillation.

Since natural oscillating systems are damped in general, why do we even talk about (undamped) simple harmonic motion? The answer is that SHM is much easier to deal with mathematically. And if the damping is not large, the oscillations can be thought of as simple harmonic motion on which the damping is super-posed, as represented by the dashed curves in Fig. 11–14. Although damping does alter the frequency of vibration, the effect can be small if the damping is small; then Eqs. 11–6 can still be useful approximations.

Sometimes the damping is so large, however, that the motion no longer resembles simple harmonic motion. Three common cases of *heavily damped* systems are shown in Fig. 11–15. Curve A represents an **underdamped** situation, in which the system makes several oscillations before coming to rest; it corresponds to a more heavily damped version of Fig. 11–14. Curve C represents the **overdamped** situation, when the damping is so large that there is no oscillation and the system takes a long time to come to rest (equilibrium). Curve B represents **critical damping**: in this case the displacement reaches zero in the shortest time. These terms all derive from the use of practical damped systems such as door-closing mechanisms and **shock absorbers** in a car (Fig. 11–16), which are usually designed to give critical damping. But as they wear out, underdamping occurs: the door of a room slams and a car bounces up and down several times when it hits a bump.

In many systems, the oscillatory motion is what counts, as in clocks and musical instruments, and damping may need to be minimized. In other systems, oscillations are the problem, such as a car's springs, so a proper amount of damping (i.e., critical) is desired. Well-designed damping is needed for all kinds of applications. Large buildings, especially in California, are now built (or retrofitted) with huge dampers to reduce possible earthquake damage (Fig. 11–17).

[†] To "damp" means to diminish, restrain, or extinguish, as to "dampen one's spirits."

**FIGURE 11–14** Damped harmonic motion.

**FIGURE 11–15** Graphs that represent (A) underdamped, (B) critically damped, and (C) overdamped oscillatory motion.

**PHYSICS APPLIED**
*Shock absorbers and building dampers*

Attached to car frame

Piston

Viscous fluid

Attached to car axle

**FIGURE 11–16** Automobile spring and shock absorber provide damping so that a car won't bounce up and down so much.

**FIGURE 11–17** These huge dampers placed in a building look a lot like huge automobile shock absorbers, and they serve a similar purpose—to reduce the amplitude and the acceleration of movement when the shock of an earthquake hits.

# 11–6 Forced Oscillations; Resonance

When an oscillating system is set into motion, it oscillates at its natural frequency (Eqs. 11–6b and 11–11b). However, a system may have an external force applied to it that has its own particular frequency. Then we have a **forced oscillation**.

For example, we might pull the mass on the spring of Fig. 11–1 back and forth at an externally applied frequency $f$. The mass then oscillates at the external frequency $f$ of the external force, even if this frequency is different from the **natural frequency** of the spring, which we will now denote by $f_0$, where (see Eq. 11–6b)

$$f_0 = \frac{1}{2\pi} \sqrt{\frac{k}{m}}.$$

For a forced oscillation with only light damping, the amplitude of oscillation is found to depend on the difference between $f$ and $f_0$, and is a maximum when the frequency of the external force equals the natural frequency of the system— that is, when $f = f_0$. The amplitude is plotted in Fig. 11–18 as a function of the external frequency $f$. Curve A represents light damping and curve B heavy damping. When the external driving frequency $f$ is near the natural frequency, $f \approx f_0$, the amplitude can become large if the damping is small. This effect of increased amplitude at $f = f_0$ is known as **resonance**. The natural oscillation frequency $f_0$ of a system is also called its **resonant frequency**.

A simple illustration of resonance is pushing a child on a swing. A swing, like any pendulum, has a natural frequency of oscillation. If you push on the swing at a random frequency, the swing bounces around and reaches no great amplitude. But if you push with a frequency equal to the natural frequency of the swing, the amplitude increases greatly. At resonance, relatively little effort is required to obtain and maintain a large amplitude.

The great tenor Enrico Caruso was said to be able to shatter a crystal goblet by singing a note of just the right frequency at full voice. This is an example of resonance, for the sound waves emitted by the voice act as a forced oscillation on the glass. At resonance, the resulting oscillation of the goblet may be large enough in amplitude that the glass exceeds its elastic limit and breaks (Fig. 11–19).

Since material objects are, in general, elastic, resonance is an important phenomenon in a variety of situations. It is particularly important in construction, although the effects are not always foreseen. For example, it has been reported that a railway bridge collapsed because a nick in one of the wheels of a crossing train set up a resonant oscillation in the bridge. Marching soldiers break step when crossing a bridge to avoid the possibility that their rhythmic march might match a resonant frequency of the bridge. The famous collapse of the Tacoma Narrows Bridge (Fig. 11–20a) in 1940 occurred as a result of strong gusting winds driving the span into large-amplitude oscillatory motion. Bridges and tall buildings are now designed with more inherent damping. The Oakland freeway collapse in the 1989 California earthquake (Fig. 11–20b) involved resonant oscillation of a section built on mudfill that readily transmitted that frequency.

Resonance can be very useful, too, and we will meet important examples later, such as in musical instruments and tuning a radio. We will also see that vibrating objects often have not one, but many resonant frequencies.

**FIGURE 11–18** Amplitude as a function of driving frequency $f$, showing resonance for lightly damped (A) and heavily damped (B) systems.

**PHYSICS APPLIED**

*Child on a swing*

**PHYSICS APPLIED**

*Shattering glass via resonance*

**FIGURE 11–19** This goblet breaks as it vibrates in resonance to a trumpet call.

**PHYSICS APPLIED**

*Resonant collapse*

**FIGURE 11–20** (a) Large-amplitude oscillations of the Tacoma Narrows Bridge, due to gusty winds, led to its collapse (November 7, 1940). (b) Collapse of a freeway in California, due to the 1989 earthquake.

(a)

(b)

**FIGURE 11–21** Water waves spreading outward from a source. In this case the source is a small spot of water oscillating up and down briefly where a rock hit (left photo).

# 11–7 Wave Motion

When you throw a stone into a lake or pool of water, circular waves form and move outward, Fig. 11–21. Waves will also travel along a rope that is stretched out straight on a table if you vibrate one end back and forth as shown in Fig. 11–22. Water waves and waves on a rope or cord are two common examples of **mechanical waves**, which propagate as oscillations of matter. We will discuss other kinds of waves in later Chapters, including electromagnetic waves and light.

**FIGURE 11–22** Wave traveling on a rope or cord. The wave travels to the right along the rope. Particles of the rope oscillate back and forth on the tabletop.

If you have ever watched ocean waves moving toward shore before they break, you may have wondered if the waves were carrying water from far out at sea onto the beach. They don't.[†] Water waves move with a recognizable velocity. But each particle (or molecule) of the water itself merely oscillates about an equilibrium point. This is clearly demonstrated by observing leaves on a pond as waves move by. The leaves (or a cork) are not carried forward by the waves, but oscillate more or less up and down about an equilibrium point because this is the motion of the water itself.

**CONCEPTUAL EXAMPLE 11–10** | **Wave vs. particle velocity.** Is the velocity of a wave moving along a rope the same as the velocity of a particle of the rope? See Fig. 11–22.

**RESPONSE** No. The two velocities are different, both in magnitude and direction. The wave on the rope of Fig. 11–22 moves to the right along the tabletop, but each piece of the rope only vibrates to and fro, perpendicular to the traveling wave. (The rope clearly does not travel in the direction that the wave on it does.)

Waves can move over large distances, but the medium (the water or the rope) itself has only a limited movement, oscillating about an equilibrium point as in simple harmonic motion. Thus, although a wave is not itself matter, the wave pattern can travel in matter. A wave consists of oscillations that move without carrying matter with them.

[†]Do not be confused by the "breaking" of ocean waves, which occurs when a wave interacts with the ground in shallow water and hence is no longer a simple wave.

**FIGURE 11–22** (Repeated.) Wave traveling on a rope or cord. The wave travels to the right along the rope. Particles of the rope oscillate back and forth on the tabletop.

**FIGURE 11–23** A wave pulse is generated by a hand holding the end of a cord and moving up and down once. Motion of the wave pulse is to the right. Arrows indicate velocity of cord particles.

(a)

(b)

(c)

(d)

Waves carry energy from one place to another. Energy is given to a water wave, for example, by a rock thrown into the water, or by wind far out at sea. The energy is transported by waves to the shore. The oscillating hand in Fig. 11–22 transfers energy to the rope, and that energy is transported down the rope and can be transferred to an object at the other end. All forms of traveling waves transport energy.

**EXERCISE G** Return to Chapter-Opening Question 2, page 292, and answer it again now. Try to explain why you may have answered differently the first time.

Let us look more closely at how a wave is formed and how it comes to "travel." We first look at a single wave bump, or **pulse**. A single pulse can be formed on a cord by a quick up-and-down motion of the hand, Fig. 11–23. The hand pulls up on one end of the cord. Because the end section is attached to adjacent sections, these also feel an upward force and they too begin to move upward. As each succeeding section of cord moves upward, the wave crest moves outward along the cord. Meanwhile, the end section of cord has been returned to its original position by the hand. As each succeeding section of cord reaches its peak position, it too is pulled back down again by tension from the adjacent section of cord. Thus the source of a traveling wave pulse is a disturbance (or vibration), and cohesive forces between adjacent sections of cord cause the pulse to travel. Waves in other media are created and propagate outward in a similar fashion. A dramatic example of a wave pulse is a tsunami or tidal wave that is created by an earthquake in the Earth's crust under the ocean. The bang you hear when a door slams is a sound wave pulse.

A **continuous** or **periodic wave**, such as that shown in Fig. 11–22, has as its source a disturbance that is continuous and oscillating; that is, the source is a *vibration* or *oscillation*. In Fig. 11–22, a hand oscillates one end of the rope. Water waves may be produced by any vibrating object at the surface, such as your hand; or the water itself is made to vibrate when wind blows across it or a rock is thrown into it. A vibrating tuning fork or drum membrane gives rise to sound waves in air. We will see later that oscillating electric charges give rise to light waves. Indeed, almost any vibrating object sends out waves.

The source of any wave, then, is a vibration. And it is a *vibration* that propagates outward and thus constitutes the wave. If the source vibrates sinusoidally in SHM, then the wave itself—if the medium is elastic—will have a sinusoidal shape both in space and in time. (1) In space: if you take a picture of the wave in space at a given instant of time, the wave will have the shape of a sine or cosine as a function of position. (2) In time: if you look at the motion of the medium at one place over a long period of time—for example, if you look between two closely spaced posts of a pier or out of a ship's porthole as water waves pass by—the up-and-down motion of that small segment of water will be simple harmonic motion. The water moves up and down sinusoidally in time.

Some of the important quantities used to describe a periodic sinusoidal wave are shown in Fig. 11–24. The high points on a wave are called *crests*; the low points, *troughs*. The **amplitude**, $A$, is the maximum height of a crest, or depth of a trough, relative to the normal (or equilibrium) level. The total swing from a crest to a trough is $2A$ (twice the amplitude). The distance between two successive crests is the **wavelength**, $\lambda$ (the Greek letter lambda). The wavelength is also equal to the distance between *any* two successive identical points on the wave. The **frequency**, $f$, is the number of crests—or complete cycles—that pass a given point per unit time. The **period**, $T$, equals $1/f$ and is the time elapsed between two successive crests passing by the same point in space.

**FIGURE 11–24** Characteristics of a single-frequency continuous wave moving through space.

The **wave speed**, $v$, is the speed at which wave crests (or any other fixed point on the wave shape) move forward. The wave speed must be distinguished from the speed of a particle of the medium itself as we saw in Example 11–10.

A wave crest travels a distance of one wavelength, $\lambda$, in a time equal to one period, $T$. Thus the wave speed is $v = \lambda/T$. Then, since $1/T = f$,

$$v = \lambda f. \qquad (11\text{–}12)$$

For example, suppose a wave has a wavelength of 5 m and a frequency of 3 Hz. Since three crests pass a given point per second, and the crests are 5 m apart, the first crest (or any other part of the wave) must travel a distance of 15 m during the 1 s. So the wave speed is 15 m/s.

**EXERCISE H** You notice a water wave pass by the end of a pier, with about 0.5 s between crests. Therefore (*a*) the frequency is 0.5 Hz; (*b*) the velocity is 0.5 m/s; (*c*) the wavelength is 0.5 m; (*d*) the period is 0.5 s.

# 11–8 Types of Waves and Their Speeds: Transverse and Longitudinal

When a wave travels down a cord—say, from left to right as in Fig. 11–22—the particles of the cord vibrate back and forth in a direction transverse (that is, perpendicular) to the motion of the wave itself. Such a wave is called a **transverse wave** (Fig. 11–25a). There exists another type of wave known as a **longitudinal wave**. In a longitudinal wave, the vibration of the particles of the medium is *along* the direction of the wave's motion. Longitudinal waves are readily formed on a stretched spring or Slinky by alternately compressing and expanding one end. This is shown in Fig. 11–25b, and can be compared to the transverse wave in Fig. 11–25a. A series of compressions and expansions travel along the spring. The *compressions* are those areas where the coils are momentarily close together. *Expansions* (sometimes called *rarefactions*) are regions where the coils are momentarily far apart. Compressions and expansions correspond to the crests and troughs of a transverse wave.

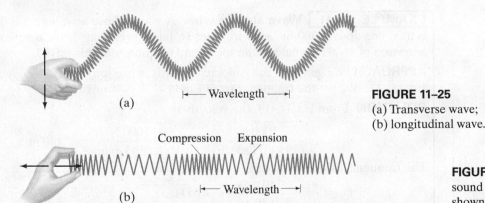

(a)

(b)

|← Wavelength →|

Compression    Expansion

|← Wavelength →|

**FIGURE 11–25**
(a) Transverse wave;
(b) longitudinal wave.

An important example of a longitudinal wave is a sound wave in air. A vibrating drumhead, for instance, alternately compresses and expands the air in contact with it, producing a longitudinal wave that travels outward in the air, as shown in Fig. 11–26.

As in the case of transverse waves, each section of the medium in which a longitudinal wave passes oscillates over a very small distance, whereas the wave itself can travel large distances. Wavelength, frequency, and wave speed all have meaning for a longitudinal wave. The wavelength is the distance between successive compressions (or between successive expansions), and frequency is the number of compressions that pass a given point per second. The wave speed is the speed with which each compression appears to move; it is equal to the product of wavelength and frequency, $v = \lambda f$ (Eq. 11–12).

**FIGURE 11–26** Production of a sound wave, which is longitudinal, shown at two moments in time about a half period $(\frac{1}{2}T)$ apart.

Drum membrane    Compression    Expansion

A longitudinal wave can be represented graphically by plotting the density of air molecules (or coils of a Slinky) versus position at a given instant, as shown in Fig. 11–27. Such a graphical representation makes it easy to illustrate what is happening. Note that the graph looks much like a transverse wave.

(a)

**FIGURE 11–27** (a) A longitudinal wave in air, with (b) its graphical representation at a particular instant in time.

(b)

## Speed of Transverse Waves

The speed of a wave depends on the properties of the medium in which it travels. The speed of a transverse wave on a stretched string or cord, for example, depends on the tension in the cord, $F_T$, and on the mass per unit length of the cord, $\mu$ (the Greek letter mu). If $m$ is the mass of a length $\ell$ of wire, $\mu = m/\ell$. For waves of small amplitude, the wave speed is

$$v = \sqrt{\frac{F_T}{\mu}}. \qquad \left[\begin{array}{c}\text{transverse wave}\\\text{on a cord}\end{array}\right] \quad \textbf{(11–13)}$$

This formula makes sense qualitatively on the basis of Newtonian mechanics. That is, we do expect the tension to be in the numerator and the mass per unit length in the denominator. Why? Because when the tension is greater, we expect the speed to be greater since each segment of cord is in tighter contact with its neighbor. Also, the greater the mass per unit length, the more inertia the cord has and the more slowly the wave would be expected to propagate.

---

**EXAMPLE 11–11** **Wave along a wire.** A wave whose wavelength is 0.30 m is traveling down a 300-m-long wire whose total mass is 15 kg. If the wire is under a tension of 1000 N, what are the speed and frequency of this wave?

**APPROACH** We assume the velocity of this wave on a wire is given by Eq. 11–13. We get the frequency from Eq. 11–12, $f = v/\lambda$.

**SOLUTION** From Eq. 11–13, the velocity is

$$v = \sqrt{\frac{1000\,\text{N}}{(15\,\text{kg})/(300\,\text{m})}} = \sqrt{\frac{1000\,\text{N}}{(0.050\,\text{kg/m})}} = 140\,\text{m/s}.$$

The frequency is

$$f = \frac{v}{\lambda} = \frac{140\,\text{m/s}}{0.30\,\text{m}} = 470\,\text{Hz}.$$

**NOTE** A higher tension would increase both $v$ and $f$, whereas a thicker, denser wire would reduce $v$ and $f$.

---

## Speed of Longitudinal Waves

The speed of a longitudinal wave has a form similar to that for a transverse wave on a cord (Eq. 11–13); that is,

$$v = \sqrt{\frac{\text{elastic force factor}}{\text{inertia factor}}}.$$

In particular, for a longitudinal wave traveling down a long solid rod,

$$v = \sqrt{\frac{E}{\rho}}, \qquad \left[\begin{array}{c}\text{longitudinal wave}\\\text{in a long rod}\end{array}\right] \quad \textbf{(11–14a)}$$

where $E$ is the elastic modulus (Section 9–5) of the material and $\rho$ is its density.

For a longitudinal wave traveling in a liquid or gas,

$$v = \sqrt{\frac{B}{\rho}}, \qquad \left[\begin{array}{l}\text{longitudinal wave} \\ \text{in a fluid}\end{array}\right] \quad \textbf{(11–14b)}$$

where $B$ is the bulk modulus (Section 9–5) and $\rho$ again is the density.

**EXAMPLE 11–12** **Echolocation.** Echolocation is a form of sensory perception used by animals such as bats, dolphins, and toothed whales (Fig. 11–28). The animal emits a pulse of sound (a longitudinal wave) which, after reflection from objects, returns and is detected by the animal. Echolocation waves can have frequencies of about 100,000 Hz. (*a*) Estimate the wavelength of a sea animal's echolocation wave. (*b*) If an obstacle is 100 m from the animal, how long after the animal emits a wave is its reflection detected?

**APPROACH** We first compute the speed of longitudinal (sound) waves in sea water, using Eq. 11–14b and Tables 9–1 and 10–1. The wavelength is $\lambda = v/f$.

**SOLUTION** (*a*) The speed of longitudinal waves in sea water, which is slightly more dense than pure water, is (Tables 9–1 and 10–1)

$$v = \sqrt{\frac{B}{\rho}} = \sqrt{\frac{2.0 \times 10^9\,\text{N/m}^2}{1.025 \times 10^3\,\text{kg/m}^3}} = 1.4 \times 10^3\,\text{m/s}.$$

Then, using Eq. 11–12, we find

$$\lambda = \frac{v}{f} = \frac{(1.4 \times 10^3\,\text{m/s})}{(1.0 \times 10^5\,\text{Hz})} = 14\,\text{mm}.$$

(*b*) The time required for the round trip between the animal and the object is

$$t = \frac{\text{distance}}{\text{speed}} = \frac{2(100\,\text{m})}{1.4 \times 10^3\,\text{m/s}} = 0.14\,\text{s}.$$

**NOTE** We shall see later that waves can be used to "resolve" (or detect) objects whose size is comparable to or larger than the wavelength. Thus, a dolphin can resolve objects on the order of a centimeter or larger in size.

**PHYSICS APPLIED**
*Space perception
by animals using sound waves*

**FIGURE 11–28** A toothed whale (Example 11–12).

## Other Waves

Both transverse and longitudinal waves are produced when an **earthquake** occurs. The transverse waves that travel through the body of the Earth are called S waves (S for shear), and the longitudinal waves are called P waves (P for pressure) or *compression* waves. Both longitudinal and transverse waves can travel through a solid since the atoms or molecules can vibrate about their relatively fixed positions in any direction. But only longitudinal waves can propagate through a fluid, because any transverse motion would not experience any restoring force since a fluid is readily deformable. This fact was used by geophysicists to infer that a portion of the Earth's core must be liquid: after an earthquake, longitudinal waves are detected diametrically across the Earth, but not transverse waves.

Besides these two types of waves that can pass through the body of the Earth (or other substance), there can also be *surface waves* that travel along the boundary between two materials. A wave on water is actually a surface wave that moves on the boundary between water and air. The motion of each particle of water at the surface is circular or elliptical (Fig. 11–29), so it is a combination of horizontal and vertical motions. Below the surface, there is also horizontal plus vertical motion, as shown. At the bottom, the motion is only horizontal. (When a wave approaches shore, the water drags at the bottom and is slowed down, while the crests move ahead at higher speed (Fig. 11–30) and "spill" over the top.)

Surface waves are also set up on the Earth when an earthquake occurs. The waves that travel along the surface are mainly responsible for the damage caused by earthquakes.

**PHYSICS APPLIED**
*Earthquake waves*

**FIGURE 11–29** A shallow water wave is an example of a *surface wave*, which is a combination of transverse and longitudinal wave motions.

**FIGURE 11–30** How a water wave breaks. The green arrows represent the local velocity of water molecules.

Waves which travel along a line in one dimension, such as transverse waves on a stretched string, or longitudinal waves in a rod or fluid-filled tube, are *linear* or *one-dimensional waves*. Surface waves, such as water waves (Fig. 11–21), are *two-dimensional waves*. Finally, waves that move out from a source in all directions, such as sound from a loudspeaker or earthquake waves through the Earth, are *three-dimensional waves*.

# 11–9 Energy Transported by Waves

Waves transport energy from one place to another. As waves travel through a medium, the energy is transferred as vibrational energy from particle to particle of the medium. For a sinusoidal wave of frequency $f$, the particles move in SHM as a wave passes, so each particle has an energy $E = \frac{1}{2}kA^2$, where $A$ is the amplitude of its motion, either transversely or longitudinally. See Eq. 11–4a.

Thus, we have the important result that the **energy transported by a wave is proportional to the square of the amplitude**. The **intensity** $I$ of a wave is defined as the power (energy per unit time) transported across unit area perpendicular to the direction of energy flow:

$$I = \frac{\text{energy/time}}{\text{area}} = \frac{\text{power}}{\text{area}}.$$

The SI unit of intensity is watts per square meter $(\text{W/m}^2)$. Since the energy is proportional to the wave amplitude squared, so too is the intensity:

$$I \propto A^2. \tag{11–15}$$

If a wave flows out from the source in all directions, it is a three-dimensional wave. Examples are sound traveling in open air, earthquake waves, and light waves. If the medium is isotropic (same in all directions), the wave is a *spherical wave* (Fig. 11–31). As the wave moves outward, the energy it carries is spread over a larger and larger area since the surface area of a sphere of radius $r$ is $4\pi r^2$. Thus the intensity of a spherical wave is

$$I = \frac{\text{power}}{\text{area}} = \frac{P}{4\pi r^2}. \qquad \text{[spherical wave]} \quad \textbf{(11–16a)}$$

If the power output $P$ of the source is constant, then the intensity decreases as the inverse square of the distance from the source:

$$I \propto \frac{1}{r^2}. \qquad \text{[spherical wave]} \quad \textbf{(11–16b)}$$

This is often called the **inverse square law**, or the "one over $r^2$ law." If we consider two points at distances $r_1$ and $r_2$ from the source, as in Fig. 11–31, then $I_1 = P/4\pi r_1^2$ and $I_2 = P/4\pi r_2^2$, so

$$\frac{I_2}{I_1} = \frac{r_1^2}{r_2^2}. \qquad \text{[spherical wave]} \quad \textbf{(11–16c)}$$

Thus, for example, when the distance doubles $(r_2/r_1 = 2)$, the intensity is reduced to $\frac{1}{4}$ its earlier value: $I_2/I_1 = \left(\frac{1}{2}\right)^2 = \frac{1}{4}$.

The amplitude of a wave also decreases with distance. Since the intensity is proportional to the square of the amplitude (Eq. 11–15), the amplitude $A$ must decrease as $1/r$ so that $I \propto A^2$ will be proportional to $1/r^2$ (as in Eq. 11–16b). Hence

$$A \propto \frac{1}{r}.$$

If we consider again two distances from the source, $r_1$ and $r_2$, then

$$\frac{A_2}{A_1} = \frac{r_1}{r_2}. \qquad \text{[spherical wave]}$$

When the wave is twice as far from the source, the amplitude is half as large, and so on (ignoring damping due to friction).

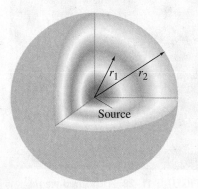

**FIGURE 11–31** A wave traveling uniformly outward in three dimensions from a source is spherical. Two crests (or compressions) are shown, of radii $r_1$ and $r_2$.

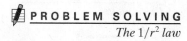

**PROBLEM SOLVING**

*The $1/r^2$ law*

**EXAMPLE 11-13** **Earthquake intensity.** The intensity of an earthquake P wave traveling through the Earth and detected 100 km from the source is $1.0 \times 10^6$ W/m². What is the intensity of that wave if detected 400 km from the source?

**APPROACH** We assume the wave is spherical, so the intensity decreases as the square of the distance from the source.

**SOLUTION** At 400 km the distance is 4 times greater than at 100 km, so the intensity will be $\left(\frac{1}{4}\right)^2 = \frac{1}{16}$ of its value at 100 km, or $(1.0 \times 10^6 \text{ W/m}^2)/16 = 6.3 \times 10^4$ W/m².

**NOTE** Using Eq. 11–16c directly gives:

$$I_2 = I_1 r_1^2 / r_2^2 = (1.0 \times 10^6 \text{ W/m}^2)(100 \text{ km})^2/(400 \text{ km})^2 = 6.3 \times 10^4 \text{ W/m}^2.$$

The situation is different for a one-dimensional wave, such as a transverse wave on a string or a longitudinal wave pulse traveling down a thin uniform metal rod. The area remains constant, so the amplitude $A$ also remains constant (ignoring friction). Thus the amplitude and the intensity do not decrease with distance.

In practice, frictional damping is generally present, and some of the energy is transformed into thermal energy. Thus the amplitude and intensity of a one-dimensional wave will decrease with distance from the source. For a three-dimensional wave, the decrease will be greater than that discussed above, more than $1/r^2$, although the effect may often be small.

## Intensity Related to Amplitude and Frequency

For a sinusoidal wave of frequency $f$, the particles move in SHM as a wave passes, so each particle has an energy $E = \frac{1}{2}kA^2$, where $A$ is the amplitude of its motion. Using Eq. 11–6b, we can write $k$ in terms of the frequency: $k = 4\pi^2 m f^2$, where $m$ is the mass of a particle (or small volume) of the medium. Then

$$E = \tfrac{1}{2}kA^2 = 2\pi^2 m f^2 A^2.$$

**FIGURE 11–32** Calculating the energy carried by a wave moving with velocity $v$.

The mass $m = \rho V$, where $\rho$ is the density of the medium and $V$ is the volume of a small slice of the medium as shown in Fig. 11–32. The volume $V = S\ell$, where $S$ is the cross-sectional surface area through which the wave travels. (We use $S$ instead of $A$ for area because we are using $A$ for amplitude.) We can write $\ell$ as the distance the wave travels in a time $t$ as $\ell = vt$, where $v$ is the speed of the wave. Thus $m = \rho V = \rho S \ell = \rho S v t$, and

$$E = 2\pi^2 \rho S v t f^2 A^2. \tag{11–17a}$$

From this equation, we see again the important result that the energy transported by a wave is proportional to the square of the amplitude. The average power transported, $\overline{P} = E/t$, is

$$\overline{P} = \frac{E}{t} = 2\pi^2 \rho S v f^2 A^2. \tag{11–17b}$$

Finally, the **intensity** $I$ of a wave is the average power transported across unit area perpendicular to the direction of energy flow:

$$I = \frac{\overline{P}}{S} = 2\pi^2 \rho v f^2 A^2. \tag{11–18}$$

This relation shows explicitly that the intensity of a wave is proportional both to the square of the wave amplitude $A$ at any point and to the square of the frequency $f$.

# 11–10 Reflection and Transmission of Waves

When a wave strikes an obstacle, or comes to the end of the medium in which it is traveling, at least a part of the wave is reflected. You have probably seen water waves reflect off a rock or the side of a swimming pool. And you may have heard a shout reflected from a distant cliff—which we call an "echo."

A wave pulse traveling along a cord is reflected as shown in Fig. 11–33 (time increases going downward in both a and b). The reflected pulse returns inverted as in Fig. 11–33a if the end of the cord is fixed; it returns right side up if the end is free as in Fig. 11–33b. When the end is fixed to a support, as in Fig. 11–33a, the pulse reaching that fixed end exerts a force (upward) on the support. The support exerts an equal but opposite force downward on the cord (Newton's third law). This downward force on the cord is what "generates" the inverted reflected pulse.

(a)            (b)

**FIGURE 11–33** Reflection of a wave pulse traveling along a cord lying on a table. (Time increases going down.) (a) The end of the cord is fixed to a peg. (b) The end of the cord is free to move.

Consider next a pulse that travels along a cord which consists of a light section and a heavy section, as shown in Fig. 11–34. When the wave pulse reaches the boundary between the two sections, part of the pulse is reflected and part is transmitted, as shown. The heavier the second section of the cord, the less the energy that is transmitted. (When the second section is a wall or rigid support, very little is transmitted and most is reflected, as in Fig. 11–33a.) For a sinusoidal wave, the frequency of the transmitted wave does not change across the boundary because the boundary point oscillates at that frequency. Thus if the transmitted wave has a lower speed, its wavelength is also less $(\lambda = v/f)$.

For a two or three dimensional wave, such as a water wave, we are concerned with **wave fronts**, by which we mean all the points along the wave forming the wave crest (what we usually refer to simply as a "wave" at the seashore). A line drawn in the direction of wave motion, perpendicular to the wave front, is called a **ray**, as shown in Fig. 11–35. Wave fronts far from the source have lost almost all their curvature (Fig. 11–35b) and are nearly straight, as ocean waves often are. They are then called **plane waves**.

**FIGURE 11–34** When a wave pulse traveling to the right along a thin cord (a) reaches a discontinuity where the cord becomes thicker and heavier, then part is reflected and part is transmitted (b).

**FIGURE 11–35** Rays, signifying the direction of wave motion, are always perpendicular to the wave fronts (wave crests). (a) Circular or spherical waves near the source. (b) Far from the source, the wave fronts are nearly straight or flat, and are called plane waves.

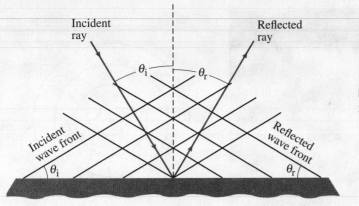

**FIGURE 11–36** Law of reflection: $\theta_r = \theta_i$.

For reflection of a two or three dimensional plane wave, as shown in Fig. 11–36, the angle that the incoming or *incident wave* makes with the reflecting surface is equal to the angle made by the reflected wave. This is the **law of reflection:**

**the angle of reflection equals the angle of incidence.**

The **angle of incidence** is defined as the angle ($\theta_i$) the incident ray makes with the perpendicular to the reflecting surface (or the wave front makes with the surface). The **angle of reflection** is the corresponding angle ($\theta_r$) for the reflected wave.

# 11–11 Interference; Principle of Superposition

**Interference** refers to what happens when two waves pass through the same region of space at the same time. Consider, for example, the two wave pulses on a cord traveling toward each other as shown in Fig. 11–37 (time increases downward in both a and b). In Fig. 11–37a the two pulses have the same amplitude, but one is a crest and the other a trough; in Fig. 11–37b they are both crests. In both cases, the waves meet and pass right by each other. However, in the region where they overlap, the resultant displacement is the *algebraic sum of their separate displacements* (a crest is considered positive and a trough negative). This is the **principle of superposition**. In Fig. 11–37a, the two waves have opposite displacements at the instant they pass one another, and they add to zero. The result is called **destructive interference**. In Fig. 11–37b, at the instant the two pulses overlap, they produce a resultant displacement that is greater than the displacement of either separate pulse, and the result is **constructive interference**.

You may wonder where the energy is at the moment of destructive interference in Fig. 11–37a; the cord may be straight at this instant, but the central parts of it are still moving up or down (kinetic energy).

(a)             (b)

Pulses far apart, approaching

Time

Pulses overlap precisely (for an instant)

Pulses far apart, receding

**FIGURE 11–37** Two wave pulses pass each other. Where they overlap, interference occurs: (a) destructive, and (b) constructive. Read (a) and (b) downward (increasing time).

(a)

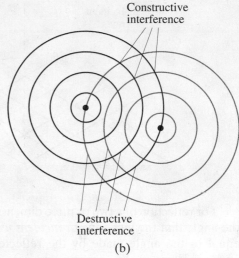

Constructive
interference

Destructive
interference

(b)

**FIGURE 11–38** (a) Interference of water waves. (b) Constructive interference occurs where one wave's maximum (a crest) meets the other's maximum. Destructive interference ("flat water") occurs where one wave's maximum (a crest) meets the other's mininum (a trough).

When two rocks are thrown into a pond simultaneously, the two sets of circular waves that move outward interfere with one another as shown in Fig. 11–38a. In some areas of overlap, crests of one wave repeatedly meet crests of the other (and troughs meet troughs), Fig. 11–38b. Constructive interference is occurring at these points, and the water continuously oscillates up and down with greater amplitude than either wave separately. In other areas, destructive interference occurs where the water does not move up and down at all over time. This is where crests of one wave meet troughs of the other, and vice versa. Figure 11–39a shows the displacement of two identical waves graphically as a function of time, as well as their sum, for the case of constructive interference. For any two such waves, we use the term **phase** to describe the relative positions of their crests. When the crests and troughs are aligned as in Fig. 11–39a, for constructive interference, the two waves are **in phase**. At points where destructive interference occurs (Fig. 11–39b), crests of one wave repeatedly meet troughs of the other wave and the two waves are said to be completely **out of phase** or, more precisely, out of phase by one-half wavelength (or 180°).[†] That is, the crests of one wave occur a half wavelength behind the crests of the other wave. The relative phase of the two water waves in Fig. 11–38 in most areas is intermediate between these two extremes, resulting in *partially* destructive interference, as illustrated in Fig. 11–39c. If the amplitudes of two interfering waves are not equal, fully destructive interference (as in Fig. 11–39b) does not occur.

[†]One wavelength, or one full oscillation, corresponds to 360°—see Section 11–3, just after Eq. 11–8c, and also Fig. 11–7.

**FIGURE 11–39** Graphs showing two identical waves, and their sum, as a function of time at three locations. In (a) the two waves interfere constructively, in (b) destructively, and in (c) partially destructively.

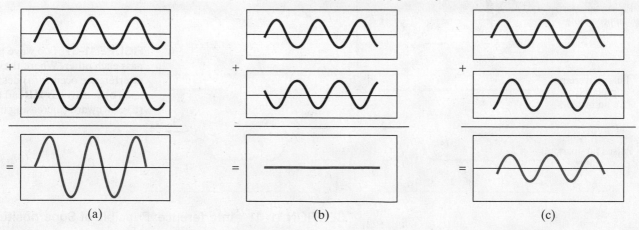

(a)

(b)

(c)

# 11–12 Standing Waves; Resonance

If you shake one end of a cord and the other end is kept fixed, a continuous wave will travel down to the fixed end and be reflected back, inverted, as we saw in Fig. 11–33a. As you continue to oscillate the cord, waves will travel in both directions, and the wave traveling along the cord, away from your hand, will interfere with the reflected wave coming back. Usually there will be quite a jumble. But if you oscillate the cord at just the right frequency, the two traveling waves will interfere in such a way that a large-amplitude **standing wave** will be produced, Fig. 11–40. It is called a "standing wave" because it does not appear to be traveling. The cord simply appears to have segments that oscillate up and down in a fixed pattern. The points of destructive interference, where the cord remains still at all times, are called **nodes**. Points of constructive interference, where the cord oscillates with maximum amplitude, are called **antinodes**. The nodes and antinodes remain in fixed positions for a particular frequency.

Standing waves can occur at more than one frequency. The lowest frequency of oscillation that produces a standing wave gives rise to the pattern shown in Fig. 11–40a. The standing waves shown in Figs. 11–40b and 11–40c are produced at precisely twice and three times the lowest frequency, respectively, assuming the tension in the cord is the same. The cord can also oscillate with four loops (four antinodes) at four times the lowest frequency, and so on.

The frequencies at which standing waves are produced are the **natural frequencies** or **resonant frequencies** of the cord, and the different standing wave patterns shown in Fig. 11–40 are different "resonant modes of vibration." A standing wave on a cord is the result of the interference of two waves traveling in opposite directions. A standing wave can also be considered a vibrating object at resonance. Standing waves represent the same phenomenon as the resonance of an oscillating spring or pendulum, which we discussed in Section 11–6. However, a spring or pendulum has only one resonant frequency, whereas the cord has an infinite number of resonant frequencies, each of which is a whole-number multiple of the lowest resonant frequency.

Consider a string stretched between two supports that is plucked like a guitar or violin string, Fig. 11–41a. Waves of a great variety of frequencies will travel in both directions along the string, will be reflected at the ends, and will travel back in the opposite direction. Most of these waves interfere with each other and quickly die out. However, those waves that correspond to the resonant frequencies of the string will persist. The ends of the string, since they are fixed, will be nodes. There may be other nodes as well. Some of the possible resonant modes of vibration (standing waves) are shown in Fig. 11–41b. Generally, the motion will be a combination of these different resonant modes, but only those frequencies that correspond to a resonant frequency will be present.

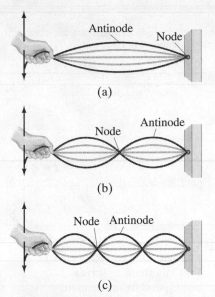

**FIGURE 11–40** Standing waves corresponding to three resonant frequencies.

**FIGURE 11–41** (a) A string is plucked. (b) Only standing waves corresponding to resonant frequencies persist for long.

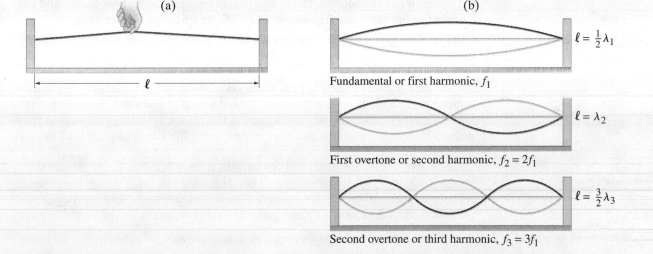

(a)

(b)

$\ell = \frac{1}{2}\lambda_1$

Fundamental or first harmonic, $f_1$

$\ell = \lambda_2$

First overtone or second harmonic, $f_2 = 2f_1$

$\ell = \frac{3}{2}\lambda_3$

Second overtone or third harmonic, $f_3 = 3f_1$

Fundamental or first harmonic, $f_1$

$\ell = \frac{1}{2}\lambda_1$

First overtone or second harmonic, $f_2 = 2f_1$

$\ell = \lambda_2$

Second overtone or third harmonic, $f_3 = 3f_1$

$\ell = \frac{3}{2}\lambda_3$

**FIGURE 11–41b** (Repeated.)
(b) Only standing waves corresponding to resonant frequencies persist for long.

To determine the resonant frequencies, we first note that the wavelengths of the standing waves bear a simple relationship to the length $\ell$ of the string. The lowest frequency, called the **fundamental frequency**, corresponds to one antinode (or loop). And as can be seen in Fig. 11–41b, the whole length corresponds to one-half wavelength. Thus $\ell = \frac{1}{2}\lambda_1$, where $\lambda_1$ stands for the wavelength of the fundamental frequency. The other natural frequencies are called **overtones**; for a vibrating string they are whole-number (integral) multiples of the fundamental, and then are also called **harmonics**, with the fundamental being referred to as the **first harmonic.**[†] The next mode of vibration after the fundamental has two loops and is called the **second harmonic** (or first overtone), Fig. 11–41b. The length of the string $\ell$ at the second harmonic corresponds to one complete wavelength: $\ell = \lambda_2$. For the third and fourth harmonics, $\ell = \frac{3}{2}\lambda_3$, and $\ell = \frac{4}{2}\lambda_4 = 2\lambda_4$, respectively, and so on. In general, we can write

$$\ell = \frac{n\lambda_n}{2}, \qquad \text{where } n = 1, 2, 3, \cdots.$$

The integer $n$ labels the number of the harmonic: $n = 1$ for the fundamental, $n = 2$ for the second harmonic, and so on. We solve for $\lambda_n$ and find

$$\lambda_n = \frac{2\ell}{n}, \qquad n = 1, 2, 3, \cdots. \qquad \begin{bmatrix} \text{string fixed} \\ \text{at both ends} \end{bmatrix} \quad \textbf{(11–19a)}$$

To find the frequency $f$ of each vibration we use Eq. 11–12, $f = v/\lambda$, and see that

$$f_n = \frac{v}{\lambda_n} = n\frac{v}{2\ell} = nf_1, \qquad n = 1, 2, 3, \cdots, \qquad \textbf{(11–19b)}$$

where $f_1 = v/\lambda_1 = v/2\ell$ is the fundamental frequency. We see that each resonant frequency is an integer multiple of the fundamental frequency on a vibrating string.

Because a standing wave is equivalent to two traveling waves moving in opposite directions, the concept of wave velocity still makes sense and is given by Eq. 11–13 in terms of the tension $F_T$ in the string and its mass per unit length ($\mu = m/\ell$). That is, $v = \sqrt{F_T/\mu}$ for waves traveling in either direction.

---

**EXAMPLE 11–14** **Piano string.** A piano string 1.10 m long has mass 9.00 g. (a) How much tension must the string be under if it is to vibrate at a fundamental frequency of 131 Hz? (b) What are the frequencies of the first four harmonics?

**APPROACH** To determine the tension, we need to find the wave speed using Eq. 11–12 ($v = \lambda f$), and then use Eq. 11–13, solving it for $F_T$.

**SOLUTION** (a) The wavelength of the fundamental is $\lambda = 2\ell = 2.20$ m (Eq. 11–19a with $n = 1$). The speed of the wave on the string is $v = \lambda f = (2.20\text{ m})(131\text{ s}^{-1}) = 288$ m/s. Then we have (Eq. 11–13)

$$F_T = \mu v^2 = \frac{m}{\ell}v^2 = \left(\frac{9.00 \times 10^{-3}\text{ kg}}{1.10\text{ m}}\right)(288\text{ m/s})^2 = 679\text{ N}.$$

(b) The first harmonic (the fundamental) has a frequency $f_1 = 131$ Hz. The frequencies of the second, third, and fourth harmonics are two, three, and four times the fundamental frequency: 262, 393, and 524 Hz, respectively.

**NOTE** The speed of the wave on the string is *not* the same as the speed of the sound wave that the piano string produces in the air (as we shall see in Chapter 12).

---

A standing wave does appear to be standing in place (and a traveling wave appears to move). The term "standing" wave is also meaningful from the point of view of energy. Since the string is at rest at the nodes, no energy flows past these points. Hence the energy is not transmitted down the string but "stands" in place in the string.

Standing waves are produced not only on strings, but also on any object that is struck, such as a drum membrane or an object made of metal or wood. The resonant frequencies depend on the dimensions of the object, just as for a string they depend on its length. Large objects have lower resonant frequencies than small objects.

[†]The term "harmonic" comes from music, because such integral multiples of frequencies "harmonize."

All musical instruments, from stringed to wind instruments (in which a column of air oscillates as a standing wave) to drums and other percussion instruments, depend on standing waves to produce their particular musical sounds, as we shall see in Chapter 12.

# *11–13 Refraction[†]

When any wave strikes a boundary, some of the energy is reflected and some is transmitted or absorbed. When a two- or three-dimensional wave traveling in one medium crosses a boundary into a medium where its speed is different, the transmitted wave may move in a different direction than the incident wave, as shown in Fig. 11–42. This phenomenon is known as **refraction**. One example is a water wave; the velocity decreases in shallow water and the waves refract, as shown in Fig. 11–43. [When the wave velocity changes gradually, as in Fig. 11–43, without a sharp boundary, the waves change direction (refract) gradually.]

In Fig. 11–42, the velocity of the wave in medium 2 is less than in medium 1. In this case, the wave front bends so that it travels more nearly parallel to the boundary. That is, the *angle of refraction*, $\theta_r$, is less than the *angle of incidence*, $\theta_i$. To see why this is so, and to help us get a quantitative relation between $\theta_r$ and $\theta_i$, let us think of each wave front as a row of soldiers. The soldiers are marching from firm ground (medium 1) into mud (medium 2) and hence are slowed down after the boundary. The soldiers that reach the mud first are slowed down first, and the row bends as shown in Fig. 11–44a. Let us consider the wave front (or row of soldiers) labeled A in Fig. 11–44b. In the same time $t$ that $A_1$ moves a distance $\ell_1 = v_1 t$, we see that $A_2$ moves a distance $\ell_2 = v_2 t$. The two right triangles in Fig. 11–44b, shaded yellow and green, have the side labeled $a$ in common. Thus

$$\sin \theta_1 = \frac{\ell_1}{a} = \frac{v_1 t}{a}$$

since $a$ is the hypotenuse, and

$$\sin \theta_2 = \frac{\ell_2}{a} = \frac{v_2 t}{a}.$$

Dividing these two equations, we obtain the **law of refraction**:

$$\frac{\sin \theta_2}{\sin \theta_1} = \frac{v_2}{v_1}. \qquad \textbf{(11–20)}$$

Since $\theta_1$ is the angle of incidence $(\theta_i)$, and $\theta_2$ is the angle of refraction $(\theta_r)$, Eq. 11–20 gives the quantitative relation between the two. If the wave were going in the opposite direction, the geometry would not change; only $\theta_1$ and $\theta_2$ would change roles: $\theta_2$ would be the angle of incidence and $\theta_1$ the angle of refraction. Thus, if the wave travels into a medium where it can move faster, it will bend the opposite way, $\theta_r > \theta_i$. We see from Eq. 11–20 that if the velocity increases, the angle increases, and vice versa.

[†]This Section and the next are covered in more detail in Chapters 23 and 24 on optics.

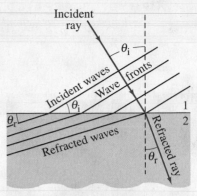

**FIGURE 11–42** Refraction of waves passing a boundary.

**FIGURE 11–43** Water waves refract gradually as they approach the shore, as their velocity decreases. There is no distinct boundary, as in Fig. 11–42, because the wave velocity changes gradually.

**FIGURE 11–44** (a) Marching soldier analogy to derive (b) law of refraction for waves.

(a)　　　　　　　　　(b)

Earthquake waves refract within the Earth as they travel through rock layers of different densities (which have different velocities) just as water waves do. Light waves refract as well, and when we discuss light, we shall find Eq. 11–20 very useful.

# *11–14 Diffraction

(a)

(b)

**FIGURE 11–45** Wave diffraction. In (a) the waves pass through a slit and into the "shadow region" behind. In (b) the waves are coming from the upper left. As they pass an obstacle, they bend around it into the shadow region behind it.

Waves spread as they travel. When waves encounter an obstacle, they bend around it somewhat and pass into the region behind it, as shown in Fig. 11–45 for water waves. This phenomenon is called **diffraction**.

The amount of diffraction depends on the wavelength of the wave and on the size of the obstacle, as shown in Fig. 11–46. If the wavelength is much larger than the object, as with the grass blades of Fig. 11–46a, the wave bends around them almost as if they are not there. For larger objects, parts (b) and (c), there is more of a "shadow" region behind the obstacle where we might not expect the waves to penetrate—but they do, at least a little. Then notice in part (d), where the obstacle is the same as in part (c) but the wavelength is longer, that there is more diffraction into the shadow region. As a rule of thumb, *only if the wavelength is smaller than the size of the object will there be a significant shadow region.* This rule applies to *reflection* from an obstacle as well. Very little of a wave is reflected unless the wavelength is smaller than the size of the obstacle.

A rough guide to the amount of diffraction is

$$\theta(\text{radians}) \approx \frac{\lambda}{\ell},$$

where $\theta$ is roughly the angular spread of waves after they have passed through an opening of width $\ell$ or around an obstacle of width $\ell$.

That waves can bend around obstacles, and thus can carry energy to areas behind obstacles, is very different from energy carried by material particles. A clear example is the following: if you are standing around a corner on one side of a building, you cannot be hit by a baseball thrown from the other side, but you can hear a shout or other sound because the sound waves diffract around the edges of the building.

(a) Water waves passing blades of grass

(b) Stick in water

(c) Short-wavelength waves passing log

(d) Long-wavelength waves passing log

**FIGURE 11–46** Water waves, coming from upper left, pass objects of various sizes. Note that the longer the wavelength compared to the size of the object, the more diffraction there is into the "shadow region."

---

**CONCEPTUAL EXAMPLE 11–15** | **Cell phones.** Cellular phones operate by radio waves with frequencies of about 1 or 2 GHz (1 gigahertz = $10^9$ Hz). These waves cannot penetrate objects that conduct electricity, such as a sheet of metal or a tree trunk. The sound quality is best if the transmitting antenna is within clear view of the handset. Yet it is possible to carry on a phone conversation even if the tower is blocked by trees, or if the handset is inside a car. Why?

**RESPONSE** If the radio waves have a frequency of about 2 GHz, and the speed of propagation is equal to the speed of light, $3 \times 10^8$ m/s (Section 1–5), then the wavelength is $\lambda = v/f = (3 \times 10^8 \text{ m/s})/(2 \times 10^9 \text{ Hz}) = 0.15$ m. The waves can diffract readily around objects 15 cm in diameter or smaller.

# *11–15 Mathematical Representation of a Traveling Wave

A simple wave with a single frequency, as in Fig. 11–47, is sinusoidal. To express such a wave mathematically, we assume it has a particular wavelength $\lambda$ and frequency $f$. At $t = 0$, the wave shape shown is

$$y = A \sin \frac{2\pi}{\lambda} x, \qquad (11–21)$$

where $y$ is the **displacement** of the wave (either longitudinal or transverse) at position $x$, $\lambda$ is the wavelength, and $A$ is the **amplitude** of the wave. [Equation 11–21 works because it repeats itself every wavelength: when $x = \lambda$, $y = \sin 2\pi = \sin 0$.]

Suppose the wave is moving to the right with speed $v$. After a time $t$, each part of the wave (indeed, the whole wave "shape") has moved to the right a distance $vt$. Figure 11–48 shows the wave at $t = 0$ as a solid curve, and at a later time $t$ as a dashed curve. Consider any point on the wave at $t = 0$: say, a crest at some position $x$. After a time $t$, that crest will have traveled a distance $vt$, so its new position is a distance $vt$ greater than its old position. To describe this crest (or other point on the wave shape), the argument of the sine function must have the same numerical value, so we replace $x$ in Eq. 11–21 by $(x - vt)$:

$$y = A \sin \left[ \frac{2\pi}{\lambda} (x - vt) \right]. \qquad (11–22)$$

Said another way, if you are on a crest, as $t$ increases, $x$ must increase at the same rate so that $(x - vt)$ remains constant.

For a wave traveling along the $x$ axis to the left, toward decreasing values of $x$, $v$ becomes $-v$, so

$$y = A \sin \left[ \frac{2\pi}{\lambda} (x + vt) \right].$$

**FIGURE 11–47** The characteristics of a single-frequency wave at $t = 0$ (just as in Fig. 11–24).

**FIGURE 11–48** A traveling wave. In time $t$, the wave moves a distance $vt$.

# Summary

An oscillating (or vibrating) object undergoes **simple harmonic motion** (SHM) if the restoring force is proportional to (the negative of) the displacement,

$$F = -kx. \qquad (11–1)$$

The maximum displacement from equilibrium is called the **amplitude**.

The **period**, $T$, is the time required for one complete cycle (back and forth), and the **frequency**, $f$, is the number of cycles per second; they are related by

$$f = \frac{1}{T}. \qquad (11–2)$$

The period of oscillation for a mass $m$ on the end of a spring is given by

$$T = 2\pi \sqrt{\frac{m}{k}}. \qquad (11–6a)$$

SHM is **sinusoidal**, which means that the displacement as a function of time follows a sine curve.

During SHM, the total energy

$$E = \tfrac{1}{2}mv^2 + \tfrac{1}{2}kx^2 \qquad (11–3)$$

is continually changing from potential to kinetic and back again.

A **simple pendulum** of length $\ell$ approximates SHM if its amplitude is small and friction can be ignored. For small amplitudes, its period is given by

$$T = 2\pi \sqrt{\frac{\ell}{g}}, \qquad (11–11a)$$

where $g$ is the acceleration of gravity.

When friction is present (for all real springs and pendulums), the motion is said to be **damped**. The maximum displacement decreases in time, and the mechanical energy is eventually all transformed to thermal energy.

If a varying force of frequency $f$ is applied to a system capable of oscillating, the amplitude of oscillation can be very large if the frequency of the applied force is near the **natural** (or **resonant**) **frequency** of the oscillator. This is called **resonance**.

Vibrating objects act as sources of waves that travel outward from the source. Waves on water and on a cord are examples. The wave may be a **pulse** (a single crest), or it may be continuous (many crests and troughs).

The **wavelength** of a continuous sinusoidal wave is the distance between two successive crests.

The **frequency** is the number of full wavelengths (or crests) that pass a given point per unit time.

The **amplitude** of a wave is the maximum height of a crest, or depth of a trough, relative to the normal (or equilibrium) level.

The **wave speed** (how fast a crest moves) is equal to the product of wavelength and frequency,

$$v = \lambda f. \tag{11–12}$$

In a **transverse wave**, the oscillations are perpendicular to the direction in which the wave travels. An example is a wave on a cord.

In a **longitudinal wave**, the oscillations are along (parallel to) the line of travel; sound is an example.

Waves carry energy from place to place without matter being carried. The **intensity** of a wave is the energy per unit time carried across unit area (in watts/m²). For three-dimensional waves traveling outward from a point source, the intensity decreases inversely as the square of the distance from the source (ignoring damping):

$$I \propto \frac{1}{r^2}. \tag{11–16b}$$

Wave intensity is proportional to the amplitude squared and to the frequency squared.

Waves reflect off objects in their path. When the **wave front** (of a two- or three-dimensional wave) strikes an object, the **angle of reflection** is equal to the **angle of incidence**. This is the **law of reflection**. When a wave strikes a boundary between two materials in which it can travel, part of the wave is reflected and part is transmitted.

When two waves pass through the same region of space at the same time, they **interfere**. The resultant displacement at any point and time is the sum of their separate displacements (= the **superposition principle**). This can result in **constructive interference**, **destructive interference**, or something in between, depending on the amplitudes and relative phases of the waves.

Waves traveling on a string of fixed length interfere with waves that have reflected off the end and are traveling back in the opposite direction. At certain frequencies, **standing waves** can be produced in which the waves seem to be standing still rather than traveling. The string (or other medium) is vibrating as a whole. This is a resonance phenomenon, and the frequencies at which standing waves occur are called **resonant frequencies**. Points of destructive interference (no oscillation) are called **nodes**. Points of constructive interference (maximum amplitude of vibration) are called **antinodes**.

[*Waves change direction, or **refract**, when traveling from one medium into a second medium where their speed is different. Waves spread, or **diffract**, as they travel and encounter obstacles. A rough guide to the amount of diffraction is $\theta \approx \lambda/\ell$, where $\lambda$ is the wavelength and $\ell$ the width of an obstacle or opening. There is a significant "shadow region" only if the wavelength $\lambda$ is smaller than the size of the obstacle.]

[*A traveling wave can be represented mathematically as $y = A \sin \{(2\pi/\lambda)(x \pm vt)\}$.]

# Questions

1. Is the acceleration of a simple harmonic oscillator ever zero? If so, where?

2. Real springs have mass. Will the true period and frequency be larger or smaller than given by the equations for a mass oscillating on the end of an idealized massless spring? Explain.

3. How could you double the maximum speed of a simple harmonic oscillator (SHO)?

4. If a pendulum clock is accurate at sea level, will it gain or lose time when taken to high altitude? Why?

5. A tire swing hanging from a branch reaches nearly to the ground (Fig. 11–49). How could you estimate the height of the branch using only a stopwatch?

**FIGURE 11–49** Question 5.

6. For a simple harmonic oscillator, when (if ever) are the displacement and velocity vectors in the same direction? When are the displacement and acceleration vectors in the same direction?

7. Two equal masses are attached to separate identical springs next to one another. One mass is pulled so its spring stretches 40 cm and the other is pulled so its spring stretches only 20 cm. The masses are released simultaneously. Which mass reaches the equilibrium point first?

8. What is the approximate period of your walking step?

9. What happens to the period of a playground swing if you rise up from sitting to a standing position?

10. Why can you make water slosh back and forth in a pan only if you shake the pan at a certain frequency?

11. Is the frequency of a simple periodic wave equal to the frequency of its source? Why or why not?

12. Explain the difference between the speed of a transverse wave traveling along a cord and the speed of a tiny piece of the cord.

13. What kind of waves do you think will travel along a horizontal metal rod if you strike its end (*a*) vertically from above and (*b*) horizontally parallel to its length?

14. Since the density of air decreases with an increase in temperature, but the bulk modulus *B* is nearly independent of temperature, how would you expect the speed of sound waves in air to vary with temperature?

15. If a rope has a free end, a pulse sent down the rope behaves differently on reflection than if the rope has that end fixed in position. What is this difference, and why does it occur?

16. How did geophysicists determine that part of the Earth's interior is liquid?

17. The speed of sound in most solids is somewhat greater than in air, yet the density of solids is much greater ($10^3$ to $10^4$ times). Explain.

18. Give two reasons why circular water waves decrease in amplitude as they travel away from the source.

19. Two linear waves have the same amplitude and speed, and otherwise are identical, except one has half the wavelength of the other. Which transmits more energy? By what factor?

20. When a sinusoidal wave crosses the boundary between two sections of cord as in Fig. 11–34, the frequency does not change (although the wavelength and velocity do change). Explain why.

21. Is energy always conserved when two waves interfere? Explain.

22. If a string is vibrating as a standing wave in three loops, are there any places you could touch it with a knife blade without disturbing the motion?

23. Why do the strings used for the lowest-frequency notes on a piano normally have wire wrapped around them?

24. When a standing wave exists on a string, the vibrations of incident and reflected waves cancel at the nodes. Does this mean that energy was destroyed? Explain.

25. Can the amplitude of the standing waves in Fig. 11–40 be greater than the amplitude of the vibrations that cause them (up and down motion of the hand)?

26. "In a round bowl of water, waves move from the center to the rim, or from the rim to the center, depending on whether you strike at the center or at the rim." So wrote Dante Alighieri 700 years ago in his great poem *Paradiso* (Canto 14), the last part of his famous *Divine Comedy*. Try this experiment and discuss your results.

*27. AM radio signals can usually be heard behind a hill, but FM often cannot. That is, AM signals bend more than FM. Explain. (Radio signals, as we shall see, are carried by electromagnetic waves whose wavelength for AM is typically 200 to 600 m and for FM about 3 m.)

# MisConceptual Questions

1. A mass on a spring in SHM (Fig. 11–1) has amplitude $A$ and period $T$. At what point in the motion is the velocity zero and the acceleration zero simultaneously?
   (a) $x = A$.
   (b) $x > 0$ but $x < A$.
   (c) $x = 0$.
   (d) $x < 0$.
   (e) None of the above.

2. An object oscillates back and forth on the end of a spring. Which of the following statements are true at some time during the course of the motion?
   (a) The object can have zero velocity and, simultaneously, nonzero acceleration.
   (b) The object can have zero velocity and, simultaneously, zero acceleration.
   (c) The object can have zero acceleration and, simultaneously, nonzero velocity.
   (d) The object can have nonzero velocity and nonzero acceleration simultaneously.

3. An object of mass $M$ oscillates on the end of a spring. To double the period, replace the object with one of mass:
   (a) $2M$.
   (b) $M/2$.
   (c) $4M$.
   (d) $M/4$.
   (e) None of the above.

4. An object of mass $m$ rests on a frictionless surface and is attached to a horizontal ideal spring with spring constant $k$. The system oscillates with amplitude $A$. The oscillation frequency of this system can be increased by
   (a) decreasing $k$.
   (b) decreasing $m$.
   (c) increasing $A$.
   (d) More than one of the above.
   (e) None of the above will work.

5. When you use the approximation $\sin \theta \approx \theta$ for a pendulum, you must specify the angle $\theta$ in
   (a) radians only.
   (b) degrees only.
   (c) revolutions or radians.
   (d) degrees or radians.

6. Suppose you pull a simple pendulum to one side by an angle of 5°, let go, and measure the period of oscillation that ensues. Then you stop the oscillation, pull the pendulum to an angle of 10°, and let go. The resulting oscillation will have a period about _____ the period of the first oscillation.
   (a) four times
   (b) twice
   (c) half
   (d) one-fourth
   (e) the same as

7. At a playground, two young children are on identical swings. One child appears to be about twice as heavy as the other. If you pull them back together the same distance and release them to start them swinging, what will you notice about the oscillations of the two children?
   (a) The heavier child swings with a period twice that of the lighter one.
   (b) The lighter child swings with a period twice that of the heavier one.
   (c) Both children swing with the same period.

8. A grandfather clock is "losing" time because its pendulum moves too slowly. Assume that the pendulum is a massive bob at the end of a string. The motion of this pendulum can be sped up by (list all that work):
   (a) shortening the string.
   (b) lengthening the string.
   (c) increasing the mass of the bob.
   (d) decreasing the mass of the bob.

9. Consider a wave traveling down a cord and the transverse motion of a small piece of the cord. Which of the following is true?
   (a) The speed of the wave must be the same as the speed of a small piece of the cord.
   (b) The frequency of the wave must be the same as the frequency of a small piece of the cord.
   (c) The amplitude of the wave must be the same as the amplitude of a small piece of the cord.
   (d) All of the above are true.
   (e) Both (b) and (c) are true.

10. Two waves are traveling toward each other along a rope. When they meet, the waves
    (a) pass through each other.
    (b) bounce off of each other.
    (c) disappear.

11. Which of the following increases the speed of waves in a stretched elastic cord? (More than one answer may apply.)
    (a) Increasing the wave amplitude.
    (b) Increasing the wave frequency.
    (c) Increasing the wavelength.
    (d) Stretching the elastic cord further.

12. Consider a wave on a string moving to the right, as shown in Fig. 11–50. What is the direction of the velocity of a particle of string at point B?

    (a) →
    (b) ↘
    (c) ↓
    (d) ↑
    (e) $\vec{v} = 0$, so no direction.

    **FIGURE 11–50**
    MisConceptual Question 12.

13. What happens when two waves, such as waves on a lake, come from different directions and run into each other?
    (a) They cancel each other out and disappear.
    (b) If they are the same size, they cancel each other out and disappear. If one wave is larger than the other, the smaller one disappears and the larger one shrinks but continues.
    (c) They get larger where they run into each other; then they continue in a direction between the direction of the two original waves and larger than either original wave.
    (d) They may have various patterns where they overlap, but each wave continues with its original pattern away from the region of overlap.
    (e) Waves cannot run into each other; they always come from the same direction and so are parallel.

14. A student attaches one end of a Slinky to the top of a table. She holds the other end in her hand, stretches it to a length $\ell$, and then moves it back and forth to send a wave down the Slinky. If she next moves her hand faster while keeping the length of the Slinky the same, how does the wavelength down the Slinky change?
    (a) It increases.
    (b) It stays the same.
    (c) It decreases.

15. A wave transports
    (a) energy but not matter.
    (b) matter but not energy.
    (c) both energy and matter.

For assigned homework and other learning materials, go to the MasteringPhysics website. (MP)

# Problems

## 11–1 to 11–3 Simple Harmonic Motion

1. (I) If a particle undergoes SHM with amplitude 0.21 m, what is the total distance it travels in one period?

2. (I) The springs of a 1700-kg car compress 5.0 mm when its 66-kg driver gets into the driver's seat. If the car goes over a bump, what will be the frequency of oscillations? Ignore damping.

3. (II) An elastic cord is 61 cm long when a weight of 75 N hangs from it but is 85 cm long when a weight of 210 N hangs from it. What is the "spring" constant $k$ of this elastic cord?

4. (II) Estimate the stiffness of the spring in a child's pogo stick if the child has a mass of 32 kg and bounces once every 2.0 seconds.

5. (II) A fisherman's scale stretches 3.6 cm when a 2.4-kg fish hangs from it. (a) What is the spring stiffness constant and (b) what will be the amplitude and frequency of oscillation if the fish is pulled down 2.1 cm more and released so that it oscillates up and down?

6. (II) A small fly of mass 0.22 g is caught in a spider's web. The web oscillates predominantly with a frequency of 4.0 Hz. (a) What is the value of the effective spring stiffness constant $k$ for the web? (b) At what frequency would you expect the web to oscillate if an insect of mass 0.44 g were trapped?

7. (II) A mass $m$ at the end of a spring oscillates with a frequency of 0.83 Hz. When an additional 780-g mass is added to $m$, the frequency is 0.60 Hz. What is the value of $m$?

8. (II) A vertical spring with spring stiffness constant 305 N/m oscillates with an amplitude of 28.0 cm when 0.235 kg hangs from it. The mass passes through the equilibrium point $(y = 0)$ with positive velocity at $t = 0$. (a) What equation describes this motion as a function of time? (b) At what times will the spring be longest and shortest?

9. (II) Figure 11–51 shows two examples of SHM, labeled A and B. For each, what is (a) the amplitude, (b) the frequency, and (c) the period?

**FIGURE 11–51**
Problem 9.

10. (II) A balsa wood block of mass 52 g floats on a lake, bobbing up and down at a frequency of 3.0 Hz. (a) What is the value of the effective spring constant of the water? (b) A partially filled water bottle of mass 0.28 kg and almost the same size and shape of the balsa block is tossed into the water. At what frequency would you expect the bottle to bob up and down? Assume SHM.

11. (II) At what displacement of a SHO is the energy half kinetic and half potential?

12. (II) An object of unknown mass $m$ is hung from a vertical spring of unknown spring constant $k$, and the object is observed to be at rest when the spring has stretched by 14 cm. The object is then given a slight push upward and executes SHM. Determine the period $T$ of this oscillation.

13. (II) A 1.65-kg mass stretches a vertical spring 0.215 m. If the spring is stretched an additional 0.130 m and released, how long does it take to reach the (new) equilibrium position again?

14. (II) A 1.15-kg mass oscillates according to the equation $x = 0.650 \cos(8.40t)$ where $x$ is in meters and $t$ in seconds. Determine (a) the amplitude, (b) the frequency, (c) the total energy, and (d) the kinetic energy and potential energy when $x = 0.360$ m.

15. (II) A 0.25-kg mass at the end of a spring oscillates 2.2 times per second with an amplitude of 0.15 m. Determine (a) the speed when it passes the equilibrium point, (b) the speed when it is 0.10 m from equilibrium, (c) the total energy of the system, and (d) the equation describing the motion of the mass, assuming that at $t = 0$, $x$ was a maximum.

16. (II) It takes a force of 91.0 N to compress the spring of a toy popgun 0.175 m to "load" a 0.160-kg ball. With what speed will the ball leave the gun if fired horizontally?

17. (II) If one oscillation has 3.0 times the energy of a second one of equal frequency and mass, what is the ratio of their amplitudes?

18. (II) A mass of 240 g oscillates on a horizontal frictionless surface at a frequency of 2.5 Hz and with amplitude of 4.5 cm. (a) What is the effective spring constant for this motion? (b) How much energy is involved in this motion?

19. (II) A mass resting on a horizontal, frictionless surface is attached to one end of a spring; the other end is fixed to a wall. It takes 3.6 J of work to compress the spring by 0.13 m. If the spring is compressed, and the mass is released from rest, it experiences a maximum acceleration of 12 m/s². Find the value of (a) the spring constant and (b) the mass.

20. (II) An object with mass 2.7 kg is executing simple harmonic motion, attached to a spring with spring constant $k = 310$ N/m. When the object is 0.020 m from its equilibrium position, it is moving with a speed of 0.55 m/s. (a) Calculate the amplitude of the motion. (b) Calculate the maximum speed attained by the object.

21. (II) At $t = 0$, an 885-g mass at rest on the end of a horizontal spring ($k = 184$ N/m) is struck by a hammer which gives it an initial speed of 2.26 m/s. Determine (a) the period and frequency of the motion, (b) the amplitude, (c) the maximum acceleration, (d) the total energy, and (e) the kinetic energy when $x = 0.40A$ where $A$ is the amplitude.

22. (III) Agent Arlene devised the following method of measuring the muzzle velocity of a rifle (Fig. 11–52). She fires a bullet into a 4.148-kg wooden block resting on a smooth surface, and attached to a spring of spring constant $k = 162.7$ N/m. The bullet, whose mass is 7.870 g, remains embedded in the wooden block. She measures the maximum distance that the block compresses the spring to be 9.460 cm. What is the speed $v$ of the bullet?

**FIGURE 11–52** Problem 22.

23. (III) A bungee jumper with mass 65.0 kg jumps from a high bridge. After arriving at his lowest point, he oscillates up and down, reaching a low point seven more times in 43.0 s. He finally comes to rest 25.0 m below the level of the bridge. Estimate the spring stiffness constant and the unstretched length of the bungee cord assuming SHM.

24. (III) A block of mass $m$ is supported by two identical parallel vertical springs, each with spring stiffness constant $k$ (Fig. 11–53). What will be the frequency of vertical oscillation?

**FIGURE 11–53**
Problem 24.

25. (III) A 1.60-kg object oscillates at the end of a vertically hanging light spring once every 0.45 s. (a) Write down the equation giving its position $y$ (+ upward) as a function of time $t$. Assume the object started by being compressed 16 cm from the equilibrium position (where $y = 0$), and released. (b) How long will it take to get to the equilibrium position for the first time? (c) What will be its maximum speed? (d) What will be the object's maximum acceleration, and where will it first be attained?

26. (III) Consider two objects, A and B, both undergoing SHM, but with different frequencies, as described by the equations $x_A = (2.0 \text{ m}) \sin(4.0t)$ and $x_B = (5.0 \text{ m}) \sin(3.0t)$, where $t$ is in seconds. After $t = 0$, find the next three times $t$ at which both objects simultaneously pass through the origin.

## 11–4 Simple Pendulum

27. (I) A pendulum has a period of 1.85 s on Earth. What is its period on Mars, where the acceleration of gravity is about 0.37 that on Earth?

28. (I) How long must a simple pendulum be if it is to make exactly one swing per second? (That is, one complete oscillation takes exactly 2.0 s.)

29. (I) A pendulum makes 28 oscillations in exactly 50 s. What is its (a) period and (b) frequency?

30. (II) What is the period of a simple pendulum 47 cm long (a) on the Earth, and (b) when it is in a freely falling elevator?

31. (II) Your grandfather clock's pendulum has a length of 0.9930 m. If the clock runs slow and loses 21 s per day, how should you adjust the length of the pendulum?

32. (II) Derive a formula for the maximum speed $v_{max}$ of a simple pendulum bob in terms of $g$, the length $\ell$, and the maximum angle of swing $\theta_{max}$.

33. (III) A simple pendulum oscillates with an amplitude of 10.0°. What fraction of the time does it spend between +5.0° and −5.0°? Assume SHM.

34. (III) A clock pendulum oscillates at a frequency of 2.5 Hz. At $t = 0$, it is released from rest starting at an angle of 12° to the vertical. Ignoring friction, what will be the position (angle in radians) of the pendulum at (a) $t = 0.25$ s, (b) $t = 1.60$ s, and (c) $t = 500$ s?

## 11–7 and 11–8 Waves

35. (I) A fisherman notices that wave crests pass the bow of his anchored boat every 3.0 s. He measures the distance between two crests to be 7.0 m. How fast are the waves traveling?

36. (I) A sound wave in air has a frequency of 282 Hz and travels with a speed of 343 m/s. How far apart are the wave crests (compressions)?

37. (I) Calculate the speed of longitudinal waves in (a) water, (b) granite, and (c) steel.

38. (I) AM radio signals have frequencies between 550 kHz and 1600 kHz (kilohertz) and travel with a speed of $3.0 \times 10^8$ m/s. What are the wavelengths of these signals? On FM the frequencies range from 88 MHz to 108 MHz (megahertz) and travel at the same speed. What are their wavelengths?

39. (II) P and S waves from an earthquake travel at different speeds, and this difference helps locate the earthquake "epicenter" (where the disturbance took place). (a) Assuming typical speeds of 8.5 km/s and 5.5 km/s for P and S waves, respectively, how far away did an earthquake occur if a particular seismic station detects the arrival of these two types of waves 1.5 min apart? (b) Is one seismic station sufficient to determine the position of the epicenter? Explain.

40. (II) A cord of mass 0.65 kg is stretched between two supports 8.0 m apart. If the tension in the cord is 120 N, how long will it take a pulse to travel from one support to the other?

41. (II) A 0.40-kg cord is stretched between two supports, 8.7 m apart. When one support is struck by a hammer, a transverse wave travels down the cord and reaches the other support in 0.85 s. What is the tension in the cord?

42. (II) A sailor strikes the side of his ship just below the surface of the sea. He hears the echo of the wave reflected from the ocean floor directly below 2.4 s later. How deep is the ocean at this point?

43. (II) Two children are sending signals along a cord of total mass 0.50 kg tied between tin cans with a tension of 35 N. It takes the vibrations in the string 0.55 s to go from one child to the other. How far apart are the children?

## 11–9 Energy Transported by Waves

44. (II) What is the ratio of (a) the intensities, and (b) the amplitudes, of an earthquake P wave passing through the Earth and detected at two points 15 km and 45 km from the source?

45. (II) The intensity of an earthquake wave passing through the Earth is measured to be $3.0 \times 10^6$ J/m²·s at a distance of 54 km from the source. (a) What was its intensity when it passed a point only 1.0 km from the source? (b) At what rate did energy pass through an area of 2.0 m² at 1.0 km?

46. (II) A bug on the surface of a pond is observed to move up and down a total vertical distance of 7.0 cm, from the lowest to the highest point, as a wave passes. If the ripples decrease to 4.5 cm, by what factor does the bug's maximum KE change?

47. (II) Two earthquake waves of the same frequency travel through the same portion of the Earth, but one is carrying 5.0 times the energy. What is the ratio of the amplitudes of the two waves?

## 11–11 Interference

48. (I) The two pulses shown in Fig. 11–54 are moving toward each other. (a) Sketch the shape of the string at the moment they directly overlap. (b) Sketch the shape of the string a few moments later. (c) In Fig. 11–37a, at the moment the pulses pass each other, the string is straight. What has happened to the energy at this moment?

**FIGURE 11–54** Problem 48.

## 11–12 Standing Waves; Resonance

49. (I) If a violin string vibrates at 440 Hz as its fundamental frequency, what are the frequencies of the first four harmonics?

50. (I) A violin string vibrates at 294 Hz when unfingered. At what frequency will it vibrate if it is fingered one-third of the way down from the end? (That is, only two-thirds of the string vibrates as a standing wave.)

51. (I) A particular string resonates in four loops at a frequency of 240 Hz. Give at least three other frequencies at which it will resonate. What is each called?

52. (II) The speed of waves on a string is 97 m/s. If the frequency of standing waves is 475 Hz, how far apart are two adjacent nodes?

**53.** (II) If two successive overtones of a vibrating string are 280 Hz and 350 Hz, what is the frequency of the fundamental?

**54.** (II) A guitar string is 92 cm long and has a mass of 3.4 g. The distance from the bridge to the support post is $\ell = 62$ cm, and the string is under a tension of 520 N. What are the frequencies of the fundamental and first two overtones?

**55.** (II) One end of a horizontal string is attached to a small-amplitude mechanical 60.0-Hz oscillator. The string's mass per unit length is $3.5 \times 10^{-4}$ kg/m. The string passes over a pulley, a distance $\ell = 1.50$ m away, and weights are hung from this end, Fig. 11–55. What mass $m$ must be hung from this end of the string to produce (a) one loop, (b) two loops, and (c) five loops of a standing wave? Assume the string at the oscillator is a node, which is nearly true.

**FIGURE 11–55** Problems 55 and 56.

**56.** (II) In Problem 55 (Fig. 11–55), the length $\ell$ of the string may be adjusted by moving the pulley. If the hanging mass $m$ is fixed at 0.080 kg, how many different standing wave patterns may be achieved by varying $\ell$ between 10 cm and 1.5 m?

**57.** (II) When you slosh the water back and forth in a tub at just the right frequency, the water alternately rises and falls at each end, remaining relatively calm at the center. Suppose the frequency to produce such a standing wave in a 75-cm-wide tub is 0.85 Hz. What is the speed of the water wave?

## *11–13 Refraction

***58.** (I) An earthquake P wave traveling at 8.0 km/s strikes a boundary within the Earth between two kinds of material. If it approaches the boundary at an incident angle of 44° and the angle of refraction is 33°, what is the speed in the second medium?

***59.** (II) A sound wave is traveling in warm air when it hits a layer of cold, dense air. If the sound wave hits the cold air interface at an angle of 25°, what is the angle of refraction? Assume that the cold air temperature is $-15°C$ and the warm air temperature is $+15°C$. The speed of sound as a function of temperature can be approximated by $v = (331 + 0.60\,T)$ m/s, where $T$ is in °C.

***60.** (III) A longitudinal earthquake wave strikes a boundary between two types of rock at a 38° angle. As the wave crosses the boundary, the specific gravity changes from 3.6 to 2.5. Assuming that the elastic modulus is the same for both types of rock, determine the angle of refraction.

## *11–14 Diffraction

***61.** (II) What frequency of sound would have a wavelength the same size as a 0.75-m-wide window? (The speed of sound is 344 m/s at 20°C.) What frequencies would diffract through the window?

# General Problems

**62.** A 62-kg person jumps from a window to a fire net 20.0 m directly below, which stretches the net 1.4 m. Assume that the net behaves like a simple spring. (a) Calculate how much it would stretch if the same person were lying in it. (b) How much would it stretch if the person jumped from 38 m?

**63.** An energy-absorbing car bumper has a spring constant of 410 kN/m. Find the maximum compression of the bumper if the car, with mass 1300 kg, collides with a wall at a speed of 2.0 m/s (approximately 5 mi/h).

**64.** The length of a simple pendulum is 0.72 m, the pendulum bob has a mass of 295 g, and it is released at an angle of 12° to the vertical. Assume SHM. (a) With what frequency does it oscillate? (b) What is the pendulum bob's speed when it passes through the lowest point of the swing? (c) What is the total energy stored in this oscillation assuming no losses?

**65.** A block of mass $M$ is suspended from a ceiling by a spring with spring stiffness constant $k$. A penny of mass $m$ is placed on top of the block. What is the maximum amplitude of oscillations that will allow the penny to just stay on top of the block? (Assume $m \ll M$.)

**66.** A block with mass $M = 6.0$ kg rests on a frictionless table and is attached by a horizontal spring ($k = 130$ N/m) to a wall. A second block, of mass $m = 1.25$ kg, rests on top of $M$. The coefficient of static friction between the two blocks is 0.30. What is the maximum possible amplitude of oscillation such that $m$ will not slip off $M$?

**67.** A simple pendulum oscillates with frequency $f$. What is its frequency if the entire pendulum accelerates at $0.35\,g$ (a) upward, and (b) downward?

**68.** A 0.650-kg mass oscillates according to the equation $x = 0.25 \sin(4.70\,t)$ where $x$ is in meters and $t$ is in seconds. Determine (a) the amplitude, (b) the frequency, (c) the period, (d) the total energy, and (e) the kinetic energy and potential energy when $x$ is 15 cm.

**69.** An oxygen atom at a particular site within a DNA molecule can be made to execute simple harmonic motion when illuminated by infrared light. The oxygen atom is bound with a spring-like chemical bond to a phosphorus atom, which is rigidly attached to the DNA backbone. The oscillation of the oxygen atom occurs with frequency $f = 3.7 \times 10^{13}$ Hz. If the oxygen atom at this site is chemically replaced with a sulfur atom, the spring constant of the bond is unchanged (sulfur is just below oxygen in the Periodic Table). Predict the frequency after the sulfur substitution.

**70.** A rectangular block of wood floats in a calm lake. Show that, if friction is ignored, when the block is pushed gently down into the water and then released, it will then oscillate with SHM. Also, determine an equation for the force constant.

71. A 320-kg wooden raft floats on a lake. When a 68-kg man stands on the raft, it sinks 3.5 cm deeper into the water. When he steps off, the raft oscillates for a while. (*a*) What is the frequency of oscillation? (*b*) What is the total energy of oscillation (ignoring damping)?

72. A diving board oscillates with simple harmonic motion of frequency 2.8 cycles per second. What is the maximum amplitude with which the end of the board can oscillate in order that a pebble placed there (Fig. 11–56) does not lose contact with the board during the oscillation?

**FIGURE 11–56** Problem 72.

73. A 950-kg car strikes a huge spring at a speed of 25 m/s (Fig. 11–57), compressing the spring 4.0 m. (*a*) What is the spring stiffness constant of the spring? (*b*) How long is the car in contact with the spring before it bounces off in the opposite direction?

**FIGURE 11–57** Problem 73.

74. A mass attached to the end of a spring is stretched a distance $x_0$ from equilibrium and released. At what distance from equilibrium will it have (*a*) velocity equal to half its maximum velocity, and (*b*) acceleration equal to half its maximum acceleration?

75. Carbon dioxide is a linear molecule. The carbon–oxygen bonds in this molecule act very much like springs. Figure 11–58 shows one possible way the oxygen atoms in this molecule can oscillate: the oxygen atoms oscillate symmetrically in and out, while the central carbon atom remains at rest. Hence each oxygen atom acts like a simple harmonic oscillator with a mass equal to the mass of an oxygen atom. It is observed that this oscillation occurs at a frequency $f = 2.83 \times 10^{13}$ Hz. What is the spring constant of the C—O bond?

**FIGURE 11–58** Problem 75, the $CO_2$ molecule.

76. A mass $m$ is gently placed on the end of a freely hanging spring. The mass then falls 27.0 cm before it stops and begins to rise. What is the frequency of the oscillation?

77. Tall buildings are designed to sway in the wind. In a 100-km/h wind, suppose the top of a 110-story building oscillates horizontally with an amplitude of 15 cm at its natural frequency, which corresponds to a period of 7.0 s. Assuming SHM, find the maximum horizontal velocity and acceleration experienced by an employee as she sits working at her desk located on the top floor. Compare the maximum acceleration (as a percentage) with the acceleration due to gravity.

78. When you walk with a cup of coffee (diameter 8 cm) at just the right pace of about one step per second, the coffee sloshes higher and higher in your cup until eventually it starts to spill over the top, Fig 11–59. Estimate the speed of the waves in the coffee.

**FIGURE 11–59**
Problem 78.

79. A bug on the surface of a pond is observed to move up and down a total vertical distance of 0.12 m, lowest to highest point, as a wave passes. (*a*) What is the amplitude of the wave? (*b*) If the amplitude increases to 0.16 m, by what factor does the bug's maximum kinetic energy change?

80. An earthquake-produced surface wave can be approximated by a sinusoidal transverse wave. Assuming a frequency of 0.60 Hz (typical of earthquakes, which actually include a mixture of frequencies), what amplitude is needed so that objects begin to leave contact with the ground? [*Hint*: Set the acceleration $a > g$.]

81. Two strings on a musical instrument are tuned to play at 392 Hz (G) and 494 Hz (B). (*a*) What are the frequencies of the first two overtones for each string? (*b*) If the two strings have the same length and are under the same tension, what must be the ratio of their masses $(m_G/m_B)$? (*c*) If the strings, instead, have the same mass per unit length and are under the same tension, what is the ratio of their lengths $(\ell_G/\ell_B)$? (*d*) If their masses and lengths are the same, what must be the ratio of the tensions in the two strings?

82. A string can have a "free" end if that end is attached to a ring that can slide without friction on a vertical pole (Fig. 11–60). Determine the wavelengths of the resonant vibrations of such a string with one end fixed and the other free.

**FIGURE 11–60**
Problem 82.

83. The ripples in a certain groove 10.2 cm from the center of a $33\frac{1}{3}$-rpm phonograph record have a wavelength of 1.55 mm. What will be the frequency of the sound emitted?

**84.** A wave with a frequency of 180 Hz and a wavelength of 10.0 cm is traveling along a cord. The maximum speed of particles on the cord is the same as the wave speed. What is the amplitude of the wave?

**85.** Estimate the average power of a moving water wave that strikes the chest of an adult standing in the water at the seashore. Assume that the amplitude of the wave is 0.50 m, the wavelength is 2.5 m, and the period is 4.0 s.

**86.** A tsunami is a sort of pulse or "wave packet" consisting of several crests and troughs that become dramatically large as they enter shallow water at the shore. Suppose a tsunami of wavelength 235 km and velocity 550 km/h travels across the Pacific Ocean. As it approaches Hawaii, people observe an unusual decrease of sea level in the harbors. Approximately how much time do they have to run to safety? (In the absence of knowledge and warning, people have died during tsunamis, some of them attracted to the shore to see stranded fishes and boats.)

**\*87.** For any type of wave that reaches a boundary beyond which its speed is increased, there is a maximum incident angle if there is to be a transmitted refracted wave. This maximum incident angle $\theta_{iM}$ corresponds to an angle of refraction equal to 90°. If $\theta_i > \theta_{iM}$, all the wave is reflected at the boundary and none is refracted, because refraction would correspond to $\sin \theta_r > 1$ (where $\theta_r$ is the angle of refraction), which is impossible. This phenomenon is referred to as *total internal reflection*. (a) Find a formula for $\theta_{iM}$ using the law of refraction, Eq. 11–20. (b) How far from the bank should a trout fisherman stand (Fig. 11–61) so trout won't be frightened by his voice (1.8 m above the ground)? The speed of sound is about 343 m/s in air and 1440 m/s in water.

**FIGURE 11–61** Problem 87b.

# Search and Learn

**1.** Describe a procedure to measure the spring constant $k$ of a car's springs. Assume that the owner's manual gives the car's mass $M$ and that the shock absorbers are worn out so that the springs are underdamped. (See Sections 11–3 and 11–5.)

**2.** A particular unbalanced wheel of a car shakes when the car moves at 90.0 km/h. The wheel plus tire has mass 17.0 kg and diameter 0.58 m. By how much will the springs of this car compress when it is loaded with 280 kg? (Assume the 280 kg is split evenly among all four springs, which are identical.) [*Hint*: Reread Sections 11–1, 11–3, 11–6, and 8–3.]

**3.** Sometimes a car develops a pronounced rattle or vibration at a particular speed, especially if the road is hot enough that the tar between concrete slabs bumps up at regularly spaced intervals. Reread Sections 11–5 and 11–6, and decide whether each of the following is a factor and, if so, how: underdamping, overdamping, critical damping, and forced resonance.

**4.** Destructive interference occurs where two overlapping waves are $\frac{1}{2}$ wavelength or 180° out of phase. Explain why 180° is equivalent to $\frac{1}{2}$ wavelength.

**5.** Estimate the effective spring constant of a trampoline. [*Hint*: Go and jump, or watch, and give your data.]

**6.** A highway overpass was observed to resonate as one full loop $\left(\frac{1}{2}\lambda\right)$ when a small earthquake shook the ground vertically at 3.0 Hz. The highway department put a support at the center of the overpass, anchoring it to the ground as shown in Fig. 11–62. What resonant frequency would you now expect for the overpass? It is noted that earthquakes rarely do significant shaking above 5 or 6 Hz. Did the modifications do any good? Explain. (See Section 11–3.)

Before modification

After modification

**FIGURE 11–62** Search and Learn 6.

"If music be the food of physics, play on."
[Paraphrase of Shakespeare's *Twelfth Night*, line 1.]

Stringed instruments depend on transverse standing waves on strings to produce their harmonious sounds. The sound of wind instruments originates in longitudinal standing waves of an air column. Percussion instruments create more complicated standing waves.

Besides examining sources of sound, we also study the decibel scale of sound level, sound wave interference and beats, the Doppler effect, shock waves and sonic booms, and ultrasound imaging.

# C H A P T E R

# 12

# Sound

**CHAPTER-OPENING QUESTION—Guess now!**

A pianist plays the note "middle C." The sound is made by the vibration of the piano string and is propagated outward as a vibration of the air (which can reach your ear). How does the vibration on the string compare to the vibration in the air?

**(a)** The vibration on the string and the vibration in the air have the same wavelength.

**(b)** They have the same frequency.

**(c)** They have the same speed.

**(d)** Neither wavelength, frequency, nor speed are the same in the air as on the string.

Sound is associated with our sense of hearing and, therefore, with the physiology of our ears and the psychology of our brain, which interprets the sensations that reach our ears. The term *sound* also refers to the physical sensation that stimulates our ears: namely, longitudinal pressure waves.

We can distinguish three aspects of any sound. First, there must be a *source* for a sound; as with any mechanical wave, the source of a sound wave is a vibrating object. Second, the energy is transferred from the source in the form of longitudinal sound *waves* in air or other material. And third, the sound is *detected*, usually by an ear or by a microphone. We start by looking at sound waves themselves.

# 12–1 Characteristics of Sound

We saw in Chapter 11, Fig. 11–26, how a vibrating drumhead produces a sound wave in air. Indeed, we usually think of sound waves traveling in the air, because normally it is the vibrations of the air that force our eardrums to vibrate. But sound waves can also travel in other materials.

Two stones struck together under water can be heard by a swimmer beneath the surface, for the vibrations are carried to the ear by the water. When you put your ear flat against the ground, you can hear an approaching train or truck. In this case the ground does not actually touch your eardrum, but the longitudinal wave transmitted by the ground is called a sound wave just the same, since its vibrations cause the outer ear and the air within it to vibrate. Sound cannot travel in the absence of matter. For example, a bell ringing inside an evacuated jar cannot be heard, and sound cannot travel through the empty reaches of outer space.

The **speed of sound** is different in different materials. In air at 0°C and 1 atm, sound travels at a speed of 331 m/s. The speed of sound in various materials is given in Table 12–1. The values depend somewhat on temperature, especially for gases. For example, in air near room temperature, the speed increases approximately 0.60 m/s for each Celsius degree increase in temperature:

$$v \approx (331 + 0.60T)\,\text{m/s}, \qquad \text{[speed of sound in air]}$$

where $T$ is the temperature in °C. Unless stated otherwise, we will assume in this Chapter that $T = 20°C$, so $v = [331 + (0.60)(20)]\,\text{m/s} = 343\,\text{m/s}$.

| TABLE 12–1 Speed of Sound in Various Materials (20°C and 1 atm) | |
| --- | --- |
| **Material** | **Speed (m/s)** |
| Air | 343 |
| Air (0°C) | 331 |
| Helium | 1005 |
| Hydrogen | 1300 |
| Water | 1440 |
| Sea water | 1560 |
| Iron and steel | $\approx 5000$ |
| Glass | $\approx 4500$ |
| Aluminum | $\approx 5100$ |
| Hardwood | $\approx 4000$ |
| Concrete | $\approx 3000$ |

| **CONCEPTUAL EXAMPLE 12–1** | **Distance from a lightning strike.** A rule of thumb that tells how close lightning has struck is "one mile for every five seconds before the thunder is heard." Explain why this works, noting that the speed of light is so high ($3 \times 10^8$ m/s, almost a million times faster than sound) that the time for light to travel to us is negligible compared to the time for the sound.

**RESPONSE** The speed of sound in air is about 340 m/s, so to travel 1 km = 1000 m takes about 3 seconds. One mile is about 1.6 kilometers, so the time for the thunder to travel a mile is about $(1.6)(3) \approx 5$ seconds.

**PHYSICS APPLIED**
*How far away is the lightning?*

Two aspects of any sound are immediately evident to a human listener: "loudness" and "pitch." Each refers to a sensation in the consciousness of the listener. But to each of these subjective sensations there corresponds a physically measurable quantity. **Loudness** is related to the intensity (energy per unit time crossing unit area) in the sound wave, and we shall discuss it in the next Section.

The **pitch** of a sound refers to whether it is high, like the sound of a piccolo or violin, or low, like the sound of a bass drum or string bass. The physical quantity that determines pitch is the frequency, as was first noted by Galileo. The lower the frequency, the lower the pitch; the higher the frequency, the higher the pitch.[†] The best human ears can respond to frequencies from about 20 Hz to almost 20,000 Hz. (Recall that 1 Hz is 1 cycle per second.) This frequency range is called the **audible range**. These limits vary somewhat from one individual to another. One general trend is that as people age, they are less able to hear high frequencies, so the high-frequency limit may be 10,000 Hz or less.

Sound waves whose frequencies are outside the audible range may reach the ear, but we are not generally aware of them. Frequencies above 20,000 Hz are called **ultrasonic** (do not confuse with *supersonic*, which is used for an object moving with a speed faster than the speed of sound). Many animals can hear ultrasonic frequencies; dogs, for example, can hear sounds as high as 50,000 Hz, and bats can detect frequencies as high as 100,000 Hz. Ultrasonic waves have many useful applications in medicine and other fields, which we will discuss later in this Chapter.

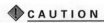

**CAUTION**
*Do not confuse
ultrasonic (high frequency)
with supersonic (high speed)*

---

[†]Although pitch is determined mainly by frequency, it also depends to a slight extent on loudness. For example, a very loud sound may seem slightly lower in pitch than a quiet sound of the same frequency.

**PHYSICS APPLIED**
*Autofocusing camera*

**EXAMPLE 12–2** | **Autofocusing with sound waves.** Autofocusing cameras emit a pulse of very high frequency (ultrasonic) sound that travels to the object being photographed, and include a sensor that detects the returning reflected sound, as shown in Fig. 12–1. To get an idea of the time sensitivity of the detector, calculate the travel time of the pulse for an object (*a*) 1.0 m away, and (*b*) 20 m away.

**APPROACH** If we assume the temperature is about 20°C, then the speed of sound is 343 m/s. Using this speed $v$ and the total distance $d$ back and forth in each case, we can obtain the time ($v = d/t$).

**SOLUTION** (*a*) The pulse travels 1.0 m to the object and 1.0 m back, for a total of 2.0 m. We solve for $t$ in $v = d/t$:

$$t = \frac{d}{v} = \frac{2.0 \text{ m}}{343 \text{ m/s}} = 0.0058 \text{ s} = 5.8 \text{ ms}.$$

(*b*) The total distance now is $2 \times 20$ m $= 40$ m, so

$$t = \frac{40 \text{ m}}{343 \text{ m/s}} = 0.12 \text{ s} = 120 \text{ ms}.$$

**NOTE** Newer autofocus cameras use infrared light ($v = 3 \times 10^8$ m/s) instead of ultrasound, and/or a digital sensor array that detects light intensity differences between adjacent receptors as the lens is automatically moved back and forth, choosing the lens position that provides maximum intensity differences (sharpest focus).

**FIGURE 12–2** The membrane of a drum, as it vibrates, alternately compresses the air and, as it recedes (moves to the left), leaves a rarefaction or expansion of air. See also Fig. 11–26.

We often describe a sound wave in terms of the vibration of the molecules of the medium in which it travels—that is, in terms of the motion or displacement of the molecules. Sound waves can also be analyzed from the point of view of pressure. Indeed, longitudinal waves are often called **pressure waves**. The pressure variation is usually easier to measure than the displacement. As Fig. 12–2 shows, in a wave "compression" (where molecules are closest together), the pressure is higher than normal, whereas in an expansion (or *rarefaction*) the pressure is less than normal. Figure 12–3 shows a graphical representation of a sound wave in air in terms of (a) displacement and (b) pressure. Note that the displacement wave is a quarter wavelength out of phase with the pressure wave: where the pressure is a maximum or minimum, the displacement from equilibrium is zero; and where the pressure variation is zero, the displacement is a maximum or minimum.

Sound waves whose frequencies are below the audible range (that is, less than 20 Hz) are called **infrasonic**. Sources of infrasonic waves include earthquakes, thunder, volcanoes, and waves produced by vibrating heavy machinery. This last source can be particularly troublesome to workers, because infrasonic waves—even though inaudible—can cause damage to the human body. These low-frequency waves act in a resonant fashion, causing motion and irritation of the body's organs.

**FIGURE 12–3** Representation of a sound wave in space at a given instant in terms of (a) displacement, and (b) pressure.

# 12–2 Intensity of Sound: Decibels

**Loudness** is a sensation in the consciousness of a human being and is related to a physically measurable quantity, the **intensity** of the wave. Intensity is defined as the energy transported by a wave per unit time across a unit area perpendicular to the energy flow. As we saw in Chapter 11, intensity is proportional to the square of the wave amplitude. Intensity has units of power per unit area, or watts/meter$^2$ ($W/m^2$).

An average human ear can detect sounds with an intensity as low as $10^{-12} W/m^2$ and as high as $1 W/m^2$ (and even higher, although above this it is painful). This is an incredibly wide range of intensity, spanning a factor of $10^{12}$ from lowest to highest. Presumably because of this wide range, what we perceive as loudness is not directly proportional to the intensity. To produce a sound that sounds about twice as loud requires a sound wave that has about 10 times the intensity. This is roughly valid at any sound level for frequencies near the middle of the audible range. For example, a sound wave of intensity $10^{-2} W/m^2$ sounds to an average human being like it is about twice as loud as one with intensity of $10^{-3} W/m^2$, and four times as loud as $10^{-4} W/m^2$.

## Sound Level

Because of this relationship between the subjective sensation of loudness and the physically measurable quantity "intensity," sound intensity levels are usually specified on a logarithmic scale. The unit on this scale is a **bel**, after the inventor Alexander Graham Bell, or much more commonly, the **decibel** (dB), which is $\frac{1}{10}$ bel (10 dB = 1 bel).[†] The **sound level**, $\beta$, of any sound is defined in terms of its intensity, $I$, as

$$\beta \ (\text{in dB}) = 10 \log \frac{I}{I_0}, \tag{12–1}$$

where $I_0$ is the intensity of a chosen reference level, and the logarithm is to the base 10. $I_0$ is usually taken as the minimum intensity audible to a good ear—the "threshold of hearing," which is $I_0 = 1.0 \times 10^{-12} W/m^2$. Thus, for example, the sound level of a sound whose intensity $I = 1.0 \times 10^{-10} W/m^2$ will be

$$\beta = 10 \log \left( \frac{1.0 \times 10^{-10} W/m^2}{1.0 \times 10^{-12} W/m^2} \right) = 10 \log 100 = 20 \, dB,$$

since log 100 is equal to 2.0. (Appendix A has a brief review of logarithms.) Notice that the sound level at the threshold of hearing is 0 dB. That is, $\beta = 10 \log 10^{-12}/10^{-12} = 10 \log 1 = 0$ since $\log 1 = 0$. Notice too that an increase in intensity by a factor of 10 corresponds to a sound level increase of 10 dB. An increase in intensity by a factor of 100 corresponds to a sound level increase of 20 dB. Thus a 50-dB sound is 100 times more intense than a 30-dB sound, and so on.

Intensities and sound levels for some common sounds are listed in Table 12–2.

---

**EXAMPLE 12–3** **Sound intensity on the street.** At a busy street corner, the sound level is 75 dB. What is the intensity of sound there?

**APPROACH** We have to solve Eq. 12–1 for intensity $I$, remembering that $I_0 = 1.0 \times 10^{-12} W/m^2$.

**SOLUTION** From Eq. 12–1

$$\log \frac{I}{I_0} = \frac{\beta}{10}.$$

Recalling that $x = \log y$ is the same as $y = 10^x$ (Appendix A–8), then

$$\frac{I}{I_0} = 10^{\beta/10}.$$

With $\beta = 75$, then

$$I = I_0 10^{\beta/10} = (1.0 \times 10^{-12} W/m^2)(10^{7.5}) = 3.2 \times 10^{-5} W/m^2.$$

---

**PHYSICS APPLIED**
*Wide range of human hearing*

**CAUTION**
*0 dB does not mean zero intensity*

**TABLE 12–2**
**Intensity of Various Sounds**

| Source of the Sound | Sound Level (dB) | Intensity ($W/m^2$) |
|---|---|---|
| Jet plane at 30 m | 140 | 100 |
| Threshold of pain | 120 | 1 |
| Loud rock concert | 120 | 1 |
| Siren at 30 m | 100 | $1 \times 10^{-2}$ |
| Busy street traffic | 80 | $1 \times 10^{-4}$ |
| Noisy restaurant | 70 | $1 \times 10^{-5}$ |
| Talk, at 50 cm | 65 | $3 \times 10^{-6}$ |
| Quiet radio | 40 | $1 \times 10^{-8}$ |
| Whisper | 30 | $1 \times 10^{-9}$ |
| Rustle of leaves | 10 | $1 \times 10^{-11}$ |
| Threshold of hearing | 0 | $1 \times 10^{-12}$ |

---

[†]The dB is dimensionless and so does not have to be included in calculations.

**EXAMPLE 12–4** **Loudspeaker response.** A high-quality loudspeaker is advertised to reproduce, at full volume, frequencies from 30 Hz to 18,000 Hz with uniform sound level $\pm 3$ dB. That is, over this frequency range, the sound level output does not vary by more than 3 dB for a given input level. By what factor does the intensity change for a 3 dB change in output level?

**APPROACH** Let us call the average intensity $I_1$ and the average sound level $\beta_1$. Then the maximum intensity, $I_2$, corresponds to a level $\beta_2 = \beta_1 + 3$ dB. We then use the relation between intensity and sound level, Eq. 12–1.

**SOLUTION** Equation 12–1 gives

$$\beta_2 - \beta_1 = 10 \log \frac{I_2}{I_0} - 10 \log \frac{I_1}{I_0}$$

$$3 \text{ dB} = 10\left(\log \frac{I_2}{I_0} - \log \frac{I_1}{I_0}\right) = 10 \log \frac{I_2}{I_1}$$

because $(\log a - \log b) = \log a/b$, as discussed in Appendix A–8. This last equation gives

$$\log \frac{I_2}{I_1} = 0.30,$$

or

$$\frac{I_2}{I_1} = 10^{0.30} = 2.0.$$

So $\pm 3$ dB corresponds to a doubling or halving of the intensity.

**NOTE** From this last equation, we see also that $\log 2.0 = 0.30$. See also the short Table of Logarithms in Appendix A–8.

**EXERCISE A** If an increase of 3 dB means "twice as intense," what does an increase of 6 dB mean?

It is worth noting that a sound-level difference of 3 dB (which corresponds to a doubled intensity, as we just saw) corresponds to only a very small change in the subjective sensation of apparent loudness. Indeed, the average human can distinguish a difference in sound level of only about 1 or 2 dB.

Normally, the loudness or intensity of a sound decreases as you get farther from the source of the sound. Indoors, this effect is altered because of reflections from the walls. However, if a source is in the open so that sound can radiate out freely in all directions, the intensity decreases as the inverse square of the distance,

$$I \propto \frac{1}{r^2},$$

as we saw in Section 11–9. Over large distances, the intensity decreases faster than $1/r^2$ because some of the energy is transferred into irregular motion of air molecules. This loss happens more for higher frequencies, so any sound of mixed frequencies will be less "bright" at a distance. Far from an outdoor band, you hear mainly the boom of the drums.

**CONCEPTUAL EXAMPLE 12–5** **Trumpet players.** A trumpeter plays at a sound level of 75 dB. Three equally loud trumpet players join in. What is the new sound level?

**RESPONSE** The intensity of four trumpets is four times the intensity of one trumpet $(= I_1)$ or $4I_1$. The sound level of the four trumpets would be

$$\beta = 10 \log \frac{4I_1}{I_0} = 10 \log 4 + 10 \log \frac{I_1}{I_0}$$

$$= 6.0 \text{ dB} + 75 \text{ dB} = 81 \text{ dB},$$

since $\log 4 = 0.60$.

**EXERCISE B** From Table 12–2, we see that ordinary conversation corresponds to a sound level of about 65 dB. If two people are talking at once, the sound level is (a) 65 dB, (b) 68 dB, (c) 75 dB, (d) 130 dB, (e) 62 dB.

**EXAMPLE 12–6** **Airplane roar.** The sound level measured 30 m from a jet plane is 140 dB. Estimate the sound level at 300 m. (Ignore reflections from the ground.)

**APPROACH** Given the sound level, we can determine the intensity at 30 m using Eq. 12–1. Because intensity decreases as the square of the distance, ignoring reflections, we can find $I$ at 300 m and again apply Eq. 12–1 to obtain the sound level.

**SOLUTION** The intensity $I$ at 30 m is

$$140 \, \text{dB} = 10 \log\left(\frac{I}{10^{-12} \, \text{W/m}^2}\right)$$

or, dividing through by 10,

$$14 = \log\left(\frac{I}{10^{-12} \, \text{W/m}^2}\right).$$

Recall that $y = \log x$ means $10^y = x$ (Appendix A–8). Then

$$10^{14} = \frac{I}{10^{-12} \, \text{W/m}^2},$$

so $I = (10^{14})(10^{-12} \, \text{W/m}^2) = 10^2 \, \text{W/m}^2$. At 300 m, 10 times as far, the intensity, which decreases as $1/r^2$, will be $\left(\frac{1}{10}\right)^2 = 1/100$ as much, or $1 \, \text{W/m}^2$. Hence, the sound level is

$$\beta = 10 \log\left(\frac{1 \, \text{W/m}^2}{10^{-12} \, \text{W/m}^2}\right) = 120 \, \text{dB}.$$

Even at 300 m, the sound is at the threshold of pain. This is why workers at airports wear ear covers to protect their ears from damage (Fig. 12–4).

**NOTE** Here is a simpler approach that avoids Eq. 12–1. Because the intensity decreases as the square of the distance, at 10 times the distance the intensity decreases by $\left(\frac{1}{10}\right)^2 = \frac{1}{100}$. We can use the result that 10 dB corresponds to an intensity change by a factor of 10 (see just before Example 12–3). Then an intensity change by a factor of 100 corresponds to a sound-level change of $(2)(10 \, \text{dB}) = 20 \, \text{dB}$. This confirms our result above: $140 \, \text{dB} - 20 \, \text{dB} = 120 \, \text{dB}$.

**PHYSICS APPLIED**
*Jet plane noise*

**FIGURE 12–4** Example 12–6. Airport worker with sound-intensity-reducing ear covers (headphones).

## Intensity Related to Amplitude

The intensity $I$ of a wave is proportional to the square of the wave amplitude, $A$, as we saw in Section 11–9. We can therefore relate the amplitude quantitatively to the intensity $I$ or sound level $\beta$, as the following Example shows.

**EXAMPLE 12–7** **How tiny the displacement is.** Calculate the displacement of air molecules for a sound of frequency 1000 Hz at the threshold of hearing.

**APPROACH** In Section 11–9 we found a relation between intensity $I$ and displacement amplitude $A$ of a wave, Eq. 11–18. The amplitude of oscillation of air molecules is what we want to solve for, given the intensity. Assume the temperature is 20°C so the speed of sound is 343 m/s.

**SOLUTION** At the threshold of hearing, $I = 1.0 \times 10^{-12} \, \text{W/m}^2$ (Table 12–2). We solve for the amplitude $A$ in Eq. 11–18:

$$A = \frac{1}{\pi f}\sqrt{\frac{I}{2\rho v}}$$

$$= \frac{1}{(3.14)(1.0 \times 10^3 \, \text{s}^{-1})}\sqrt{\frac{1.0 \times 10^{-12} \, \text{W/m}^2}{(2)(1.29 \, \text{kg/m}^3)(343 \, \text{m/s})}} = 1.1 \times 10^{-11} \, \text{m},$$

where we have taken the density of air to be $1.29 \, \text{kg/m}^3$ (Table 10–1).

**NOTE** We see how incredibly sensitive the human ear is: it can detect displacements of air molecules which are less than the diameter of atoms (about $10^{-10} \, \text{m}$).

**PHYSICS APPLIED**
*Incredible sensitivity of the ear*

# *12–3 The Ear and Its Response; Loudness

The human ear is a remarkably sensitive detector of sound. Mechanical detectors of sound (microphones) can barely match the ear in detecting low-intensity sounds.

The function of the ear is to transform the vibrational energy of waves into electrical signals which are carried to the brain by way of nerves. A microphone performs a similar task. Sound waves striking the diaphragm of a microphone set it into vibration, and these vibrations are transformed into an electrical signal (Chapter 21) with the same frequencies, which can then be amplified and sent to a loudspeaker or recorder.

Figure 12–5 is a diagram of the human ear. The ear consists of three main divisions: the outer ear, middle ear, and inner ear. In the outer ear, sound waves from the outside travel down the ear canal to the eardrum (the tympanum), which vibrates in response to the impinging waves. The middle ear consists of three small bones known as the hammer, anvil, and stirrup, which transfer the vibrations of the eardrum to the inner ear at the oval window. This delicate system of levers, coupled with the relatively large area of the eardrum compared to the area of the oval window, results in the pressure being amplified by a factor of about 20. The inner ear consists of the semicircular canals, which are important for controlling balance, and the liquid-filled cochlea where the vibrational energy of sound waves is transformed into electrical energy and sent to the brain.

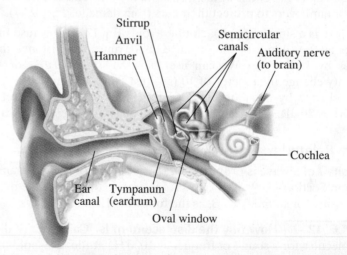

**FIGURE 12–5** Diagram of the human ear.

## *The Ear's Response

The ear is not equally sensitive to all frequencies. To hear the same loudness for sounds of different frequencies requires different intensities. Studies on large numbers of people have produced the averaged curves shown in Fig. 12–6 (top of next page). On this graph, each curve represents sounds that seemed to be equally loud. The number labeling each curve represents the **loudness level** (the units are called *phons*), which is numerically equal to the sound level in dB at 1000 Hz. For example, the curve labeled 40 represents sounds that are heard by an average person to have the same loudness as a 1000-Hz sound with a sound level of 40 dB. From this 40-phon curve, we see that a 100-Hz tone must be at a level of about 62 dB to be perceived as loud as a 1000-Hz tone of only 40 dB.

**FIGURE 12–6** Sensitivity of the human ear as a function of frequency (see text). Note that the frequency scale is "logarithmic" in order to cover a wide range of frequencies.

The lowest curve in Fig. 12–6 (labeled 0) represents the sound level, as a function of frequency, for the *threshold of hearing*, the softest sound that is just audible by a very good ear. Note that the ear is most sensitive to sounds of frequency between 2000 and 4000 Hz, which are common in speech and music. Note too that whereas a 1000-Hz sound is audible at a level of 0 dB, a 100-Hz sound must be nearly 40 dB to be heard. The top curve in Fig. 12–6, labeled 120 phons, represents the *threshold of pain*. Sounds above this level can actually be felt and cause pain.

Figure 12–6 shows that at lower sound levels, our ears are less sensitive to the high and low frequencies relative to middle frequencies. The "loudness" control on some stereo systems is intended to compensate for this low-volume insensitivity. As the volume is turned down, the loudness control boosts the high and low frequencies relative to the middle frequencies so that the sound will have a more "normal-sounding" frequency balance. Many listeners, however, find the sound more pleasing or natural without the loudness control.

# 12–4 Sources of Sound: Vibrating Strings and Air Columns

The source of any sound is a vibrating object. Almost any object can vibrate and hence be a source of sound. We now discuss some simple sources of sound, particularly musical instruments. In musical instruments, the source is set into vibration by striking, plucking, bowing, or blowing. Standing waves are produced and the source vibrates at its natural resonant frequencies. The vibrating source is in contact with the air (or other medium) and pushes on it to produce sound waves that travel outward. The frequencies of the waves are the same as those of the source, but the speed and wavelengths can be different. A drum has a stretched membrane that vibrates. Xylophones and marimbas have metal or wood bars that can be set into vibration. Bells, cymbals, and gongs also make use of a vibrating metal. Many instruments make use of vibrating strings, such as the violin, guitar, and piano, or make use of vibrating columns of air, such as the flute, trumpet, and pipe organ. We have already seen that the pitch of a pure sound is determined by the frequency. Typical frequencies for musical notes on the "equally tempered chromatic scale" are given in Table 12–3 for the *octave* beginning with middle C. Note that one **octave** corresponds to a doubling of frequency. For example, middle C has frequency of 262 Hz whereas C′ (C above middle C) has twice that frequency, 524 Hz. [Middle C is the C or "do" note at the middle of a piano keyboard.]

**TABLE 12–3 Equally Tempered Chromatic Scale[†]**

| Note | Name | Frequency (Hz) |
|---|---|---|
| C | do | 262 |
| C# or D♭ | | 277 |
| D | re | 294 |
| D# or E♭ | | 311 |
| E | mi | 330 |
| F | fa | 349 |
| F# or G♭ | | 370 |
| G | sol | 392 |
| G# or A♭ | | 415 |
| A | la | 440 |
| A# or B♭ | | 466 |
| B | ti | 494 |
| C′ | do | 524 |

[†] Only one octave is included.

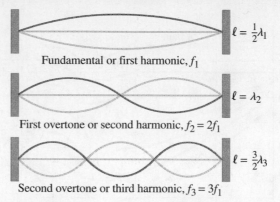

$\ell = \frac{1}{2}\lambda_1$

Fundamental or first harmonic, $f_1$

**FIGURE 12–7** Standing waves on a string—only the lowest three frequencies are shown.

$\ell = \lambda_2$

First overtone or second harmonic, $f_2 = 2f_1$

$\ell = \frac{3}{2}\lambda_3$

Second overtone or third harmonic, $f_3 = 3f_1$

## Stringed Instruments

We saw in Chapter 11, Fig. 11–41, how standing waves are established on a string, and we show this again here in Fig. 12–7. Such standing waves are the basis for all stringed instruments. The pitch is normally determined by the lowest resonant frequency, the **fundamental**, which corresponds to nodes occurring only at the ends. The string vibrating up and down as a whole corresponds to a half wavelength as shown at the top of Fig. 12–7; so the wavelength of the fundamental on the string is equal to twice the length $\ell$ of the string. Therefore, the fundamental frequency is $f_1 = v/\lambda = v/2\ell$, where $v$ is the velocity of the wave on the string (*not* in the air). The possible frequencies for standing waves on a stretched string are whole-number multiples of the fundamental frequency:

$$f_n = nf_1 = n\frac{v}{2\ell}, \qquad n = 1, 2, 3, \cdots$$

(just as in Eq. 11–19b), where $n = 1$ refers to the fundamental and $n = 2, 3, \cdots$ are the overtones. All of the standing waves, $n = 1, 2, 3, \cdots$, are called harmonics[†], as we saw in Section 11–12.

When a finger is placed on the string of a guitar or violin, the effective length of the vibrating string is shortened. So its fundamental frequency, and pitch, is higher since the wavelength of the fundamental is shorter (Fig. 12–8). The strings on a guitar or violin are all the same length. They sound at a different pitch because the strings have different mass per unit length, $\mu = m/\ell$, which affects the velocity on the string, Eq. 11–13,

$$v = \sqrt{F_T/\mu}. \qquad \text{[stretched string]}$$

Thus the velocity on a heavier string is lower and the frequency will be lower for the same wavelength. The tension $F_T$ also has an effect: indeed, adjusting the tension is the means for tuning the pitch of each string. In pianos and harps the strings are of different lengths. For the lower notes the strings are not only longer, but heavier as well, and the reason is illustrated in the following Example.

**FIGURE 12–8** The wavelength of (a) an unfingered string is longer than that of (b) a fingered string. Hence, the frequency of the fingered string is higher. Only one string is shown on this guitar, and only the simplest standing wave, the fundamental, is shown.

(a)

(b)

---

**EXAMPLE 12–8** **Piano strings.** The highest key on a piano corresponds to a frequency about 150 times that of the lowest key. If the string for the highest note is 5.0 cm long, how long would the string for the lowest note have to be if it had the same mass per unit length and was under the same tension?

**APPROACH** Since $v = \sqrt{F_T/\mu}$, the velocity would be the same on each string. So the frequency is inversely proportional to the length $\ell$ of the string ($f = v/\lambda = v/2\ell$).

**SOLUTION** We can write, for the fundamental frequencies of each string, the ratio

$$\ell_L/\ell_H = f_H/f_L,$$

where the subscripts L and H refer to the lowest and highest notes, respectively. Thus $\ell_L = \ell_H(f_H/f_L) = (5.0\,\text{cm})(150) = 750\,\text{cm}$, or 7.5 m. This would be ridiculously long ($\approx 25\,\text{ft}$) for a piano.

**NOTE** The longer strings of lower frequency are made heavier (higher mass per unit length), so even on grand pianos the strings are less than 3 m long.

---

[†]When the resonant frequencies above the fundamental (that is, the overtones) are integral multiples of the fundamental, as here, they are called harmonics. But if the overtones are not integral multiples of the fundamental, as is the case for a vibrating drumhead, for example, they are not harmonics.

**EXAMPLE 12–9** | **Frequencies and wavelengths in the violin.** A 0.32-m-long violin string is tuned to play A above middle C at 440 Hz. (*a*) What is the wavelength of the fundamental string vibration, and (*b*) what are the frequency and wavelength of the sound wave produced? (*c*) Why is there a difference?

**APPROACH** The wavelength of the fundamental string vibration equals twice the length of the string (Fig. 12–7). As the string vibrates, it pushes on the air, which is thus forced to oscillate at the same frequency as the string.

**SOLUTION** (*a*) From Fig. 12–7 the wavelength of the fundamental is

$$\lambda = 2\ell = 2(0.32\,\text{m}) = 0.64\,\text{m} = 64\,\text{cm}.$$

This is the wavelength of the standing wave on the string.

(*b*) The sound wave that travels outward in the air (to reach our ears) has the same frequency, 440 Hz. Its wavelength is

$$\lambda = \frac{v}{f} = \frac{343\,\text{m/s}}{440\,\text{Hz}} = 0.78\,\text{m} = 78\,\text{cm},$$

where $v$ is the speed of sound in air (assumed at 20°C), Section 12–1.

(*c*) The wavelength of the sound wave is different from that of the standing wave on the string because the speed of sound in air (343 m/s at 20°C) is different from the speed of the wave on the string ($= f\lambda = 440\,\text{Hz} \times 0.64\,\text{m} = 280\,\text{m/s}$) which depends on the tension in the string and its mass per unit length.

**NOTE** The frequencies on the string and in the air are the same: the string and air are in contact, and the string "forces" the air to vibrate at the same frequency. But the wavelengths are different because the wave speed on the string is different than that in air.

| **EXERCISE D** Return to the Chapter-Opening Question, page 328, and answer it again now. Try to explain why you may have answered differently the first time.

Stringed instruments would not be very loud if they relied on their vibrating strings to produce the sound waves since the strings are too thin to compress and expand much air. Stringed instruments therefore make use of a kind of mechanical amplifier known as a *sounding board* (piano) or *sounding box* (guitar, violin), which acts to amplify the sound by putting a greater surface area in contact with the air (Fig. 12–9). When the strings are set into vibration, the sounding board or box is set into vibration as well. Since it has much greater area in contact with the air, it can produce a more intense sound wave. On an electric guitar, the sounding box is not so important since the vibrations of the strings are amplified electronically.

## Wind Instruments

Instruments such as woodwinds, the brasses, and the pipe organ produce sound from the vibrations of standing waves in a column of air within a tube (Fig. 12–10). Standing waves can occur in the air of any cavity, including the human throat, but the frequencies present are complicated for any but very simple shapes such as the uniform, narrow tube of a flute or an organ pipe. In some instruments, a vibrating reed or the vibrating lip of the player helps to set up vibrations of the air column. In others, a stream of air is directed against one edge of the opening or mouthpiece, leading to turbulence which sets up the vibrations. Because of the disturbance, whatever its source, the air within the tube vibrates with a variety of frequencies, but only frequencies that correspond to standing waves will persist.

For a string fixed at both ends, Fig. 12–7, the standing waves have nodes (no movement) at the two ends, and one or more antinodes (large amplitude of vibration) in between. A node separates successive antinodes. The lowest-frequency standing wave, the *fundamental*, corresponds to a single antinode. The higher-frequency standing waves are called **overtones** or **harmonics**, as we saw in Section 11–12. Specifically, the first harmonic is the fundamental, the second harmonic (= first overtone) has twice the frequency of the fundamental, and so on.

(a)

(b)

**FIGURE 12–9** (a) Piano, showing sounding board to which strings are attached; (b) sounding box (guitar).

**FIGURE 12–10** Wind instruments: flute (left) and clarinet.

# TUBE OPEN AT BOTH ENDS

(a) Displacement of air

First harmonic = fundamental
$$\ell = \tfrac{1}{2}\lambda_1$$
$$f_1 = \frac{v}{2\ell}$$

←[motion of air molecules]

Second harmonic
$$\ell = \lambda_2$$
$$f_2 = \frac{v}{\ell} = 2f_1$$

Overtones

Third harmonic
$$\ell = \tfrac{3}{2}\lambda_3$$
$$f_3 = \frac{3v}{2\ell} = 3f_1$$

(b) Pressure variation in the air

**FIGURE 12–11** Graphs of the three simplest modes of vibration (standing waves) for a uniform tube open at both ends ("open tube"). These simplest modes of vibration are graphed in (a), on the left, in terms of the motion of the air (displacement), and in (b), on the right, in terms of air pressure. Each graph shows the wave format at two times, A and B, a half period apart. The actual motion of molecules for one case, the fundamental, is shown just below the tube at top left.

**PHYSICS APPLIED**

*Wind instruments*

The situation is similar for a column of air in a tube of uniform diameter, but we must remember that it is now air itself that is vibrating. We can describe the waves either in terms of the flow of the air—that is, in terms of the *displacement* of air—or in terms of the *pressure* in the air (see Figs. 12–2 and 12–3). In terms of displacement, the air at the closed end of a tube is a displacement node since the air is not free to move there, whereas near the open end of a tube there will be an antinode because the air can move freely in and out. The air within the tube vibrates in the form of longitudinal standing waves. A few of the possible modes of vibration for a tube open at both ends (called an **open tube**) are shown graphically in Fig. 12–11. Possible vibration modes for a tube that is open at one end but closed at the other (called a **closed tube**) are shown in Fig. 12–12. [A tube closed at *both* ends, having no connection to the outside air, would be useless as an instrument.] The graphs in part (a) of each Figure (left-hand sides) represent the displacement amplitude of the vibrating air in the tube. Note that these are graphs, and that the air molecules themselves oscillate *horizontally*, parallel to the tube length, as shown by the small arrows in the top diagram of Fig. 12–11a (on the left). The exact position of the antinode near the open end of a tube depends on the diameter of the tube, but if the diameter is small compared to the length, which is the usual case, the antinode occurs very close to the end as shown.[†] We assume this is the case in what follows. (The position of the antinode may also depend slightly on the wavelength and other factors.)

Let us look in detail at the open tube, in Fig. 12–11a, which might be an organ pipe or a flute. An open tube has displacement antinodes at both ends since the air is free to move at open ends. There must be at least one node within an open tube if there is to be a standing wave at all. A single node corresponds to the *fundamental frequency* of the tube. Since the distance between two successive nodes, or between two successive antinodes, is $\tfrac{1}{2}\lambda$, there is one-half of a wavelength within the length of the tube for the simplest case of the fundamental (top diagram in Fig. 12–11a): $\ell = \tfrac{1}{2}\lambda$, or $\lambda = 2\ell$. So the fundamental frequency is $f_1 = v/\lambda = v/2\ell$, where $v$ is the velocity of sound in air (the air in the tube). The standing wave with two nodes is the *first overtone* or *second harmonic* and has half the wavelength ($\ell = \lambda$) and twice the frequency of the fundamental. Indeed, in a uniform tube open at both ends, the frequency of each overtone is an integral multiple of the fundamental frequency, as shown in Fig. 12–11a. This is just what is found for a string.

---

[†]The diameter $D$ of a tube does affect the node at the open end of a tube. The end correction can be roughly approximated as adding $D/3$ to $\ell$ to give us an effective length for the tube in calculations.

**(a)** Displacement of air    **(b)** Pressure variation in the air

First harmonic = fundamental
$$\ell = \tfrac{1}{4}\lambda_1$$
$$f_1 = \frac{v}{4\ell}$$

Third harmonic
$$\ell = \tfrac{3}{4}\lambda_3$$
$$f_3 = \frac{3v}{4\ell} = 3f_1$$

⎫
⎬ Overtones
⎭

Fifth harmonic
$$\ell = \tfrac{5}{4}\lambda_5$$
$$f_5 = \frac{5v}{4\ell} = 5f_1$$

**FIGURE 12–12** Modes of vibration (standing waves) for a tube closed at one end ("closed tube"). See caption for Fig. 12–11.

For a closed tube, shown in Fig. 12–12a, which could be an organ pipe, there is always a displacement node at the closed end (because the air is not free to move) and an antinode at the open end (where the air can move freely). Since the distance between a node and the nearest antinode is $\tfrac{1}{4}\lambda$, we see that the fundamental in a closed tube corresponds to only one-fourth of a wavelength within the length of the tube: $\ell = \lambda/4$, and $\lambda = 4\ell$. The fundamental frequency is thus $f_1 = v/4\ell$, or half that for an open pipe of the same length. There is another difference, for as we can see from Fig. 12–12a, only the odd harmonics are present in a closed tube: the overtones have frequencies equal to $3, 5, 7, \cdots$ times the fundamental frequency. There is no way for waves with $2, 4, 6, \cdots$ times the fundamental frequency to have a node at one end and an antinode at the other, and thus they cannot exist as standing waves in a closed tube.

Another way to analyze the vibrations in a uniform tube is to consider a description in terms of the *pressure* in the air, shown in part (b) of Figs. 12–11 and 12–12 (right-hand sides). Where the air in a wave is compressed, the pressure is higher, whereas in a wave expansion (or rarefaction), the pressure is less than normal. The open end of a tube is open to the atmosphere. Hence the pressure variation at an open end must be a *node*: the pressure does not alternate, but remains at the outside atmospheric pressure. If a tube has a closed end, the pressure at that closed end can readily alternate to be above or below atmospheric pressure. Hence there is a pressure *antinode* at a closed end of a tube. There can be pressure nodes and antinodes within the tube. Some of the possible vibrational modes in terms of pressure are shown in Fig. 12–11b for an open tube, and in Fig. 12–12b for a closed tube.

**EXAMPLE 12–10** | **Organ pipes.** What will be the fundamental frequency and first three overtones for a 26-cm-long organ pipe at 20°C if it is (a) open, (b) closed?

**APPROACH** All our calculations can be based on Figs. 12–11a and 12–12a.

**SOLUTION** (a) For the open pipe, Fig. 12–11a, the fundamental frequency is

$$f_1 = \frac{v}{2\ell} = \frac{343 \text{ m/s}}{2(0.26 \text{ m})} = 660 \text{ Hz}.$$

The speed $v$ is the speed of sound in air (the air vibrating in the pipe). The overtones include all harmonics: 1320 Hz, 1980 Hz, 2640 Hz, and so on.

(b) For a closed pipe, Fig. 12–12a, the fundamental frequency is

$$f_1 = \frac{v}{4\ell} = \frac{343 \text{ m/s}}{4(0.26 \text{ m})} = 330 \text{ Hz}.$$

Only odd harmonics are present: the first three overtones are 990 Hz, 1650 Hz, and 2310 Hz.

**NOTE** The closed pipe plays 330 Hz, which, from Table 12–3, is E above middle C, whereas the open pipe of the same length plays 660 Hz, an octave higher.

Pipe organs use both open and closed pipes, with lengths from a few centimeters to 5 m or more. A flute acts as an open tube, for it is open not only where you blow into it, but is open also at the opposite end. The different notes on a flute are obtained by shortening the length of the vibrating air column, by uncovering holes along the tube (so a displacement antinode can occur at the hole). The shorter the length of the vibrating air column, the higher the fundamental frequency.

**EXAMPLE 12–11** **Flute.** A flute is designed to play middle C (262 Hz) as the fundamental frequency when all the holes are covered. Approximately how long should the distance be from the mouthpiece to the far end of the flute? (This is only approximate since the antinode does not occur precisely at the mouthpiece.) Assume the temperature is 20°C.

**APPROACH** When all holes are covered, the length of the vibrating air column is the full length. The speed of sound in air at 20°C is 343 m/s. Because a flute is open at both ends, we use Fig. 12–11: the fundamental frequency $f_1$ is related to the length $\ell$ of the vibrating air column by $f = v/2\ell$.

**SOLUTION** Solving for $\ell$, we find

$$\ell = \frac{v}{2f} = \frac{343 \text{ m/s}}{2(262 \text{ s}^{-1})} = 0.655 \text{ m}.$$

**EXERCISE E** To see why players of wind instruments "warm up" their instruments (so they will be in tune), determine the fundamental frequency of the flute of Example 12–11 when all holes are covered and the temperature is 10°C instead of 20°C.

**EXAMPLE 12–12** **ESTIMATE** **Wind noise frequencies.** Wind can be noisy—it can "howl" in trees; it can "moan" in chimneys. What is causing the noise, and about what range of frequencies would you expect to hear?

**APPROACH** Gusts of air in the wind cause vibrations or oscillations of the tree limb (or air column in the chimney), which produce sound waves of the same frequency. The end of a tree limb fixed to the tree trunk is a node, whereas the other end is free to move and therefore is an antinode; the tree limb is thus about $\frac{1}{4}\lambda$ (Fig. 12–13).

**SOLUTION** We estimate $v \approx 4000$ m/s for the speed of sound in wood (Table 12–1). Suppose that a tree limb has length $\ell \approx 2$ m; then $\lambda = 4\ell = 8$ m and $f = v/\lambda = (4000 \text{ m/s})/(8 \text{ m}) \approx 500$ Hz.

**NOTE** Wind can excite air oscillations in a chimney, much like in an organ pipe or flute. A chimney is a fairly long tube, perhaps 3 m in length, acting like a tube open at either one end or even both ends. If open at both ends ($\lambda = 2\ell$), with $v \approx 340$ m/s, we find $f_1 \approx v/2\ell \approx 57$ Hz, which is a fairly low note—no wonder chimneys "moan"!

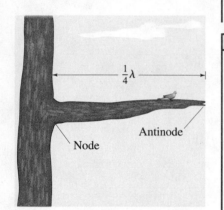

**FIGURE 12–13** Example 12–12.

**FIGURE 12–14** The amplitudes of the fundamental and first two overtones are added at each point to get the "sum," or composite waveform.

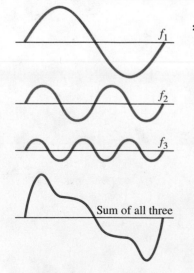

# *12–5 Quality of Sound, and Noise; Superposition

Whenever we hear a sound, particularly a musical sound, we are aware of its loudness, its pitch, and also of a third aspect called "quality." For example, when a piano and then a flute play a note of the same loudness and pitch (say, middle C), there is a clear difference in the overall sound. We would never mistake a piano for a flute. This is what is meant by the **quality** of a sound. For musical instruments, the terms *timbre* and *tone color* are also used.

Just as loudness and pitch can be related to physically measurable quantities, so too can quality. The quality of a sound depends on the presence of overtones—their number and their relative amplitudes. Generally, when a note is played on a musical instrument, the fundamental as well as overtones are present simultaneously. Figure 12–14 illustrates how the *principle of superposition* (Section 11–11) applies to three wave forms, in this case the fundamental and first two overtones (with particular amplitudes): they add together at each point to give a composite **waveform**.

By "waveform" we mean the shape of the wave in space at a given moment. Normally, more than two overtones are present. [Any complex wave can be analyzed into a superposition of sinusoidal waves of appropriate amplitudes, wavelengths, and frequencies. Such an analysis is called a *Fourier analysis*.]

The relative amplitudes of the overtones for a given note are different for different musical instruments, which is what gives each instrument its characteristic quality or timbre. A bar graph showing the relative amplitudes of the harmonics for a given note produced by an instrument is called a *sound spectrum*. Several typical examples for different musical instruments are shown in Fig. 12–15. The fundamental usually has the greatest amplitude, and its frequency is what is heard as the pitch.

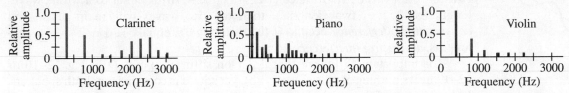

**FIGURE 12–15** Sound spectra for different instruments. The spectra change when the instruments play different notes. The clarinet is a bit complicated: it acts like a closed tube at lower frequencies, having only odd harmonics, but at higher frequencies all harmonics occur as for an open tube.

The manner in which an instrument is played strongly influences the sound quality. Plucking a violin string, for example, makes a very different sound than pulling a bow across it. The sound spectrum at the very start (or end) of a note (as when a hammer strikes a piano string) can be very different from the subsequent sustained tone. This too affects the subjective tone quality of an instrument.

An ordinary sound, like that made by striking two stones together, is a noise that has a certain quality, but a clear pitch is not discernible. Such a noise is a mixture of many frequencies which bear little relation to one another. A sound spectrum made of that noise would not show discrete lines like those of Fig. 12–15. Instead it would show a continuous, or nearly continuous, spectrum of frequencies. Such a sound we call "noise" in comparison with the more harmonious sounds which contain frequencies that are simple multiples of the fundamental.

# 12–6 Interference of Sound Waves; Beats

## Interference in Space

We saw in Section 11–11 that when two waves simultaneously pass through the same region of space, they interfere with one another. Interference also occurs with sound waves.

Consider two large loudspeakers, A and B, a distance *d* apart on the stage of an auditorium as shown in Fig. 12–16. Let us assume the two speakers are emitting sound waves of the same single frequency and that they are in phase: that is, when one speaker is forming a compression, so is the other. (We ignore reflections from walls, floor, etc.) The curved lines in the diagram represent the crests of sound waves from each speaker at one instant in time. We must remember that for a sound wave, a crest is a compression in the air whereas a trough—which falls between two crests—is a rarefaction. A human ear or detector at a point such as C, which is the same distance from each speaker, will experience a loud sound because the interference will be constructive—two crests reach it at one moment, two troughs reach it a moment later. On the other hand, at a point such as D in the diagram, little if any sound will be heard because destructive interference occurs—compressions of one wave meet rarefactions of the other and vice versa (see Fig. 11–38 and the related discussion on water waves in Section 11–11).

**FIGURE 12–16** Sound waves from two loudspeakers interfere.

(a)

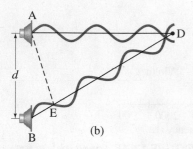

(b)

**FIGURE 12–17** Sound waves of a single frequency from loudspeakers A and B (see Fig. 12–16) constructively interfere at C and destructively interfere at D. [Shown here are graphical representations, not the actual longitudinal sound waves.]

An analysis of this situation is perhaps clearer if we graphically represent the waveforms as in Fig. 12–17. In Fig. 12–17a it can be seen that at point C, constructive interference occurs because both waves simultaneously have crests or simultaneously have troughs when they arrive at C. In Fig. 12–17b we see that, to reach point D, the wave from speaker B must travel a greater distance than the wave from A. Thus the wave from B lags behind that from A. In this diagram, point E is chosen so that the distance ED is equal to AD. Thus we see that if the distance BE is equal to precisely one-half the wavelength of the sound, the two waves will be exactly out of phase when they reach D, and destructive interference occurs. This then is the criterion for determining at what points destructive interference occurs: destructive interference occurs at any point whose distance from one speaker is one-half wavelength greater than its distance from the other speaker. Notice that if this extra distance (BE in Fig. 12–17b) is equal to a whole wavelength (or 2, 3, ··· wavelengths), then the two waves will be in phase and *constructive interference* occurs. If the distance BE equals $\frac{1}{2}$, $1\frac{1}{2}$, $2\frac{1}{2}$, ··· wavelengths, *destructive interference* occurs.

It is important to realize that a person sitting at point D in Fig. 12–16 or 12–17 hears nothing at all (or nearly so), yet sound is coming from both speakers. Indeed, if one of the speakers is turned off, the sound from the other speaker will be clearly heard.

If a loudspeaker emits a whole range of frequencies, only specific wavelengths will destructively interfere completely at a given point.

---

**EXAMPLE 12–13** **Loudspeakers' interference.** Two loudspeakers are 1.00 m apart. A person stands 4.00 m from one speaker. How far should this person be from the second speaker to detect destructive interference when the speakers emit an 1150-Hz sound? Assume the temperature is 20°C.

**APPROACH** To sense destructive interference, the person should be one-half wavelength closer to or farther from one speaker than from the other—that is, at a distance = 4.00 m ± $\lambda/2$. We can determine $\lambda$ because we know $f$ and $v$.

**SOLUTION** The speed of sound at 20°C is 343 m/s, so the wavelength of this sound is (Eq. 11–12)

$$\lambda = \frac{v}{f}$$

$$= \frac{343 \text{ m/s}}{1150 \text{ Hz}} = 0.30 \text{ m}.$$

For destructive interference to occur, the person must be one-half wavelength farther from one loudspeaker than from the other, or 0.15 m. Thus the person must be 3.85 m or 4.15 m from the second speaker.

**NOTE** If the speakers are less than 0.15 m apart, there will be no location that is 0.15 m farther from one speaker than the other, and there will be no place where destructive interference could occur.

---

## Beats—Interference in Time

We have been discussing interference of sound waves that takes place in space. An interesting and important example of interference that occurs in time is the phenomenon known as **beats**: if two sources of sound—say, two tuning forks—are close in frequency but not exactly the same, sound waves from the two sources interfere with each other. The sound level at a given position alternately rises and falls in time, because the two waves are sometimes in phase and sometimes out of phase due to their different wavelengths. The regularly spaced intensity changes are called beats.

**FIGURE 12–18** Beats occur as a result of the superposition of two sound waves of slightly different frequency.

To see how beats arise, consider two equal-amplitude sound waves of frequency $f_A = 50$ Hz and $f_B = 60$ Hz, respectively. In 1.00 s, the first source makes 50 vibrations whereas the second makes 60. We now examine the waves at one point in space equidistant from the two sources. The waveforms for each wave as a function of time, at a fixed position, are shown on the top graph of Fig. 12–18; the red line represents the 50-Hz wave, and the blue line represents the 60-Hz wave. The lower graph in Fig. 12–18 shows the sum of the two waves as a function of time. At time $t = 0$ the two waves are shown to be in phase and interfere constructively. Because the two waves vibrate at different rates, at time $t = 0.05$ s they are completely out of phase and interfere destructively. At $t = 0.10$ s, they are again in phase and the resultant amplitude again is large. Thus the resultant amplitude is large every 0.10 s and drops drastically in between. This rising and falling of the intensity is what is heard as beats.[†] In this case the beats are 0.10 s apart. That is, the **beat frequency** is ten per second, or 10 Hz. This result, that the beat frequency equals the difference in frequency of the two waves, is valid in general.

The phenomenon of beats can occur with any kind of wave and is a very sensitive method for comparing frequencies. For example, to tune a piano, a piano tuner listens for beats produced between his standard tuning fork and the frequency of a particular string on the piano, and knows it is in tune when the beats disappear. The members of an orchestra tune up by listening for beats between their instruments and the frequency of a standard tone (usually A above middle C at 440 Hz) produced by a piano or an oboe. Humans hear the individual beats if they are only a few per second. For higher beat frequencies, they run together and up to about 20 Hz or so you hear an intensity modulation (a wavering between loud and soft); above 20 Hz you hear a separate low tone (audible if the tones are strong enough).

**PHYSICS APPLIED**
*Tuning a piano*

---

**EXAMPLE 12–14** **Beats.** A tuning fork produces a steady 400-Hz tone. When this tuning fork is struck and held near a vibrating guitar string, twenty beats are counted in five seconds. What are the possible frequencies produced by the guitar string?

**APPROACH** For beats to occur, the string must vibrate at a frequency different from 400 Hz by whatever the beat frequency is.

**SOLUTION** The beat frequency is

$$f_{beat} = 20\text{ vibrations}/5\text{ s} = 4\text{ Hz}.$$

This is the difference of the frequencies of the two waves. Because one wave is known to be 400 Hz, the other must be either 404 Hz or 396 Hz.

---

[†]Beats will be heard even if the amplitudes are not equal, as long as the difference in amplitude is not great.

# 12–7 Doppler Effect

You may have noticed that you hear the pitch of the siren on a speeding fire truck drop abruptly as it passes you. Or you may have noticed the change in pitch of a blaring horn on a fast-moving car as it passes by you. The pitch of the engine noise of a race car changes as the car passes an observer. When a source of sound is moving toward an observer, the pitch the observer hears is higher than when the source is at rest; and when the source is traveling away from the observer, the pitch is lower. This phenomenon is known as the **Doppler effect**[†] and occurs for all types of waves. Let us now see why it occurs, and calculate the difference between the perceived and source sound frequencies when there is relative motion between source and observer.

(a) At rest

(b) Fire truck moving

**FIGURE 12–19** (a) Both observers on the sidewalk hear the same frequency from the fire truck at rest. (b) Doppler effect: observer toward whom the fire truck moves hears a higher-frequency sound, and observer behind the fire truck hears a lower-frequency sound.

**FIGURE 12–20** Determination of the frequency shift in the Doppler effect (see text). The red dot is the source.

Source $\bullet$ $\qquad d = \lambda \qquad$

(a) Source fixed

Crest emitted when source was at point 1.

Next crest emitted when source was at point 2.

1 $\vec{v}_{source}$ 2 $\qquad \lambda'$

$d_{source} = v_{source} T$

(b) Source moving

Consider the siren of a fire truck at rest, which is emitting sound of a particular frequency in all directions as shown in Fig. 12–19a. The sound waves are moving at the speed of sound in air, $v_{snd}$, which is independent of the velocity of the source or observer. If our source, the fire truck, is moving, the siren emits sound at the same frequency as it does at rest. But the sound wavefronts it emits forward, in front of it, are closer together than when the fire truck is at rest, as shown in Fig. 12–19b. This is because the fire truck, as it moves, is partly "catching up" to the previously emitted wavefronts, and emits each crest closer to the previous one. Thus an observer on the sidewalk in front of the truck will detect more wave crests passing per second, so the frequency heard is higher. The wavefronts emitted behind the truck, on the other hand, are farther apart than when the truck is at rest because the truck is speeding away from them. Each new wavefront emitted is farther from the preceding one than when the truck is at rest. Hence, fewer wave crests per second pass by an observer behind the moving truck (Fig. 12–19b) and the perceived pitch is lower.

We can calculate the frequency shift by making use of Fig. 12–20. We assume the air (or other medium) is at rest in our reference frame. We consider first the stationary observer off to the right in Fig. 12–19. In Fig. 12–20a, the source of the sound is shown as a red dot, and is at rest. Two successive wave crests are shown, the second of which has just been emitted and is still near the source. The distance between these crests is $\lambda$, the wavelength. If the frequency of the source is $f$, then the time between emissions of wave crests is

$$T = \frac{1}{f} = \frac{\lambda}{v_{snd}}.$$

In Fig. 12–20b, the source is moving with a velocity $v_{source}$ toward the observer.

[†]After J. C. Doppler (1803–1853).

In a time $T$ (just defined), the first wave crest has moved a distance $d = v_{snd} T = \lambda$, where $v_{snd}$ is the velocity of the sound wave in air (which is the same whether the source is moving or not). In this same time, the source has moved a distance $d_{source} = v_{source} T$. Then the distance between successive wave crests, which is the wavelength $\lambda'$ the observer on the right will perceive, is

$$\begin{aligned} \lambda' &= d - d_{source} \\ &= \lambda - v_{source} T \\ &= \lambda - v_{source} \frac{\lambda}{v_{snd}} \\ &= \lambda \left( 1 - \frac{v_{source}}{v_{snd}} \right). \end{aligned}$$

We subtract $\lambda$ from both sides of this equation and find that the shift in wavelength, $\Delta\lambda$, is

$$\Delta\lambda = \lambda' - \lambda = -\lambda \frac{v_{source}}{v_{snd}}.$$

So the shift in wavelength is directly proportional to the source speed $v_{source}$. The frequency $f'$ that will be perceived by our stationary observer on the ground is given by (Eq. 11–12)

$$f' = \frac{v_{snd}}{\lambda'} = \frac{v_{snd}}{\lambda \left( 1 - \frac{v_{source}}{v_{snd}} \right)}.$$

Because $v_{snd}/\lambda = f$, then

$$f' = \frac{f}{\left( 1 - \frac{v_{source}}{v_{snd}} \right)}. \qquad \begin{bmatrix} \text{source moving toward} \\ \text{stationary observer} \end{bmatrix} \quad \textbf{(12–2a)}$$

Because the denominator is less than 1, the observed frequency $f'$ is higher than the source frequency $f$. That is, $f' > f$. For example, if a source emits a sound of frequency 400 Hz when at rest, then when the source moves toward a fixed observer with a speed of 30 m/s, the observer hears a frequency (at 20°C) of

$$f' = \frac{400 \text{ Hz}}{1 - \frac{30 \text{ m/s}}{343 \text{ m/s}}} = 438 \text{ Hz}.$$

Now consider a source moving *away* from a stationary observer at a speed $v_{source}$ (observer on the left in Fig. 12–19). Using the same arguments as above, the wavelength $\lambda'$ perceived by our observer will have the minus sign on $d_{source}$ (first equation on this page) changed to plus:

$$\begin{aligned} \lambda' &= d + d_{source} \\ &= \lambda \left( 1 + \frac{v_{source}}{v_{snd}} \right). \end{aligned}$$

The difference between the observed and emitted wavelengths will be $\Delta\lambda = \lambda' - \lambda = +\lambda(v_{source}/v_{snd})$. The observed frequency of the wave is $f' = v_{snd}/\lambda'$, which equals

$$f' = \frac{f}{\left( 1 + \frac{v_{source}}{v_{snd}} \right)}. \qquad \begin{bmatrix} \text{source moving away from} \\ \text{stationary observer} \end{bmatrix} \quad \textbf{(12–2b)}$$

If a source emitting at 400 Hz is moving away from a fixed observer at 30 m/s, the observer hears a frequency $f' = (400 \text{ Hz})/\left[ 1 + (30 \text{ m/s})/(343 \text{ m/s}) \right] = 368 \text{ Hz}$.

**FIGURE 12–21** Observer moving with speed $v_{obs}$ toward a stationary source detects wave crests passing at speed $v' = v_{snd} + v_{obs}$ where $v_{snd}$ is the speed of the sound waves in air.

The Doppler effect also occurs when the source is at rest and the observer is in motion. If the observer is traveling *toward* the source, the pitch heard is higher than that of the emitted source frequency. If the observer is traveling *away* from the source, the pitch heard is lower. Quantitatively the change in frequency is different than for the case of a moving source. With a fixed source and a moving observer, the distance between wave crests, the wavelength $\lambda$, is not changed. But the velocity of the crests with respect to the observer *is* changed. If the observer is moving toward the source, Fig. 12–21, the speed $v'$ of the waves relative to the observer is a simple addition of velocities: $v' = v_{snd} + v_{obs}$, where $v_{snd}$ is the velocity of sound in air (we assume the air is still) and $v_{obs}$ is the velocity of the observer. Hence, the frequency heard is

$$f' = \frac{v'}{\lambda} = \frac{v_{snd} + v_{obs}}{\lambda}.$$

Because $\lambda = v_{snd}/f$, then

$$f' = \frac{(v_{snd} + v_{obs})f}{v_{snd}},$$

or

$$f' = \left(1 + \frac{v_{obs}}{v_{snd}}\right)f. \qquad \begin{bmatrix} \text{observer moving toward} \\ \text{stationary source} \end{bmatrix} \quad \textbf{(12–3a)}$$

If the observer is moving away from the source, the relative velocity is $v' = v_{snd} - v_{obs}$, so

$$f' = \left(1 - \frac{v_{obs}}{v_{snd}}\right)f. \qquad \begin{bmatrix} \text{observer moving away} \\ \text{from stationary source} \end{bmatrix} \quad \textbf{(12–3b)}$$

---

**EXAMPLE 12–15** **A moving siren.** The siren of a police car at rest emits at a predominant frequency of 1600 Hz. What frequency will you hear if you are at rest and the police car moves at 25.0 m/s (*a*) toward you, and (*b*) away from you?

**APPROACH** The observer is fixed, and the source moves, so we use Eqs. 12–2. The frequency you (the observer) hear is the emitted frequency $f$ divided by the factor $\left(1 \pm v_{source}/v_{snd}\right)$ where $v_{source}$ is the speed of the police car. Use the minus sign when the car moves toward you (giving a higher frequency); use the plus sign when the car moves away from you (lower frequency).

**SOLUTION** (*a*) The car is moving toward you, so (Eq. 12–2a)

$$f' = \frac{f}{\left(1 - \dfrac{v_{source}}{v_{snd}}\right)} = \frac{1600\ \text{Hz}}{\left(1 - \dfrac{25.0\ \text{m/s}}{343\ \text{m/s}}\right)} = 1726\ \text{Hz} \approx 1730\ \text{Hz}.$$

(*b*) The car is moving away from you, so (Eq. 12–2b)

$$f' = \frac{f}{\left(1 + \dfrac{v_{source}}{v_{snd}}\right)} = \frac{1600\ \text{Hz}}{\left(1 + \dfrac{25.0\ \text{m/s}}{343\ \text{m/s}}\right)} = 1491\ \text{Hz} \approx 1490\ \text{Hz}.$$

---

**EXERCISE F** Suppose the police car of Example 12–15 is at rest and emits at 1600 Hz. What frequency would you hear if you were moving at 25.0 m/s (*a*) toward it, and (*b*) away from it?

When a sound wave is reflected from a moving obstacle, the frequency of the reflected wave will, because of the Doppler effect, be different from that of the incident wave. This is illustrated in the following Example.

**EXAMPLE 12–16** | **Two Doppler shifts.** A 5000-Hz sound wave is emitted by a stationary source. This sound wave reflects from an object moving 3.50 m/s toward the source (Fig. 12–22). What is the frequency of the wave reflected by the moving object as detected by a detector at rest near the source?

**APPROACH** There are actually two Doppler shifts in this situation. First, the moving object acts like an observer moving toward the source with speed $v_{obs} = 3.50$ m/s (Fig. 12–22a) and so "detects" a sound wave of frequency (Eq. 12–3a) $f' = f[1 + (v_{obs}/v_{snd})]$. Second, reflection of the wave from the moving object is equivalent to the object reemitting the wave at the same frequency, and thus acting effectively as a moving source with speed $v_{source} = 3.50$ m/s (Fig. 12–22b). The final frequency detected, $f''$, is given by $f'' = f'/[1 - (v_{source}/v_{snd})]$, Eq. 12–2a.

**SOLUTION** The frequency $f'$ that is "detected" by the moving object is (Eq. 12–3a):

$$f' = \left(1 + \frac{v_{obs}}{v_{snd}}\right)f = \left(1 + \frac{3.50 \text{ m/s}}{343 \text{ m/s}}\right)(5000 \text{ Hz}) = 5051 \text{ Hz}.$$

**FIGURE 12–22** Example 12–16.

The moving object now "emits" (reflects) a sound of frequency (Eq. 12–2a)

$$f'' = \frac{f'}{\left(1 - \dfrac{v_{source}}{v_{snd}}\right)} = \frac{5051 \text{ Hz}}{\left(1 - \dfrac{3.50 \text{ m/s}}{343 \text{ m/s}}\right)} = 5103 \text{ Hz}.$$

Thus the frequency shifts by 103 Hz.

**NOTE** Bats use this technique to be aware of their surroundings. This is also the principle behind Doppler radar as speed-measuring devices for vehicles, baseball pitches, tennis serves, storms such as tornadoes, and other objects.

The emitted wave and the reflected wave in Example 12–16, when mixed together (say, electronically), interfere with one another and beats are produced. The beat frequency is equal to the difference in the two frequencies, 103 Hz. This Doppler technique is used in a variety of medical applications, usually with ultrasonic waves in the megahertz frequency range. For example, ultrasonic waves reflected from red blood cells can be used to determine the velocity of blood flow. Similarly, the technique can be used to detect the movement of the chest of a young fetus and to monitor its heartbeat.

**PHYSICS APPLIED**
*Doppler blood-flow meter and other medical uses*

For convenience, we can write Eqs. 12–2 and 12–3 as a single equation that covers all cases of both source and observer in motion:

$$f' = f\left(\frac{v_{snd} \pm v_{obs}}{v_{snd} \mp v_{source}}\right). \qquad \left[\begin{array}{l} \text{source and} \\ \text{observer moving} \end{array}\right] \quad \textbf{(12–4)}$$

**PROBLEM SOLVING**
*Getting the signs right*

To get the signs right, recall from your own experience that the frequency is higher when observer and source approach each other, and lower when they move apart. Thus the upper signs in numerator and denominator apply if source and/or observer move toward each other; the lower signs apply if they are moving apart.

**EXERCISE G** How fast would a source have to approach an observer at rest for the observed frequency to be one octave above the produced frequency (frequency doubled)? (a) $\frac{1}{2}v_{snd}$, (b) $v_{snd}$, (c) $2v_{snd}$, (d) $4v_{snd}$.

## Doppler Effect for Light

**PHYSICS APPLIED**
*Doppler effect for EM waves
and weather forecasting*

The Doppler effect occurs for waves other than sound. Light and other types of electromagnetic waves (such as for radar) exhibit the Doppler effect: although the formulas for the frequency shift are not identical to Eqs. 12–2 and 12–3, the effect is similar (see Chapter 33). One important application is for weather forecasting using radar. The time delay between the emission of a radar pulse and its reception after being reflected off raindrops gives the position of precipitation. Measuring the Doppler shift in frequency (as in Example 12–16) tells how fast the storm is moving and in which direction. "Radar guns" used by police work similarly, measuring a car's speed by the Doppler shift of electromagnetic waves.

**PHYSICS APPLIED**
*Redshift in cosmology*

Another important application is to astronomy, where the velocities of galaxies can be estimated from the Doppler shift. Light from distant galaxies is shifted toward lower frequencies, indicating that the galaxies are moving away from us. This is called the **redshift** since red has the lowest frequency of visible light. The greater the frequency shift, the greater the velocity of recession. It is found that the farther the galaxies are from us, the faster they move away. This observation is the basis for the idea that the universe is expanding, and is one basis for the idea that the universe had a beginning affectionately called the "Big Bang" (Chapter 33).

# *12–8 Shock Waves and the Sonic Boom

An object such as an airplane traveling faster than the speed of sound is said to have a **supersonic speed**. Such a speed is often given as a **Mach**[†] **number**, which is defined as the ratio of the speed of the object to the speed of sound in the surrounding medium. For example, a plane traveling 600 m/s high in the atmosphere, where the speed of sound is only 300 m/s, has a speed of Mach 2.

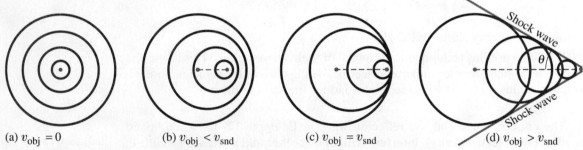

(a) $v_{obj} = 0$  (b) $v_{obj} < v_{snd}$  (c) $v_{obj} = v_{snd}$  (d) $v_{obj} > v_{snd}$

**FIGURE 12–23** Sound waves emitted by an object (a) at rest or (b, c, and d) moving. (b) If the object's velocity is less than the velocity of sound, the Doppler effect occurs. (c) At $v_{obj} = v_{snd}$, the waves pile up in front, forming a "sound barrier." (d) If the object's velocity is greater than the velocity of sound, a shock wave is produced.

**FIGURE 12–24** Bow waves produced by (a) a boat, (b) a duck.

(a)

(b)

When a source of sound moves at subsonic speeds (less than the speed of sound), the pitch of the sound is altered as we have seen (the Doppler effect); see also Figs. 12–23a and b. But if a source of sound moves faster than the speed of sound, a more dramatic effect known as a **shock wave** occurs. In this case, the source is actually "outrunning" the waves it produces. As shown in Fig. 12–23c, when the source is traveling *at* the speed of sound, the wave fronts it emits in the forward direction "pile up" directly in front of it. When the object moves faster, at a supersonic speed, the wave fronts pile up on one another along the sides, as shown in Fig. 12–23d. The different wave crests overlap one another and form a single very large crest which is the shock wave. Behind this very large crest there is usually a very large trough. A shock wave is essentially the result of constructive interference of a large number of wave fronts. A shock wave in air is analogous to the bow wave of a boat traveling faster than the speed of the water waves it produces, Fig. 12–24.

[†]After the Austrian physicist Ernst Mach (1838–1916).

(a)                 (b)

**FIGURE 12–25** (a) The (double) sonic boom has already been heard by person A on the left. The front shock wave is just being heard by person B in the center. And both will shortly be heard by person C on the right. (b) Special photo of supersonic aircraft showing shock waves produced in the air. (Several closely spaced shock waves are produced by different parts of the aircraft.)

When an airplane travels at supersonic speeds, greater than the speed of sound in the air, the noise it makes and its disturbance of the air form into a shock wave containing a tremendous amount of sound energy. When the shock wave passes a listener, it is heard as a loud **sonic boom**. A sonic boom lasts only a fraction of a second, but the energy it contains is sometimes sufficient to break windows and cause other damage. Actually, a sonic boom is made up of two or more booms since major shock waves can form at the front and the rear of the aircraft (Fig. 12–25), as well as at the wings and other parts. Bow waves of a boat are also multiple, as can be seen in Fig. 12–24a.

When an aircraft accelerates toward the speed of sound, it encounters a barrier of sound waves in front of it (see Fig. 12–23c). To exceed the speed of sound, the aircraft needs extra thrust to pass through this **sound barrier**. This is called "breaking the sound barrier." Once a supersonic speed is attained, this barrier no longer impedes the motion. It is sometimes erroneously thought that a sonic boom is produced only at the moment an aircraft is breaking through the sound barrier. Actually, a shock wave follows the aircraft at all times it is traveling at supersonic speed. A series of observers on the ground will each hear a loud "boom" as the shock wave passes, Fig. 12–25. The shock wave consists of a cone whose apex is at the aircraft. The angle of this cone, $\theta$ (see Fig. 12–23d), is given by

$$\sin \theta = \frac{v_{snd}}{v_{obj}}, \tag{12–5}$$

where $v_{obj}$ is the velocity of the object (the aircraft) and $v_{snd}$ is the velocity of sound in the medium (air for an airplane).

**PHYSICS APPLIED**
*Sonic boom*

## Applications: Sonar, Ultrasound, and Medical Imaging

**PHYSICS APPLIED**
*Sonar: depth finding, Earth soundings*

... is used in many applications to determine distance. The ... technique is used to locate underwater objects. A transmitter ... through the water, and a detector receives its reflection, ... time interval is carefully measured, and from it ... can be determined since the speed of sound ... the location of reefs, sunken ships, sub- ... in this way. The interior structure of ... reflections of waves traveling ... explosion (called "soundings"). ... structures and boundaries within the ... also useful in the exploration for oil ... aircraft involves a similar pulse-echo ... magnetic (EM) waves, which, like visible light,

... ag."

Sonar generally makes use of **ultrasonic** frequencies: that is, sound waves whose frequencies are above 20 kHz, beyond the range of human detection. For sonar, the frequencies are typically in the range 20 kHz to 100 kHz. One reason for using ultrasound waves, other than the fact that they are inaudible, is that for shorter wavelengths there is less diffraction (Section 11–14) so the beam spreads less and smaller objects can be detected.

## *Ultrasound Medical Imaging

The diagnostic use of ultrasound in medicine, in the form of images (sometimes called *sonograms*), is an important and interesting application of physical principles. A **pulse-echo technique** is used, much like sonar, except that the frequencies used are in the range of 1 to 10 MHz ($1 \text{ MHz} = 10^6 \text{ Hz}$). A high-frequency sound pulse is directed into the body, and its reflections from boundaries or interfaces between organs and other structures and lesions in the body are then detected. Tumors and other abnormal growths, or pockets of fluid, can be distinguished; the action of heart valves and the development of a fetus (Fig. 12–26) can be examined; and information about various organs of the body, such as the brain, heart, liver, and kidneys, can be obtained. Although ultrasound does not replace X-rays, for certain kinds of diagnosis it is more helpful. Some kinds of tissue or fluid are not detected in X-ray photographs, but ultrasound waves are reflected from their boundaries. "Real-time" ultrasound images are like a movie of a section of the interior of the body.

**FIGURE 12–26** Ultrasound image of a human fetus within the uterus.

**PHYSICS APPLIED**
*Ultrasound medical imaging*

The pulse-echo technique for medical imaging works as follows. A brief pulse of ultrasound is emitted by a transducer that transforms an electrical pulse into a sound-wave pulse. Part of the pulse is reflected as echoes at each interface in the body, and most of the pulse (usually) continues on, Fig. 12–27a.

**FIGURE 12–27** (a) Ultrasound pulse passes through the abdomen, reflecting from surfaces in its path. (b) Reflected pulses plotted as a function of time when received by transducer. The vertical dashed lines point out which reflected pulse goes with which surface. (c) "Dot display" for the same echoes: brightness of each dot is related to signal strength.

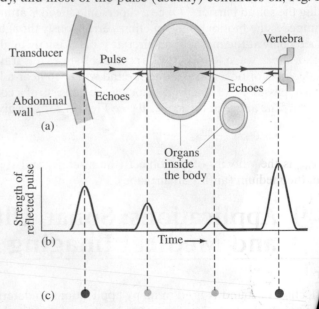

The detection of reflected pulses by the same transducer can then be displ[...]
the screen of a display monitor. The time elapsed from when the [...]
emitted to when each reflection (echo) is received is proportional to the [...]
to the reflecting surface. For example, if the distance from the transd[...]
vertebra is 25 cm, the pulse travels a round-trip distance of $2 \times 25 \text{ cm}$[...]
The speed of sound in human tissue is about 1540 m/s (close to that of [...]
so the time taken is

$$t = \frac{d}{v} = \frac{(0.50 \text{ m})}{(1540 \text{ m/s})} = 320 \text{ } \mu s.$$

The *strength* of a reflected pulse depends mainly on the diffe[...]
of the two materials on either side of the interface and can be di[...]
or as a dot (Figs. 12–27b and c). Each echo dot (Fig. 12–27c) [...]
as a point whose position is given by the time delay and whose [...]

(a)

(b)

Body organs

**FIGURE 12–28** (a) Ten traces are made across the abdomen by moving the transducer, or by using an array of transducers. (b) The echoes from interfaces or boundaries (of organs) are plotted as dots to produce the image. More closely spaced traces would give a more detailed image.

on the strength of the echo. A two-dimensional image can then be formed out of these dots from a series of scans. The transducer is moved, or an array of transducers is used, each of which sends out a pulse at each position and receives echoes as shown in Fig. 12–28a. Each trace can be plotted, spaced appropriately one below the other, to form an image on a monitor screen as shown in Fig. 12–28b. Only 10 lines are shown in Fig. 12–28, so the image is crude. More lines give a more precise image.

# Summary

Sound travels as a longitudinal wave in air and other materials. In air, the speed of sound increases with temperature. At 20°C, it is about 343 m/s.

The **pitch** of a sound is determined by the frequency; the higher the frequency, the higher the pitch.

The **audible range** of frequencies for humans is roughly 20 Hz to 20,000 Hz (1 Hz = 1 cycle per second).

The **loudness** or **intensity** of a sound is related to the amplitude squared of the wave. Because the human ear can detect sound intensities from $10^{-12}$ W/m$^2$ to over 1 W/m$^2$, sound levels are specified on a logarithmic scale. The **sound level** $\beta$, specified in decibels, is defined in terms of intensity $I$ as

$$\beta \text{ (in dB)} = 10 \log\left(\frac{I}{I_0}\right), \qquad \textbf{(12–1)}$$

where the reference intensity $I_0$ is usually taken to be $10^{-12}$ W/m$^2$.

Musical instruments are simple sources of sound in which *standing waves* are produced.

The strings of a stringed instrument may vibrate as a whole with nodes only at the ends; the frequency at which this standing wave occurs is called the **fundamental**. The fundamental frequency corresponds to a wavelength equal to twice the length of the string, $\lambda_1 = 2\ell$. The string can also vibrate at higher frequencies, called **overtones** or **harmonics**, in which there are one or more additional nodes. The frequency of each harmonic is a whole-number multiple of the fundamental.

In wind instruments, standing waves are set up in the column of air within the tube.

The vibrating air in an **open tube** (open at both ends) has displacement antinodes at both ends. The fundamental frequency corresponds to a wavelength equal to twice the tube length: $\lambda_1 = 2\ell$. The harmonics have frequencies that are 1, 2, 3, 4, $\cdots$ times the fundamental frequency, just as for strings.

For a **closed tube** (closed at one end), the fundamental corresponds to a wavelength four times the length of the tube: $\lambda_1 = 4\ell$. Only the odd harmonics are present, equal to 1, 3, 5, 7, $\cdots$ times the fundamental frequency.

Sound waves from different sources can interfere with each other. If two sounds are at slightly different frequencies, **beats** can be heard at a frequency equal to the difference in frequency of the two sources.

The **Doppler effect** refers to the change in pitch of a sound due to the motion either of the source or of the observer. If source and observer are approaching each other, the perceived pitch is higher. If they are moving apart, the perceived pitch is lower.

[*Shock waves and a sonic boom occur when an object moves at a **supersonic** speed—faster than the speed of sound. **Ultrasonic**-frequency (higher than 20 kHz) sound waves are used in many applications, including sonar and medical imaging.]

# Questions

1. What is the evidence that sound travels as a wave?

2. What is the evidence that sound is a form of energy?

3. Children sometimes play with a homemade "telephone" by attaching a string to the bottoms of two paper cups. When the string is stretched and a child speaks into one cup, the sound can be heard at the other cup (Fig. 12–29). Explain clearly how the sound wave travels from one cup to the other.

**FIGURE 12–29** Question 3.

4. When a sound wave passes from air into water, do you expect the frequency or wavelength to change?

5. What evidence can you give that the speed of sound in air does not depend significantly on frequency?

6. The voice of a person who has inhaled helium sounds very high-pitched. Why?

7. How will the air temperature in a room affect the pitch of organ pipes?

8. Explain how a tube might be used as a filter to reduce the amplitude of sounds in various frequency ranges. (An example is a car muffler.)

9. Why are the frets on a guitar (Fig. 12–30) spaced closer together as you move up the fingerboard toward the bridge?

**FIGURE 12–30**
Question 9.

10. A noisy truck approaches you from behind a building. Initially you hear it but cannot see it. When it emerges and you do see it, its sound is suddenly "brighter"—you hear more of the high-frequency noise. Explain. [*Hint*: See Section 11–14 on diffraction.]

11. Standing waves can be said to be due to "interference in space," whereas beats can be said to be due to "interference in time." Explain.

12. In Fig. 12–16, if the frequency of the speakers is lowered, would the points D and C (where destructive and constructive interference occur) move farther apart or closer together? Explain.

13. Traditional methods of protecting the hearing of people who work in areas with very high noise levels have consisted mainly of efforts to block or reduce noise levels. With a relatively new technology, headphones are worn that do not block the ambient noise. Instead, a device is used which detects the noise, inverts it electronically, then feeds it to the headphones *in addition to* the ambient noise. How could adding *more* noise reduce the sound levels reaching the ears?

14. Consider the two waves shown in Fig. 12–31. Each wave can be thought of as a superposition of two sound waves with slightly different frequencies, as in Fig. 12–18. In which of the waves, (a) or (b), are the two component frequencies farther apart? Explain.

(a)

(b)

**FIGURE 12–31** Question 14.

15. Is there a Doppler shift if the source and observer move in the same direction, with the same velocity? Explain.

16. If a wind is blowing, will this alter the frequency of the sound heard by a person at rest with respect to the source? Is the wavelength or velocity changed?

17. Figure 12–32 shows various positions of a child on a swing moving toward a person on the ground who is blowing a whistle. At which position, A through E, will the child hear the highest frequency for the sound of the whistle? Explain your reasoning.

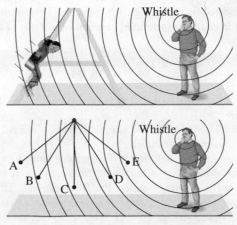

**FIGURE 12–32** Question 17.

# MisConceptual Questions

1. Do you expect an echo to return to you more quickly on a hot day or a cold day?
   (a) Hot day.
   (b) Cold day.
   (c) Same on both days.

2. Sound waves are
   (a) transverse waves characterized by the displacement of air molecules.
   (b) longitudinal waves characterized by the displacement of air molecules.
   (c) longitudinal waves characterized by pressure differences.
   (d) Both (b) and (c).
   (e) (a), (b), and (c).

3. The sound level near a noisy air conditioner is 70 dB. If two such units operate side by side, the sound level near them would be
   (a) 70 dB.
   (b) 73 dB.
   (c) 105 dB.
   (d) 140 dB.

4. To make a given sound seem twice as loud, how should a musician change the intensity of the sound?
   (a) Double the intensity.
   (b) Halve the intensity.
   (c) Quadruple the intensity.
   (d) Quarter the intensity.
   (e) Increase the intensity by a factor of 10.

5. A musical note that is two octaves higher than a second note
   (a) has twice the frequency of the second note.
   (b) has four times the frequency of the second note.
   (c) has twice the amplitude of the second note.
   (d) is 3 dB louder than the second note.
   (e) None of the above.

6. In which of the following is the wavelength of the lowest vibration mode the same as the length of the string or tube?
   (a) A string.
   (b) An open tube.
   (c) A tube closed at one end.
   (d) All of the above.
   (e) None of the above.

7. When a sound wave passes from air into water, what properties of the wave will change?
   (a) Frequency.
   (b) Wavelength.
   (c) Wave speed.
   (d) Both frequency and wavelength.
   (e) Both wave speed and wavelength.

8. A guitar string vibrates at a frequency of 330 Hz with wavelength 1.40 m. The frequency and wavelength of this sound in air (20°C) as it reaches our ears is
   (a) same frequency, same wavelength.
   (b) higher frequency, same wavelength.
   (c) lower frequency, same wavelength.
   (d) same frequency, longer wavelength.
   (e) same frequency, shorter wavelength.

9. A guitar player shortens the length of a guitar's vibrating string by pressing the string straight down onto a fret. The guitar then emits a higher-pitched note, because
   (a) the string's tension has been dramatically increased.
   (b) the string can vibrate with a much larger amplitude.
   (c) the string vibrates at a higher frequency.

10. An organ pipe with a fundamental frequency $f$ is open at both ends. If one end is closed off, the fundamental frequency will
    (a) drop by half.
    (b) not change.
    (c) double.

11. Two loudspeakers are about 10 m apart in the front of a large classroom. If either speaker plays a pure tone at a single frequency of 400 Hz, the loudness seems pretty even as you wander around the room, and gradually decreases in volume as you move farther from the speaker. If both speakers then play the same tone together, what do you hear as you wander around the room?
    (a) The pitch of the sound increases to 800 Hz, and the sound is louder but not twice as loud. It is louder closer to the speakers and gradually decreases as you move away from the speakers—except near the back wall, where a slight echo makes the sound louder.
    (b) The sound is louder but maintains the same relative spatial pattern of gradually decreasing volume as you move away from the speakers.
    (c) As you move around the room, some areas seem to be dead spots with very little sound, whereas other spots seem to be louder than with only one speaker.
    (d) The sound is twice as loud—so loud that you cannot hear any difference as you move around the room.
    (e) At points equidistant from both speakers, the sound is twice as loud. In the rest of the room, the sound is the same as if a single speaker were playing.

12. You are driving at 75 km/h. Your sister follows in the car behind at 75 km/h. When you honk your horn, your sister hears a frequency
    (a) higher than the frequency you hear.
    (b) lower than the frequency you hear.
    (c) the same as the frequency you hear.
    (d) You cannot tell without knowing the horn's frequency.

13. A guitar string is vibrating at its fundamental frequency $f$. Which of the following is *not* true?
    (a) Each small section of the guitar string oscillates up and down at a frequency $f$.
    (b) The wavelength of the standing wave on the guitar string is $\lambda = v/f$, where $v$ is the velocity of the wave on the string.
    (c) A sound wave created by this vibrating string propagates through the air with frequency $f$.
    (d) A sound wave created by this vibrating string propagates through the air with wavelength $\lambda = v/f$, where $v$ is the velocity of sound in air.
    (e) The wavelength of the standing wave on the guitar string is $\lambda = \ell$, where $\ell$ is the length of the string.

For assigned homework and other learning materials, go to the MasteringPhysics website.

# Problems

[Unless stated otherwise, assume $T = 20°C$ and $v_{sound} = 343$ m/s in air.]

## 12–1 Characteristics of Sound

1. (I) A hiker determines the length of a lake by listening for the echo of her shout reflected by a cliff at the far end of the lake. She hears the echo 2.5 s after shouting. Estimate the length of the lake.

2. (I) A sailor strikes the side of his ship just below the water-line. He hears the echo of the sound reflected from the ocean floor directly below 2.0 s later. How deep is the ocean at this point? Assume the speed of sound in sea water is 1560 m/s (Table 12–1) and does not vary significantly with depth.

3. (I) (a) Calculate the wavelengths in air at 20°C for sounds in the maximum range of human hearing, 20 Hz to 20,000 Hz. (b) What is the wavelength of an 18-MHz ultrasonic wave?

4. (II) On a warm summer day (31°C), it takes 4.80 s for an echo to return from a cliff across a lake. On a winter day, it takes 5.20 s. What is the temperature on the winter day?

5. (II) An ocean fishing boat is drifting just above a school of tuna on a foggy day. Without warning, an engine backfire occurs on another boat 1.55 km away (Fig. 12–33). How much time elapses before the backfire is heard (a) by the fish, and (b) by the fishermen?

**FIGURE 12–33**
Problem 5.

6. (II) A person, with his ear to the ground, sees a huge stone strike the concrete pavement. A moment later two sounds are heard from the impact: one travels in the air and the other in the concrete, and they are 0.80 s apart. How far away did the impact occur? See Table 12–1.

7. (III) A stone is dropped from the top of a cliff. The splash it makes when striking the water below is heard 2.7 s later. How high is the cliff?

## 12–2 Intensity of Sound; Decibels

8. (I) What is the intensity of a sound at the pain level of 120 dB? Compare it to that of a whisper at 20 dB.

9. (I) What is the sound level of a sound whose intensity is $1.5 \times 10^{-6}$ W/m²?

10. (II) You are trying to decide between two new stereo amplifiers. One is rated at 75 W per channel and the other is rated at 120 W per channel. In terms of dB, how much louder will the more powerful amplifier be when both are producing sound at their maximum levels?

11. (II) If two firecrackers produce a combined sound level of 85 dB when fired simultaneously at a certain place, what will be the sound level if only one is exploded? [Hint: Add intensities, not dBs.]

12. (II) A person standing a certain distance from an airplane with four equally noisy jet engines is experiencing a sound level of 140 dB. What sound level would this person experience if the captain shut down all but one engine? [Hint: Add intensities, not dBs.]

13. (II) One CD player is said to have a signal-to-noise ratio of 82 dB, whereas for a second CD player it is 98 dB. What is the ratio of intensities of the signal and the background noise for each device?

14. (II) A 55-dB sound wave strikes an eardrum whose area is $5.0 \times 10^{-5}$ m². (a) How much energy is received by the eardrum per second? (b) At this rate, how long would it take your eardrum to receive a total energy of 1.0 J?

15. (II) At a rock concert, a dB meter registered 130 dB when placed 2.5 m in front of a loudspeaker on stage. (a) What was the power output of the speaker, assuming uniform spherical spreading of the sound and neglecting absorption in the air? (b) How far away would the sound level be 85 dB?

16. (II) A fireworks shell explodes 100 m above the ground, creating colorful sparks. How much greater is the sound level of the explosion for a person at a point directly below the explosion than for a person a horizontal distance of 200 m away (Fig. 12–34)?

**FIGURE 12–34** Problem 16.

17. (II) If the amplitude of a sound wave is made 3.5 times greater, (a) by what factor will the intensity increase? (b) By how many dB will the sound level increase?

18. (II) Two sound waves have equal displacement amplitudes, but one has 2.2 times the frequency of the other. What is the ratio of their intensities?

19. (II) What would be the sound level (in dB) of a sound wave in air that corresponds to a displacement amplitude of vibrating air molecules of 0.13 mm at 440 Hz?

20. (II) (a) Estimate the power output of sound from a person speaking in normal conversation. Use Table 12–2. Assume the sound spreads roughly uniformly over a sphere centered on the mouth. (b) How many people would it take to produce a total sound output of 60 W of ordinary conversation? [Hint: Add intensities, not dBs.]

21. (III) Expensive amplifier A is rated at 220 W, while the more modest amplifier B is rated at 45 W. (a) Estimate the sound level in decibels you would expect at a point 3.5 m from a loudspeaker connected in turn to each amp. (b) Will the expensive amp sound twice as loud as the cheaper one?

## *12–3 Loudness

*22. (I) A 5000-Hz tone must have what sound level to seem as loud as a 100-Hz tone that has a 50-dB sound level? (See Fig. 12–6.)

*23. (I) What are the lowest and highest frequencies that an ear can detect when the sound level is 40 dB? (See Fig. 12–6.)

*24. (II) Your ears can accommodate a huge range of sound levels. What is the ratio of highest to lowest intensity at (a) 100 Hz, (b) 5000 Hz? (See Fig. 12–6.)

## 12–4 Sources of Sound: Strings and Air Columns

25. (I) Estimate the number of octaves in the human audible range, 20 Hz to 20 kHz.

26. (I) What would you estimate for the length of a bass clarinet, assuming that it is modeled as a closed tube and that the lowest note that it can play is a D♭ whose frequency is 69 Hz?

27. (I) The A string on a violin has a fundamental frequency of 440 Hz. The length of the vibrating portion is 32 cm, and it has mass 0.35 g. Under what tension must the string be placed?

28. (I) An organ pipe is 116 cm long. Determine the fundamental and first three audible overtones if the pipe is (a) closed at one end, and (b) open at both ends.

29. (I) (a) What resonant frequency would you expect from blowing across the top of an empty soda bottle that is 24 cm deep, if you assumed it was a closed tube? (b) How would that change if it was one-third full of soda?

30. (I) If you were to build a pipe organ with open-tube pipes spanning the range of human hearing (20 Hz to 20 kHz), what would be the range of the lengths of pipes required?

31. (II) A tight guitar string has a frequency of 540 Hz as its third harmonic. What will be its fundamental frequency if it is fingered at a length of only 70% of its original length?

32. (II) Estimate the frequency of the "sound of the ocean" when you put your ear very near a 15-cm-diameter seashell (Fig. 12–35).

**FIGURE 12–35**
Problem 32.

33. (II) An unfingered guitar string is 0.68 m long and is tuned to play E above middle C (330 Hz). (a) How far from the end of this string must a fret (and your finger) be placed to play A above middle C (440 Hz)? (b) What is the wavelength on the string of this 440-Hz wave? (c) What are the frequency and wavelength of the sound wave produced in air at 22°C by this fingered string?

34. (II) (a) Determine the length of an open organ pipe that emits middle C (262 Hz) when the temperature is 18°C. (b) What are the wavelength and frequency of the fundamental standing wave in the tube? (c) What are $\lambda$ and $f$ in the traveling sound wave produced in the outside air?

35. (II) An organ is in tune at 22.0°C. By what percent will the frequency be off at 11°C?

36. (II) How far from the mouthpiece of the flute in Example 12–11 should the hole be that must be uncovered to play F above middle C at 349 Hz?

37. (II) (a) At $T = 22°C$, how long must an open organ pipe be to have a fundamental frequency of 294 Hz? (b) If this pipe is filled with helium, what is its fundamental frequency?

38. (II) A particular organ pipe can resonate at 264 Hz, 440 Hz, and 616 Hz, but not at any other frequencies in between. (a) Show why this is an open or a closed pipe. (b) What is the fundamental frequency of this pipe?

39. (II) A uniform narrow tube 1.70 m long is open at both ends. It resonates at two successive harmonics of frequencies 275 Hz and 330 Hz. What is (a) the fundamental frequency, and (b) the speed of sound in the gas in the tube?

40. (II) A pipe in air at 23.0°C is to be designed to produce two successive harmonics at 280 Hz and 320 Hz. How long must the pipe be, and is it open or closed?

41. (II) How many overtones are present within the audible range for a 2.18-m-long organ pipe at 20°C (a) if it is open, and (b) if it is closed?

42. (II) Determine the fundamental and first overtone frequencies when you are in a 9.0-m-long hallway with all doors closed. Model the hallway as a tube closed at both ends.

43. (III) When a player's finger presses a guitar string down onto a fret, the length of the vibrating portion of the string is shortened, thereby increasing the string's fundamental frequency (see Fig. 12–36). The string's tension and mass per unit length remain unchanged. If the unfingered length of the string is $\ell = 75.0$ cm, determine the positions $x$ of the first six frets, if each fret raises the pitch of the fundamental by one musical note compared to the neighboring fret. On the equally tempered chromatic scale, the ratio of frequencies of neighboring notes is $2^{1/12}$.

**FIGURE 12–36**
Problem 43.

44. (III) The human ear canal is approximately 2.5 cm long. It is open to the outside and is closed at the other end by the eardrum. Estimate the frequencies (in the audible range) of the standing waves in the ear canal. What is the relationship of your answer to the information in the graph of Fig. 12–6?

## *12–5 Quality of Sound, Superposition

*45. (II) Approximately what are the intensities of the first two overtones of a violin compared to the fundamental? How many decibels softer than the fundamental are the first and second overtones? (See Fig. 12–15.)

## 12–6 Interference; Beats

46. (I) A piano tuner hears one beat every 2.0 s when trying to adjust two strings, one of which is sounding 350 Hz. How far off in frequency is the other string?

47. (I) A certain dog whistle operates at 23.5 kHz, while another (brand X) operates at an unknown frequency. If humans can hear neither whistle when played separately, but a shrill whine of frequency 5000 Hz occurs when they are played simultaneously, estimate the operating frequency of brand X.

48. (II) What is the beat frequency if middle C (262 Hz) and C♯ (277 Hz) are played together? What if each is played two octaves lower (each frequency reduced by a factor of 4)?

49. (II) A guitar string produces 3 beats/s when sounded with a 350-Hz tuning fork and 8 beats/s when sounded with a 355-Hz tuning fork. What is the vibrational frequency of the string? Explain your reasoning.

50. (II) Two violin strings are tuned to the same frequency, 294 Hz. The tension in one string is then decreased by 2.5%. What will be the beat frequency heard when the two strings are played together? [*Hint*: Recall Eq. 11–13.]

51. (II) The two sources of sound in Fig. 12–16 face each other and emit sounds of equal amplitude and equal frequency (305 Hz) but 180° out of phase. For what minimum separation of the two speakers will there be some point at which (a) complete constructive interference occurs and (b) complete destructive interference occurs. (Assume $T = 20°C$.)

**52.** (II) Two piano strings are supposed to be vibrating at 220 Hz, but a piano tuner hears three beats every 2.5 s when they are played together. (a) If one is vibrating at 220.0 Hz, what must be the frequency of the other (is there only one answer)? (b) By how much (in percent) must the tension be increased or decreased to bring them in tune?

**53.** (III) Two loudspeakers are 1.60 m apart. A person stands 3.00 m from one speaker and 3.50 m from the other. (a) What is the lowest frequency at which destructive interference will occur at this point if the speakers are in phase? (b) Calculate two other frequencies that also result in destructive interference at this point (give the next two highest). Let $T = 20°C$.

**54.** (III) Two loudspeakers are placed 3.00 m apart, as shown in Fig. 12–37. They emit 474-Hz sounds, in phase. A microphone is placed 3.20 m distant from a point midway between the two speakers, where an intensity maximum is recorded. (a) How far must the microphone be moved to the right to find the first intensity minimum? (b) Suppose the speakers are reconnected so that the 474-Hz sounds they emit are exactly out of phase. At what positions are the intensity maximum and minimum now?

**FIGURE 12–37**
Problem 54.

**55.** (III) A source emits sound of wavelengths 2.54 m and 2.72 m in air. (a) How many beats per second will be heard? (Assume $T = 20°C$.) (b) How far apart in space are the regions of maximum intensity?

## 12–7 Doppler Effect

**56.** (I) The predominant frequency of a certain fire truck's siren is 1650 Hz when at rest. What frequency do you detect if you move with a speed of 30.0 m/s (a) toward the fire truck, and (b) away from it?

**57.** (II) A bat at rest sends out ultrasonic sound waves at 50.0 kHz and receives them returned from an object moving directly away from it at 27.5 m/s. What is the received sound frequency?

**58.** (II) Two automobiles are equipped with the same single-frequency horn. When one is at rest and the other is moving toward the first at 18 m/s, the driver at rest hears a beat frequency of 4.5 Hz. What is the frequency the horns emit? Assume $T = 20°C$.

**59.** (II) As a bat flies toward a wall at a speed of 6.0 m/s, the bat emits an ultrasonic sound wave with frequency 30.0 kHz. What frequency does the bat hear in the reflected wave?

**60.** (II) In one of the original Doppler experiments, a tuba was played at a frequency of 75 Hz on a moving flat train car, and a second identical tuba played the same tone while at rest in the railway station. What beat frequency was heard in the station if the train car approached the station at a speed of 14.0 m/s?

**61.** (II) A wave on the ocean surface with wavelength 44 m travels east at a speed of 18 m/s relative to the ocean floor. If, on this stretch of ocean, a powerboat is moving at 14 m/s (relative to the ocean floor), how often does the boat encounter a wave crest, if the boat is traveling (a) west, and (b) east?

**62.** (III) A police car sounding a siren with a frequency of 1580 Hz is traveling at 120.0 km/h. (a) What frequencies does an observer standing next to the road hear as the car approaches and as it recedes? (b) What frequencies are heard in a car traveling at 90.0 km/h in the opposite direction before and after passing the police car? (c) The police car passes a car traveling in the same direction at 80.0 km/h. What two frequencies are heard in this car?

**63.** (III) The Doppler effect using ultrasonic waves of frequency $2.25 \times 10^6$ Hz is used to monitor the heartbeat of a fetus. A (maximum) beat frequency of 240 Hz is observed. Assuming that the speed of sound in tissue is 1540 m/s, calculate the maximum velocity of the surface of the beating heart.

## *12–8 Shock Waves; Sonic Boom

**\*64.** (I) (a) How fast is an object moving on land if its speed at 24°C is Mach 0.33? (b) A high-flying jet cruising at 3000 km/h displays a Mach number of 3.1 on a screen. What is the speed of sound at that altitude?

**\*65.** (I) The wake of a speedboat is 12° in a lake where the speed of the water wave is 2.2 km/h. What is the speed of the boat?

**\*66.** (II) An airplane travels at Mach 2.1 where the speed of sound is 310 m/s. (a) What is the angle the shock wave makes with the direction of the airplane's motion? (b) If the plane is flying at a height of 6500 m, how long after it is directly overhead will a person on the ground hear the shock wave?

**\*67.** (II) A space probe enters the thin atmosphere of a planet where the speed of sound is only about 42 m/s. (a) What is the probe's Mach number if its initial speed is 15,000 km/h? (b) What is the angle of the shock wave relative to the direction of motion?

**\*68.** (II) A meteorite traveling 9200 m/s strikes the ocean. Determine the shock wave angle it produces (a) in the air just before entering the ocean, and (b) in the water just after entering. Assume $T = 20°C$.

**\*69.** (III) You look directly overhead and see a plane exactly 1.45 km above the ground flying faster than the speed of sound. By the time you hear the sonic boom, the plane has traveled a horizontal distance of 2.0 km. See Fig. 12–38. Determine (a) the angle of the shock cone, $\theta$, and (b) the speed of the plane and its Mach number. Assume the speed of sound is 330 m/s.

**FIGURE 12–38**
Problem 69.

**\*70.** (III) A supersonic jet traveling at Mach 2.0 at an altitude of 9500 m passes directly over an observer on the ground. Where will the plane be relative to the observer when the latter hears the sonic boom? (See Fig. 12–39.)

**FIGURE 12–39** Problem 70.

# General Problems

71. A fish finder uses a sonar device that sends 20,000-Hz sound pulses downward from the bottom of the boat, and then detects echoes. If the maximum depth for which it is designed to work is 85 m, what is the minimum time between pulses (in fresh water)?

72. A single mosquito 5.0 m from a person makes a sound close to the threshold of human hearing (0 dB). What will be the sound level of 200 such mosquitoes?

73. What is the resultant sound level when an 81-dB sound and an 87-dB sound are heard simultaneously?

74. The sound level 8.25 m from a loudspeaker, placed in the open, is 115 dB. What is the acoustic power output (W) of the speaker, assuming it radiates equally in all directions?

75. A stereo amplifier is rated at 225 W output at 1000 Hz. The power output drops by 12 dB at 15 kHz. What is the power output in watts at 15 kHz?

76. Workers around jet aircraft typically wear protective devices over their ears. Assume that the sound level of a jet airplane engine, at a distance of 30 m, is 130 dB, and that the average human ear has an effective radius of 2.0 cm. What would be the power intercepted by an unprotected ear at a distance of 30 m from a jet airplane engine?

77. In audio and communications systems, the **gain**, $\beta$, in decibels is defined for an amplifier as

$$\beta = 10 \log\left(\frac{P_{out}}{P_{in}}\right),$$

where $P_{in}$ is the power input to the system and $P_{out}$ is the power output. (a) A particular amplifier puts out 135 W of power for an input of 1.0 mW. What is its gain in dB? (b) If a signal-to-noise ratio of 93 dB is specified, what is the noise power if the output signal is 10 W?

78. Manufacturers typically offer a particular guitar string in a choice of diameters so that players can tune their instruments with a preferred string tension. For example, a nylon high-E string is available in a low- and high-tension model with diameter 0.699 mm and 0.724 mm, respectively. Assuming the density $\rho$ of nylon is the same for each model, compare (as a ratio) the tension in a tuned high- and low-tension string.

79. A tuning fork is set into vibration above a vertical open tube filled with water (Fig. 12–40). The water level is allowed to drop slowly. As it does so, the air in the tube above the water level is heard to resonate with the tuning fork when the distance from the tube opening to the water level is 0.125 m and again at 0.395 m. What is the frequency of the tuning fork?

**FIGURE 12–40**
Problem 79.

80. Two identical tubes, each closed at one end, have a fundamental frequency of 349 Hz at 25.0°C. The air temperature is increased to 31.0°C in one tube. If the two pipes are now sounded together, what beat frequency results?

81. Each string on a violin is tuned to a frequency $1\frac{1}{2}$ times that of its neighbor. The four equal-length strings are to be placed under the same tension; what must be the mass per unit length of each string relative to that of the lowest string?

82. A particular whistle produces sound by setting up the fundamental standing wave in an air column 7.10 cm long. The tube is closed at one end. The whistle blower is riding in a car moving away from you at 25 m/s. What frequency do you hear?

83. The diameter $D$ of a tube does affect the node at the open end of a tube. The end correction can be roughly approximated as adding $D/3$ to $\ell$ to give us an effective length for the tube in calculations. For a closed tube of length 0.55 m and diameter 3.0 cm, what are the frequencies of the first four harmonics, taking the end correction into consideration?

84. The frequency of a steam train whistle as it approaches you is 565 Hz. After it passes you, its frequency is measured as 486 Hz. How fast was the train moving (assume constant velocity)?

85. Two trains emit 508-Hz whistles. One train is stationary. The conductor on the stationary train hears a 3.5-Hz beat frequency when the other train approaches. What is the speed of the moving train?

86. Two loudspeakers are at opposite ends of a railroad car as it moves past a stationary observer at 12.0 m/s, as shown in Fig. 12–41. If the speakers have identical sound frequencies of 348 Hz, what is the beat frequency heard by the observer when (a) he listens from position A, in front of the car, (b) he is between the speakers, at B, and (c) he hears the speakers after they have passed him, at C?

**FIGURE 12–41** Problem 86.

87. Two open organ pipes, sounding together, produce a beat frequency of 6.0 Hz. The shorter one is 2.40 m long. How long is the other?

88. A bat flies toward a moth at speed 7.8 m/s while the moth is flying toward the bat at speed 5.0 m/s. The bat emits a sound wave of 51.35 kHz. What is the frequency of the wave detected by the bat after that wave reflects off the moth?

89. A bat emits a series of high-frequency sound pulses as it approaches a moth. The pulses are approximately 70.0 ms apart, and each is about 3.0 ms long. How far away can the moth be detected by the bat so that the echo from one pulse returns before the next pulse is emitted?

90. Two loudspeakers face each other at opposite ends of a long corridor. They are connected to the same source which produces a pure tone of 282 Hz. A person walks from one speaker toward the other at a speed of 1.6 m/s. What "beat" frequency does the person hear?

91. A sound-insulating door reduces the sound level by 30 dB. What fraction of the sound intensity passes through this door?

**92.** The "alpenhorn" (Fig. 12–42) was once used to send signals from one Alpine village to another. Since lower frequency sounds are less susceptible to intensity loss, long horns were used to create deep sounds. When played as a musical instrument, the alpenhorn must be blown in such a way that only one of the overtones is resonating. The most popular alpenhorn is about 3.4 m long, and it is called the F# (or G♭) horn. What is the fundamental frequency of this horn, and which overtone is close to F#? (See Table 12–3.) Model as a tube open at both ends.

**FIGURE 12–42** Problem 92.

# Search and Learn

**1.** At a painfully loud concert, a 120-dB sound wave travels away from a loudspeaker at 343 m/s. How much sound wave energy is contained in each 1.0-cm³ volume of air in the region near this loudspeaker? (See Sections 12–2 and 11–9.)

**2.** At a race track, you can estimate the speed of cars just by listening to the difference in pitch of the engine noise between approaching and receding cars. Suppose the sound of a certain car drops by a full octave (frequency halved) as it goes by on the straightaway. How fast is it going?

**3.** A person hears a pure tone in the 500 to 1000-Hz range coming from two sources. The sound is loudest at points equidistant from the two sources. To determine exactly what the frequency is, the person moves about and finds that the sound level is minimal at a point 0.25 m farther from one source than the other. What is the frequency of the sound?

**93.** Room acoustics for stereo listening can be compromised by the presence of standing waves, which can cause acoustic "dead spots" at the locations of the pressure nodes. Consider a living room 4.7 m long, 3.6 m wide, and 2.8 m high. Calculate the fundamental frequencies for the standing waves in this room.

**94.** A dramatic demonstration, called "singing rods," involves a long, slender aluminum rod held in the hand near the rod's midpoint. The rod is stroked with the other hand. With a little practice, the rod can be made to "sing," or emit a clear, loud, ringing sound. For an 80-cm-long rod, (a) what is the fundamental frequency of the sound? (b) What is its wavelength in the rod, and (c) what is the traveling wavelength of the sound in air at 20°C?

**\*95.** The intensity at the threshold of hearing for the human ear at a frequency of about 1000 Hz is $I_0 = 1.0 \times 10^{-12} \text{ W/m}^2$, for which $\beta$, the sound level, is 0 dB. The threshold of pain at the same frequency is about 120 dB, or $I = 1.0 \text{ W/m}^2$, corresponding to an increase of intensity by a factor of $10^{12}$. By what factor does the displacement amplitude, $A$, vary?

**96.** A **Doppler flow meter** uses ultrasound waves to measure blood-flow speeds. Suppose the device emits sound at 3.5 MHz, and the speed of sound in human tissue is about 1540 m/s. What is the expected beat frequency if blood is flowing in large leg arteries at 3.0 cm/s directly away from the sound source?

**4.** A factory whistle emits sound of frequency 770 Hz. The wind velocity is 15.0 m/s from the north (heading south). What frequency will observers hear who are located, at rest, (a) due north, (b) due south, (c) due east, and (d) due west, of the whistle? What frequency is heard by a cyclist heading (e) north or (f) west, toward the whistle at 12.0 m/s? Assume $T = 20°C$.

**5.** A bugle is a tube of fixed length that behaves as if it is open at both ends. A bugler, by adjusting his lips correctly and blowing with proper air pressure, can cause a harmonic (usually other than the fundamental) of the air column within the tube to sound loudly. Standard military tunes like *Taps* and *Reveille* require only four musical notes: G4 (392 Hz), C5 (523 Hz), E5 (659 Hz), and G5 (784 Hz). (a) For a certain length ℓ, a bugle will have a sequence of four consecutive harmonics whose frequencies very nearly equal those associated with the notes G4, C5, E5, and G5. Determine this ℓ. (b) Which harmonic is each of the (approximate) notes G4, C5, E5, and G5 for the bugle?

## ANSWERS TO EXERCISES

**A:** 4 times as intense.
**B:** (b).
**C:** 7 octaves, plus. [Note: This is like counting in binary, $2^7 = 128$; for more see Section 17–10.]
**D:** (b).
**E:** 257 Hz.
**F:** (a) 1717 Hz, (b) 1483 Hz.
**G:** (a).

Heating the air inside a "hot-air" balloon raises the air's temperature, causing it to expand, and forces some of the air out the opening at the bottom. The reduced amount of air inside means its density is lower than the outside air, so there is a net buoyant force upward on the balloon (Chapter 10). In this Chapter we study temperature and its effects on matter: thermal expansion and the gas laws. We examine the microscopic theory of matter as atoms or molecules that are continuously in motion, which we call kinetic theory. The temperature of a gas is directly related to the average translational kinetic energy of its molecules. We will consider ideal gases, but will also look at real gases and how they change phase, including evaporation, vapor pressure, and humidity.

# Temperature and Kinetic Theory

## CHAPTER 13

**CHAPTER-OPENING QUESTION—Guess now!**

A hot-air balloon, open at one end (see photos above), rises when the air inside is heated by a flame. For the following properties, is the air inside the balloon higher, lower, or the same as for the air outside the balloon?
(i) Temperature.     (ii) Pressure.     (iii) Density.

This Chapter is the first of three (Chapters 13, 14, and 15) devoted to temperature, heat, and thermodynamics. Much of this Chapter discusses the theory that matter is made up of atoms and that these atoms are in continuous random motion. This theory is called the *kinetic theory*. ("Kinetic," you may recall from Chapter 6, is Greek for "moving.")

We also discuss the concept of temperature and how it is measured, as well as the measured properties of gases which serve as a foundation for kinetic theory.

## 13–1 Atomic Theory of Matter

The idea that matter is made up of atoms dates back to the ancient Greeks. According to the Greek philosopher Democritus, if a pure substance—say, a piece of iron—were cut into smaller and smaller bits, eventually a smallest piece of that substance would be obtained which could not be divided further. This smallest piece was called an **atom**, which in Greek means "indivisible." Today an atom is still the smallest piece of a substance, but we do not consider it indivisible. Rather it is viewed as consisting of a central nucleus (containing protons and neutrons) surrounded by electrons, Chapter 27.

### CONTENTS

(a)

(b)

(c)

Today the atomic theory is universally accepted. The experimental evidence in its favor, however, came mainly in the eighteenth, nineteenth, and twentieth centuries, and much of it was obtained from the analysis of chemical reactions.

We will often speak of the relative masses of individual atoms and molecules—what we call the **atomic mass** or **molecular mass**, respectively. (The terms *atomic weight* and *molecular weight* are sometimes used.) These masses are based on arbitrarily assigning the most abundant form of carbon atom, $^{12}C$, the atomic mass of exactly 12.0000 **unified atomic mass units** (u). In terms of kilograms,

$$1\,u\ =\ 1.6605 \times 10^{-27}\,kg.$$

The average atomic mass of hydrogen is 1.0079 u, and the values for other atoms are as listed in the Periodic Table inside the back cover of this book, and also in Appendix B.[†] The molecular mass of a compound is the sum of atomic masses of the atoms making up the molecules of that compound.

[An **element** is a substance, such as neon, gold, iron, or copper, that cannot be broken down into simpler substances by chemical means. **Compounds** are substances made up of elements, and can be broken down into them; examples are carbon dioxide and water. The smallest piece of an element is an atom; the smallest piece of a compound is a **molecule**. Molecules are made up of atoms; a molecule of water, for example, is made up of two atoms of hydrogen and one of oxygen; its chemical formula is $H_2O$.]

An important piece of evidence for the atomic theory is called **Brownian motion**, named after the biologist Robert Brown, who is credited with its discovery in 1827. While he was observing tiny pollen grains suspended in water under his microscope, Brown noticed that the tiny grains moved about in erratic paths (Fig. 13–1), even though the water appeared to be perfectly still. The atomic theory easily explains Brownian motion if we assume that the atoms of any substance are continually in motion. Then Brown's tiny pollen grains are jostled about by the vigorous barrage of rapidly moving molecules of water.

In 1905, Albert Einstein examined Brownian motion from a theoretical point of view and was able to calculate from the experimental data the approximate size and mass of atoms and molecules. His calculations showed that the diameter of a typical atom is about $10^{-10}$ m.

At the start of Chapter 10, we distinguished the three common phases (or states) of matter—solid, liquid, gas—based on **macroscopic**, or "large-scale," properties. Now let us see how these three phases of matter differ, from the atomic or **microscopic** point of view. First of all, atoms and molecules must exert attractive forces on each other, because only this explains why a brick or a block of aluminum holds together in one piece. The attractive forces between molecules are of an electrical nature (more on this in later Chapters). When molecules come too close together, the force between them must become repulsive (electric repulsion between their outer electrons). We need this assumption to explain that matter takes up space. Thus molecules maintain a minimum distance from each other. In a solid material, the attractive forces are strong enough that the atoms or molecules move only slightly (oscillate) about relatively fixed positions, often in an array known as a crystal lattice, as shown in Fig. 13–2a. In a liquid, the atoms or molecules are moving more rapidly, or the forces between them are weaker, so that they are sufficiently free to pass around one another, as in Fig. 13–2b. In a gas, the forces are so weak, or

[†] The relative masses of different atoms came from analysis of chemical reactions, and the **law of definite proportions**. It states that when two or more elements combine to form a compound, they always do so in the same proportions by mass. For example, table salt is always formed from 23 parts sodium and 35 parts chlorine; and water from one part hydrogen and eight parts oxygen. A continuous theory of matter could not account for the law of definite proportions but atomic theory does: the proportions of each element that form a compound correspond to the relative masses of the combining atoms. One atom of sodium (Na) combines with one atom of chlorine (Cl) to form one molecule of salt (NaCl), and one atom of sodium has a mass 23/35 times as large as one of chlorine. Hydrogen, the lightest atom, was arbitrarily assigned the relative mass of 1. On this scale, carbon was about 12, oxygen 16, sodium 23, and so on. It was sometimes more complicated. For example, from the various compounds oxygen formed, its relative mass was judged to be 16; but this was inconsistent with the mass ratio in water of oxygen to hydrogen, only 8 to 1. This was explained by assuming two H atoms combine with one O atom to form a water molecule.

**FIGURE 13–1** Path of a tiny particle (pollen grain, for example) suspended in water. The straight lines connect observed positions of the particle at equal time intervals.

**FIGURE 13–2** Atomic arrangements in (a) a crystalline solid, (b) a liquid, and (c) a gas.

the speeds so high, that the molecules do not even stay close together. They move rapidly every which way, Fig. 13–2c, filling any container and occasionally colliding with one another. On average, the speeds are sufficiently high in a gas that when two molecules collide, the force of attraction is not strong enough to keep them close together and they fly off in new directions.

---

**EXAMPLE 13–1** | **ESTIMATE** | **Distance between atoms.** The density of copper is $8.9 \times 10^3 \, \text{kg/m}^3$, and each copper atom has a mass of 63 u. Estimate the average distance between the centers of neighboring copper atoms.

**APPROACH** We consider a cube of copper 1 m on a side. From the given density $\rho$ we can calculate the mass $m$ of a cube of volume $V = 1 \, \text{m}^3$ ($m = \rho V$). We divide this mass $m$ by the mass of one atom (63 u) to obtain the number of atoms in $1 \, \text{m}^3$. We assume the atoms are in a uniform array, and we let $N$ be the number of atoms in a 1-m length; then $(N)(N)(N) = N^3$ equals the total number of atoms in $1 \, \text{m}^3$.

**SOLUTION** The mass of 1 copper atom is $63 \, \text{u} = 63 \times 1.66 \times 10^{-27} \, \text{kg} = 1.05 \times 10^{-25} \, \text{kg}$. This means that in a cube of copper 1 m on a side (volume $= 1 \, \text{m}^3$), there are

$$\frac{8.9 \times 10^3 \, \text{kg}}{1.05 \times 10^{-25} \, \text{kg/atom}} = 8.5 \times 10^{28} \, \text{atoms}.$$

The volume of a cube of side $\ell$ is $V = \ell^3$, so on one edge of the 1-m-long cube there are $(8.5 \times 10^{28})^{\frac{1}{3}}$ atoms $= 4.4 \times 10^9$ atoms. Hence the distance between neighboring atoms is

$$\frac{1 \, \text{m}}{4.4 \times 10^9 \, \text{atoms}} = 2.3 \times 10^{-10} \, \text{m}.$$

**NOTE** Watch out for units. Even though "atoms" is not a unit, it is helpful to include it to make sure you calculate correctly.

**NOTE** The distance between atoms is essentially what we mean when we speak of the size or diameter of an atom. So we have calculated the size of a copper atom.

# 13–2 Temperature and Thermometers

In everyday life, **temperature** is a measure of how hot or cold something is. A hot oven is said to have a high temperature, whereas the ice of a frozen lake is said to have a low temperature.

Many properties of matter change with temperature. For example, most materials expand when their temperature is increased.[†] An iron beam is longer when hot than when cold. Concrete roads and sidewalks expand and contract slightly according to temperature, which is why compressible spacers or expansion joints (Fig. 13–3) are placed at regular intervals. The electrical resistance of matter changes with temperature (Chapter 18). So too does the color radiated by objects, at least at high temperatures: you may have noticed that the heating element of an electric stove glows with a red color when hot. At higher temperatures, solids such as iron glow orange or even white. The white light from an incandescent lightbulb comes from an extremely hot tungsten wire. The surface temperatures of the Sun and other stars can be measured by the predominant color (more precisely, wavelengths) of light they emit.

Instruments designed to measure temperature are called **thermometers**. There are many kinds of thermometers, but their operation always depends on some property of matter that changes with temperature. Many common thermometers rely on the expansion of a material with an increase in temperature. The first idea for a thermometer, by Galileo, made use of the expansion of a gas. Common thermometers today consist of a hollow glass tube filled with mercury or with alcohol colored with a red dye, as were the earliest usable thermometers (Fig. 13–4).

[†] Most materials expand when their temperature is raised, but not all. Water, for example, in the range 0°C to 4°C contracts with an increase in temperature (see Section 13–4).

**FIGURE 13–3** Expansion joint on a bridge. Note center white line of highway.

**FIGURE 13–4** Thermometers built by the Accademia del Cimento (1657–1667) in Florence, Italy, are among the earliest known. These sensitive and exquisite instruments contained alcohol, sometimes colored, like many thermometers today.

Tube

Bulb (acts as
a reservoir)

(a)                        (b)

**FIGURE 13–5** (a) Mercury- or
alcohol-in-glass thermometer;
(b) bimetallic strip.

**FIGURE 13–6** Photograph of a
thermometer using a coiled
bimetallic strip.

**FIGURE 13–7** Celsius and
Fahrenheit scales compared.

Inside a common liquid-in-glass thermometer, the liquid expands more than the glass when the temperature is increased, so the liquid level rises in the tube (Fig. 13–5a). Although metals also expand with temperature, the change in length of a metal rod, say, is generally too small to measure accurately for ordinary changes in temperature. However, a useful thermometer can be made by bonding together two different metals with different rates of expansion (Fig. 13–5b). When the temperature is increased, the different amounts of expansion cause the **bimetallic strip** to bend. Often the bimetallic strip is in the form of a coil, one end of which is fixed while the other is attached to a pointer, Fig. 13–6. Such thermometers are used as ordinary air thermometers, oven thermometers, automatic off switches in electric coffeepots, and in room thermostats for determining when the heater or air conditioner should go on or off. Very precise thermometers make use of electrical properties (Chapter 18), such as resistance thermometers, thermocouples, and thermistors, often with a digital readout.

## Temperature Scales

In order to measure temperature quantitatively, some sort of numerical scale must be defined. The most common scale today is the **Celsius** or **centigrade** scale. In the United States, the **Fahrenheit** scale is common. The most important scale in scientific work is the absolute, or Kelvin, scale, and it will be discussed later in this Chapter.

One way to define a temperature scale is to assign arbitrary values to two readily reproducible temperatures. For both the Celsius and Fahrenheit scales these two fixed points are chosen to be the freezing point and the boiling point[†] of water, both taken at standard atmospheric pressure. On the Celsius scale, the freezing point of water is chosen to be 0°C ("zero degrees Celsius") and the boiling point 100°C. On the Fahrenheit scale, the freezing point is defined as 32°F and the boiling point 212°F. A practical thermometer is calibrated by placing it in carefully prepared environments at each of the two temperatures and marking the position of the liquid or pointer. For a Celsius scale, the distance between the two marks is divided into one hundred equal intervals representing each degree between 0°C and 100°C (hence the name "centigrade scale" meaning "hundred steps"). For the Fahrenheit scale, the two points are labeled 32°F and 212°F and the distance between them is divided into 180 equal intervals. For temperatures below the freezing point of water and above the boiling point of water, the scales may be extended using the same equally spaced intervals. However, thermometers can be used only over a limited temperature range because of their own limitations—for example, an alcohol-in-glass thermometer is rendered useless above temperatures where the alcohol vaporizes. For very low or very high temperatures, specialized thermometers are required, some of which we will mention later.

Every temperature on the Celsius scale corresponds to a particular temperature on the Fahrenheit scale, Fig. 13–7. To convert from one to the other, remember that 0°C corresponds to 32°F and that a range of 100° on the Celsius scale corresponds to a range of 180° on the Fahrenheit scale. Thus, one Fahrenheit degree (1 F°) corresponds to $100/180 = \frac{5}{9}$ of a Celsius degree (1 C°). That is, $1\,F° = \frac{5}{9}C°$. (Notice that when we refer to a specific temperature, we say "degrees Celsius," as in 20°C; but when we refer to a *change* in temperature or a temperature *interval*, we say "Celsius degrees," as in "2 C°.") The conversion between the two temperature scales can be written

$$T(°C) = \tfrac{5}{9}\left[T(°F) - 32\right]$$

or

$$T(°F) = \tfrac{9}{5}T(°C) + 32.$$

Rather than memorizing these relations, it may be simpler to remember that 0°C = 32°F and that a change of 5 C° = a change of 9 F°.

[†] The freezing point of a substance is defined as that temperature at which the solid and liquid phases coexist in equilibrium—that is, without any net liquid changing into the solid or vice versa. Experimentally, this is found to occur at only one definite temperature, for a given pressure. Similarly, the boiling point is defined as that temperature at which the liquid and gas coexist in equilibrium. Since these points vary with pressure, the pressure must be specified (usually it is 1 atm).

**EXAMPLE 13–2** **Taking your temperature.** Normal body temperature is 98.6°F. What is this on the Celsius scale?

**APPROACH** We recall that 0°C = 32°F and that a change of 5 C° = 9 F°.

**SOLUTION** First we relate the given temperature to the freezing point of water (0°C). That is, 98.6°F is 98.6 − 32.0 = 66.6 F° above the freezing point of water. Since each F° is equal to $\frac{5}{9}$ C°, this corresponds to 66.6 × $\frac{5}{9}$ = 37.0 Celsius degrees above the freezing point. The freezing point of water is 0°C, so normal body temperature is 37.0°C.

**⚠ CAUTION**

*Convert temperature by remembering 0°C = 32°F and a change of 5 C° = 9 F°*

## *Standard Temperature Scale

Different materials do not expand in quite the same way over a wide temperature range. Consequently, if we calibrate different kinds of thermometers exactly as described above, they will not usually agree precisely.

Because of such discrepancies, some standard kind of thermometer must be chosen so that all temperatures can be precisely defined. The chosen standard for this purpose is the **constant-volume gas thermometer**. As shown in the simplified diagram of Fig. 13–8, this thermometer consists of a bulb filled with a low-pressure gas connected by a thin tube to a mercury manometer (Section 10–6). The volume of the gas is kept constant by raising or lowering the right-hand tube of the manometer so that the mercury in the left-hand tube coincides with the reference mark. An increase in temperature causes a proportional increase in pressure in the bulb. Thus the tube must be lifted higher to keep the gas volume constant. The height of the mercury in the right-hand column is then a measure of the temperature. This thermometer gives the same results for all gases in the limit of reducing the gas pressure in the bulb toward zero. The resulting scale serves as a basis for the **standard temperature scale**.

**FIGURE 13–8** Constant-volume gas thermometer.

# 13–3 Thermal Equilibrium and the Zeroth Law of Thermodynamics

If two objects at different temperatures are placed in thermal contact (meaning thermal energy can transfer from one to the other), the two objects will eventually reach the same temperature. They are then said to be in **thermal equilibrium**. For example, you leave a fever thermometer in your mouth until it comes into thermal equilibrium with that environment; then you read it. Two objects are defined to be in thermal equilibrium if, when placed in thermal contact, no net energy flows from one to the other, and their temperatures don't change.

## *The Zeroth Law of Thermodynamics

Experiments indicate that

> **if two systems are in thermal equilibrium with a third system, then they are in thermal equilibrium with each other.**

This postulate is called the **zeroth law of thermodynamics**. It has this unusual name because it was not until after the first and second laws of thermodynamics (Chapter 15) were worked out that scientists realized that this apparently obvious postulate needed to be stated first.

Temperature is a property of a system that determines whether the system will be in thermal equilibrium with other systems. When two systems are in thermal equilibrium, their temperatures are (by definition) equal, and no net thermal energy is exchanged between them. This is consistent with our everyday notion of temperature: when a hot object and a cold one are put into contact, they eventually come to the same temperature. Thus the importance of the zeroth law is that it allows a useful definition of temperature.

# 13–4 Thermal Expansion

Most substances expand when heated and contract when cooled. However, the amount of expansion or contraction varies, depending on the material.

## Linear Expansion

Experiments indicate that the change in length $\Delta\ell$ of almost all solids is, to a good approximation, directly proportional to the change in temperature $\Delta T$, as long as $\Delta T$ is not too large. The change in length is also proportional to the original length of the object, $\ell_0$. That is, for the same temperature increase, a 4-m-long iron rod will increase in length twice as much as a 2-m-long iron rod. We can write this proportionality as an equation:

$$\Delta\ell = \alpha\ell_0\,\Delta T, \tag{13–1a}$$

where $\alpha$, the proportionality constant, is called the **coefficient of linear expansion** for the particular material and has units of $(\text{C}°)^{-1}$. We write $\ell = \ell_0 + \Delta\ell$, Fig. 13–9, and rewrite this equation as

$$\ell = \ell_0(1 + \alpha\,\Delta T), \tag{13–1b}$$

where $\ell_0$ is the length initially, at temperature $T_0$, and $\ell$ is the length after heating or cooling to a temperature $T$. If the temperature change $\Delta T = T - T_0$ is negative, then $\Delta\ell = \ell - \ell_0$ is also negative; the length shortens as the temperature decreases.

The values of $\alpha$ for various materials at 20°C are listed in Table 13–1. Actually, $\alpha$ does vary slightly with temperature (which is why thermometers made of different materials do not agree precisely). However, if the temperature range is not too great, the variation can usually be ignored.

**FIGURE 13–9** A thin rod of length $\ell_0$ at temperature $T_0$ is heated to a new uniform temperature $T$ and acquires length $\ell$, where $\ell = \ell_0 + \Delta\ell$.

### TABLE 13–1 Coefficients of Expansion, near 20°C

| Material | Coefficient of Linear Expansion, $\alpha$ $(\text{C}°)^{-1}$ | Coefficient of Volume Expansion, $\beta$ $(\text{C}°)^{-1}$ |
|---|---|---|
| *Solids* | | |
| Aluminum | $25 \times 10^{-6}$ | $75 \times 10^{-6}$ |
| Brass | $19 \times 10^{-6}$ | $56 \times 10^{-6}$ |
| Copper | $17 \times 10^{-6}$ | $50 \times 10^{-6}$ |
| Gold | $14 \times 10^{-6}$ | $42 \times 10^{-6}$ |
| Iron or steel | $12 \times 10^{-6}$ | $35 \times 10^{-6}$ |
| Lead | $29 \times 10^{-6}$ | $87 \times 10^{-6}$ |
| Glass (Pyrex®) | $3 \times 10^{-6}$ | $9 \times 10^{-6}$ |
| Glass (ordinary) | $9 \times 10^{-6}$ | $27 \times 10^{-6}$ |
| Quartz | $0.4 \times 10^{-6}$ | $1 \times 10^{-6}$ |
| Concrete and brick | $\approx 12 \times 10^{-6}$ | $\approx 36 \times 10^{-6}$ |
| Marble | $1.4\text{–}3.5 \times 10^{-6}$ | $4\text{–}10 \times 10^{-6}$ |
| *Liquids* | | |
| Gasoline | | $950 \times 10^{-6}$ |
| Mercury | | $180 \times 10^{-6}$ |
| Ethyl alcohol | | $1100 \times 10^{-6}$ |
| Glycerin | | $500 \times 10^{-6}$ |
| Water | | $210 \times 10^{-6}$ |
| *Gases* | | |
| Air (and most other gases at atmospheric pressure) | | $3400 \times 10^{-6}$ |

**EXAMPLE 13–3** **Bridge expansion.** The steel bed of a suspension bridge is 200 m long at 20°C. If the extremes of temperature to which it might be exposed are −30°C to +40°C, how much will it contract and expand?

**PHYSICS APPLIED**
*Expansion in structures*

**APPROACH** We assume the bridge bed will expand and contract linearly with temperature, as given by Eq. 13–1a.

**SOLUTION** From Table 13–1, we find that $\alpha = 12 \times 10^{-6}(\text{C}°)^{-1}$ for steel. The increase in length when it is at 40°C will be

$$\Delta\ell = \alpha\ell_0\,\Delta T = (12 \times 10^{-6}/\text{C}°)(200\,\text{m})(40°\text{C} - 20°\text{C}) = 4.8 \times 10^{-2}\,\text{m},$$

or 4.8 cm. When the temperature decreases to −30°C, $\Delta T = -50\,\text{C}°$. Then

$$\Delta\ell = (12 \times 10^{-6}/\text{C}°)(200\,\text{m})(-50\,\text{C}°) = -12.0 \times 10^{-2}\,\text{m},$$

or a decrease in length of 12 cm. The total range the expansion joints must accommodate is 12 cm + 4.8 cm ≈ 17 cm (Fig. 13–3).

**CONCEPTUAL EXAMPLE 13–4** **Do holes expand or contract?** If you heat a thin, circular ring (Fig. 13–10a) in the oven, does the ring's hole get larger or smaller?

**RESPONSE** If you guessed that the metal expands into the hole, making the hole smaller, it is not so. Imagine the ring is solid, like a coin (Fig. 13–10b). Draw a circle on it with a pen as shown. When the metal expands, the material inside the circle will expand along with the rest of the metal; so the dashed circle expands. Cutting the metal where the circle is shows that the hole in Fig. 13–10a increases in diameter.

(a)          (b)

**FIGURE 13–10** Example 13–4.

←6.420 cm→

**FIGURE 13–11**
Example 13–5.

←6.445 cm→

**EXAMPLE 13–5** **Ring on a rod.** An iron ring is to fit snugly on a cylindrical iron rod (Fig. 13–11). At 20°C, the diameter of the rod is 6.445 cm and the inside diameter of the ring is 6.420 cm. To slip over the rod, the ring must be slightly larger than the rod diameter by about 0.008 cm. To what temperature must the ring be brought if its hole is to be large enough so it will slip over the rod?

**APPROACH** The hole in the ring must be increased from a diameter of 6.420 cm to 6.445 cm + 0.008 cm = 6.453 cm. The ring must be heated since the hole diameter will increase linearly with temperature (Example 13–4).

**FIGURE 13–12** Example 13–6.

**SOLUTION** We solve for $\Delta T$ in Eq. 13–1a and find

$$\Delta T = \frac{\Delta\ell}{\alpha\ell_0} = \frac{6.453\,\text{cm} - 6.420\,\text{cm}}{(12 \times 10^{-6}/\text{C}°)(6.420\,\text{cm})} = 430\,\text{C}°.$$

So the ring must be raised at least to $T = (20°\text{C} + 430\,\text{C}°) = 450°\text{C}$.

**NOTE** In doing Problems, do not forget the last step, adding in the initial temperature (20°C here).

**CONCEPTUAL EXAMPLE 13–6** **Opening a tight jar lid.** When the lid of a glass jar is tight, holding the lid under hot water for a short time will often make it easier to open (Fig. 13–12). Why?

**RESPONSE** The lid may be struck by the hot water more directly than the glass and so expand sooner. But even if not, metals generally expand more than glass for the same temperature change ($\alpha$ is greater—see Table 13–1).

**PHYSICS APPLIED**
*Opening a tight lid*

## Volume Expansion

The change in *volume* of a material which undergoes a temperature change is given by a relation similar to Eq. 13–1a, namely,

$$\Delta V = \beta V_0 \Delta T, \tag{13–2}$$

where $V_0$ is the original volume, $\Delta V$ is the change in volume when the temperature changes by $\Delta T$, and $\beta$ is the **coefficient of volume expansion**. The units of $\beta$ are $(C°)^{-1}$.

Values of $\beta$ for various materials are given in Table 13–1. Notice that for solids, $\beta$ is normally equal to approximately $3\alpha$. Note also that linear expansion has no meaning for liquids and gases because they do not have fixed shapes.

Equations 13–1 and 13–2 are accurate only if $\Delta \ell$ (or $\Delta V$) is small compared to $\ell_0$ (or $V_0$). This is of particular concern for liquids and even more so for gases because of the large values of $\beta$. Furthermore, $\beta$ itself varies substantially with temperature for gases. Therefore, a more convenient way of dealing with gases is needed, and will be discussed starting in Section 13–5.

**EXAMPLE 13–7** | **Gas tank in the Sun.** The 70-liter (L) steel gas tank of a car is filled to the top with gasoline at 20°C. The car sits in the Sun and the tank reaches a temperature of 40°C (104°F). How much gasoline do you expect to overflow from the tank?

**APPROACH** Both the gasoline and the tank expand as the temperature increases, and we assume they do so linearly as described by Eq. 13–2. The volume of overflowing gasoline equals the volume increase of the gasoline minus the increase in volume of the tank.

**SOLUTION** The gasoline expands by

$$\Delta V = \beta V_0 \Delta T = (950 \times 10^{-6}/C°)(70\,L)(40°C - 20°C)$$
$$= 1.3\,L.$$

The tank also expands. We can think of it as a steel shell that undergoes volume expansion $(\beta = 35 \times 10^{-6}/C° \approx 3\alpha)$. If the tank were solid, the surface layer (the shell) would expand just the same (as in Example 13–4). Thus the tank increases in volume by

$$\Delta V = (35 \times 10^{-6}/C°)(70\,L)(40°C - 20°C) = 0.049\,L,$$

so the tank expansion has little effect. More than a liter of gas could spill out.

*PHYSICS APPLIED*
*Gas tank overflow*

**FIGURE 13–13** Behavior of water as a function of temperature near 4°C. (a) Volume of 1.00000 gram of water as a function of temperature. (b) Density vs. temperature. [Note the break in each axis.]

(a)

(b)

## Anomalous Behavior of Water Below 4°C

Most substances expand more or less uniformly with an increase in temperature, as long as no phase change occurs. Water, however, does not follow the usual pattern. If water at 0°C is heated, it actually *decreases* in volume until it reaches 4°C. Above 4°C water behaves normally and expands in volume as the temperature is increased, Fig. 13–13. Water thus has its greatest density at 4°C. This anomalous behavior of water is of great importance for the survival of aquatic life during cold winters. When water in a lake (or river) is above 4°C and begins to cool by contact with cold air, the water at the surface sinks because it is denser. It is replaced by warmer water from below. This mixing continues until the temperature of the entire lake reaches 4°C. As the surface water cools further, it remains on the surface because it is less dense than the 4°C water below. Water thus freezes first at the surface, and the ice remains on the surface since ice (specific gravity = 0.917) is less dense than water. The water at the bottom remains liquid unless it is so cold that the whole body of water freezes. If water were like most substances, becoming more dense as it cools, the water at the bottom of a lake would be frozen first.

Lakes would freeze solid more easily because circulation would bring the warmer water to the surface to be efficiently cooled. The complete freezing of a lake would cause severe damage to its plant and animal life. Because of the unusual behavior of water below 4°C, it is rare for any large and deep body of water to freeze completely, and this is helped by the layer of ice on the surface which acts as an insulator to reduce the flow of heat out of the water into the cold air above. Without this peculiar but wonderful property of water, life on this planet as we know it might not have been possible.

Not only does water expand as it cools from 4°C to 0°C, it expands even more as it freezes to ice. This is why ice cubes float in water and pipes break when water inside them freezes.

## *Thermal Stresses

In many situations, such as in buildings and roads, the ends of a beam or slab of material are rigidly fixed, which greatly limits expansion or contraction. If the temperature should change, large compressive or tensile stresses, called **thermal stresses**, will occur. The magnitude of such stresses can be calculated using the concept of elastic modulus developed in Chapter 9. To calculate the internal stress in a beam, we can think of this process as occurring in two steps: (1) the beam tries to expand (or contract) by an amount $\Delta\ell$ given by Eq. 13–1; (2) the solid in contact with the beam exerts a force to compress (or expand) it, keeping it at its original length. The force $F$ required is given by Eq. 9–4:

$$\Delta\ell = \frac{1}{E}\frac{F}{A}\ell_0,$$

where $E$ is Young's modulus for the material. To calculate the internal stress, $F/A$, we then set $\Delta\ell$ in Eq. 13–1a equal to $\Delta\ell$ in the equation above and find

$$\alpha\ell_0\,\Delta T = \frac{1}{E}\frac{F}{A}\ell_0.$$

Hence, the stress is

$$\frac{F}{A} = \alpha E\,\Delta T.$$

For example, if 10-m-long concrete slabs are placed touching each other in a new park you are designing, a 30°C increase in temperature would produce a stress $F/A = \alpha E\,\Delta T = (12 \times 10^{-6}/\text{C}°)(20 \times 10^{9}\,\text{N/m}^2)(30\,\text{C}°) = 7.2 \times 10^{6}\,\text{N/m}^2$. That stress would exceed the shear strength of concrete (Table 9–2), no doubt causing fracture and cracks. This is why soft spacers (or expansion joints) are placed between slabs on sidewalks and highways.

# 13–5 The Gas Laws and Absolute Temperature

Equation 13–2 is not useful for describing the expansion of a gas, partly because the expansion can be so great, and partly because gases generally expand to fill whatever container they are in. Indeed, Eq. 13–2 is meaningful only if the pressure is kept constant. The volume of a gas depends very much on the pressure as well as on the temperature. It is therefore valuable to determine a relation between the volume, the pressure, the temperature, and the quantity of a gas. Such a relation is called an **equation of state**. (By the word *state*, we mean the physical condition of the system.)

If the state of a system is changed, we will always wait until the pressure and temperature have reached the same values throughout. We thus consider only **equilibrium states** of a system—when the variables that describe it (such as temperature and pressure) are the same throughout the system and are not changing in time. We also note that the results of this Section are accurate only for gases that are not too dense (the pressure is not too high, on the order of an atmosphere or less) and not close to the liquefaction (boiling) point.

**FIGURE 13–14** Pressure vs. volume of a fixed amount of gas at a constant temperature, showing the inverse relationship as given by Boyle's law: as the pressure decreases, the volume increases.

**FIGURE 13–15** Volume of a fixed amount of gas as a function of (a) Celsius temperature, and (b) Kelvin temperature, when the pressure is kept constant.

Temperature (°C)

(a)

Temperature (kelvins, or K)

(b)

For a given quantity of gas it is found experimentally that, to a good approximation, *the volume of a gas is inversely proportional to the absolute pressure applied to it when the temperature is kept constant.* That is,

$$V \propto \frac{1}{P},$$ [constant $T$]

where $P$ is the absolute pressure (*not* "gauge pressure"—see Section 10–4). For example, if the pressure on a gas is doubled, the volume is reduced to half its original volume. This relation is known as **Boyle's law**, after the Englishman Robert Boyle (1627–1691), who first stated it on the basis of his own experiments. A graph of $P$ vs. $V$ for a fixed temperature is shown in Fig. 13–14. Boyle's law can also be written

$$PV = \text{constant}$$ [constant $T$]

for a fixed quantity of a gas kept at constant temperature. If either the pressure or volume of a fixed amount of gas is allowed to vary, the other variable also changes so that the product $PV$ remains constant.

Temperature also affects the volume of a gas, but a quantitative relationship between $V$ and $T$ was not found until more than a century after Boyle's work. The Frenchman Jacques Charles (1746–1823) found that when the pressure is not too high and is kept constant, the volume of a gas increases with temperature at a nearly linear rate, as shown in Fig. 13–15a. However, all gases liquefy at low temperatures (for example, oxygen liquefies at −183°C), so the graph cannot be extended below the liquefaction point. Nonetheless, the graph is essentially a straight line and if projected to lower temperatures, as shown by the dashed line, it crosses the axis at about −273°C.

Such a graph can be drawn for any gas, and a straight line results which always projects back to −273°C at zero volume. This seems to imply that if a gas could be cooled to −273°C, it would have zero volume, and at lower temperatures a negative volume, which makes no sense. It could be argued that −273°C is the lowest temperature possible; indeed, many other more recent experiments indicate that this is so. This temperature is called the **absolute zero** of temperature. Its value has been determined to be −273.15°C.

Absolute zero forms the basis of a temperature scale known as the **absolute scale** or **Kelvin scale**, and it is used extensively in scientific work. On this scale the temperature is specified as degrees Kelvin or, preferably, simply as **kelvins** (K) without the degree sign. The intervals are the same as for the Celsius scale, but the zero on this scale (0 K) is chosen as absolute zero. Thus the freezing point of water (0°C) is 273.15 K, and the boiling point of water is 373.15 K. Indeed, any temperature on the Celsius scale can be changed to kelvins by adding 273.15 to it:

$$T(\text{K}) = T(°\text{C}) + 273.15.$$

Now let us look at Fig. 13–15b, where the graph of the volume of a gas versus absolute temperature is a straight line that passes through the origin. Thus, to a good approximation, *the volume of a fixed quantity of gas is directly proportional to the absolute temperature when the pressure is kept constant.* This is known as **Charles's law**, and is written

$$V \propto T.$$ [constant $P$]

A third gas law, known as **Gay-Lussac's law**, after Joseph Gay-Lussac (1778–1850), states that *at constant volume, the absolute pressure of a fixed quantity of a gas is directly proportional to the absolute temperature*:

$$P \propto T.$$ [constant $V$]

The laws of Boyle, Charles, and Gay-Lussac are not really laws in the sense that we use this term today (precise, deep, wide-ranging validity). They are really only approximations that are accurate for real gases only as long as the pressure and density of the gas are not too high, and the gas is not too close to liquefaction (condensation). The term *law* applied to these three relationships has become traditional, however, so we have stuck with that usage.

| CONCEPTUAL EXAMPLE 13–8 | Why you should not put a closed glass |

**CONCEPTUAL EXAMPLE 13–8** | **Why you should not put a closed glass jar into a campfire.** What could happen if you tossed an empty glass jar, with the lid on tight, into a fire, and why?

**RESPONSE** The inside of the jar is not empty. It is filled with air. As the fire heats the air inside, its temperature rises. The volume of the glass jar changes only slightly due to the heating. According to Gay-Lussac's law the pressure $P$ of the air inside the jar can increase enough to cause the jar to explode, throwing glass pieces outward.

**PHYSICS APPLIED**
*Throwing a jar into a campfire*

# 13–6 The Ideal Gas Law

The gas laws of Boyle, Charles, and Gay-Lussac were obtained by means of an important scientific technique: namely, *considering one quantity and how it is affected by changing only one other variable, keeping all other variables constant.* These laws can now be combined into a single more general relation among all three variables—absolute pressure, volume, and absolute temperature of a fixed amount of gas:

$$PV \propto T.$$

This relation indicates how any of the quantities $P$, $V$, or $T$ will vary when the other two quantities change. This relation reduces to Boyle's, Charles's, or Gay-Lussac's law when either $T$, $P$, or $V$, respectively, is held constant.

Finally, we must incorporate the effect of the amount of gas present. For example, when more air is forced into a balloon, the balloon gets bigger (Fig. 13–16). Indeed, careful experiments show that at constant temperature and pressure, the volume $V$ of an enclosed gas increases in direct proportion to the mass $m$ of gas present. Hence we write

$$PV \propto mT.$$

**FIGURE 13–16** Blowing up a balloon means putting more air (more air molecules) into the balloon, which increases its volume. The pressure is nearly constant, at atmospheric pressure, except for the small effect of the balloon's elasticity.

This proportion can be made into an equation by inserting a constant of proportionality. Experiment shows that this constant has a different value for different gases. However, the constant of proportionality turns out to be the same for all gases if, instead of the mass $m$, we use the number of *moles*.

The "mole" is an official SI unit for the amount of substance. One **mole** (abbreviated mol) is the amount of substance that contains $6.02 \times 10^{23}$ objects (usually atoms, molecules, or ions, etc.). This number is called *Avogadro's number*, as discussed in Section 13–8. Its value comes from measurements. The mole's precise definition is the number of atoms in exactly 12 grams of carbon-12 (page 360).

Equivalently, 1 mol is that amount of substance whose mass in grams is numerically equal to the molecular mass of the substance (Section 13–1). For example, the mass of 1 mole of $CO_2$ is $[12 + (2 \times 16)] = 44$ g because carbon has atomic mass of 12 and oxygen 16 (see Periodic Table inside the rear cover).

In general, the number of moles, $n$, in a given sample of a pure substance is equal to the mass of the sample in grams divided by the molecular mass specified as grams per mole:

$$n \, (\text{mole}) = \frac{\text{mass (grams)}}{\text{molecular mass (g/mol)}}.$$

For example, the number of moles in 132 g of $CO_2$ (molecular mass 44 u) is

$$n = \frac{132 \, \text{g}}{44 \, \text{g/mol}} = 3.0 \, \text{mol}.$$

We can now write the proportion above ($PV \propto mT$) as an equation:

$$PV = nRT, \qquad\qquad (13\text{–}3)$$

where $n$ represents the number of moles and $R$ is the constant of proportionality. $R$ is called the **universal gas constant** because its value is found experimentally to be the same for all gases. The value of $R$, in several sets of units (only the first is the proper SI unit), is

$$R = 8.314 \, \text{J}/(\text{mol} \cdot \text{K}) \qquad\qquad [\text{SI units}]$$
$$= 0.0821 \, (\text{L} \cdot \text{atm})/(\text{mol} \cdot \text{K})$$
$$= 1.99 \, \text{calories}/(\text{mol} \cdot \text{K}).^{\dagger}$$

Equation 13–3 is called the **ideal gas law**, or the **equation of state for an ideal gas**. We use the term "ideal" because real gases do not follow Eq. 13–3 precisely, particularly at high pressure (and density) or when the gas is near the liquefaction point (= boiling point). However, at pressures less than an atmosphere or so, and when $T$ is not close to the liquefaction point of the gas, Eq. 13–3 is quite accurate and useful for real gases.

Always remember, when using the ideal gas law, that temperatures must be given in kelvins (K) and that the pressure $P$ must always be *absolute* pressure, not gauge pressure (Section 10–4).

⚠ **CAUTION**
*Always give T in kelvins and P as absolute (not gauge) pressure*

**EXERCISE A** Return to the Chapter-Opening Question, page 359, and answer it again now. Try to explain why you may have answered differently the first time.

**EXERCISE B** An ideal gas is contained in a steel sphere at 27.0°C and 1.00 atm absolute pressure. If no gas is allowed to escape and the temperature is raised to 127°C, what will be the new pressure? (*a*) 0.21 atm; (*b*) 0.75 atm; (*c*) 1.00 atm; (*d*) 1.33 atm; (*e*) 4.7 atm.

# 13–7 Problem Solving with the Ideal Gas Law

The ideal gas law is an extremely useful tool, and we now consider some Examples. We will often refer to "standard conditions" or **standard temperature and pressure** (**STP**), which means:

$$T = 273 \, \text{K} \;\; (0°\text{C}) \qquad \text{and} \qquad P = 1.00 \, \text{atm} = 1.013 \times 10^5 \, \text{N/m}^2 = 101.3 \, \text{kPa}.$$

**EXAMPLE 13–9** | **Volume of one mole at STP.** Determine the volume of 1.00 mol of any gas, assuming it behaves like an ideal gas, at STP.

**APPROACH** We use the ideal gas law, solving for $V$ with $n = 1.00$ mol.

**SOLUTION** We solve for $V$ in Eq. 13–3:

$$V = \frac{nRT}{P} = \frac{(1.00 \, \text{mol})(8.314 \, \text{J/mol} \cdot \text{K})(273 \, \text{K})}{(1.013 \times 10^5 \, \text{N/m}^2)} = 22.4 \times 10^{-3} \, \text{m}^3.$$

✎ **PROBLEM SOLVING**
*1 mol of gas at STP has V = 22.4 L*

Since 1 liter (L) is $1000 \, \text{cm}^3 = 1.00 \times 10^{-3} \, \text{m}^3$, 1.00 mol of any (ideal) gas has volume $V = 22.4 \, \text{L}$ at STP.

The value of 22.4 L for the volume of 1 mol of an ideal gas at STP is worth remembering, for it sometimes makes calculation simpler.

**EXERCISE C** What is the volume of 1.00 mol of ideal gas at 546 K (= 2 × 273 K) and 2.0 atm absolute pressure? (*a*) 11.2 L; (*b*) 22.4 L; (*c*) 44.8 L; (*d*) 67.2 L; (*e*) 89.6 L.

$^{\dagger}$ Sometimes it is useful to use $R$ as given in terms of calories; calories will be defined in Section 14–1.

**EXAMPLE 13–10** **Helium balloon.** A helium party balloon, assumed to be a perfect sphere, has a radius of 18.0 cm. At room temperature (20°C), its internal pressure is 1.05 atm. Find the number of moles of helium in the balloon and the mass of helium needed to inflate the balloon to these values.

**APPROACH** We can use the ideal gas law to find $n$, since we are given $P$ and $T$, and can find $V$ from the given radius.

**SOLUTION** We get the volume $V$ from the formula for a sphere:

$$V = \tfrac{4}{3}\pi r^3$$
$$= \tfrac{4}{3}\pi (0.180 \text{ m})^3 = 0.0244 \text{ m}^3.$$

The pressure is given as $1.05 \text{ atm} = 1.064 \times 10^5 \text{ N/m}^2$. The temperature must be expressed in kelvins, so we change 20°C to $(20 + 273)\text{K} = 293 \text{ K}$. Finally, we use the value $R = 8.314 \text{ J/(mol·K)}$ because we are using SI units. Thus

$$n = \frac{PV}{RT} = \frac{(1.064 \times 10^5 \text{ N/m}^2)(0.0244 \text{ m}^3)}{(8.314 \text{ J/mol·K})(293 \text{ K})} = 1.066 \text{ mol}.$$

The mass of helium (atomic mass = 4.00 g/mol as given in the Periodic Table or Appendix B) can be obtained from

$$\text{mass} = n \times \text{molecular mass} = (1.066 \text{ mol})(4.00 \text{ g/mol}) = 4.26 \text{ g}$$

or $4.26 \times 10^{-3}$ kg.

**EXAMPLE 13–11** **ESTIMATE** **Mass of air in a room.** Estimate the mass of air in a room whose dimensions are 5 m × 3 m × 2.5 m high, at STP.

**APPROACH** First we determine the number of moles $n$ using the given volume. Then we can multiply by the mass of one mole to get the total mass.

**SOLUTION** Example 13–9 told us that 1 mol of a gas at 0°C has a volume of $22.4 \text{ L} = 22.4 \times 10^{-3} \text{ m}^3$. The room's volume is 5 m × 3 m × 2.5 m, so

$$n = \frac{(5 \text{ m})(3 \text{ m})(2.5 \text{ m})}{22.4 \times 10^{-3} \text{ m}^3/\text{mol}} \approx 1700 \text{ mol}.$$

Air is a mixture of about 20% oxygen $(O_2)$ and 80% nitrogen $(N_2)$. The molecular masses are $2 \times 16 \text{ u} = 32 \text{ u}$ and $2 \times 14 \text{ u} = 28 \text{ u}$, respectively, for an average of about 29 u. Thus, 1 mol of air has a mass of about $29 \text{ g} = 0.029 \text{ kg}$, so our room has a mass of air

$$m \approx (1700 \text{ mol})(0.029 \text{ kg/mol}) \approx 50 \text{ kg}.$$

**NOTE** That is roughly 100 lb of air!

**PHYSICS APPLIED**
*The mass of the air
in a room is significant*

**EXERCISE D** At 20°C, would there be (*a*) more, (*b*) less, or (*c*) the same mass of air in a room than at 0°C?

Frequently, volume is specified in liters and pressure in atmospheres. Rather than convert these to SI units, we can instead use the value of $R$ given in Section 13–6 as $0.0821 \text{ L·atm/mol·K}$.

In many situations it is not necessary to use the value of $R$ at all. For example, many problems involve a change in the pressure, temperature, and volume of a fixed amount of gas. In this case, $PV/T = nR = $ constant, since $n$ and $R$ remain constant. If we now let $P_1$, $V_1$, and $T_1$ represent the appropriate variables initially, and $P_2$, $V_2$, $T_2$ represent the variables after the change is made, then we can write

$$\frac{P_1 V_1}{T_1} = \frac{P_2 V_2}{T_2}.$$    [fixed $n$]

**PROBLEM SOLVING**
*Using the ideal gas law as a ratio*

If we know any five of the quantities in this equation, we can solve for the sixth. Or, if one of the three variables is constant ($V_1 = V_2$, or $P_1 = P_2$, or $T_1 = T_2$) then we can use this equation to solve for one unknown when given the other three quantities.

**FIGURE 13–17** Example 13–12.

**EXAMPLE 13–12** **Check tires cold.** An automobile tire is filled (Fig. 13–17) to a gauge pressure of 210 kPa (= 30 psi) at 10°C. After a drive of 100 km, the temperature within the tire rises to 40°C. What is the pressure within the tire now?

**APPROACH** We do not know the number of moles of gas, or the volume of the tire, but we assume they are constant. We use the ratio form of the ideal gas law.

**SOLUTION** Since $V_1 = V_2$, then

$$\frac{P_1}{T_1} = \frac{P_2}{T_2}.$$

This is, incidentally, a statement of Gay-Lussac's law. Since the pressure given is the gauge pressure (Section 10–4), we must add atmospheric pressure (= 101 kPa) to get the absolute pressure $P_1 = (210\,\text{kPa} + 101\,\text{kPa}) = 311\,\text{kPa}$. We convert temperatures to kelvins by adding 273 and solve for $P_2$:

$$P_2 = P_1\left(\frac{T_2}{T_1}\right) = (3.11 \times 10^5\,\text{Pa})\left(\frac{313\,\text{K}}{283\,\text{K}}\right) = 344\,\text{kPa}.$$

Subtracting atmospheric pressure, we find the resulting gauge pressure to be 243 kPa, which is a 16% increase (= 35 psi).

**NOTE** This Example shows why car manuals emphasize checking tire pressure when the tires are cold.

# 13–8 Ideal Gas Law in Terms of Molecules: Avogadro's Number

The fact that the gas constant, $R$, has the same value for all gases is a remarkable reflection of simplicity in nature. It was first recognized, although in a slightly different form, by the Italian scientist Amedeo Avogadro (1776–1856). Avogadro stated that *equal volumes of gas at the same pressure and temperature contain equal numbers of molecules.* This is sometimes called **Avogadro's hypothesis**. That this is consistent with $R$ being the same for all gases can be seen as follows. From Eq. 13–3, $PV = nRT$, we see that for the same number of moles, $n$, and the same pressure and temperature, the volume will be the same for all gases as long as $R$ is the same. Second, the number of molecules in 1 mole is the same for all gases (see page 369). Thus Avogadro's hypothesis is equivalent to $R$ being the same for all gases.

The number of molecules in one mole of any pure substance is known as **Avogadro's number**, $N_A$. Although Avogadro conceived the notion, he was not able to actually determine the value of $N_A$. Indeed, precise measurements were not done until the twentieth century.

A number of methods have been devised to measure $N_A$, and the accepted value today is (see inside front cover for more precise value)

*Avogadro's number*

$$N_A = 6.02 \times 10^{23}. \qquad\qquad\text{[molecules/mole]}$$

Since the total number of molecules, $N$, in a gas is equal to $N_A$ times the number of moles $(N = nN_A)$, then the ideal gas law, Eq. 13–3, can be written in terms of the number of molecules present:

$$PV = nRT = \frac{N}{N_A}RT,$$

or

*IDEAL GAS LAW (in terms of molecules)*

$$PV = NkT, \qquad\qquad\qquad (13\text{–}4)$$

where $k = R/N_A$ is called the **Boltzmann constant** and has the value

$$k = \frac{R}{N_A} = \frac{8.314\,\text{J/mol}\cdot\text{K}}{6.02 \times 10^{23}/\text{mol}} = 1.38 \times 10^{-23}\,\text{J/K}.$$

**EXAMPLE 13–13** | **Hydrogen atom mass.** Use Avogadro's number to determine the mass of a hydrogen atom.

**APPROACH** The mass of one atom equals the mass of 1 mol divided by the number of atoms in 1 mol, $N_A$.

**SOLUTION** One mole of hydrogen atoms (atomic mass = 1.008 u, Section 13–1 or Appendix B) has a mass of $1.008 \times 10^{-3}$ kg and contains $6.02 \times 10^{23}$ atoms. Thus one atom has a mass

$$m = \frac{1.008 \times 10^{-3} \text{ kg}}{6.02 \times 10^{23}} = 1.67 \times 10^{-27} \text{ kg}.$$

**NOTE** Historically, the reverse was done: a precise value of $N_A$ was obtained from a precise measurement of the mass of the hydrogen atom.

**EXAMPLE 13–14** | **ESTIMATE** | **How many molecules in one breath?** Estimate how many molecules you breathe in with a 1.0-L breath of air.

**APPROACH** We determine what fraction of a mole 1.0 L is using the result of Example 13–9 that 1 mole has a volume of 22.4 L at STP, and then multiply that by $N_A$ to get the number of molecules in this number of moles.

**SOLUTION** One mole corresponds to 22.4 L at STP, so 1.0 L of air is $(1.0 \text{ L})/(22.4 \text{ L/mol}) = 0.045$ mol. Then 1.0 L of air contains

$$(0.045 \text{ mol})(6.02 \times 10^{23} \text{ molecules/mole}) \approx 3 \times 10^{22} \text{ molecules}.$$

**PHYSICS APPLIED**
*Molecules in a breath*

# 13–9 Kinetic Theory and the Molecular Interpretation of Temperature

The analysis of matter in terms of atoms in continuous random motion is called **kinetic theory**. We now investigate the properties of a gas from the point of view of kinetic theory, which is based on the laws of classical mechanics. But to apply Newton's laws to each one of the vast number of molecules in a gas ($>10^{25}/\text{m}^3$ at STP) is far beyond the capability of any present computer. Instead we take a statistical approach and determine averages of certain quantities, and connect these averages to macroscopic variables. We will demand that our microscopic description correspond to the macroscopic properties of gases; otherwise our theory would be of little value. Most importantly, we will arrive at an important relation between the average kinetic energy of molecules in a gas and the absolute temperature.

We make the following assumptions about the molecules in a gas. These assumptions reflect a simple view of a gas, but nonetheless the results they predict correspond well to the essential features of real gases that are at low pressure and are far from the liquefaction point. Under these conditions real gases follow the ideal gas law quite closely, and indeed the gas we now describe is referred to as an **ideal gas**. The assumptions representing the basic postulates of kinetic theory for an ideal gas are:

1. There are a large number of molecules, $N$, each of mass $m$, moving in random directions with a variety of speeds. This assumption agrees with our observation that a gas fills its container and, in the case of air on Earth, is kept from escaping only by the force of gravity.

2. The molecules are, on average, far apart from one another. That is, their average separation is much greater than the diameter of each molecule.

3. The molecules are assumed to obey the laws of classical mechanics, and are assumed to interact with one another only when they collide. Although molecules exert weak attractive forces on each other between collisions, the potential energy associated with these forces is small compared to the kinetic energy.

4. Collisions with another molecule or the wall of the vessel are assumed to be perfectly elastic, like the collisions of perfectly elastic billiard balls (Chapter 7). We assume the collisions are of very short duration compared to the time between collisions. Then we can ignore the potential energy associated with collisions in comparison to the kinetic energy between collisions.

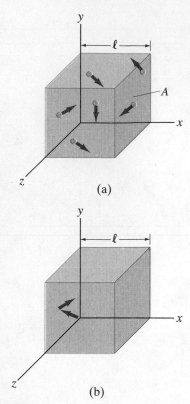

**FIGURE 13–18** (a) Molecules of a gas moving about in a rectangular container. (b) Arrows indicate the momentum of one molecule as it rebounds from the end wall.

We can see how this kinetic view of a gas can explain Boyle's law (Section 13–5). The pressure exerted on a wall of a container of gas is due to the constant bombardment of molecules. If the volume is reduced by (say) half, the molecules are closer together and twice as many will be striking a given area of the wall per second. Hence we expect the pressure to be twice as great, in agreement with Boyle's law.

Now let us calculate quantitatively the pressure a gas exerts on its container in terms of microscopic quantities. We imagine that the molecules are inside a rectangular container (at rest) whose ends have area $A$ and whose length is $\ell$, as shown in Fig. 13–18a. The pressure exerted by the gas on the walls of its container is, according to our model, due to the collisions of the molecules with the walls. Let us focus our attention on the wall, of area $A$, at the left end of the container and examine what happens when one molecule strikes this wall, as shown in Fig. 13–18b. This molecule exerts a force on the wall, and according to Newton's third law the wall exerts an equal and opposite force back on the molecule. The magnitude of this force on the molecule, according to Newton's second law, is equal to the molecule's rate of change of momentum, $F = \Delta(mv)/\Delta t$ (Eq. 7–2). Assuming the collision is elastic, only the $x$ component of the molecule's momentum changes, and it changes from $-mv_x$ (it is moving in the negative $x$ direction) to $+mv_x$. Thus the change in the molecule's momentum, $\Delta(mv)$, which is the final momentum minus the initial momentum, is

$$\Delta(mv) = mv_x - (-mv_x) = 2mv_x$$

for one collision. This molecule will make many collisions with the wall, each separated by a time $\Delta t$, which is the time it takes the molecule to travel across the container and back again, a distance ($x$ component) equal to $2\ell$. Thus $2\ell = v_x \Delta t$, or

$$\Delta t = \frac{2\ell}{v_x}.$$

The time $\Delta t$ between collisions with a wall is very small, so the number of collisions per second is very large. Thus the average force—averaged over many collisions—will be equal to the momentum change during one collision divided by the time between collisions (Newton's second law, Eq. 7–2):

$$F = \frac{\Delta(mv)}{\Delta t} = \frac{2mv_x}{2\ell/v_x} = \frac{mv_x^2}{\ell}. \quad \text{[due to one molecule]}$$

During its passage back and forth across the container, the molecule may collide with the tops and sides of the container, but this does not alter its $x$ component of momentum and thus does not alter our result. [It may also collide with other molecules, which may change its $v_x$. However, any loss (or gain) of momentum is acquired by other molecules, and because we will eventually sum over all the molecules, this effect will be included. So our result above is not altered.]

The actual force due to one molecule is intermittent, but because a huge number of molecules are striking the wall per second, the force is, on average, nearly constant. To calculate the force due to *all* the molecules in the container, we have to add the contributions of each. If all $N$ molecules have the same mass $m$, the net force on the wall is

$$F = \frac{m}{\ell}\left(v_{x1}^2 + v_{x2}^2 + \cdots + v_{xN}^2\right),$$

where $v_{x1}$ means $v_x$ for molecule number 1 (we arbitrarily assign each molecule a number) and the sum extends over the total number of molecules $N$ in the container. The average value of the square of the $x$ component of velocity is

$$\overline{v_x^2} = \frac{v_{x1}^2 + v_{x2}^2 + \cdots + v_{xN}^2}{N}, \tag{13–5}$$

where the overbar ($^-$) means "average." Thus we can write the force as

$$F = \frac{m}{\ell} N \overline{v_x^2}. \tag{i}$$

We know that the square of any vector is equal to the sum of the squares of its components (theorem of Pythagoras). Thus $v^2 = v_x^2 + v_y^2 + v_z^2$ for any velocity $v$.

Taking averages, we obtain

$$\overline{v^2} = \overline{v_x^2} + \overline{v_y^2} + \overline{v_z^2}. \qquad\qquad\text{(ii)}$$

Since the velocities of the molecules in our gas are assumed to be random, there is no preference to one direction or another. Hence

$$\overline{v_x^2} = \overline{v_y^2} = \overline{v_z^2}. \qquad\qquad\text{(iii)}$$

Combining Eqs. (iii) and (ii), we get

$$\overline{v^2} = 3\overline{v_x^2}. \qquad\qquad\text{(iv)}$$

We substitute Eq. (iv) into Eq. (i) for net force $F$ (bottom of previous page):

$$F = \frac{m}{\ell} N \frac{\overline{v^2}}{3}.$$

The pressure on the wall is then

$$P = \frac{F}{A} = \frac{1}{3} \frac{Nm\overline{v^2}}{A\ell}$$

or

$$P = \frac{1}{3} \frac{Nm\overline{v^2}}{V}, \qquad \begin{bmatrix}\text{pressure in an}\\ \text{ideal gas}\end{bmatrix} \quad \text{(13–6)}$$

where $V = \ell A$ is the volume of the container. This is the result we wanted, the pressure exerted by a gas on its container expressed in terms of molecular properties.

Equation 13–6 can be rewritten in a clearer form by multiplying both sides by $V$ and rearranging the right-hand side:

$$PV = \frac{2}{3} N\left(\frac{1}{2} m\overline{v^2}\right). \qquad\qquad\text{(13–7)}$$

The quantity $\frac{1}{2} m\overline{v^2}$ is the average translational kinetic energy ($\overline{KE}$) of the molecules in the gas. If we compare Eq. 13–7 with Eq. 13–4, the ideal gas law $PV = NkT$, we see that the two agree if

$$\frac{2}{3}\left(\frac{1}{2} m\overline{v^2}\right) = kT,$$

or

$$\boxed{\overline{KE} = \frac{1}{2} m\overline{v^2} = \frac{3}{2} kT.} \qquad \text{[ideal gas]} \quad \text{(13–8)}$$

This equation tells us that

> **the average translational kinetic energy of molecules in random motion in an ideal gas is directly proportional to the absolute temperature of the gas.**

The higher the temperature, according to kinetic theory, the faster the molecules are moving on average. This relation is one of the triumphs of kinetic theory.

---

**EXAMPLE 13–15** | **Molecular kinetic energy.** What is the average translational kinetic energy of molecules in an ideal gas at 37°C?

**APPROACH** We use the absolute temperature in Eq. 13–8.

**SOLUTION** We change 37°C to 310 K and insert into Eq. 13–8:

$$\overline{KE} = \frac{3}{2} kT = \frac{3}{2}(1.38 \times 10^{-23}\,\text{J/K})(310\,\text{K}) = 6.42 \times 10^{-21}\,\text{J}.$$

**NOTE** A mole of molecules would have a total translational kinetic energy equal to $(6.42 \times 10^{-21}\,\text{J})(6.02 \times 10^{23}) = 3860\,\text{J}$, which equals the kinetic energy of a 1-kg stone traveling almost 90 m/s.

---

**EXERCISE E** If molecules of hydrogen gas and oxygen gas were placed in the same balloon at room temperature, how would the average kinetic energies of the molecules compare? (*a*) They would be the same. (*b*) The hydrogen molecules would have greater kinetic energy. (*c*) The oxygen molecules would have greater kinetic energy. (*d*) Need more information.

Equation 13–8 holds not only for gases, but also applies reasonably accurately to liquids and solids. Thus the result of Example 13–15 would apply to molecules within living cells at body temperature (37°C).

We can use Eq. 13–8 to calculate how fast molecules are moving on average. Notice that the average in Eqs. 13–5 through 13–8 is over the *square* of the speed. The square root of $\overline{v^2}$ is called the **root-mean-square** speed, $v_{rms}$ (since we are taking the square *root* of the *mean* of the *square* of the speed):

$$v_{rms} = \sqrt{\overline{v^2}} = \sqrt{\frac{3kT}{m}}. \qquad \textbf{(13–9)}$$

---

**EXAMPLE 13–16** **Speeds of air molecules.** What is the rms speed of air molecules ($O_2$ and $N_2$) at room temperature (20°C)?

**APPROACH** To obtain $v_{rms}$, we need the masses of $O_2$ and $N_2$ molecules and then apply Eq. 13–9 to oxygen and nitrogen separately, since they have different masses.

**SOLUTION** The masses of one molecule of $O_2$ (molecular mass = 32 u) and $N_2$ (molecular mass = 28 u) are (where 1 u = $1.66 \times 10^{-27}$ kg)

$$m(O_2) = (32)(1.66 \times 10^{-27}\,\text{kg}) = 5.3 \times 10^{-26}\,\text{kg},$$
$$m(N_2) = (28)(1.66 \times 10^{-27}\,\text{kg}) = 4.6 \times 10^{-26}\,\text{kg}.$$

Thus, for oxygen

$$v_{rms} = \sqrt{\frac{3kT}{m}} = \sqrt{\frac{(3)(1.38 \times 10^{-23}\,\text{J/K})(293\,\text{K})}{(5.3 \times 10^{-26}\,\text{kg})}} = 480\,\text{m/s},$$

and for nitrogen the result is $v_{rms} = 510$ m/s.

**NOTE** These speeds are more than 1700 km/h or 1000 mi/h, and are greater than the speed of sound, $\approx 340$ m/s at 20°C (Chapter 12).

---

**EXERCISE F** By what factor must the absolute temperature change to double $v_{rms}$? (*a*) $\sqrt{2}$; (*b*) 2; (*c*) $2\sqrt{2}$; (*d*) 4; (*e*) 16.

## *Kinetic Energy Near Absolute Zero

Equation 13–8, $\overline{KE} = \frac{3}{2}kT$, implies that as the temperature approaches absolute zero, the kinetic energy of molecules approaches zero. Modern quantum theory, however, tells us this is not quite so. Instead, as absolute zero is approached, the kinetic energy approaches a very small nonzero minimum value. Even though all real gases become liquid or solid near 0 K, molecular motion does not cease, even at absolute zero.

# 13–10 Distribution of Molecular Speeds

FIGURE 13–19 Distribution of speeds of molecules in an ideal gas. Note that $v_{rms}$ is not at the peak of the curve (that speed is called the "most probable speed," $v_P$). This is because the curve is skewed to the right: it is not symmetrical.

The molecules in a gas are assumed to be in random motion, which means that many molecules have speeds less than the rms speed and others have greater speeds. In 1859, James Clerk Maxwell (1831–1879) derived, on the basis of kinetic theory, that the speeds of molecules in a gas are distributed according to the graph shown in Fig. 13–19. This is known as the **Maxwell distribution of speeds**.[†] The speeds vary from zero to many times the rms speed, but as the graph shows, most molecules have speeds that are not far from the average. Less than 1% of the molecules exceed four times $v_{rms}$.

Experiments to determine the distribution of molecular speeds in real gases, starting in the 1920s, confirmed with considerable accuracy the Maxwell distribution and the direct proportion between average kinetic energy and absolute temperature, Eq. 13–8.

[†] Mathematically, the distribution is given by $\Delta N = Cv^2 \exp(-\frac{1}{2}mv^2/kT)\Delta v$, where $\Delta N$ is the number of molecules with speed between $v$ and $v + \Delta v$, $C$ is a constant, and exp means the expression in parentheses is an exponent on the "natural number" $e = 2.718\ldots$.

Figure 13–20 shows the Maxwell distribution for two different temperatures. Just as $v_{rms}$ increases with temperature, so the whole distribution curve shifts to the right at higher temperatures. Kinetic theory can be applied approximately to liquids and solutions. Figure 13–20 illustrates how kinetic theory can explain why many chemical reactions, including those in biological cells, take place more rapidly as the temperature increases. Most chemical reactions occur in a liquid solution, and the molecules have a speed distribution close to the Maxwell distribution. Two molecules may chemically react only if their kinetic energy is above some particular minimum value (called the *activation energy*), $E_A$, so that when they collide, they penetrate into each other somewhat. Figure 13–20 shows that at a higher temperature, many more molecules have a speed and kinetic energy KE above the needed threshold $E_A$.

**FIGURE 13–20** Distribution of molecular speeds for two different temperatures. Color shading shows proportions of molecules above a certain speed (corresponding to an activation energy $E_A = \frac{1}{2}mv^2$).

# 13–11 Real Gases and Changes of Phase

The ideal gas law, $PV = NkT$, is an accurate description of the behavior of a real gas as long as the pressure is not too high and the temperature is far from the liquefaction point. But what happens when these two criteria are not satisfied? First we discuss real gas behavior, and then we examine how kinetic theory can help us understand this behavior.

Let us look at a graph of pressure plotted against volume for a given amount of gas. On such a **PV diagram**, Fig. 13–21, each point represents the pressure and volume of an equilibrium state of the given substance. The various curves (labeled A, B, C, and D) show how the pressure varies as a function of volume for four different values of constant temperature $T_A$, $T_B$, $T_C$, and $T_D$. The red dashed curve A′ represents the behavior of a gas as predicted by the ideal gas law; that is, $PV$ = constant. The solid curve A represents the behavior of a real gas at the same temperature. Notice that at high pressure, the volume of a real gas is less than that predicted by the ideal gas law. The curves B and C in Fig. 13–21 represent the gas at successively lower temperatures, and we see that the behavior deviates even more from the curves predicted by the ideal gas law (for example, B′), and the deviation is greater the closer the gas is to liquefying.

To explain this behavior, note that at higher pressure we expect the molecules to be closer together. And at lower temperatures, the potential energy associated with attractive forces between the molecules (which we ignored before) is no longer negligible. These attractive forces tend to pull the molecules closer together so the volume is less than expected from the ideal gas law. At still lower temperatures, these forces cause liquefaction, and the molecules become very close together.

Curve D represents the situation when liquefaction occurs. At low pressure on curve D (on the right in Fig. 13–21), the substance is a gas and occupies a large volume. As the pressure is increased, the volume decreases until point b is reached. From point b to point a, the volume decreases with no change in pressure; the substance is gradually changing from the gas to the liquid phase. At point a, all of the substance has changed to liquid. Further increase in pressure reduces the volume only slightly—liquids are nearly incompressible—so on the left the curve is very steep as shown. The shaded area under the gold dashed line represents the region where the gas and liquid phases exist together in equilibrium.

Curve C in Fig. 13–21 represents the behavior of the substance at its **critical temperature**; the point c (the one point where curve C is horizontal) is called the **critical point**. At temperatures less than the critical temperature, a gas will change to the liquid phase if sufficient pressure is applied. Above the critical temperature (and this is the definition of the term), no amount of pressure can cause a gas to change phase and become a liquid. (Thus curves A and B represent the substance at temperatures where it can only be a gas.) The critical temperatures for various gases are given in Table 13–2. Scientists tried for many years to liquefy oxygen without success, which led to the idea that there must be a critical point. Oxygen can be liquefied only if first cooled below its critical temperature of −118°C.

**FIGURE 13–21** *PV* diagram for a real substance. Curves A, B, C, and D represent the same substance at different temperatures ($T_A > T_B > T_C > T_D$).

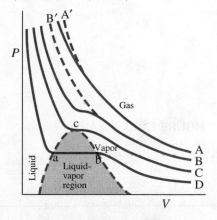

**TABLE 13–2 Critical Temperatures and Pressures**

| Substance | Critical Temperature °C | K | Critical Pressure (atm) |
|---|---|---|---|
| Water | 374 | 647 | 218 |
| $CO_2$ | 31 | 304 | 72.8 |
| Oxygen | −118 | 155 | 50 |
| Nitrogen | −147 | 126 | 33.5 |
| Hydrogen | −239.9 | 33.3 | 12.8 |
| Helium | −267.9 | 5.3 | 2.3 |

Often a distinction is made between the terms "gas" and "vapor": a substance below its critical temperature in the gaseous state is called a **vapor**; above the critical temperature, it is called a **gas**.

The behavior of a substance can be diagrammed not only on a *PV* diagram but also on a *PT* diagram. A *PT* **diagram**, often called a **phase diagram**, is particularly convenient for comparing the different phases of a substance. Figure 13–22 is the phase diagram for water. The curve labeled *ℓ-v* represents those points where the liquid and vapor phases are in equilibrium—it is thus a graph of the boiling point versus pressure. Note that the curve correctly shows that at a pressure of 1 atm the boiling point is 100°C and that the boiling point is lowered for a decreased pressure. The curve *s-ℓ* represents points where solid and liquid exist in equilibrium and thus is a graph of the freezing point versus pressure.

**FIGURE 13–22** Phase diagram for water (note that the scales are not linear).

At 1 atm, the freezing point of water is 0°C, as shown. Notice also in Fig. 13–22 that at a pressure of 1 atm, the substance is in the liquid phase if the temperature is between 0°C and 100°C, but is in the solid or vapor phase if the temperature is below 0°C or above 100°C. The curve labeled *s-v* is the *sublimation point* versus pressure curve. **Sublimation** refers to the process whereby at low pressures a solid changes directly into the vapor phase without passing through the liquid phase. For water, sublimation occurs if the pressure of the water vapor is less than 0.0060 atm. Carbon dioxide, which in the solid phase is called dry ice, sublimates even at atmospheric pressure (Fig. 13–23).

**FIGURE 13–23** Phase diagram for carbon dioxide.

The intersection of the three curves (in Fig. 13–22) is the **triple point**. For water this occurs at $T = 273.16$ K and $P = 6.03 \times 10^{-3}$ atm. It is only at the triple point that the three phases can exist together in equilibrium. Because the triple point corresponds to a unique value of temperature and pressure, it is precisely reproducible and is often used as a point of reference. For example, the standard of temperature is usually specified as exactly 273.16 K at the triple point of water, rather than 273.15 K at the freezing point of water at 1 atm.

Notice that the solid-liquid (*s-ℓ*) curve for water (Fig. 13–22) slopes upward to the left. This is true only of substances that *expand* upon freezing: at a higher pressure, a lower temperature is needed to cause the liquid to freeze. More commonly, substances contract upon freezing and the *s-ℓ* curve slopes upward to the right, as shown for carbon dioxide $(CO_2)$ in Fig. 13–23.

The phase transitions we have been discussing are the common ones. Some substances, however, can exist in several forms in the solid phase. A transition from one phase to another occurs at a particular temperature and pressure, just like ordinary phase changes. For example, ice has been observed in at least eight forms at very high pressure. Ordinary helium has two distinct liquid phases, called helium I and II. They exist only at temperatures within a few degrees of absolute zero. Helium II exhibits very unusual properties referred to as **superfluidity**. It has essentially zero viscosity and exhibits strange properties such as climbing up the sides of an open container. Also interesting are **liquid crystals** (used for computer and TV monitors, Section 24–11) which can be considered to be in a phase between liquid and solid.

# 13–12 Vapor Pressure and Humidity

## Evaporation

If a glass of water is left out overnight, the water level will have dropped by morning. We say the water has evaporated, meaning that some of the water has changed to the vapor or gas phase.

This process of **evaporation** can be explained on the basis of kinetic theory. The molecules in a liquid move past one another with a variety of speeds that follow, approximately, the Maxwell distribution. There are strong attractive forces between these molecules, which is what keeps them close together in the liquid phase. A molecule near the surface of the liquid may, because of its speed, leave the liquid momentarily. But just as a rock thrown into the air returns to the Earth, so the attractive forces of the other molecules can pull the vagabond molecule back to the liquid surface—that is, if its velocity is not too large. A molecule with a high enough velocity, however, will escape the liquid entirely, like a rocket escaping the Earth, and become part of the gas phase. Only those molecules that have kinetic energy above a particular value can escape to the gas phase. We have already seen that kinetic theory predicts that the relative number of molecules with kinetic energy above a particular value (such as $E_A$ in Fig. 13–20) increases with temperature. This is in accord with the well-known observation that the evaporation rate is greater at higher temperatures.

Because it is the fastest molecules that escape from the surface, the average speed of those remaining is less. When the average speed is less, the absolute temperature is less. Thus kinetic theory predicts that *evaporation is a cooling process*. You may have noticed this effect when you stepped out of a warm shower and felt cold as the water on your body began to evaporate; and after working up a sweat on a hot day, even a slight breeze makes you feel cool through evaporation. Try licking your finger and then blow on it.

## Vapor Pressure

Air normally contains water vapor (water in the gas phase), and it comes mainly from evaporation. To look at this process in a little more detail, consider a closed container that is partially filled with water (or another liquid) and from which the air has been removed (Fig. 13–24). The fastest moving molecules quickly evaporate into the empty space above the liquid's surface. As they move about, some of these molecules strike the liquid surface and again become part of the liquid phase: this is called **condensation**. The number of molecules in the vapor increases until the number of molecules returning to the liquid equals the number leaving in the same time interval. Equilibrium then exists, and the space above the liquid surface is said to be *saturated*. The pressure of the vapor when it is saturated is called the **saturated vapor pressure** (or simply the vapor pressure).

The saturated vapor pressure of any substance depends on the temperature. At higher temperatures, more molecules have sufficient kinetic energy to break from the liquid surface into the vapor phase. Hence equilibrium will be reached at a higher vapor pressure. The saturated vapor pressure of water at various temperatures is given in Table 13–3. Notice that even solids—for example, ice—have a measurable saturated vapor pressure.

In everyday situations, evaporation from a liquid takes place into the air above it rather than into a vacuum. This does not materially alter the discussion above relating to Fig. 13–24. Equilibrium will still be reached when there are sufficient molecules in the gas phase that the number reentering the liquid equals the number leaving. The concentration of particular molecules (such as water) in the gas phase is not affected by the presence of air, although collisions with air molecules may lengthen the time needed to reach equilibrium. Thus equilibrium occurs at the same value of the saturated vapor pressure as if air were not there.

If the container is large or is not closed, all the liquid may evaporate before saturation is reached. And if the container is not sealed—as, for example, a room in your house—it is not likely that the air will become saturated with water vapor (unless it is raining outside).

**FIGURE 13–24** Vapor appears above a liquid in a closed container.

**PHYSICS APPLIED**
*Evaporation cools*

**TABLE 13–3 Saturated Vapor Pressure of Water**

| Temp-erature (°C) | Saturated Vapor Pressure | |
|---|---|---|
| | torr (= mm-Hg) | Pa (= N/m²) |
| −50 | 0.030 | 4.0 |
| −10 | 1.95 | $2.60 \times 10^2$ |
| 0 | 4.58 | $6.11 \times 10^2$ |
| 5 | 6.54 | $8.72 \times 10^2$ |
| 10 | 9.21 | $1.23 \times 10^3$ |
| 15 | 12.8 | $1.71 \times 10^3$ |
| 20 | 17.5 | $2.33 \times 10^3$ |
| 25 | 23.8 | $3.17 \times 10^3$ |
| 30 | 31.8 | $4.24 \times 10^3$ |
| 40 | 55.3 | $7.37 \times 10^3$ |
| 50 | 92.5 | $1.23 \times 10^4$ |
| 60 | 149 | $1.99 \times 10^4$ |
| 70[†] | 234 | $3.12 \times 10^4$ |
| 80 | 355 | $4.73 \times 10^4$ |
| 90 | 526 | $7.01 \times 10^4$ |
| 100[‡] | 760 | $1.01 \times 10^5$ |
| 120 | 1489 | $1.99 \times 10^5$ |
| 150 | 3570 | $4.76 \times 10^5$ |

[†] Boiling point on summit of Mt. Everest.
[‡] Boiling point at sea level.

## Boiling

The saturated vapor pressure of a liquid increases with temperature. When the temperature is raised to the point where the saturated vapor pressure at that temperature equals the external pressure, **boiling** occurs (Fig. 13–25). As the boiling point is approached, tiny bubbles tend to form in the liquid, which indicate a change from the liquid to the gas phase. However, if the vapor pressure inside the bubbles is less than the external pressure, the bubbles immediately are crushed. As the temperature is increased, the saturated vapor pressure inside a bubble eventually becomes equal to or exceeds the external air pressure. The bubble will then not collapse but can rise to the surface. Boiling has then begun. *A liquid boils when its saturated vapor pressure equals the external pressure.* This occurs for water at a pressure of 1 atm (760 torr) at 100°C, as can be seen from Table 13–3.

The boiling point of a liquid depends on the external pressure. At high elevations, the boiling point of water is somewhat less than at sea level because the air pressure is less up there. For example, on the summit of Mt. Everest (8850 m) the air pressure is about one-third of what it is at sea level, and from Table 13–3 we can see that water will boil at about 70°C. Cooking food by boiling takes longer at high elevations, because the boiling water is cooking at a lower temperature. Pressure cookers reduce cooking time because they build up a pressure as high as 2 atm, allowing a higher boiling (and cooking) temperature to be attained (Problem 64 and Fig. 13–32).

## Partial Pressure and Humidity

When we refer to the weather as being dry or humid, we are referring to the water vapor content of the air. In a gas such as air, which is a mixture of several types of gases, the total pressure is the sum of the *partial pressures* of each gas present.[†] By **partial pressure**, we mean the pressure each gas would exert if it alone were present. The partial pressure of water in the air can be as low as zero and can vary up to a maximum equal to the saturated vapor pressure of water at the given temperature. Thus, at 20°C, the partial pressure of water cannot exceed 17.5 torr (see Table 13–3) or about 0.02 atm. The **relative humidity** is defined as the ratio of the partial pressure of water vapor to the saturated vapor pressure at a given temperature. It is usually expressed as a percentage:

$$\text{Relative humidity} = \frac{\text{partial pressure of } H_2O}{\text{saturated vapor pressure of } H_2O} \times 100\%.$$

Thus, when the humidity is close to 100%, the air holds nearly all the water vapor it can.

**EXAMPLE 13–17** **Relative humidity.** On a particular hot day, the temperature is 30°C and the partial pressure of water vapor in the air is 21.0 torr. What is the relative humidity?

**APPROACH** From Table 13–3, we see that the saturated vapor pressure of water at 30°C is 31.8 torr.

**SOLUTION** The relative humidity is thus

$$\frac{21.0 \text{ torr}}{31.8 \text{ torr}} \times 100\% = 66\%.$$

Humans are sensitive to humidity. A relative humidity of 40–50% is generally optimum for both health and comfort. High humidity, particularly on a hot day, reduces the evaporation of moisture from the skin, which is one of the body's vital mechanisms for regulating body temperature. Very low humidity, on the other hand, can dry the skin and mucous membranes.

[†] For example, 78% (by volume) of air molecules are nitrogen and 21% oxygen, with much smaller amounts of water vapor, argon, carbon dioxide, and other gases. At an air pressure of 1 atm, oxygen exerts a partial pressure of 0.21 atm and nitrogen 0.78 atm.

**FIGURE 13–25** Boiling: bubbles of water vapor float upward from the bottom (where the temperature is highest).

**PHYSICS APPLIED**
*Humidity and comfort*

Air is saturated with water vapor when the partial pressure of water in the air is equal to the saturated vapor pressure at that temperature. If the partial pressure of water exceeds the saturated vapor pressure, the air is said to be **supersaturated**. This situation can occur when a temperature decrease occurs. For example, suppose the temperature is 30°C and the partial pressure of water is 21 torr, which represents a humidity of 66% as we saw in Example 13–17. Suppose now that the temperature falls to, say, 20°C, as might happen at nightfall. From Table 13–3 we see that the saturated vapor pressure of water at 20°C is 17.5 torr. Hence the relative humidity would be greater than 100%, and the supersaturated air cannot hold this much water vapor. The excess water may condense and appear as dew, clouds, or as fog or rain (Fig. 13–26).

When air containing a given amount of water is cooled, a temperature is reached where the partial pressure of water equals the saturated vapor pressure. This is called the **dew point**. Measurement of the dew point is the most accurate means of determining the relative humidity. One method uses a polished metal surface which is gradually cooled down while in contact with air. The temperature at which moisture begins to appear on the surface is the dew point, and the partial pressure of water can then be obtained from saturated vapor pressure Tables. If, for example, on a given day the temperature is 20°C and the dew point is 5°C, then the partial pressure of water (Table 13–3) in the 20°C air is 6.54 torr, whereas its saturated vapor pressure is 17.5 torr; hence the relative humidity is 6.54/17.5 = 37%.

(a)

(b)

(c)

**FIGURE 13–26** (*a*) Fog or mist settling in a valley where the temperature has dropped below the dew point. (*b*) Dew drops on a leaf. (*c*) Clouds form on a sunny day at the beach due to air, nearly saturated with water vapor, rising to an altitude where the cooler temperature is at the dew point.

---

| CONCEPTUAL EXAMPLE 13–18 | **Dryness in winter.** Why does the air inside heated buildings seem very dry on a cold winter day?

**RESPONSE** Suppose the relative humidity outside on a −10°C day is 50%. Table 13–3 tells us the partial pressure of water in the air is about 1.0 torr. If this air is brought indoors and heated to +20°C, the relative humidity is (1.0 torr)/(17.5 torr) = 5.7%. Even if the outside air were saturated at a partial pressure of 1.95 torr, the inside relative humidity would still be at a low 11%.

# *13–13 Diffusion

If you carefully place a few drops of food coloring in a glass of water as in Fig. 13–27, you will find that the color spreads throughout the water. The process may take some time (assuming you do not shake the glass), but eventually the color will become uniform. This mixing, known as **diffusion**, is readily explained by kinetic theory as due to the random movement of the molecules. Diffusion occurs in gases too. Common examples include perfume or smoke (or the odor of something cooking on a stove) diffusing in air, although convection (moving air currents) often plays a greater role in spreading odors than does diffusion. Diffusion depends on concentration, by which we mean the number of molecules or moles per unit volume. In general, *the diffusing substance moves from a region where its concentration is high to a region where its concentration is low.*

(a)

(b)

(c)

**FIGURE 13–27** A few drops of food coloring (a) dropped into water, (b) spreads slowly throughout the water, eventually (c) becoming uniform.

Region 1; | $\leftarrow\!\Delta x\!\rightarrow$ | Region 2;
concentration | | concentration
$= C_1$ | | $= C_2$

**FIGURE 13–28** Diffusion occurs from a region of high concentration to one of lower concentration (only one type of molecule is shown).

### TABLE 13–4 Diffusion Constants, $D$ (20°C, 1 atm)

| Diffusing Molecules | Medium | $D$ (m²/s) |
|---|---|---|
| $H_2$ | Air | $6.3 \times 10^{-5}$ |
| $O_2$ | Air | $1.8 \times 10^{-5}$ |
| $O_2$ | Water | $100 \times 10^{-11}$ |
| Glycine (an amino acid) | Water | $95 \times 10^{-11}$ |
| Blood hemoglobin | Water | $6.9 \times 10^{-11}$ |
| DNA (mass $6 \times 10^6$ u) | Water | $0.13 \times 10^{-11}$ |

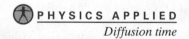

PHYSICS APPLIED

*Diffusion time*

Diffusion can be readily understood on the basis of kinetic theory and the random motion of molecules. Consider a tube of cross-sectional area $A$ containing molecules in a higher concentration on the left than on the right, Fig. 13–28. We assume the molecules are in random motion. Yet there will be a net flow of molecules to the right. To see why, let us consider the small section of tube of length $\Delta x$ as shown. Molecules from both regions 1 and 2 cross into this central section as a result of their random motion. The more molecules there are in a region, the more will strike a given area or cross a boundary. Since there is a greater concentration of molecules in region 1 than in region 2, more molecules cross into the central section from region 1 than from region 2. There is, then, a net flow of molecules from left to right, from high concentration $(C_1)$ toward low concentration $(C_2)$. The net flow becomes zero only when the concentrations become equal.

We might expect that the greater the difference in concentration, the greater the flow rate. Indeed, the rate of diffusion, $J$ (number of molecules or moles or kg per second), is directly proportional to the difference in concentration per unit distance, $(C_1 - C_2)/\Delta x$ (which is called the **concentration gradient**), and to the cross-sectional area $A$ (see Fig. 13–28):

$$J = DA \frac{C_1 - C_2}{\Delta x}. \tag{13–10}$$

$D$ is a constant of proportionality called the **diffusion constant**. Equation 13–10 is known as the **diffusion equation**, or **Fick's law**. If the concentrations are given in mol/m³, then $J$ is the number of moles passing a given point per second. If the concentrations are given in kg/m³, then $J$ is the mass movement per second (kg/s). The length $\Delta x$ is given in meters, and area $A$ in m². The values of $D$ for a variety of substances diffusing in a particular medium are given in Table 13–4.

**EXAMPLE 13–19 | ESTIMATE | Diffusion of ammonia in air.** To get an idea of the time required for diffusion, estimate how long it might take for ammonia $(NH_3)$ to be detected 10 cm from a bottle after it is opened, assuming only diffusion is occurring.

**APPROACH** This will be an order-of-magnitude calculation. The rate of diffusion $J$ can be set equal to the number of molecules $N$ diffusing across area $A$ in a time $t$: $J = N/t$. Then the time $t = N/J$, where $J$ is given by Eq. 13–10. We will have to make some assumptions and rough approximations about concentrations to use Eq. 13–10.

**SOLUTION** Using Eq. 13–10, we find

$$t = \frac{N}{J} = \frac{N}{DA} \frac{\Delta x}{\Delta C}.$$

The average concentration (midway between bottle and nose) can be approximated by $\overline{C} \approx N/V$, where $V$ is the volume over which the molecules move and is roughly on the order of $V \approx A \Delta x$, where $\Delta x$ is 10 cm = 0.10 m. We substitute $N = \overline{C}V = \overline{C}A \Delta x$ into the above equation:

$$t \approx \frac{(\overline{C}A \Delta x) \Delta x}{DA \Delta C} = \frac{\overline{C}}{\Delta C} \frac{(\Delta x)^2}{D}.$$

The concentration of ammonia is high near the bottle $(C)$ and low near the detecting nose $(\approx 0)$, so $\overline{C} \approx C/2 \approx \Delta C/2$, or $(\overline{C}/\Delta C) \approx \frac{1}{2}$. Since $NH_3$ molecules have a size somewhere between $H_2$ and $O_2$, from Table 13–4 we can estimate $D \approx 4 \times 10^{-5}$ m²/s. Then

$$t \approx \frac{1}{2} \frac{(0.10 \text{ m})^2}{(4 \times 10^{-5} \text{ m}^2/\text{s})} \approx 100 \text{ s},$$

or about a minute or two.

**NOTE** This result seems rather long from experience, suggesting that air currents (convection) are more important than diffusion for transmitting odors.

Diffusion is extremely important for living organisms. For example, molecules produced in certain chemical reactions within cells diffuse to other areas where they take part in other reactions.

Gas diffusion is important too. Plants require carbon dioxide for photosynthesis. The $CO_2$ diffuses into leaves from the outside air through tiny openings (stomata). As $CO_2$ is utilized by the cells, its concentration drops below that in the air outside, and more diffuses inward. Water vapor and oxygen produced by the cells diffuse outward into the air.

Animals also exchange oxygen and $CO_2$ with the environment. Oxygen is required for energy-producing reactions and must diffuse into cells. $CO_2$ is produced as an end product of many metabolic reactions and must diffuse out of cells. But diffusion is slow over longer distances, so only the smallest organisms in the animal world could survive without having developed complex respiratory and circulatory systems. In humans, oxygen is taken into the lungs, where it diffuses short distances across lung tissue and into the blood. Then the blood circulates it to cells throughout the body. The blood also carries $CO_2$ produced by the cells back to the lungs, where it diffuses outward.

# Summary

The atomic theory of matter postulates that all matter is made up of tiny entities called **atoms**, which are typically $10^{-10}$ m in diameter.

**Atomic** and **molecular masses** are specified on a scale where the most common form of carbon ($^{12}C$) is arbitrarily given the value 12.0000 u (atomic mass units), exactly.

The distinction between solids, liquids, and gases can be attributed to the strength of the attractive forces between the atoms or molecules and to their average speed.

**Temperature** is a measure of how hot or cold something is. **Thermometers** are used to measure temperature on the **Celsius** (°C), **Fahrenheit** (°F), and **Kelvin** (K) scales. Two standard points on each scale are the freezing point of water (0°C, 32°F, 273.15 K) and the boiling point of water (100°C, 212°F, 373.15 K). A one-kelvin change in temperature equals a change of one Celsius degree or $\frac{9}{5}$ Fahrenheit degrees. Kelvins are related to °C by

$$T(\text{K}) = T(\text{°C}) + 273.15.$$

When two objects at different temperatures are placed in contact, they eventually reach the same temperature and are then said to be in **thermal equilibrium**.

The change in length, $\Delta\ell$, of a solid, when its temperature changes by an amount $\Delta T$, is directly proportional to the temperature change and to its original length $\ell_0$. That is,

$$\Delta\ell = \alpha\ell_0\,\Delta T, \tag{13-1a}$$

where $\alpha$ is the *coefficient of linear expansion*.

The change in volume of most solids, liquids, and gases is proportional to the temperature change and to the original volume $V_0$:

$$\Delta V = \beta V_0\,\Delta T. \tag{13-2}$$

The *coefficient of volume expansion*, $\beta$, is approximately equal to $3\alpha$ for uniform solids.

Water is unusual because, unlike most materials whose volume increases with temperature, its volume in the range from 0°C to 4°C actually decreases as the temperature increases.

The **ideal gas law**, or **equation of state for an ideal gas**, relates the pressure $P$, volume $V$, and temperature $T$ (in kelvins) of $n$ moles of gas by the equation

$$PV = nRT, \tag{13-3}$$

where $R = 8.314\,\text{J/mol·K}$ for all gases. Real gases obey the ideal gas law quite accurately if they are not at too high a pressure or near their liquefaction point.

One **mole** is that amount of a substance whose mass in grams is numerically equal to the atomic or molecular mass of that substance.

**Avogadro's number**, $N_A = 6.02 \times 10^{23}$, is the number of atoms or molecules in 1 mol of any pure substance.

The ideal gas law can be written in terms of the number of molecules $N$ in the gas as

$$PV = NkT, \tag{13-4}$$

where $k = R/N_A = 1.38 \times 10^{-23}\,\text{J/K}$ is Boltzmann's constant.

According to the **kinetic theory** of gases, which is based on the idea that a gas is made up of molecules that are moving rapidly and randomly, the average translational kinetic energy of molecules is proportional to the Kelvin temperature $T$:

$$\overline{\text{KE}} = \tfrac{1}{2}m\overline{v^2} = \tfrac{3}{2}kT, \tag{13-8}$$

where $k$ is Boltzmann's constant. At any moment, there exists a wide distribution of molecular speeds within a gas.

The behavior of real gases at high pressure, and/or when near their liquefaction point, deviates from the ideal gas law due to the attractive forces between molecules. Below the **critical temperature**, a gas can change to a liquid if sufficient pressure is applied; but if the temperature is higher than the critical temperature, no amount of pressure will cause a liquid surface to form. The **triple point** of a substance is that unique temperature and pressure at which all three phases—solid, liquid, and gas—can coexist in equilibrium.

**Evaporation** of a liquid is the result of the fastest moving molecules escaping from the surface. **Saturated vapor pressure** refers to the pressure of the vapor above a liquid when the two phases are in equilibrium. The vapor pressure of a substance (such as water) depends strongly on temperature, and at the boiling point is equal to atmospheric pressure. **Relative humidity** of air is the ratio of the actual partial pressure of water vapor in the air to the saturated vapor pressure at that temperature; it is usually expressed as a percentage.

[***Diffusion** is the process whereby molecules of a substance move (on average) from one area to another because of a difference in that substance's concentration.]

# Questions

1. Which has more atoms: 1 kg of lead or 1 kg of copper? (See the Periodic Table or Appendix B.) Explain why.

2. Name several properties of materials that could be used to make a thermometer.

3. Which is larger, 1 C° or 1 F°? Explain why.

4. In the relation $\Delta \ell = \alpha \ell_0 \Delta T$, should $\ell_0$ be the initial length, the final length, or does it matter?

5. A flat bimetallic strip consists of a strip of aluminum riveted to a strip of iron. When heated, the strip will bend. Which metal will be on the outside of the curve? Why? [*Hint*: See Table 13–1.]

6. Long steam pipes that are fixed at the ends often have a section in the shape of a ∪. Why?

7. Figure 13–29 shows a diagram of a simple bimetallic **thermostat** used to control a furnace (or other heating or cooling system). The electric switch (attached to the bimetallic strip) is a glass vessel containing liquid mercury that conducts electricity when it touches both contact wires. Explain how this device controls the furnace and how it can be set at different temperatures.

Temperature setting lever

Bimetallic strip

Liquid mercury switch

Liquid mercury

Wires to heater

**FIGURE 13–29** A thermostat (Question 7).

8. A glass container may break if one part of it is heated or cooled more rapidly than adjacent parts. Explain.

9. Explain why it is advisable to add water to an overheated automobile engine only slowly, and only with the engine running.

10. The units for the coefficient of linear expansion $\alpha$ are $(C°)^{-1}$, and there is no mention of a length unit such as meters. Would the expansion coefficient change if we used feet or millimeters instead of meters? Explain.

11. When a cold alcohol-in-glass thermometer is first placed in a hot tub of water, the alcohol initially descends a bit and then rises. Explain.

12. The principal virtue of Pyrex glass is that its coefficient of linear expansion is much smaller than that for ordinary glass (Table 13–1). Explain why this gives rise to the higher heat resistance of Pyrex.

13. Will a clock using a pendulum supported on a long thin brass rod that is accurate at 20°C run fast or slow on a hot day (30°C)? Explain.

14. Freezing a can of soda will cause its bottom and top to bulge so badly the can will not stand up. What has happened?

15. Will the buoyant force on an aluminum sphere submerged in water increase, decrease, or remain the same, if the temperature is increased from 20°C to 40°C? Explain.

16. Can you determine the temperature of a vacuum? Explain.

17. Escape velocity from the Earth refers to the minimum speed an object must have to leave the Earth and never return. (See also Fig 5-22, page 122). (*a*) The escape velocity from the Moon is about one-fifth what it is for the Earth, due to the Moon's smaller mass. Explain why the Moon has practically no atmosphere. (*b*) If hydrogen was once in the Earth's atmosphere, why would it have probably escaped?

18. What exactly does it mean when we say that oxygen boils at −183°C?

19. A length of thin wire is placed over a block of ice (or an ice cube) at 0°C. The wire hangs down both sides of the ice, and weights are hung from the ends of the wire. It is found that the wire cuts its way through the ice cube, but leaves a solid block of ice behind it. This process is called *regelation*. Explain how this happens by inferring how the freezing point of water depends on pressure.

20. (*a*) Why does food cook faster in a pressure cooker? (*b*) Why does pasta or rice need to boil longer at high altitudes? (*c*) Is it harder to boil water at high altitudes?

21. Is it possible to boil water at room temperature (20°C) without heating it? Explain.

22. Why does exhaled air appear as a little white cloud in the winter (Fig. 13–30)?

**FIGURE 13–30** Question 22.

23. Explain why it is dangerous to open the radiator cap of an overheated automobile engine.

# MisConceptual Questions

1. Rod A has twice the diameter of rod B, but both are made of iron and have the same initial length. Both rods are now subjected to the same change in temperature (but remain solid). How would the change in the rods' lengths compare?
(*a*) Rod A > rod B.
(*b*) Rod B > rod A.
(*c*) Rod A = rod B.
(*d*) Need to know whether the rods were cooled or heated.

2. The linear expansion of a material depends on which of the following?
(*a*) The length of the material.
(*b*) The change in temperature of the material.
(*c*) The type of material.
(*d*) All of the above.
(*e*) Both (*b*) and (*c*).

3. A steel plate has a hole in it with a diameter of exactly 1.0 cm when the plate is at a temperature of 20°C. A steel ring has an inner diameter of exactly 1.0 cm at 20°C. Both the plate and the ring are heated to 100°C. Which statement is true?
   (a) The hole in the plate gets smaller, and the opening in the ring gets larger.
   (b) The opening in the ring gets larger, but we need the relative size of the plate and the hole to know what happens to the hole.
   (c) The hole in the plate and the opening in the ring get larger.
   (d) The hole in the plate and the opening in the ring get smaller.
   (e) The hole in the plate gets larger, and the opening in the ring gets smaller.

4. One mole of an ideal gas in a sealed rigid container is initially at a temperature of 100°C. The temperature is then increased to 200°C. The pressure in the gas
   (a) remains constant.
   (b) increases by about 25%.
   (c) doubles.
   (d) triples.

5. When an ideal gas is warmed from 20°C to 40°C, the gas's temperature $T$ that appears in the ideal gas law increases by a factor
   (a) of 2.
   (b) of 1.07.
   (c) that depends on the temperature scale you use.

6. Two identical bottles at the same temperature contain the same gas. If bottle B has twice the volume and contains half the number of moles of gas as bottle A, how does the pressure in B compare with the pressure in A?
   (a) $P_B = \frac{1}{2} P_A$.
   (b) $P_B = 2 P_A$.
   (c) $P_B = \frac{1}{4} P_A$.
   (d) $P_B = 4 P_A$.
   (e) $P_B = P_A$.

7. The temperature of an ideal gas increases. Which of the following is true?
   (a) The pressure must decrease.
   (b) The pressure must increase.
   (c) The pressure must increase while the volume decreases.
   (d) The volume must increase while the pressure decreases.
   (e) The pressure, the volume, or both, may increase.

8. An ideal gas is in a sealed rigid container. The average kinetic energy of the gas molecules depends most on
   (a) the size of the container.
   (b) the number of molecules in the container.
   (c) the temperature of the gas.
   (d) the mass of the molecules.

9. Two ideal gases, A and B, are at the same temperature. If the molecular mass of the molecules in gas A is twice that of the molecules in gas B, the molecules' root-mean-square speed is
   (a) the same in both gases.     (d) twice as great in B.
   (b) twice as great in A.        (e) 1.4 times greater in B.
   (c) 1.4 times greater in A.

10. In a mixture of the gases oxygen and helium, which statement is valid?
    (a) The helium atoms will be moving faster than the oxygen molecules, on average.
    (b) Both will be moving at the same speed.
    (c) The oxygen molecules will, on average, be moving more rapidly than the helium atoms.
    (d) The kinetic energy of helium atoms will exceed that of oxygen molecules.
    (e) None of the above.

11. Which of the following is *not* true about an ideal gas?
    (a) The average kinetic energy of the gas molecules increases as the temperature increases.
    (b) The volume of an ideal gas increases with temperature if the pressure is held constant.
    (c) The pressure of an ideal gas increases with temperature if the volume is held constant.
    (d) All gas molecules have the same speed at a particular temperature.
    (e) The molecules are assumed to be far apart compared to their size.

12. When using the ideal gas law, which of the following rules must be obeyed?
    (a) Always use temperature in kelvins and absolute pressure.
    (b) Always use volume in m³ and temperature in kelvins.
    (c) Always use gauge pressure and temperature in degrees Celsius.
    (d) Always use gauge pressure and temperature in kelvins.
    (e) Always use volume in m³ and gauge pressure.

13. The rms speed of the molecules of an ideal gas
    (a) is the same as the most probable speed of the molecules.
    (b) is always equal to $\sqrt{2}$ times the maximum molecular speed.
    (c) will increase as the temperature of a gas increases.
    (d) All of the above.

For assigned homework and other learning materials, go to the MasteringPhysics website.

# Problems

## 13–1 Atomic Theory

1. (I) How does the number of atoms in a 27.5-gram gold ring compare to the number in a silver ring of the same mass?

2. (I) How many atoms are there in a 3.4-g copper coin?

## 13–2 Temperature and Thermometers

3. (I) (a) "Room temperature" is often taken to be 68°F. What is this on the Celsius scale? (b) The temperature of the filament in a lightbulb is about 1900°C. What is this on the Fahrenheit scale?

4. (I) Among the highest and lowest natural air temperatures claimed are 136°F in the Libyan desert and −129°F in Antarctica. What are these temperatures on the Celsius scale?

5. (I) A thermometer tells you that you have a fever of 38.9°C. What is this in Fahrenheit?

6. (I) (a) 18° below zero on the Celsius scale is what Fahrenheit temperature? (b) 18° below zero on the Fahrenheit scale is what Celsius temperature?

**7.** (II) Determine the temperature at which the Celsius and Fahrenheit scales give the same numerical reading $(T_C = T_F)$.

**8.** (II) In an alcohol-in-glass thermometer, the alcohol column has length 12.61 cm at 0.0°C and length 22.79 cm at 100.0°C. What is the temperature if the column has length (a) 18.70 cm, and (b) 14.60 cm?

## 13–4 Thermal Expansion

**9.** (I) The Eiffel Tower (Fig. 13–31) is built of wrought iron approximately 300 m tall. Estimate how much its height changes between January (average temperature of 2°C) and July (average temperature of 25°C). Ignore the angles of the iron beams and treat the tower as a vertical beam.

**FIGURE 13–31** Problem 9. The Eiffel Tower in Paris.

**10.** (I) A concrete highway is built of slabs 12 m long (15°C). How wide should the expansion cracks between the slabs be (at 15°C) to prevent buckling if the range of temperature is −30°C to +50°C?

**11.** (I) Super Invar™, an alloy of iron and nickel, is a strong material with a very low coefficient of thermal expansion $(0.20 \times 10^{-6}/\text{C}°)$. A 1.8-m-long tabletop made of this alloy is used for sensitive laser measurements where extremely high tolerances are required. How much will this alloy table expand along its length if the temperature increases 6.0 C°? Compare to tabletops made of steel.

**12.** (II) To what temperature would you have to heat a brass rod for it to be 1.5% longer than it is at 25°C?

**13.** (II) To make a secure fit, rivets that are larger than the rivet hole are often used and the rivet is cooled (usually in dry ice) before it is placed in the hole. A steel rivet 1.872 cm in diameter is to be placed in a hole 1.870 cm in diameter in a metal at 22°C. To what temperature must the rivet be cooled if it is to fit in the hole?

**14.** (II) An ordinary glass is filled to the brim with 450.0 mL of water at 100.0°C. If the temperature of glass and water is decreased to 20.0°C, how much water could be added to the glass?

**15.** (II) An aluminum sphere is 8.75 cm in diameter. What will be its % change in volume if it is heated from 30°C to 160°C?

**16.** (II) It is observed that 55.50 mL of water at 20°C completely fills a container to the brim. When the container and the water are heated to 60°C, 0.35 g of water is lost. (a) What is the coefficient of volume expansion of the container? (b) What is the most likely material of the container? Density of water at 60°C is 0.98324 g/mL.

**17.** (II) A brass plug is to be placed in a ring made of iron. At 15°C, the diameter of the plug is 8.755 cm and that of the inside of the ring is 8.741 cm. They must both be brought to what common temperature in order to fit?

**18.** (II) A certain car has 14.0 L of liquid coolant circulating at a temperature of 93°C through the engine's cooling system. Assume that, in this normal condition, the coolant completely fills the 3.5-L volume of the aluminum radiator and the 10.5-L internal cavities within the aluminum engine. When a car overheats, the radiator, engine, and coolant expand and a small reservoir connected to the radiator catches any resultant coolant overflow. Estimate how much coolant overflows to the reservoir if the system goes from 93°C to 105°C. Model the radiator and engine as hollow shells of aluminum. The coefficient of volume expansion for coolant is $\beta = 410 \times 10^{-6}/\text{C}°$.

***19.** (II) An aluminum bar has the desired length when at 12°C. How much stress is required to keep it at this length if the temperature increases to 35°C? [See Table 9–1.]

**20.** (III) The pendulum in a grandfather clock is made of brass and keeps perfect time at 17°C. How much time is gained or lost in a year if the clock is kept at 29°C? (Assume the frequency dependence on length for a simple pendulum applies.) [*Hint*: See Chapter 8.]

## 13–5 Gas Laws; Absolute Temperature

**21.** (I) Absolute zero is what temperature on the Fahrenheit scale?

**22.** (II) Typical temperatures in the interior of the Earth and Sun are about 4000°C and $15 \times 10^6 \, °\text{C}$, respectively. (a) What are these temperatures in kelvins? (b) What percent error is made in each case if a person forgets to change °C to K?

## 13–6 and 13–7 Ideal Gas Law

**23.** (I) If $3.50 \, \text{m}^3$ of a gas initially at STP is placed under a pressure of 3.20 atm, the temperature of the gas rises to 38.0°C. What is the volume?

**24.** (I) In an internal combustion engine, air at atmospheric pressure and a temperature of about 20°C is compressed in the cylinder by a piston to $\frac{1}{9}$ of its original volume (compression ratio = 9.0). Estimate the temperature of the compressed air, assuming the pressure reaches 40 atm.

**25.** (II) If 16.00 mol of helium gas is at 10.0°C and a gauge pressure of 0.350 atm, calculate (a) the volume of the helium gas under these conditions, and (b) the temperature if the gas is compressed to precisely half the volume at a gauge pressure of 1.00 atm.

**26.** (II) A storage tank contains 21.6 kg of nitrogen $(N_2)$ at an absolute pressure of 3.45 atm. What will the pressure be if the nitrogen is replaced by an equal mass of $CO_2$ at the same temperature?

**27.** (II) A storage tank at STP contains 28.5 kg of nitrogen $(N_2)$. (a) What is the volume of the tank? (b) What is the pressure if an additional 32.2 kg of nitrogen is added without changing the temperature?

**28.** (II) A scuba tank is filled with air to a gauge pressure of 204 atm when the air temperature is 29°C. A diver then jumps into the ocean and, after a short time on the ocean surface, checks the tank's gauge pressure and finds that it is only 191 atm. Assuming the diver has inhaled a negligible amount of air from the tank, what is the temperature of the ocean water?

**29.** (II) What is the pressure inside a 38.0-L container holding 105.0 kg of argon gas at 21.6°C?

**30.** (II) A sealed metal container contains a gas at 20.0°C and 1.00 atm. To what temperature must the gas be heated for the pressure to double to 2.00 atm? (Ignore expansion of the container.)

**31.** (II) A tire is filled with air at 15°C to a gauge pressure of 230 kPa. If the tire reaches a temperature of 38°C, what fraction of the original air must be removed if the original pressure of 230 kPa is to be maintained?

**32.** (II) If 61.5 L of oxygen at 18.0°C and an absolute pressure of 2.45 atm are compressed to 38.8 L and at the same time the temperature is raised to 56.0°C, what will the new pressure be?

**33.** (II) You buy an "airtight" bag of potato chips packaged at sea level, and take the chips on an airplane flight. When you take the potato chips out of your "carry-on" bag, you notice it has noticeably "puffed up." Airplane cabins are typically pressurized at 0.75 atm, and assuming the temperature inside an airplane is about the same as inside a potato chip processing plant, by what percentage has the bag "puffed up" in comparison to when it was packaged?

**34.** (II) A helium-filled balloon escapes a child's hand at sea level and 20.0°C. When it reaches an altitude of 3600 m, where the temperature is 5.0°C and the pressure only 0.68 atm, how will its volume compare to that at sea level?

**35.** (II) Compare the value for the density of water vapor at exactly 100°C and 1 atm (Table 10–1) with the value predicted from the ideal gas law. Why would you expect a difference?

**36.** (III) A sealed test tube traps $25.0 \text{ cm}^3$ of air at a pressure of 1.00 atm and temperature of 18°C. The test tube's stopper has a diameter of 1.50 cm and will "pop off" the test tube if a net upward force of 10.0 N is applied to it. To what temperature would you have to heat the trapped air in order to "pop off" the stopper? Assume the air surrounding the test tube is always at a pressure of 1.00 atm.

**37.** (III) An air bubble at the bottom of a lake 41.0 m deep has a volume of $1.00 \text{ cm}^3$. If the temperature at the bottom is 5.5°C and at the top 18.5°C, what is the radius of the bubble just before it reaches the surface?

### 13–8 Ideal Gas Law in Terms of Molecules; Avogadro's Number

**38.** (I) Calculate the number of molecules/$\text{m}^3$ in an ideal gas at STP.

**39.** (I) How many moles of water are there in 1.000 L at STP? How many molecules?

**40.** (II) Estimate the number of (a) moles and (b) molecules of water in all the Earth's oceans. Assume water covers 75% of the Earth to an average depth of 3 km.

**41.** (II) The lowest pressure attainable using the best available vacuum techniques is about $10^{-12} \text{ N/m}^2$. At such a pressure, how many molecules are there per $\text{cm}^3$ at 0°C?

**42.** (II) Is a gas mostly empty space? Check by assuming that the spatial extent of the gas molecules in air is about $\ell_0 = 0.3 \text{ nm}$ so one gas molecule occupies an approximate volume equal to $\ell_0^3$. Assume STP.

### 13–9 Molecular Interpretation of Temperature

**43.** (I) (a) What is the average translational kinetic energy of a nitrogen molecule at STP? (b) What is the total translational kinetic energy of 1.0 mol of $N_2$ molecules at 25°C?

**44.** (I) Calculate the rms speed of helium atoms near the surface of the Sun at a temperature of about 6000 K.

**45.** (I) By what factor will the rms speed of gas molecules increase if the temperature is increased from 20°C to 160°C?

**46.** (I) A gas is at 20°C. To what temperature must it be raised to triple the rms speed of its molecules?

**47.** (I) What speed would a 1.0-g paper clip have if it had the same kinetic energy as a molecule at 22°C?

**48.** (II) The rms speed of molecules in a gas at 20.0°C is to be increased by 4.0%. To what temperature must it be raised?

**49.** (II) If the pressure in a gas is tripled while its volume is held constant, by what factor does $v_{\text{rms}}$ change?

**50.** (II) Show that the rms speed of molecules in a gas is given by $v_{\text{rms}} = \sqrt{3P/\rho}$, where $P$ is the pressure in the gas and $\rho$ is the gas density.

**51.** (II) Show that for a mixture of two gases at the same temperature, the ratio of their rms speeds is equal to the inverse ratio of the square roots of their molecular masses, $v_1/v_2 = \sqrt{M_2/M_1}$.

**52.** (II) What is the rms speed of nitrogen molecules contained in an $8.5\text{-m}^3$ volume at 2.9 atm if the total amount of nitrogen is 2100 mol?

**53.** (II) Two isotopes of uranium, $^{235}U$ and $^{238}U$ (the superscripts refer to their atomic masses), can be separated by a gas diffusion process by combining them with fluorine to make the gaseous compound $UF_6$. Calculate the ratio of the rms speeds of these molecules for the two isotopes, at constant $T$. Use Appendix B for masses.

**54.** (III) Calculate (a) the rms speed of an oxygen molecule at 0°C and (b) determine how many times per second it would move back and forth across a 5.0-m-long room on average, assuming it made no collisions with other molecules.

### 13–11 Real Gases; Phase Changes

**55.** (I) $CO_2$ exists in what phase when the pressure is 35 atm and the temperature is 35°C (Fig. 13–23)?

**56.** (I) (a) At atmospheric pressure, in what phases can $CO_2$ exist? (b) For what range of pressures and temperatures can $CO_2$ be a liquid? Refer to Fig. 13–23.

**57.** (I) Water is in which phase when the pressure is 0.01 atm and the temperature is (a) 90°C, (b) −20°C?

**58.** (II) You have a sample of water and are able to control temperature and pressure arbitrarily. (a) Using Fig. 13–22, describe the phase changes you would see if you started at a temperature of 85°C, a pressure of 180 atm, and decreased the pressure down to 0.004 atm while keeping the temperature fixed. (b) Repeat part (a) with the temperature at 0.0°C. Assume that you held the system at the starting conditions long enough for the system to stabilize before making further changes.

### 13–12 Vapor Pressure and Humidity

**59.** (I) What is the partial pressure of water vapor at 30°C if the humidity is 75%?

**60.** (I) What is the air pressure at a place where water boils at 80°C?

**61.** (II) What is the dew point if the humidity is 65% on a day when the temperature is 25°C?

**62.** (II) If the air pressure at a particular place in the mountains is 0.80 atm, estimate the temperature at which water boils.

63. (II) What is the mass of water in a closed room 5.0 m × 6.0 m × 2.4 m when the temperature is 25°C and the relative humidity is 55%?

64. (II) A **pressure cooker** is a sealed pot designed to cook food with the steam produced by boiling water somewhat above 100°C. The pressure cooker in Fig. 13–32 uses a weight of mass $m$ to allow steam to escape at a certain pressure through a small hole (diameter $d$) in the cooker's lid. If $d = 3.0$ mm, what should $m$ be in order to cook food at 120°C? Assume that atmospheric pressure outside the cooker is $1.01 \times 10^5$ Pa.

Weight (mass $m$)

Diameter $d$

Steam

Water

**FIGURE 13–32**
Problem 64.

65. (II) If the humidity in a sealed room of volume 420 m³ at 20°C is 65%, what mass of water can still evaporate from an open pan?

66. (III) Air that is at its dew point of 5°C is drawn into a building where it is heated to 22°C. What will be the relative humidity at this temperature? Assume constant pressure of 1.0 atm. Take into account the expansion of the air.

67. (III) When using a mercury barometer (Section 10–6), the vapor pressure of mercury is usually assumed to be zero. At room temperature mercury's vapor pressure is about 0.0015 mm-Hg. At sea level, the height $h$ of mercury in a barometer is about 760 mm. (a) If the vapor pressure of mercury is neglected, is the true atmospheric pressure greater or less than the value read from the barometer? (b) What is the percent error? (c) What is the percent error if you use a water barometer and ignore water's saturated vapor pressure at STP?

## *13–13 Diffusion

*68. (II) Estimate the time needed for a glycine molecule (see Table 13–4) to diffuse a distance of 25 $\mu$m in water at 20°C if its concentration varies over that distance from 1.00 mol/m³ to 0.50 mol/m³? Compare this "speed" to its rms (thermal) speed. The molecular mass of glycine is about 75 u.

*69. (II) Oxygen diffuses from the surface of insects to the interior through tiny tubes called tracheae. An average trachea is about 2 mm long and has cross-sectional area of $2 \times 10^{-9}$ m². Assuming the concentration of oxygen inside is half what it is outside in the atmosphere, (a) show that the concentration of oxygen in the air (assume 21% is oxygen) at 20°C is about 8.7 mol/m³, then (b) calculate the diffusion rate $J$, and (c) estimate the average time for a molecule to diffuse in. Assume the diffusion constant is $1 \times 10^{-5}$ m²/s.

# General Problems

70. A Pyrex measuring cup was calibrated at normal room temperature. How much error will be made in a recipe calling for 375 mL of cool water, if the water and the cup are hot, at 95°C, instead of at room temperature? Neglect the glass expansion.

71. A precise steel tape measure has been calibrated at 14°C. At 37°C, (a) will it read high or low, and (b) what will be the percentage error?

72. A cubic box of volume $6.15 \times 10^{-2}$ m³ is filled with air at atmospheric pressure at 15°C. The box is closed and heated to 165°C. What is the net force on each side of the box?

73. The gauge pressure in a helium gas cylinder is initially 32 atm. After many balloons have been blown up, the gauge pressure has decreased to 5 atm. What fraction of the original gas remains in the cylinder?

74. If a scuba diver fills his lungs to full capacity of 5.5 L when 9.0 m below the surface, to what volume would his lungs expand if he quickly rose to the surface? Is this advisable?

75. A house has a volume of 1200 m³. (a) What is the total mass of air inside the house at 15°C? (b) If the temperature drops to −15°C, what mass of air enters or leaves the house?

76. Estimate the number of air molecules in a room of length 6.0 m, width 3.0 m, and height 2.5 m. Assume the temperature is 22°C. How many moles does that correspond to?

77. An iron cube floats in a bowl of liquid mercury at 0°C. (a) If the temperature is raised to 25°C, will the cube float higher or lower in the mercury? (b) By what percent will the fraction of volume submerged change? [*Hint*: See Chapter 10.]

78. A helium balloon, assumed to be a perfect sphere, has a radius of 24.0 cm. At room temperature (20°C), its internal pressure is 1.08 atm. Determine the number of moles of helium in the balloon, and the mass of helium needed to inflate the balloon to these values.

79. A standard cylinder of oxygen used in a hospital has gauge pressure = 2000 psi (13,800 kPa) and volume = 14 L (0.014 m³) at $T = 295$ K. How long will the cylinder last if the flow rate, measured at atmospheric pressure, is constant at 2.1 L/min?

80. A brass lid screws tightly onto a glass jar at 15°C. To help open the jar, it can be placed into a bath of hot water. After this treatment, the temperatures of the lid and the jar are both 55°C. The inside diameter of the lid is 8.0 cm. Find the size of the gap (difference in radius) that develops by this procedure.

81. The density of gasoline at 0°C is $0.68 \times 10^3$ kg/m³. (a) What is the density on a hot day, when the temperature is 33°C? (b) What is the percent change in density?

82. The first real length standard, adopted more than 200 years ago, was a platinum bar with two very fine marks separated by what was defined to be exactly one meter. If this standard bar was to be accurate to within $\pm 1.0$ $\mu$m, how carefully would the trustees have needed to control the temperature? The coefficient of linear expansion is $9 \times 10^{-6}$/C°.

83. If a steel band were to fit snugly around the Earth's equator at 25°C, but then was heated to 55°C, how high above the Earth would the band be (assume equal everywhere)?

84. In outer space the density of matter is about one atom per $cm^3$, mainly hydrogen atoms, and the temperature is about 2.7 K. Calculate the rms speed of these hydrogen atoms, and the pressure (in atmospheres).

85. (a) Estimate the rms speed of an amino acid, whose molecular mass is 89 u, in a living cell at 37°C. (b) What would be the rms speed of a protein of molecular mass 85,000 u at 37°C?

86. The escape speed from the Earth is $1.12 \times 10^4$ m/s, so that a gas molecule traveling away from Earth near the outer boundary of the Earth's atmosphere would, at this speed, be able to escape from the Earth's gravitational field and be lost to the atmosphere. At what temperature is the rms speed of (a) oxygen molecules, and (b) helium atoms equal to $1.12 \times 10^4$ m/s? (c) Can you explain why our atmosphere contains oxygen but not helium?

87. Consider a container of oxygen gas at a temperature of 23°C that is 1.00 m tall. Compare the gravitational potential energy of a molecule at the top of the container (assuming the potential energy is zero at the bottom) with the average kinetic energy of the molecules. Is it reasonable to neglect the potential energy?

88. A space vehicle returning from the Moon enters the Earth's atmosphere at a speed of about 42,000 km/h. Molecules (assume nitrogen) striking the nose of the vehicle with this speed correspond to what temperature? (Because of this high temperature, the nose of a space vehicle must be made of special materials; indeed, part of it does vaporize, and this is seen as a bright blaze upon reentry.)

89. A sauna has 8.5 $m^3$ of air volume, and the temperature is 85°C. The air is perfectly dry. How much water (in kg) should be evaporated if we want to increase the relative humidity from 0% to 10%? (See Table 13–3.)

90. A 0.50-kg trash-can lid is suspended against gravity by tennis balls thrown vertically upward at it. How many tennis balls per second must rebound from the lid elastically, assuming they have a mass of 0.060 kg and are thrown at 15 m/s?

91. In humid climates, people constantly **dehumidify** their cellars to prevent rot and mildew. If the cellar in a house (kept at 20°C) has 105 $m^2$ of floor space and a ceiling height of 2.4 m, what is the mass of water that must be removed from it in order to drop the humidity from 95% to a more reasonable 40%?

# Search and Learn

1. This Chapter gives two ways to calculate the thermal expansion of a gas at a constant pressure of 1.0 atm. Use both methods to calculate the volume change of 1000 L of an ideal gas as it goes from −100°C to 0°C and from 0°C to 100°C. Why are the answers different?

2. A scuba tank when fully charged has a pressure of 180 atm at 18°C. The volume of the tank is 11.3 L. (a) What would the volume of the air be at 1.00 atm and at the same temperature? (b) Before entering the water, a person consumes 2.0 L of air in each breath, and breathes 12 times a minute. At this rate, how long would the tank last? (c) At a depth of 23.0 m in sea water at a temperature of 10°C, how long would the same tank last assuming the breathing rate does not change?

3. A 28.4-kg solid aluminum cylindrical wheel of radius 0.41 m is rotating about its axle in frictionless bearings with angular velocity $\omega$ = 32.8 rad/s. If its temperature is then raised from 15.0°C to 95.0°C, what is the fractional change in $\omega$?

4. A hot-air balloon achieves its buoyant lift by heating the air inside the balloon, which makes it less dense than the air outside. Suppose the volume of a balloon is 1800 $m^3$ and the required lift is 3300 N (rough estimate of the weight of the equipment and passenger). Calculate the temperature of the air inside the balloon which will produce the required lift. Assume the outside air is an ideal gas at 0°C. What factors limit the maximum altitude attainable by this method for a given load? [*Hint*: See Chapter 10.]

5. Estimate how many molecules of air are in each 2.0-L breath you inhale that were also in the last breath Galileo took. Assume the atmosphere is about 10 km high and of constant density. What other assumptions did you make?

6. (a) The second postulate of kinetic theory is that the molecules are, on average, far apart from one another. That is, their average separation is much greater than the diameter of each molecule. Is this assumption reasonable? To check, calculate the average distance between molecules of a gas at STP, and compare it to the diameter of a typical gas molecule, about 0.3 nm. (b) If the molecules were the diameter of ping-pong balls, say 4 cm, how far away would the next ping-pong ball be on average? (c) Repeat part a, but now assume the gas has been compressed so that the pressure is now 3 atm but still at 273 K. (d) Estimate what % of the total volume of gas is taken up by molecules themselves in parts a and c. [Note that the volume of the molecules themselves can become a significant part of the total volume at lower temperatures and higher pressures. Hence the actual volume the molecules have to bounce around in is less than the total volume. This contributes to the effect shown in Fig. 13–21 at high pressures where real gases (solid red lines) deviate from ideal gas behavior (dashed lines A' and B').]

## ANSWERS TO EXERCISES

A: (i) Higher, (ii) same, (iii) lower.
B: (d).
C: (b).

D: (b) Less.
E: (a).
F: (d).

When it is cold, warm clothes act as insulators to reduce heat loss from the body to the environment by conduction and convection. Heat radiation from a campfire can warm you and your clothes. The fire can also transfer energy directly by heat convection and conduction to what you are cooking. Heat, like work, represents a transfer of energy. Heat is defined as a transfer of energy due to a difference of temperature. Internal energy $U$ is the sum total of all the energy of all the molecules of the system.

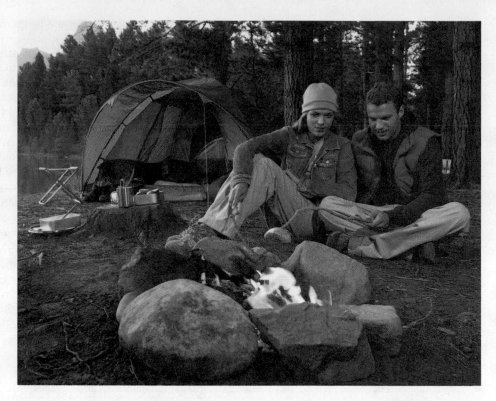

# 14

# Heat

## CONTENTS

**CHAPTER-OPENING QUESTION—Guess now!**

A 5-kg cube of warm iron (60°C) is put in thermal contact with a 10-kg cube of cold iron (15°C). Which statement is valid?

**(a)** Heat flows spontaneously from the warm cube to the cold cube until both cubes have the same heat content.

**(b)** Heat flows spontaneously from the warm cube to the cold cube until both cubes have the same temperature.

**(c)** Heat can flow spontaneously from the warm cube to the cold cube, but can also flow spontaneously from the cold cube to the warm cube.

**(d)** Heat flows from the larger cube to the smaller one because the larger one has more internal energy.

When a pot of cold water is placed on a hot burner of a stove, the temperature of the water increases. We say that heat "flows" from the hot burner to the cold water. When two objects at different temperatures are put in contact, heat spontaneously flows from the hotter one to the colder one. The spontaneous flow of heat is in the direction tending to equalize the temperature. If the two objects are kept in contact long enough for their temperatures to become equal, the objects are said to be in thermal equilibrium, and there is no further heat flow between them. For example, when a fever thermometer is first placed in your mouth, heat flows from your mouth to the thermometer. When the thermometer reaches the same temperature as the inside of your mouth, the thermometer and your mouth are then in equilibrium, and no more heat flows.

Heat and temperature are often confused. They are very different concepts, and in this Chapter we will make a clear distinction between them. We begin by defining and using the concept of heat. We also discuss how heat is used in calorimetry, how it is involved in changes of state of matter, and the processes of heat transfer—conduction, convection, and radiation.

# 14–1 Heat as Energy Transfer

We use the term "heat" in everyday life as if we knew what we meant. But the term is often used inconsistently, so it is important for us to define heat clearly, and to clarify the phenomena and concepts related to heat.

We commonly speak of the flow of heat—heat flows from a stove burner to a pot of soup, from the Sun to the Earth, from a person's mouth into a fever thermometer. Heat flows spontaneously from an object at higher temperature to one at lower temperature. Indeed, an eighteenth-century model of heat pictured heat flow as movement of a fluid substance called *caloric*. However, the caloric fluid could never be detected. In the nineteenth century, it was found that the various phenomena associated with heat could be described consistently using a new model that views heat as being akin to work, as we will discuss in a moment. First we note that a common unit for heat, still in use today, is named after caloric. It is called the **calorie** (cal) and is defined as *the amount of heat necessary to raise the temperature of 1 gram of water by 1 Celsius degree.* [To be precise, the particular temperature range from 14.5°C to 15.5°C is specified because the heat required is very slightly different at different temperatures. The difference is less than 1% over the range 0 to 100°C, and we will ignore it for most purposes.] More often used than the calorie is the **kilocalorie** (kcal), which is 1000 calories. Thus *1 kcal is the heat needed to raise the temperature of 1 kg of water by 1 C°.* Often a kilocalorie is called a **Calorie** (with a capital C), and this Calorie (or the kJ) is used to specify the energy value of food. In the British system of units, heat is measured in British thermal units (Btu). One **Btu** is defined as the heat needed to raise the temperature of 1 lb of water by 1 F°. It can be shown (Problem 5) that 1 Btu = 0.252 kcal = 1056 J. Also, one **therm** is $10^5$ Btu.

The idea that heat is related to energy transfer was pursued by a number of scientists in the 1800s, particularly by an English brewer, James Prescott Joule (1818–1889). Joule and others performed a number of experiments that were crucial for establishing our present-day view that heat, like work, represents a transfer of energy. One of Joule's experiments is shown (simplified) in Fig. 14–1. The falling weight causes the paddle wheel to turn. The friction between the water and the paddle wheel causes the temperature of the water to rise slightly (barely measurable, in fact, by Joule). The same temperature rise could also be obtained by heating the water on a hot stove. In this and many other experiments (some involving electrical energy), Joule determined that a given amount of work done was always equivalent to a particular amount of heat input. Quantitatively, 4.186 joules (J) of work was found to be equivalent to 1 calorie (cal) of heat. This is known as the **mechanical equivalent of heat**:

$$4.186 \text{ J} = 1 \text{ cal};$$

$$4.186 \text{ kJ} = 1 \text{ kcal}.$$

As a result of these and other experiments, scientists came to interpret heat not as a substance, and not exactly as a form of energy. Rather, heat refers to a *transfer of energy*: when heat flows from a hot object to a cooler one, it is energy that is being transferred from the hot to the cold object. Thus, **heat** is *energy transferred from one object to another because of a difference in temperature.* In SI units, the unit for heat, as for any form of energy, is the joule. Nonetheless, calories and kcal are still sometimes used. Today the calorie is *defined* in terms of the joule (via the mechanical equivalent of heat, above), rather than in terms of the properties of water, as given previously. The latter is still handy to remember: 1 cal raises 1 g of water by 1 C°, or 1 kcal raises 1 kg of water by 1 C°.

**CAUTION**
*Heat is not a fluid*

**FIGURE 14–1** Joule's experiment on the mechanical equivalent of heat. [The work is transformed into internal energy (Section 14–2).]

**CAUTION**
*Heat is energy transferred because of a $\Delta T$*

**PHYSICS APPLIED**

*Working off Calories*

---

**EXAMPLE 14–1** | **ESTIMATE** | **Working off the extra Calories.** Suppose you throw caution to the wind and eat 500 Calories of ice cream and cake. To compensate, you want to do an equivalent amount of work climbing stairs or a mountain. How much total height must you climb?

**APPROACH** The work $W$ you need to do in climbing stairs equals the change in gravitational potential energy: $W = \Delta PE = mgh$, where $h$ is the vertical height climbed. For this estimate, let us approximate your mass as $m \approx 60$ kg.

**SOLUTION** 500 food Calories is 500 kcal, which in joules is

$$(500 \text{ kcal})(4.186 \times 10^3 \text{ J/kcal}) = 2.1 \times 10^6 \text{ J}.$$

The work done to climb a vertical height $h$ is $W = mgh$. We solve for $h$:

$$h = \frac{W}{mg} = \frac{2.1 \times 10^6 \text{ J}}{(60 \text{ kg})(9.80 \text{ m/s}^2)} = 3600 \text{ m}.$$

This is a huge elevation change (over 11,000 ft).

**NOTE** The human body does not transform food energy with 100% efficiency—it is more like 20% efficient. As we shall discuss in the next Chapter, some energy is always "wasted," so you would actually have to climb only about $(0.2)(3600 \text{ m}) \approx 700$ m, which is more reasonable (about 2300 ft of elevation gain).

# 14–2 Internal Energy

The sum total of all the energy of all the molecules in an object is called its **internal energy**. (Sometimes **thermal energy** is used to mean the same thing.) We introduce the concept of internal energy now since it will help clarify ideas about heat.

### Distinguishing Temperature, Heat, and Internal Energy

**CAUTION**

*Distinguish heat from internal energy and from temperature*

Using the kinetic theory, we can make a clear distinction between temperature, heat, and internal energy. Temperature (in kelvins) is a measure of the *average* kinetic energy of individual molecules (Eq. 13–8). Internal energy refers to the *total* energy of all the molecules within the object. (Thus two equal-mass hot ingots of iron may have the same temperature, but two of them have twice as much internal energy as one does.) Heat, finally, refers to a *transfer* of energy from one object to another because of a difference in temperature.

**CAUTION**

*Direction of heat flow depends on temperatures (not on amount of internal energy)*

Notice that the direction of heat flow between two objects depends on their temperatures, not on how much internal energy each has. Thus, if 50 g of water at 30°C is placed in contact (or mixed) with 200 g of water at 25°C, heat flows *from* the water at 30°C *to* the water at 25°C even though the internal energy of the 25°C water is much greater because there is so much more of it.

### Internal Energy of an Ideal Gas

Let us calculate the internal energy of $n$ moles of an ideal monatomic (one atom per molecule) gas. The internal energy, $U$, is the sum of the translational kinetic energies of all the atoms. This sum is equal to the average kinetic energy per molecule times the total number of molecules, $N$:

$$U = N\left(\tfrac{1}{2}m\overline{v^2}\right).$$

Using Eq. 13–8, $\overline{KE} = \tfrac{1}{2}m\overline{v^2} = \tfrac{3}{2}kT$, we can write this as

$$U = \tfrac{3}{2}NkT$$

or (recall Section 13–8)

$$U = \tfrac{3}{2}nRT, \qquad \begin{bmatrix} \text{internal energy of} \\ \text{ideal monatomic gas} \end{bmatrix} \quad \textbf{(14–1)}$$

where $n$ is the number of moles. Thus, the internal energy of an ideal gas depends only on temperature and the number of moles (or molecules) of gas.

If the gas molecules contain more than one atom, then the rotational and vibrational energy of the molecules (Fig. 14–2) must also be taken into account. The internal energy will be greater at a given temperature than for a monatomic gas, but it will still be a function only of temperature for an ideal gas.

The internal energy of real gases also depends mainly on temperature, but where real gases deviate from ideal gas behavior, their internal energy depends also somewhat on pressure and volume (due to atomic potential energy).

The internal energy of liquids and solids is quite complicated, for it includes electrical potential energy associated with the forces (or "chemical" bonds) between atoms and molecules.

# 14–3 Specific Heat

If heat flows into an object, the object's temperature rises (assuming no phase change). But how much does the temperature rise? That depends. As early as the eighteenth century, experimenters had recognized that the amount of heat $Q$ required to change the temperature of a given material is proportional to the mass $m$ of the material present and to the temperature change $\Delta T$. This remarkable simplicity in nature can be expressed in the equation

$$Q = mc\,\Delta T, \tag{14–2}$$

where $c$ is a quantity characteristic of the material called its **specific heat**. Because $c = Q/(m\,\Delta T)$, specific heat is specified in units[†] of J/kg·C° (the proper SI unit) or kcal/kg·C°. For water at 15°C and a constant pressure of 1 atm, $c = 4.186 \times 10^3$ J/kg·C° or 1.00 kcal/kg·C°, since, by definition of the cal and the joule, it takes 1 kcal of heat to raise the temperature of 1 kg of water by 1 C°. Table 14–1 gives the values of specific heat for other substances at 20°C. The values of $c$ depend to some extent on temperature (as well as slightly on pressure), but for temperature changes that are not too great, $c$ can often be considered constant.

---

**EXAMPLE 14–2** **How heat transferred depends on specific heat.** (*a*) How much heat input is needed to raise the temperature of an empty 20-kg vat made of iron from 10°C to 90°C? (*b*) What if the vat is filled with 20 kg of water?

**APPROACH** We apply Eq. 14–2 to the different materials involved.

**SOLUTION** (*a*) Our system is the iron vat alone. From Table 14–1, the specific heat of iron is 450 J/kg·C°. The change in temperature is (90°C − 10°C) = 80 C°. Thus,

$$Q_{\text{vat}} = mc\,\Delta T = (20\,\text{kg})(450\,\text{J/kg·C°})(80\,\text{C°}) = 7.2 \times 10^5\,\text{J} = 720\,\text{kJ}.$$

(*b*) Our system is the vat plus the water. The water alone would require

$$Q_{\text{w}} = mc\,\Delta T = (20\,\text{kg})(4186\,\text{J/kg·C°})(80\,\text{C°}) = 6.7 \times 10^6\,\text{J} = 6700\,\text{kJ},$$

or almost 10 times what an equal mass of iron requires. The total, for the vat plus the water, is 720 kJ + 6700 kJ = 7400 kJ.

**NOTE** In (*b*), the iron vat and the water underwent the same temperature change, $\Delta T = 80$ C°, but their specific heats are different.

---

If the iron vat in part (*a*) of Example 14–2 had been *cooled* from 90°C to 10°C, 720 kJ of heat would have flowed *out* of the iron. In other words, Eq. 14–2 is valid for heat flow either in or out, with a corresponding increase or decrease in temperature.

We saw in part (*b*) of Example 14–2 that water requires almost 10 times as much heat as an equal mass of iron to make the same temperature change. Water has one of the highest specific heats of all substances, which makes it an ideal substance for hot-water space-heating systems and other uses that require a minimal drop in temperature for a given amount of heat transfer. It is the water content, too, that causes the apples rather than the crust in hot apple pie to burn our tongues, through heat transfer.

[†]Note that J/kg·C° means $\dfrac{\text{J}}{\text{kg·C°}}$ and *not* (J/kg)·C° = J·C°/kg (otherwise we would have written it that way).

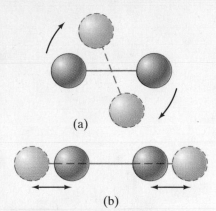

**FIGURE 14–2** Besides translational kinetic energy, molecules can have (a) rotational kinetic energy, and (b) vibrational energy (both kinetic and potential).

**TABLE 14–1 Specific Heats**
(at 1 atm constant pressure and 20°C unless otherwise stated)

| Substance | Specific Heat, $c$ | |
| --- | --- | --- |
| | J/kg·C° | kcal/kg·C° (= cal/g·C°) |
| Aluminum | 900 | 0.22 |
| Alcohol (ethyl) | 2400 | 0.58 |
| Copper | 390 | 0.093 |
| Glass | 840 | 0.20 |
| Iron or steel | 450 | 0.11 |
| Lead | 130 | 0.031 |
| Marble | 860 | 0.21 |
| Mercury | 140 | 0.033 |
| Silver | 230 | 0.056 |
| Wood | 1700 | 0.4 |
| Water | | |
|   Ice (−5°C) | 2100 | 0.50 |
|   Liquid (15°C) | 4186 | 1.00 |
|   Steam (110°C) | 2010 | 0.48 |
| Human body (average) | 3470 | 0.83 |
| Protein | 1700 | 0.4 |

**PHYSICS APPLIED**
*Practical effects of water's high specific heat*

**A very hot frying pan.** You accidentally let an empty iron frying pan get very hot on the stove (200°C or even more). What happens when you dunk it into a few inches of cool water in the bottom of the sink? Will the final temperature be midway between the initial temperatures of the water and pan? Will the water start boiling? Assume the mass of water is roughly the same as the mass of the frying pan.

**RESPONSE** Experience may tell you that the water warms up—perhaps by as much as 10 or 20 degrees. The water doesn't come close to boiling. The water's temperature increase is a lot less than the frying pan's temperature decrease. Why? Because the mass of water is roughly equal to that of the pan, and iron has a specific heat nearly 10 times smaller than that of water (Table 14–1). As heat leaves the frying pan and enters the water, the iron pan's temperature change will be about 10 times greater than that of the water. If, instead, you let a few drops of water fall onto the hot pan, that very small mass of water will sizzle and boil away (the pan's mass may be hundreds of times larger than that of the water drops).

**EXERCISE A** Return to the Chapter-Opening Question, page 390, and answer it again now. Try to explain why you may have answered differently the first time.

## *Specific Heats for Gases

Specific heats for gases are more complicated than for solids and liquids, which change in volume only slightly with a change in temperature. Gases change strongly in volume with a change in temperature at constant pressure, as we saw in Chapter 13 with the gas laws; or, if kept at constant volume, the pressure in a gas changes strongly with temperature. The specific heat of a gas depends very much on how the process of changing its temperature is carried out. Most commonly, we deal with the specific heats of gases kept (a) at constant pressure $(c_P)$ or (b) at constant volume $(c_V)$. Some values are given in Table 14–2, where we see that $c_P$ is always greater than $c_V$. For liquids and solids, this distinction is usually negligible. More details are given in Appendix D on molecular specific heats and the equipartition of energy.

**TABLE 14–2**
**Specific Heats of Gases**
(kcal/kg · C°)

| Gas | $c_P$ (constant pressure) | $c_V$ (constant volume) |
|---|---|---|
| Steam (100°C) | 0.482 | 0.350 |
| Oxygen | 0.218 | 0.155 |
| Helium | 1.15 | 0.75 |
| Carbon dioxide | 0.199 | 0.153 |
| Nitrogen | 0.248 | 0.177 |

# 14–4 Calorimetry—Solving Problems

In discussing heat and thermodynamics, we often consider a particular **system**, which is any object or set of objects we choose to consider. Everything else in the universe is its "environment" (or the "surroundings"). There are several categories of systems. A **closed system** is one for which no mass enters or leaves (but energy may be exchanged with the environment). In an **open system**, mass may enter or leave (as may energy). Many (idealized) systems we study in physics are closed systems. But many systems, including plants and animals, are open systems since they exchange materials (food, oxygen, waste products) with the environment. A closed system is said to be **isolated** if no energy in any form (as well as no mass) passes across its boundaries; otherwise it is not isolated.

A perfectly isolated system is an ideal, but we often try to set up a system that can be closely approximated as an isolated system (preferably one we can deal with fairly easily). When different parts of an isolated system are at different temperatures, heat will flow (energy is transferred) from the part at higher temperature to the part at lower temperature—that is, within the system—until **thermal equilibrium** is reached, meaning the entire system is at the same temperature. For an isolated system, no energy is transferred into or out of it. So we can apply *conservation of energy* to such an isolated system. A simple intuitive way to set up a conservation of energy equation is to write that the heat lost by one part of the system is equal to the heat gained by the other part:

$$\text{heat lost} = \text{heat gained} \qquad \text{[isolated system]}$$

or

$$\text{energy } out \text{ of one part} = \text{energy } into \text{ another part.} \qquad \text{[isolated system]}$$

**EXAMPLE 14–4** **The cup cools the tea.** If 200 cm³ of tea at 95°C is poured into a 150-g glass cup initially at 25°C (Fig. 14–3), what will be the common final temperature $T$ of the tea and cup when equilibrium is reached, assuming no heat flows to the surroundings?

**APPROACH** We apply conservation of energy to our system of tea plus cup, which we are assuming is isolated: all of the heat that leaves the tea flows into the cup. We can use the specific heat equation, Eq. 14–2, to determine how the heat flow is related to the temperature changes.

FIGURE 14–3 Example 14–4.

**SOLUTION** Because tea is mainly water, we can take its specific heat as 4186 J/kg·C° (Table 14–1), and its mass $m$ is its density times its volume $(V = 200\,\text{cm}^3 = 200 \times 10^{-6}\,\text{m}^3)$: $m = \rho V = (1.0 \times 10^3\,\text{kg/m}^3)(200 \times 10^{-6}\,\text{m}^3) = 0.20\,\text{kg}$. We use Eq. 14–2, apply conservation of energy, and let $T$ be the as yet unknown final temperature:

$$\text{heat lost by tea} = \text{heat gained by cup}$$
$$m_{\text{tea}}c_{\text{tea}}(95°\text{C} - T) = m_{\text{cup}}c_{\text{cup}}(T - 25°\text{C}).$$

Putting in numbers and using Table 14–1 ($c_{\text{cup}} = 840\,\text{J/kg·C}°$ for glass), we solve for $T$, and find

$$(0.20\,\text{kg})(4186\,\text{J/kg·C}°)(95°\text{C} - T) = (0.15\,\text{kg})(840\,\text{J/kg·C}°)(T - 25°\text{C})$$
$$79{,}500\,\text{J} - (837\,\text{J/C}°)T = (126\,\text{J/C}°)T - 3150\,\text{J}$$
$$T = 86°\text{C}.$$

The tea drops in temperature by 9 C° by coming into equilibrium with the cup.

**NOTE** The cup increases in temperature by $86°\text{C} - 25°\text{C} = 61\,\text{C}°$. Its much greater change in temperature (compared with that of the tea water) is due to its much smaller specific heat compared to that of water.

**NOTE** In this calculation, the $\Delta T$ (of Eq. 14–2, $Q = mc\,\Delta T$) is a positive quantity on both sides of our conservation of energy equation. On the left is "heat lost" and $\Delta T$ is the initial minus the final temperature ($95°\text{C} - T$), whereas on the right is "heat gained" and $\Delta T$ is the final minus the initial temperature.

Another, perhaps more general, way to set up the energy conservation equation for heat transfer within an isolated system is to write that the sum of all internal heat transfers within the system adds up to zero:

$$\Sigma Q = 0. \qquad \text{[isolated system]} \quad \textbf{(14–3)}$$

Each $Q$ represents the heat entering or leaving one part of the system. Each term is written as $Q = mc(T_f - T_i)$, and $\Delta T = T_f - T_i$ is always the final minus the initial temperature. $\Delta T$ can be either positive or negative, depending on whether heat flows into or out of that part. Let us redo Example 14–4 using $\Sigma Q = 0$.

**EXAMPLE 14–4′** **Alternate Solution, $\Sigma Q = 0$.**

**APPROACH** We use Eq. 14–3, $\Sigma Q = 0$.

**SOLUTION** For each term, $\Delta T = T_f - T_i$. If $T_f < T_i$, then $\Delta T < 0$. Equation 14–3, $\Sigma Q = 0$, becomes

$$m_{\text{cup}}c_{\text{cup}}(T - 25°\text{C}) + m_{\text{tea}}c_{\text{tea}}(T - 95°\text{C}) = 0.$$

The second term is negative because $T$ will be less than 95°C. Solving the algebra gives the same result, $T = 86°\text{C}$.

You are free to use either approach. They are entirely equivalent, algebraically. For example, if you move the first term in the displayed equation of the alternate Example 14–4 over to the other side of the equals sign, you obtain the "heat lost = heat gained" equation in the first version of Example 14–4.

**FIGURE 14–4** Simple water calorimeter.

The exchange of energy, as exemplified in Example 14–4, is the basis for a technique known as **calorimetry**, which is the quantitative measurement of heat exchange. To make such measurements, a **calorimeter** is used; a simple water calorimeter is shown in Fig. 14–4. It is very important that the calorimeter be well insulated so that almost no heat is exchanged with the surroundings. One important use of the calorimeter is in the determination of specific heats of substances. In the technique known as the "method of mixtures," a sample of a substance is heated to a high temperature, which is accurately measured, and then quickly placed in the cool water of the calorimeter. The heat lost by the sample will be gained by the water and the calorimeter cup. By measuring the final temperature of the mixture, the specific heat can be calculated, as illustrated in the following Example.

| **EXAMPLE 14–5** | **Unknown specific heat determined by calorimetry.** An engineer wishes to determine the specific heat of a new metal alloy. A 0.150-kg sample of the alloy is heated to 540°C. It is then quickly placed in 0.400 kg of water at 10.0°C, which is contained in a 0.200-kg aluminum calorimeter cup. The final temperature of the system is 30.5°C. Calculate the specific heat of the alloy.

**APPROACH** We apply conservation of energy to our system, which we take to be the alloy sample, the water, and the calorimeter cup. We assume this system is isolated, and apply Eq. 14–3, $\Sigma Q = 0$.

**SOLUTION** Each term is of the form $Q = mc(T_f - T_i)$. Thus $\Sigma Q = 0$ gives

$$m_a c_a \, \Delta T_a + m_w c_w \, \Delta T_w + m_{cal} c_{cal} \, \Delta T_{cal} = 0$$

where the subscripts a, w, and cal refer to the alloy, water, and calorimeter, respectively, and each $\Delta T$ is the final temperature (30.5°C) minus the initial temperature for each object. When we put in values and use Table 14–1, this equation becomes

$$(0.150 \, \text{kg})(c_a)(30.5°C - 540°C) + (0.400 \, \text{kg})(4186 \, \text{J/kg} \cdot \text{C°})(30.5°C - 10.0°C)$$
$$+ (0.200 \, \text{kg})(900 \, \text{J/kg} \cdot \text{C°})(30.5°C - 10.0°C) = 0,$$

or

$$-(76.4 \, \text{kg} \cdot \text{C°})c_a + 34{,}300 \, \text{J} + 3690 \, \text{J} = 0.$$

Solving for $c_a$ we obtain:

$$c_a = 497 \, \text{J/kg} \cdot \text{C°} \approx 500 \, \text{J/kg} \cdot \text{C°}.$$

**NOTE** We rounded off because we have ignored any heat transferred to the thermometer and the stirrer (which is used to quicken the heat transfer process and thus reduce heat loss to the outside). To take them into account we would have to add (small) additional terms to the equation.

**PROBLEM SOLVING**

*Be sure to consider all possible sources of energy transfer*

In all Examples and Problems of this sort, be sure to include *all* objects that gain or lose heat (within reason). For simplicity, we have ignored very small masses, such as the thermometer and the stirrer, which will affect the energy balance only very slightly.

## Bomb Calorimeter

**PHYSICS APPLIED**

*Measuring Calorie content*

A **bomb calorimeter** is used to measure the thermal energy released when a substance burns (including foods) to determine their Calorie content. A carefully weighed sample of the substance, with an excess amount of oxygen, is placed in a sealed container (the "bomb"). The bomb is placed in the water of the calorimeter and a fine wire passing into the bomb is then heated to ignite the mixture. The Calorie content of foods determined in this way can be unreliable because our bodies may not metabolize all the available energy (which would be excreted). Careful measurements and calculations need to take this into account.

**FIGURE 14–5** Temperature as a function of the heat added to bring 1.0 kg of ice at −40°C to steam above 100°C.

# 14–5 Latent Heat

When a material changes phase from solid to liquid, or from liquid to gas (see also Section 13–11), a certain amount of energy is involved in this **change of phase**. For example, let us trace what happens when a 1.0-kg block of ice at −40°C is heated at a slow steady rate until all the ice has changed to water, then the (liquid) water is heated to 100°C and changed to steam, and heated further above 100°C, all at 1 atm pressure. As shown in the graph of Fig. 14–5, as the ice is heated starting at −40°C, its temperature rises at a rate of about 2 C°/kcal of heat added (since for ice, $c \approx 0.50 \, \text{kcal/kg·C°}$). However, when 0°C is reached, the temperature stops increasing even though heat is still being added. The ice gradually changes to water in the liquid state, with no change in temperature. After about 40 kcal has been added at 0°C, half the ice remains and half has changed to water. After about 80 kcal, or 330 kJ, has been added, all the ice has changed to water, still at 0°C. Continued addition of heat causes the water's temperature to again increase, now at a rate of 1 C°/kcal. When 100°C is reached, the temperature again remains constant as the heat added changes the liquid water to vapor (steam). About 540 kcal (2260 kJ) is required to change the 1.0 kg of water completely to steam, after which the graph rises again, indicating that the temperature of the steam rises as heat is added.

The heat required to change 1.0 kg of a substance from the solid to the liquid state is called the **heat of fusion**; it is denoted by $L_F$. The heat of fusion of water is 79.7 kcal/kg or, in proper SI units, 333 kJ/kg ($= 3.33 \times 10^5 \, \text{J/kg}$). The heat required to change a substance from the liquid to the vapor phase is called the **heat of vaporization**, $L_V$. For water it is 539 kcal/kg or 2260 kJ/kg. Other substances follow graphs similar to Fig. 14–5, although the melting-point and boiling-point temperatures are different, as are the specific heats and heats of fusion and vaporization. Values for the heats of fusion and vaporization, which are also called the **latent heats**, are given in Table 14–3 for a number of substances.

**TABLE 14–3 Latent Heats** (at 1 atm)

| Substance | Melting Point (°C) | Heat of Fusion | | Boiling Point (°C) | Heat of Vaporization | |
|---|---|---|---|---|---|---|
| | | kJ/kg | kcal/kg[†] | | kJ/kg | kcal/kg[†] |
| Oxygen | −218.8 | 14 | 3.3 | −183 | 210 | 51 |
| Nitrogen | −210.0 | 26 | 6.1 | −195.8 | 200 | 48 |
| Ethyl alcohol | −114 | 104 | 25 | 78 | 850 | 204 |
| Ammonia | −77.8 | 33 | 8.0 | −33.4 | 137 | 33 |
| Water | 0 | 333 | 79.7 | 100 | 2260 | 539 |
| Lead | 327 | 25 | 5.9 | 1750 | 870 | 208 |
| Silver | 961 | 88 | 21 | 2193 | 2300 | 558 |
| Iron | 1538 | 289 | 69.1 | 3023 | 6340 | 1520 |
| Tungsten | 3410 | 184 | 44 | 5900 | 4800 | 1150 |

[†]Numerical values in kcal/kg are the same in cal/g.

The heats of vaporization and fusion also refer to the amount of heat *released* by a substance when it changes from a gas to a liquid, or from a liquid to a solid. Thus, steam releases 2260 kJ/kg when it changes to water, and water releases 333 kJ/kg when it becomes ice. [In these cases of heat release, $Q < 0$ when using the $\Sigma Q = 0$ approach, Eq. 14–3.]

The heat involved in a change of phase depends not only on the latent heat but also on the total mass of the substance. That is,

$$Q = mL, \tag{14–4}$$

where $L$ is the latent heat of the particular process and substance, $m$ is the mass of the substance, and $Q$ is the heat added or released during the phase change. For example, when 5.00 kg of water freezes at 0°C, $(5.00\,\text{kg})(3.33 \times 10^5\,\text{J/kg}) = 1.67 \times 10^6\,\text{J}$ of energy is released.

**EXERCISE B** A pot of water is boiling on a gas stove, and then you turn up the heat. What happens? (*a*) The temperature of the water starts increasing. (*b*) There is a tiny decrease in the rate of water loss by evaporation. (*c*) The rate of water loss by evaporation increases. (*d*) There is an appreciable increase in both the rate of boiling and the temperature of the water. (*e*) None of these.

Calorimetry sometimes involves a change of state, as the following Examples show. Indeed, latent heats are often measured using calorimetry.

---

**EXAMPLE 14–6** **Making ice.** How much energy does a freezer have to remove from 1.5 kg of water at 20°C to make ice at −12°C?

**APPROACH** We need to calculate the total energy removed by adding the heat outflow (1) to reduce the water temperature from 20°C to 0°C, (2) to change the liquid water to solid ice at 0°C, and (3) to lower the ice temperature from 0°C to −12°C.

**SOLUTION** The heat $Q$ that needs to be removed from the 1.5 kg of water is

$$
\begin{aligned}
Q &= mc_\text{w}(20°C - 0°C) + mL_\text{F} + mc_\text{ice}\big[0° - (-12°C)\big] \\
&= (1.5\,\text{kg})(4186\,\text{J/kg}\cdot\text{C}°)(20\,\text{C}°) + (1.5\,\text{kg})(3.33 \times 10^5\,\text{J/kg}) \\
&\quad + (1.5\,\text{kg})(2100\,\text{J/kg}\cdot\text{C}°)(12\,\text{C}°) \\
&= 6.6 \times 10^5\,\text{J} = 660\,\text{kJ}.
\end{aligned}
$$

---

**| EXERCISE C** Which process in Example 14–6 required the greatest heat loss?

---

**EXAMPLE 14–7** **ESTIMATE** **Will all the ice melt?** At a reception, a 0.50-kg chunk of ice at −10°C is placed in 3.0 kg of "iced" tea at 20°C. At what temperature and in what phase will the final mixture be? The tea can be considered as water. Ignore any heat flow to the surroundings, including the container.

**PROBLEM SOLVING**
*First determine (or estimate) the final state*

**APPROACH** Before we can write down an equation applying conservation of energy, we must first check to see if the final state will be all ice, a mixture of ice and water at 0°C, or all water. To bring the 3.0 kg of water at 20°C down to 0°C would require an energy release of

$$m_\text{w}\,c_\text{w}(20°C - 0°C) = (3.0\,\text{kg})(4186\,\text{J/kg}\cdot\text{C}°)(20\,\text{C}°) = 250{,}000\,\text{J}.$$

On the other hand, to raise the ice from −10°C to 0°C would require

$$
\begin{aligned}
m_\text{ice}\,c_\text{ice}\big[0°C - (-10°C)\big] &= (0.50\,\text{kg})(2100\,\text{J/kg}\cdot\text{C}°)(10\,\text{C}°) \\
&= 10{,}500\,\text{J},
\end{aligned}
$$

and to change the ice to water at 0°C would require

$$m_\text{ice}\,L_\text{F} = (0.50\,\text{kg})(333\,\text{kJ/kg}) = 167{,}000\,\text{J}.$$

The sum of the last two quantities is 10.5 kJ + 167 kJ = 177 kJ. This is not enough energy to bring the 3.0 kg of water at 20°C down to 0°C, so we see that the mixture must end up all water, somewhere between 0°C and 20°C.

**SOLUTION** To determine the final temperature $T$, we apply conservation of energy. We present both of the techniques discussed in Section 14–4.

**PROBLEM SOLVING**

*Then determine the final temperature*

*Method 1*: "$\Sigma Q = 0$" gives

$$\begin{pmatrix} \text{heat to raise} \\ \text{0.50 kg of ice} \\ \text{from } -10°C \\ \text{to } 0°C \end{pmatrix} + \begin{pmatrix} \text{heat to change} \\ \text{0.50 kg} \\ \text{of ice} \\ \text{to water} \end{pmatrix} + \begin{pmatrix} \text{heat to raise} \\ \text{0.50 kg of water} \\ \text{from } 0°C \\ \text{to } T \end{pmatrix} + \begin{pmatrix} \text{heat lost by} \\ \text{3.0 kg of} \\ \text{water cooling} \\ \text{from } 20°C \text{ to } T \end{pmatrix} = 0.$$

Using some of the results from the "Approach" above, we obtain

$$10{,}500 \text{ J} + 167{,}000 \text{ J} + (0.50 \text{ kg})(4186 \text{ J/kg} \cdot C°)(T - 0°C)$$
$$+ (3.0 \text{ kg})(4186 \text{ J/kg} \cdot C°)(T - 20°C) = 0.$$

Solving for $T$ we obtain

$$T = 5.0°C.$$

*Method 2*: "heat gained = heat lost" produces a word equation like the one above (the last plus sign becomes an equals sign and we lose the " = 0"). That is,

$$10{,}500 \text{ J} + 167{,}000 \text{ J} + (0.50 \text{ kg})(4186 \text{ J/kg} \cdot C°)(T - 0°C)$$
$$= (3.0 \text{ kg})(4186 \text{ J/kg} \cdot C°)(20°C - T).$$

The term on the right is heat lost by 3.0 kg of water cooling from 20°C to $T$; here $\Delta T = T_i - T_f = 20°C - T$ for this approach. Algebraically, this equation is identical to the one above in the first method.[†]

**EXERCISE D** How much more ice at $-10°C$ would be needed in Example 14–7 to bring the tea down to 0°C, while just melting all the ice?

## Calorimetry

1. Be sure you have sufficient information to apply energy conservation. Ask yourself: **is the system isolated** (or nearly so, enough to get a good estimate)? Do we know or can we calculate all significant sources of energy transfer?

2. Apply **conservation of energy**. Either write

   $$\text{heat gained} = \text{heat lost},$$

   or use

   $$\Sigma Q = 0.$$

3. If **no phase changes** occur, each term in the energy conservation equation will have the form

   $$Q = mc \, \Delta T.$$

4. If **phase changes** do or might occur, there may be terms in the energy conservation equation of the form $Q = mL$, where $L$ is the latent heat. But *before* applying energy conservation, determine (or estimate) in which phase the final state will be, as we did in Example 14–7 by calculating the different contributing values for heat $Q$.

5. Note that when the system reaches thermal **equilibrium**, the final **temperature** of each substance will have the *same* value. There is only one $T_f$.

6. **Solve** your energy equation for the unknown.

### Evaporation

The latent heat to change a liquid to a gas is needed not only at the boiling point. Water can change from the liquid to the gas phase even at room temperature. This process is called **evaporation** (see previous Chapter, Section 13–12). The value of the heat of vaporization of water increases slightly with a decrease in temperature: at 20°C, for example, it is 2450 kJ/kg (585 kcal/kg) compared to 2260 kJ/kg (= 539 kcal/kg) at 100°C. When water evaporates, the remaining liquid cools, because the energy required (the latent heat of vaporization) comes from the water itself; so its internal energy, and therefore its temperature, must drop.[‡]

---

[†]For any algebraic equation $A = B$, if you subtract $B$ from both sides you get $A - B = B - B = 0$, or $A - B = 0$. Thus, if you move a term from one side of an equals sign to the other, the sign (+ or −) changes.

[‡]Also from the point of view of kinetic theory, evaporation is a cooling process because it is the fastest-moving molecules that escape from the surface (Section 13–12). Hence the average speed of the remaining molecules is less, so by Eq. 13–8 the temperature is less.

Evaporation of water from skin is one of the most important ways the body uses to control its temperature. When the temperature of the blood rises slightly above normal, the hypothalamus region of the brain detects this temperature increase and sends a signal to the sweat glands to increase their production. The energy (latent heat) needed to vaporize this water comes from the body, so the body cools.

### Kinetic Theory of Latent Heats

We can make use of kinetic theory to see why energy is needed to melt or vaporize a substance. At the melting point, the latent heat of fusion does not act to increase the average kinetic energy (and the temperature) of the molecules in the solid, but instead is used to overcome the potential energy associated with the forces between the molecules. That is, work must be done against these attractive forces to break the molecules loose from their relatively fixed positions in the solid so they can freely roll over one another in the liquid phase. Similarly, energy is required for molecules held close together in the liquid phase to escape into the gaseous phase where they are far apart. This process is a more energetic reorganization of the molecules than is melting (the average distance between the molecules is greatly increased), and hence the heat of vaporization is generally much greater than the heat of fusion for a given substance.

# 14–6 Heat Transfer: Conduction

Heat is transferred from one place or object to another in three different ways: by *conduction,* by *convection,* and by *radiation.* We now discuss each of these in turn; but in practical situations, any two or all three may be operating at the same time. This Section deals with conduction.

When a metal poker is put in a hot fire, or a silver spoon is placed in a hot bowl of soup, the exposed end of the poker or spoon soon becomes hot as well, even though it is not directly in contact with the source of heat. We say that heat has been *conducted* from the hot end to the cold end.

Heat **conduction** in many materials can be visualized as being carried out via molecular collisions. As one end of an object is heated, the molecules there move faster and faster (= higher temperature). As these faster molecules collide with slower-moving neighbors, they transfer some of their kinetic energy to them, which in turn transfer some energy by collision with molecules still farther along the object. Thus the kinetic energy of thermal motion is transferred by molecular collision along the object. In metals, collisions of free electrons are mainly responsible for conduction. Conduction between objects in physical contact occurs similarly.

Heat conduction from one point to another takes place only if there is a difference in temperature between the two points. Indeed, it is found experimentally that the rate of heat flow through a substance is proportional to the difference in temperature between its ends. The rate of heat flow also depends on the size and shape of the object. To investigate this quantitatively, let us consider the heat flow through a uniform cylinder, as illustrated in Fig. 14–6. It is found experimentally that the heat flow $Q$ over a time interval $t$ is given by the relation

$$\frac{Q}{t} = kA\frac{T_1 - T_2}{\ell} \tag{14–5}$$

where $A$ is the cross-sectional area of the object, $\ell$ is the distance between the two ends, which are at temperatures $T_1$ and $T_2$, and $k$ is a proportionality constant called the **thermal conductivity** which is characteristic of the material. From Eq. 14–5, we see that the rate of heat flow (units of J/s) is directly proportional to the cross-sectional area and to the temperature gradient[†] $(T_1 - T_2)/\ell$.

**FIGURE 14–6** Heat conduction between regions at temperatures $T_1$ and $T_2$. If $T_1$ is greater than $T_2$, the heat flows to the right; the rate is given by Eq. 14–5.

[†]Equation 14–5 is quite similar to the relations describing diffusion (Section 13–13) and the flow of fluids through a pipe (Section 10–12). In those cases, the flow of matter was found to be proportional to the concentration gradient $(C_1 - C_2)/\ell$, or to the pressure gradient $(P_1 - P_2)/\ell$. This close similarity is one reason we speak of the "flow" of heat. Yet we must keep in mind that no substance is flowing in the case of heat—it is energy that is being transferred.

The thermal conductivities, $k$, for a variety of substances are given in Table 14–4. Substances for which $k$ is large conduct heat rapidly and are said to be good **thermal conductors**. Most metals fall in this category, although there is a wide range even among them, as you may observe by holding the ends of a silver spoon and a stainless-steel spoon immersed in the same hot cup of soup. Substances for which $k$ is small, such as wool, fiberglass, polyurethane, and goose down, are poor conductors of heat and are therefore good **thermal insulators**. The relative magnitudes of $k$ can explain simple phenomena such as why a tile floor is much colder on the feet than a rug-covered floor at the same temperature. Tile is a better conductor of heat than the rug. Heat that flows from your foot to the rug is not conducted away rapidly, so the rug's surface quickly warms up to the temperature of your foot and feels good. But the tile conducts the heat away rapidly and thus can take more heat from your foot quickly, so your foot's surface temperature drops.

**TABLE 14–4**
**Thermal Conductivities**

| Substance | Thermal Conductivity, $k$ | |
|---|---|---|
| | $\dfrac{J}{(s \cdot m \cdot C°)}$ | $\dfrac{kcal}{(s \cdot m \cdot C°)}$ |
| Silver | 420 | $10 \times 10^{-2}$ |
| Copper | 380 | $9.2 \times 10^{-2}$ |
| Aluminum | 200 | $5.0 \times 10^{-2}$ |
| Steel | 40 | $1.1 \times 10^{-2}$ |
| Ice | 2 | $5 \times 10^{-4}$ |
| Glass | 0.84 | $2.0 \times 10^{-4}$ |
| Brick | 0.84 | $2.0 \times 10^{-4}$ |
| Concrete | 0.84 | $2.0 \times 10^{-4}$ |
| Water | 0.56 | $1.4 \times 10^{-4}$ |
| Human tissue | 0.2 | $0.5 \times 10^{-4}$ |
| Wood | 0.1 | $0.3 \times 10^{-4}$ |
| Fiberglass | 0.048 | $0.12 \times 10^{-4}$ |
| Cork | 0.042 | $0.10 \times 10^{-4}$ |
| Wool | 0.040 | $0.10 \times 10^{-4}$ |
| Goose down | 0.025 | $0.060 \times 10^{-4}$ |
| Polyurethane | 0.024 | $0.057 \times 10^{-4}$ |
| Air | 0.023 | $0.055 \times 10^{-4}$ |

**EXAMPLE 14–8** | **Heat loss through windows.** A major source of heat loss from a house in cold weather is through the windows. Calculate the rate of heat flow through a glass window 2.0 m $\times$ 1.5 m in area and 3.2 mm thick, if the temperatures at the inner and outer surfaces are 15.0°C and 14.0°C, respectively (Fig. 14–7).

**APPROACH** Heat flows by conduction through the 3.2-mm thickness of glass from the higher inside temperature to the lower outside temperature. We use the heat conduction equation, Eq. 14–5.

**SOLUTION** Here $A = (2.0\,\text{m})(1.5\,\text{m}) = 3.0\,\text{m}^2$ and $\ell = 3.2 \times 10^{-3}\,\text{m}$. Using Table 14–4 to get $k$, we have

$$\frac{Q}{t} = kA\frac{T_1 - T_2}{\ell}$$

$$= \frac{(0.84\,\text{J/s}\cdot\text{m}\cdot\text{C}°)(3.0\,\text{m}^2)(15.0°C - 14.0°C)}{(3.2 \times 10^{-3}\,\text{m})}$$

$$= 790\,\text{J/s}.$$

**NOTE** This rate of heat flow is equivalent to $(790\,\text{J/s})/(4.19 \times 10^3\,\text{J/kcal}) = 0.19\,\text{kcal/s}$, or $(0.19\,\text{kcal/s}) \times (3600\,\text{s/h}) = 680\,\text{kcal/h}$.

You might notice in Example 14–8 that 15°C is not very warm for the living room of a house. The room itself may indeed be much warmer, and the outside might be colder than 14°C. But the temperatures of 15°C and 14°C were specified as those at the window surfaces, and there is usually a considerable drop in temperature of the air in the vicinity of the window both on the inside and the outside. That is, the layer of air on either side of the window acts as an insulator, and normally the major part of the temperature drop between the inside and outside of the house takes place across the air layer. If there is a heavy wind, the air outside a window will constantly be replaced with cold air; the temperature gradient across the glass will be greater and there will be a much greater rate of heat loss. Increasing the width of the air layer, such as using two panes of glass separated by an air gap, will reduce the heat loss more than simply increasing the glass thickness, since the thermal conductivity of air is much less than that for glass. Such "double-pane windows" are often called *thermal windows*.

The insulating properties of clothing come from the insulating properties of air. Without clothes, our bodies in still air would heat the air in contact with the skin and would soon become reasonably comfortable because air is a very good insulator. But since air moves—there are breezes and drafts, and people move about—the warm air would be replaced by cold air, thus increasing the temperature difference and the heat loss from the body. Clothes keep us warm by trapping air so it cannot move readily. It is not the cloth that insulates us, but the air that the cloth traps. Goose down is a very good insulator because even a small amount of it fluffs up and traps a great amount of air.

| **EXERCISE E** Explain why drapes in front of a window reduce heat loss from a house.

**PHYSICS APPLIED**
*Heat loss through windows*

**FIGURE 14–7** Example 14–8.

**PHYSICS APPLIED**
*Thermal windows*

**PHYSICS APPLIED**
*Clothes insulate by trapping an air layer*

**TABLE 14–5  R-values**

| Material | Thickness | R-value (ft²·h·F°/Btu) |
|---|---|---|
| Glass | $\frac{1}{8}$ inch | 1 |
| Brick | $3\frac{1}{2}$ inches | 0.6–1 |
| Plywood | $\frac{1}{2}$ inch | 0.6 |
| Fiberglass insulation | 4 inches | 12 |

Cooler water

Hotter water

**FIGURE 14–8** Convection currents in a pot of water being heated on a stove.

**PHYSICS APPLIED**

*Ocean currents and wind*

**PHYSICS APPLIED**

*Convective home heating*

**FIGURE 14–9** Convection plays a role in heating a house. The circular arrows show convective air currents in the rooms.

Radiator

Furnace  Hot water

Cool

**PHYSICS APPLIED**

*Body heat: convection by blood*

## *R-values for Building Materials

The insulating properties of building materials are often specified by *R*-values (or "thermal resistance"), defined for a given thickness $\ell$ of material as:

$$R = \frac{\ell}{k}.$$

The *R*-value of a given piece of material combines the thickness $\ell$ and the thermal conductivity $k$ in one number. Larger *R* means better insulation from heat or cold. In the United States, *R*-values are given in British units as ft²·h·F°/Btu (for example, *R*-19 means $R = 19$ ft²·h·F°/Btu). Table 14–5 gives *R*-values for some common building materials. *R*-values increase directly with material thickness: for example, 2 inches of fiberglass is *R*-6, whereas 4 inches is *R*-12.

# 14–7  Heat Transfer: Convection

Although liquids and gases are generally not very good conductors of heat, they can transfer heat rapidly by convection. **Convection** is the process whereby heat flows by the bulk movement of molecules from one place to another. Whereas conduction involves molecules (and/or electrons) moving only over small distances and colliding, convection involves the movement of large numbers of molecules over large distances.

A forced-air furnace, in which air is heated and then blown by a fan into a room, is an example of *forced convection. Natural convection* occurs as well, and one familiar example is that hot air rises. For instance, the air above a radiator (or other type of heater) expands as it is heated (Chapter 13), and hence its density decreases. Because its density is less than that of the surrounding cooler air, it rises via buoyancy, just as a log submerged in water floats upward because its density is less than that of water. Warm or cold ocean currents, such as the balmy Gulf Stream, represent natural convection on a global scale. Wind is another example of convection, and weather in general is strongly influenced by convective air currents.

When a pot of water is heated (Fig. 14–8), convection currents are set up as the heated water at the bottom of the pot rises because of its reduced density. That heated water is replaced by cooler water from above. This principle is used in many heating systems, such as the hot-water radiator system shown in Fig. 14–9. Water is heated in the furnace, and as its temperature increases, it expands and rises as shown. This causes the water to circulate in the heating system. Hot water then enters the radiators, heat is transferred by conduction to the air, and the cooled water returns to the furnace. Thus, the water circulates because of convection; pumps are sometimes used to improve circulation. The air throughout the room also becomes heated as a result of convection. The air heated by the radiators rises and is replaced by cooler air, resulting in convective air currents, as shown by the green arrows in Fig. 14–9.

Other types of furnaces also depend on convection. Hot-air furnaces with registers (openings) near the floor often do not have fans but depend on natural convection, which can be appreciable. In other systems, a fan is used. In either case, it is important that cold air can return to the furnace so that convective currents circulate throughout the room if the room is to be uniformly heated.

The human body produces a great deal of thermal energy. Of the food energy transformed within the body, at best 20% is used to do work, so over 80% appears as thermal energy. During light activity, for example, if this thermal energy were not dissipated, the body temperature would rise about 3 C° per hour. Clearly, the heat generated by the body must be transferred to the outside. Is the heat transferred by conduction? The temperature of the skin in a comfortable environment is 33 to 35°C, whereas the interior of the body is at 37°C. Calculation shows (Problem 52) that, because of this small temperature difference, plus the low thermal conductivity of tissue, direct conduction is responsible for very little of the heat that must leave the body. Instead, the heat is carried to the skin by the blood. In addition to all its other important responsibilities, blood acts as a convective fluid to transfer heat to just beneath the surface of the skin. It is then conducted (over a very short distance) to the surface. Once at the surface, the heat is transferred to the environment by convection, evaporation, and radiation (Section 14–8).

# 14–8 Heat Transfer: Radiation

Convection and conduction require the presence of matter as a medium to carry the heat from the hotter to the colder region. But a third type of heat transfer occurs without any medium at all. All life on Earth depends on the transfer of energy from the Sun, and this energy is transferred to the Earth over empty (or nearly empty) space. This form of energy transfer is heat—since the Sun's surface temperature (6000 K) is much higher than Earth's ($\approx 300$ K)—and is referred to as **radiation** (Fig. 14–10). The warmth we receive from a fire is mainly radiant energy.

As we shall see in later Chapters, radiation consists essentially of electromagnetic waves. Suffice it to say for now that radiation from the Sun consists of visible light plus many other wavelengths that the eye is not sensitive to, including infrared (IR) radiation, which is mainly responsible for heating the Earth.

The rate at which an object radiates energy has been found to be proportional to the fourth power of the Kelvin temperature, $T$. That is, an object at 2000 K, as compared to one at 1000 K, radiates energy at a rate $2^4 = 16$ times as much. The rate of radiation is also proportional to the area $A$ of the emitting object, so the rate at which energy leaves the object, $Q/t$, is

$$\frac{Q}{t} = \epsilon \sigma A T^4. \tag{14–6}$$

This is called the **Stefan-Boltzmann equation**, and $\sigma$ is a universal constant called the **Stefan-Boltzmann constant** which has the value

$$\sigma = 5.67 \times 10^{-8} \, \text{W/m}^2 \cdot \text{K}^4.$$

The factor $\epsilon$, called the **emissivity**, is a number[†] between 0 and 1 that is characteristic of the surface of the radiating material. Very black surfaces, such as charcoal, have emissivity close to 1, whereas shiny metal surfaces have $\epsilon$ close to zero and thus emit correspondingly less radiation. The value of $\epsilon$ depends somewhat on the temperature of the material.

Not only do shiny surfaces emit less radiation, but they absorb little of the radiation that falls upon them (most is reflected). Black and very dark objects are good emitters ($\epsilon \approx 1$); they also absorb nearly all the radiation that falls on them—which is why light-colored clothing is usually preferable to dark clothing on a hot day. Thus, **a good absorber is also a good emitter**.

Any object not only emits energy by radiation but also absorbs energy radiated by other objects. If an object of emissivity $\epsilon$ and area $A$ is at a temperature $T_1$, it radiates energy at a rate $\epsilon \sigma A T_1^4$. If the object is surrounded by an environment at temperature $T_2$, the rate at which the surroundings radiate energy is proportional to $T_2^4$; then the rate that energy is *absorbed* by the object is proportional to $T_2^4$. The *net* rate of radiant heat flow from the object is given by the equation

$$\frac{Q}{t} = \epsilon \sigma A (T_1^4 - T_2^4), \tag{14–7}$$

where $A$ is the surface area of the object, $T_1$ its temperature and $\epsilon$ its emissivity (at temperature $T_1$), and $T_2$ is the temperature of the surroundings. This equation is consistent with the experimental fact that equilibrium between the object and its surroundings is reached when they come to the same temperature. That is, $Q/t$ must equal zero when $T_1 = T_2$, so $\epsilon$ must be the same for emission and absorption. This confirms the idea that a good emitter is a good absorber. Because both the object and its surroundings radiate energy, there is a net transfer of energy from one to the other unless everything is at the same temperature. From Eq. 14–7 it is clear that if $T_1 > T_2$, the net flow of heat is from the object to the surroundings, so the object cools. But if $T_1 < T_2$, the net heat flow is from the surroundings into the object, and the object's temperature rises. If different parts of the surroundings are at different temperatures, Eq. 14–7 becomes more complicated.

[†]$\epsilon$ is the Greek letter epsilon.

**FIGURE 14–10** The Sun's surface radiates at 6000 K—much higher than the Earth's surface.

**PHYSICS APPLIED**
*Dark vs. light clothing*

**EXAMPLE 14–9 | ESTIMATE | Cooling by radiation.** An athlete is sitting unclothed in a locker room whose dark walls are at a temperature of 15°C. Estimate the body's rate of heat loss by radiation, assuming a skin temperature of 34°C and $\epsilon = 0.70$. Take the surface area of the body not in contact with the chair to be $1.5 \text{ m}^2$.

**APPROACH** We use Eq. 14–7, which requires Kelvin temperatures.

**SOLUTION** We have

$$\frac{Q}{t} = \epsilon \sigma A (T_1^4 - T_2^4)$$
$$= (0.70)(5.67 \times 10^{-8} \text{ W/m}^2 \cdot \text{K}^4)(1.5 \text{ m}^2)[(307 \text{ K})^4 - (288 \text{ K})^4] = 120 \text{ W}.$$

**NOTE** This person's "output" is a bit more than what a 100-W bulb uses.

**NOTE** Avoid a common error: $(T_1^4 - T_2^4) \neq (T_1 - T_2)^4$.

A resting person naturally produces heat internally at a rate of about 100 W, as we will see in Section 15–3, less than the heat loss by radiation as calculated in Example 14–9. Hence, our person's temperature would drop, causing considerable discomfort. The body responds to excessive heat loss by increasing its metabolic rate, and shivering is one method by which the body increases its metabolism. Naturally, clothes help a lot. Example 14–9 illustrates that a person may be uncomfortable even if the temperature of the air is, say, 25°C, which is quite a warm room. If the walls or floor are cold, radiation to them occurs no matter how warm the air is. Indeed, it is estimated that radiation accounts for about 50% of the heat loss from a sedentary person in a normal room. Rooms are most comfortable when the walls and floor are warm and the air is not so warm. Floors and walls can be heated by means of hot-water conduits or electric heating elements. Such first-rate heating systems are becoming more common today, and it is interesting to note that 2000 years ago the Romans, even in houses in the remote province of Great Britain, made use of hot-water and steam conduits in the floor to heat their houses.

**EXAMPLE 14–10 | ESTIMATE | Two teapots.** A ceramic teapot ($\epsilon = 0.70$) and a shiny one ($\epsilon = 0.10$) each hold 0.75 L of tea at 95°C. (*a*) Estimate the rate of heat loss from each, and (*b*) estimate the temperature drop after 30 min for each. Consider only radiation, and assume the surroundings are at 20°C.

**APPROACH** We are given all the information necessary to calculate the heat loss due to radiation, except for the area. The teapot holds 0.75 L, and we can approximate it as a cube 10 cm on a side (volume = 1.0 L), with five sides exposed. To estimate the temperature drop in (*b*), we use the concept of specific heat and ignore the contribution of the pots compared to that of the water.

**SOLUTION** (*a*) The teapot, approximated by a cube 10 cm on a side with five sides exposed, has a surface area of about $5 \times (0.1 \text{ m})^2 = 5 \times 10^{-2} \text{ m}^2$. The rate of heat loss would be about

$$\frac{Q}{t} = \epsilon \sigma A (T_1^4 - T_2^4)$$
$$= \epsilon (5.67 \times 10^{-8} \text{ W/m}^2 \cdot \text{K}^4)(5 \times 10^{-2} \text{ m}^2)[(368 \text{ K})^4 - (293 \text{ K})^4] \approx \epsilon (30) \text{ W},$$

or about 20 W for the ceramic pot ($\epsilon = 0.70$) and 3 W for the shiny one ($\epsilon = 0.10$).

(*b*) To estimate the temperature drop, we use the specific heat of water and ignore the contribution of the pots. The mass of 0.75 L of water is 0.75 kg. (Recall that $1.0 \text{ L} = 1000 \text{ cm}^3 = 1 \times 10^{-3} \text{ m}^3$ and $\rho = 1000 \text{ kg/m}^3$.) Using Eq. 14–2 and Table 14–1, we get

$$\frac{Q}{t} = mc \frac{\Delta T}{t}.$$

Then

$$\frac{\Delta T}{t} = \frac{Q/t}{mc} \approx \frac{\epsilon (30) \text{ J/s}}{(0.75 \text{ kg})(4186 \text{ J/kg} \cdot \text{C}°)} \approx \epsilon (0.01) \text{ C}°/\text{s}.$$

After 30 min (1800 s), $\Delta T = \epsilon (0.01 \text{ C}°/\text{s})t = \epsilon (0.01 \text{ C}°/\text{s})(1800 \text{ s}) \approx 18\epsilon \text{ C}°$, or about 12 C° for the ceramic pot ($\epsilon = 0.70$) and about 2 C° for the shiny one ($\epsilon = 0.10$). The shiny one clearly has an advantage, at least in terms of radiation.

**NOTE** Convection and conduction could play a greater role than radiation.

Heating of an object by radiation from the Sun cannot be calculated using Eq. 14–7 since this equation assumes a uniform temperature, $T_2$, of the environment surrounding the object, whereas the Sun is essentially a point source. Hence the Sun must be treated as a separate source of energy. About 1350 J of energy from the Sun strikes Earth's atmosphere per second per square meter of area at right angles to the Sun's rays. This number, 1350 W/m², is called the **solar constant**. The atmosphere may absorb as much as 70% of this energy before it reaches the ground, depending on the cloud cover. On a clear day, about 1000 W/m² reaches the Earth's surface. An object of emissivity $\epsilon$ with area $A$ facing the Sun absorbs energy from the Sun at a rate, in watts, of about

$$\frac{Q}{t} = (1000 \text{ W/m}^2)\epsilon A \cos \theta, \qquad (14\text{–}8)$$

where $\theta$ is the angle between the Sun's rays and a line perpendicular to the area $A$ (Fig. 14–11). That is, $A \cos \theta$ is the "effective" area, at right angles to the Sun's rays.

The explanation for the **seasons** and the polar ice caps (see Fig. 14–12) depends on this $\cos \theta$ factor in Eq. 14–8. The seasons are *not* a result of how close the Earth is to the Sun—in fact, in the Northern Hemisphere, summer occurs when the Earth is farthest from the Sun. It is the angle (i.e., $\cos \theta$) that really matters. Furthermore, the reason the Sun heats the Earth more at midday than at sunrise or sunset is also related to this $\cos \theta$ factor.

---

**EXAMPLE 14–11** | **ESTIMATE** | **Getting a tan—energy absorption.** What is the rate of energy absorption from the Sun by a person lying flat on the beach on a clear day if the Sun makes a 30° angle with the vertical? Assume that $\epsilon = 0.70$ and that 1000 W/m² reaches the Earth's surface.

**APPROACH** We use Eq. 14–8 and estimate a typical human to be roughly 2 m tall by 0.4 m wide, so $A \approx (2 \text{ m})(0.4 \text{ m}) = 0.8 \text{ m}^2$. See Fig. 14–11.

**SOLUTION** Since $\cos 30° = 0.866$, we have

$$\frac{Q}{t} = (1000 \text{ W/m}^2)\epsilon A \cos \theta$$
$$= (1000 \text{ W/m}^2)(0.70)(0.8 \text{ m}^2)(0.866) = 500 \text{ W}.$$

**NOTE** If a person wears light-colored clothing, $\epsilon$ is much smaller, so the energy absorbed is less.

---

An interesting application of thermal radiation to diagnostic medicine is **thermography**. A special instrument, the thermograph, scans the body, measuring the intensity of infrared[†] radiation from many points and forming a picture that resembles an X-ray (Fig. 14–13). Areas where metabolic activity is high, such as in tumors, can often be detected on a thermogram as a result of their higher temperature and consequent increased radiation.

[†]Infrared radiation is light whose wavelengths are longer than visible light (see Fig. 22–8).

(a)

(b)

**PHYSICS APPLIED**
*Radiation from the Sun*

**FIGURE 14–11** Radiant energy striking a body at an angle $\theta$.

**FIGURE 14–12** (a) Earth's seasons arise from the tilt of Earth's axis relative to its orbit around the Sun. (b) June sunlight makes an angle of about $23\frac{1}{2}°$ with the equator. Thus $\theta$ in the southern U.S. (label A) is near 0° (direct summer sunlight), but in the Southern Hemisphere (B), $\theta$ is 50° or 60°, and less heat can be absorbed—hence it is winter. Near the poles (C), there is never strong direct sunlight: $\cos \theta$ varies from about $\frac{1}{2}$ in summer to 0 in winter; so with little heating, ice can form.

(a)

(b)

**PHYSICS APPLIED**
*Thermography*

**FIGURE 14–13** Thermograms of a healthy person's arms and hands (a) before and (b) after smoking a cigarette, showing a temperature decrease due to impaired blood circulation associated with smoking. The thermograms have been color-coded according to temperature; the scale on the right goes from blue (cold) to white (hot).

**EXAMPLE 14–12** **ESTIMATE** **Star radius.** The giant star Betelgeuse emits radiant energy at a rate $10^4$ times greater than our Sun, whereas its surface temperature is only half (2900 K) that of our Sun. Estimate the radius of Betelgeuse, assuming $\epsilon = 1$. The Sun's radius is $r_S = 7 \times 10^8$ m.

**APPROACH** We assume both stars are spherical, with surface area $4\pi r^2$.

**SOLUTION** We solve Eq. 14–6 for $A$:

$$4\pi r^2 = A = \frac{(Q/t)}{\epsilon \sigma T^4}.$$

Then

$$\frac{r_B^2}{r_S^2} = \frac{(Q/t)_B}{(Q/t)_S} \cdot \frac{T_S^4}{T_B^4} = (10^4)(2^4) = 16 \times 10^4.$$

Hence $r_B = \sqrt{16 \times 10^4}\, r_S = (400)(7 \times 10^8 \text{ m}) \approx 3 \times 10^{11}$ m. If Betelgeuse were our Sun, it would envelop us (Earth is $1.5 \times 10^{11}$ m from the Sun).

**EXERCISE F** Fanning yourself on a hot day cools you by (*a*) increasing the radiation rate of the skin; (*b*) increasing conductivity; (*c*) decreasing the mean free path of air; (*d*) increasing the evaporation of perspiration; (*e*) none of these.

# Summary

**Internal energy**, $U$, refers to the total energy of all the molecules in an object. For an ideal monatomic gas,

$$U = \tfrac{3}{2}NkT = \tfrac{3}{2}nRT \qquad \textbf{(14–1)}$$

where $N$ is the number of molecules or $n$ is the number of moles.

**Heat** refers to the transfer of energy from one object to another because of a difference of temperature. Heat is thus measured in energy units, such as joules.

Heat and internal energy are also sometimes specified in calories or kilocalories (kcal), where

$$1 \text{ kcal} = 4.186 \text{ kJ}$$

is the amount of heat needed to raise the temperature of 1 kg of water by 1 C°.

The **specific heat**, $c$, of a substance is defined as the energy (or heat) required to change the temperature of unit mass of the substance by 1 degree; as an equation,

$$Q = mc\,\Delta T, \qquad \textbf{(14–2)}$$

where $Q$ is the heat absorbed or given off, $\Delta T$ is the temperature increase or decrease, and $m$ is the mass of the substance.

When heat flows between parts of an isolated system, conservation of energy tells us that the heat gained by one part of the system is equal to the heat lost by the other part of the system. This is the basis of **calorimetry**, which is the quantitative measurement of heat exchange.

Exchange of energy occurs, without a change in temperature, whenever a substance changes phase. The **heat of fusion** is the heat required to melt 1 kg of a solid into the liquid phase; it is also equal to the heat given off when the substance changes from liquid to solid. The **heat of vaporization** is the energy required to change 1 kg of a substance from the liquid to the vapor phase; it is also the energy given off when the substance changes from vapor to liquid.

Heat is transferred from one place (or object) to another in three different ways: conduction, convection, and radiation.

In **conduction**, energy is transferred through a substance by means of collisions between hotter (faster) molecules or electrons with their slower moving neighbors.

**Convection** is the transfer of energy by the mass movement of molecules even over considerable distances.

**Radiation**, which does not require the presence of matter, is energy transfer by electromagnetic waves, such as from the Sun. All objects radiate energy in an amount that is proportional to the fourth power of their Kelvin temperature $(T^4)$ and to their surface area. The energy radiated (or absorbed) also depends on the nature of the surface (dark surfaces absorb and radiate more than do bright shiny ones), which is characterized by the emissivity, $\epsilon$.

Radiation from the Sun arrives at the surface of the Earth on a clear day at a rate of about 1000 W/m².

# Questions

1. What happens to the work done on a jar of orange juice when it is vigorously shaken?

2. When a hot object warms a cooler object, does temperature flow between them? Are the temperature changes of the two objects equal? Explain.

3. (*a*) If two objects of different temperatures are placed in contact, will heat naturally flow from the object with higher internal energy to the object with lower internal energy? (*b*) Is it possible for heat to flow even if the internal energies of the two objects are the same? Explain.

4. In warm regions where tropical plants grow but the temperature may drop below freezing a few times in the winter, the destruction of sensitive plants due to freezing can be reduced by watering them in the evening. Explain.

5. The specific heat of water is quite large. Explain why this fact makes water particularly good for heating systems (that is, hot-water radiators).

6. Why does water in a metal canteen stay cooler if the cloth jacket surrounding the canteen is kept moist?

7. Explain why burns caused by steam at 100°C on the skin are often more severe than burns caused by water at 100°C.

8. Explain why water cools (its temperature drops) when it evaporates, using the concepts of latent heat and internal energy.

9. Will pasta cook faster if the water boils more vigorously? Explain.

10. Very high in the Earth's atmosphere, the temperature can be 700°C. Yet an animal there would freeze to death rather than roast. Explain.

11. Explorers on failed Arctic expeditions have survived by covering themselves with snow. Why would they do that?

12. Why is wet sand at a beach cooler to walk on than dry sand?

13. If you hear that an object has "high heat content," does that mean that its temperature is high? Explain.

14. When hot-air furnaces are used to heat a house, why is it important that there be a vent for air to return to the furnace? What happens if this vent is blocked by a bookcase?

15. Ceiling fans are sometimes reversible, so that they drive the air down in one season and pull it up in another season. Explain which way you should set the fan (a) for summer, (b) for winter.

16. Goose down sleeping bags and parkas are often specified as so many inches or centimeters of *loft*, the actual thickness of the garment when it is fluffed up. Explain.

17. Microprocessor chips have a "heat sink" glued on top that looks like a series of fins. Why are they shaped like that?

18. Sea breezes are often encountered on sunny days at the shore of a large body of water. Explain, noting that the temperature of the land rises more rapidly than that of the nearby water.

19. The floor of a house on a foundation under which the air can flow is often cooler than a floor that rests directly on the ground (such as a concrete slab foundation). Explain.

20. A 22°C day is warm, while a swimming pool at 22°C feels cool. Why?

21. Explain why air temperature readings are always taken with the thermometer in the shade.

22. A premature baby in an incubator can be dangerously cooled even when the air temperature in the incubator is warm. Explain.

23. Does an ordinary electric fan cool the air? Why or why not? If not, why use it?

24. Heat loss occurs through windows by the following processes: (1) through the glass panes; (2) through the frame, particularly if it is metal; (3) ventilation around edges; and (4) radiation. (a) For the first three, what is (are) the mechanism(s): conduction, convection, or radiation? (b) Heavy curtains reduce which of these heat losses? Explain in detail.

25. A piece of wood lying in the Sun absorbs more heat than a piece of shiny metal. Yet the metal feels hotter than the wood when you pick it up. Explain.

26. The Earth cools off at night much more quickly when the weather is clear than when cloudy. Why?

27. An "emergency blanket" is a thin shiny (metal-coated) plastic foil. Explain how it can help to keep an immobile person warm.

28. Explain why cities situated by the ocean tend to have less extreme temperatures than inland cities at the same latitude.

29. A paper cup placed among hot coals will burn if empty (note burn spots at top of cup in Fig. 14–14), but won't burn if filled with water. Explain. Forget the marshmallows.

**FIGURE 14–14** Question 29.

30. On a cold windy day, a window will feel colder than on an equally cold day with no wind. This is true even if no air leaks in near the window. Why?

# MisConceptual Questions

1. When you put an ice cube in a glass of warm tea, which of the following happens?
   (a) Cold flows from the ice cube into the tea.
   (b) Cold flows from the ice cube into the tea and heat flows from the tea into the ice cube.
   (c) Heat flows from the tea into the ice cube.
   (d) Neither heat nor cold flows. Only temperature flows between the ice and the tea.

2. Both beakers A and B in Fig. 14–15 contain a mixture of ice and water at equilibrium. Which beaker is the coldest, or are they equal in temperature?
   (a) Beaker A.
   (b) Beaker B.
   (c) Equal.

**FIGURE 14–15**
MisConceptual
Question 2.        A        B

3. For objects at thermal equilibrium, which of the following is true?
   (a) Each is at the same temperature.
   (b) Each has the same internal energy.
   (c) Each has the same heat.
   (d) All of the above.
   (e) None of the above.

4. Which of the following happens when a material undergoes a phase change?
   (a) The temperature changes.
   (b) The chemical composition changes.
   (c) Heat flows into or out of the material.
   (d) The molecules break apart into atoms.

5. As heat is added to water, is it possible for the temperature measured by a thermometer in the water to remain constant?
   (a) Yes, the water could be changing phase.
   (b) No, adding heat will always change the temperature.
   (c) Maybe; it depends on the rate at which the heat is added.
   (d) Maybe; it depends on the initial water temperature.

6. A typical thermos bottle has a thin vacuum space between the shiny inner flask (which holds a liquid) and the shiny protective outer flask, often stainless steel. The vacuum space is excellent at preventing
   (a) conduction.
   (b) convection.
   (c) radiation.
   (d) conduction and convection.
   (e) conduction, convection, and radiation.

7. Heat is
   (a) a fluid called caloric.
   (b) a measure of the average kinetic energy of atoms.
   (c) the amount of energy transferred between objects as a result of a difference in temperature.
   (d) an invisible, odorless, weightless substance.
   (e) the total kinetic energy of an ideal gas.

8. Radiation is emitted
   (a) only by glowing objects such as the Sun.
   (b) only by objects whose temperature is greater than the temperature of the surroundings.
   (c) only by objects with more caloric than their surroundings.
   (d) by any object not at 0 K.
   (e) only by objects that have a large specific heat.

9. Ten grams of water is added to ten grams of ice in an insulated container. Will all of the ice melt?
   (a) Yes.    (b) No.    (c) More information is needed.

10. Two objects are made of the same material, but they have different masses and temperatures. If the objects are brought into thermal contact, which one will have the greater temperature change?
    (a) The one with the higher initial temperature.
    (b) The one with the lower initial temperature.
    (c) The one with the greater mass.
    (d) The one with the lesser mass.
    (e) The one with the higher specific heat.
    (f) Not enough information.

11. It has been a hot summer, so when you arrive at a lake, you decide to go for a swim even though it is nighttime. The water is cold! The next day, you go swimming again during the hottest part of the day, and even though the air is warmer the water is still almost as cold. Why?
    (a) Water is fairly dense compared with many other liquids.
    (b) Water remains in a liquid state for a wide range of temperatures.
    (c) Water has a high bulk modulus.
    (d) Water has a high specific heat.

12. Two equal-mass liquids, initially at the same temperature, are heated for the same time over the same stove. You measure the temperatures and find that one liquid has a higher temperature than the other. Which liquid has the higher specific heat?
    (a) The cooler one.
    (b) The hotter one.
    (c) Both are the same.

For assigned homework and other learning materials, go to the MasteringPhysics website.

# Problems

## 14–1 Heat as Energy Transfer

1. (I) To what temperature will 8200 J of heat raise 3.0 kg of water that is initially at 10.0°C?

2. (I) How much heat (in joules) is required to raise the temperature of 34.0 kg of water from 15°C to 95°C?

3. (II) When a diver jumps into the ocean, water leaks into the gap region between the diver's skin and her wetsuit, forming a water layer about 0.5 mm thick. Assuming the total surface area of the wetsuit covering the diver is about 1.0 m², and that ocean water enters the suit at 10°C and is warmed by the diver to skin temperature of 35°C, estimate how much energy (in units of candy bars = 300 kcal) is required by this heating process.

4. (II) An average active person consumes about 2500 Cal a day. (a) What is this in joules? (b) What is this in kilowatt-hours? (c) If your power company charges about 10 ¢ per kilowatt-hour, how much would your energy cost per day if you bought it from the power company? Could you feed yourself on this much money per day?

5. (II) A British thermal unit (Btu) is a unit of heat in the British system of units. One Btu is defined as the heat needed to raise 1 lb of water by 1 F°. Show that

   $$1 \text{ Btu} = 0.252 \text{ kcal} = 1056 \text{ J.}$$

6. (II) How many joules and kilocalories are generated when the brakes are used to bring a 1300-kg car to rest from a speed of 95 km/h?

7. (II) A water heater can generate 32,000 kJ/h. How much water can it heat from 12°C to 42°C per hour?

8. (II) A small immersion heater is rated at 375 W. Estimate how long it will take to heat a cup of soup (assume this is 250 mL of water) from 15°C to 75°C.

## 14–3 and 14–4 Specific Heat; Calorimetry

9. (I) An automobile cooling system holds 18 L of water. How much heat does it absorb if its temperature rises from 15°C to 95°C?

10. (I) What is the specific heat of a metal substance if 135 kJ of heat is needed to raise 4.1 kg of the metal from 18.0°C to 37.2°C?

11. (II) (a) How much energy is required to bring a 1.0-L pot of water at 20°C to 100°C? (b) For how long could this amount of energy run a 60-W lightbulb?

12. (II) Samples of copper, aluminum, and water experience the same temperature rise when they absorb the same amount of heat. What is the ratio of their masses? [*Hint*: See Table 14–1.]

13. (II) How long does it take a 750-W coffeepot to bring to a boil 0.75 L of water initially at 11°C? Assume that the part of the pot which is heated with the water is made of 280 g of aluminum, and that no water boils away.

14. (II) What will be the equilibrium temperature when a 265-g block of copper at 245°C is placed in a 145-g aluminum calorimeter cup containing 825 g of water at 12.0°C?

15. (II) A 31.5-g glass thermometer reads 23.6°C before it is placed in 135 mL of water. When the water and thermometer come to equilibrium, the thermometer reads 41.8°C. What was the original temperature of the water? Ignore the mass of fluid inside the glass thermometer.

16. (II) A 0.40-kg iron horseshoe, just forged and very hot (Fig. 14–16), is dropped into 1.25 L of water in a 0.30-kg iron pot initially at 20.0°C. If the final equilibrium temperature is 25.0°C, estimate the initial temperature of the hot horseshoe.

**FIGURE 14–16**
Problem 16.

17. (II) When a 290-g piece of iron at 180°C is placed in a 95-g aluminum calorimeter cup containing 250 g of glycerin at 10°C, the final temperature is observed to be 38°C. Estimate the specific heat of glycerin.

18. (II) The *heat capacity, C,* of an object is defined as the amount of heat needed to raise its temperature by 1 C°. Thus, to raise the temperature by $\Delta T$ requires heat $Q$ given by
$$Q = C \, \Delta T.$$
(*a*) Write the heat capacity $C$ in terms of the specific heat, $c$, of the material. (*b*) What is the heat capacity of 1.0 kg of water? (*c*) Of 45 kg of water?

19. (II) The 1.20-kg head of a hammer has a speed of 7.5 m/s just before it strikes a nail (Fig. 14–17) and is brought to rest. Estimate the temperature rise of a 14-g iron nail generated by eight such hammer blows done in quick succession. Assume the nail absorbs all the energy.

**FIGURE 14–17**
Problem 19.

20. (II) A 215-g sample of a substance is heated to 330°C and then plunged into a 105-g aluminum calorimeter cup containing 185 g of water and a 17-g glass thermometer at 10.5°C. The final temperature is 35.0°C. What is the specific heat of the substance? (Assume no water boils away.)

21. (II) A 0.095-kg aluminium sphere is dropped from the roof of a 55-m-high building. If 65% of the thermal energy produced when it hits the ground is absorbed by the sphere, what is its temperature increase?

22. (II) Estimate the Calorie content of 65 g of candy from the following measurements. A 15-g sample of the candy is placed in a small aluminum container of mass 0.325 kg filled with oxygen. This container is placed in 1.75 kg of water in an aluminum calorimeter cup of mass 0.624 kg at an initial temperature of 15.0°C. The oxygen–candy mixture in the small container (a "bomb calorimeter") is ignited, and the final temperature of the whole system is 53.5°C.

23. (II) Determine the energy content of 100 g of Karen's fudge cookies from the following measurements. A 10-g sample of a cookie is allowed to dry before putting it in a *bomb calorimeter* (page 396). The aluminum bomb has a mass of 0.615 kg and is placed in 2.00 kg of water contained in an aluminum calorimeter cup of mass 0.524 kg. The initial temperature of the system is 15.0°C, and its temperature after ignition is 36.0°C.

## 14–5 Latent Heat

24. (I) If $3.40 \times 10^5$ J of energy is supplied to a container of liquid oxygen at −183°C, how much oxygen can evaporate?

25. (II) How much heat is needed to melt 23.50 kg of silver that is initially at 25°C?

26. (II) During exercise, a person may give off 185 kcal of heat in 25 min by evaporation of water (at 20°C) from the skin. How much water has been lost? [*Hint*: See page 399.]

27. (II) What mass of steam at 100°C must be added to 1.00 kg of ice at 0°C to yield liquid water at 30°C?

28. (II) A 28-g ice cube at its melting point is dropped into an insulated container of liquid nitrogen. How much nitrogen evaporates if it is at its boiling point of 77 K and has a latent heat of vaporization of 200 kJ/kg? Assume for simplicity that the specific heat of ice is a constant and is equal to its value near its melting point.

29. (II) High-altitude mountain climbers do not eat snow, but always melt it first with a stove. To see why, calculate the energy absorbed from your body if you: (*a*) eat 1.0 kg of −15°C snow which your body warms to body temperature of 37°C; (*b*) melt 1.0 kg of −15°C snow using a stove and drink the resulting 1.0 kg of water at 2°C, which your body has to warm to 37°C.

30. (II) An iron boiler of mass 180 kg contains 730 kg of water at 18°C. A heater supplies energy at the rate of 58,000 kJ/h. How long does it take for the water (*a*) to reach the boiling point, and (*b*) to all have changed to steam?

31. (II) Determine the latent heat of fusion of mercury using the following calorimeter data: 1.00 kg of solid Hg at its melting point of −39.0°C is placed in a 0.620-kg aluminum calorimeter with 0.400 kg of water at 12.80°C; the resulting equilibrium temperature is 5.06°C.

32. (II) At a crime scene, the forensic investigator notes that the 6.2-g lead bullet that was stopped in a doorframe apparently melted completely on impact. Assuming the bullet was shot at room temperature (20°C), what does the investigator calculate as the minimum muzzle velocity of the gun?

33. (II) A 64-kg ice-skater moving at 7.5 m/s glides to a stop. Assuming the ice is at 0°C and that 50% of the heat generated by friction is absorbed by the ice, how much ice melts?

34. (II) A cube of ice is taken from the freezer at −8.5°C and placed in an 85-g aluminum calorimeter filled with 310 g of water at room temperature of 20.0°C. The final situation is all water at 17.0°C. What was the mass of the ice cube?

35. (II) A 55-g bullet traveling at 250 m/s penetrates a block of ice at 0°C and comes to rest within the ice. Assuming that the temperature of the bullet doesn't change appreciably, how much ice is melted as a result of the collision?

## 14–6 to 14–8 Conduction, Convection, Radiation

36. (I) Calculate the rate of heat flow by conduction through the windows of Example 14–8, assuming that there are strong gusty winds and the external temperature is −5°C.

37. (I) One end of a 56-cm-long copper rod with a diameter of 2.0 cm is kept at 460°C, and the other is immersed in water at 22°C. Calculate the heat conduction rate along the rod.

38. (II) (*a*) How much power is radiated by a tungsten sphere (emissivity $\epsilon = 0.35$) of radius 19 cm at a temperature of 25°C? (*b*) If the sphere is enclosed in a room whose walls are kept at −5°C, what is the *net* flow rate of energy out of the sphere?

**39.** (II) How long does it take the Sun to melt a block of ice at 0°C with a flat horizontal area 1.0 m² and thickness 1.0 cm? Assume that the Sun's rays make an angle of 35° with the vertical and that the emissivity of ice is 0.050.

**40.** (II) *Heat conduction to skin.* Suppose 150 W of heat flows by conduction from the blood capillaries beneath the skin to the body's surface area of 1.5 m². If the temperature difference is 0.50 C°, estimate the average distance of capillaries below the skin surface.

**41.** (II) Two rooms, each a cube 4.0 m per side, share a 14-cm-thick brick wall. Because of a number of 100-W lightbulbs in one room, the air is at 30°C, while in the other room it is at 10°C. How many of the 100-W bulbs are needed to maintain the temperature difference across the wall?

**42.** (II) A 100-W lightbulb generates 95 W of heat, which is dissipated through a glass bulb that has a radius of 3.0 cm and is 0.50 mm thick. What is the difference in temperature between the inner and outer surfaces of the glass?

**43.** (III) Approximately how long should it take 8.2 kg of ice at 0°C to melt when it is placed in a carefully sealed Styrofoam ice chest of dimensions 25 cm × 35 cm × 55 cm whose walls are 1.5 cm thick? Assume that the conductivity of Styrofoam is double that of air and that the outside temperature is 34°C.

**44.** (III) A copper rod and an aluminum rod of the same length and cross-sectional area are attached end to end (Fig. 14–18). The copper end is placed in a furnace maintained at a constant temperature of 205°C. The aluminum end is placed in an ice bath held at a constant temperature of 0.0°C. Calculate the temperature at the point where the two rods are joined.

**FIGURE 14–18**
Problem 44.  205°C   $T = ?$   0.0°C

**\*45.** (III) Suppose the insulating qualities of the wall of a house come mainly from a 4.0-in. layer of brick and an *R*-19 layer of insulation, as shown in Fig. 14–19. What is the total rate of heat loss through such a wall, if its total area is 195 ft² and the temperature difference across it is 35 F°?

**FIGURE 14–19**
Problem 45. Two layers insulating a wall.

# General Problems

**46.** A soft-drink can contains about 0.35 kg of liquid at 5°C. Drinking this liquid can actually consume some of the fat in the body, since energy is needed to warm the liquid to body temperature (37°C). How many food Calories should the drink have so that it is in perfect balance with the heat needed to warm the liquid (essentially water)?

**47.** (*a*) Estimate the total power radiated into space by the Sun, assuming it to be a perfect emitter at $T = 5500$ K. The Sun's radius is $7.0 \times 10^8$ m. (*b*) From this, determine the power per unit area arriving at the Earth, $1.5 \times 10^{11}$ m away (Fig. 14–20).

**FIGURE 14–20**
Problem 47.

$r = 1.5 \times 10^{11}$ m

Sun — Earth

**48.** To get an idea of how much thermal energy is contained in the world's oceans, estimate the heat liberated when a cube of ocean water, 1 km on each side, is cooled by 1 K. (Approximate the ocean water as pure water for this estimate.)

**49.** What will be the final result when equal masses of ice at 0°C and steam at 100°C are mixed together?

**50.** A mountain climber wears a goose-down jacket 3.5 cm thick with total surface area 0.95 m². The temperature at the surface of the clothing is −18°C and at the skin is 34°C. Determine the rate of heat flow by conduction through the jacket assuming (*a*) it is dry and the thermal conductivity *k* is that of goose down, and (*b*) the jacket is wet, so *k* is that of water and the jacket has matted to 0.50 cm thickness.

**51.** During light activity, a 70-kg person may generate 200 kcal/h. Assuming that 20% of this goes into useful work and the other 80% is converted to heat, estimate the temperature rise of the body after 45 min if none of this heat is transferred to the environment.

**52.** Estimate the rate at which heat can be conducted from the interior of the body to the surface. As a model, assume that the thickness of tissue is 4.0 cm, that the skin is at 34°C and the interior at 37°C, and that the surface area is 1.5 m². Compare this to the measured value of about 230 W that must be dissipated by a person working lightly. This clearly shows the necessity of convective cooling by the blood.

**53.** A bicyclist consumes 9.0 L of water over the span of 3.5 hours during a race. Making the approximation that 80% of the cyclist's energy goes into evaporating this water (at 20°C) as sweat, how much energy in kcal did the rider use during the ride? [*Hint*: See page 399.]

**54.** If coal gives off 30 MJ/kg when burned, how much coal is needed to heat a house requiring $2.0 \times 10^5$ MJ for the whole winter? Assume that 30% of the heat is lost up the chimney.

**55.** A 15-g lead bullet is tested by firing it into a fixed block of wood with a mass of 35 kg. The block and imbedded bullet together absorb all the heat generated. After thermal equilibrium has been reached, the system has a temperature rise measured as 0.020 C°. Estimate the bullet's entering speed.

**56.** A 310-kg marble boulder rolls off the top of a cliff and falls a vertical height of 120 m before striking the ground. Estimate the temperature rise of the rock if 50% of the heat generated remains in the rock.

**57.** A 2.3-kg lead ball is placed in a 2.5-L insulated pail of water initially at 20.0°C. If the final temperature of the water–lead combination is 32.0°C, what was the initial temperature of the lead ball?

**58.** A microwave oven is used to heat 250 g of water. On its maximum setting, the oven can raise the temperature of the liquid water from 20°C to 100°C in 1 min 45 s (= 105 s). (*a*) At what rate does the oven put energy into the liquid water? (*b*) If the power input from the oven to the water remains constant, determine how many grams of water will boil away if the oven is operated for 2 min (rather than just 1 min 45 s).

59. In a typical squash game (Fig. 14–21), two people hit a soft rubber ball at a wall. Assume that the ball hits the wall at a velocity of 22 m/s and bounces back at a velocity of 12 m/s, and that the kinetic energy lost in the process heats the ball. What will be the temperature increase of the ball after one bounce? (The specific heat of rubber is about 1200 J/kg·C°.)

**FIGURE 14–21**
Problem 59.

60. The temperature within the Earth's crust increases about 1.0 C° for each 30 m of depth. The thermal conductivity of the crust is 0.80 J/s·C°·m. (a) Determine the heat transferred from the interior to the surface for the entire Earth in 1.0 h. (b) Compare this heat to the 1000 W/m² that reaches the Earth's surface in 1.0 h from the Sun.

61. An iron meteorite melts when it enters the Earth's atmosphere. If its initial temperature was −105°C outside of Earth's atmosphere, calculate the minimum velocity the meteorite must have had before it entered Earth's atmosphere.

62. The temperature of the glass surface of a 75-W lightbulb is 75°C when the room temperature is 18°C. Estimate the temperature of a 150-W lightbulb with a glass bulb the same size. Consider only radiation, and assume that 90% of the energy is emitted as heat.

63. In a cold environment, a person can lose heat by conduction and radiation at a rate of about 200 W. Estimate how long it would take for the body temperature to drop from 36.6°C to 35.6°C if metabolism were nearly to stop. Assume a mass of 65 kg. (See Table 14–1.)

64. A 12-g lead bullet traveling at 220 m/s passes through a thin wall and emerges at a speed of 160 m/s. If the bullet absorbs 50% of the heat generated, (a) what will be the temperature rise of the bullet? (b) If the bullet's initial temperature was 20°C, will any of the bullet melt, and if so, how much?

65. A leaf of area 40 cm² and mass $4.5 \times 10^{-4}$ kg directly faces the Sun on a clear day. The leaf has an emissivity of 0.85 and a specific heat of 0.80 kcal/kg·K. (a) Estimate the energy absorbed per second by the leaf from the Sun, and then (b) estimate the rate of rise of the leaf's temperature. (c) Will the temperature rise continue for hours? Why or why not? (d) Calculate the temperature the leaf would reach if it lost all its heat by radiation to the surroundings at 24°C. (e) In what other ways can the heat be dissipated by the leaf?

66. Using the result of part (a) in Problem 65, take into account radiation from the leaf to calculate how much water must be transpired (evaporated) by the leaf per hour to maintain a temperature of 35°C.

67. After a hot shower and dishwashing, there seems to be no hot water left in the 65-gal (245-L) water heater. This suggests that the tank has emptied and refilled with water at roughly 10°C. (a) How much energy does it take to reheat the water to 45°C? (b) How long would it take if the heater output is 9500 W?

68. A house thermostat is normally set to 22°C, but at night it is turned down to 16°C for 9.0 h. Estimate how much more heat would be needed (state as a percentage of daily usage) if the thermostat were not turned down at night. Assume that the outside temperature averages 0°C for the 9.0 h at night and 8°C for the remainder of the day, and that the heat loss from the house is proportional to the temperature difference inside and out. To obtain an estimate from the data, you must make other simplifying assumptions; state what these are.

# Search and Learn

1. Create graphs similar to Fig. 14–5, but for lead and ethyl alcohol. Compare and contrast them with each other and with the graph for water. Are there any temperature ranges for which all three substances are liquids? All vapors? All solids? For convenience, use the specific heats given in Table 14–1 for all states of lead and ethyl alcohol.

2. (a) Using the solar constant, estimate the rate at which the whole Earth receives energy from the Sun. (b) Assume the Earth radiates an equal amount back into space (that is, the Earth is in equilibrium). Then, assuming the Earth is a perfect emitter ($\epsilon = 1.0$), estimate its average surface temperature. [Hint: Discuss why you use area $A = \pi r_E^2$ or $A = 4\pi r_E^2$ in each part.]

3. Calculate what will happen when 1000 J of heat is added to 100 grams of (a) ice at −20°C, (b) ice at 0°C, (c) water at 10°C, (d) water at 100°C, and (e) steam at 110°C.

4. A house has well-insulated walls 19.5 cm thick (assume conductivity of air) and area 410 m², a roof of wood 5.5 cm thick and area 250 m², and uncovered windows 0.65 cm thick and total area 33 m². (a) Assuming that heat is lost only by conduction, calculate the rate at which heat must be supplied to this house to maintain its inside temperature at 23°C if the outside temperature is −15°C. (b) If the house is initially at 15°C, estimate how much heat must be supplied to raise the temperature to 23°C within 30 min. Assume that only the air needs to be heated and that its volume is 750 m³. (c) If natural gas costs $0.080/kg and its heat of combustion is $5.4 \times 10^7$ J/kg, what is the monthly cost to maintain the house as in part (a) for 24 h each day, assuming 90% of the heat produced is used to heat the house? Take the specific heat of air to be 0.24 kcal/kg·C°.

## ANSWERS TO EXERCISES

A: (b).
B: (c).
C: The phase change, liquid to ice (second process).
D: 0.21 kg.

E: The drapes trap a layer of air between the inside of the window and the room, which acts as an excellent insulator.
F: (d).

Thermodynamics is the study of heat and work. Heat is a transfer of energy due to a difference of temperature. Work is a transfer of energy by mechanical means, not due to a temperature difference. The first law of thermodynamics links the two in a general statement of energy conservation: the heat $Q$ added to a system minus the net work $W$ done by the system equals the change in internal energy $\Delta U$ of the system: $\Delta U = Q - W$.

There are many uses for a heat engine such as a modern coal-burning power plant, or a steam locomotive. The photograph shows a steam locomotive which produces steam that does work on a piston that moves linkage to turn locomotive wheels. The efficiency of any engine is limited by nature as described in the second law of thermodynamics. This great law is best stated in terms of a quantity called entropy, which is *not* conserved, but instead is constrained always to increase in any real process. Entropy is a measure of disorder. The second law of thermodynamics tells us that as time moves forward, the disorder in the universe increases. We also discuss practical matters such as heat engines, heat pumps, refrigerators, and air conditioners.

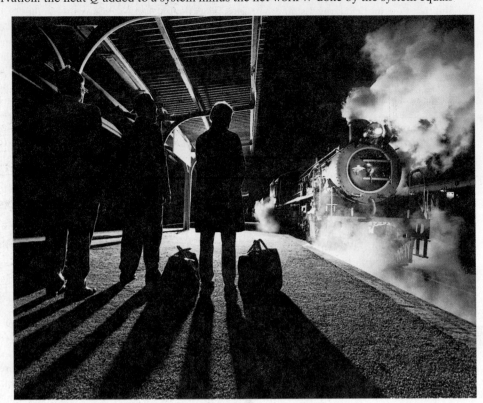

## CHAPTER 15

# The Laws of Thermodynamics

**CHAPTER-OPENING QUESTION—Guess now!**

Fossil-fuel electric generating plants produce "thermal pollution." Part of the heat produced by the burning fuel is not converted to electric energy. The reason for this waste is

(a) the efficiency is higher if some heat is allowed to escape.
(b) engineering technology has not yet reached the point where 100% waste heat recovery is possible.
(c) some waste heat *must* be produced: this is a fundamental property of nature when converting heat to useful work.
(d) the plants rely on fossil fuels, not nuclear fuel.
(e) None of the above.

**T**hermodynamics is the name we give to the study of processes in which energy is transferred as heat and as work.

In Chapter 6 we saw that work is done when energy is transferred from one object to another by mechanical means. In Chapter 14 we saw that heat is a transfer of energy from one object to a second one at a lower temperature. Thus, heat is much like work. To distinguish them, *heat* is defined as a *transfer of energy due to a difference in temperature*, whereas work is a transfer of energy that is not due to a temperature difference.

In discussing thermodynamics, we often refer to particular systems. A **system** is any object or set of objects that we wish to consider (see Section 14–4). Everything else in the universe is referred to as its "environment" or the "surroundings."

In this Chapter, we examine the two great laws of thermodynamics. The first law of thermodynamics relates work and heat transfers to the change in internal energy of a system, and is a general statement of the conservation of energy. The second law of thermodynamics expresses limits on the ability to do useful work, and is often stated in terms of *entropy*, which is a measure of disorder. Besides these two great laws, we also discuss some important related practical devices: heat engines, refrigerators, heat pumps, and air conditioners.

# 15–1 The First Law of Thermodynamics

In Section 14–2, we defined the internal energy of a system as the sum total of all the energy of the molecules within the system. Then the internal energy of a system should increase if work is done on the system, or if heat is added to it. Similarly the internal energy should decrease if heat flows out of the system or if work is done by the system on something in the surroundings.

Thus it is reasonable to extend conservation of energy and propose an important law: the change in internal energy of a closed system, $\Delta U$, will be equal to the energy added to the system by heating minus the work done by the system on the surroundings. In equation form we write

$$\Delta U = Q - W \qquad (15\text{–}1)$$

FIRST LAW OF
THERMODYNAMICS

where $Q$ is the net heat *added* to the system and $W$ is the net work done *by* the system. We must be careful and consistent in following the sign conventions for $Q$ and $W$. Because $W$ in Eq. 15–1 is the work done *by* the system, then if work is done *on* the system, $W$ will be negative and $U$ will increase. Similarly, $Q$ is positive for heat added to the system, so if heat leaves the system, $Q$ is negative. [Caution: Elsewhere you may sometimes encounter the opposite convention for $W$ where $W$ is defined as the work done *on* the system; in that case Eq. 15–1 is written as $\Delta U = Q + W$.]

Equation 15–1 is known as the **first law of thermodynamics**. It is one of the great laws of physics, and its validity rests on experiments (such as Joule's) to which no exceptions have been seen. Since $Q$ and $W$ represent energy transferred into or out of the system, the internal energy changes accordingly. Thus, the first law of thermodynamics is a general statement of the *law of conservation of energy*.

Note that the conservation of energy law was not able to be formulated until the 1800s, because it depended on the interpretation of heat as a transfer of energy.

A given system does not "have" a certain amount of heat or work. Rather, work and heat are involved in *thermodynamic processes* that can change the system from one state to another; they are not characteristic of the state itself. Quantities which describe the state of a system, such as internal energy $U$, pressure $P$, volume $V$, temperature $T$, and mass $m$ or number of moles $n$, are called **state variables**. $Q$ and $W$ are *not* state variables.

> **⚠ CAUTION**
> *Heat added is +*
> *Heat lost is −*
> *Work on system is −*
> *Work by system is +*

> **⚠ CAUTION**
> *P, V, T, U, m, n are state variables. W and Q are not: a system does not have an amount of heat or work*

---

**EXAMPLE 15–1** **Using the first law.** 2500 J of heat is added to a system, and 1800 J of work is done on the system. What is the change in internal energy of the system?

**APPROACH** We apply the first law of thermodynamics, Eq. 15–1, to our system.

**SOLUTION** The heat added to the system is $Q = 2500$ J. The work $W$ done *by* the system is $-1800$ J. Why the minus sign? Because 1800 J done *on* the system (as given) equals $-1800$ J done *by* the system, and it is the latter we need for the sign conventions we used for Eq. 15–1. Hence

$$\Delta U = 2500\,\text{J} - (-1800\,\text{J}) = 2500\,\text{J} + 1800\,\text{J} = 4300\,\text{J}.$$

**NOTE** We did this calculation in detail to emphasize the importance of keeping careful track of signs. Both the heat and the work are inputs to the system, so we expect $\Delta U$ to be increased by both.

## *The First Law of Thermodynamics Extended

To write the first law of thermodynamics in a more complete form, consider a system that is moving so it has kinetic energy KE, and suppose there is also potential energy PE. Then the first law of thermodynamics would have to include these terms and would be written as

$$\Delta\text{KE} + \Delta\text{PE} + \Delta U = Q - W. \qquad (15\text{–}2)$$

---

**EXAMPLE 15–2** **Kinetic energy transformed to thermal energy.** A 3.0-g bullet traveling at a speed of 400 m/s enters a tree and exits the other side with a speed of 200 m/s. Where did the bullet's lost KE go, and how much energy was transferred?

**APPROACH** Take the bullet and tree as our system. No potential energy is involved. No work is done on (or by) the system by outside forces, nor is any heat added because no energy was transferred to or from the system due to a temperature difference. Thus the kinetic energy gets transformed into internal energy of the bullet and tree. This answers the first question.

**SOLUTION** In the first law of thermodynamics as written in Eq. 15–2, we are given $Q = W = \Delta\text{PE} = 0$, so we have

$$\Delta\text{KE} + \Delta U = 0$$

or, using subscripts i and f for initial and final velocities,

$$\Delta U = -\Delta\text{KE} = -(\text{KE}_f - \text{KE}_i) = \tfrac{1}{2}m(v_i^2 - v_f^2)$$
$$= \tfrac{1}{2}(3.0 \times 10^{-3}\,\text{kg})\big[(400\,\text{m/s})^2 - (200\,\text{m/s})^2\big] = 180\,\text{J}.$$

**NOTE** The internal energy of the bullet and tree both increase, as both experience a rise in temperature. If we had chosen the bullet alone as our system, work would be done on it and heat transfer would occur.

---

# 15–2 Thermodynamic Processes and the First Law

Let us analyze some thermodynamic processes in light of the first law of thermodynamics.

## Isothermal Processes ($\Delta T = 0$)

To begin, we choose a very simple system: a fixed mass of an ideal gas enclosed in a container fitted with a movable piston as shown in Fig. 15–1.

First we consider an idealized process, such as adding heat or doing work, that is carried out at constant temperature. Such a process is called an **isothermal** process (from the Greek meaning "same temperature"). If the system is an ideal gas, then $PV = nRT$ (Eq. 13–3), so for a fixed amount of gas kept at constant temperature, $PV = $ constant. Thus a graph of pressure $P$ vs. volume $V$, a **PV diagram**, would follow a curve like AB in Fig. 15–2 for an isothermal process. Each point on the curve, such as point A, represents the state of the system at a given moment—that is, its pressure $P$ and volume $V$. At a lower temperature, another isothermal process would be represented by a curve like A′ B′ in Fig. 15–2 (the product $PV = nRT = $ constant is less when $T$ is less). The curves shown in Fig. 15–2 are referred to as *isotherms*.

We assume the gas is in contact with a **heat reservoir** (a body whose mass is so large that, ideally, its temperature does not change significantly when heat is exchanged with our system). We also assume that a process of compression (volume decrease) or expansion (volume increase) is done very slowly, so that the process can be considered a *series of equilibrium states* all at the same constant temperature.[†]

**FIGURE 15–1** An ideal gas in a cylinder fitted with a movable piston.

**FIGURE 15–2** *PV* diagram for an ideal gas undergoing isothermal processes at two different temperatures.

---

[†]If a gas expands or is compressed quickly, there is turbulence and different parts of the gas in its container would be at different pressures and temperatures.

If the gas is initially in a state represented by point A in Fig. 15–2, and an amount of heat $Q$ is added to the system, the pressure and volume will change and the state of the system will be represented by another point, B, on the diagram. If the temperature is to remain constant, the gas will expand and do an amount of work $W$ on the environment (it exerts a force on the piston in Fig. 15–1 and moves it through a distance). The temperature and mass are kept constant so, from Eq. 14–1, the internal energy does not change: $\Delta U = \frac{3}{2}nR\,\Delta T = 0$. Hence, by the first law of thermodynamics, Eq. 15–1, $\Delta U = Q - W = 0$, so $W = Q$: the work done by the gas in an isothermal process equals the heat added to the gas.

## Adiabatic Processes ($Q = 0$)

An **adiabatic** process is one in which no heat is allowed to flow into or out of the system: $Q = 0$. This situation can occur if the system is extremely well insulated, or the process happens so quickly that heat—which flows slowly—has no time to flow in or out. The very rapid expansion of gases in an internal combustion engine is one example of a process that is very nearly adiabatic. An adiabatic expansion of an ideal gas done very slowly can be represented by a curve like that labeled AC in Fig. 15–3. Since $Q = 0$, we have from Eq. 15–1 that $\Delta U = -W$. When a gas expands, it does work and $W$ is positive, so the internal energy decreases; hence the temperature decreases as well (because $\Delta U = \frac{3}{2}nR\,\Delta T$). This is seen in Fig. 15–3 where the product $PV$ ($= nRT$) is less at point C than at point B. (Compare to curve AB for an isothermal process, in which $\Delta U = 0$ and $\Delta T = 0$.) In the reverse operation, an adiabatic compression (going from C to A, for example), work is done *on* the gas, and hence the internal energy increases and the temperature rises. In a diesel engine, the fuel–air mixture is rapidly compressed adiabatically by a factor of 15 or more; the temperature rise is so great that the mixture ignites spontaneously, without spark plugs.

## Isobaric and Isovolumetric Processes

Isothermal and adiabatic processes are just two possible processes that can occur. Two other simple thermodynamic processes are illustrated on the $PV$ diagrams of Fig. 15–4: (a) an **isobaric** process is one in which the pressure is kept constant, so the process is represented by a straight horizontal line on the $PV$ diagram (Fig. 15–4a); (b) an **isovolumetric** (or *isochoric*) process is one in which the volume does not change (Fig. 15–4b). In these, and in all other processes, the first law of thermodynamics holds.

## Work Done in Volume Changes

It is often valuable to calculate the work done in a process. If the pressure is kept constant during a process (isobaric), the work done is easily calculated. For example, if the gas in Fig. 15–5 expands very slowly against the piston, the work done by the gas to raise the piston is the force $F$ times the distance $d$. But the force is just the pressure $P$ of the gas times the area $A$ of the piston, $F = PA$. Thus,

$$W = Fd = PAd.$$

Note that $Ad = \Delta V$, the change in volume of the gas, so

$$W = P\,\Delta V. \qquad \text{[constant pressure]} \quad \textbf{(15–3)}$$

Equation 15–3 also holds if the gas is *compressed* at constant pressure, in which case $\Delta V$ is negative (since $V$ decreases); $W$ is then negative, which indicates that work is done *on* the gas. Equation 15–3 is also valid for liquids and solids, as long as the pressure is constant during the process.

In an isovolumetric process (Fig. 15–4b) the volume does not change, so no work is done, $W = 0$.

**FIGURE 15–3** $PV$ diagram for adiabatic (AC) and isothermal (AB) processes on an ideal gas.

**FIGURE 15–4** (a) Isobaric ("same pressure") process. (b) Isovolumetric ("same volume") process.

(a) Isobaric      (b) Isovolumetric

**FIGURE 15–5** Work is done on the piston when the gas expands, moving the piston a distance $d$.

area $A$

$P$

$d$

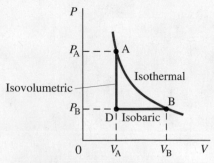

**FIGURE 15–6** *PV* diagram for different processes (see the text), where the system changes from A to B.

Figure 15–6 shows the isotherm AB we saw in Fig. 15–2 as well as another possible process represented by the path ADB. In going from A to D, the gas does no work since the volume does not change. But in going from D to B, the gas does work equal to $P_B(V_B - V_A)$, and this is the total work done in the process ADB.

If the pressure varies during a process, such as for the isothermal process AB in Fig. 15–6 (and Fig. 15–2), Eq. 15–3 cannot be used directly to determine the work. A rough estimate can be obtained, however, by using an "average" value for $P$ in Eq. 15–3. More accurately, the work done is equal to the area under the *PV* curve. This is obvious when the pressure is constant: as Fig. 15–7a shows, the shaded area is just $P_B(V_B - V_A)$, and this is the work done. Similarly, the work done during an isothermal process is equal to the shaded area shown in Fig. 15–7b. The calculation of work done in this case can be carried out using calculus, or by estimating the area on graph paper.

(a)  (b)

**FIGURE 15–7** Work done by a gas is equal to the area under the *PV* curve.

---

**CONCEPTUAL EXAMPLE 15–3** | **Work in isothermal and adiabatic processes.** In Fig. 15–3 we saw the *PV* diagrams for a gas expanding in two ways, isothermally and adiabatically. The initial volume $V_A$ was the same in each case, and the final volumes were the same $(V_B = V_C)$. In which process was more work done by the gas?

**RESPONSE** Our system is the gas. More work was done by the gas in the isothermal process, which we can see in two simple ways by looking at Fig. 15–3. First, the "average" pressure was higher during the isothermal process AB, so $W = P_{av} \Delta V$ was greater ($\Delta V$ is the same for both processes). Second, we can look at the area under each curve as we showed in Fig. 15–7b: the area under curve AB, which represents the work done, is greater (because curve AB is higher) than the area under AC in Fig. 15–3.

---

**EXERCISE B** Is the work done by the gas in process ADB of Fig. 15–6 greater than, less than, or equal to the work done in the isothermal process AB?

Table 15–1 gives a brief summary of the processes we have discussed. Many other types of processes can occur, but these "simple" ones are useful and can be dealt with by fairly simple means.

**TABLE 15–1  Simple Thermodynamic Processes and the First Law**

| Process | What is constant: | The first law, $\Delta U = Q - W$, predicts: |
|---|---|---|
| Isothermal | $T = $ constant | $\Delta T = 0$ makes $\Delta U = 0$, so $Q = W$ |
| Isobaric | $P = $ constant | $Q = \Delta U + W = \Delta U + P \Delta V$ |
| Isovolumetric | $V = $ constant | $\Delta V = 0$ makes $W = 0$, so $Q = \Delta U$ |
| Adiabatic | $Q = 0$ | $\Delta U = -W$ |

**EXAMPLE 15-4** **First law in isobaric and isovolumetric processes.** An ideal gas is slowly compressed at a constant pressure of 2.0 atm from 10.0 L to 2.0 L. This process is represented in Fig. 15–8 as the path B to D. (In this process, some heat flows out of the gas and the temperature drops.) Heat is then added to the gas, holding the volume constant, and the pressure and temperature are allowed to rise (line DA) until the temperature reaches its original value $(T_A = T_B)$. In the process BDA, calculate (a) the total work done by the gas, and (b) the total heat flow into the gas.

**FIGURE 15–8** Example 15–4.

**APPROACH** (a) Work is done only in the compression process BD. In process DA, the volume is constant so $\Delta V = 0$ and no work is done (Eq. 15–3). (b) We use the first law of thermodynamics, Eq. 15–1.

**SOLUTION** (a) During the compression BD, the pressure is 2.0 atm $= 2(1.01 \times 10^5 \text{ N/m}^2)$ and the work done is $(1 \text{ L} = 10^3 \text{ cm}^3 = 10^{-3} \text{ m}^3)$

$$
\begin{aligned}
W &= P \Delta V \\
&= (2.02 \times 10^5 \text{ N/m}^2)[(2.0 \times 10^{-3} \text{ m}^3) - (10.0 \times 10^{-3} \text{ m}^3)] \\
&= -1.6 \times 10^3 \text{ J}.
\end{aligned}
$$

The total work done *by* the gas is $-1.6 \times 10^3$ J, where the minus sign means that $+1.6 \times 10^3$ J of work is done *on* the gas.

(b) Because the temperature at the beginning and at the end of process BDA is the same, there is no change in internal energy: $\Delta U = 0$. From the first law of thermodynamics we have

$$0 = \Delta U = Q - W,$$

so

$$Q = W = -1.6 \times 10^3 \text{ J}.$$

Because $Q$ is negative, 1600 J of heat flows out of the gas for the whole process, BDA.

**EXERCISE C** In Example 15–4, if the heat lost from the gas in the process BD is $8.4 \times 10^3$ J, what is the change in internal energy of the gas during process BD?

**EXAMPLE 15-5** **Work done in an engine.** In an engine, 0.25 mol of an ideal monatomic gas in the cylinder expands rapidly and adiabatically against the piston. In the process, the temperature of the gas drops from 1150 K to 400 K. How much work does the gas do?

**APPROACH** We take the gas as our system (the piston is part of the surroundings). The pressure is not constant, so we can't use Eq. 15–3 $(W = P \Delta V)$. Instead, we can use the first law of thermodynamics to find $W$ because we can determine $\Delta U$ (from $\Delta T$) and $Q = 0$ (the process is adiabatic).

**SOLUTION** We determine $\Delta U$ from Eq. 14–1 for the internal energy of an ideal monatomic gas, using subscripts f and i for final and initial states:

$$
\begin{aligned}
\Delta U &= U_f - U_i = \tfrac{3}{2}nR(T_f - T_i) \\
&= \tfrac{3}{2}(0.25 \text{ mol})(8.314 \text{ J/mol·K})(400 \text{ K} - 1150 \text{ K}) \\
&= -2300 \text{ J}.
\end{aligned}
$$

Then, from the first law of thermodynamics, Eq. 15–1, the work done by the gas is

$$W = Q - \Delta U = 0 - (-2300 \text{ J}) = 2300 \text{ J}.$$

**EXAMPLE 15–6** $\Delta U$ **for boiling water to steam.** Determine the change in internal energy of 1.00 liter of water (mass 1.00 kg) at 100°C when it is fully boiled from liquid to gas, which results in 1671 liters of steam at 100°C. Assume the process is done at atmospheric pressure.

**APPROACH** Our system is the water. The heat required here does not result in a temperature change; rather, a change in phase occurs. We can determine the heat $Q$ required using the latent heat of water, as in Section 14–5. Work too will be done: $W = P \Delta V$. The first law of thermodynamics will then give us $\Delta U$.

**SOLUTION** The latent heat of vaporization of water (Table 14–3) is $L_V = 22.6 \times 10^5$ J/kg. So the heat input required for this process is (Eq. 14–4)

$$Q = mL = (1.00 \text{ kg})(22.6 \times 10^5 \text{ J/kg}) = 22.6 \times 10^5 \text{ J}.$$

The work done by the water is (Eq. 15–3 since $P$ is constant)

$$W = P \Delta V = (1.01 \times 10^5 \text{ N/m}^2)[(1671 \times 10^{-3} \text{ m}^3) - (1 \times 10^{-3} \text{ m}^3)]$$
$$= 1.69 \times 10^5 \text{ J},$$

where we used 1 atm $= 1.01 \times 10^5$ N/m² and 1 L $= 10^3$ cm³ $= 10^{-3}$ m³. Then

$$\Delta U = Q - W = (22.6 \times 10^5 \text{ J}) - (1.7 \times 10^5 \text{ J}) = 20.9 \times 10^5 \text{ J}.$$

**NOTE** Most of the heat added goes to increasing the internal energy of the water (increasing molecular energy to overcome the attraction that held the molecules close together in the liquid state). Only a small part ($< 10\%$) goes into doing work.

**NOTE** Equation 14–1, $U = \frac{3}{2}nRT$, would tell us that $\Delta U = 0$ here because $\Delta T = 0$. Yet we determined that $\Delta U = 21 \times 10^5$ J. What is wrong? Equation 14–1 applies only to an ideal monatomic gas, not to liquid water.

# *15–3 Human Metabolism and the First Law

**PHYSICS APPLIED**
*Energy in the human body*

**FIGURE 15–9** Bike rider getting an input of energy.

Human beings and other animals do work. Work is done when a person walks or runs, or lifts a heavy object. Work requires energy. Energy is also needed for growth—to make new cells, and to replace old cells that have died. A great many energy-transforming processes occur within an organism, and they are referred to as *metabolism*.

We can apply the first law of thermodynamics,

$$\Delta U = Q - W,$$

to an organism: say, the human body. Work $W$ is done by the body in its various activities; if this is not to result in a decrease in the body's internal energy (and temperature), energy must somehow be added to compensate. The body's internal energy is not maintained by a flow of heat $Q$ into the body, however. Normally, the body is at a higher temperature than its surroundings, so heat usually flows *out* of the body. Even on a very hot day when heat is absorbed, the body has no way of utilizing this heat to support its vital processes. What then is the source of energy that allows us to do work? It is the internal energy (chemical potential energy) stored in foods (Fig. 15–9). In a closed system, the internal energy changes only as a result of heat flow or work done. In an open system, such as a human, internal energy itself can flow into or out of the system. When we eat food, we are bringing internal energy into our bodies directly, which thus increases the total internal energy $U$ in our bodies. This energy eventually goes into work and heat flow from the body according to the first law.

The metabolic rate is the rate at which internal energy is transformed within the body. It is usually specified in kcal/h or in watts. Typical metabolic rates for a variety of human activities are given in Table 15–2 (top of next page) for an "average" 65-kg adult.

**EXAMPLE 15–7** **Energy transformation in the body.** How much energy is transformed in 24 h by a 65-kg person who spends 8.0 h sleeping, 1.0 h at moderate physical labor, 4.0 h in light activity, and 11.0 h working at a desk or relaxing?

**APPROACH** The energy transformed during each activity equals the metabolic rate (Table 15–2) multiplied by the time.

**SOLUTION** Table 15–2 gives the metabolic rate in watts (J/s). Since there are 3600 s in an hour, the total energy transformed is

$$\left[ \begin{array}{l} (8.0\,\text{h})(70\,\text{J/s}) + (1.0\,\text{h})(460\,\text{J/s}) \\ + (4.0\,\text{h})(230\,\text{J/s}) + (11.0\,\text{h})(115\,\text{J/s}) \end{array} \right](3600\,\text{s/h}) = 1.15 \times 10^7\,\text{J}.$$

**NOTE** Since $4.186 \times 10^3\,\text{J} = 1\,\text{kcal}$, this is equivalent to 2800 kcal; a food intake of 2800 Cal would compensate for this energy output. A 65-kg person who wanted to lose weight would have to eat less than 2800 Cal a day, or increase his or her level of activity. Exercise beats *any* diet technique.

**TABLE 15–2 Metabolic Rates**
(65-kg human)

| Activity | Metabolic Rate (approximate) | |
| --- | --- | --- |
| | kcal/h | watts |
| Sleeping | 60 | 70 |
| Sitting upright | 100 | 115 |
| Light activity (eating, dressing, household chores) | 200 | 230 |
| Moderate work (tennis, walking) | 400 | 460 |
| Running (15 km/h) | 1000 | 1150 |
| Bicycling (race) | 1100 | 1270 |

# 15–4 The Second Law of Thermodynamics—Introduction

The first law of thermodynamics states that energy is conserved. There are, however, many processes we can imagine that conserve energy but are not observed to occur in nature. For example, when a hot object is placed in contact with a cold object, heat flows from the hotter one to the colder one, never spontaneously from colder to hotter. If heat were to leave the colder object and pass to the hotter one, energy could still be conserved. Yet it does not happen spontaneously.[†] As a second example, consider what happens when you drop a rock and it hits the ground. The initial potential energy of the rock changes to kinetic energy as the rock falls. When the rock hits the ground, this energy in turn is transformed into internal energy of the rock and the ground in the vicinity of the impact; the molecules move faster and the temperature rises slightly. But have you seen the reverse happen—a rock at rest on the ground suddenly rise up in the air because the thermal energy of molecules is transformed into kinetic energy of the rock as a whole? Energy could be conserved in this process, yet we never see it happen.

There are many other examples of processes that occur in nature but whose reverse does not. Here are two more. (1) If you put a layer of salt in a jar and cover it with a layer of similar-sized grains of pepper, when you shake it you get a thorough mixture. But no matter how long you shake it, the mixture does not separate into two layers again. (2) Coffee cups and glasses break spontaneously if you drop them. But they do not go back together spontaneously (Fig. 15–10).

The first law of thermodynamics (conservation of energy) would not be violated if any of these processes occurred in reverse. To explain this lack of reversibility, scientists in the latter half of the nineteenth century formulated a new principle known as the second law of thermodynamics.

[†]By spontaneously, we mean by itself without input of work of some sort. (A refrigerator does move heat from a cold environment to a warmer one, but only because its motor does work—Section 15–6.)

(a) Initial state.  (b) Later: cup reassembles and rises up.  (c) Later still: cup lands on table.

**FIGURE 15–10** Have you ever observed this process, a broken cup spontaneously reassembling and rising up onto a table? This process could conserve energy. But it never happens.

The **second law of thermodynamics** is a statement about which processes occur in nature and which do not. It can be stated in a variety of ways, all of which are equivalent. One statement, due to R. J. E. Clausius (1822–1888), is that

**heat can flow spontaneously from a hot object to a cold object; heat will not flow spontaneously from a cold object to a hot object.**

Since this statement applies to one particular process, it is not obvious how it applies to other processes. A more general statement is needed that will include other possible processes in a more obvious way.

The development of a general statement of the second law of thermodynamics was based partly on the study of heat engines. A **heat engine** is any device that changes thermal energy into mechanical work, such as a steam engine or an automobile engine. We now examine heat engines, both from a practical point of view and to show their importance in developing the second law of thermodynamics.

# 15–5 Heat Engines

It is easy to produce thermal energy by doing work—for example, by simply rubbing your hands together briskly, or indeed by any frictional process. But to get work from thermal energy is more difficult, and a practical device to do so was invented only about 1700 with the development of the steam engine.

The basic idea behind any heat engine is that mechanical energy can be obtained from thermal energy only when heat is allowed to flow from a high temperature to a low temperature. In the process, some of the heat can then be transformed to mechanical work, as diagrammed schematically in Fig. 15–11. Useful heat engines run in a repeating *cycle*: that is, the system returns repeatedly to its starting point, and thus can run continuously. In each cycle the change in internal energy of the system is $\Delta U = 0$ because it returns to the starting state. Thus a heat input $Q_H$ at a high temperature $T_H$ is partly transformed into work $W$ and partly exhausted as heat $Q_L$ at a lower temperature $T_L$ (Fig. 15–11). By conservation of energy, $Q_H = W + Q_L$. The high and low temperatures, $T_H$ and $T_L$, are called the **operating temperatures** of the engine. Note carefully that we are now using a new (and intuitive) sign convention for heat engines: we take $Q_H$, $Q_L$, and $W$ as always positive. The direction of each energy transfer is shown by the arrow on the applicable diagram, such as Fig. 15–11.

**FIGURE 15–11** Schematic diagram of energy transfers for a heat engine.

**◆ CAUTION**

*Sign convention for heat engines:*
$Q_H > 0$, $Q_L > 0$, $W > 0$

**⊕ PHYSICS APPLIED**

*Engines*

## Steam Engine and Internal Combustion Engine

The operation of a steam engine is shown in Fig. 15–12. Steam engines are of two main types, each using steam heated by combustion of coal, oil, or gas, or by nuclear energy.

**FIGURE 15–12**
Steam engines.

(a) Reciprocating type

(b) Turbine (boiler and condenser not shown)

In a reciprocating engine, Fig. 15–12a, the heated steam passes through the intake valve and expands against a piston, forcing it to move. As the piston returns to its original position, it forces the gases out the exhaust valve which opens. A steam turbine, Fig. 15–12b, is very similar except that the reciprocating piston is replaced by a rotating turbine that resembles a paddlewheel with many sets of blades. Most of our electricity today is generated using steam turbines.[†] The material that is heated and cooled, steam in this case, is called the **working substance**. In an old-time steam engine (see page 412), the high temperature is obtained by burning coal, oil, or other fuel to heat the steam.

In an internal combustion engine (used in most automobiles), the high temperature is achieved by burning the gasoline–air mixture in the cylinder itself (ignited by the spark plug), as described in Fig. 15–13.

**FIGURE 15–13** Four-stroke-cycle internal combustion engine: (a) the gasoline–air mixture flows into the cylinder as the piston moves down; (b) the piston moves upward and compresses the gas; (c) the brief instant when firing of the spark plug ignites the highly compressed gasoline–air mixture, raising it to a high temperature; (d) the gases, now at high temperature and pressure, expand against the piston in this, the power stroke; (e) the burned gases are pushed out to the exhaust pipe. When the piston reaches the top, the exhaust valve closes and the intake valve opens, and the whole cycle repeats. (a), (b), (d), and (e) are the four strokes of the cycle.

## *Why a $\Delta T$ Is Needed to Drive a Heat Engine

To see why a *temperature difference* is required to run an engine, consider a steam engine. In the reciprocating engine, for example, suppose there were no condenser or pump (Fig. 15–12a), and that the steam was at the same temperature throughout the system. Then the pressure of the gas being exhausted would be the same as on intake. The work done by the gas *on* the piston when it expanded would equal the amount of work done *by* the piston to force the steam out the exhaust; hence, no net work would be done. In a real engine, the exhausted gas is cooled to a lower temperature and condensed so that the exhaust pressure is less than the intake pressure. Thus, the work the piston must do on the gas to expel it on the exhaust stroke is less than the work done by the gas on the piston during the intake. So a net amount of work can be obtained—but only if there is this difference of temperature. Similarly, in the gas turbine if the gas isn't cooled, the pressure on each side of the blades would be the same. By cooling the gas on the exhaust side, the pressure on the back side of the blade is less and hence the turbine turns.

[†]Even nuclear power plants utilize steam turbines; the nuclear fuel—uranium—serves as fuel to heat the steam.

**FIGURE 15–11** (Repeated.) Schematic diagram of energy transfers for a heat engine.

## Efficiency

The **efficiency**, $e$, of any heat engine can be defined as the ratio of the work it does, $W$, to the heat input at the high temperature, $Q_H$ (Fig. 15–11):

$$e = \frac{W}{Q_H}. \tag{15–4a}$$

This is a sensible definition since $W$ is the output (what you get from the engine), whereas $Q_H$ is what you put in and pay for in burned fuel. Since energy is conserved, the heat input $Q_H$ must equal the work done plus the heat that flows out at the low temperature $(Q_L)$:

$$Q_H = W + Q_L.$$

Thus $W = Q_H - Q_L$, and the efficiency of an engine is

$$e = \frac{W}{Q_H} = \frac{Q_H - Q_L}{Q_H}$$

or

$$e = 1 - \frac{Q_L}{Q_H}. \tag{15–4b}$$

To give the efficiency as a percent, we multiply Eq. 15–4 by 100. Note that $e$ could be 1.0 (or 100%) only if $Q_L$ were zero—that is, only if no heat were exhausted to the environment (which we will see shortly never happens).

---

**EXAMPLE 15–8** **Car efficiency.** An automobile engine has an efficiency of 20% and produces an average of 23,000 J of mechanical work per second during operation. (*a*) How much heat input is required, and (*b*) how much heat is discharged as waste heat from this engine, per second?

**APPROACH** We want to find the heat input $Q_H$ as well as the heat output $Q_L$, given $W = 23,000$ J each second and an efficiency $e = 0.20$. We can use the definition of efficiency, Eq. 15–4 in its various forms, to find first $Q_H$ and then $Q_L$.

**SOLUTION** (*a*) From Eq. 15–4a, $e = W/Q_H$, we solve for $Q_H$:

$$Q_H = \frac{W}{e} = \frac{23,000 \text{ J}}{0.20}$$
$$= 1.15 \times 10^5 \text{ J} = 115 \text{ kJ}.$$

The engine requires 115 kJ/s = 115 kW of heat input.

(*b*) Now we use Eq. 15–4b $(e = 1 - Q_L/Q_H)$ and solve for $Q_L$:

$$\frac{Q_L}{Q_H} = 1 - e$$

so

$$Q_L = (1 - e)Q_H = (0.80)115 \text{ kJ}$$
$$= 92 \text{ kJ}.$$

The engine discharges heat to the environment at a rate of 92 kJ/s = 92 kW.

**NOTE** Of the 115 kJ that enters the engine per second, only 23 kJ (20%) does useful work whereas 92 kJ (80%) is wasted as heat output.

**NOTE** The problem was stated in terms of energy per unit time. We could just as well have stated it in terms of power, since 1 J/s = 1 watt.

---

## Carnot Engine

To see how to increase efficiency, the French scientist Sadi Carnot (1796–1832) examined the characteristics of an ideal engine, now called a **Carnot engine**. No Carnot engine actually exists, but as a theoretical idea it played an important role in the development and understanding of the second law of thermodynamics.

**FIGURE 15–14** The Carnot cycle. Heat engines work in a cycle, and the cycle for the theoretical Carnot engine begins at point "a" on this $PV$ diagram for an ideal gas. (1) The gas is first expanded isothermally, with the addition of heat $Q_H$, along the path "ab" at temperature $T_H$. (2) Next the gas expands adiabatically from "b" to "c"— no heat is exchanged, but the temperature drops to $T_L$. (3) The gas is then compressed at constant temperature $T_L$, path cd, and heat $Q_L$ flows out. (4) Finally, the gas is compressed adiabatically, path da, back to its original state.

The idealized Carnot engine consisted of four processes done in a cycle, two of which are adiabatic ($Q = 0$) and two are isothermal ($\Delta T = 0$). This idealized cycle is shown in Fig. 15–14. Each of the processes was considered to be done **reversibly**. That is, each of the processes (say, during expansion of the gases against a piston) was done so slowly that the process could be considered a series of equilibrium states, and the whole process could be done in reverse with no change in the magnitude of work done or heat exchanged. A real process, on the other hand, would occur more quickly; there would be turbulence in the gas, friction would be present, and so on. Because of these factors, a real process cannot be done precisely in reverse—the turbulence would be different and the heat lost to friction would not reverse itself. Thus, real processes are **irreversible**.

The isothermal processes of a Carnot engine, where heats $Q_H$ and $Q_L$ are transferred, are assumed to be done at constant temperatures $T_H$ and $T_L$. That is, the system is assumed to be in contact with idealized *heat reservoirs* (page 414) which are so large their temperatures don't change significantly when $Q_H$ and $Q_L$ are transferred.

Carnot showed that for an ideal reversible engine, the heats $Q_H$ and $Q_L$ are proportional to the operating temperatures $T_H$ and $T_L$ (in kelvins): $Q_H/Q_L = T_H/T_L$. So the efficiency can be written as

$$e_{ideal} = \frac{T_H - T_L}{T_H} = 1 - \frac{T_L}{T_H}. \qquad \begin{bmatrix} \text{Carnot (ideal)} \\ \text{efficiency} \end{bmatrix} \quad \textbf{(15–5)}$$

Equation 15–5 expresses the fundamental upper limit to the efficiency of any heat engine. A higher efficiency would violate the second law of thermodynamics.[†] Real engines always have an efficiency lower than this because of losses due to friction and the like. Real engines that are well designed reach 60 to 80% of the Carnot efficiency.

[†]If an engine had a higher efficiency than Eq. 15–5, it could be used in conjunction with a Carnot engine that is made to work in reverse as a refrigerator. If $W$ was the same for both, the net result would be a flow of heat at a low temperature to a high temperature without work being done. That would violate the Clausius statement of the second law.

**EXAMPLE 15–9** **Steam engine efficiency.** A steam engine operates between 500°C and 270°C. What is the maximum possible efficiency of this engine?

**APPROACH** The maximum possible efficiency is the idealized Carnot efficiency, Eq. 15–5. We must use kelvin temperatures.

**SOLUTION** We first change the temperature to kelvins by adding 273 to the given Celsius temperatures: $T_H = 773$ K and $T_L = 543$ K. Then

$$e_{ideal} = 1 - \frac{543}{773} = 0.30.$$

To get the efficiency in percent, we multiply by 100. Thus, the maximum (or Carnot) efficiency is 30%. Realistically, an engine might attain 0.70 of this value, or 21%.

**NOTE** In this Example the exhaust temperature is still rather high, 270°C. Steam engines are often arranged in series so that the exhaust of one engine is used as intake by a second or third engine.

**EXAMPLE 15–10** **A phony claim?** An engine manufacturer makes the following claims: An engine's heat input per second is 9.0 kJ at 435 K. The heat output per second is 4.0 kJ at 285 K. Do you believe these claims?

**APPROACH** The engine's efficiency can be calculated from the definition, Eq. 15–4. It must be less than the maximum possible, Eq. 15–5.

**SOLUTION** The claimed efficiency of the engine is (Eq. 15–4)

$$e = \frac{Q_H - Q_L}{Q_H}$$

$$= \frac{9.0\,\text{kJ} - 4.0\,\text{kJ}}{9.0\,\text{kJ}} = 0.56,$$

or 56%. The maximum possible efficiency is given by the Carnot efficiency, Eq. 15–5:

$$e_{ideal} = \frac{T_H - T_L}{T_H}$$

$$= \frac{435\,\text{K} - 285\,\text{K}}{435\,\text{K}} = 0.34,$$

or 34%. The manufacturer's claims violate the second law of thermodynamics and cannot be believed.

**EXERCISE D** A motor is running with an intake temperature $T_H = 400$ K and an exhaust temperature $T_L = 300$ K. Which of the following are *not* possible efficiencies for the engine? (*a*) 0.10; (*b*) 0.16; (*c*) 0.24; (*d*) 0.30; (*e*) 0.33.

We can see from Eq. 15–5 that at normal temperatures, a 100% efficient engine is not possible. Only if the exhaust temperature, $T_L$, were at absolute zero would 100% efficiency be reachable. But getting to absolute zero is a practical (as well as theoretical) impossibility. [Careful experimentation suggests that absolute zero is unattainable. This result is known as the **third law of thermodynamics**.] Because no engine can be 100% efficient, it can be stated that

**no device is possible whose sole effect is to transform a given amount of heat completely into work.**

This is known as the **Kelvin-Planck statement of the second law of thermodynamics**. Figure 15–15 diagrams the ideal perfect heat engine, which can not exist.

**FIGURE 15–15** Diagram of an impossible perfect heat engine in which all heat input is used to do work.

SECOND LAW OF THERMODYNAMICS
*(Kelvin-Planck statement)*

If the second law were not true, so that a perfect engine could be built, rather remarkable things could happen. For example, if the engine of a ship did not need a low-temperature reservoir to exhaust heat into, the ship could sail across the ocean using the vast resources of the internal energy of the ocean water. Indeed, we would have no fuel problems at all!

**EXERCISE E** Return to the Chapter-Opening Question, page 412, and answer it again now. Try to explain why you may have answered differently the first time.

# 15–6 Refrigerators, Air Conditioners, and Heat Pumps

The operating principle of refrigerators, air conditioners, and heat pumps is just the reverse of a heat engine. Each operates to transfer heat *out* of a cool environment into a warm environment. As diagrammed in Fig. 15–16, by doing work $W$, heat is taken from a low-temperature region, $T_L$ (such as inside a refrigerator), and a greater amount of heat is exhausted at a high temperature, $T_H$ (the room). Heat $Q_L$ is removed from cooling coils *inside* the refrigerator and heat $Q_H$ is given off by coils *outside* the rear of the refrigerator, Fig. 15–17. You can often feel this heated air coming out beneath the refrigerator. The work $W$ is usually done by an electric motor which compresses a fluid, as illustrated in Fig. 15–17. (We assume $Q_L$, $Q_H$, and $W$ are all positive, as in Section 15–5.)

Low pressure

Expansion valve

High pressure

Cooling coils (inside refrigerator)

Condenser coils (outside refrigerator)

$Q_H$ (to outside)

$Q_L$ (from inside refrigerator to coils)

Low pressure vapor

High pressure vapor

$W$

(a)

Compressor motor          Plug

$Q_L$          $Q_H$

$W$

(b)

Motor          Plug

**FIGURE 15–16** Schematic diagram of energy transfers for a refrigerator or air conditioner (AC).

**FIGURE 15–17** (a) Typical refrigerator system. The electric compressor motor forces a gas at high pressure through a heat exchanger (condenser) on the rear *outside* wall of the refrigerator, where $Q_H$ is given off and the gas cools to become liquid. The liquid passes from a high-pressure region, via a valve, to low-pressure tubes on the *inside* walls of the refrigerator; the liquid evaporates at this lower pressure and thus absorbs heat ($Q_L$) from the inside of the refrigerator. The fluid returns to the compressor, where the cycle begins again. (b) Schematic diagram, like Fig. 15–16.

A perfect refrigerator—one in which no work is required to take heat from the low-temperature region to the high-temperature region—is not possible. This is the **Clausius statement of the second law of thermodynamics**, already mentioned in Section 15–4: it can be stated formally as

**no device is possible whose sole effect is to transfer heat from one system at a temperature $T_L$ into a second system at a higher temperature $T_H$.**

To make heat flow from a low-temperature object (or system) to one at a higher temperature, work must be done. Thus, *there can be no perfect refrigerator.*

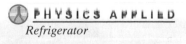

**PHYSICS APPLIED**
*Refrigerator*

SECOND LAW OF THERMODYNAMICS
(Clausius statement)

FIGURE 15–16 (Repeated.)
Schematic diagram of energy
transfers for a refrigerator or air
conditioner (AC).

PHYSICS APPLIED
*Air conditioner*

FIGURE 15–18 A heat pump uses
an electric motor to "pump" heat
from the cold outside to the warm
inside of a house.

PHYSICS APPLIED
*Heat pump*

The **coefficient of performance** (COP) of a refrigerator is defined as the heat $Q_L$ removed from the low-temperature area (inside a refrigerator) divided by the work $W$ done to remove the heat (Fig. 15–16):

$$\text{COP} = \frac{Q_L}{W}. \qquad \begin{bmatrix} \text{refrigerator and} \\ \text{air conditioner} \end{bmatrix} \quad \textbf{(15–6a)}$$

We use $Q_L$ because it is the heat removed from inside that matters from a practical point of view. This makes sense because the more heat $Q_L$ that can be removed from inside the refrigerator for a given amount of work, the better (more efficient) the refrigerator is. Energy is conserved, so from the first law of thermodynamics we can write $Q_L + W = Q_H$, or $W = Q_H - Q_L$ (see Fig. 15–16). Then Eq. 15–6a becomes

$$\text{COP} = \frac{Q_L}{W} = \frac{Q_L}{Q_H - Q_L}. \qquad \begin{bmatrix} \text{refrigerator and} \\ \text{air conditioner} \end{bmatrix} \quad \textbf{(15–6b)}$$

For an ideal refrigerator (not a perfect one, which is impossible), the best we could do would be

$$\text{COP}_{\text{ideal}} = \frac{T_L}{T_H - T_L}, \qquad \begin{bmatrix} \text{refrigerator and} \\ \text{air conditioner} \end{bmatrix} \quad \textbf{(15–6c)}$$

analagous to an ideal (Carnot) engine (Eq. 15–5).

An **air conditioner** works very much like a refrigerator, although the actual construction details are different: an air conditioner takes heat $Q_L$ from inside a room or building at a low temperature, and deposits heat $Q_H$ outside to the environment at a higher temperature. Equations 15–6 also describe the coefficient of performance for an air conditioner.

---

**EXAMPLE 15–11** **Making ice.** A freezer has a COP of 2.8 and uses 200 watts of power. How long would it take to freeze an ice-cube tray that contains 600 g of water at 0°C?

**APPROACH** In Eq. 15–6b, $Q_L$ is the heat that must be transferred out of the water so it will become ice. To determine $Q_L$, we use the latent heat of fusion $L$ of water and Eq. 14–4, $Q = mL$.

**SOLUTION** From Table 14–3, $L = 333$ kJ/kg for water. Hence

$$Q_L = mL = (0.600 \text{ kg})(3.33 \times 10^5 \text{ J/kg}) = 2.0 \times 10^5 \text{ J}$$

is the total energy that needs to be removed from the water. The freezer does work at the rate of 200 watts $= 200$ J/s $= W/t$, which is the work $W$ it can do in $t$ seconds. We solve for $t$: $t = W/(200 \text{ J/s})$. For $W$, we can also use Eq. 15–6a: $W = Q_L/\text{COP}$. Thus

$$t = \frac{W}{200 \text{ J/s}} = \frac{Q_L/\text{COP}}{200 \text{ J/s}}$$

$$= \frac{2.0 \times 10^5 \text{ J}}{(2.8)(200 \text{ J/s})} = 360 \text{ s},$$

or about 6 min.

---

Heat naturally flows from high temperature to low temperature. Refrigerators and air conditioners do work to accomplish the opposite: to make heat flow from cold to hot. We might say they "pump" heat from cold areas to hotter areas, against the natural tendency of heat to flow from hot to cold, just as water can be pumped uphill, against the natural tendency to flow downhill. The term **heat pump** is usually reserved for a device that can heat a house in winter by using an electric motor that does work $W$ to take heat $Q_L$ from the outside at low temperature and delivers heat $Q_H$ to the warmer inside of the house; see Fig. 15–18.

As in a refrigerator, there is an indoor and an outdoor heat exchanger (coils of the refrigerator) and an electric compressor motor. The operating principle is like that for a refrigerator or air conditioner; but the objective of a heat pump is to heat (deliver $Q_H$), rather than to cool (remove $Q_L$). Thus, the coefficient of performance of a heat pump is defined differently than for an air conditioner because it is the heat $Q_H$ delivered to the inside of the house that is important now:

$$\text{COP} = \frac{Q_H}{W}. \qquad \text{[heat pump]} \quad \textbf{(15–7)}$$

The COP is necessarily greater than 1. Typical heat pumps today have COP $\approx$ 2.5 to 3. Most heat pumps can be "turned around" and used as air conditioners in the summer.

**EXAMPLE 15–12** **Heat pump.** A heat pump has a coefficient of performance of 3.0 and is rated to do work at 1500 watts. (*a*) How much heat can it add to a room per second? (*b*) If the heat pump were turned around to act as an air conditioner in the summer, what would you expect its coefficient of performance to be, assuming all else stays the same?

**APPROACH** We use the definitions of coefficient of performance, which are different for the two devices in (*a*) and (*b*).

**SOLUTION** (*a*) We use Eq. 15–7 for the heat pump, and, since our device does 1500 J of work per second, it can pour heat into the room at a rate of

$$Q_H = \text{COP} \times W = 3.0 \times 1500\,\text{J} = 4500\,\text{J}$$

per second, or at a rate of 4500 W. [$\approx 4$ Btu/s.]

(*b*) If our device is turned around in summer, it can take heat $Q_L$ from inside the house, doing 1500 J of work per second to then dump $Q_H = 4500$ J per second to the hot outside. Energy is conserved, so $Q_L + W = Q_H$ (see Fig. 15–18, but reverse the inside and outside of the house). Then

$$Q_L = Q_H - W = 4500\,\text{J} - 1500\,\text{J} = 3000\,\text{J}.$$

The coefficient of performance as an air conditioner would thus be (Eq. 15–6a)

$$\text{COP} = \frac{Q_L}{W} = \frac{3000\,\text{J}}{1500\,\text{J}} = 2.0.$$

**NOTE** The coefficients of performance are defined differently for heat pumps and air conditioners.

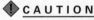

**CAUTION**

*Heat pumps and air conditioners have different COP definitions*

**EXERCISE F** The heat pump of Example 15–12 uses 1500 W of electric power to deliver 4500 W of heat. Does this sound like we're getting something for nothing? (*a*) Explain why we aren't. (*b*) Compare to the refrigerator of Example 15–11: the motor uses 200 W to extract how much heat?

A good heat pump can sometimes be a money saver and an energy saver, depending on the cost of the unit and installation, etc. Compare, for example, our heat pump in Example 15–12 to, say, a 1500-W electric heater. We plug the heater into the wall, it draws 1500 W of electricity, and it delivers 1500 W of heat to the room. Our heat pump when plugged into the wall also draws 1500 W of electricity (which is what we pay for), but it delivers 4500 W of heat!

## *SEER Rating

Cooling devices such as refrigerators and air conditioners are often given a rating known as SEER (Seasonal Energy Efficiency Ratio), which is defined as

$$\text{SEER} = \frac{(\text{heat removed in Btu})}{(\text{electrical input in watt-hours})},$$

**PHYSICS APPLIED**

*SEER rating*

as measured by averaging over varying (seasonal) conditions. The definition of the SEER is basically the same as the COP except for the (unfortunate) mixed units. Given that 1 Btu = 1056 J (see Section 14–1 and Problem 5 in Chapter 14), then a SEER = 1 is a COP equal to (1 Btu/1 W·h) = (1056 J)/(1 J/s × 3600 s) = 0.29. A COP = 1 is a SEER = 1/0.29 = 3.4.

# 15–7 Entropy and the Second Law of Thermodynamics

Thus far we have stated the second law of thermodynamics for specific situations. What we really need is a general statement of the second law of thermodynamics that will cover all situations, including processes discussed earlier in this Chapter that are not observed in nature even though they would not violate the first law of thermodynamics. It was not until the latter half of the nineteenth century that the second law of thermodynamics was finally stated in a general way—namely, in terms of a quantity called **entropy**, introduced by Clausius in the 1860s. Entropy, unlike heat, is a function of the state of a system. That is, a system in a given state has a temperature, a volume, a pressure, a mass, and also has a particular value of entropy. In the next Section, we will see that entropy can be interpreted as a measure of the order or disorder of a system.

When we deal with entropy—as with potential energy—it is the *change* in entropy during a process that is important, not the absolute amount. According to Clausius, the change in entropy $S$ of a system, when an amount of heat $Q$ is *added* to it by a reversible[†] process at constant temperature, is given by

$$\Delta S = \frac{Q}{T},\tag{15–8}$$

where $T$ is the kelvin temperature. (If heat is lost, $Q$ is negative in this equation, as per our original sign conventions on page 413.)

---

**EXAMPLE 15–13** **Entropy change in melting.** An ice cube of mass 56 g is taken from a storage compartment at 0°C and placed in a paper cup. After a few minutes, exactly half of the mass of the ice cube has melted, becoming water at 0°C. Find the change in entropy of the ice/water.

**APPROACH** We consider the 56 g of water, initially in the form of ice, as our system. To determine the entropy change, we first must find the heat needed to melt the ice, which we do using the latent heat of fusion of water, $L = 333$ kJ/kg (Section 14–5). The heat $Q$ required comes from the surroundings.

**SOLUTION** The heat required to melt 28 g of ice (half of the 56-g ice cube) is

$$Q = mL = (0.028 \text{ kg})(333 \text{ kJ/kg}) = 9.3 \text{ kJ}.$$

The temperature remains constant in our process, so we can find the change in entropy from Eq. 15–8:

$$\Delta S = \frac{Q}{T} = \frac{9.3 \text{ kJ}}{273 \text{ K}} = 34 \text{ J/K}.$$

**NOTE** The change in entropy of the surroundings (cup, air) has not been computed.

---

The temperature in Example 15–13 was constant, so the calculation was short. If the temperature varies during a process, a summation of the heat flow over the changing temperature can often be calculated using calculus or a computer. However, if the temperature change is not too great, a reasonable approximation can be made using the average value of the temperature, as indicated in the next Example.

---

**EXAMPLE 15–14 | ESTIMATE | Entropy change when water samples are mixed.** A sample of 50.0 kg of water at 20.00°C is mixed with 50.0 kg of water at 24.00°C. Estimate the change in entropy.

**APPROACH** The final temperature of the mixture will be 22.00°C, since we started with equal amounts of water. We use the specific heat of water and the methods of calorimetry (Sections 14–3 and 14–4) to determine the heat transferred. Then we use the average temperature of each sample of water to estimate the entropy change ($\Delta Q/T$).

---

[†]Real processes are irreversible. Because entropy is a state variable, the change in entropy $\Delta S$ for an irreversible process can be determined by calculating $\Delta S$ for a reversible process between the same two states.

**SOLUTION** A quantity of heat,

$$Q = mc\,\Delta T = (50.0\,\text{kg})(4186\,\text{J/kg}\cdot\text{C}°)(2.00\,\text{C}°) = 4.186 \times 10^5\,\text{J},$$

flows out of the hot water as it cools down from 24°C to 22°C, and this heat flows into the cold water as it warms from 20°C to 22°C. The total change in entropy, $\Delta S$, will be the sum of the changes in entropy of the hot water, $\Delta S_H$, and that of the cold water, $\Delta S_C$:

$$\Delta S = \Delta S_H + \Delta S_C.$$

We estimate entropy changes by writing $\Delta S = Q/T_{av}$, where $T_{av}$ is an "average" temperature for each process, which ought to give a reasonable estimate since the temperature change is small. For the hot water we use an average temperature of 23°C (296 K), and for the cold water an average temperature of 21°C (294 K). Thus

$$\Delta S_H \approx -\frac{4.186 \times 10^5\,\text{J}}{296\,\text{K}} = -1414\,\text{J/K}$$

which is negative because this heat flows out (sign conventions, page 413), whereas heat is added to the cold water:

$$\Delta S_C \approx \frac{4.186 \times 10^5\,\text{J}}{294\,\text{K}} = 1424\,\text{J/K}.$$

The entropy of the hot water $(S_H)$ decreases because heat flows out of the hot water. But the entropy of the cold water $(S_C)$ increases by a greater amount. The total change in entropy is

$$\Delta S = \Delta S_H + \Delta S_C \approx -1414\,\text{J/K} + 1424\,\text{J/K} \approx 10\,\text{J/K}.$$

In Example 15–14, we saw that although the entropy of one part of the system decreased, the entropy of the other part increased by a greater amount; the net change in entropy of the whole system was positive. This result, which we have calculated for a specific case in Example 15–14, has been found to hold in all other cases tested. That is, the total entropy of an isolated system is found to increase in all natural processes. The second law of thermodynamics can be stated in terms of entropy as follows: *The entropy of an isolated system never decreases. It can only stay the same or increase.* Entropy can remain the same only for an idealized (reversible) process. For any real process, the change in entropy $\Delta S$ is greater than zero:

$$\Delta S > 0. \qquad\qquad \text{[real process]} \quad \textbf{(15–9)}$$

If the system is not isolated, then the change in entropy of the system, $\Delta S_{sys}$, plus the change in entropy of the environment, $\Delta S_{env}$, must be greater than or equal to zero:

$$\Delta S = \Delta S_{sys} + \Delta S_{env} \geq 0. \qquad\qquad \textbf{(15–10)}$$

Only idealized processes can have $\Delta S = 0$. Real processes always have $\Delta S > 0$. This, then, is the *general statement of the second law of thermodynamics*:

**the total entropy of any system plus that of its environment increases as a result of any natural process.**

*SECOND LAW OF THERMODYNAMICS (general statement)*

Although the entropy of one part of the universe may decrease in any natural process (see Example 15–14), the entropy of some other part of the universe always increases by a greater amount, so the total entropy always increases.

Now that we finally have a quantitative general statement of the second law of thermodynamics, we can see that it is an unusual law. It differs considerably from other laws of physics, which are typically equalities (such as $F = ma$) or conservation laws (such as for energy and momentum). The second law of thermodynamics introduces a new quantity, the entropy $S$, but does not tell us it is conserved. Quite the opposite. Entropy is *not* conserved in natural processes. Entropy always increases in time for real processes.

# 15–8 Order to Disorder

The concept of entropy, as we have discussed it so far, may seem rather abstract. But we can relate it to the more ordinary concepts of *order* and *disorder*. In fact, the entropy of a system can be considered a *measure of the disorder of the system*. Then the second law of thermodynamics can be stated simply as:

**natural processes tend to move toward a state of greater disorder.**

Exactly what we mean by **disorder** may not always be clear, so we now consider a few examples. Some of these will show us how this very general statement of the second law applies beyond what we usually consider as thermodynamics.

Let us look at the simple processes mentioned in Section 15–4. First, a jar containing separate layers of salt and pepper is more orderly than a jar in which the salt and pepper are all mixed up. Shaking a jar containing separate layers results in a mixture, and no amount of shaking brings the orderly layers back again. The natural process is from a state of relative order (layers) to one of relative disorder (a mixture), not the reverse. That is, disorder increases. Second, a solid coffee cup is a more "orderly" and useful object than the pieces of a broken cup. Cups break when they fall, but they do not spontaneously mend themselves (as faked in Fig. 15–10). Again, the normal course of events is an increase of disorder.

Let us consider some processes for which we have actually calculated the entropy change, and see that an increase in entropy results in an increase in disorder (or vice versa). When ice melts to water at 0°C, the entropy of the water increases (Example 15–13). Intuitively, we can think of solid water, ice, as being more ordered than the less orderly fluid state which can flow all over the place. This change from order to disorder can be seen more clearly from the molecular point of view: the orderly arrangement of water molecules in an ice crystal has changed to the disorderly and somewhat random motion of the molecules in the fluid state.

When a hot substance is put in contact with a cold substance, heat flows from the high temperature to the low until the two substances reach the same intermediate temperature. Entropy increases, as we saw in Example 15–14. At the beginning of the process we can distinguish two classes of molecules: those with a high average kinetic energy (the hot object), and those with a low average kinetic energy (the cooler object). After the process in which heat flows, all the molecules are in one class with the same average kinetic energy; we no longer have the more orderly arrangement of molecules in two classes. Order has gone to disorder. Furthermore, the separate hot and cold objects could serve as the hot- and cold-temperature regions of a heat engine, and thus could be used to obtain useful work. But once the two objects are put in contact and reach the same temperature, no work can be obtained. Disorder has increased, because a system that has the ability to perform work must surely be considered to have a higher order than a system no longer able to do work.

When a stone falls to the ground, its macroscopic kinetic energy is transformed to thermal energy. (We noted earlier that the reverse never happens: a stone never absorbs thermal energy and rises into the air of its own accord.) This is another example of order changing to disorder. Thermal energy is associated with the disorderly random motion of molecules, but the molecules in the falling stone all have the same velocity downward in addition to their own random velocities. Thus, the more orderly kinetic energy of the stone as a whole (which could do useful work) is changed to disordered thermal energy when the stone strikes the ground. Disorder increases in this process, as it does in all processes that occur in nature.

## Biological Development

**PHYSICS APPLIED**
*Biological development*

An interesting example of the increase in entropy relates to the biological development and growth of organisms. Clearly, a human being is a highly ordered organism. The development of an individual from a single cell to a grown person is a process of increasing order. Evolution too might be seen as an increase in order.

Do these processes violate the second law of thermodynamics? No, they do not. In the processes of growth and evolution, and even during the mature life of an individual, waste products are eliminated. These small molecules that remain as a result of metabolism are simple molecules without much order. Thus they represent relatively higher disorder or entropy. Indeed, the total entropy of the molecules cast aside by organisms during the processes of development and growth is greater than the decrease in entropy associated with the order of the growing individual or evolving species.

### "Time's Arrow"

Another aspect of the second law of thermodynamics is that it tells us in which *direction* processes go. If you were to see a film being run backward, you would undoubtedly be able to tell that it *was* run backward. For you would see odd occurrences, such as a broken coffee cup rising from the floor and reassembling on a table, or a torn balloon suddenly becoming whole again and filled with air. We know these things don't happen in real life; they are processes in which order increases—or entropy decreases. They violate the second law of thermodynamics. When watching a movie (or imagining that time could go backward), we are tipped off to a reversal of time by observing whether entropy (and disorder) is increasing or decreasing. Hence, entropy has been called **time's arrow**, because it can tell us in which direction time is going.

# 15–9 Unavailability of Energy; Heat Death

In the process of heat conduction from a hot object to a cold one, we have seen that entropy increases and that order goes to disorder. The separate hot and cold objects could serve as the high- and low-temperature regions for a heat engine and thus could be used to obtain useful work. But after the two objects are put in contact with each other and reach the same uniform temperature, no work can be obtained from them. With regard to being able to do useful work, order has gone to disorder in this process.

The same can be said about a falling rock that comes to rest upon striking the ground. Before hitting the ground, all the kinetic energy of the rock could have been used to do useful work. But once the rock's mechanical kinetic energy becomes thermal energy, doing useful work is no longer possible.

Both these examples illustrate another important aspect of the second law of thermodynamics:

**in any natural process, some energy becomes unavailable to do useful work.**

In any process, no energy is ever lost (it is always conserved). Rather, energy becomes less useful—it can do less useful work. As time goes on, **energy is degraded**, in a sense. It goes from more orderly forms (such as mechanical) eventually to the least orderly form: internal, or thermal, energy. Entropy is a factor here because the amount of energy that becomes unavailable to do work is proportional to the change in entropy during any process.

A natural outcome of the degradation of energy is the prediction that as time goes on, the universe should approach a state of maximum disorder. Matter would become a uniform mixture, and heat would have flowed from high-temperature regions to low-temperature regions until the whole universe is at one temperature. No work could then be done. All the energy of the universe would have degraded to thermal energy. This prediction, called the **heat death** of the universe, has been much discussed, but it would lie *very* far in the future. It is a complicated subject, and some scientists question whether thermodynamics, as we now understand it, actually applies to the universe as a whole, which is at a much larger scale.[†]

[†]When a star, like our Sun, loses energy by radiation, it becomes *hotter* (not cooler). By losing energy, gravity is able to compress the gas of which the Sun is made—becoming smaller and denser means the Sun gets hotter. For astronomical objects, when heat flows from a hot object to a cooler object, the hot object gets hotter and the cool one cooler. That is, the temperature difference *increases*.

# *15–10 Statistical Interpretation of Entropy and the Second Law

The ideas of entropy and disorder are made clearer with the use of a statistical or probabilistic analysis of the molecular state of a system. This statistical approach, which was first applied toward the end of the nineteenth century by Ludwig Boltzmann (1844–1906), makes a clear distinction between the "macrostate" and the "microstate" of a system. The **microstate** of a system would be specified by giving the position and velocity of every particle (or molecule). The **macrostate** of a system is specified by giving the far fewer macroscopic properties of the system—the temperature, pressure, number of moles, and so on. In reality, we can know only the macrostate of a system. There are generally far too many molecules in a system to be able to know the velocity and position of every one at a given moment. Nonetheless, we can hypothesize a great many different microstates that can correspond to the *same* macrostate.

Let us take a very simple example. Suppose you repeatedly shake four coins in your hand and drop them on a table. Specifying the *number* of heads and the number of tails that appear on a given throw is the *macrostate* of this system. Specifying *each coin* as being a head or a tail is the *microstate* of the system. In the following Table we see how many microstates correspond to each macrostate:

| Macrostate | Possible Microstates (H = heads, T = tails) | Number of Microstates |
|---|---|---|
| 4 heads | HHHH | 1 |
| 3 heads, 1 tail | HHHT, HHTH, HTHH, THHH | 4 |
| 2 heads, 2 tails | HHTT, HTHT, THHT, HTTH, THTH, TTHH | 6 |
| 1 head, 3 tails | TTTH, TTHT, THTT, HTTT | 4 |
| 4 tails | TTTT | 1 |

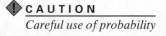

**CAUTION**

*Careful use of probability*

A basic assumption behind the statistical approach is that *each microstate is equally probable*. Thus the number of microstates that give the same macrostate corresponds to the relative probability of that macrostate occurring. The macrostate of two heads and two tails is the most probable one in our case of tossing four coins; out of the total of 16 possible microstates, six correspond to two heads and two tails, so the probability of throwing two heads and two tails is 6 out of 16, or 38%. The probability of throwing one head and three tails is 4 out of 16, or 25%. The probability of four heads is only 1 in 16, or 6%. If you threw the coins 16 times, you might not find that two heads and two tails appear exactly 6 times, or four tails exactly once. These are only probabilities or averages. But if you made 1600 throws, very nearly 38% of them would be two heads and two tails. The greater the number of tries, the closer the percentages are to the calculated probabilities.

**EXERCISE G** In the Table above, what is the probability that there will be at least two heads? (a) $\frac{1}{2}$; (b) $\frac{1}{16}$; (c) $\frac{1}{8}$; (d) $\frac{3}{8}$; (e) $\frac{11}{16}$.

If we toss more coins—say, 100 all at the same time—the relative probability of throwing all heads (or all tails) is greatly reduced. There is only one microstate corresponding to all heads. For 99 heads and 1 tail, there are 100 microstates because each of the coins could be the one tail. The relative probabilities for other macrostates are given in Table 15–3 (top of next page). About $1.3 \times 10^{30}$ microstates are possible.[†] Thus the relative probability of finding all heads is about 1 in $10^{30}$, an incredibly unlikely event! The probability of obtaining 50 heads and 50 tails (see Table 15–3) is $(1.0 \times 10^{29})/(1.3 \times 10^{30}) = 0.08$ or 8%. The probability of obtaining anything between 45 and 55 heads is about 70%.

[†]Each coin has two possibilities, heads or tails. Then the possible number of microstates is $2 \times 2 \times 2 \times \cdots = 2^{100} = 1.3 \times 10^{30}$ (using a calculator or logarithms).

| TABLE 15–3 Probabilities of Various Macrostates for 100 Coin Tosses | | | |
|---|---|---|---|
| Macrostate | | Number of Microstates | Probability |
| Heads | Tails | | |
| 100 | 0 | 1 | $7.9 \times 10^{-31}$ |
| 99 | 1 | $1.0 \times 10^2$ | $7.9 \times 10^{-29}$ |
| 90 | 10 | $1.7 \times 10^{13}$ | $1.4 \times 10^{-17}$ |
| 80 | 20 | $5.4 \times 10^{20}$ | $4.2 \times 10^{-10}$ |
| 60 | 40 | $1.4 \times 10^{28}$ | 0.011 |
| 55 | 45 | $6.1 \times 10^{28}$ | 0.047 |
| 50 | 50 | $1.0 \times 10^{29}$ | 0.077 |
| 45 | 55 | $6.1 \times 10^{28}$ | 0.047 |
| 40 | 60 | $1.4 \times 10^{28}$ | 0.011 |
| 20 | 80 | $5.4 \times 10^{20}$ | $4.2 \times 10^{-10}$ |
| 10 | 90 | $1.7 \times 10^{13}$ | $1.4 \times 10^{-17}$ |
| 1 | 99 | $1.0 \times 10^2$ | $7.9 \times 10^{-29}$ |
| 0 | 100 | 1 | $7.9 \times 10^{-31}$ |

Thus we see that as the number of coins increases, the probability of obtaining the most orderly arrangement (all heads or all tails) becomes extremely unlikely. The least orderly arrangement (half heads, half tails) is the most probable, and the probability of being within, say, 5% of the most probable arrangement greatly increases as the number of coins increases. These same ideas can be applied to the molecules of a system. For example, the most probable state of a gas (say, the air in a room) is one in which the molecules take up the whole space and move about randomly; this corresponds to the Maxwellian distribution, Fig. 15–19a (and see Section 13–10). On the other hand, the very orderly arrangement of all the molecules located in one corner of the room and all moving with the same velocity (Fig. 15–19b) is extremely unlikely.

From these examples, we can see that probability is directly related to disorder and hence to entropy. That is, the most probable state is the one with greatest entropy, or greatest disorder and randomness. The processes we actually observe are those where the entropy increase is greatest.

In terms of probability, the second law of thermodynamics—which tells us that entropy increases in any process—reduces to the statement that **those processes occur which are most probable**.

There is an additional element, however. The second law in terms of probability does not *forbid* a decrease in entropy. Rather, it says the probability is extremely low. It is not impossible that salt and pepper could separate spontaneously into layers, or that a broken cup could mend itself (Fig. 15–10). It is even possible that a lake could freeze over on a hot summer day (that is, for heat to flow out of the cold lake into the warmer surroundings). But the probability for such events occurring is almost always miniscule. In our coin examples, we saw that increasing the number of coins from 4 to 100 drastically reduced the probability of large deviations from the average, or most probable, arrangement. In ordinary systems we are not dealing with only 100 molecules, but with extremely large numbers of molecules: in 1 mole alone there are $6 \times 10^{23}$ molecules. Hence the probability of deviation far from the average is incredibly tiny. For example, it has been calculated that the probability that a stone resting on the ground could transform 1 cal of thermal energy into mechanical energy and rise up into the air is much less likely than the probability that a group of monkeys typing randomly would by chance produce the complete works of Shakespeare. Put another way, the probability is less than once in the entire age of the universe; this could be what "never" means.

(a)

(b)

**FIGURE 15–19** (a) Most probable distribution of molecular speeds in a gas (Maxwellian, or random). (b) Orderly, but highly unlikely, distribution of speeds in which all molecules have nearly the same speed.

(a)

(b)

(c)

**FIGURE 15–20** (a) A fossil-fuel steam plant (this one uses forest waste products = biomass). (b) An array of mirrors focuses sunlight on a boiler to produce steam at a solar energy installation. (c) Large cooling towers at an electric generating plant.

**FIGURE 15–21** Mechanical energy is transformed to electric energy with a turbine and generator.

Source of energy: water, steam, wind

Electric generator

Turbine

Electric energy

**PHYSICS APPLIED**

*Heat engines and thermal pollution*

# *15–11 Thermal Pollution, Global Warming, and Energy Resources

Much of the energy we use in everyday life—from motor vehicles to most of the electricity produced by power plants—uses a heat engine. More than $\frac{2}{3}$ of the electric energy produced in the U.S. is generated at fossil-fuel steam plants (coal, oil, or gas—see Fig. 15–20a), and they make use of a heat engine (steam engines). In electric power plants, the steam drives the turbines and generators (Fig. 15–21) whose output is electric energy. The various means to turn the turbine are shown in Table 15–4, along with some of the advantages and disadvantages of each. Even nuclear power plants use a steam engine, run on nuclear fuel. Electricity produced by falling water at dams, by windmills, or by solar cells (Fig. 15–20b) does not involve a heat engine.

The heat output $Q_L$ from every heat engine, from power plants to cars, is referred to as **thermal pollution** because this heat $(Q_L)$ must be absorbed by the environment—such as by water from rivers or lakes, or by the air using large cooling towers (Fig. 15–20c). When water is the coolant, this heat raises the temperature of the water before returning it to its source, altering the natural ecology of aquatic life (largely because warmer water holds less oxygen). In the case of air cooling towers, the output heat $Q_L$ raises the temperature of the atmosphere, which affects weather.

**Air pollution**—by which we mean the chemicals released in the burning of fossil fuels in cars, power plants, and industrial furnaces—gives rise to smog and other problems. Another issue is the buildup of $CO_2$ in the Earth's atmosphere due to the burning of fossil fuels; the carbon in them combines with $O_2$ of the air, forming $CO_2$. This $CO_2$ absorbs some of the infrared radiation that the Earth naturally emits (Section 14–8), and thus contributes to **global warming**. The **carbon footprint** of any activity (home appliances, steel production, manufacture of goods, transportation) refers to the negative impact on the environment due to the burning of fossil fuels which release $CO_2$ (and other noxious products) into the atmosphere, contributing to climate change. A carbon footprint can be expressed in kg or tons of carbon (in the $CO_2$). Limiting the burning of fossil fuels can help reduce these problems.

Thermal pollution, however, is unavoidable. Engineers can try to design and build engines that are more efficient, but they cannot surpass the Carnot efficiency and must live with $T_L$ being at best the ambient temperature of water or air. The second law of thermodynamics tells us the limit imposed by nature. Even alternative energy sources like wind and solar cells can contribute to global warming (windmills slow down cooling winds, solar cells are "dark" and absorb more of the incident energy). What we can do, in light of the second law of thermodynamics, is use less energy and conserve our fuel resources. There is no other solution.

## TABLE 15–4 Electric Energy Resources

| Form of Electric Energy Production | % of Production (approx.) | | Advantages | Disadvantages |
| --- | --- | --- | --- | --- |
| | U.S. | World | | |
| **Fossil-fuel steam plants:** burn coal, oil, or natural gas to boil water, producing high-pressure steam that turns a turbine of a generator (Figs. 15–12b, 15–21); uses heat engine. | 72 | 66 | We know how to build them; for now relatively inexpensive. | Air pollution; thermal pollution; limited efficiency; land devastation from extraction of raw materials (mining); global warming; accidents such as oil spills at sea; limited fuel supply (estimates range from a couple of decades to a few centuries). |
| **Nuclear energy:** | | | | |
| **Fission:** nuclei of uranium or plutonium atoms split ("fission") with release of energy (Chapter 31) that heats steam; uses heat engine. | 20 | 17 | Normally almost no air pollution; less contribution to global warming; relatively inexpensive. | Thermal pollution; accidents can release damaging radioactivity; difficult disposal of radioactive by-products; possible diversion of nuclear material by terrorists; limited fuel supply. |
| **Fusion:** energy released when isotopes of hydrogen (or other small nuclei) combine or "fuse" (Chapter 31). | 0 | 0 | Relatively "clean"; vast fuel supply (hydrogen in water molecules in oceans); less contribution to global warming. | Not yet workable. |
| **Hydroelectric:** falling water turns turbines at the base of a dam. | 7 | 16 | No heat engine needed; no air, water, or thermal pollution; relatively inexpensive; high efficiency; dams can control flooding. | Reservoirs behind dams inundate scenic or inhabited land; dams block upstream migration of salmon and other fish for reproduction; few locations remain for new dams; drought. |
| **Geothermal:** natural steam from inside the Earth comes to the surface (hot springs, geysers, steam vents); or cold water passed down into contact with hot, dry rock is heated to steam. | <1 | <1 | No heat engine needed; little air pollution; good efficiency; relatively inexpensive and "clean." | Few appropriate sites; small production; mineral content of spent hot water can pollute. |
| **Wind power:** 3-kW to 5-MW windmills (vanes up to 50 m long) turn a generator. | ≈1 | <1 | No heat engine; no air, water, or thermal pollution; relatively inexpensive. | Large array of big windmills might affect weather and be eyesores; slows down cooling winds; hazardous to migratory birds; winds not always strong. |
| **Solar energy:** | <1 | <1 | | |
| **Active solar heating:** rooftop solar panels absorb the Sun's rays, which heat water in tubes for space heating and hot water supply. | | | No heat engine needed; no air or thermal pollution; unlimited fuel supply. | Space limitations; may require back-up; relatively expensive; less effective when cloudy. |
| **Passive solar heating:** architectural devices—windows along southern exposure, sunshade over windows to keep Sun's rays out in summer. | | | No heat engine needed; no air or thermal pollution; relatively inexpensive. | Almost none, but other methods needed too. |
| **Solar cells (photovoltaic cells):** convert sunlight directly into electricity without use of heat engine. | | | No heat engine; thermal, air, and water pollution very low; good efficiency (>30% and improving). | Expensive; chemical pollution at manufacture; much land needed as Sun's energy not concentrated; absorption (dark color) without reemission = global warming. |

## Thermodynamics

1. Define the **system** you are dealing with; distinguish the system under study from its surroundings.

2. When applying the first law of thermodynamics, be careful of **signs** associated with **work** and **heat**. In the first law, work done *by* the system is positive; work done *on* the system is negative. Heat *added* to the system is positive, but heat *removed* from it is negative. With heat engines, we usually consider the heat intake, the heat exhausted, and the work done as positive.

3. Watch the **units** used for work and heat; work is most often expressed in joules, and heat can be in calories, kilocalories, or joules. Be consistent: choose only one unit for use throughout a given problem.

4. **Temperatures** must generally be expressed in kelvins; temperature *differences* may be expressed in C° or K.

5. **Efficiency** (or coefficient of performance) is a ratio of two energy transfers: useful output divided by required input. Efficiency (but *not* coefficient of performance) is always less than 1 in value, and hence is often stated as a percentage.

6. The **entropy** of a system increases when heat is added to the system, and decreases when heat is removed. If heat is transferred from system A to system B, the change in entropy of A is negative and the change in entropy of B is positive.

# Summary

The **first law of thermodynamics** states that the change in internal energy $\Delta U$ of a system is equal to the heat *added* to the system, $Q$, minus the work done *by* the system, $W$:

$$\Delta U = Q - W. \quad (15\text{--}1)$$

This important law is a statement of the conservation of energy, and is found to hold for all processes.

An **isothermal** process is a process carried out at constant temperature.

In an **adiabatic** process, no heat is exchanged ($Q = 0$).

The work $W$ done by a gas at constant pressure $P$ is given by

$$W = P\,\Delta V, \quad (15\text{--}3)$$

where $\Delta V$ is the change in volume of the gas.

A **heat engine** is a device for changing thermal energy, by means of heat flow, into useful work.

The **efficiency** $e$ of a heat engine is defined as the ratio of the work $W$ done by the engine to the high temperature heat input $Q_H$. Because of conservation of energy, the work output equals $Q_H - Q_L$, where $Q_L$ is the heat exhausted at low temperature to the environment; hence

$$e = \frac{W}{Q_H} = \frac{Q_H - Q_L}{Q_H} = 1 - \frac{Q_L}{Q_H}. \quad (15\text{--}4)$$

$Q_H$, $Q_L$, and $W$, as defined for heat engines, are positive.

The *upper limit* on the efficiency (the *Carnot efficiency*) can be written in terms of the higher and lower operating temperatures (in kelvins) of the engine, $T_H$ and $T_L$, as

$$e_{ideal} = 1 - \frac{T_L}{T_H}. \quad (15\text{--}5)$$

Real (irreversible) engines always have an efficiency less than this.

The operation of **refrigerators** and **air conditioners** is the reverse of a heat engine: work is done to extract heat $Q_L$ from a cool region and exhaust it to a region at a higher temperature. The coefficient of performance (COP) for either is

$$COP = \frac{Q_L}{W}, \quad \begin{bmatrix} \text{refrigerator or} \\ \text{air conditioner} \end{bmatrix} \quad (15\text{--}6a)$$

where $W$ is the work needed to remove heat $Q_L$ from the area with the low temperature.

A **heat pump** uses work $W$ to bring heat $Q_L$ from the cold outside and deliver heat $Q_H$ to warm the interior. The coefficient of performance of a heat pump is

$$COP = \frac{Q_H}{W}, \quad [\text{heat pump}] \quad (15\text{--}7)$$

because it is the heat $Q_H$ delivered inside the building that counts.

The **second law of thermodynamics** can be stated in several equivalent ways:

(a) heat flows spontaneously from a hot object to a cold one, but not the reverse;

(b) there can be no 100% efficient heat engine—that is, one that can change a given amount of heat completely into work;

(c) natural processes tend to move toward a state of greater disorder or greater **entropy**.

Statement (c) is the most general statement of the second law of thermodynamics, and can be restated as: the total entropy, $S$, of any system plus that of its environment increases as a result of any natural process:

$$\Delta S > 0. \quad (15\text{--}9)$$

The change in entropy in a process that transfers heat $Q$ at a constant temperature $T$ is

$$\Delta S = \frac{Q}{T}. \quad (15\text{--}8)$$

Entropy is a quantitative measure of the **disorder** of a system. The second law of thermodynamics also indicates that as time goes on, energy is **degraded** to less useful forms—that is, it is less available to do useful work.

The second law of thermodynamics tells us in which direction processes tend to proceed; hence entropy is called "time's arrow."

[*Entropy can be examined from a statistical point of view, considering **macrostates** (for example, $P$, $V$, $T$) and **microstates** (state of each molecule). The most probable processes are the ones we observe. They are the ones that increase entropy the most. Processes that violate the second law "could" occur, but only with *extremely* low probability.]

[*All heat engines give rise to **thermal pollution** because they exhaust heat to the environment.]

# Questions

1. In an isothermal process, 3700 J of work is done by an ideal gas. Is this enough information to tell how much heat has been added to the system? If so, how much? If not, why not?

2. Can mechanical energy ever be transformed completely into heat or internal energy? Can the reverse happen? In each case, if your answer is no, explain why not; if yes, give one or two examples.

3. Can the temperature of a system remain constant even though heat flows into or out of it? If so, give examples.

4. Explain why the temperature of a gas increases when it is compressed adiabatically.

5. An ideal monatomic gas expands slowly to twice its volume (1) isothermally; (2) adiabatically; (3) isobarically. Plot each on a $PV$ diagram. In which process is $\Delta U$ the greatest, and in which is $\Delta U$ the least? In which is $W$ the greatest and the least? In which is $Q$ the greatest and the least?

6. (a) What happens if you remove the lid of a bottle containing chlorine gas? (b) Does the reverse process ever happen? Why or why not? (c) Can you think of two other examples of irreversibility?

7. Would a definition of heat engine efficiency as $e = W/Q_L$ be useful? Explain.

8. What are the high-temperature and the low-temperature areas for (a) an internal combustion engine, and (b) a steam engine? Are they, strictly speaking, heat reservoirs?

9. The oceans contain a tremendous amount of thermal (internal) energy. Why, in general, is it not possible to put this energy to useful work?

10. Can you warm a kitchen in winter by leaving the oven door open? Can you cool the kitchen on a hot summer day by leaving the refrigerator door open? Explain.

11. The COPs are defined differently for heat pumps and air conditioners. Explain why.

12. You are asked to test a machine that the inventor calls an "in-room air conditioner": a big box, standing in the middle of the room, with a cable that plugs into a power outlet. When the machine is switched on, you feel a stream of cold air coming out of it. How do you know that this machine cannot cool the room?

13. Think up several processes (other than those already mentioned) that would obey the first law of thermodynamics, but, if they actually occurred, would violate the second law.

14. Suppose a lot of papers are strewn all over the floor; then you stack them neatly. Does this violate the second law of thermodynamics? Explain.

15. The first law of thermodynamics is sometimes whimsically stated as, "You can't get something for nothing," and the second law as, "You can't even break even." Explain how these statements could be equivalent to the formal statements.

16. A gas is allowed to expand (a) adiabatically and (b) isothermally. In each process, does the entropy increase, decrease, or stay the same? Explain.

17. Which do you think has the greater entropy, 1 kg of solid iron or 1 kg of liquid iron? Why?

18. Give three examples, other than those mentioned in this Chapter, of naturally occurring processes in which order goes to disorder. Discuss the observability of the reverse process.

19. Entropy is often called "time's arrow" because it tells us in which direction natural processes occur. If a movie were run backward, name some processes that you might see that would tell you that time was "running backward."

20. Living organisms, as they grow, convert relatively simple food molecules into a complex structure. Is this a violation of the second law of thermodynamics? Explain your answer.

# MisConceptual Questions

1. In an isobaric compression of an ideal gas,
   (a) no heat flows into the gas.
   (b) the internal energy of the gas remains constant.
   (c) no work is done on the gas.
   (d) work is done on the gas.
   (e) work is done by the gas.

2. Which is possible: converting (i) 100 J of work entirely into 100 J of heat, (ii) 100 J of heat entirely into 100 J of work?
   (a) Only (i) is possible.
   (b) Only (ii) is possible.
   (c) Both (i) and (ii) are possible.
   (d) Neither (i) nor (ii) is possible.

3. An ideal gas undergoes an isobaric compression and then an isovolumetric process that brings it back to its initial temperature. Had the gas undergone one isothermal process instead,
   (a) the work done on the gas would be the same.
   (b) the work done on the gas would be less.
   (c) the work done on the gas would be greater.
   (d) Need to know the temperature of the isothermal process.

4. An ideal gas undergoes an isothermal expansion from state A to state B. In this process (use sign conventions, page 413),
   (a) $Q = 0$, $\Delta U = 0$, $W > 0$.
   (b) $Q > 0$, $\Delta U = 0$, $W < 0$.
   (c) $Q = 0$, $\Delta U > 0$, $W > 0$.
   (d) $Q > 0$, $\Delta U = 0$, $W > 0$.
   (e) $Q = 0$, $\Delta U < 0$, $W < 0$.

5. An ideal gas undergoes an isothermal process. Which of the following statements are true? (i) No heat is added to or removed from the gas. (ii) The internal energy of the gas does not change. (iii) The average kinetic energy of the molecules does not change.
   (a) (i) only.
   (b) (i) and (ii) only.
   (c) (i) and (iii) only.
   (d) (ii) and (iii) only.
   (e) (i), (ii), and (iii).
   (f) None of the above.

6. An ideal gas undergoes an adiabatic expansion, a process in which no heat flows into or out of the gas. As a result,
   (a) the temperature of the gas remains constant and the pressure decreases.
   (b) both the temperature and pressure of the gas decrease.
   (c) the temperature of the gas decreases and the pressure increases.
   (d) both the temperature and volume of the gas increase.
   (e) both the temperature and pressure of the gas increase.

7. A heat engine operates between a high temperature of about 600°C and a low temperature of about 300°C. What is the maximum theoretical efficiency for this engine?
   (a) = 100%.    (b) ≈ 66%.    (c) ≈ 50%.    (d) ≈ 34%.
   (e) Cannot be determined from the given information.

8. On a very hot day, could you cool your kitchen by leaving the refrigerator door open?
   (a) Yes, but it would be very expensive.
   (b) Yes, but only if the humidity is below 50%.
   (c) No, the refrigerator would exhaust the same amount of heat into the room as it takes out of the room.
   (d) No, the heat exhausted by the refrigerator into the room is more than the heat the refrigerator takes out of the room.

9. Which of the following possibilities could increase the efficiency of a heat engine or an internal combustion engine?
   (a) Increase the temperature of the hot part of the system and reduce the temperature of the exhaust.
   (b) Increase the temperatures of both the hot part and the exhaust part of the system by the same amount.
   (c) Decrease the temperatures of both the hot part and the exhaust part of the system by the same amount.
   (d) Decrease the temperature of the hot part and increase the temperature of the exhaust part by the same amount.
   (e) None of the above; only redesigning the engine or using better gas could improve the engine's efficiency.

10. About what percentage of the heat produced by burning gasoline is turned into useful work by a typical automobile?
    (a) 20%.   (b) 50%.   (c) 80%.   (d) 90%.   (e) Nearly 100%.

11. Which statement is true regarding the entropy change of an ice cube that melts?
    (a) Since melting occurs at the melting point temperature, there is no temperature change so there is no entropy change.
    (b) Entropy increases.
    (c) Entropy decreases.

For assigned homework and other learning materials, go to the MasteringPhysics website.

# Problems

## 15–1 and 15–2 First Law of Thermodynamics
[Recall sign conventions, page 413.]

1. (I) An ideal gas expands isothermally, performing $4.30 \times 10^3$ J of work in the process. Calculate (a) the change in internal energy of the gas, and (b) the heat absorbed during this expansion.

2. (I) One liter of air is cooled at constant pressure until its volume is halved, and then it is allowed to expand isothermally back to its original volume. Draw the process on a PV diagram.

3. (II) Sketch a PV diagram of the following process: 2.5 L of ideal gas at atmospheric pressure is cooled at constant pressure to a volume of 1.0 L, and then expanded isothermally back to 2.5 L, whereupon the pressure is increased at constant volume until the original pressure is reached.

4. (II) A gas is enclosed in a cylinder fitted with a light frictionless piston and maintained at atmospheric pressure. When 254 kcal of heat is added to the gas, the volume is observed to increase slowly from 12.0 m³ to 16.2 m³. Calculate (a) the work done by the gas and (b) the change in internal energy of the gas.

5. (II) A 1.0-L volume of air initially at 3.5 atm of (gauge) pressure is allowed to expand isothermally until the pressure is 1.0 atm. It is then compressed at constant pressure to its initial volume, and lastly is brought back to its original pressure by heating at constant volume. Draw the process on a PV diagram, including numbers and labels for the axes.

6. (II) The pressure in an ideal gas is cut in half slowly, while being kept in a container with rigid walls. In the process, 465 kJ of heat left the gas. (a) How much work was done during this process? (b) What was the change in internal energy of the gas during this process?

7. (II) In an engine, an almost ideal gas is compressed adiabatically to half its volume. In doing so, 2630 J of work is done on the gas. (a) How much heat flows into or out of the gas? (b) What is the change in internal energy of the gas? (c) Does its temperature rise or fall?

8. (II) An ideal gas expands at a constant total pressure of 3.0 atm from 410 mL to 690 mL. Heat then flows out of the gas at constant volume, and the pressure and temperature are allowed to drop until the temperature reaches its original value. Calculate (a) the total work done by the gas in the process, and (b) the total heat flow into the gas.

9. (II) 8.5 moles of an ideal monatomic gas expand adiabatically, performing 8300 J of work in the process. What is the change in temperature of the gas during this expansion?

10. (II) Consider the following two-step process. Heat is allowed to flow out of an ideal gas at constant volume so that its pressure drops from 2.2 atm to 1.4 atm. Then the gas expands at constant pressure, from a volume of 5.9 L to 9.3 L, where the temperature reaches its original value. See Fig. 15–22. Calculate (a) the total work done by the gas in the process, (b) the change in internal energy of the gas in the process, and (c) the total heat flow into or out of the gas.

**FIGURE 15–22**
Problem 10.

11. (II) Use the conservation of energy to explain why the temperature of a well-insulated gas increases when it is compressed—say, by pushing down on a piston—whereas the temperature decreases when the gas expands. Show your reasoning.

12. (III) The $PV$ diagram in Fig. 15–23 shows two possible states of a system containing 1.75 moles of a monatomic ideal gas. ($P_1 = P_2 = 425\,\text{N/m}^2$, $V_1 = 2.00\,\text{m}^3$, $V_2 = 8.00\,\text{m}^3$.) (a) Draw the process which depicts an isobaric expansion from state 1 to state 2, and label this process A. (b) Find the work done by the gas and the change in internal energy of the gas in process A. (c) Draw the two-step process which depicts an isothermal expansion from state 1 to the volume $V_2$, followed by an isovolumetric increase in temperature to state 2, and label this process B. (d) Find the change in internal energy of the gas for the two-step process B.

**FIGURE 15–23** Problem 12.

13. (III) When a gas is taken from a to c along the curved path in Fig. 15–24, the work done by the gas is $W = -35\,\text{J}$ and the heat added to the gas is $Q = -175\,\text{J}$. Along path abc, the work done by the gas is $W = -56\,\text{J}$. (That is, 56 J of work is done on the gas.) (a) What is $Q$ for path abc? (b) If $P_c = \frac{1}{2}P_b$, what is $W$ for path cda? (c) What is $Q$ for path cda? (d) What is $U_a - U_c$? (e) If $U_d - U_c = 42\,\text{J}$, what is $Q$ for path da?

**FIGURE 15–24** Problem 13.

## *15–3 Human Metabolism

*14. (I) How much energy would the person of Example 15–7 transform if instead of working 11.0 h she took a noontime break and ran at 15 km/h for 1.0 h?

*15. (I) Calculate the average metabolic rate of a 65-kg person who sleeps 8.0 h, sits at a desk 6.0 h, engages in light activity 6.0 h, watches TV 2.0 h, plays tennis 1.5 h, and runs 0.50 h daily.

*16. (II) A 65-kg person decides to lose weight by sleeping one hour less per day, using the time for light activity. How much weight (or mass) can this person expect to lose in 1 year, assuming no change in food intake? Assume that 1 kg of fat stores about 40,000 kJ of energy.

*17. (II) (a) How much energy is transformed by a typical 65-kg person who runs at 15 km/h for 30 min/day in one week (Table 15–2)? (b) How many food calories would the person have to eat to make up for this energy loss?

## 15–5 Heat Engines

18. (I) A heat engine exhausts 8200 J of heat while performing 2600 J of useful work. What is the efficiency of this engine?

19. (I) What is the maximum efficiency of a heat engine whose operating temperatures are 560°C and 345°C?

20. (I) The exhaust temperature of a heat engine is 230°C. What is the high temperature if the Carnot efficiency is 34%?

21. (I) A heat engine does 9200 J of work per cycle while absorbing 25.0 kcal of heat from a high-temperature reservoir. What is the efficiency of this engine?

22. (I) A heat engine's high temperature $T_H$ could be ambient temperature, because liquid nitrogen at 77 K could be $T_L$ and is cheap. What would be the efficiency of a Carnot engine that made use of heat transferred from air at room temperature (293 K) to the liquid nitrogen "fuel" (Fig. 15–25)?

**FIGURE 15–25** Problem 22.

23. (II) Which will improve the efficiency of a Carnot engine more: a 10 C° increase in the high-temperature reservoir, or a 10 C° decrease in the low-temperature reservoir? Give detailed results. Can you state a generalization?

24. (II) A certain power plant puts out 580 MW of electric power. Estimate the heat discharged per second, assuming that the plant has an efficiency of 32%.

25. (II) A nuclear power plant operates at 65% of its maximum theoretical (Carnot) efficiency between temperatures of 660°C and 330°C. If the plant produces electric energy at the rate of 1.4 GW, how much exhaust heat is discharged per hour?

26. (II) A heat engine exhausts its heat at 340°C and has a Carnot efficiency of 36%. What exhaust temperature would enable it to achieve a Carnot efficiency of 42%?

27. (II) A Carnot engine's operating temperatures are 210°C and 45°C. The engine's power output is 910 W. Calculate the rate of heat output.

28. (II) A four-cylinder gasoline engine has an efficiency of 0.22 and delivers 180 J of work per cycle per cylinder. If the engine runs at 25 cycles per second (1500 rpm), determine (a) the work done per second, and (b) the total heat input per second from the gasoline. (c) If the energy content of gasoline is 130 MJ per gallon, how long does one gallon last?

29. (II) A Carnot engine performs work at the rate of 520 kW with an input of 950 kcal of heat per second. If the temperature of the heat source is 520°C, at what temperature is the waste heat exhausted?

30. (II) A heat engine uses a heat source at 580°C and has an ideal (Carnot) efficiency of 22%. To increase the ideal efficiency to 42%, what must be the temperature of the heat source?

31. (III) A typical compact car experiences a total drag force of about 350 N at 55 mi/h. If this car gets 32 miles per gallon of gasoline at this speed, and a liter of gasoline (1 gal = 3.8 L) releases about $3.2 \times 10^7\,\text{J}$ when burned, what is the car's efficiency?

## 15–6 Refrigerators, Air Conditioners, Heat Pumps

32. (I) If an ideal refrigerator keeps its contents at 2.5°C when the house temperature is 22°C, what is its COP?

33. (I) The low temperature of a freezer cooling coil is −8°C and the discharge temperature is 33°C. What is the maximum theoretical coefficient of performance?

34. (II) What is the temperature inside an ideal refrigerator-freezer that operates with a COP = 7.0 in a 22°C room?

35. (II) A heat pump is used to keep a house warm at 22°C. How much work is required of the pump to deliver 3100 J of heat into the house if the outdoor temperature is (a) 0°C, (b) −15°C? Assume a COP of 3.0. (c) Redo for both temperatures, assuming an ideal (Carnot) coefficient of performance $COP = T_H/(T_H - T_L)$.

36. (II) (a) What is the coefficient of performance of an ideal heat pump that extracts heat from 6°C air outside and deposits heat inside a house at 24°C? (b) If this heat pump operates on 1200 W of electrical power, what is the maximum heat it can deliver into the house each hour? See Problem 35.

37. (II) What volume of water at 0°C can a freezer make into ice cubes in 1.0 h, if the coefficient of performance of the cooling unit is 6.0 and the power input is 1.2 kilowatt?

38. (II) How much less per year would it cost a family to operate a heat pump that has a coefficient of performance of 2.9 than an electric heater that costs $2000 to heat their home for a year? If the conversion to the heat pump costs $15,000, how long would it take the family to break even on heating costs? How much would the family save in 20 years?

## 15–7 Entropy

39. (I) What is the change in entropy of 320 g of steam at 100°C when it is condensed to water at 100°C?

40. (I) 1.0 kg of water is heated from 0°C to 100°C. Estimate the change in entropy of the water.

41. (I) What is the change in entropy of $1.00 \, m^3$ of water at 0°C when it is frozen to ice at 0°C?

42. (II) A 5.8-kg box having an initial speed of 4.0 m/s slides along a rough table and comes to rest. Estimate the total change in entropy of the universe. Assume all objects are at room temperature (293 K).

43. (II) If $1.00 \, m^3$ of water at 0°C is frozen and cooled to −8.0°C by being in contact with a great deal of ice at −8.0°C, estimate the total change in entropy of the process.

44. (II) An aluminum rod conducts 8.40 cal/s from a heat source maintained at 225°C to a large body of water at 22°C. Calculate the rate at which entropy increases in this process.

45. (II) A 2.8-kg piece of aluminum at 28.5°C is placed in 1.0 kg of water in a Styrofoam container at room temperature (20.0°C). Estimate the net change in entropy of the system.

46. (II) A falling rock has kinetic energy KE just before striking the ground and coming to rest. What is the total change in entropy of rock plus environment as a result of this collision?

47. (II) 1.0 kg of water at 35°C is mixed with 1.0 kg of water at 45°C in a well-insulated container. Estimate the net change in entropy of the system.

48. (III) A real heat engine working between heat reservoirs at 970 K and 650 K produces 550 J of work per cycle for a heat input of 2500 J. (a) Compare the efficiency of this real engine to that of an ideal (Carnot) engine. (b) Calculate the total entropy change of the universe per cycle of the real engine, and (c) also if the engine is ideal (Carnot).

## *15–10 Statistical Interpretation

*49. (II) Calculate the probabilities, when you throw two dice, of obtaining (a) a 4, and (b) a 10.

*50. (II) Suppose that you repeatedly shake six coins in your hand and drop them on the floor. Construct a table showing the number of microstates that correspond to each macrostate. What is the probability of obtaining (a) three heads and three tails, and (b) six heads?

*51. (III) A bowl contains many red, orange, and green jelly beans, in equal numbers. You are to make a line of 3 jelly beans by randomly taking 3 beans from the bowl. (a) Construct a table showing the number of microstates that correspond to each macrostate. Then determine the probability of (b) all 3 beans red, and (c) 2 greens, 1 orange.

*52. (III) Rank the following five-card hands in order of increasing probability: (a) four aces and a king; (b) six of hearts, eight of diamonds, queen of clubs, three of hearts, jack of spades; (c) two jacks, two queens, and an ace; and (d) any hand having no two equal-value cards (no pairs, etc.). Discuss your ranking in terms of microstates and macrostates.

## *15–11 Energy Resources

*53. (I) Solar cells (Fig. 15–26) can produce about 40 W of electricity per square meter of surface area if directly facing the Sun. How large an area is required to supply the needs of a house that requires 24 kWh/day? Would this fit on the roof of an average house? (Assume the Sun shines about 9 h/day.)

**FIGURE 15–26**
Problem 53.

*54. (II) Energy may be stored by pumping water to a high reservoir when demand is low and then releasing it to drive turbines during peak demand. Suppose water is pumped to a lake 115 m above the turbines at a rate of $1.00 \times 10^5 \, kg/s$ for 10.0 h at night. (a) How much energy (kWh) is needed to do this each night? (b) If all this energy is released during a 14-h day, at 75% efficiency, what is the average power output?

*55. (II) Water is stored in an artificial lake created by a dam (Fig. 15–27). The water depth is 48 m at the dam, and a steady flow rate of $32 \, m^3/s$ is maintained through hydroelectric turbines installed near the base of the dam. How much electrical power can be produced?

**FIGURE 15–27**
Problem 55:
Flaming Gorge
Dam on the
Green River in
Utah.

# General Problems

**56.** An inventor claims to have built an engine that produces 2.00 MW of usable work while taking in 3.00 MW of thermal energy at 425 K, and rejecting 1.00 MW of thermal energy at 215 K. Is there anything fishy about his claim? Explain.

**57.** When $5.80 \times 10^5$ J of heat is added to a gas enclosed in a cylinder fitted with a light frictionless piston maintained at atmospheric pressure, the volume is observed to increase from $1.9 \, \text{m}^3$ to $4.1 \, \text{m}^3$. Calculate (a) the work done by the gas, and (b) the change in internal energy of the gas. (c) Graph this process on a PV diagram.

**58.** A restaurant refrigerator has a coefficient of performance of 4.6. If the temperature in the kitchen outside the refrigerator is 32°C, what is the lowest temperature that could be obtained inside the refrigerator if it were ideal?

**59.** A particular car does work at the rate of about 7.0 kJ/s when traveling at a steady 21.8 m/s along a level road. This is the work done against friction. The car can travel 17 km on 1.0 L of gasoline at this speed (about 40 mi/gal). What is the minimum value for $T_H$ if $T_L$ is 25°C? The energy available from 1.0 L of gas is $3.2 \times 10^7$ J.

**60.** A "Carnot" refrigerator (the reverse of a Carnot engine) absorbs heat from the freezer compartment at a temperature of −17°C and exhausts it into the room at 25°C. (a) How much work would the refrigerator do to change 0.65 kg of water at 25°C into ice at −17°C? (b) If the compressor output is 105 W and runs 25% of the time, how long will this take?

**61.** It has been suggested that a heat engine could be developed that made use of the temperature difference between water at the surface of the ocean and water several hundred meters deep. In the tropics, the temperatures may be 27°C and 4°C, respectively. (a) What is the maximum efficiency such an engine could have? (b) Why might such an engine be feasible in spite of the low efficiency? (c) Can you imagine any adverse environmental effects that might occur?

**62.** A cooling unit for a new freezer has an inner surface area of $8.0 \, \text{m}^2$, and is bounded by walls 12 cm thick with a thermal conductivity of $0.050 \, \text{W/m} \cdot \text{K}$. The inside must be kept at −15°C in a room that is at 22°C. The motor for the cooling unit must run no more than 15% of the time. What is the minimum power requirement of the cooling motor?

**63.** Refrigeration units can be rated in "tons." A 1-ton air conditioning system can remove sufficient energy to freeze 1 ton (2000 pounds = 909 kg) of 0°C water into 0°C ice in one 24-h day. If, on a 35°C day, the interior of a house is maintained at 22°C by the continuous operation of a 5-ton air conditioning system, how much does this cooling cost the homeowner per hour? Assume the work done by the refrigeration unit is powered by electricity that costs $0.10 per kWh and that the unit's coefficient of performance is 18% that of an ideal refrigerator. 1 kWh = $3.60 \times 10^6$ J.

**64.** Two 1100 kg cars are traveling 85 km/h in opposite directions when they collide and are brought to rest. Estimate the change in entropy of the universe as a result of this collision. Assume $T = 20°C$.

**65.** A 110-g insulated aluminum cup at 35°C is filled with 150 g of water at 45°C. After a few minutes, equilibrium is reached. (a) Determine the final temperature, and (b) estimate the total change in entropy.

**66.** The burning of gasoline in a car releases about $3.0 \times 10^4$ kcal/gal. If a car averages 41 km/gal when driving 110 km/h, which requires 25 hp, what is the efficiency of the engine under those conditions?

**67.** A Carnot engine operates with $T_L = 20°C$ and has an efficiency of 25%. By how many kelvins should the high operating temperature $T_H$ be increased to achieve an efficiency of 35%?

**68.** Calculate the work done by an ideal gas in going from state A to state C in Fig. 15–28 for each of the following processes: (a) ADC, (b) ABC, and (c) AC directly.

**FIGURE 15–28**
Problem 68.

**69.** A 38% efficient power plant puts out 850 MW of electrical power. Cooling towers take away the exhaust heat. (a) If the air temperature is allowed to rise 7.0 C°, estimate what volume of air $(\text{km}^3)$ is heated per day. Will the local climate be heated significantly? (b) If the heated air were to form a layer 180 m thick, estimate how large an area it would cover for 24 h of operation. Assume the air has density 1.3 kg/m³ and has specific heat of about 1.0 kJ/kg·C° at constant pressure.

**70.** Suppose a power plant delivers energy at 880 MW using steam turbines. The steam goes into the turbines superheated at 625 K and deposits its unused heat in river water at 285 K. Assume that the turbine operates as an ideal Carnot engine. (a) If the river flow rate is 37 m³/s, estimate the average temperature increase of the river water immediately downstream from the power plant. (b) What is the entropy increase per kilogram of the downstream river water in J/kg·K?

**71.** A car engine whose output power is 135 hp operates at about 15% efficiency. Assume the engine's water temperature of 85°C is its cold-temperature (exhaust) reservoir and 495°C is its thermal "intake" temperature (the temperature of the exploding gas–air mixture). (a) What is the ratio of its efficiency relative to its maximum possible (Carnot) efficiency? (b) Estimate how much power (in watts) goes into moving the car, and how much heat, in joules and in kcal, is exhausted to the air in 1.0 h.

**72.** An ideal monatomic gas is contained in a tall cylindrical jar of cross-sectional area 0.080 m² fitted with an airtight frictionless 0.15-kg movable piston. When the gas is heated (at constant pressure) from 25°C to 55°C, the piston rises 1.0 cm. How much heat was required for this process? Assume atmospheric pressure outside. [*Hint*: See Section 14–2.]

**73.** Metabolizing 1.0 kg of fat results in about $3.7 \times 10^7$ J of internal energy in the body. (a) In one day, how much fat does the body burn to maintain the body temperature of a person staying in bed and metabolizing at an average rate of 95 W? (b) How long would it take to burn 1.0 kg of fat this way assuming there is no food intake?

**74.** (*a*) At a steam power plant, steam engines work in pairs, the heat output of the first one being the approximate heat input of the second. The operating temperatures of the first are 750°C and 440°C, and of the second 415°C and 270°C. If the heat of combustion of coal is $2.8 \times 10^7$ J/kg, at what rate must coal be burned if the plant is to put out 950 MW of power? Assume the efficiency of the engines is 65% of the ideal (Carnot) efficiency. (*b*) Water is used to cool the power plant. If the water temperature is allowed to increase by no more than 4.5 C°, estimate how much water must pass through the plant per hour.

**75.** Suppose a heat pump has a stationary bicycle attachment that allows *you* to provide the work instead of using an electrical wall outlet. If your heat pump has a coefficient of performance of 2.0 and you can cycle at a racing pace (Table 15–2) for a half hour, how much heat can you provide?

**76.** An ideal air conditioner keeps the temperature inside a room at 21°C when the outside temperature is 32°C. If 4.8 kW of power enters a room through the windows in the form of direct radiation from the Sun, how much electrical power would be saved if the windows were shaded so only 500 W enters?

**77.** An ideal heat pump is used to maintain the inside temperature of a house at $T_{in} = 22$°C when the outside temperature is $T_{out}$. Assume that when it is operating, the heat pump does work at a rate of 1500 W. Also assume that the house loses heat via conduction through its walls and other surfaces at a rate given by $(650 \text{ W/C°})(T_{in} - T_{out})$. (*a*) For what outside temperature would the heat pump have to operate all the time in order to maintain the house at an inside temperature of 22°C? (*b*) If the outside temperature is 8°C, what percentage of the time does the heat pump have to operate in order to maintain the house at an inside temperature of 22°C?

# Search and Learn

**1.** What happens to the internal energy of water vapor in the air that condenses on the outside of a cold glass of water? Is work done or heat exchanged? Explain in detail.

**2.** Draw a *PV* diagram for an ideal gas which undergoes a three-step cyclic thermodynamic process in which the first step has $\Delta U = 0$ and $W > 0$, the second step has $W = 0$, and the third step has $Q = 0$ and $W < 0$.

**3.** What exactly is a Carnot engine and why is it important? How practical is it?

**4.** (*a*) Make up an advertisement for a refrigerator or air conditioner that violates the first law of thermodynamics (see Section 15–6). (*b*) Make up an ad for a car engine that violates the second law of thermodynamics (see Section 15–5).

**\*5.** One day a person sleeps for 7.0 h, goes running for an hour, gets dressed and eats breakfast for an hour, sits at work for the next 9.0 h, does household chores for a couple of hours, eats dinner for an hour, surfs the Internet and watches TV for a couple of hours, and finally takes an hour for a bath and getting ready for bed. If all of the energy associated with these activities is considered as heat that the body outputs to the environment, estimate the change in entropy the person has provided. (See Sections 15–3 and 15–7.)

**6.** A particular 1.5-m² photovoltaic panel operating in direct sunlight produces electricity at 20% efficiency. The resulting electricity is used to operate an electric stove that can be used to heat water. A second system uses a 1.5-m² curved mirror to concentrate the Sun's energy directly onto a container of water. Estimate how long it takes each system to heat 1.0 kg of water from 25°C to 95°C. (See also Chapter 14.)

**7.** A dehumidifier removes water vapor from air and has been referred to as a "refrigerator with an open door." The humid air is pulled in by a fan and passes over a cold coil, whose temperature is less than the dew point, and some of the air's water condenses. After this water is extracted, the air is warmed back to its original temperature and sent into the room. In a well-designed dehumidifier, the heat that is removed by the cooling coil mostly comes from the condensation of water vapor to liquid, and this heat is used to re-warm the air. Estimate how much water is removed in 1.0 h by an ideal dehumidifier, if the temperature of the room is 25°C, the water condenses at 8°C, and the dehumidifier does work at the rate of 600 W of electrical power. (See Sections 15–6, 13–12, and 14–5.)

**\*8.** **Trees offsetting $CO_2$.** Trees can help offset the buildup of $CO_2$ due to burning coal and other fossil fuels. $CO_2$ can be absorbed by tree foliage. Trees use the carbon to grow, and release $O_2$ into the atmosphere. Suppose a refrigerator uses 600 kWh of electricity per year (about $2 \times 10^9$ J) from a 33% efficient coal-fired power plant. Burning 1 kg of coal releases about $2 \times 10^7$ J of energy. Assume coal is all carbon, which when burned in air becomes $CO_2$. (*a*) How much coal is burned per year to run this refrigerator? (*b*) Assuming a forest can capture 1700 kg of carbon per hectare (= 10,000 m²) per year, estimate how many square meters of forest are needed to capture the carbon (in the form now of $CO_2$) emitted in (*a*).

## ANSWERS TO EXERCISES

**A:** 700 J.

**B:** Less.

**C:** $-6.8 \times 10^3$ J.

**D:** (*d*), (*e*).

**E:** (*c*).

**F:** (*a*) Heat $Q_L$ comes from outside to conserve energy; (*b*) 560 W.

**G:** (*e*).

# Mathematical Review

## A–1 Relationships, Proportionality, and Equations

One of the important aspects of physics is the search for relationships between different quantities—that is, determining how one quantity affects another.

As a simple example, the ancients found that if one circle has twice the diameter of a second circle, the first also has twice the circumference. If the diameter is three times as large, the circumference is also three times as large. In other words, an increase in the diameter results in a proportional increase in the circumference. We say that the circumference is **directly proportional** to the diameter. This can be written in symbols as $C \propto D$, where "$\propto$" means "is proportional to," and $C$ and $D$ refer to the circumference and diameter of a circle, respectively. The next step is to change this proportionality to an equation, which will make it possible to link the two quantities numerically. This means inserting a proportionality constant, which in many cases is determined by measurement. The ancients found that the ratio of the circumference to the diameter of any circle was 3.1416 (to keep only the first few decimal places). This number is designated by the Greek letter $\pi$. It is the constant of proportionality for the relationship $C \propto D$. To obtain an equation, we insert $\pi$ into the proportion and change the $\propto$ to $=$. Thus,

$$C = \pi D.$$

Other kinds of proportionality occur as well. For example, the area of a circle is proportional to the *square* of its radius. That is, if the radius is doubled, the area becomes four times as large; and so on. In this case we can write $A \propto r^2$, where $A$ stands for the area and $r$ for the radius of the circle. The constant of proportionality is found to be $\pi$ again: $A = \pi r^2$.

Sometimes two quantities are related in such a way that an increase in one leads to a proportional *decrease* in the other. This is called **inverse proportion**. For example, the time required to travel a given distance is inversely proportional to the speed of travel. The greater the speed, the less time it takes. We can write this inverse proportion as

time $\propto$ 1/speed.

The larger the denominator of a fraction, the lower the value of the fraction is as a whole. For example, $\frac{1}{4}$ is less than $\frac{1}{2}$. Thus, if the speed is doubled, the time is halved, which is what we want to express by this inverse proportionality relationship.

If you suspect that a relationship exists between two or more quantities, you can try to determine the precise nature of this relationship by varying one of the quantities and measuring how the other varies as a result. Sometimes a given quantity is affected by two or more quantities; for instance, the acceleration of an object is related to both its mass and the applied force. In such a case, only one quantity is varied at a time, while the others are held constant.

When one quantity affects another, we often use the expression **is a function of** to indicate this dependence; for example, we say that the pressure in a tire is a function of the temperature.

Whatever kind of proportion is found to hold, it can be changed to an equality by finding the proper proportionality constant. Quantitative statements or predictions about the physical world can then be made with the equation.

# A–2 Exponents

When we write $10^4$, we mean that you multiply 10 by itself four times: $10^4 = 10 \times 10 \times 10 \times 10 = 10{,}000$. The superscript 4 is called an **exponent**, and 10 is said to be raised to the fourth power. Any number or symbol can be raised to a power. Special names are used when the exponent is 2 ($a^2$ is "$a$ squared") or 3 ($a^3$ is "$a$ cubed"). For any other power, we say $a^n$ is "$a$ to the $n$th power." If the exponent is 1, it is usually dropped: $a^1 = a$, since no multiplication is involved.

The rules for multiplying numbers expressed as powers are as follows: first,

$$(a^n)(a^m) = a^{n+m}. \tag{A–1}$$

That is, the exponents are added. To see why, consider the result of the multiplication of $3^3$ by $3^4$:

$$(3^3)(3^4) = (3)(3)(3) \times (3)(3)(3)(3) = (3)^7.$$

Here the sum of the exponents is $3 + 4 = 7$, so rule A–1 works. Notice that this rule works only if the base numbers ($a$ in Eq. A–1) are the same. Thus we *cannot* use the rule of summing exponents for $(6^3)(5^2)$; these numbers would have to be written out. However, if the base numbers are different but the exponents are the same, we can write a second rule:

$$(a^n)(b^n) = (ab)^n. \tag{A–2}$$

For example, $(5^3)(6^3) = (30)^3$ since

$$(5)(5)(5)(6)(6)(6) = (30)(30)(30).$$

The third rule involves a power raised to another power: $(a^3)^2$ means $(a^3)(a^3)$, which is equal to $a^{3+3} = a^6$. The general rule is then

$$(a^n)^m = a^{nm}. \tag{A–3}$$

In this case, the exponents are multiplied.

**Negative exponents** are used for reciprocals. Thus,

$$\frac{1}{a} = a^{-1}, \qquad \frac{1}{a^3} = a^{-3},$$

and so on. The reason for using negative exponents is to allow us to use the multiplication rules given above. For example, $(a^5)(a^{-3})$ means

$$\frac{(a)(a)(a)(a)(a)}{(a)(a)(a)} = a^2,$$

after canceling 3 of the $a$'s. Rule A–1 gives us the same result:

$$(a^5)(a^{-3}) = a^{5-3} = a^2.$$

What does an exponent of zero mean? That is, what is $a^0$? Any number raised to the zeroth power is defined as being equal to 1:

$$a^0 = 1.$$

This definition is used because it follows from the rules for adding exponents. For example,

$$a^3 a^{-3} = a^{3-3} = a^0 = 1.$$

But *does* $a^3 a^{-3}$ actually equal 1? Yes, because

$$a^3 a^{-3} = \frac{a^3}{a^3} = 1.$$

**Fractional exponents** are used to represent *roots*. For example, $a^{\frac{1}{2}}$ means the square root of $a$; that is, $a^{\frac{1}{2}} = \sqrt{a}$. Similarly, $a^{\frac{1}{3}}$ means the cube root of $a$, and so on. The fourth root of $a$ means that if you multiply the fourth root of $a$ by itself four times, you again get $a$:

$$\left(a^{\frac{1}{4}}\right)^4 = a.$$

This is consistent with rule A–3 since $\left(a^{\frac{1}{4}}\right)^4 = a^{\frac{4}{4}} = a^1 = a$.

# A–3 Powers of 10, or Exponential Notation

Writing out very large and very small numbers such as the distance of Neptune from the Sun, 4,500,000,000 km, or the diameter of a typical atom, 0.00000001 cm, is inconvenient and prone to error. It also leaves in question (see Section 1–4) the number of significant figures. (How many of the zeros are significant in the number 4,500,000,000 km?)

For these reasons we make use of the "powers of 10," or exponential notation. The distance from Neptune to the Sun is then expressed as $4.50 \times 10^9$ km (assuming that the value is significant to three digits), and the diameter of an atom $1.0 \times 10^{-8}$ cm. This way of writing numbers is based on the use of exponents, where $a^n$ signifies $a$ multiplied by itself $n$ times. For example, $10^4 = 10 \times 10 \times 10 \times 10 = 10,000$. Thus, $4.50 \times 10^9 = 4.50 \times 1,000,000,000 = 4,500,000,000$. Notice that the exponent (9 in this case) is just the number of places the decimal point is moved to the right to obtain the fully written-out number (4.500,000,000.)

When two numbers are multiplied (or divided), you first multiply (or divide) the simple parts and then the powers of 10. Thus, $2.0 \times 10^3$ multiplied by $5.5 \times 10^4$ equals $(2.0 \times 5.5) \times (10^3 \times 10^4) = 11 \times 10^7$, where we have used the rule for adding exponents (Appendix A–2). Similarly, $8.2 \times 10^5$ divided by $2.0 \times 10^2$ equals

$$\frac{8.2 \times 10^5}{2.0 \times 10^2} = \frac{8.2}{2.0} \times \frac{10^5}{10^2} = 4.1 \times 10^3.$$

For numbers less than 1, say 0.01, the exponent power of 10 is written with a negative sign (see previous page): $0.01 = 1/100 = 1/10^2 = 1 \times 10^{-2}$. Similarly, $0.002 = 2 \times 10^{-3}$. The decimal point has again been moved the number of places expressed in the exponent. For example, $0.020 \times 3600 = 72$, or in exponential notation $(2.0 \times 10^{-2}) \times (3.6 \times 10^3) = 7.2 \times 10^1 = 72$.

Notice also that $10^1 \times 10^{-1} = 10 \times 0.1 = 1$, and by the law of exponents, $10^1 \times 10^{-1} = 10^0$. Therefore, $10^0 = 1$.

When writing a number in exponential notation, it is usual to make the simple number be between 1 and 10. Thus it is conventional to write $4.5 \times 10^9$ rather than $45 \times 10^8$, although they are the same number.[†] This notation also allows the number of *significant figures* to be clearly expressed. We write $4.50 \times 10^9$ if this value is accurate to three significant figures, but $4.5 \times 10^9$ if it is accurate to only two.

# A–4 Algebra

Physical relationships between quantities can be represented as equations involving symbols (usually letters of the Greek or Roman alphabet) that represent the quantities. The manipulation of such equations is the field of algebra, and it is used a great deal in physics. An equation involves an equals sign, which tells us that the quantities on either side of the equals sign have the same value. Examples of equations are

$$3 + 8 = 11$$
$$2x + 7 = 15$$
$$a^2b + c = 6.$$

The first equation involves only numbers, so is called an arithmetic equation. The other two equations are algebraic since they involve symbols. In the third equation, the quantity $a^2b$ means the product of $a$ times $a$ times $b$: $a^2b = a \times a \times b$.

---

[†] Another convention used, particularly with computers, is that the simple number be between 0.1 and 1. Thus we could write 4,500,000,000 as $0.450 \times 10^{10}$. This is slightly less compact.

## Solving for an Unknown

Often we wish to solve for one (or more) symbols, and we treat it as an *unknown*. For example, in the equation $2x + 7 = 15$, $x$ is the unknown; this equation is true, however, only when $x = 4$. Determining what value (or values) the unknown can have to satisfy the equation is called *solving the equation*. To solve an equation, the following rule can be used:

> *An equation will remain true if any operation performed on one side is also performed on the other side.* For example: (*a*) addition or subtraction of a number or symbol; (*b*) multiplication or division by a number or symbol; (*c*) raising each side of the equation to the same power, or taking the same root (such as square root).

**EXAMPLE A–1** Solve for $x$ in the equation

$$2x + 7 = 15.$$

**APPROACH** We perform the same operations on both sides of the equation to isolate $x$ as the only variable on the left side of the equals sign.

**SOLUTION** We first subtract 7 from both sides:

$$2x + 7 - 7 = 15 - 7$$

or

$$2x = 8.$$

Then we divide both sides by 2 to get

$$\frac{2x}{2} = \frac{8}{2},$$

or, carrying out the divisions,

$$x = 4,$$

and this solves the equation.

**EXAMPLE A–2** (*a*) Solve the equation

$$a^2 b + c = 24$$

for the unknown $a$ in terms of $b$ and $c$. (*b*) Solve for $a$ assuming that $b = 2$ and $c = 6$.

**APPROACH** We perform operations to isolate $a$ as the only variable on the left side of the equals sign.

**SOLUTION** (*a*) We are trying to solve for $a$, so we first subtract $c$ from both sides:

$$a^2 b = 24 - c,$$

then divide by $b$:

$$a^2 = \frac{24 - c}{b},$$

and finally take square roots:

$$a = \sqrt{\frac{24 - c}{b}}.$$

(*b*) If we are given that $b = 2$ and $c = 6$, then

$$a = \sqrt{\frac{24 - 6}{2}} = 3.$$

But this is not the only answer. Whenever we take a square root, the number can be either positive or negative. Thus $a = -3$ is also a solution. Why? Because $(-3)^2 = 9$, just as $(+3)^2 = 9$. So we actually get two solutions: $a = +3$ and $a = -3$.

**NOTE** When an unknown appears squared in an equation, there are generally two solutions for that unknown.

To check a solution, we put it back into the original equation (this is really a check that we did all the manipulations correctly). In the equation

$$a^2b + c = 24,$$

we put in $a = 3$, $b = 2$, $c = 6$ and find

$$(3)^2(2) + (6) \stackrel{?}{=} 24$$
$$24 = 24,$$

which checks.

**EXERCISE A** Put $a = -3$ into the equation of Example A–2 and show that it works too.

## Two or More Unknowns

If we have two or more unknowns, one equation is not sufficient to find them. In general, if there are $n$ unknowns, $n$ independent equations are needed. For example, if there are two unknowns, we need two equations. If the unknowns are called $x$ and $y$, a typical procedure is to solve one equation for $x$ in terms of $y$, and substitute this into the second equation.

**EXAMPLE A–3** Solve the following pair of equations for $x$ and $y$:

$$3x - 2y = 19$$
$$x + 4y = -3.$$

**APPROACH** We have two unknowns and two equations; we can start by solving the second equation for $x$ in terms of $y$. Then we substitute this result for $x$ into the first equation.

**SOLUTION** We subtract $4y$ from both sides of the second equation:

$$x = -3 - 4y.$$

We substitute this expression for $x$ into the first equation, and simplify:

$$3(-3 - 4y) - 2y = 19$$
$$-9 - 12y - 2y = 19 \quad \text{(carried out the multiplication by 3)}$$
$$-14y = 28 \quad \text{(added 9 to both sides)}$$
$$y = -2. \quad \text{(divided both sides by } -14)$$

Now that we know $y = -2$, we substitute this into the expression for $x$:

$$x = -3 - 4y$$
$$= -3 - 4(-2) = -3 + 8 = 5.$$

Our solution is $x = 5$, $y = -2$. We check this solution by putting these values back into the original equations:

$$3x - 2y \stackrel{?}{=} 19$$
$$3(5) - 2(-2) \stackrel{?}{=} 19$$
$$15 + 4 \stackrel{?}{=} 19$$
$$19 = 19 \quad \text{(it checks)}$$

and

$$x + 4y \stackrel{?}{=} -3$$
$$5 + 4(-2) \stackrel{?}{=} -3$$
$$-3 = -3. \quad \text{(it checks)}$$

Other methods for solving two or more equations, such as the method of determinants, can be found in an algebra textbook.

## The Quadratic Formula

We sometimes encounter equations that involve an unknown, say $x$, that appears not only to the first power, but squared as well. Such a **quadratic equation** can be written in the general form

$$ax^2 + bx + c = 0.$$

The quantities $a$, $b$, and $c$ are typically numbers or constants that are given.[†] The general solutions to such an equation are given by the **quadratic formula**:

$$x = \frac{-b \pm \sqrt{b^2 - 4ac}}{2a}.$$  (A–4)

The $\pm$ sign indicates that there are two solutions for $x$: one where the plus sign is used, the other where the minus sign is used.

---

**EXAMPLE A–4** Find the solutions for $x$ in the equation

$$3x^2 - 5x = 2.$$

**APPROACH** Here $x$ appears both to the first power and squared, so we use the quadratic equation.

**SOLUTION** First we write this equation in the standard form

$$ax^2 + bx + c = 0$$

by subtracting 2 from both sides:

$$3x^2 - 5x - 2 = 0.$$

In this case, $a$, $b$, and $c$ in the standard formula take the values $a = 3$, $b = -5$, and $c = -2$. The two solutions for $x$ are, using Eq. A–4,

$$x = \frac{+5 + \sqrt{25 - (4)(3)(-2)}}{(2)(3)} = \frac{5 + 7}{6} = 2$$

and

$$x = \frac{+5 - \sqrt{25 - (4)(3)(-2)}}{(2)(3)} = \frac{5 - 7}{6} = -\frac{1}{3}.$$

---

In this Example, the two solutions are $x = 2$ and $x = -\frac{1}{3}$. In physics problems, it sometimes happens that only one of the solutions corresponds to a real-life situation; in this case, the other solution is discarded. In other cases, both solutions may correspond to physical reality.

Notice, incidentally, that $b^2$ must be greater than $4ac$, so that $\sqrt{b^2 - 4ac}$ yields a real number. If $(b^2 - 4ac)$ is less than zero (negative), there is no real solution. The square root of a negative number is called **imaginary**.

A second-order equation—one in which the highest power of $x$ is 2—has two solutions; a third-order equation—involving $x^3$—has three solutions; and so on.

# A–5 The Binomial Expansion

Sometimes we end up with a quantity of the form $(1 + x)^n$. That is, the quantity $(1 + x)$ is raised to the $n$th power. This can be written as an infinite sum of terms known as the **binomial expansion**:

$$(1 + x)^n = 1 + nx + \frac{n(n - 1)}{2} x^2 + \cdots.$$  (A–5)

This formula is useful for us mainly when $x$ is very small compared to one ($x \ll 1$). In this case, each successive term is much smaller than the preceding

---

[†]Or one or more of them could be variables, in which case additional equations are needed.

term. For example, let $x = 0.01$, and $n = 2$. Then whereas the first term equals 1, the second term is $nx = (2)(0.01) = 0.02$, and the third term is $[(2)(1)/2](0.01)^2 = 0.0001$, and so on. Thus, when $x$ is small, we can ignore all but the first two (or three) terms and can write

$$(1 + x)^n \approx 1 + nx. \tag{A–6}$$

This approximation often allows us to solve an equation easily that otherwise might be very difficult. Some examples of the binomial expansion are

$$(1 + x)^2 \approx 1 + 2x,$$

$$\frac{1}{1 + x} = (1 + x)^{-1} \approx 1 - x,$$

$$\sqrt{1 + x} = (1 + x)^{\frac{1}{2}} \approx 1 + \tfrac{1}{2}x,$$

$$\frac{1}{\sqrt{1 + x}} = (1 + x)^{-\frac{1}{2}} \approx 1 - \tfrac{1}{2}x,$$

where $x \ll 1$.

As a numerical example, let us evaluate $\sqrt{1.02}$ using the binomial expansion since $x = 0.02$ is much smaller than 1:

$$\sqrt{1.02} = (1.02)^{\frac{1}{2}} = (1 + 0.02)^{\frac{1}{2}} \approx 1 + \tfrac{1}{2}(0.02) = 1.01.$$

You can check with a calculator (and maybe not even more quickly) that $\sqrt{1.02} \approx 1.01$.

# A–6 Plane Geometry

We review here a number of theorems involving angles and triangles that are useful in physics.

**FIGURE A–1**

**FIGURE A–2**

1. **Equal angles.** Two angles are equal if any of the following conditions are true:
   (*a*) They are vertical angles (Fig. A–1); *or*
   (*b*) the left side of one is parallel to the left side of the other, and the right side of one is parallel to the right side of the other (Fig. A–2; the left and right sides are as seen from the vertex, where the two sides meet); *or*
   (*c*) the left side of one is perpendicular to the left side of the other, and the right sides are likewise perpendicular (Fig. A–3).

**FIGURE A–3**

2. *The sum of the angles* in any plane triangle is 180°.

3. **Similar triangles.** Two triangles are said to be similar if all three of their angles are equal (in Fig. A–4, $\theta_1 = \phi_1$, $\theta_2 = \phi_2$, and $\theta_3 = \phi_3$). Similar triangles thus have the same basic shape but may be different sizes and have different orientations. Two useful theorems about similar triangles are:
   (*a*) Two triangles are similar if any two of their angles are equal. (This follows because the third angles must also be equal since the sum of the angles of a triangle is 180°.)
   (*b*) The ratios of corresponding sides of two similar triangles are equal. That is (Fig. A–4),

$$\frac{a_1}{b_1} = \frac{a_2}{b_2} = \frac{a_3}{b_3}.$$

**FIGURE A–4**

4. **Congruent triangles.** Two triangles are congruent if one can be placed precisely on top of the other. That is, they are similar triangles and they have the same size. Two triangles are congruent if any of the following holds:
   (*a*) The three corresponding sides are equal.
   (*b*) Two sides and the enclosed angle are equal ("side-angle-side").
   (*c*) Two angles and the enclosed side are equal ("angle-side-angle").

5. **Right triangles.** A right triangle has one angle that is 90° (a **right angle**); that is, the two sides that meet at the right angle are perpendicular (Fig. A–5). The two other (acute) angles in the right triangle add up to 90°.

6. **Pythagorean theorem.** In any right triangle, the square of the length of the hypotenuse (the side opposite the right angle) is equal to the sum of the squares of the lengths of the other two sides. In Fig. A–5,

$$c^2 = a^2 + b^2.$$

# A–7 Trigonometric Functions and Identities

Trigonometric functions for any angle $\theta$ are defined by constructing a right triangle about that angle as shown in Fig. A–6; opp and adj are the lengths of the sides opposite and adjacent to the angle $\theta$, and hyp is the length of the hypotenuse:

**FIGURE A–6**

$$\sin \theta = \frac{\text{opp}}{\text{hyp}} \qquad\qquad \csc \theta = \frac{1}{\sin \theta} = \frac{\text{hyp}}{\text{opp}}$$

$$\cos \theta = \frac{\text{adj}}{\text{hyp}} \qquad\qquad \sec \theta = \frac{1}{\cos \theta} = \frac{\text{hyp}}{\text{adj}}$$

$$\tan \theta = \frac{\text{opp}}{\text{adj}} = \frac{\sin \theta}{\cos \theta} \qquad \cot \theta = \frac{1}{\tan \theta} = \frac{\text{adj}}{\text{opp}}$$

$$\text{adj}^2 + \text{opp}^2 = \text{hyp}^2 \qquad \text{(Pythagorean theorem)}.$$

Figure A–7 shows the signs (+ or −) that cosine, sine, and tangent take on for angles $\theta$ in the four quadrants (0° to 360°). Note that angles are measured counterclockwise from the $x$ axis as shown; negative angles are measured from *below* the $x$ axis, clockwise: for example, $-30° = +330°$, and so on.

The following are some useful identities among the trigonometric functions:

$$\sin^2 \theta + \cos^2 \theta = 1$$

$$\sin 2\theta = 2 \sin \theta \cos \theta$$

$$\cos 2\theta = \cos^2 \theta - \sin^2 \theta = 2 \cos^2 \theta - 1 = 1 - 2 \sin^2 \theta$$

$$\tan 2\theta = \frac{2 \tan \theta}{1 - \tan^2 \theta}$$

$$\sin(A \pm B) = \sin A \cos B \pm \cos A \sin B$$

$$\cos(A \pm B) = \cos A \cos B \mp \sin A \sin B$$

$$\tan(A \pm B) = \frac{\tan A \pm \tan B}{1 \mp \tan A \tan B}$$

$$\sin(180° - \theta) = \sin \theta$$

$$\cos(180° - \theta) = -\cos \theta$$

$$\sin(90° - \theta) = \cos \theta$$

$$\cos(90° - \theta) = \sin \theta$$

$$\sin \tfrac{1}{2}\theta = \sqrt{\frac{1 - \cos \theta}{2}}$$

$$\cos \tfrac{1}{2}\theta = \sqrt{\frac{1 + \cos \theta}{2}}$$

$$\tan \tfrac{1}{2}\theta = \sqrt{\frac{1 - \cos \theta}{1 + \cos \theta}}$$

$$\sin A \pm \sin B = 2 \sin\left(\frac{A \pm B}{2}\right) \cos\left(\frac{A \mp B}{2}\right).$$

**FIGURE A–7**

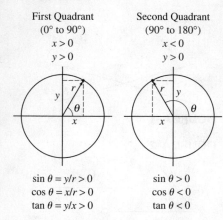

First Quadrant
(0° to 90°)
$x > 0$
$y > 0$

$\sin \theta = y/r > 0$
$\cos \theta = x/r > 0$
$\tan \theta = y/x > 0$

Second Quadrant
(90° to 180°)
$x < 0$
$y > 0$

$\sin \theta > 0$
$\cos \theta < 0$
$\tan \theta < 0$

Third Quadrant
(180° to 270°)
$x < 0$
$y < 0$

$\sin \theta < 0$
$\cos \theta < 0$
$\tan \theta > 0$

Fourth Quadrant
(270° to 360°)
$x > 0$
$y < 0$

$\sin \theta < 0$
$\cos \theta > 0$
$\tan \theta < 0$

**FIGURE A–8**

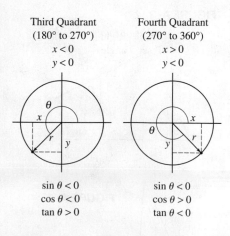

For any triangle (see Fig. A–8):

$$\frac{\sin \alpha}{a} = \frac{\sin \beta}{b} = \frac{\sin \gamma}{c} \qquad \text{(law of sines)}$$

$$c^2 = a^2 + b^2 - 2ab \cos \gamma. \qquad \text{(law of cosines)}$$

## Trigonometric Table: Numerical Values of Sin, Cos, Tan

| Angle in Degrees | Angle in Radians | Sine | Cosine | Tangent | Angle in Degrees | Angle in Radians | Sine | Cosine | Tangent |
|---|---|---|---|---|---|---|---|---|---|
| 0° | 0.000 | 0.000 | 1.000 | 0.000 | | | | | |
| 1° | 0.017 | 0.017 | 1.000 | 0.017 | 46° | 0.803 | 0.719 | 0.695 | 1.036 |
| 2° | 0.035 | 0.035 | 0.999 | 0.035 | 47° | 0.820 | 0.731 | 0.682 | 1.072 |
| 3° | 0.052 | 0.052 | 0.999 | 0.052 | 48° | 0.838 | 0.743 | 0.669 | 1.111 |
| 4° | 0.070 | 0.070 | 0.998 | 0.070 | 49° | 0.855 | 0.755 | 0.656 | 1.150 |
| 5° | 0.087 | 0.087 | 0.996 | 0.087 | 50° | 0.873 | 0.766 | 0.643 | 1.192 |
| 6° | 0.105 | 0.105 | 0.995 | 0.105 | 51° | 0.890 | 0.777 | 0.629 | 1.235 |
| 7° | 0.122 | 0.122 | 0.993 | 0.123 | 52° | 0.908 | 0.788 | 0.616 | 1.280 |
| 8° | 0.140 | 0.139 | 0.990 | 0.141 | 53° | 0.925 | 0.799 | 0.602 | 1.327 |
| 9° | 0.157 | 0.156 | 0.988 | 0.158 | 54° | 0.942 | 0.809 | 0.588 | 1.376 |
| 10° | 0.175 | 0.174 | 0.985 | 0.176 | 55° | 0.960 | 0.819 | 0.574 | 1.428 |
| 11° | 0.192 | 0.191 | 0.982 | 0.194 | 56° | 0.977 | 0.829 | 0.559 | 1.483 |
| 12° | 0.209 | 0.208 | 0.978 | 0.213 | 57° | 0.995 | 0.839 | 0.545 | 1.540 |
| 13° | 0.227 | 0.225 | 0.974 | 0.231 | 58° | 1.012 | 0.848 | 0.530 | 1.600 |
| 14° | 0.244 | 0.242 | 0.970 | 0.249 | 59° | 1.030 | 0.857 | 0.515 | 1.664 |
| 15° | 0.262 | 0.259 | 0.966 | 0.268 | 60° | 1.047 | 0.866 | 0.500 | 1.732 |
| 16° | 0.279 | 0.276 | 0.961 | 0.287 | 61° | 1.065 | 0.875 | 0.485 | 1.804 |
| 17° | 0.297 | 0.292 | 0.956 | 0.306 | 62° | 1.082 | 0.883 | 0.469 | 1.881 |
| 18° | 0.314 | 0.309 | 0.951 | 0.325 | 63° | 1.100 | 0.891 | 0.454 | 1.963 |
| 19° | 0.332 | 0.326 | 0.946 | 0.344 | 64° | 1.117 | 0.899 | 0.438 | 2.050 |
| 20° | 0.349 | 0.342 | 0.940 | 0.364 | 65° | 1.134 | 0.906 | 0.423 | 2.145 |
| 21° | 0.367 | 0.358 | 0.934 | 0.384 | 66° | 1.152 | 0.914 | 0.407 | 2.246 |
| 22° | 0.384 | 0.375 | 0.927 | 0.404 | 67° | 1.169 | 0.921 | 0.391 | 2.356 |
| 23° | 0.401 | 0.391 | 0.921 | 0.424 | 68° | 1.187 | 0.927 | 0.375 | 2.475 |
| 24° | 0.419 | 0.407 | 0.914 | 0.445 | 69° | 1.204 | 0.934 | 0.358 | 2.605 |
| 25° | 0.436 | 0.423 | 0.906 | 0.466 | 70° | 1.222 | 0.940 | 0.342 | 2.747 |
| 26° | 0.454 | 0.438 | 0.899 | 0.488 | 71° | 1.239 | 0.946 | 0.326 | 2.904 |
| 27° | 0.471 | 0.454 | 0.891 | 0.510 | 72° | 1.257 | 0.951 | 0.309 | 3.078 |
| 28° | 0.489 | 0.469 | 0.883 | 0.532 | 73° | 1.274 | 0.956 | 0.292 | 3.271 |
| 29° | 0.506 | 0.485 | 0.875 | 0.554 | 74° | 1.292 | 0.961 | 0.276 | 3.487 |
| 30° | 0.524 | 0.500 | 0.866 | 0.577 | 75° | 1.309 | 0.966 | 0.259 | 3.732 |
| 31° | 0.541 | 0.515 | 0.857 | 0.601 | 76° | 1.326 | 0.970 | 0.242 | 4.011 |
| 32° | 0.559 | 0.530 | 0.848 | 0.625 | 77° | 1.344 | 0.974 | 0.225 | 4.331 |
| 33° | 0.576 | 0.545 | 0.839 | 0.649 | 78° | 1.361 | 0.978 | 0.208 | 4.705 |
| 34° | 0.593 | 0.559 | 0.829 | 0.675 | 79° | 1.379 | 0.982 | 0.191 | 5.145 |
| 35° | 0.611 | 0.574 | 0.819 | 0.700 | 80° | 1.396 | 0.985 | 0.174 | 5.671 |
| 36° | 0.628 | 0.588 | 0.809 | 0.727 | 81° | 1.414 | 0.988 | 0.156 | 6.314 |
| 37° | 0.646 | 0.602 | 0.799 | 0.754 | 82° | 1.431 | 0.990 | 0.139 | 7.115 |
| 38° | 0.663 | 0.616 | 0.788 | 0.781 | 83° | 1.449 | 0.993 | 0.122 | 8.144 |
| 39° | 0.681 | 0.629 | 0.777 | 0.810 | 84° | 1.466 | 0.995 | 0.105 | 9.514 |
| 40° | 0.698 | 0.643 | 0.766 | 0.839 | 85° | 1.484 | 0.996 | 0.087 | 11.43 |
| 41° | 0.716 | 0.656 | 0.755 | 0.869 | 86° | 1.501 | 0.998 | 0.070 | 14.301 |
| 42° | 0.733 | 0.669 | 0.743 | 0.900 | 87° | 1.518 | 0.999 | 0.052 | 19.081 |
| 43° | 0.750 | 0.682 | 0.731 | 0.933 | 88° | 1.536 | 0.999 | 0.035 | 28.636 |
| 44° | 0.768 | 0.695 | 0.719 | 0.966 | 89° | 1.553 | 1.000 | 0.017 | 57.290 |
| 45° | 0.785 | 0.707 | 0.707 | 1.000 | 90° | 1.571 | 1.000 | 0.000 | ∞ |

# A–8 Logarithms

Logarithms are defined in the following way:

$$\text{if } y = A^x, \quad \text{then } x = \log_A y.$$

That is, the logarithm of a number $y$ to the base $A$ is that number which, as the exponent of $A$, gives back the number $y$. For **common logarithms**, the base is 10, so

$$\text{if } y = 10^x, \quad \text{then } x = \log y.$$

The subscript 10 on $\log_{10}$ is usually omitted when dealing with common logs. Another base sometimes used is the exponential base $e = 2.718\cdots$, a natural number.[†] Such logarithms are called **natural logarithms** and are written "ln". Thus,

$$\text{if } y = e^x, \quad \text{then } x = \ln y.$$

For any number $y$, the two types of logarithm are related by

$$\ln y = 2.3026 \log y.$$

Some simple rules for logarithms include:

$$\log(ab) = \log a + \log b. \tag{A–7}$$

This is true because if $a = 10^n$ and $b = 10^m$, then $ab = 10^{n+m}$. From the definition of logarithm, $\log a = n$, $\log b = m$, and $\log(ab) = n + m$; hence, $\log(ab) = n + m = \log a + \log b$. In a similar way, we can show the rules

$$\log\!\left(\frac{a}{b}\right) = \log a - \log b \tag{A–8}$$

and

$$\log a^n = n \log a. \tag{A–9}$$

These three rules apply not only to common logs but to natural or any other kind of logarithm.

Logs were once used as a technique for simplifying certain types of calculation. Because of the advent of electronic calculators and computers, they are not often used any more for that purpose. However, logs do appear in certain physical equations, so it is helpful to know how to deal with them. If you do not have a calculator that calculates logs, you can use a **log table**, such as the small one shown here (Table A–1). The number $N$ is given to two digits (some tables give $N$ to three or more digits); the first digit is in the vertical column to the left, the second digit is in the horizontal row across the top. For example, the Table tells us that $\log 1.0 = 0.000$, $\log 1.1 = 0.041$, and $\log 4.1 = 0.613$. The Table gives logs for numbers between 1.0 and 9.9; for larger or smaller numbers, we use rule A–7:

$$\log(ab) = \log a + \log b.$$

For example,

$$\log(380) = \log(3.8 \times 10^2) = \log(3.8) + \log(10^2).$$

From the Table, $\log 3.8 = 0.580$; and from rule A–9,

$$\log(10^2) = 2\log(10) = 2,$$

since $\log(10) = 1$. [This follows from the definition of the logarithm: if

---

[†]The exponential base $e$ can be written as an infinite series:

$$e = 1 + \frac{1}{1} + \frac{1}{1\cdot 2} + \frac{1}{1\cdot 2\cdot 3} + \frac{1}{1\cdot 2\cdot 3\cdot 4} + \cdots.$$

## TABLE A–1 Short Table of Common Logarithms

| N | 0.0 | 0.1 | 0.2 | 0.3 | 0.4 | 0.5 | 0.6 | 0.7 | 0.8 | 0.9 |
|---|-----|-----|-----|-----|-----|-----|-----|-----|-----|-----|
| 1 | .000 | .041 | .079 | .114 | .146 | .176 | .204 | .230 | .255 | .279 |
| 2 | .301 | .322 | .342 | .362 | .380 | .398 | .415 | .431 | .447 | .462 |
| 3 | .477 | .491 | .505 | .519 | .531 | .544 | .556 | .568 | .580 | .591 |
| 4 | .602 | .613 | .623 | .633 | .643 | .653 | .663 | .672 | .681 | .690 |
| 5 | .699 | .708 | .716 | .724 | .732 | .740 | .748 | .756 | .763 | .771 |
| 6 | .778 | .785 | .792 | .799 | .806 | .813 | .820 | .826 | .833 | .839 |
| 7 | .845 | .851 | .857 | .863 | .869 | .875 | .881 | .886 | .892 | .898 |
| 8 | .903 | .908 | .914 | .919 | .924 | .929 | .935 | .940 | .944 | .949 |
| 9 | .954 | .959 | .964 | .968 | .973 | .978 | .982 | .987 | .991 | .996 |

$10 = 10^1$, then $1 = \log(10)$.] Thus,

$$\log(380) = \log(3.8) + \log(10^2)$$
$$= 0.580 + 2$$
$$= 2.580.$$

Similarly,

$$\log(0.081) = \log(8.1) + \log(10^{-2})$$
$$= 0.908 - 2 = -1.092.$$

Sometimes we need to do the reverse process: find the number $N$ whose log is, say, 2.670. This is called "taking the **antilogarithm**." To do so, we separate our number 2.670 into two parts, making the separation at the decimal point:

$$\log N = 2.670 = 2 + 0.670$$
$$= \log 10^2 + 0.670.$$

We now look at Table A–1 to see what number has its log equal to 0.670; none does, so we must **interpolate**: we see that $\log 4.6 = 0.663$ and $\log 4.7 = 0.672$. So the number we want is between 4.6 and 4.7, and closer to the latter by $\frac{7}{9}$. Approximately we can say that $\log 4.68 = 0.670$. Thus

$$\log N = 2 + 0.670$$
$$= \log(10^2) + \log(4.68) = \log(4.68 \times 10^2),$$

so $N = 4.68 \times 10^2 = 468$.

If the given logarithm is negative, say, $-2.180$, we proceed as follows:

$$\log N = -2.180 = -3 + 0.820$$
$$= \log 10^{-3} + \log 6.6 = \log 6.6 \times 10^{-3},$$

so $N = 6.6 \times 10^{-3}$. Notice that we added to our given logarithm the next largest integer (3 in this case) so that we have an integer, plus a decimal number between 0 and 1.0 whose antilogarithm can be looked up in the Table.

# B Selected Isotopes

| (1) Atomic Number Z | (2) Element | (3) Symbol | (4) Mass Number A | (5) Atomic Mass† | (6) % Abundance (or Radioactive Decay‡ Mode) | (7) Half-life (if radioactive) |
|---|---|---|---|---|---|---|
| 0 | (Neutron) | n | 1 | 1.008665 | $\beta^-$ | 10.183 min |
| 1 | Hydrogen | H | 1 | 1.007825 | 99.9885% | |
| | [proton | p | 1 | 1.007276] | | |
| | Deuterium | $^2_1$H | 2 | 2.014102 | 0.0115% | |
| | [deuteron | d or D | 2 | 2.013553] | | |
| | Tritium | $^3_1$H | 3 | 3.016049 | $\beta^-$ | 12.32 yr |
| | [triton | t or T | 3 | 3.015500] | | |
| 2 | Helium | He | 3 | 3.016029 | 0.000137% | |
| | | | 4 | 4.002603 | 99.999863% | |
| 3 | Lithium | Li | 6 | 6.015123 | 7.59% | |
| | | | 7 | 7.016003 | 92.41% | |
| 4 | Beryllium | Be | 7 | 7.016929 | EC, $\gamma$ | 53.24 days |
| | | | 9 | 9.012183 | 100% | |
| 5 | Boron | B | 10 | 10.012937 | 19.9% | |
| | | | 11 | 11.009305 | 80.1% | |
| 6 | Carbon | C | 11 | 11.011434 | $\beta^+$, EC | 20.334 min |
| | | | 12 | 12.000000 | 98.93% | |
| | | | 13 | 13.003355 | 1.07% | |
| | | | 14 | 14.003242 | $\beta^-$ | 5730 yr |
| 7 | Nitrogen | N | 13 | 13.005739 | $\beta^+$, EC | 9.965 min |
| | | | 14 | 14.003074 | 99.632% | |
| | | | 15 | 15.000109 | 0.368% | |
| 8 | Oxygen | O | 15 | 15.003066 | $\beta^+$, EC | 122.24 s |
| | | | 16 | 15.994915 | 99.757% | |
| | | | 18 | 17.999160 | 0.205% | |
| 9 | Fluorine | F | 19 | 18.998403 | 100% | |
| 10 | Neon | Ne | 20 | 19.992440 | 90.48% | |
| | | | 22 | 21.991385 | 9.25% | |
| 11 | Sodium | Na | 22 | 21.994437 | $\beta^+$, EC, $\gamma$ | 2.6027 yr |
| | | | 23 | 22.989769 | 100% | |
| | | | 24 | 23.990963 | $\beta^-$, $\gamma$ | 14.997 h |
| 12 | Magnesium | Mg | 24 | 23.985042 | 78.99% | |
| 13 | Aluminum | Al | 27 | 26.981539 | 100% | |
| 14 | Silicon | Si | 28 | 27.976927 | 92.223% | |
| | | | 31 | 30.975363 | $\beta^-$, $\gamma$ | 157.3 min |
| 15 | Phosphorus | P | 31 | 30.973762 | 100% | |
| | | | 32 | 31.973908 | $\beta^-$ | 14.262 days |

† The masses (atomic mass units) given in column (5) are those for the neutral atom, including the Z electrons (except for the proton, deuteron, triton).

‡ Chapter 30; EC = electron capture.

| (1) Atomic Number Z | (2) Element | (3) Symbol | (4) Mass Number A | (5) Atomic Mass | (6) % Abundance (or Radioactive Decay Mode) | (7) Half-life (if radioactive) |
|---|---|---|---|---|---|---|
| 16 | Sulfur | S | 32 | 31.972071 | 94.99% | |
| | | | 35 | 34.969032 | $\beta^-$ | 87.37 days |
| 17 | Chlorine | Cl | 35 | 34.968853 | 75.76% | |
| | | | 37 | 36.965903 | 24.24% | |
| 18 | Argon | Ar | 40 | 39.962383 | 99.6035% | |
| 19 | Potassium | K | 39 | 38.963706 | 93.2581% | |
| | | | 40 | 39.963998 | 0.0117% | |
| | | | | | $\beta^-$, EC, $\gamma$, $\beta^+$ | $1.248 \times 10^9$ yr |
| 20 | Calcium | Ca | 40 | 39.962591 | 96.94% | |
| 21 | Scandium | Sc | 45 | 44.955908 | 100% | |
| 22 | Titanium | Ti | 48 | 47.947942 | 73.72% | |
| 23 | Vanadium | V | 51 | 50.943957 | 99.750% | |
| 24 | Chromium | Cr | 52 | 51.940506 | 83.789% | |
| 25 | Manganese | Mn | 55 | 54.938044 | 100% | |
| 26 | Iron | Fe | 56 | 55.934936 | 91.754% | |
| 27 | Cobalt | Co | 59 | 58.933194 | 100% | |
| | | | 60 | 59.933816 | $\beta^-$, $\gamma$ | 5.2713 yr |
| 28 | Nickel | Ni | 58 | 57.935342 | 68.077% | |
| | | | 60 | 59.930786 | 26.223% | |
| 29 | Copper | Cu | 63 | 62.929598 | 69.15% | |
| | | | 65 | 64.927790 | 30.85% | |
| 30 | Zinc | Zn | 64 | 63.929142 | 49.17% | |
| | | | 66 | 65.926034 | 27.73% | |
| 31 | Gallium | Ga | 69 | 68.925574 | 60.108% | |
| 32 | Germanium | Ge | 72 | 71.922076 | 27.45% | |
| | | | 74 | 73.921178 | 36.50% | |
| 33 | Arsenic | As | 75 | 74.921595 | 100% | |
| 34 | Selenium | Se | 80 | 79.916522 | 49.61% | |
| 35 | Bromine | Br | 79 | 78.918338 | 50.69% | |
| 36 | Krypton | Kr | 84 | 83.911498 | 56.987% | |
| 37 | Rubidium | Rb | 85 | 84.911790 | 72.17% | |
| 38 | Strontium | Sr | 86 | 85.909261 | 9.86% | |
| | | | 88 | 87.905612 | 82.58% | |
| | | | 90 | 89.907730 | $\beta^-$ | 28.90 yr |
| 39 | Yttrium | Y | 89 | 88.905840 | 100% | |
| 40 | Zirconium | Zr | 90 | 89.904698 | 51.45% | |
| 41 | Niobium | Nb | 93 | 92.906373 | 100% | |
| 42 | Molybdenum | Mo | 98 | 97.905405 | 24.39% | |
| 43 | Technetium | Tc | 98 | 97.907212 | $\beta^-$, $\gamma$ | $4.2 \times 10^6$ yr |
| 44 | Ruthenium | Ru | 102 | 101.904344 | 31.55% | |
| 45 | Rhodium | Rh | 103 | 102.905498 | 100% | |
| 46 | Palladium | Pd | 106 | 105.903480 | 27.33% | |
| 47 | Silver | Ag | 107 | 106.905092 | 51.839% | |
| | | | 109 | 108.904755 | 48.161% | |
| 48 | Cadmium | Cd | 114 | 113.903365 | 28.73% | |
| 49 | Indium | In | 115 | 114.903879 | 95.71%; $\beta^-$ | $4.41 \times 10^{14}$ yr |
| 50 | Tin | Sn | 120 | 119.902202 | 32.58% | |
| 51 | Antimony | Sb | 121 | 120.903812 | 57.21% | |

| (1) Atomic Number Z | (2) Element | (3) Symbol | (4) Mass Number A | (5) Atomic Mass | (6) % Abundance (or Radioactive Decay Mode) | (7) Half-life (if radioactive) |
|---|---|---|---|---|---|---|
| 52 | Tellurium | Te | 130 | 129.906223 | 34.08%; $\beta^-\beta^-$ | $>3.0 \times 10^{24}$ yr |
| 53 | Iodine | I | 127 | 126.904472 | 100% | |
| | | | 131 | 130.906126 | $\beta^-, \gamma$ | 8.0252 days |
| 54 | Xenon | Xe | 132 | 131.904155 | 26.9086% | |
| | | | 136 | 135.907214 | 8.8573%; $\beta^-\beta^-$ | $>2.4 \times 10^{21}$ yr |
| 55 | Cesium | Cs | 133 | 132.905452 | 100% | |
| 56 | Barium | Ba | 137 | 136.905827 | 11.232% | |
| | | | 138 | 137.905247 | 71.698% | |
| 57 | Lanthanum | La | 139 | 138.906356 | 99.9119% | |
| 58 | Cerium | Ce | 140 | 139.905443 | 88.450% | |
| 59 | Praseodymium | Pr | 141 | 140.907658 | 100% | |
| 60 | Neodymium | Nd | 142 | 141.907729 | 27.152% | |
| 61 | Promethium | Pm | 145 | 144.912756 | EC, $\alpha$ | 17.7 yr |
| 62 | Samarium | Sm | 152 | 151.919740 | 26.75% | |
| 63 | Europium | Eu | 153 | 152.921238 | 52.19% | |
| 64 | Gadolinium | Gd | 158 | 157.924112 | 24.84% | |
| 65 | Terbium | Tb | 159 | 158.925355 | 100% | |
| 66 | Dysprosium | Dy | 164 | 163.929182 | 28.260% | |
| 67 | Holmium | Ho | 165 | 164.930329 | 100% | |
| 68 | Erbium | Er | 166 | 165.930300 | 33.503% | |
| 69 | Thulium | Tm | 169 | 168.934218 | 100% | |
| 70 | Ytterbium | Yb | 174 | 173.938866 | 31.026% | |
| 71 | Lutetium | Lu | 175 | 174.940775 | 97.401% | |
| 72 | Hafnium | Hf | 180 | 179.946557 | 35.08% | |
| 73 | Tantalum | Ta | 181 | 180.947996 | 99.98799% | |
| 74 | Tungsten (wolfram) | W | 184 | 183.950931 | 30.64%; $\alpha$ | $>8.9 \times 10^{21}$ yr |
| 75 | Rhenium | Re | 187 | 186.955750 | 62.60%; $\beta^-$ | $4.33 \times 10^{10}$ yr |
| 76 | Osmium | Os | 191 | 190.960926 | $\beta^-, \gamma$ | 15.4 days |
| | | | 192 | 191.961477 | 40.78% | |
| 77 | Iridium | Ir | 191 | 190.960589 | 37.3% | |
| | | | 193 | 192.962922 | 62.7% | |
| 78 | Platinum | Pt | 195 | 194.964792 | 33.78% | |
| 79 | Gold | Au | 197 | 196.966569 | 100% | |
| 80 | Mercury | Hg | 199 | 198.968281 | 16.87% | |
| | | | 202 | 201.970643 | 29.86% | |
| 81 | Thallium | Tl | 205 | 204.974428 | 70.48% | |
| 82 | Lead | Pb | 206 | 205.974466 | 24.1% | |
| | | | 207 | 206.975897 | 22.1% | |
| | | | 208 | 207.976652 | 52.4% | |
| | | | 210 | 209.984189 | $\beta^-, \gamma, \alpha$ | 22.20 yr |
| | | | 211 | 210.988737 | $\beta^-, \gamma$ | 36.1 min |
| | | | 212 | 211.991898 | $\beta^-, \gamma$ | 10.64 h |
| | | | 214 | 213.999806 | $\beta^-, \gamma$ | 26.8 min |
| 83 | Bismuth | Bi | 209 | 208.980399 | 100% | |
| | | | 211 | 210.987270 | $\alpha, \gamma, \beta^-$ | 2.14 min |
| 84 | Polonium | Po | 210 | 209.982874 | $\alpha, \gamma$, EC | 138.376 days |
| | | | 214 | 213.995202 | $\alpha, \gamma$ | 164.3 $\mu$s |
| 85 | Astatine | At | 218 | 218.008695 | $\alpha, \beta^-$ | 1.5 s |

| (1) Atomic Number Z | (2) Element | (3) Symbol | (4) Mass Number A | (5) Atomic Mass | (6) % Abundance (or Radioactive Decay Mode) | (7) Half-life (if radioactive) |
|---|---|---|---|---|---|---|
| 86 | Radon | Rn | 222 | 222.017578 | $\alpha, \gamma$ | 3.8235 days |
| 87 | Francium | Fr | 223 | 223.019736 | $\beta^-, \gamma, \alpha$ | 22.00 min |
| 88 | Radium | Ra | 226 | 226.025410 | $\alpha, \gamma$ | 1600 yr |
| 89 | Actinium | Ac | 227 | 227.027752 | $\beta^-, \gamma, \alpha$ | 21.772 yr |
| 90 | Thorium | Th | 228 | 228.028741 | $\alpha, \gamma$ | 1.9116 yr |
|  |  |  | 232 | 232.038056 | 100%; $\alpha, \gamma$ | $1.40 \times 10^{10}$ yr |
| 91 | Protactinium | Pa | 231 | 231.035884 | $\alpha, \gamma$ | $3.276 \times 10^4$ yr |
| 92 | Uranium | U | 232 | 232.037156 | $\alpha, \gamma$ | 68.9 yr |
|  |  |  | 233 | 233.039636 | $\alpha, \gamma$ | $1.592 \times 10^5$ yr |
|  |  |  | 235 | 235.043930 | 0.7204%; $\alpha, \gamma$ | $7.04 \times 10^8$ yr |
|  |  |  | 236 | 236.045568 | $\alpha, \gamma$ | $2.342 \times 10^7$ yr |
|  |  |  | 238 | 238.050788 | 99.2742%; $\alpha, \gamma$ | $4.468 \times 10^9$ yr |
|  |  |  | 239 | 239.054294 | $\beta^-, \gamma$ | 23.45 min |
| 93 | Neptunium | Np | 237 | 237.048174 | $\alpha, \gamma$ | $2.144 \times 10^6$ yr |
|  |  |  | 239 | 239.052939 | $\beta^-, \gamma$ | 2.356 days |
| 94 | Plutonium | Pu | 239 | 239.052164 | $\alpha, \gamma$ | 24,110 yr |
|  |  |  | 244 | 244.064205 | $\alpha$ | $8.00 \times 10^7$ yr |
| 95 | Americium | Am | 243 | 243.061381 | $\alpha, \gamma$ | 7370 yr |
| 96 | Curium | Cm | 247 | 247.070354 | $\alpha, \gamma$ | $1.56 \times 10^7$ yr |
| 97 | Berkelium | Bk | 247 | 247.070307 | $\alpha, \gamma$ | 1380 yr |
| 98 | Californium | Cf | 251 | 251.079589 | $\alpha, \gamma$ | 898 yr |
| 99 | Einsteinium | Es | 252 | 252.082980 | $\alpha$, EC, $\gamma$ | 471.7 days |
| 100 | Fermium | Fm | 257 | 257.095106 | $\alpha, \gamma$ | 100.5 days |
| 101 | Mendelevium | Md | 258 | 258.098431 | $\alpha, \gamma$ | 51.5 days |
| 102 | Nobelium | No | 259 | 259.101030 | $\alpha$, EC | 58 min |
| 103 | Lawrencium | Lr | 262 | 262.109610 | $\alpha$, EC, fission | $\approx 4$ h |
| 104 | Rutherfordium | Rf | 263 | 263.112500 | fission | 10 min |
| 105 | Dubnium | Db | 268 | 268.125670 | fission | 32 h |
| 106 | Seaborgium | Sg | 271 | 271.133930 | $\alpha$, fission | 2.4 min |
| 107 | Bohrium | Bh | 274 | 274.143550 | $\alpha$, fission | 0.9 min |
| 108 | Hassium | Hs | 270 | 270.134290 | $\alpha$ | 22 s |
| 109 | Meitnerium | Mt | 278 | 278.156310 | $\alpha$, fission | 8 s |
| 110 | Darmstadtium | Ds | 281 | 281.164510 | $\alpha$, fission | 20 s |
| 111 | Roentgenium | Rg | 281 | 281.166360 | $\alpha$, fission | 26 s |
| 112 | Copernicium | Cn | 285 | 285.177120 | $\alpha$ | 30 s |
| 113† |  |  | 286 | 286.18210 | $\alpha$, fission | 20 s |
| 114 | Flerovium | Fl | 289 | 289.190420 | $\alpha$ | 2.7 s |
| 115† |  |  | 289 | 289.193630 | $\alpha$, fission | 0.22 s |
| 116 | Livermorium | Lv | 293 | 293.204490 | $\alpha$ | 53 ms |
| 117† |  |  | 294 | 294.210460 | $\alpha$ | 0.08 s |
| 118† |  |  | 294 | 294.213920 | $\alpha$, fission | 0.9 ms |

† Preliminary evidence (unconfirmed) has been reported for elements 113, 115, 117, and 118.

# Rotating Frames of Reference; Inertial Forces; Coriolis Effect

APPENDIX C

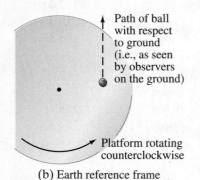

People on ground appear to move this way

Path of ball with respect to rotating platform (i.e., as seen by observer on platform)

(a) Rotating reference frame

Path of ball with respect to ground (i.e., as seen by observers on the ground)

Platform rotating counterclockwise

(b) Earth reference frame

**FIGURE C–1** Path of a ball released on a rotating merry-go-round (a) in the reference frame of the merry-go-round, and (b) in a reference frame fixed on the ground.

## Inertial and Noninertial Reference Frames

In Chapters 5 and 8 we examined the motion of objects, including circular and rotational motion, from the outside, as observers fixed on the Earth. Sometimes it is convenient to place ourselves (in theory, if not physically) into a reference frame that is rotating. Let us examine the motion of objects from the point of view, or frame of reference, of persons seated on a rotating platform such as a merry-go-round. It looks to them as if the rest of the world is going around *them*. But let us focus on what they observe when they place a tennis ball on the floor of the rotating platform, which we assume is frictionless. If they put the ball down gently, without giving it any push, they will observe that it accelerates from rest and moves outward as shown in Fig. C–1a. According to Newton's first law, an object initially at rest should stay at rest if no net force acts on it. But, according to the observers on the rotating platform, the ball starts moving even though there is no net force acting on it. To observers on the ground, this is all very clear: the ball has an initial velocity when it is released (because the platform is moving), and it simply continues moving in a straight-line path as shown in Fig. C–1b, in accordance with Newton's first law.

But what shall we do about the frame of reference of the observers on the rotating platform? Newton's first law, the law of inertia, does not hold in this rotating frame of reference since the ball starts moving with no net force on it. For this reason, such a frame is called a **noninertial reference frame**. An **inertial reference frame** (as discussed in Chapter 4) is one in which the law of inertia—Newton's first law—does hold, and so do Newton's second and third laws. In a noninertial reference frame, such as our rotating platform, Newton's second law also does not hold. For instance in the situation described above, there is no net force on the ball; yet, with respect to the rotating platform, the ball accelerates.

## Fictitious (Inertial) Forces

Because Newton's laws do not hold when observations are made with respect to a rotating frame of reference, calculation of motion can be complicated. However, we can still apply Newton's laws in such a reference frame if we make use of a trick. The ball on the rotating platform of Fig. C–1a flies outward when released (even though no force is actually acting on it). So the trick we use is to write down the equation $\Sigma F = ma$ as if a force equal to $mv^2/r$ (or $m\omega^2 r$) were acting radially outward on the object in addition to any other forces that may be acting. This extra force, which might be designated as "centrifugal force" since it *seems* to act outward, is called a **fictitious force** or **pseudoforce**. It is a pseudoforce ("pseudo" means "false") because there is no object that exerts this force. Furthermore, when viewed from an inertial reference frame, the effect doesn't exist at all. We have made up this pseudoforce so that we can make calculations in a noninertial frame using Newton's second law, $\Sigma F = ma$. Thus the observer in the noninertial frame of Fig. C–1a uses Newton's second law for the ball's outward motion by assuming that a force equal to $mv^2/r$ acts on it. Such pseudoforces are also called **inertial forces** since they arise only because the reference frame is not an inertial one.

In Section 5–3 we discussed the forces on a person in a car going around a curve (Fig. 5–11) from the point of view of an inertial frame. The car, on the other hand, is not an inertial frame. Passengers in such a car could interpret this being pressed outward as the effect of a "centrifugal" force. But they need to recognize that it is a pseudoforce because there is no identifiable object exerting it. It is an effect of being in a noninertial frame of reference.

The Earth itself is rotating on its axis. Thus, strictly speaking, Newton's laws are not valid on the Earth. However, the effect of the Earth's rotation is usually so small that it can be ignored, although it does influence the movement of large air masses and ocean currents. Because of the Earth's rotation, the material of the Earth is concentrated slightly more at the equator. The Earth is thus not a perfect sphere but is slightly fatter at the equator than at the poles.

## Coriolis Effect

In a reference frame that rotates at a constant angular speed $\omega$ (relative to an inertial frame), there exists another pseudoforce known as the *Coriolis force*. It appears to act on an object in a rotating reference frame only if the object is moving relative to that rotating reference frame, and it acts to deflect the object sideways. It, too, is an effect of the rotating reference frame being noninertial and hence is referred to as an *inertial force*. It also affects the weather.

To see how the Coriolis force arises, consider two people, A and B, at rest on a platform rotating with angular speed $\omega$, as shown in Fig. C–2a. They are situated at distances $r_A$ and $r_B$ from the axis of rotation (at O). The woman at A throws a ball with a horizontal velocity $\vec{v}$ (in her reference frame) radially outward toward the man at B on the outer edge of the platform. In Fig. C–2a, we view the situation from an inertial reference frame. The ball initially has not only the velocity $\vec{v}$ radially outward, but also a tangential velocity $\vec{v}_A$ due to the rotation of the platform. Now Eq. 8–4 tells us that $v_A = r_A \omega$, where $r_A$ is the woman's radial distance from the axis of rotation at O. If the man at B had this same velocity $v_A$, the ball would reach him perfectly. But his speed is $v_B = r_B \omega$, which is greater than $v_A$ because $r_B > r_A$. Thus, when the ball reaches the outer edge of the platform, it passes a point that the man at B has already gone by because his speed in that direction is greater than the ball's. So the ball passes behind him.

Figure C–2b shows the situation as seen from the rotating platform as frame of reference. Both A and B are at rest, and the ball is thrown with velocity $\vec{v}$ toward B, but the ball deflects to the right as shown and passes behind B as previously described. This is not a centrifugal-force effect, because that would act radially outward. Instead, this effect acts sideways, perpendicular to $\vec{v}$, and is called a **Coriolis acceleration**; it is said to be due to the Coriolis force, which is a fictitious inertial force. Its explanation as seen from an inertial system was given above: it is an effect of being in a rotating system, for which a point farther from the rotation axis has a higher linear speed. On the other hand, when viewed from the rotating system, the motion can be described using Newton's second law, $\Sigma\vec{F} = m\vec{a}$, if we add a "pseudoforce" term corresponding to this Coriolis effect.

(a) Inertial reference frame

(b) Rotating reference frame

**FIGURE C–2** The origin of the Coriolis effect. Looking down on a rotating platform, (a) as seen from a nonrotating inertial reference frame, and (b) as seen from the rotating platform as frame of reference.

**FIGURE C–2** (Repeated.)
The origin of the Coriolis effect. Looking down on a rotating platform, (a) as seen from a nonrotating inertial reference frame, and (b) as seen from the rotating platform as frame of reference.

(a) Inertial reference frame

(b) Rotating reference frame

Let us determine the magnitude of the Coriolis acceleration for the simple case described above. (We assume $v$ is large and distances short, so we can ignore gravity.) We do the calculation from the inertial reference frame (Fig. C–2a). The ball moves radially outward a distance $r_B - r_A$ at speed $v$ in a short time $t$ given by

$$r_B - r_A = vt.$$

During this time, the ball moves to the side a distance $s_A$ given by

$$s_A = v_A t.$$

The man at B, in this time $t$, moves a distance

$$s_B = v_B t.$$

The ball therefore passes behind him a distance $s$ (Fig. C–2a) given by

$$s = s_B - s_A = (v_B - v_A)t.$$

We saw earlier that $v_A = r_A \omega$ and $v_B = r_B \omega$, so

$$s = (r_B - r_A)\omega t.$$

We substitute $r_B - r_A = vt$ (see above) and get

$$s = \omega v t^2. \tag{C–1}$$

This same $s$ equals the sideways displacement as seen from the noninertial rotating system (Fig. C–2b).

Equation C–1 corresponds to motion at constant acceleration, because as we saw in Chapter 2 (Eq. 2–11b), $y = \frac{1}{2}at^2$ for a constant acceleration (with zero initial velocity in the $y$ direction). Thus, if we write Eq. C–1 in the form $s = \frac{1}{2}a_{Cor}t^2$, we see that the Coriolis acceleration $a_{Cor}$ is

$$a_{Cor} = 2\omega v. \tag{C–2}$$

This relation is valid for any velocity in the plane of rotation perpendicular to the axis of rotation (in Fig. C–2, the axis through point O perpendicular to the page).

Because the Earth rotates, the Coriolis effect has some interesting manifestations on the Earth. It affects the movement of air masses and thus has an influence on weather. In the absence of the Coriolis effect, air would rush directly into a region of low pressure, as shown in Fig. C–3a. But because of the Coriolis effect, the winds are deflected to the right in the Northern Hemisphere (Fig. C–3b), since the Earth rotates from west to east. So there tends to be a counterclockwise wind pattern around a low-pressure area. The reverse is true in the Southern Hemisphere. Thus cyclones rotate counterclockwise in the Northern Hemisphere and clockwise in the Southern Hemisphere. The same effect explains the easterly trade winds near the equator: any winds heading south toward the equator will be deflected toward the west (that is, as if coming from the east).

The Coriolis effect also acts on a falling object. An object released from the top of a high tower will not hit the ground directly below the release point, but will be deflected slightly to the east. Viewed from an inertial frame, this happens because the top of the tower revolves with the Earth at a slightly higher speed than the bottom of the tower.

(a)

(b)

(c)

**FIGURE C–3** (a) Winds (moving air masses) would flow directly toward a low-pressure area if the Earth did not rotate; (b) and (c): because of the Earth's rotation, the winds are deflected to the right in the Northern Hemisphere (as in Fig. C–2) as if a fictitious (Coriolis) force were acting.

# Molar Specific Heats for Gases, and the Equipartition of Energy

## Molar Specific Heats for Gases

The values of the specific heats for gases depend on how the thermodynamic process is carried out. Two important processes are those in which either the volume or the pressure is kept constant, and Table D–1 shows how different they can be.

The difference in specific heats for gases is nicely explained in terms of the first law of thermodynamics and kinetic theory. For gases we usually use **molar specific heats**, $C_V$ and $C_P$, which are defined as the heat required to raise 1 mol of a gas by 1 K (or 1 C°) at constant volume and at constant pressure, respectively. In analogy to Eq. 14–2, the heat $Q$ needed to raise the temperature of $n$ moles of gas by $\Delta T$ is

$$Q = nC_V \Delta T \qquad \text{[volume constant]} \quad \textbf{(D–1a)}$$

$$Q = nC_P \Delta T. \qquad \text{[pressure constant]} \quad \textbf{(D–1b)}$$

We can see from the definition of molar specific heat (compare Eqs. 14–2 and D–1) that

$$C_V = Mc_V$$

$$C_P = Mc_P,$$

where $M$ is the molecular mass of the gas ($M = m/n$ in grams/mol).[†] The values for molar specific heats are included in Table D–1. These values are nearly the same for different gases that have the same number of atoms per molecule.

Now we use kinetic theory and imagine that an ideal gas is slowly heated via two different processes—first at constant volume, and then at constant pressure. In both processes, we let the temperature increase by the same amount, $\Delta T$.

[†]For example, $M = 2\,\text{g/mol}$ for He, and $M = 32\,\text{g/mol}$ for $O_2$.

| TABLE D–1 Specific Heats of Gases at 15°C | | | | | |
|---|---|---|---|---|---|
| | Specific Heats (kcal/kg · K) | | Molar Specific Heats (cal/mol · K) | | $C_P - C_V$ (cal/mol · K) |
| Gas | $c_V$ | $c_P$ | $C_V$ | $C_P$ | |
| Monatomic | | | | | |
| He | 0.75 | 1.15 | 2.98 | 4.97 | 1.99 |
| Ne | 0.148 | 0.246 | 2.98 | 4.97 | 1.99 |
| Diatomic | | | | | |
| $N_2$ | 0.177 | 0.248 | 4.96 | 6.95 | 1.99 |
| $O_2$ | 0.155 | 0.218 | 5.03 | 7.03 | 2.00 |
| Triatomic | | | | | |
| $CO_2$ | 0.153 | 0.199 | 6.80 | 8.83 | 2.03 |
| $H_2O$ (100°C) | 0.350 | 0.482 | 6.20 | 8.20 | 2.00 |
| Polyatomic | | | | | |
| $C_2H_6$ | 0.343 | 0.412 | 10.30 | 12.35 | 2.05 |

In the constant-volume process, no work is done since $\Delta V = 0$. Thus, according to the first law of thermodynamics ($Q = \Delta U + W$, Section 15–1), the heat added (which we denote by $Q_V$) all goes into increasing the internal energy of the gas:

$$Q_V = \Delta U.$$

In the constant-pressure process, work *is* done. Hence the heat added, $Q_P$, must not only increase the internal energy but also is used to do work $W = P \Delta V$. Thus, to increase the temperature by the same $\Delta T$, more heat must be added in the process at constant pressure than in the process at constant volume. For the process at constant pressure, the first law of thermodynamics gives

$$Q_P = \Delta U + P \Delta V.$$

Since $\Delta U$ is the same in the two processes (we chose $\Delta T$ to be the same), we can combine the two above equations:

$$Q_P - Q_V = P \Delta V.$$

From the ideal gas law, $V = nRT/P$, so for a process at constant pressure $\Delta V = nR \Delta T/P$. Putting this into the last equation and using Eqs. D–1, we find

$$nC_P \Delta T - nC_V \Delta T = P\left(\frac{nR \Delta T}{P}\right)$$

or, after cancellations,

$$C_P - C_V = R. \tag{D–2}$$

Since the gas constant $R = 8.314\,\text{J/mol}\cdot\text{K} = 1.99\,\text{cal/mol}\cdot\text{K}$, our prediction is that $C_P$ will be larger than $C_V$ by about $1.99\,\text{cal/mol}\cdot\text{K}$. Indeed, this is very close to what is obtained experimentally, as shown in the last column in Table D–1.

Now we calculate the molar specific heat of a monatomic gas using kinetic theory. For a process carried out at constant volume, no work is done, so the first law of thermodynamics tells us that

$$\Delta U = Q_V.$$

For an ideal monatomic gas, the internal energy, $U$, is the total kinetic energy of all the molecules,

$$U = N(\tfrac{1}{2}m\overline{v^2}) = \tfrac{3}{2}nRT$$

as we saw in Section 14–2. Then, using Eq. D–1a, we write $\Delta U = Q_V$ as

$$\Delta U = \tfrac{3}{2}nR \Delta T = nC_V \Delta T \tag{D–3}$$

or

$$C_V = \tfrac{3}{2}R. \tag{D–4}$$

Since $R = 8.314\,\text{J/mol}\cdot\text{K} = 1.99\,\text{cal/mol}\cdot\text{K}$, kinetic theory predicts that $C_V = 2.98\,\text{cal/mol}\cdot\text{K}$ for an ideal monatomic gas. This is very close to the experimental values for monatomic gases such as helium and neon (Table D–1). From Eq. D–2, $C_P$ is predicted to be $R + C_V = (1.99 + 2.98)\,\text{cal/mol}\cdot\text{K} = 4.97\,\text{cal/mol}\cdot\text{K}$, also in agreement with experiment (Table D–1).

## Equipartition of Energy

The measured molar specific heats for more complex gases (Table D–1), such as diatomic (two atoms) and triatomic (three atoms) gases, increase with the increased number of atoms per molecule. We can explain this by assuming that the internal energy includes not only translational kinetic energy but other forms of energy as well. For example, in a diatomic gas (Fig. D–1), the two atoms can rotate about two different axes (but rotation about a third axis passing through the two atoms would give rise to very little energy since the moment of inertia is so small). The molecules can have rotational as well as translational kinetic energy.

It is useful to introduce the idea of **degrees of freedom**, by which we mean the number of independent ways molecules can possess energy. For example, a monatomic gas has three degrees of freedom, because an atom can have velocity along the $x$, $y$, and $z$ axes. These are considered to be three independent motions because a change in any one of the components would not affect the others. A diatomic molecule has the same three degrees of freedom associated with translational kinetic energy plus two more degrees of freedom associated with rotational kinetic energy (Fig. D–1), for a total of five degrees of freedom.

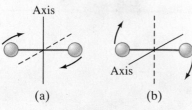

**FIGURE D–1** A diatomic molecule can rotate about two different axes.

Axis

(a)          (b)

Axis

Table D–1 indicates that the $C_V$ for diatomic gases is about $\frac{5}{3}$ times as great as for a monatomic gas—that is, in the same ratio as their degrees of freedom. This result led nineteenth-century physicists to the **principle of equipartition of energy**. This principle states that energy is shared equally among the active degrees of freedom, and each active degree of freedom of a molecule has on average an energy equal to $\frac{1}{2}kT$. Thus, the average energy for a molecule of a monatomic gas would be $\frac{3}{2}kT$ (which we already knew) and of a diatomic gas $\frac{5}{2}kT$. Hence the internal energy of a diatomic gas would be $U = N(\frac{5}{2}kT) = \frac{5}{2}nRT$, where $n$ is the number of moles. Using the same argument we did for monatomic gases, we see that for diatomic gases the molar specific heat at constant volume would be $\frac{5}{2}R = 4.97$ cal/mol·K, close to measured values (Table D–1). More complex molecules have even more degrees of freedom and thus greater molar specific heats.

However, measurements showed that for diatomic gases at very low temperatures, $C_V$ has a value of only $\frac{3}{2}R$, as if it had only three degrees of freedom. And at very high temperatures, $C_V$ was about $\frac{7}{2}R$, as if there were seven degrees of freedom. The explanation is that at low temperatures, nearly all molecules have only translational kinetic energy; that is, no energy goes into rotational energy and only three degrees of freedom are "active." At very high temperatures, all five degrees of freedom are active plus two additional ones. We interpret the two new degrees of freedom as being associated with the two atoms vibrating, as if they were connected by a spring (Fig. D–2). One degree of freedom comes from the kinetic energy of the vibrational motion, and the second comes from the potential energy of vibrational motion ($\frac{1}{2}kx^2$). At room temperature, these two degrees of freedom are apparently not active (Fig. D–3). Why fewer degrees of freedom are "active" at lower temperatures was eventually explained by Einstein using quantum theory.

**FIGURE D–2** A diatomic molecule can vibrate, as if the two atoms were connected by a spring. They aren't, but rather they exert forces on each other that are electrical in nature—of a form that resembles a spring force.

**FIGURE D–3** Molar specific heat $C_V$ as a function of temperature for hydrogen molecules ($H_2$). As the temperature is increased, some of the translational kinetic energy can be transferred in collisions into rotational kinetic energy and, at still higher temperature, into vibrational kinetic and potential energy. [Note: $H_2$ dissociates into two atoms at about 3200 K, so the last part of the curve is shown dashed.]

## Solids

The principle of equipartition of energy can be applied to solids as well. The molar specific heat of any solid at high temperature is close to $3R$ (6.0 cal/mol·K), Fig. D–4. This is called the *Dulong and Petit value* after the scientists who first measured it in 1819. (Note that Table 14–1 gave the specific heats per kilogram, not per mole.) At high temperatures, each atom apparently has six degrees of freedom, although some are not active at low temperatures. Each atom in a crystalline solid can vibrate about its equilibrium position as if it were connected by springs to each of its neighbors (Fig. D–5). Thus it can have three degrees of freedom for kinetic energy and three more associated with potential energy of vibration in each of the $x$, $y$, and $z$ directions, which is in accord with measured values.

**FIGURE D–4** Molar specific heats of solids as a function of temperature.

**FIGURE D–5** The atoms in a crystalline solid can vibrate about their equilibrium positions as if they were connected to their neighbors by springs. (The forces between atoms are actually electrical in nature.)

# Galilean and Lorentz Transformations

We examine in detail the mathematics of relating quantities in one inertial reference frame to the equivalent quantities in another. In particular, we will see how positions and velocities *transform* (that is, change) from one frame of reference to the other.

We begin with the classical, or Galilean, viewpoint. Consider two inertial reference frames S and S′ which are each characterized by a set of coordinate axes, Fig. E–1. The axes $x$ and $y$ ($z$ is not shown) refer to S, and $x'$ and $y'$ refer to S′. The $x'$ and $x$ axes overlap one another, and we assume that frame S′ moves to the right (in the $x$ direction) at speed $v$ with respect to S. For simplicity let us assume the origins 0 and 0′ of the two reference frames are superimposed at time $t = 0$.

**FIGURE E–1** Inertial reference frame S′ moves to the right at speed $v$ with respect to inertial frame S.

Now consider an event that occurs at some point P (Fig. E–1) represented by the coordinates $x'$, $y'$, $z'$ in reference frame S′ at the time $t'$. What will be the coordinates of P in S? Since S and S′ overlap precisely initially, after a time $t'$, S′ will have moved a distance $vt'$. Therefore, at time $t'$, $x = x' + vt'$. The $y$ and $z$ coordinates, on the other hand, are not altered by motion along the $x$ axis; thus $y = y'$ and $z = z'$. Finally, since time is assumed to be absolute in Galilean–Newtonian physics, clocks in the two frames will agree with each other; so $t = t'$. We summarize these in the following **Galilean transformation equations**:

$$
\begin{aligned}
x &= x' + vt' \\
y &= y' \\
z &= z' \\
t &= t'.
\end{aligned}
$$

[Galilean] **(E–1)**

These equations give the coordinates of an event in the S frame when those in the S′ frame are known. If those in the S frame are known, then the S′ coordinates are obtained from

$$ x' = x - vt, \qquad y' = y, \qquad z' = z, \qquad t' = t. \qquad \text{[Galilean]} $$

These four equations are the "inverse" transformation, and are obtained from Eqs. E–1 by exchanging primed and unprimed quantities and replacing $v$ by $-v$. This makes sense because, as seen from the S′ frame, S moves to the left (negative $x$ direction) with speed $v$.

Now suppose that the point P in Fig. E–1 represents an object that is moving. Let the components of its velocity vector in S′ be $u'_x$, $u'_y$, and $u'_z$. (We use $u$ to distinguish it from the relative velocity of the two frames, $v$.) Now $u'_x = \Delta x'/\Delta t'$, $u'_y = \Delta y'/\Delta t'$, and $u'_z = \Delta z'/\Delta t'$, where all quantities are as measured in the S′ frame. For example, if at time $t'_1$ the particle is at $x'_1$ and a short time later, $t'_2$, it is at $x'_2$, then

$$u'_x = \frac{x'_2 - x'_1}{t'_2 - t'_1} = \frac{\Delta x'}{\Delta t'}.$$

The velocity of P as seen from S will have components $u_x$, $u_y$, and $u_z$. We can show how these are related to the velocity components in S′ by using Eqs. E–1. For example,

$$u_x = \frac{\Delta x}{\Delta t} = \frac{x_2 - x_1}{t_2 - t_1} = \frac{(x'_2 + vt'_2) - (x'_1 + vt'_1)}{t'_2 - t'_1}$$

$$= \frac{(x'_2 - x'_1) + v(t'_2 - t'_1)}{t'_2 - t'_1}$$

$$= \frac{\Delta x'}{\Delta t'} + v = u'_x + v.$$

For the other components, $u'_y = u_y$ and $u'_z = u_z$, so we have

$$u_x = u'_x + v$$
$$u_y = u'_y \qquad\qquad \text{[Galilean]} \quad \textbf{(E–2)}$$
$$u_z = u'_z.$$

These are known as the **Galilean velocity transformation equations**. We see that the $y$ and $z$ components of velocity are unchanged, but the $x$ components differ by $v$. This is just what we have used before when dealing with relative velocity (Section 3–8). For example, if S′ is a train and S the Earth, and the train moves with speed $v$ with respect to Earth, a person walking toward the front of the train with speed $u'_x$ will have a speed with respect to the Earth of $u_x = u'_x + v$.

The Galilean transformations, Eqs. E–1 and E–2, are accurate only when the velocities involved are not relativistic (Chapter 26)—that is, much less than the speed of light, $c$. We can see, for example, that the first of Eqs. E–2 will not work for the speed of light, $c$, which is the same in all inertial reference frames (a basic postulate in the theory of relativity). That is, light traveling in S′ with speed $u'_x = c$ will have speed $c + v$ in S, according to Eq. E–2, whereas the theory of relativity insists it must be $c$ in S. Clearly, then, a new set of transformation equations is needed to deal with relativistic velocities.

We will derive the required equations, again looking at Fig. E–1. We assume the transformation is linear and for $x$ is of the form

$$x = \gamma(x' + vt'). \qquad\qquad \textbf{(i)}$$

That is, we modify the first of Eqs. E–1 by multiplying by a factor $\gamma$ which is yet to be determined.[†] We assume the $y$ and $z$ equations are unchanged

$$y = y', \qquad z = z'$$

because there is no length contraction in these directions. We will not assume a form for $t$, but will derive it. The inverse equations must have the same form with $v$ replaced by $-v$. (The principle of relativity demands it, since S′ moving to the right with respect to S is equivalent to S moving to the left with respect to S′.) Therefore

$$x' = \gamma(x - vt). \qquad\qquad \textbf{(ii)}$$

Suppose a light pulse leaves the common origin of S and S′ at time $t = t' = 0$.

[†] We are NOT assuming $\gamma$ is $1/\sqrt{1 - v^2/c^2}$, as in Chapter 26. Our minds are open. Let's see.

Then after a time $t$ it will have traveled along the $x$ axis a distance $x = ct$ (in S), or $x' = ct'$ (in S'). Therefore, from Eqs. (i) and (ii) above,

$$ct = x = \gamma(ct' + vt') = \gamma(c + v)t', \qquad \text{(iii)}$$
$$ct' = x' = \gamma(ct - vt) = \gamma(c - v)t. \qquad \text{(iv)}$$

From Eq. (iv), $t' = \gamma(c - v)(t/c)$, and we substitute this into Eq. (iii) and find $ct = \gamma(c + v)\gamma(c - v)(t/c) = \gamma^2(c^2 - v^2)t/c$. We cancel out the $t$ on each side and solve for $\gamma$ to find

$$\gamma = \frac{1}{\sqrt{1 - v^2/c^2}}.$$

We have found that $\gamma$ is, in fact, the value for $\gamma$ we used in Chapter 26, Eq. 26–2.

Now that we have found $\gamma$, we need only find the relation between $t$ and $t'$. To do so, we combine $x' = \gamma(x - vt)$ with $x = \gamma(x' + vt')$:

$$x' = \gamma(x - vt) = \gamma[\gamma(x' + vt') - vt].$$

We solve for $t$, doing some algebra, and find $t = \gamma(t' + vx'/c^2)$. In summary,

*LORENTZ TRANSFORMATIONS*

$$
\begin{aligned}
x &= \frac{1}{\sqrt{1 - v^2/c^2}}(x' + vt') \\
y &= y' \\
z &= z' \\
t &= \frac{1}{\sqrt{1 - v^2/c^2}}\left(t' + \frac{vx'}{c^2}\right).
\end{aligned}
\qquad \text{(E–3)}
$$

These are called the **Lorentz transformation equations**. They were first proposed, in a slightly different form, by Lorentz in 1904 to explain the null result of the Michelson–Morley experiment and to make Maxwell's equations take the same form in all inertial reference frames. A year later, Einstein derived them independently based on his theory of relativity. Notice that not only is the $x$ equation modified as compared to the Galilean transformation, but so is the $t$ equation. Indeed, we see directly in this last equation how the space and time coordinates mix.

The relativistically correct velocity equations are obtained using Eqs. E–3 (we let $\gamma = 1/\sqrt{1 - v^2/c^2}$) and $u_x = \Delta x/\Delta t$, $u_x' = \Delta x'/\Delta t'$:

$$u_x = \frac{\Delta x}{\Delta t} = \frac{\gamma(\Delta x' + v\,\Delta t')}{\gamma(\Delta t' + v\,\Delta x'/c^2)} = \frac{(\Delta x'/\Delta t') + v}{1 + (v/c^2)(\Delta x'/\Delta t')}$$
$$= \frac{u_x' + v}{1 + vu_x'/c^2}.$$

The others are obtained in the same way, and we collect them here:

*RELATIVISTIC VELOCITY TRANSFORMATIONS*

$$
\begin{aligned}
u_x &= \frac{u_x' + v}{1 + vu_x'/c^2} \\
u_y &= \frac{u_y'\sqrt{1 - v^2/c^2}}{1 + vu_x'/c^2} \\
u_z &= \frac{u_z'\sqrt{1 - v^2/c^2}}{1 + vu_x'/c^2}.
\end{aligned}
\qquad \text{(E–4)}
$$

The first of these equations is Eq. 26–11, which we used in Section 26–10 where we discussed how velocities do not add in our commonsense (Galilean) way, because of the denominator $(1 + vu_x'/c^2)$. We can now also see that the $y$ and $z$ components of velocity are also altered and that they depend on the $x'$ component of velocity.

**EXAMPLE E-1** **Length contraction.** Derive the length contraction formula, Eq. 26–3, from the Lorentz transformation equations.

**APPROACH** We consider measurements in two reference frames, S and S', that move with speed $v$ relative to each other, as in Fig. E–1.

**SOLUTION** Let an object of length $\ell_0$ be at rest on the $x$ axis in S. The coordinates of its two end points are $x_1$ and $x_2$, so that $x_2 - x_1 = \ell_0$. At any instant in S', the end points will be at $x_1'$ and $x_2'$ as given by the Lorentz transformation equations. The length measured in S' is $\ell = x_2' - x_1'$. An observer in S' measures this length by measuring $x_2'$ and $x_1'$ at the same time (in the S' frame), so $t_2' = t_1'$. Then, from the first of Eqs. E–3,

$$\ell_0 = x_2 - x_1 = \frac{1}{\sqrt{1 - v^2/c^2}}(x_2' + vt_2' - x_1' - vt_1').$$

Since $t_2' = t_1'$, we have

$$\ell_0 = \frac{1}{\sqrt{1 - v^2/c^2}}(x_2' - x_1') = \frac{\ell}{\sqrt{1 - v^2/c^2}},$$

or

$$\ell = \ell_0\sqrt{1 - v^2/c^2},$$

which is Eq. 26–3a: the length contraction formula.

**EXAMPLE E-2** **Time dilation.** Derive the time dilation formula, Eq. 26–1, from the Lorentz transformation equations.

**APPROACH** Again we compare measurements in two reference frames, S and S', that move with speed $v$ relative to each other, Fig. E–1.

**SOLUTION** The time $\Delta t_0$ between two events that occur at the same place $(x_2' = x_1')$ in S' is measured to be $\Delta t_0 = t_2' - t_1'$. Since $x_2' = x_1'$, then from the last of Eqs. E–3, the time $\Delta t$ between the events as measured in S is

$$\Delta t = t_2 - t_1 = \frac{1}{\sqrt{1 - v^2/c^2}}\left(t_2' + \frac{vx_2'}{c^2} - t_1' - \frac{vx_1'}{c^2}\right)$$

$$= \frac{1}{\sqrt{1 - v^2/c^2}}(t_2' - t_1')$$

$$= \frac{\Delta t_0}{\sqrt{1 - v^2/c^2}},$$

which is the time dilation formula, Eq. 26–1. Notice that we chose S' to be the frame in which the two events occur at the same place, so that $x_1' = x_2'$, and then the terms containing $x_1'$ and $x_2'$ cancel out.

# Answers to Odd-Numbered Problems

## Chapter 1

1. (a) 3;
   (b) 4;
   (c) 3;
   (d) 1;
   (e) 2;
   (f) 4;
   (g) 2.
3. (a) 86,900;
   (b) 9100;
   (c) 0.88;
   (d) 476;
   (e) 0.0000362.
5. 4.6%.
7. $1.00 \times 10^5$ s.
9. 1%.
11. $(3.0 \pm 0.2) \times 10^9$ cm$^2$.
13. (a) 1 megavolt;
   (b) 2 micrometers;
   (c) 6 kilodays;
   (d) 18 hectobucks;
   (e) 700 nanoseconds.
15. (a) $1.5 \times 10^{11}$ m;
   (b) $1.5 \times 10^8$ km.
17. (a) $3.9 \times 10^{-9}$ in.;
   (b) $1.0 \times 10^8$ atoms.
19. (a) $9.46 \times 10^{15}$ m;
   (b) $6.31 \times 10^4$ AU.
21. Soccer; 9.4 yd, 8.6 m, 9.4%.
23. (a) $10^{12}$ protons or neutrons;
   (b) $10^{10}$ protons or neutrons;
   (c) $10^{29}$ protons or neutrons;
   (d) $10^{68}$ protons or neutrons.
25. (a) $10^3$;
   (b) $10^5$;
   (c) $10^{-2}$;
   (d) $10^9$.
27. 500 hr.
29. 2.5 hr.
31. (a) 700;
   (b) answers vary.
33. Second method.
35. 8.8 s.
37. (a) $L/T^4$, $L/T^2$;
   (b) m/s$^4$, m/s$^2$.
39. $10^{-35}$ m.
41. $3.3 \times 10^5$ chips/cylinder.

43. 46,000 years.
45. 400 jelly beans.
47. 75 minutes.
49. $5 \times 10^5$ metric tons, $1 \times 10^8$ gal.
51. 3000 m.
53. (a) 0.10 nm;
   (b) $1.0 \times 10^5$ fm;
   (c) $1.0 \times 10^{10}$ angstroms;
   (d) $9.5 \times 10^{25}$ angstroms.
55. (a) 3%, 3%;
   (b) 0.7%, 0.2%.
57. $8 \times 10^{-2}$ m$^3$.
59. $1.18 \times 10^9$ atoms/m$^2$.
61. $4 \times 10^{51}$ kg.

## Chapter 2

1. 53 m.
3. 0.57 cm/s, no, we need the distance traveled.
5. 0.14 h.
7. (a) 350 km;
   (b) 78 km/h.
9. (a) 3.68 m/s;
   (b) 0.
11. 38 s.
13. 1.6 min.
15. 6.00 m/s.
17. 6.1 m/s$^2$.
19. 6.0 m/s$^2$, 0.61$g$'s.
21. (a) 21.2 m/s;
   (b) 2.00 m/s$^2$.
23. 1 m/s$^2$, 110 m.
25. 260 m/s$^2$.
27. 112 m.
29. 44$g$'s.
31. (a) 130 m;
   (b) 69 m.
33. 21 m/s.
35. 6.3 s.
37. 0.70 m/s$^2$.
39. 61.8 m.
41. 17 m/s, 14 m.
43. 1.09 s.

45. (a)

(b)

49. 5.21 s.
51. 12 m/s.
53. 1.6 m.
55. (a) 48 s;
   (b) 90 s to 108 s;
   (c) 0 s to 42 s, 65 s to 83 s, 90 s to 108 s;
   (d) 65 s to 83 s.
57. (a) 0.3 m/s;
   (b) 1.2 m/s;
   (c) 0.30 m/s;
   (d) 1.4 m/s;
   (e) −0.95 m/s.
59.

61. 1.2 m.

**63.** 3.1 m.

**65.** (a) 14.4 s, no;
(b) no, 4.6 s.

**67.** (b) 4.8 m;
(c) 36 m.

**69.** −20 m/s².

**71.** (a) 5.80 s;
(b) 41.4 m/s;
(c) 99.5 m.

**73.** She should try to stop the car.

**75.** 1.5 poles.

**77.** 30%.

**79.** 245.0 km/h.

**81.** 23.7 s, 840 km/h.

**83.** (a) 4.3 × 10⁶ bits;
(b) 67%.

## Chapter 3

**1.** 302 km, 13° south of west.

**3.** 11.70 units, −33.1°.

**5.** (a)

(b) −22.8 units, 9.85 units;
(c) 24.8 units, 23.4° above the −x axis.

**7.** (a) 1.3 units, positive x direction;
(b) 12.3 units, positive x direction;
(c) 12.3 units, negative x direction.

**9.** (a) x component 24.0,
y component 11.6;
(b) 26.7 units, 25.8°.

**11.** 64.6, 53.1°.

**13.** (a) 62.6, −31.0°;
(b) 77.5, 71.9°;
(c) 77.5, 251.9°.

**15.** (−2845 m, 3589 m, 2450 m),
5190 m.

**17.** 3.7 m.

**19.** 6 times farther.

**21.** 14.5 m.

**23.** 18°, 72°.

**25.** 2.0 m/s.

**27.** (a) 30.8 m;
(b) 5.02 s;
(c) 136 m;
(d) 28.9 m/s.

**29.** 22.3 m.

**31.** 481 m.

**33.** (a) 4.0 m.s, 55° above the horizontal;
(b) 4.6 m;
(c) 9.7 m/s, 76° below the horizontal.

**35.** No, 0.81 m too low; 4.6 m to 34.7 m.

**37.** (a) 4.0 × 10¹ m/s;
(b) 24 m/s.

**39.** 1.66 m/s, 25°.

**41.** 23.1 s.

**43.** (a) 10.4 m/s, 17° above the horizontal;
(b) 10.4 m/s, 17° below the horizontal.

**45.** 5.31° west of south.

**47.** (a) 56°;
(b) 140 s.

**49.** 23 s, 23 m.

**51.** 65 km/h, 58° west of north;
65 km/h, 32° south of east.

**53.** Horizontal: 3.9 m/s² leftward;
vertical: 1.9 m/s² downward.

**55.** 0.88 s, 0.95 m.

**57.** 1.7 m/s².

**59.** (a) 9.96 s;
(b) 531 m;
(c) 53.2 m/s, −60.4 m/s;
(d) 80.5 m/s;
(e) 48.6° below the horizon;
(f) 70.9 m.

**61.** $v_T/\tan\theta$.

**63.** (a) 13.3 m;
(b) 22.1°.

**65.** 33 m/s.

**67.** 54°.

**69.** (a) 2.51 m/s, 61.4°;
(b) 3.60 m downstream, 6.60 m across the river.

**71.** (a) 13 m;
(b) 31° below the horizontal.

**73.** (a) 68 m;
(b) 7.3 m/s.

**75.** 0.51%.

## Chapter 4

**1.** 77 N.

**3.** 1450 N.

**5.** −1.3 × 10⁶ N, 39%, 1.3 × 10⁶ N.

**7.** −3100 N.

**9.** 780 N, backward.

**11.** (a) 196 N, 196 N;
(b) 98.0 N, 294 N.

**13.** Descend with downward
$a \geq 2.2$ m/s².

**15.** −2800 m/s², 280g's, 1.9 × 10⁵ N.

**17.** (a) −7.35 m/s²;
(b) 1290 N.

**19.** (a) 7.4 m/s, downward;
(b) 2100 N, upward.

**21.** (a)                    (b)

**23.** 1410 N.

**25.** (a) 31 N (lower cord), 63 N (upper cord);
(b) 35 N (lower cord), 71 N (upper cord).

**27.** $F_{T1}/F_{T2} = 2$.

**29.** (a) 1.0 × 10¹ m/s²;
(b) 3.3 m/s.

**31.** (a) 23°;
(b) toward the windshield.

**33.** (a) 2.7 m/s²;
(b) 0.96 s;
(c) 99 kg.

**35.** 34 N.

**37.** (a) 0.60;
(b) 0.53.
**39.** 42°.
**41.** 0.46.
**43.** 1200 N.
**45.** 1.4.
**47.** (a) 5.0 kg;
(b) $1.0 \times 10^1$ kg.
**49.** (a) 1.7 m/s$^2$;
(b) 430 N;
(c) 1.7 m/s$^2$, 220 N.
**51.** $1.20 \times 10^2$ N, in the direction opposite to the child's velocity.
**53.** 4.8 s.
**55.** $4.0 \times 10^2$ m.
**57.** (a) 3.67 m/s$^2$;
(b) 9.39 m/s.
**59.** (a) 2.5 m/s$^2$;
(b) 6.3 m/s.
**61.** $-5.3$ m/s$^2$.
**63.** $4.0 \times 10^1$ N.
**65.** $-2.2$ m/s$^2$.
**67.** (a) 1.6 m/s$^2$;
(b) 0.53.
**69.** 0.86 m/s$^2$.
**71.** 9.9°.
**73.** 73 m/s.
**75.** (a) 0.67;
(b) 6.1 m/s;
(c) 15 m/s.
**77.** 4.2°.
**79.** (a) $8.76 \times 10^4$ N, upward;
(b) $1.14 \times 10^4$ N;
(c) $1.14 \times 10^4$ N, downward.
**81.** (a) 45 N (10 lb);
(b) 37 N (8.4 lb);
(c) no, the fish cannot be lifted vertically by a 10-lb force.
**83.** 380 N (between second and last climbers), 760 N (between first and second climbers).
**85.** (a) 3.0 times her weight;
(b) 7.7 times his weight; Jim.
**87.** 23 m/s (85 km/h).
**89.** $4.90 \times 10^2$ N.
**91.** (a) 0.9 m/s$^2$;
(b) 0.98 m/s$^2$.
**93.** (b) Yes.
**95.** (a) 16 m/s;
(b) 13 m/s.
**97.** 1100 N, opposite to the velocity.

## Chapter 5

**1.** (a) 1.01 m/s$^2$;
(b) 22.7 N.
**3.** 12 m/s.
**5.** 13 m/s.
**7.** 34 m/s.
**9.** 24 m/s, yes.
**11.** 8.5 rpm.
**13.** 1700 rev/day.
**15.** 0.210.
**17.** $F_A = 4\pi^2 f^2(m_A r_A + m_B r_B)$,
$F_B = 4\pi^2 f^2 m_B r_B$.
**19.** (a) 5970 N;
(b) 379 N;
(c) 29.4 m/s.
**21.** 930 m.
**23.** 59 km/h to 110 km/h.
**25.** $a_{tan} = 4.1$ m/s$^2$, $a_{rad} = 13$ m/s$^2$;
1.4.
**27.** (a) 0.930 m/s;
(b) 2.83 m/s.
**29.** (a) 24.0 kg on both;
(b) $w_{Earth} = 235$ N,
$w_{Planet} = 288$ N.
**31.** 3.94 kg, 0.06 kg.
**33.** 1.62 m/s$^2$.
**35.** $6.5 \times 10^{23}$ kg.
**37.** 27.4 m/s$^2$.
**39.** (a) 9.78 m/s$^2$;
(b) 2.44 m/s$^2$.
**41.** $9.6 \times 10^{17}$ N; $2.7 \times 10^{-5}$.
**43.** $2.02 \times 10^7$ m.
**45.** 7460 m/s.
**47.** 2.4 m/s$^2$ upward.
**49.** $7.05 \times 10^3$ s.
**51.** (a) 568 N;
(b) 568 N;
(c) 699 N;
(d) 440 N;
(e) 0.
**53.** (a) 59 N, away from the Moon;
(b) 76 N, toward the Moon.
**55.** 9.6 s.
**57.** 160 yr.
**59.** 84.5 min.
**61.** $2 \times 10^8$ yr.
**63.** Europa: $671 \times 10^3$ km,
Ganymede: $1070 \times 10^3$ km,
Callisto: $1880 \times 10^3$ km.
**65.** $5.4 \times 10^{12}$ m; yes; Pluto.

**67.** $5.97 \times 10^{-3}$ m/s$^2$, $3.56 \times 10^{22}$ N; the Sun.
**69.** 28.3 m/s, 0.410 rev/s.
**71.** 0.18; no; the wall pushes against the riders, so by Newton's third law, they push against the wall.
**73.** 9.2 m/s.
**75.** (a) In circular motion, they accelerate toward each other without moving toward each other.
(b) $9.6 \times 10^{29}$ kg.
**77.** Yes; $\sqrt{rg}$, where $r$ is the radius of the vertical circle.
**79.** $T_{inner} = 2.0 \times 10^4$ s,
$T_{outer} = 7.1 \times 10^4$ s.
**81.** (a) 3900 m/s;
(b) $4.4 \times 10^4$ s.
**83.** 25.0 m/s.
**85.** (a) $7.6 \times 10^6$ m;
(b) $3.8 \times 10^4$ N;
(c) $1.2 \times 10^6$ m.
**87.** $1.21 \times 10^6$ m.
**89.** $0.44r$.

## Chapter 6

**1.** $2.06 \times 10^4$ J.
**3.** 2300 J.
**5.** $1.0 \times 10^6$ J.
**7.** 1960 J.
**9.** 390 J.
**11.** 2 m.
**13.** (a) 2800 J;
(b) 2100 J.
**15.** 484 m/s.
**17.** $-5.51 \times 10^{-19}$ J.
**19.** The lighter one, $\sqrt{2}$; both the same.
**21.** 43 m/s.
**23.** 21 m/s.
**25.** (a) 3010 N;
(b) 7480 J;
(c) $5.42 \times 10^4$ J;
(d) $-4.67 \times 10^4$ J;
(e) 7.51 m/s.
**27.** 1.01 m.
**29.** (a) $9.00 \times 10^5$ J,
(b) $9.06 \times 10^5$ J;
(c) yes.
**31.** 45.4 m/s.
**33.** 4.89 m/s.
**35.** 74 cm.
**37.** $1.4 \times 10^5$ N/m.
**39.** (a) 7.47 m/s;
(b) 3.01 m.

**41.** 52 m.

**43.** 12 $Mg/h$.

**45.** (a) $9.19 \times 10^4$ J;

   (b) 433 N.

**47.** 332 J.

**49.** (a) 15.3 m/s;

   (b) 1.03 N, upward.

**51.** 0.091.

**53.** $1.4 \times 10^5$ J.

**55.** (a) 2.8 m;

   (b) 1.5 m;

   (c) 1.5 m.

**57.** 22.0 s.

**59.** (a) 1100 J;

   (b) 1100 W.

**61.** 2700 N.

**63.** $2.9 \times 10^4$ W, 38 hp.

**65.** $5.3 \times 10^4$ W.

**67.** 15.4 W.

**69.** 33 hp.

**71.** 610 W.

**73.** 14.9 m/s.

**75.** (a) $\sqrt{Fx/m}$;

   (b) $\sqrt{3Fx/4m}$.

**77.** (a) 0.014 J;

   (b) 0.039 J.

**79.** (a) $-9.0 \times 10^4$ J;

   (b) $8.2 \times 10^4$ N;

   (c) $-2.3 \times 10^5$ J.

**81.** 340 W.

**83.** (a) $1.0 \times 10^4$ J;

   (b) 16 m/s.

**85.** (a) 42 m/s;

   (b) $3.2 \times 10^5$ W.

**87.** (b) 420 kWh;

   (c) $1.5 \times 10^9$ J;

   (d) $50, no.

**89.** (a) $8.9 \times 10^5$ J;

   (b) 54 W, 0.072 hp;

   (c) 360 W, 0.48 hp.

**91.** (a) 0.39 m;

   (b) $\mu_s < 0.53$;

   (c) 1.4 m/s.

**93.** $1.7 \times 10^5$ m³.

## Chapter 7

**1.** 0.24 kg·m/s.

**3.** 10.2 m/s.

**5.** $5.9 \times 10^7$ N, opposite the gas velocity.

**7.** $-0.898$ m/s.

**9.** 2500 m/s.

**11.** 0.99 m/s.

**13.** $4.9 \times 10^6$ N.

**15.** 2230 N, toward the pitcher.

**17.** (a) $9.0 \times 10^1$ kg·m/s;

   (b) $1.1 \times 10^4$ N.

**19.** (a) $-0.16$ m/s;

   (b) 521 N;

   (c) astronaut: 391 J; capsule: 26 J.

**21.** (a) 290 kg·m/s eastward;

   (b) 290 kg·m/s westward;

   (c) 290 kg·m/s eastward;

   (d) 340 N eastward.

**23.** (a) 5 N·s;

   (b) 80 m/s.

**25.** 0.440-kg ball: 1.27 m/s, east;
   0.220-kg ball: 5.07 m/s, east.

**27.** Tennis ball: 2.50 m/s; other ball: 5.00 m/s; both in direction of tennis ball's initial motion.

**29.** (a) 0.840 kg;

   (b) 0.75.

**31.** (a) 1.7 m/s, in direction of initial incoming velocity;

   (b) 1.2 kg.

**33.** $\sqrt{2}$.

**35.** Vertical: 0.15 m; horizontal: 0.90 m.

**37.** 21 m/s.

**39.** 0.42.

**41.** (a) 12.1 m/s;

   (b) 56.4 J before, 13.7 J after.

**43.** (a) 920 m/s;

   (b) 0.999.

**45.** $1.14 \times 10^{-22}$ kg·m/s, 147° from the electron's momentum, 123° from the neutrino's momentum.

**47.** (a) 30°;

   (b) $v_{\text{nucleus}} = v_{\text{target}} = v/\sqrt{3}$;

   (c) 2/3.

**49.** $6.5 \times 10^{-11}$ m.

**51.** 2.62 m.

**53.** $(1.2\ell,\ 0.9\ell)$ relative to back left corner.

**55.** $0.27R$ to the left of C.

**57.** 19% of the person's height along the line from shoulder to hand.

**59.** 4.3% of their height; no.

**61.** (a) $4.66 \times 10^6$ m from center of Earth.

**63.** (a) 5.8 m;

   (b) 4.0 m;

   (c) 4.2 m.

**65.** 0.45 m toward initial position of 85-kg person.

**67.** $2.0 \times 10^1$ m.

**69.** 8.

**71.** (a) $v'_A = 3.65$ m/s,
   $v'_B = 4.45$ m/s;

   (b) $\Delta p_A = -370$ kg·m/s,
   $\Delta p_B = 370$ kg·m/s.

**73.** 110 km/h ≈ 70 mi/h.

**75.** 340 m/s.

**77.** (a) 8.6 m;

   (b) 38 m.

**79.** (a) $v'_m = 3.98$ m/s, $v'_M = 4.42$ m/s;

   (b) 1.62 m.

**81.** (a) $1.5 \times 10^{21}$ J;

   (b) 38,000.

**83.** (a) No;

   (b) $m_B/m_A$;

   (c) $m_B/m_A$;

   (d) stays at rest.

**85.** $v_m = 3D\sqrt{\dfrac{k}{12m}}$, $v_{3m} = D\sqrt{\dfrac{k}{12m}}$.

## Chapter 8

**1.** (a) 0.785 rad, $\pi/4$ rad;

   (b) 1.05 rad, $\pi/3$ rad;

   (c) 1.57 rad, $\pi/2$ rad;

   (d) 6.28 rad, $2\pi$ rad;

   (e) 7.77 rad, $89\pi/36$ rad.

**3.** $5.3 \times 10^3$ m.

**5.** (a) 750 rad/s;

   (b) 23 m/s;

   (c) $4.5 \times 10^7$ bit/s.

**7.** (a) 230 rad/s;

   (b) $4.0 \times 10^1$ m/s, $9.3 \times 10^3$ m/s².

**9.** (a) 0.105 rad/s;

   (b) $1.75 \times 10^{-3}$ rad/s;

   (c) $1.45 \times 10^{-4}$ rad/s;

   (d) 0.

**11.** (a) 464 m/s;

   (b) 185 m/s;

   (c) 345 m/s.

**13.** $3.3 \times 10^4$ rpm.

**15.** (a) $1.5 \times 10^{-4}$ rad/s²;

   (b) $a_{\text{rad}} = 1.2 \times 10^{-2}$ m/s²,
   $a_{\text{tan}} = 6.2 \times 10^{-4}$ m/s².

**17.** (a) $-96$ rad/s²;

   (b) 98 rev.

**19.** (a) 46 rev/min²;

   (b) 46 rpm.

**21.** 33 m.

**23.** (a) 0.53 rad/s²;

   (b) 13 s.

**25.** 1.2 m·N, clockwise.

**27.** $mg(\ell_2 - \ell_1)$, clockwise.

**29.** (*a*) 14 m·N;
    (*b*) −13 m·N.
**31.** 0.12 kg·m².
**33.** 1.2 × 10⁻¹⁰ m.
**35.** (*a*) 7.8 m·N;
    (*b*) 310 N.
**37.** 22 m·N.
**39.** (*a*) 7.0 kg·m²;
    (*b*) 0.70 kg·m²;
    (*c*) *y* axis.
**41.** 320 m·N; 130 N.
**43.** (*a*) $1.90 \times 10^3$ kg·m²;
    (*b*) $8.9 \times 10^3$ m·N.
**45.** 31 N.
**47.** (*a*) $a_A = 0.69$ m/s², upward;
    $a_B = 0.69$ m/s², downward;
    (*b*) 2%.
**49.** 125 hp.
**51.** 9.70 m/s.
**53.** (*a*) $2.6 \times 10^{29}$ J;
    (*b*) $2.7 \times 10^{33}$ J.
**55.** $1.63 \times 10^4$ J.
**57.** $\sqrt{\frac{10}{7} g (R_0 - r_0)}$.
**59.** 7.27 m/s.
**61.** (*a*) 15 kg·m²/s;
    (*b*) −2.5 m·N.
**63.** $\frac{1}{2}\omega$.
**65.** 1.2 kg·m²; by pulling her arms in toward the center of her body.
**67.** (*a*) 0.52 rad/s;
    (*b*) $KE_{before} = 370$ J,
    $KE_{after} = 2.0 \times 10^2$ J.
**69.** (*a*) 0.43 rad/s;
    (*b*) 0.80 rad/s.
**71.** (*a*) $5 \times 10^{-2}$ rad/s;
    (*b*) $KE_f = 2 \times 10^4$ $KE_i$.
**73.** $(3.2 \times 10^{-16})$%.
**75.** 52 kg.
**77.** $f_{R_1} = 480$ rpm, $f_{R_2} = 210$ rpm.
**79.** 4.50 m/s.
**81.** (*a*) $\omega_R/\omega_F = N_F/N_R$;
    (*b*) 4.0;
    (*c*) 1.5.
**83.** (*a*) 3.5 m;
    (*b*) 4.7 s.
**85.** $\dfrac{Mg\sqrt{2Rh - h^2}}{R - h}$.
**87.** (*a*) 4.84 J;
    (*b*) $F_{4\,kg} = 26.3$ N, $F_{3\,kg} = 19.8$ N.
**89.** (*a*) $3g/2\ell$;
    (*b*) $\frac{3}{2}g$.
**91.** $\ell/2$; $\ell/2$.
**93.** 27 h.

**95.** (*a*) 820 kg·m²/s²;
    (*b*) 820 m·N;
    (*c*) 930 W.

## Chapter 9

**1.** 528 N, 120° clockwise from $\vec{F}_A$.
**3.** (*a*) 2.3 m from vertical support;
    (*b*) 4200 kg.
**5.** (*a*) $F_A = 1.5 \times 10^3$ N, down;
    $F_B = 2.0 \times 10^3$ N, up;
    (*b*) $F_A = 1.8 \times 10^3$ N, down;
    $F_B = 2.6 \times 10^3$ N, up.
**7.** 1200 N.
**9.** $F_{closer} = 2900$ N, down;
    $F_{farther} = 1300$ N, down.
**11.** (*a*) 2.3 m from the adult;
    (*b*) 2.5 m from the adult.
**13.** $F_{left} = 260$ N, $F_{right} = 190$ N.
**15.** 20 N to 50 N.
**17.** 0.64 m to right of fulcrum rock.
**19.** (*a*) 410 N;
    (*b*) horizontal: 410 N;
    vertical: 328 N.
**21.** $F_A = 1.7 \times 10^4$ N,
    $F_B = 7.7 \times 10^3$ N.
**23.** $6.0 \times 10^1$ N; the angle is small.
**25.** (*a*) 230 N;
    (*b*) $1.0 \times 10^2$ N.
**27.** (*a*)

    (*b*) $F_{A_{vertical}} = 9$ N,
    $F_{A_{horizontal}} = 51$ N;
    (*c*) 2.4 m.
**29.** $F_{top_{horizontal}} = 55.2$ N, right;
    $F_{top_{vertical}} = 63.7$ N, up;
    $F_{bottom_{horizontal}} = 55.2$ N, left;
    $F_{bottom_{vertical}} = 63.7$ N, up;
**31.** 7.0 kg.
**33.** 2.4*w*.
**35.** 1600 N.
**37.** 1800 N.
**39.** (*b*) Yes, by $\frac{1}{24}$ of a brick length;
    (*c*) $D = \sum_{i=1}^{n} \dfrac{\ell}{2i}$;
    (*d*) 35 bricks;
**41.** (*a*) $1.8 \times 10^5$ N/m²;
    (*b*) $3.5 \times 10^{-6}$.

**43.** (*a*) $1.4 \times 10^6$ N/m²;
    (*b*) $6.9 \times 10^{-6}$;
    (*c*) $6.6 \times 10^{-5}$ m.
**45.** $9.0 \times 10^7$ N/m², $9.0 \times 10^2$ atm.
**47.** 25 kg.
**49.** $1.7 \times 10^{-2}$ J.
**51.** (*a*) 393 N;
    (*b*) thicker.
**53.** (*a*) $3.7 \times 10^{-5}$ m²;
    (*b*) $2.7 \times 10^{-3}$ m.
**55.** 1.3 cm.
**57.** 12 m.
**59.** (*a*) $F_{left} = 310$ N, up;
    $F_{right} = 210$ N, down;
    (*b*) 0.65 m from right hand;
    (*c*) 1.2 m from right hand.
**61.** $2.9 \times 10^9$ m·N, clockwise; no.
**63.** (*a*) 0.78 N;
    (*b*) 0.98 N.
**65.** $3.5 \times 10^{-4}$ m.
**67.** A: 230 N; B: 110 N.
**69.** 2.51 m.
**71.** 45°.
**73.** (*a*) 2100 N;
    (*b*) 1.3.
**75.** 2500 N, no.
**77.** $2.6 \times 10^{-4}$ m².
**79.** (*a*) 1.6 m;
    (*b*) $1.2 \times 10^4$ N, no.
**81.** $(4.0 \times 10^1)$°.

## Chapter 10

**1.** $3 \times 10^{11}$ kg.
**3.** 710 kg.
**5.** 0.8547.
**7.** (*a*) 5501 kg/m³.
    (*b*) 5497 kg/m³, −0.07%.
**9.** (*a*) $6.1 \times 10^6$ N/m²;
    (*b*) $1.7 \times 10^5$ N/m².
**11.** (*a*) $4.5 \times 10^5$ N;
    (*b*) $4.5 \times 10^5$ N.
**13.** 1.2 m.
**15.** 1900 kg.
**17.** (*a*) $7.0 \times 10^5$ N/m²;
    (*b*) 72 m.
**19.** $1.60 \times 10^4$ m.
**21.** $4.0 \times 10^7$ N/m².
**23.** 0.57.
**25.** (*a*) $1.5 \times 10^5$ N;
    (*b*) $1.8 \times 10^5$ N.
**27.** Iron or steel.
**29.** $9.9 \times 10^{-3}$ m³.

31. 10.5%.

33. 32 bottles.

35. 0.88.

37. (a) $6.68 \times 10^{-2} \, m^3$;

(b) 1.07;

(c) 12%.

39. 9 N, down, 21 N, up.

41. 4.4 m/s.

43. 9.6 m/s.

45. $4.12 \times 10^{-3} \, m^3/s$.

47. $1.6 \times 10^5 \, N/m^2$.

49. $1.2 \times 10^5 \, N$.

51. $9.7 \times 10^4 \, Pa$.

53. 2.5 m/s, 2.2 atm.

57. 1100 N.

59. $8.2 \times 10^3 \, Pa$.

61. 0.094 m.

63. (a) Laminar;

(b) 2940, turbulent.

65. 0.89 Pa/cm.

67. $2.4 \times 10^{-2} \, N/m$.

69. (a) $\gamma = F/4\pi r$;

(b) $1.7 \times 10^{-2} \, N/m$.

71. 1.5 mm.

73. (a) $7.6 \times 10^{-4} \, N$;

(b) 1.3 N.

75. $F_{T_{string}} = 0.71 \, N$, 984.2 g.

77. (a) $1.0 \times 10^{-3} \, m^2$;

(b) $4.0 \times 10^3 \, J$;

(c) $5.3 \times 10^{-3} \, m$;

(d) 80 strokes.

79. 0.6 atm.

81. 1.0 m.

83. $2 \times 10^7 \, Pa$.

85. $1.89 \times 10^4 \, m^3$.

87. $5.29 \times 10^{18} \, kg$.

89. (a) 7.9 m/s;

(b) 0.22 L/s;

(c) 0.79 m/s.

91. 130 N.

93. $1.2 \times 10^4 \, N/m^2$.

95. $3.5 \times 10^{-3} \, Pa \cdot s$.

97. 0.27 kg.

99. 68%.

## Chapter 11

1. 0.84 m.

3. 560 N/m.

5. (a) 650 N/m;

(b) 2.1 cm, 2.6 Hz.

7. 0.85 kg.

9. (a) $A_A = 2.5 \, m$, $A_B = 3.5 \, m$;

(b) $f_A = 0.25 \, Hz$, $f_B = 0.50 \, Hz$;

(c) $T_A = 4.0 \, s$, $T_B = 2.0 \, s$.

11. $\pm 70.7\%$ of the amplitude.

13. 0.233 s.

15. (a) 2.1 m/s;

(b) 1.5 m/s;

(c) 0.54 J;

(d) $x = (0.15 \, m) \cos(4.4\pi t)$.

17. $\sqrt{3} : 1$.

19. (a) 430 N/m;

(b) 4.6 kg.

21. (a) 0.436 s, 2.29 Hz;

(b) 0.157 m;

(c) $32.6 \, m/s^2$;

(d) 2.26 J;

(e) 1.90 J.

23. 68.0 N/m, 15.6 m.

25. (a) $y = (0.16 \, m) \cos(14 \, t)$;

(b) 0.11 s;

(c) 2.2 m/s;

(d) $31 \, m/s^2$, at the release point.

27. 3.0 s.

29. (a) 1.8 s;

(b) 0.56 Hz.

31. Shorten it by 0.5 mm.

33. $\frac{1}{3}$.

35. 2.3 m/s.

37. (a) 1400 m/s;

(b) 4100 m/s;

(c) 5100 m/s.

39. (a) 1400 km;

(b) No; need readings from at least two other stations.

41. 4.8 N.

43. 21 m.

45. (a) $8.7 \times 10^9 \, J/m^2 \cdot s$;

(b) $1.7 \times 10^{10} \, W$.

47. $\sqrt{5} : 1$.

49. 440 Hz, 880 Hz, 1320 Hz, 1760 Hz.

51. 60 Hz, fundamental or first harmonic; 120 Hz, first overtone or second harmonic; 180 Hz, second overtone or third harmonic.

53. 70 Hz.

55. (a) 1.2 kg;

(b) 0.29 kg;

(c) $4.6 \times 10^{-2} \, kg$.

57. 1.3 m/s.

59. 24°.

61. 460 Hz; $f < 460$ Hz.

63. 0.11 m.

65. $Mg/k$.

67. (a) $1.16 \, f$;

(b) $0.81 \, f$.

69. $2.6 \times 10^{13} \, Hz$.

71. (a) 1.2 Hz;

(b) 12 J.

73. (a) $3.7 \times 10^4 \, N/m$.

(b) 0.50 s.

75. $8.40 \times 10^2 \, N/m$.

77. 0.13 m/s, $0.12 \, m/s^2$; 1.2%.

79. (a) 0.06 m;

(b) 7.1.

81. (a) G: 784 Hz, 1180 Hz; B: 988 Hz, 1480 Hz;

(b) $1.59 : 1$;

(c) $1.26 : 1$;

(d) $0.630 : 1$.

83. $2.30 \times 10^2 \, Hz$.

85. 18 W.

87. (a) $\theta_{iM} = \sin^{-1}(v_{air}/v_{water})$
$= \sin^{-1}(v_i/v_r)$;

(b) 0.44 m.

## Chapter 12

1. 430 m.

3. (a) 1.7 cm to 17 m;

(b) $1.9 \times 10^{-5} \, m$.

5. (a) 0.994 s;

(b) 4.52 s.

7. 33 m.

9. 62 dB.

11. 82 dB.

13. 82-dB player: $1.6 \times 10^8$; 98-dB player: $6.3 \times 10^9$.

15. (a) 790 W;

(b) 440 m.

17. (a) 12;

(b) 11 dB.

19. 130 dB.

21. (a) 220-W: 122 dB; 45-W: 115 dB;

(b) no.

23. 80 Hz, 15,000 Hz.

25. 10 octaves.

27. 87 N.

29. (a) 360 Hz;

(b) 540 Hz.

31. 260 Hz.

33. (a) 17 cm;

(b) 1.02 m;

(c) $f = 440$ Hz, $\lambda = 0.78$ m.

35. 1.9%.

37. (a) 0.585 m;
    (b) 858 Hz.
39. (a) 55 Hz;
    (b) 190 m/s.
41. (a) 253 overtones;
    (b) 253 overtones.
43. 4.2 cm, 8.2 cm, 11.9 cm, 15.5 cm, 18.8 cm, 22.0 cm.
45. $I_2/I_1 = 0.64$;  $I_3/I_1 = 0.20$;
    $\beta_{2-1} = -2$ dB;  $\beta_{3-1} = -7$ dB.
47. 28.5 kHz.
49. 347 Hz.
51. (a) 0.562 m;
    (b) 0.
53. (a) 343 Hz;
    (b) 1000 Hz, 1700 Hz.
55. (a) 8.9 beats per second;
    (b) 38 m.
57. $4.27 \times 10^4$ Hz.
59. $3.11 \times 10^4$ Hz.
61. (a) Every 1.4 s;
    (b) every 11 s.
63. 0.0821 m/s.
65. 11 km/h.
67. (a) 99;
    (b) 0.58°.
69. (a) 36°;
    (b) 560 m/s, 1.7.
71. 0.12 s.
73. 88 dB.
75. 14 W.
77. (a) 51 dB;
    (b) $5 \times 10^{-9}$ W.
79. 635 Hz.
81. $\mu_{2nd} = 0.44\mu_{lowest}$,
    $\mu_{3rd} = 0.20\mu_{lowest}$,
    $\mu_{4th} = 0.088\mu_{lowest}$.
83. 150 Hz, 460 Hz, 770 Hz, 1100 Hz.
85. 2.35 m/s.
87. 2.62 m.
89. 11.5 m.
91. $\frac{1}{1000}$.
93. 36 Hz, 48 Hz, 61 Hz.
95. $10^6$.

## Chapter 13

1. $N_{gold} = 0.548N_{silver}$.
3. (a) 20°C;
   (b) 3500°F.
5. 102.0°F.
7. $-40°C = -40°F$.
9. 0.08 m.

11. $2.2 \times 10^{-6}$ m; $\frac{1}{60}$ of the change for steel.
13. $-70°C$.
15. 0.98%.
17. $-210°C$.
19. $4.0 \times 10^7$ N/m².
21. $-459.67°F$.
23. 1.25 m³.
25. (a) 0.2754 m³;
    (b) $-63°C$.
27. (a) 22.8 m³;
    (b) $2.16 \times 10^5$ Pa.
29. $1.69 \times 10^8$ Pa.
31. 7.4%.
33. 33%.
35. Actual: 0.598 kg/m³, ideal: 0.588 kg/m³; near a phase change.
37. 1.07 cm.
39. 55.51 mol, $3.343 \times 10^{25}$ molecules.
41. 300 molecules/cm³.
43. (a) $5.65 \times 10^{-21}$ J;
    (b) 3700 J.
45. 1.22.
47. $3.5 \times 10^{-9}$ m/s.
49. $\sqrt{3}$.
53. $\dfrac{(v_{rms})_{^{235}UF_6}}{(v_{rms})_{^{238}UF_6}} = 1.004$.
55. Vapor.
57. (a) Vapor;
    (b) solid.
59. 3200 Pa.
61. 18°C.
63. 0.91 kg.
65. 2.5 kg.
67. (a) Greater than;
    (b) $(-2.0 \times 10^{-4})$%;
    (c) 0.603%.
69. (b) $4 \times 10^{-11}$ mol/s;
    (c) 0.6 s.
71. (a) Low;
    (b) $(2.8 \times 10^{-2})$%.
73. 18%.
75. (a) 1500 kg;
    (b) 200 kg enters.
77. (a) Lower;
    (b) 0.36%.
79. 910 min.
81. (a) $0.66 \times 10^3$ kg/m³;
    (b) $-3.0$%.
83. 2300 m.

85. (a) 290 m/s;
    (b) 9.5 m/s.
87. PE/KE = $8.50 \times 10^{-5}$,  yes.
89. 0.30 kg.
91. 2.4 kg.

## Chapter 14

1. 10.7°C.
3. 0.04 candy bars.
7. 250 kg/h.
9. $6.0 \times 10^6$ J.
11. (a) $3.3 \times 10^5$ J;
    (b) 5600 s.
13. $4.0 \times 10^2$ s.
15. 42.6°C.
17. $2.3 \times 10^3$ J/kg·C°.
19. 43 C°.
21. 0.39 C°.
23. 473 kcal.
25. $7.1 \times 10^6$ J.
27. 0.18 kg.
29. (a) $5.2 \times 10^5$ J;
    (b) $1.5 \times 10^5$ J.
31. 11.2 kJ/kg.
33. 2.7 g.
35. 5.2 g.
37. 93 J/s = 93 W.
39. $7.5 \times 10^4$ s.
41. 20 bulbs.
43. $3.1 \times 10^4$ s.
45. 350 Btu/h.
47. (a) $3.2 \times 10^{26}$ W;
    (b) $1.1 \times 10^3$ W/m².
49. A mixture of $\frac{1}{3}$ steam and $\frac{2}{3}$ liquid water at 100°C.
51. 2 C°.
53. $6.6 \times 10^3$ kcal.
55. $4.0 \times 10^2$ m/s.
57. 450°C.
59. 0.14 C°.
61. $1.43 \times 10^3$ m/s, toward the Earth.
63. 19 min.
65. (a) 3.4 W;
    (b) 2.3 C°/s;
    (c) no, $T > 8000$°C in less than an hour;
    (d) 86°C;
    (e) conduction, convection, evaporation.
67. (a) $3.6 \times 10^7$ J;
    (b) 63 min.

# Chapter 15

**1.** (a) 0;
   (b) $4.30 \times 10^3$ J.

**3.**

P (atm)

**5.**

P (atm)

**7.** (a) 0;
   (b) 2630 J;
   (c) rise.

**9.** −78 K.

**13.** (a) −196 J;
   (b) 28 J;
   (c) 168 J;
   (d) 140 J;
   (e) 98 J.

**15.** 170 W.

**17.** (a) $1.4 \times 10^7$ J;
   (b) 3500 Cal.

**19.** 25.8%.

**21.** 8.8%.

**23.** 10°C decrease in the low-temperature reservoir.

**25.** $1.7 \times 10^{13}$ J/h.

**27.** 1800 W.

**29.** 420°C.

**31.** 0.15.

**33.** 6.5.

**35.** (a) $1.0 \times 10^3$ J;
   (b) $1.0 \times 10^3$ J;
   (c) 230 J at 0°C, 390 J at −15°C.

**37.** 78 L.

**39.** $-1.9 \times 10^3$ J/K.

**41.** $-1.22 \times 10^6$ J/K.

**43.** $4 \times 10^4$ J/K.

**45.** 0.64 J/K.

**47.** 1.1 J/K.

**49.** (a) $\frac{1}{12}$;
   (b) $\frac{1}{12}$.

**51.** (a)

(r = red, o = orange, g = green)

| Macrostate | Microstates | Number of microstates |
|---|---|---|
| 3 r, 0 o, 0 g | r r r | 1 |
| 2 r, 1 o, 0 g | r r o   r o r   o r r | 3 |
| 2 r, 0 o, 1 g | r r g   r g r   g r r | 3 |
| 1 r, 2 o, 0 g | r o o   o r o   o o r | 3 |
| 1 r, 0 o, 2 g | r g g   g r g   g g r | 3 |
| 1 r, 1 o, 1 g | r o g   r g o   o r g  o g r   g r o   g o r | 6 |
| 0 r, 3 o, 0 g | o o o | 1 |
| 0 r, 2 o, 1 g | g o o   o g o   o o g | 3 |
| 0 r, 1 o, 2 g | o g g   g o g   g g o | 3 |
| 0 r, 0 o, 3 g | g g g | 1 |

   (b) $\frac{1}{27}$;
   (c) $\frac{1}{9}$.

**53.** 70 m², yes.

**55.** $1.5 \times 10^7$ W.

**57.** (a) $2.2 \times 10^5$ J;
   (b) $3.6 \times 10^5$ J;
   (c) P (atm)

**59.** 86°C.

**61.** (a) 7.7%;
   (b) the large volume of "fuel" (ocean water) available.

**63.** $0.43/h.

**65.** (a) 44°C;
   (b) $4.3 \times 10^{-2}$ J/K.

**67.** 60 K.

**69.** (a) 13 km³/day, possibly;
   (b) 73 km².

**71.** (a) 0.281;
   (b) $1.01 \times 10^5$ W, $2.1 \times 10^9$ J, $4.9 \times 10^5$ kcal.

**73.** (a) 0.22 kg;
   (b) 4.5 days.

**75.** $4.6 \times 10^6$ J.

**77.** (a) −4°C;
   (b) 29%.

# Chapter 16

**1.** $2.7 \times 10^{-3}$ N.

**3.** $2.2 \times 10^4$ N.

**5.** $(1.9 \times 10^{-13})$%.

**7.** 3.76 cm.

**9.** $-4.6 \times 10^8$ C, 0.

**11.** $F_{\text{left}} = 120$ N, to the left;
   $F_{\text{center}} = 560$ N, to the right;
   $F_{\text{right}} = 450$ N, to the left.

**13.** $2.1 \times 10^{12}$ electrons.

**15.** $10.1 \dfrac{kQ^2}{\ell^2}$, at 61°.

**17.** (a) $88.8 \times 10^{-6}$ C, $1.2 \times 10^{-6}$ C;
   (b) $91.1 \times 10^{-6}$ C, $-1.1 \times 10^{-6}$ C.

**19.** $3.94 \times 10^{-16}$ N, west.

**21.** $6.30 \times 10^6$ N/C, upward.

**23.** $1.33 \times 10^{14}$ m/s², opposite to the field.

**25.**

$+Q$   $-3Q$

**27.** $5.97 \times 10^{-10}$ N/C, south.

**29.** Upper right corner, $E = 3.76 \times 10^4$ N/C, at 45.0°.

**31.** $\dfrac{4kQxa}{(x^2 - a^2)^2}$, to the left.

**33.** $3.7 \times 10^7$ N/C, 330°.

**35.** $E_A = 3.0 \times 10^6$ N/C, at 90°;
   $E_B = 7.8 \times 10^7$ N/C, at 56°; yes.

**37.** (a) $5 \times 10^{-10}$ N;
   (b) $7 \times 10^{-10}$ N;
   (c) $6 \times 10^{-5}$ N.

**39.** (a) $-1.1 \times 10^5$ N·m²/C;
   (b) 0.

**41.** $8.3 \times 10^{-10}$ C.

**43.** (a) $k\dfrac{Q}{r^2}$;
   (b) 0;
   (c) $k\dfrac{Q}{r^2}$;
   (d) The shell causes the field to be 0 in the shell material. The charge polarizes the shell.

**45.** $4.0 \times 10^9$ C.

**47.** $6.8 \times 10^5$ C, negative.

**49.** $1.0 \times 10^7$ electron charges.

**51.** $5.2 \times 10^{-11}$ m.

**53.** 4.3 m.

**55.** 0.14 N, rightward.

**57.** $8.2 \times 10^{-7}$ C, positive.

**59.** (a) $4 \times 10^{10}$ particles;
   (b) $4 \times 10^{-5}$ kg.

**61.** $9.90 \times 10^6$ N/C, downward.

**63.** $x = d\left(\sqrt{2} + 1\right) \approx 2.41d$.

**65.** $QE\ell$, counterclockwise.

**67.** $8.94 \times 10^{-19}$.

## Chapter 17

**1.** $5.0 \times 10^{-4}$ J.

**3.** $-1.0$ V.

**5.** 4030 V, plate B.

**7.** 5.78 V.

**9.** $-4.25 \times 10^4$ V.

**11.** $-157$ V.

**13.** $9.0 \times 10^5$ m/s.

**15.** 3000 V; only a small amount of charge was transferred.

**17.**

**19.** $2.8 \times 10^{-9}$ C.

**21.** (a) $5.8 \times 10^5$ V;
(b) $9.2 \times 10^{-14}$ J.

**23.** $9.15 \times 10^6$ m/s.

**25.** (a) 18 cm from − charge, on opposite side from + charge;
(b) 1.6 cm from − charge, toward + charge, and 8.0 cm from − charge, away from + charge.

**27.** (a) $1.6 \times 10^4$ V;
(b) $9.9 \times 10^4$ V/m, 64°.

**29.** $4.2 \times 10^6$ V.

**31.** (a) 27 V;
(b) $2.2 \times 10^{-18}$ J, or 14 eV;
(c) $-2.2 \times 10^{-18}$ J, or $-14$ eV;
(d) $2.2 \times 10^{-18}$ J, or 14 eV.

**33.** (a) $6.6 \times 10^{-3}$ V;
(b) $4.6 \times 10^{-3}$ V;
(c) $-4.6 \times 10^{-3}$ V.

**35.** $2.6 \times 10^{-6}$ F.

**37.** $6.00 \times 10^{-5}$ C.

**39.** $6.3 \times 10^{-7}$ F.

**41.** $0.24$ m².

**43.** $9 \times 10^{-16}$ m, no.

**45.** $V_{2.50\,\mu\text{F}} = V_{6.80\,\mu\text{F}} = 611$ V,
$Q_{2.50\,\mu\text{F}} = 1.53 \times 10^{-3}$ C,
$Q_{6.80\,\mu\text{F}} = 4.16 \times 10^{-3}$ C.

**47.** $4.7 \times 10^{-11}$ F.

**49.** 9.5 V.

**51.** $4.20 \times 10^{-9}$ F, 0.247 m².

**53.** $9.6 \times 10^{-5}$ F.

**55.** (a) $9 \times 10^{-12}$ F;
(b) $8 \times 10^{-11}$ C;
(c) 200 V/m;
(d) $4 \times 10^{-10}$ J;
(e) capacitance, charge, work done.

**57.** $1.0 \times 10^{-7}$ J/m³.

**59.** 1110100.

**61.** 43,690.

**63.** (a) 65,536;
(b) 16,777,216;
(c) 16,777,216.

**65.** (b) 56 Hz.

**67.** $+2.0 \times 10^5$ V/m to $-2.0 \times 10^5$ V/m.

**69.** Yes, $1.3 \times 10^{-12}$ V.

**71.** (a) Multiplied by 2;
(b) multiplied by 2.

**73.** Alpha particle, 2.

**75.** Left: $-6.85kQ/\ell$, top: $-3.46kQ/\ell$, right: $-5.15kQ/\ell$.

**77.** (a) 17 cm from − charge, on opposite side from + charge;
(b) 1.1 cm from − charge, toward + charge, and 8.1 cm from − charge, away from + charge.

**79.** (a) 31 J;
(b) $5.9 \times 10^5$ W.

**81.** 1.8 J.

**83.** $3.7 \times 10^{-10}$ C.

**85.** (a) $6.4 \times 10^{-11}$ C;
(b) $6.4 \times 10^{-11}$ C;
(c) 18 V;
(d) $2 \times 10^{-10}$ J.

**87.** (a) $3.6 \times 10^3$ m/s;
(b) $2.8 \times 10^3$ m/s.

**89.** $1.7 \times 10^6$ V.

**91.** $1.3 \times 10^{-6}$ C.

**93.** 16°.

**95.** (a) 0.32 $\mu$m²;
(b) 59 megabytes.

## Chapter 18

**1.** $1.00 \times 10^{19}$ electrons/s.

**3.** $6.2 \times 10^{-11}$ A.

**5.** 1200 V.

**7.** (a) 28 A;
(b) $8.4 \times 10^4$ C.

**9.** (a) 8.9 $\Omega$;
(b) $1.2 \times 10^4$ C.

**11.** (a) 4.8 A;
(b) 6.6 A.

**13.** $5.1 \times 10^{-2}$ $\Omega$.

**15.** Yes, for length 4.0 mm.

**17.** 2.0 V.

**19.** (a) $3.8 \times 10^{-4}$ $\Omega$;
(b) $1.5 \times 10^{-3}$ $\Omega$;
(c) $6.0 \times 10^{-3}$ $\Omega$.

**21.** 18C°.

**23.** 2400°C.

**25.** $R_{\text{carbon}} = 1.42$ k$\Omega$,
$R_{\text{Nichrome}} = 1.78$ k$\Omega$.

**27.** 0.72 W.

**29.** 31 V.

**31.** $1.7 \times 10^5$ C.

**33.** (a) 950 W;
(b) 15 $\Omega$;
(c) 9.9 $\Omega$.

**35.** (a) 1.1 A;
(b) 110 $\Omega$.

**37.** 0.046 kWh; 6.6 cents per month.

**39.** $2.8 \times 10^6$ J.

**41.** 24 bulbs.

**43.** 1.5 m; power increases 36× and could start a fire.

**45.** (a) 7.2 A;
(b) 1.7 $\Omega$.

**47.** 0.12 A.

**49.** (a) Infinite resistance;
(b) 96 $\Omega$.

**51.** (a) 930 V;
(b) 3.9 A.

**53.** (a) 3300 W;
(b) 9.7 A.

**55.** $6.0 \times 10^{-10}$ m/s.

**57.** 2.2 A/m², north.

**59.** 32 m/s (possible delay between nerve stimulation and generation of action potential).

**61.** 9.8 h.

**63.** 6.22 A.

**65.** $2.4 \times 10^{-4}$ m.

**67.** \$3200 per hour per meter.

**69.** $4.2 \times 10^{-3}$ m.

**71.** (a) 33 Hz;
(b) 0.990 A;
(c) $V = (33.6 \sin 210t)$ V.

**73.** 2.25 $\Omega$.

**75.** (b) As large as possible.

**77.** (a) 7.4 hp;
(b) 220 km.

**79.** $1.7 \times 10^{-4}$ m.

**81.** 32% increase.

**83.** (*a*) $I_A = 0.33$ A, $I_B = 3.3$ A;
 (*b*) $R_A = 360\ \Omega$, $R_B = 3.6\ \Omega$;
 (*c*) $Q_A = 1.2 \times 10^3$ C,
  $Q_B = 1.2 \times 10^4$ C;
 (*d*) $E_A = E_B = 1.4 \times 10^5$ J;
 (*e*) Bulb B.

**85.** (*a*) $4 \times 10^6$ J;
 (*b*) $2 \times 10^4$ m.

**87.** (*a*) 12 W;
 (*b*) 4.6 W.

**89.** $1.34 \times 10^{-4}\ \Omega$.

**91.** $f = 1 - \dfrac{V}{V_0}$.

# Chapter 19

**1.** (*a*) 5.92 V;
 (*b*) 5.99 V.

**3.** $0.034\ \Omega$; $0.093\ \Omega$.

**5.** (*a*) $330\ \Omega$;
 (*b*) $8.9\ \Omega$.

**7.** 2.

**9.** Connect 18 resistors in series; then measure voltage across 7 consecutive series resistors.

**11.** $0.3\ \Omega$.

**13.** $560\ \Omega$, 0.020.

**15.** $32\ \Omega$.

**17.** $140\ \Omega$.

**19.** $\frac{13}{8} R$.

**21.** $4.8\ \text{k}\Omega$.

**23.** 55 V.

**25.** 0.35 A.

**27.** 0.

**29.** (*a*) 34 V;
 (*b*) 85-V battery: 82 V;
  45-V battery: 43 V.

**31.** $I_1 = 0.68$ A, left; $I_2 = 0.33$ A, left.

**33.** (*a*) $\mathscr{E}/R$;
 (*b*) $R$.

**35.** 0.71 A.

**37.** 3 parallel sets, each with 100 cells in series.

**39.** $3.71 \times 10^{-6}$ F.

**41.** $2.0 \times 10^{-9}$ F, yes.

**43.** $1.90 \times 10^{-8}$ F in parallel,
 $1.7 \times 10^{-9}$ F in series.

**45.** 2:1.

**47.** In parallel, 750 pF.

**49.** $29.3\ \mu$F, $5.7\ \mu$F.

**51.** (*a*) $\frac{3}{5} C$;
 (*b*) $Q_1 = Q_2 = \frac{1}{5} CV$, $Q_3 = \frac{2}{5} CV$,
  $Q_4 = \frac{3}{5} CV$; $V_1 = V_2 = \frac{1}{5} V$,
  $V_3 = \frac{2}{5} V$, $V_4 = \frac{3}{5} V$.

**53.** $1.0 \times 10^6\ \Omega$.

**55.** $7.4 \times 10^{-3}$ s.

**57.** (*a*) $I_1 = \dfrac{2\mathscr{E}}{3R}$, $I_2 = I_3 = \dfrac{\mathscr{E}}{3R}$

 (*b*) $I_1 = I_2 = \dfrac{\mathscr{E}}{2R}$, $I_3 = 0$;

 (*c*) $\frac{1}{2} \mathscr{E}$.

**59.** (*a*) $2.9 \times 10^{-5}$ A;
 (*b*) $8.8 \times 10^6\ \Omega$.

**61.** Add $710\ \Omega$ in series with ammeter, $29\ \Omega/$V.

**63.** $9.60 \times 10^{-4}$ A, 4.8 V;
 current: $+20\%$, voltage: $-20\%$.

**65.** 9.8 V.

**67.** Put $9.0\ \text{k}\Omega$ in series with the body.

**69.** $\frac{1}{4} C$, $\frac{2}{5} C$, $\frac{3}{5} C$, $\frac{3}{4} C$, $C$, $\frac{4}{3} C$, $\frac{5}{3} C$,
 $\frac{5}{2} C$, $4C$.

**71.** $9.2 \times 10^4\ \Omega$.

**73.** (*a*) $3.6\ \Omega$;
 (*b*) 14 W.

**77.** 600 cells; $0.54\ \text{m}^2$, 4 banks in parallel, each containing 150 cells in series.

**79.** (*a*) $6.0\ \Omega$;
 (*b*) 2.2 V.

**81.** 11 V.

**83.** $100\ \Omega$.

**87.** $9.0\ \Omega$.

**89.** $Q_{12\ \mu\text{F}} = 1.0 \times 10^{-4}$ C,
 $Q_{48\ \mu\text{F}} = 4.1 \times 10^{-4}$ C.

**91.** (*a*) $1.9 \times 10^{-4}$ J;
 (*b*) $4.0 \times 10^{-5}$ J;
 (*c*) $Q_a = 16\ \mu$C; $Q_b = 3.3\ \mu$C.

**93.** $Q_1 = \dfrac{C_1 C_2}{C_2 + C_1} V_0$,
 $Q_2 = \dfrac{C_2^2}{C_2 + C_1} V$.

**95.** (*a*) In parallel;
 (*b*) 7.7 pF to 35 pF.

**97.** $Q_1 = 11\ \mu$C, $V_1 = 11$ V;
 $Q_2 = 13\ \mu$C, $V_2 = 6.3$ V;
 $Q_3 = 13\ \mu$C, $V_3 = 5.2$ V.

# Chapter 20

**1.** (*a*) 5.8 N/m;
 (*b*) 3.3 N/m.

**3.** 1.3 N.

**5.** $27°$.

**7.** (*a*) South pole;
 (*b*) 3.86 A;
 (*c*) $8.50 \times 10^{-2}$ N.

**9.** $5.6 \times 10^{-14}$ N, north.

**11.** 0.24 T.

**13.** (*a*) To the right;
 (*b*) downward;
 (*c*) into the page.

**15.** (*a*) $6.0 \times 10^5$ m/s;
 (*b*) $3.6 \times 10^{-2}$ m;
 (*c*) $3.8 \times 10^{-7}$ s.

**17.** 0.59 m.

**19.** $r_{\text{proton}}/r_{\text{electron}} = 42.8$.

**21.** $1.97 \times 10^{-6}$ m.

**23.** (*a*) Sign determines polarity but not magnitude of Hall emf.
 (*b*) 0.56 m/s.

**25.** $2.9 \times 10^{-4}$ T, about 5.8 times larger.

**27.** $7.8 \times 10^{-2}$ N, toward other wire.

**29.**

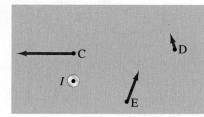

**31.** $5.1 \times 10^{-6}$ N, toward wire.

**33.** $3.8 \times 10^{-5}$ T, $17°$ below horizontal.

**35.** (*a*) $(2.0 \times 10^{-5}\ \text{T/A})(I - 25\ \text{A})$;
 (*b*) $(2.0 \times 10^{-5}\ \text{T/A})(I + 25\ \text{A})$.

**37.**

**39.** 15 A, downward.

**41.** Closer wire: $4.4 \times 10^{-2}$ N/m, attract; farther wire: $2.2 \times 10^{-2}$ N/m, repel.

**43.** $4.66 \times 10^{-5}$ T.

**45.** 1.19 A.

**47.** 0.12 N, south.

**49.** (c) No; inversely as distance from center of toroid: $B \propto 1/R$.

**51.** 1.18 T.

**53.** 69.7 $\mu$A.

**55.** $1.87 \times 10^{6}$ V/m; perpendicular to velocity and magnetic field, and in opposite direction to magnetic force on protons.

**57.** $1.3 \times 10^{-3}$ m; $6.5 \times 10^{-4}$ m.

**59.** $_1^2$H nucleus or $_2^4$He nucleus.

**61.** 0.5 T.

**63.**

**65.** $2.7 \times 10^{-2}$ T, upward.

**67.** 0.30 N, northerly, 68° above horizontal.

**69.** $7.7 \times 10^{-6}$ N.

**71.** $1.1 \times 10^{-6}$ m/s, west.

**73.** $\dfrac{2\,\mu_0 I}{\pi \ell}$, to the left.

**75.** They will exit above or below second tube; 52°.

**77.** $-2.1 \times 10^{-20}$ J.

**79.** $r = 5.3 \times 10^{-5}$ m, $p = 3.3 \times 10^{-4}$ m.

**81.** $1.9 \times 10^{-3}$ T.

**83.** $\dfrac{5.0mg}{\ell B}$, to the left.

**85.** (a) Negative; (b) $\dfrac{qB_0\left(d^2 + \ell^2\right)}{2d}$.

**87.** 1.2 A; downward.

**89.** (a) M: $5.8 \times 10^{-4}$ N/m, upward; N: $3.4 \times 10^{-4}$ N/m, at 300°; P: $3.4 \times 10^{-4}$ N/m, at 240°; (b) $1.75 \times 10^{-4}$ T, at $-14°$.

**91.** $2.9 \times 10^{-6}$ T, $B_{\text{wire}} \approx 0.06 B_{\text{Earth}}$.

## Chapter 21

**1.** 560 V.

**3.** Counterclockwise.

**5.** 0.20 V.

**7.** (a) $1.0 \times 10^{-2}$ Wb; (b) 48°; (c) $6.7 \times 10^{-3}$ Wb.

**9.** (a) $1.73 \times 10^{-2}$ V; (b) 0.114 V/m, downward.

**11.** (a) 0; (b) clockwise; (c) counterclockwise; (d) clockwise.

**13.** 0.65 mV, east or west.

**15.** (a) Magnetic force on current in moving bar, $B^2\ell^2 v/R$; (b) $B^2\ell^2 v^2/R$.

**17.** (a) 0.17 V; (b) $7.1 \times 10^{-3}$ A; (c) $7.5 \times 10^{-4}$ N, to the right.

**19.** 5.23 C.

**21.** (a) 810 V; (b) double the rotation frequency.

**23.** 17 rotations per second.

**25.** 92 V.

**27.** $1.71 \times 10^{4}$ turns.

**29.** $I_S = 0.21 I_P$.

**31.** (a) 6.2 V; (b) step-down.

**33.** 450 V, 56 A.

**35.** $6 \times 10^{9}$ m.

**37.** 55 MW.

**39.** 6.9 V.

**41.** 0.10 H.

**43.** (a) $1.5 \times 10^{-2}$ H; (b) 75 turns.

**45.** 46 m, 21 km; 0.70 kΩ.

**47.** 23 J.

**49.** $5 \times 10^{15}$ J.

**51.** 3.7.

**53.** (a) 2.3; (b) 4.6; (c) 6.9.

**55.** 3300 Hz.

**57.** $1.6 \times 10^{4}$ Ω, $1.47 \times 10^{-2}$ A.

**59.** (a) 7400 Ω; (b) 0.38 A.

**61.** (a) $3.6 \times 10^{4}$ Ω; (b) $3.7 \times 10^{4}$ Ω.

**63.** 205 Ω.

**65.** 270 Hz; the voltages are out of phase.

**67.** (a) $1.77 \times 10^{-2}$ A; (b) $-12.6°$; (c) $V_R = 117$ V, $V_C = 26.1$ V.

**69.** $3.6 \times 10^{5}$ Hz.

**71.** (a) $1.3 \times 10^{-7}$ F; (b) 37 A.

**73.** (a) 0.032 H; (b) 0.032 A; (c) 16 $\mu$J.

**75.** $6.01 \times 10^{-3}$ J.

**77.** Coil radius = 1.5 cm, 10,000 turns.

**79.** 200 kV.

**81.** (a) 41 kV; (b) 31 MW; (c) 1.0 MW; (d) 30 MW.

**85.** Put a 98-mH inductor in series with it.

**87.** 82 V.

**89.** 93 mH.

**91.** 2.

## Chapter 22

**1.** $1.1 \times 10^{5}$ V/m/s.

**3.** $1.7 \times 10^{15}$ V/m/s.

**5.** $2.4 \times 10^{-13}$ T.

**7.** 90.0 kHz, 2.33 V/m, along the horizontal north–south line.

**9.** 1.25 s.

**11.** $4.20 \times 10^{-7}$ m, violet visible light.

**13.** $2.00 \times 10^{10}$ Hz.

**15.** (a) $1.319 \times 10^{-2}$ m; (b) $2.5 \times 10^{18}$ Hz.

**17.** $4.0 \times 10^{16}$ m.

**19.** 9600 wavelengths; $3.54 \times 10^{-15}$ s.

**21.** $1.6 \times 10^{6}$ revolutions/s.

**23.** $3.02 \times 10^{7}$ s.

**25.** 5.7 W/m², 46 V/m.

**27.** $4.51 \times 10^{-6}$ J.

**29.** $3.80 \times 10^{26}$ W.

**31.** $7.3 \times 10^{-7}$ N/m²; $7.3 \times 10^{-11}$ N away from bulb.

**33.** 400 m².

**35.** 0.16 m.

**37.** Channel 2: 5.56 m, Channel 51: 0.434 m.

**39.** $1.5 \times 10^{-12}$ F.

**41.** (a) $1.3 \times 10^{-6}$ H; (b) $9.9 \times 10^{-11}$ F.

**43.** $6.25 \times 10^{-4}$ V/m; $1.04 \times 10^{-9}$ W/m².

**45.** 10 ns.

**47.** 1.36 s; inside.

**49.** $5.00 \times 10^{2}$ s.

**51.** 13 V/m, $4.4 \times 10^{-8}$ T.

**53.** (a) $1.2 \times 10^{-10}$ J; (b) $8.7 \times 10^{-6}$ V/m, $2.9 \times 10^{-14}$ T.

**55.** 61 km.

**57.** (a) $2.8 \times 10^{-3}$ J/s;

    (b) 1.0 V/m;

    (c) 1.0 V;

    (d) $2.0 \times 10^{-2}$ V.

**59.** $6 \times 10^{10}$ W.

**61.** 35 kW.

# Chapter 23

**1.**

**3.** 11°.

**5.**

**7.** $4.0 \times 10^{-6}$ m$^2$.

**9.** 10.5 cm.

**11.** (a) $d_i \approx -5$ cm;

    (b) $d_i = -5.3$ cm;

    (c) 1.0 mm.

**13.** $-6.8$ m.

**15.** 5.7 m.

**17.** 2.0 cm behind ball's front surface; virtual; upright.

**19.** 1.0 m.

**21.** (a) Concave;

    (b) upright, virtual, and magnified;

    (c) 1.40 m.

**23.**

**25.** 1.31.

**27.** 1.62.

**29.** 50.1°.

**31.** 38.6°.

**33.** 81.9°.

**35.** 61.0°, crown glass.

**37.** At least 93.5 cm away.

**39.** (a) 1.4;

    (b) no;

    (c) 1.9.

**41.** (a) ~500 mm;

    (b) 478 mm.

**43.** (a) 3.08 D, converging;

    (b) $-0.148$ m, diverging.

**45.** (a) 106 mm;

    (b) 109 mm;

    (c) 117 mm;

    (d) 513 mm.

**47.** (a) 37 cm behind lens;

    (b) $+2.3\times$.

**49.** $d_i = -6.67$ cm behind lens, virtual and upright, $h_i = 0.534$ mm.

**51.** (a) 70.0 mm;

    (b) 30.0 mm.

**53.** 64 cm.

**55.** 21.3 cm or 64.7 cm from object.

**57.** 0.105 m; 5.8 m.

**59.** 18.5 cm beyond second lens; $-0.651\times$ (inverted).

**61.** (a) 10 cm beyond second lens;

    (b) $-1.0\times$;

    (c)

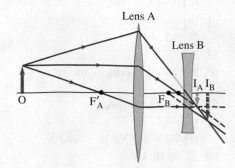

**63.** $-29.7$ cm.

**65.** 9.0 cm.

**67.** 0.34 m.

**69.** 1.25 s.

**71.** 6.04 m.

**73.** 1.58, light flint glass.

**75.** (a) Convex;

    (b) 25 cm behind mirror;

    (c) $-110$ cm;

    (d) $-220$ cm.

**77.** 67°.

**79.** 9 cm, 12 cm.

**81.** $n \geq 1.60$.

**83.** (a) $-0.33$ mm;

    (b) $-0.47$ mm;

    (c) $-0.98$ mm.

**85.** Left: converging; right: diverging.

**89.** 5.7 cm from object, between it and lens.

# Chapter 24

**1.** $5.4 \times 10^{-7}$ m.

**3.** $6.3 \times 10^{-7}$ m, $4.8 \times 10^{14}$ Hz.

**5.** 17° and 64°.

**7.** $1.5 \times 10^{-4}$ m.

**9.** 3.1 cm.

**11.**

**13.** $2.2 \times 10^{-3}$ m.

**15.** 14 mm.

**17.** 570 nm.

**19.** 0.23°.

**21.** 1.8°.

**23.** $1.0 \times 10^{-6}$ m.

**25.** $7.38 \times 10^{-2}$ m.

**27.** (a) $\lambda$;

    (b) 400 nm.

**29.** $3.4 \times 10^{-6}$ m.

**31.** Entire pattern is shifted, with central maximum at 28.0° to the normal.

**33.** 330 nm.

**35.** 556 nm.

**37.** 490 nm, 610 nm, 640 nm, 670 nm.

**39.** $(3.0 \times 10^1)°$.

**41.** Second order.

**43.** 7140 slits/cm.

**45.** 0.878 cm.

**47.** 230 nm.

**49.** 9 lenses.

**51.** 33 dark bands.

**53.** 110 nm; 230 nm.

**55.** 482 nm.

**57.** 691 nm.

**59.** 1.004328.

**61.** 57.3°.

**63.** 48.1%.

**65.** No; for diamond, $\theta_p = 59.4°$.

**67.** 61.2°.

**69.** 36.9°; 48.8°; 53.1°.

**71.** 28° relative to first polarizer.

**73.** (a) $1.3 \times 10^{-4}$ m;
    (b) $3.9 \times 10^{-7}$ m.

**75.** H: 13.7°; Ne: 13.5°; Ar: 14.5°.

**77.** 480 nm.

**79.** $\lambda_2 > 600$ nm overlaps with
    $\lambda_3 < 467$ nm.

**81.** (a) 82 nm;
    (b) 130 nm.

**83.** $4.8 \times 10^4$ m.

**85.** $3.19 \times 10^{-5}$ m.

**87.** 580 nm.

**89.** 0.6 m.

**91.** 658 nm; 782 slits/cm.

**93.** 400 nm, 600 nm.

**95.** (a) 0;
    (b) 0.11;
    (c) 0.

## Chapter 25

**1.** $f/8$.

**3.** 3.00 mm to 46 mm.

**5.** 110 mm; 220 mm.

**7.** 5.1 m to infinity.

**9.** 16 mm.

**11.** $f/2.5$.

**13.** 1.1 D.

**15.** (a) $-1.2$ D;
    (b) 35 cm.

**17.** $-8.3$ D; $-7.1$ D.

**19.** (a) 2.0 cm;
    (b) 1.9 cm.

**21.** 18.4 cm, 1.00 m.

**23.** 1.6×.

**25.** 6.2 cm from lens, 4.0×.

**27.** (a) 3.6×;
    (b) 14 mm;
    (c) 6.9 cm.

**29.** 4 cm toward contract.

**31.** $-29\times$; 85 cm.

**33.** 25 cm.

**35.** $-22\times$.

**37.** $-110\times$.

**39.** Objective: 1.09 m;
    eyepiece: 9.09 mm.

**41.** 3.0 m.

**43.** $(4.0 \times 10^2)\times$.

**45.** 520×.

**47.** (a) $(9.00 \times 10^2)\times$;
    (b) eyepiece: 1.8 cm;
        objective, 0.300 cm;
    (c) 0.306 cm from objective.

**49.** (a) 14 cm;
    (b) 130×.

**51.** (a) Converging;
    (b) 281 cm.

**53.** $(1.54 \times 10^{-5})°$.

**55.** $1.7 \times 10^{11}$ m.

**57.** $1.0 \times 10^4$ m.

**59.** 8.5 cm, $6.2 \times 10^{-6}$ rad (distance of
    2.4 km).

**61.** (a) 53.8°;
    (b) 0.19 nm.

**63.** (a) 1×;
    (b) 1× (at back of body) to
        2.7× (at front of body).

**65.** $\frac{1}{6}$ s.

**67.** (a) $1.8 \times 10^4$ m;
    (b) 23″; atmospheric effects and
        aberrations in the eye.

**69.** 0.82 mm.

**71.** 8.2.

**73.** $-0.75$ D (upper part),
    $+2.0$ D (lower part).

**75.** (a) $-2.8\times$;
    (b) $+5.5$-D lens.

**77.** $+3.9$ D.

**79.** $4.8 \times 10^3$ C°/min.

**81.** $-15\times$.

**83.** 0.4 m.

**85.** 110 m.

**87.** 2 m.

**89.** $-8.4$ cm.

## Chapter 26

**1.** 83.9 m.

**3.** 35 ly.

**5.** $0.70c$.

**7.** (a) $0.141c$;
    (b) $0.140c$.

**9.** (a) 6.92 m, 1.35 m;
    (b) 13.9 s;
    (c) $0.720c$;
    (d) 13.9 s.

**11.** (a) 3.6 yr;
    (b) 7.0 yr.

**13.** $(6.97 \times 10^{-8})\%$.

**15.** $0.9716c$.

**17.** (a) $-1.1\%$;
    (b) $-34\%$.

**19.** $0.95c$.

**21.** $8.209 \times 10^{-14}$ J, 0.512 MeV.

**23.** 1000 kg.

**25.** $5.36 \times 10^{-13}$ kg.

**27.** 1.6 GeV/$c$.

**29.** (a) 4.5 GeV;
    (b) 5.4 GeV/$c$.

**31.** $0.866c$, 0.886 MeV/$c$.

**33.** $9.0 \times 10^{13}$ J; $9.2 \times 10^9$ kg.

**35.** 1.30 MeV/$c^2$.

**37.** $0.333c$.

**39.** (a) $0.866c$;
    (b) $0.745c$.

**41.** (a) $1.8 \times 10^{19}$ J;
    (b) $-1.7\%$.

**43.** $\dfrac{2m}{\sqrt{1 - v^2/c^2}}$; 0;
    $\left( \dfrac{1}{\sqrt{1 - v^2/c^2}} - 1 \right) 2mc^2$.

**47.** $0.69c$.

**49.** (a) $0.959c$;
    (b) $0.36c$.

**51.** $0.962c$.

**53.** $0.3c$.

**55.** (a) $0.66c$;
    (b) 6.5 yr.

**57.** $0.91c$.

**59.** 1.022 MeV.

**61.** 0.79 MeV.

**63.** (a) $4 \times 10^9$ kg/s;
    (b) $4 \times 10^7$ yr;
    (c) $1 \times 10^{13}$ yr.

**67.** $5.38 \times 10^{-12}$ kg; $(1.5 \times 10^{-8})\%$.

**69.** (*a*)

(*b*)

**71.** $0.981c$.

**73.** $1.7 \times 10^{21}$ J; $\sim 20\times$ greater.

**75.** (*a*) 15 m;

(*b*) 42 min.

**77.** (*a*) $0.986c$;

(*b*) $\left(1 - 5 \times 10^{-7}\right)c$.

**79.** Yes, in barn's reference frame, if his speed is $\geq 0.639c$; no, in boy's reference frame.

**81.** $8.0 \times 10^{-8}$ s.

## Chapter 27

**1.** $6.2 \times 10^4$ C/kg.

**3.** 5 electrons.

**5.** (*a*) 10.6 $\mu$m, far infrared;

(*b*) 940 nm, near infrared;

(*c*) 0.7 mm, microwave.

**7.** $5.4 \times 10^{-20}$ J, 0.34 eV.

**9.** $9.35 \times 10^{-6}$ m.

**11.** $2.7 \times 10^{-19}$ J to $5.0 \times 10^{-19}$ J, 1.7 eV to 3.1 eV.

**13.** $1.14 \times 10^{-27}$ kg$\cdot$m/s.

**17.** $7.2 \times 10^{14}$ Hz.

**19.** 429 nm.

**21.** (*a*) 2.3 eV;

(*b*) 0.85 V.

**23.** 0.92 eV; $5.7 \times 10^5$ m/s.

**25.** (*a*) 0.78 eV;

(*b*) no ejected electrons.

**27.** 3.32 eV.

**31.** (*a*) $2.43 \times 10^{-12}$ m;

(*b*) $1.32 \times 10^{-15}$ m.

**33.** 2.62 MeV.

**35.** 212 MeV; $5.86 \times 10^{-15}$ m.

**37.** 1.592 MeV, $7.81 \times 10^{-13}$ m.

**39.** $4.7 \times 10^{-12}$ m.

**41.** 1840.

**43.** (*a*) $4 \times 10^{-10}$ m;

(*b*) $1 \times 10^{-10}$ m;

(*c*) $3.9 \times 10^{-11}$ m.

**45.** $6.4 \times 10^{-12}$ m; yes; much less than 5 cm; no.

**47.** 22 V.

**49.** (*a*) Absorption; final state; largest energy photon;

(*b*) emission, initial state;

(*c*) absorption, final state.

**51.** $n = 5$ to $n' = 3$.

**53.** (*a*) 486 nm;

(*b*) 103 nm;

(*c*) 434 nm.

**55.** 91.2 nm.

**57.** Yes; from $n = 1$ to $n = 3$.

**59.**

**61.** $-0.544$ eV.

**63.** Yes; $\left(7.30 \times 10^{-3}\right)c$, $1/\gamma = 0.99997$.

**65.** $1.20 \times 10^{29}$ m, $-4.22 \times 10^{-97}$ J.

**67.** $3.28 \times 10^{15}$ Hz.

**69.** $4.4 \times 10^{26}$ photons/s.

**71.** $5.5 \times 10^{18}$ photons/s.

**73.** (*a*) 1.7 eV;

(*b*) 3.0 eV.

**75.** 0.34 MeV for both.

**77.** $4.0 \times 10^{-14}$ m.

**79.** $4.40 \times 10^{-40}$; yes.

**81.** 0.32 V.

**83.** $4.0 \times 10^{-7}$ W/m$^2$; $1.7 \times 10^{-2}$ V/m.

**85.** $8.2 \times 10^{-10}$ m.

**87.** $3.5 \times 10^{-12}$ m.

**89.** 1200 m/s.

**91.** Paschen series, level 4.

**93.** $1.8 \times 10^{11}$ C/kg.

**95.** $E_n = -\dfrac{2.84 \times 10^{165} \text{ J}}{n^2}$, $r_n = n^2(5.16 \times 10^{-129}$ m); not apparent.

## Chapter 28

**1.** $4.5 \times 10^{-7}$ m.

**3.** $5.3 \times 10^{-11}$ m.

**5.** $7 \times 10^{-8}$ eV.

**7.** $3.00 \times 10^{-10}$ eV/$c^2$.

**9.** Electron: $1.5 \times 10^{-3}$ m; baseball: $9.7 \times 10^{-33}$ m; $\Delta x_{\text{electron}}/\Delta x_{\text{baseball}} = 1.5 \times 10^{29}$.

**11.** 7700 m/s; $1.7 \times 10^{-4}$ eV.

**13.** $5.53 \times 10^{-10}$ m.

**15.** $\ell = 0, 1, 2, 3, 4, 5$.

**17.** 14 electrons.

**19.** (*a*) $\left(1, 0, 0, -\frac{1}{2}\right)$, $\left(1, 0, 0, +\frac{1}{2}\right)$, $\left(2, 0, 0, -\frac{1}{2}\right)$, $\left(2, 0, 0, +\frac{1}{2}\right)$, $\left(2, 1, -1, -\frac{1}{2}\right)$, $\left(2, 1, -1, +\frac{1}{2}\right)$;

(*b*) $\left(1, 0, 0, -\frac{1}{2}\right)$, $\left(1, 0, 0, +\frac{1}{2}\right)$, $\left(2, 0, 0, -\frac{1}{2}\right)$, $\left(2, 0, 0, +\frac{1}{2}\right)$, $\left(2, 1, -1, -\frac{1}{2}\right)$, $\left(2, 1, -1, +\frac{1}{2}\right)$, $\left(2, 1, 0, -\frac{1}{2}\right)$, $\left(2, 1, 0, +\frac{1}{2}\right)$, $\left(2, 1, 1, -\frac{1}{2}\right)$, $\left(2, 1, 1, +\frac{1}{2}\right)$, $\left(3, 0, 0, -\frac{1}{2}\right)$, $\left(3, 0, 0, +\frac{1}{2}\right)$, $\left(3, 1, -1, -\frac{1}{2}\right)$.

**21.** $12\hbar$, or $3.65 \times 10^{-34}$ J$\cdot$s.

**23.** $n \geq 4$; $3 \leq \ell \leq n - 1$; $m_s = -\frac{1}{2}, +\frac{1}{2}$.

**25.** (*a*) $1s^2 2s^2 2p^6 3s^2 3p^6 3d^6 4s^2$;

(*b*) $1s^2 2s^2 2p^6 3s^2 3p^6 3d^{10} 4s^2 4p^4$;

(*c*) $1s^2 2s^2 2p^6 3s^2 3p^6 3d^{10} 4s^2 4p^6 5s^2$.

**27.** (*a*) 5;

(*b*) $-0.544$ eV;

(*c*) $\sqrt{6}\,\hbar$, 2;

(*d*) $-2, -1, 0, 1, 2$.

**31.** $n = 3$, $\ell = 2$.

**33.** $4.36 \times 10^{-11}$ m; $1 \times 10^{-9}$ m.

**37.** $1.798 \times 10^{-10}$ m.

**39.** $6.12 \times 10^{-11}$ m; partial shielding by $n = 2$ shell.

**41.** 0.017 J, $5.5 \times 10^{16}$ photons.

**43.** 634 nm.

**45.** $n \geq 7$, $\ell = 6$, $m_\ell = 2$.

**47.** $L_{\text{max}} = \sqrt{30}\,\hbar$, $L_{\text{min}} = 0$.

**49.** $5.3 \times 10^{-27}$ kg$\cdot$m/s, $1.3 \times 10^{-7}$ m.

**51.** $5.75 \times 10^{-13}$ m, 115 keV.

**53.** 22 electrons.

**57.** $L_{\text{Bohr}} = 2\hbar$; $L_{\text{qm}} = 0$ or $\sqrt{2}\,\hbar$.

**59.** $3.7 \times 10^{-37}$ m.

**61.** $\Delta p_{\text{electron}}/\Delta p_{\text{proton}} = 0.0234$.

**63.** $2.0 \times 10^{-35}$ m/s; yes; $10^{34}$ s.

**65.** Copper.

## Chapter 29

**1.** 5.1 eV.

**3.** HN: 110 pm, CN: 150 pm, NO: 133 pm.

**5.** 4.6 eV.

**9.** $1.10 \times 10^{-10}$ m.

**11.** $5.22 \times 10^{-4}$ m.

**13.** (a) 680 N/m;
(b) $2.1 \times 10^{-6}$ m.

**15.** 0.315 nm.

**17.** 2.0 eV.

**19.** $\lambda \leq 1.7 \ \mu$m.

**21.** $1.2 \times 10^{6}$ electrons.

**23.** $3 \times 10^{6}$.

**25.** 1.7 eV.

**27.** 14 V.

**29.** 7.3 mA.

**31.** (a) 5.1 mA;
(b) 3.6 mA.

**33.**

**35.** 0.65 V.

**37.** (a) 130;
(b) 8700.

**39.** (a) $3.1 \times 10^{4}$ K;
(b) 930 K.

**41.** (a) $-5.3$ eV;
(b) 4.4 eV.

**43.** $1.94 \times 10^{-46}$ kg·m$^2$.

**45.** $6.03 \times 10^{-4}$ eV.

**47.** $\lambda \leq 3.5 \times 10^{-7}$ m.

**49.** Yes; $1.11 \times 10^{-6}$ m.

**51.** 980 J/mol.

## Chapter 30

**1.** 0.149 u.

**3.** 0.855%.

**5.** 3727 MeV/$c^2$.

**7.** (a) 180 m;
(b) $3.5 \times 10^{4}$.

**9.** 30 MeV.

**11.** $6.0 \times 10^{26}$ nucleons; no; all nucleons have about the same mass.

**13.** 550 MeV.

**15.** 7.699 MeV/nucleon.

**17.** 12.42 MeV.

**19.** $^{23}_{11}$Na: 8.113 MeV/nucleon;
$^{24}_{11}$Na: 8.063 MeV/nucleon.

**21.** (b) Yes, binding energy is positive.

**25.** 0.782 MeV.

**27.** (a) $\beta^-$ emitter;
(b) $^{24}_{11}$Na $\rightarrow$ $^{24}_{12}$Mg + $\beta^-$ + $\bar{\nu}$, 5.515 MeV.

**29.** 2.822 MeV.

**31.** $\alpha$: 6.114 MeV; $\beta^-$: 0.259 MeV.

**33.** 10.8 MeV.

**35.** For both: KE$_{max}$ = 0.9612 MeV, KE$_{min}$ = 0.

**37.** KE$_{recoil}$ = 0.0718 MeV, $Q$ = 4.27 MeV.

**39.** 1.2 h.

**41.** $2.5 \times 10^{9}$ decays/s.

**43.** (a) $3.60 \times 10^{12}$ decays/s;
(b) $3.58 \times 10^{12}$ decays/s;
(c) $1.34 \times 10^{9}$ decays/s.

**45.** 7 $\alpha$ particles; 4 $\beta^-$ particles.

**47.** 0.91 g.

**49.** (a) $1.38 \times 10^{-13}$ s$^{-1}$;
(b) $3.73 \times 10^{7}$ decays/min.

**51.** 86 decays/s.

**53.** $1.78 \times 10^{9}$ yr.

**55.** 7900 yr.

**57.** 15.8 d.

**59.** $N_{\mathrm{D}} = N_0 \left( 1 - e^{-\lambda t} \right)$.

**61.** $2.6 \times 10^{4}$ yr.

**63.** 41 yr.

**65.** $3.0 \times 10^{-14}$ m, 4.2× nuclear radius.

**67.** $^{40}_{19}$K: 0.16 mg; $^{39}_{19}$K: 1.2 g.

**69.** $2.71 \times 10^{-11}$ g.

**71.** (a) $4.6 \times 10^{15}$ decays/s;
(b) $1.2 \times 10^{4}$ decays/s.

**73.** (a) 0.002603 u, 2.425 MeV/$c^2$;
(b) 0, 0;
(c) $-0.090739$ u, $-84.52$ MeV/$c^2$;
(d) 0.043930 u, 40.92 MeV/$c^2$;
(e) $\Delta \geq 0$ for $0 \leq Z \leq 8$ and $Z \geq 85$;
$\Delta < 0$ for $9 \leq Z \leq 84$;
$\Delta \geq 0$ for $0 \leq A \leq 15$ and $A \geq 218$;
$\Delta < 0$ for $16 \leq A < 218$.

**75.** 0.2 decays/s.

**77.** (a) 180 m;
(b) 13 km.

**79.** (a) There would be no atoms— just neutrons;
(b) 0.083%.

**81.** $1.1 \times 10^{11}$ yr.

**83.** 13 decays/s.

**85.** (b) 98.2%.

## Chapter 31

**1.** $^{28}_{13}$Al; $\beta^-$; $^{28}_{14}$Si.

**3.** Yes, $Q > 0$.

**5.** 18.000937 u.

**7.** (a) Yes;
(b) 20.4 MeV.

**9.** 4.730 MeV.

**11.** n + $^{14}_{7}$N $\rightarrow$ $^{14}_{6}$C + p, 0.626 MeV.

**13.** (a) He picks up a neutron from C;
(b) $^{11}_{6}$C;
(c) 1.856 MeV; exothermic.

**15.** 5.702 MeV released.

**17.** 126.5 MeV.

**19.** 1/1100.

**21.** (a) 5 neutrons;
(b) 171.1 MeV.

**23.** 1100 kg.

**25.** 260 MeV; about 30% > fission energy released.

**27.** 3000 eV.

**31.** (a) a: $6.03 \times 10^{23}$ MeV/g;
b: $4.89 \times 10^{23}$ MeV/g;
c: $2.11 \times 10^{24}$ MeV/g;
(b) $5.13 \times 10^{23}$ MeV/g;
a: 0.851; b: 1.05; c: 0.243.

**33.** 0.39 g.

**35.** $5.6 \times 10^{3}$ kg/h.

**37.** $2.46 \times 10^{9}$ J; 50× > gasoline.

**39.** 7000 rads.

**41.** 144 rads.

**43.** (a) 1.0 rad, or 0.010 Gy;
(b) $1.0 \times 10^{10}$ protons.

**47.** $5.61 \times 10^{-10}$ kg.

**49.** $9 \times 10^{16}$ e$^-$.

**51.** (a) $2 \times 10^{-7}$ Sv/yr, $2 \times 10^{-5}$ rem/yr;
($2 \times 10^{-1}$) × allowed dose;
(b) $2 \times 10^{-6}$ Sv/yr, $2 \times 10^{-4}$ rem/yr;
($2 \times 10^{-3}$) × allowed dose.

**53.** 7.041 m; radio wave.

**55.** (a) $^{12}_{6}$C;
(b) 5.702 MeV.

**57.** 1.0043 : 1.

**59.** $6.6 \times 10^{-2}$ rem/yr.

**61.** (a) $^{226}_{88}\text{Ra} \rightarrow \, ^{4}_{2}\text{He} + \, ^{222}_{86}\text{Rn}$;
   (b) 4.871 MeV;
   (c) 190.6 MeV/$c$ for both;
   (d) $8.78 \times 10^{-2}$ MeV.

**63.** (a) $3.7 \times 10^{26}$ W;
   (b) $3.5 \times 10^{38}$ protons/s;
   (c) $1.1 \times 10^{11}$ yr.

**65.** $8 \times 10^{12}$ J.

**67.** (a) 3700 decays/s;
   (b) $4.0 \times 10^{-4}$ Sv/yr;
      11% of background.

**69.** 7.274 MeV.

**71.** 990 kg.

**73.** (a) $^{141}_{54}\text{Xe}$;
   (b) 2 neutrons escape or are absorbed, 1 causes another fission;
   (c) 176.0 MeV.

**75.** Most to least dangerous:
   C > B > A.

# Chapter 32

**1.** 5.59 GeV.

**3.** $1.3 \times 10^{7}$ Hz.

**5.** 2.0 T.

**7.** Alpha particles;
   $\lambda_\alpha \approx d_{\text{nucleon}}$, $\lambda_p \approx 2d_{\text{nucleon}}$.

**9.** 0.9 MeV/rev.

**13.** 33.9 MeV.

**15.** 1879.2 MeV.

**17.** 67.5 MeV.

**19.** $1.32 \times 10^{-15}$ m.

**21.** $1.3 \times 10^{-12}$ m.

**23.** First, second, and fourth will not happen.

**25.** (a) 37.8 MeV;
   (b) $\text{KE}_p = 5.4$ MeV,
      $\text{KE}_{\pi^-} = 32.4$ MeV.

**27.** $9 \times 10^{-3}$ MeV.

**29.** $7.5 \times 10^{-21}$ s.

**31.** (a) 1300 eV;
   (b) 150 MeV.

**33.** $8 \times 10^{-27}$ s.

**35.** (a) udd;
   (b) $\bar{u}\bar{d}\bar{d}$;
   (c) uds;
   (d) $\bar{u}\bar{d}\bar{s}$.

**37.** $c\bar{u}$.

**39.**

**41.** 16 GeV; $7.8 \times 10^{-17}$ m.

**43.** (a) 1.022 MeV;
   (b) 1876.6 MeV.

**45.** $1.2 \times 10^{10}$ m, $4.0 \times 10^{1}$ s.

**47.** (a) Possible, strong interaction;
   (b) possible, strong interaction;
   (c) possible, strong interaction;
   (d) not possible; charge is not conserved;
   (e) possible, weak interaction.

**49.** $10^{-18}$ m corresponds to 200 GeV.

**51.** 64 fundamental fermions.

**53.** (b) $10^{29}$ K.

**55.** 798.7 MeV.

**57.** $9.3 \times 10^{10}$ eV;
   $1.3 \times 10^{-17}$ m, $0.99995c$.

**59.** (a) $\pi^0$;
   (b) $\bar{\nu}_\mu$.

**63.** $10^{-25}$ s.

# Chapter 33

**1.** 3.1 ly.

**3.** Less than; by a factor of 2.

**5.** $2 \times 10^{-3}$ kg/m$^3$.

**7.** $4.2 \times 10^{-2}$ rad, or 2.4°;
   about 4.5× Moon's width.

**9.** $1.83 \times 10^{9}$ kg/m$^3$;
   $3.33 \times 10^{5}$ times larger.

**11.** (a) 0.018″;
   (b) $(5.0 \times 10^{-6})°$.

**13.** 280 yr.

**15.** $D_1/D_2 = 0.13$.

**17.** $3 \times 10^{14}$ m.

**19.** 540°.

**21.** $3.1 \times 10^{-16}$ m.

**23.** $2.1 \times 10^{8}$ ly.

**25.** (a) 0.3 nm;
   (b) 3.2 nm.

**27.** $0.058c$.

**29.** $9 \times 10^{9}$ ly.

**31.** $6.8 \times 10^{7}$ ly.

**33.** $1.1 \times 10^{-3}$ m.

**35.** 6 nucleons/m$^3$.

**37.** (a) $10^{-5}$ s;
   (b) $10^{-7}$ s;
   (c) $10^{-4}$ s.

**39.** 0.2 rev/s.

**41.** A: $T$ increases, $L$ doesn't change, size decreases.
   B: $T$ unchanged, $L$ decreases, size decreases.
   C: $T$ decreases, $L$ increases, size increases.

**43.** $1.7 \times 10^{25}$ W.

**45.** 400 ly; $r_{\text{Sun}}/d_{\text{Earth–Sun}} = 2 \times 10^{7}$, $r_{\text{Sun}}/d_{\text{Galaxy}} = 4 \times 10^{-3}$.

**47.** (a) $(9 \times 10^{-6})°$;
   (b) 4 m.

**49.** (a) 13.93 MeV;
   (b) 4.71 MeV;
   (c) $5.46 \times 10^{10}$ K.

**51.** $t_P = 5.38 \times 10^{-44}$ s.

# Index

Note: The abbreviation *defn* means the page cited gives the definition of the term; *fn* means the reference is in a footnote; *pr* means it is found in a Problem or Question; *ff* means "also the following pages."

# Photo Credits

**CO–11 left** Ford Motor Company **CO–11 right** Jonathan Nourok/PhotoEdit **11–4** Ford Motor Company **11–6** Paul Springett A/Alamy **11–11** Berenice Abbott/Photo Researchers, Inc. **11–13** D. Giancoli **11–17** Taylor Devices **11–19** Martin Bough/Fundamental Photographs **11–20a** AP Wide World Photos **11–20b** Paul X. Scott **11–28** Art Wolfe/Getty Images **11–21a/b/c/d, 11–38** D. Giancoli **11–43** Jacynthroode/iStockphoto **11–45a/b** Richard Megna/Fundamental Photographs **11–49** Andre Gallant/Getty Images **11–59** Richard Megna/Fundamental Photographs

**CO–12** Scala/Art Resource **12–4** Willie Maldonado/Getty Images **12–9a** Ben Clark/Getty Images **12–9b** Tony Cenicola/The New York Times **12–10** Bob Daemmrich/The Image Works **12–24** Prof. Vickie Frohne **12–25** Gary S. Settles/Photo Researchers, Inc. **12–26** GE Medical Systems/Photo Researchers, Inc. **12–28a** Visions of America/SuperStock **12–35** Nation Wong/Corbis **12–42** D. Giancoli

**CO–13 left** Niall Edwards/Alamy **CO–13 right** Richard Price/Getty **13–3** Ingolfson **13–4b** Franca Principe/Istituto e Museo di Storia della Scienza, Florence **13–6** Leonard Lessin/Photolibrary **13–12** Pearson Education/Eric Schrader **13–16** D-BASE/Getty Images **13–17** Michael Newman/PhotoEdit **13–25** Paul Silverman/Fundamental Photographs **13–26a** Norbert Rosing/National Geographic Stock **13–26b** American Association of Physics Teachers/Mrinalini Modak **13–26c** Prof. Gary Wysin **13–27a/b/c** Mary Teresa Giancoli **13–30** Kennan Harvey/Getty Images **13–31** Reed Kaestner/Corbis

**CO–14** Pixland/age fotostock **14–10** Ernst Haas/Hulton Archive/Getty Images **14–13a/b** Science Photo Library/Photo Researchers, Inc. **14–14** Prof. Vickie Frohne **14–16** Stacey Bates/Shutterstock **14–21** Tobias Titz/Getty Images

**CO–15** Taxi/Getty Images **15–9** Will Hart **15–10 left, middle, right** Leonard Lessin/Photolibrary **15–20a** Carsten Koall/Getty Images **15–20b** Kevin Burke/Corbis **15–20c** David Woodfall **Table 15–4 top left** Michael Collier **Table 15–4 bottom left** Michel de Nijs/iStockphoto **Table 15–4 mid-top** Larry Lee Photography/Corbis **Table 15–4 mid-bottom** Richard Schmidt-Zuper/iStockphoto **15–25** Patrick Landmann/Photo Researchers, Inc. **15–26** Richard Schmidt-Zuper/iStockphoto **15–27** Michael Collier

**CO–16** Mike Dunning/Dorling Kindersley **16–37** Peter Menzel/Photo Researchers, Inc. **16–38** Dr. Gopal Murti/Science Source/Photo Researchers, Inc. **16–48** American Association of Physics Teachers/Matthew Claspill

**CO–17** Emily Michot/Miami Herald/MCT/Newscom **17–8** D. Giancoli **17–13c, 17–18, 17–20** Eric Schrader/Pearson Education **17–21** tunart/iStockphoto **17–31 left** Eric Schrader/Pearson Education **17–31 right** Robnil **17–36** beerkoff/Shutterstock **17–44** Andrea Sordini

**CO–18 left** Mahaux Photography **CO–18 right** Eric Schrader/Pearson Education **18–1** Jean-Loup Charmet/Science Photo Library/Photo Researchers, Inc. **18–2** The Burndy Library Collection/Huntington Library **18–6a** Richard Megna/Fundamental Photographs **18–11** T. J. Florian/Rainbow Image Library **18–15** Richard Megna/Fundamental Photographs **18–16** Tony Freeman/PhotoEdit **18–18** Clint Spencer/iStockphoto **18–32** Alexandra Truitt & Jerry Marshall **18–34** Scott T. Smith/Corbis **18–36** Jim Wehtje/Getty Images

**CO–19** Patrik Stoffarz/AFP/Getty Images/Newscom **19–15** David R. Frazier/Photolibrary, Inc./Alamy **19–24** Apogee/Photo Researchers, Inc. **19–27a** Photodisc/Getty Images **19–27b, 19–28, 19–30a** Eric Schrader/Pearson Education **19–30b** Olaf Doring/Imagebroker/AGE Fotostock **19–67** Raymond Forbes/AGE fotostock **19–74** Eric Schrader/Pearson Education

**CO–20** Richard Megna/Fundamental Photographs **20–1** Dorling Kindersley **20–4a** Stephen Oliver/Dorling Kindersley **20–6** Mary Teresa Giancoli **20–8a/b, 20–18** Richard Megna/Fundamental Photographs **20–20b** Jack Finch/Science Photo Library/Photo Researchers, Inc. **20–28b, 20–43** Richard Megna/Fundamental Photographs **20–50** Clive Streeter/Dorling Kindersley, Courtesy of The Science Museum, London

**CO–21** Richard Megna/Fundamental Photographs **21–7** Photo Courtesy of Diva de Provence, Toronto, ON, Canada **21–12** Jeff Hunter/Getty Images **21–21** Associated Press Photo/Robert F. Bukaty **21–22** Photograph by Robert Fenton Houser **21–27a** Terence Kearey **21–34a** 4kodiak/iStockphoto **21–34b** Eric Schrader/Pearson Education

**CO–22** NASA **22–1** Original photograph in the possession of Sir Henry Roscoe, courtesy AIP Emilio Segrè Visual Archives **22–9** The Image Works Archives **22–12** Time Life Pictures/Getty Images **22–19** David J. Green/Alamy Images **22–20** Don Baida **22–22** NASA

**CO–23, 23–6** D. Giancoli **23–11a** Mary Teresa Giancoli and Suzanne Saylor **23–11b** Paul Silverman/Fundamental Photographs **23–20** John Lawrence/Travel Pix Ltd. **23–22a** Shannon Fagan/age fotostock **23–23** Giuseppe Molesini, Istituto Nazionale di Ottica, Florence **23–30b** Garo/Phanie/Photo Researchers, Inc. **23–31c/d** D. Giancoli **23–32** D. Giancoli and Howard Shugat **23–34** Kari Erik Marttila Photography **23–38a/b, 23–47** D. Giancoli **23–50** Mary Teresa Giancoli **23–65** American Association of Physics Teachers/Annacy Wilson **23–66** American Association of Physics Teachers/Matt Buck **23–69** American Association of Physics Teachers/Sarah Lampen **23–70a/b** Scott Dudley

**CO–24** Giuseppe Molesini, Istituto Nazionale di Ottica, Florence  **24–4a** Kent Wood/Photo Researchers, Inc.  **24–9a** Bausch & Lomb Incorporated  **24–13** David Parker/Science Photo Library/Photo Researchers, Inc.  **24–16b** Lewis Kemper Photography  **24–17** George Diebold  **24–19a** P. M. Rinard/American Journal of Physics  **24–19b** Richard Megna/Fundamental Photographs  **24–19c** Ken Kay/Fundamental Photographs  **24–28** Wabash Instrument Corp./Fundamental Photographs  **24–29a** Giuseppe Molesini, Istituto Nazionale di Ottica, Florence  **24–29b** Richard Megna/Fundamental Photographs  **24–29c** Paul Silverman/ Fundamental Photographs  **24–31b** Ken Kay/Fundamental Photographs  **24–33b/c** Bausch & Lomb Incorporated  **24–35** D. Hurst/Alamy  **24–44** Diane Schiumo/Fundamental Photographs  **24–47a/b** JiarenLau Photography, http://creativecommons.org/ licenses/by/2.0/deed.en  **24–52** Suunto  **24–53** Daniel Rutter/Dan's Data

**CO–25, 25–5, 25–6a/b** Mary Teresa Giancoli  **25–08** Leonard Lessin/Photolibrary  **25–19a/b** Museo Galileo - Istituto e Museo di Storia della Scienza  **25–21** Yerkes Observatory, University of Chicago  **25–22c** Sandy Huffaker/Stringer/Getty Images  **25–22d** Inter-University Centre for Astronomy and Astrophysics/Laurie Hatch  **25–24b** Leica Microsystems  **25–29a/b** Reproduced by permission from M. Cagnet, M. Francon, and J. Thrier, The Atlas of Optical Phenomena. Berlin: Springer-Verlag, 1962.  **25–32** Space Telescope Science Institute  **25–33** David Parker/Photo Researchers, Inc.  **25–37** The Burndy Library Collection/Huntington Library  **25–40** Rosalind Franklin/Photo Researchers, Inc.  **25–44a** Martin M. Rotker  **25–44b** Scott Camazine/Alamy  **25–47** Ron Chapple/Ron Chapple Photography  **25–48** NOAA Space Environment Center

**CO–26** "The City Blocks Became Still Shorter" from page 4 of the book "Mr Tompkins in Paperback" by George Gamow. Reprinted with the permission of Cambridge University Press.  **26–1** Bettmann/Corbis  **26–7** D. Giancoli  **26–12** "Unbelievably Shortened" from page 3 of "Mr Tompkins in Paperback" by George Gamow. Reprinted with the permission of Cambridge University Press.

**CO–27 left** Paul Gunning/Photo Researchers, Inc.  **CO–27 right** Professors Pietro M. Motta & Silvia Correr/Photo Researchers, Inc.  **27–11** Samuel Goudsmit/AIP Emilio Segrè Visual Archives, Goudsmit Collection  **27–12** Education Development Center, Inc.  **27–16a** Lee D. Simon/Science Source/Photo Researchers, Inc.  **27–16b** Oliver Meckes/Max-Planck-Institut-Tubingen/Photo Researchers, Inc.  **27–16c** Science Photo Library/Alamy  **27–22b** Richard Megna/Fundamental Photographs  **27–23a/b/c** Wabash Instrument Corp./Fundamental Photographs

**CO–28** Richard Cummins/Corbis  **28–1** Niels Bohr Institute, courtesy AIP Emilio Segrè Visual Archives  **28–2** F. D. Rosetti/ American Institute of Physics/Emilio Segrè Visual Archives  **28–4** Hitachi, Ltd., Advanced Research Laboratory  **28–15** Paul Silverman/Fundamental Photographs  **28–22** NIH/Photo Researchers, Inc.  **28–23** AP Photo/Dusan Vranic

**CO–29** Intel  **29–35** Jeff J. Daly/Fundamental Photographs  **29–36** D. Giancoli  **29–39** General Motors

**CO–30** D. Giancoli  **30–3** French Government Tourist Office  **30–7** Enrico Fermi Stamp Design/2001 United States Postal Service. All Rights Reserved. Used with Permission from the US Postal Service and Rachel Fermi.  **30–16** Fermilab Visual Media Services

**CO–31** Peter Beck/Corbis  **31–6** Archival Photofiles, [apf2-00502], Special Collections Research Center, University of Chicago Library  **31–9** Tokyo Electric Power Company/Jana Press/ZUMAPRESS.com/Newscom  **31–10** LeRoy N. Sanchez/Los Alamos National Laboratory  **31–11** Corbis/Bettmann  **31–15a** UPPA/Photoshot/Newscom  **31–15b** National Ignition Facility/Lawrence Livermore National Laboratory  **31–19a** Robert Turgeon, Cornell University  **31–19b** Jack Van't Hof/Brookhaven National Laboratory  **31–20b** Needell M.D./Custom Medical Stock Photography  **31–22a/b** National Institute of Health  **31–25b** Slaven/Custom Medical Stock Photo  **31–27** Scott Camazine/Photo Researchers, Inc.

**CO–32** CERN  **32–1** Lawrence Berkeley National Laboratory  **32–3** Reidar Hahn/Fermilab Visual Media Services  **32–5a/b, 32–6** CERN  **32–10a** David Parker/Science Photo Library/Photo Researchers, Inc.  **32–10b** Science Photo Library/Photo Researchers, Inc.  **32–11** Brookhaven National Laboratory  **32–12** Lawrence Berkeley National Laboratory  **32–15** Fermilab/ Science Photo Library/Photo Researchers, Inc.  **32–20** Denis Balibouse/AFP/Getty Images

**CO–33** NASA, ESA, R. Ellis (Caltech), and the UDF 2012 Team  **33–1a** NASA  **33–1b** Allan Morton/Dennis Milon/Science Photo Library/Photo Researchers, Inc.  **33–2c** NASA  **33–3** U.S. Naval Observatory/NASA  **33–4** National Optical Astronomy Observatories  **33–5a** Reginald J. Dofour, Rice University  **33–5b** U.S. Naval Observatory/NASA  **33–5c** National Optical Astronomy Observatories  **33–9a/b** Australian Astronomical Observatory/David Main Images  **33–15a** NASA  **33–23** Alcatel-Lucent  **33–25** AP Photo/Fredrik Persson  **33–26** WMAP Science Team/NASA  **33–32** Bertil Ericson/SCANPIX/AP/Corbis

## Useful Geometry Formulas — Areas, Volumes

Circumference of circle   $C = \pi d = 2\pi r$

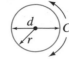

Area of circle   $A = \pi r^2 = \dfrac{\pi d^2}{4}$

Area of rectangle   $A = \ell w$

Area of parallelogram   $A = bh$

Area of triangle   $A = \frac{1}{2} hb$

Right triangle (Pythagoras)   $c^2 = a^2 + b^2$

Sphere: surface area   $A = 4\pi r^2$
　　　　volume   $V = \frac{4}{3}\pi r^3$

Rectangular solid: volume   $V = \ell w h$

Cylinder (right):
　　surface area   $A = 2\pi r\ell + 2\pi r^2$
　　volume   $V = \pi r^2 \ell$

Right circular cone:
　　surface area   $A = \pi r^2 + \pi r \sqrt{r^2 + h^2}$
　　volume   $V = \frac{1}{3}\pi r^2 h$

## Exponents [See Appendix A–2 for details]

$(a^n)(a^m) = a^{n+m}$   [Example: $(a^3)(a^2) = a^5$]
$(a^n)(b^n) = (ab)^n$   [Example: $(a^3)(b^3) = (ab)^3$]
$(a^n)^m = a^{nm}$   $\left[\begin{array}{l}\text{Example: } (a^3)^2 = a^6 \\ \text{Example: } (a^{\frac{1}{4}})^4 = a\end{array}\right]$

$a^{-1} = \dfrac{1}{a}$ 　　$a^{-n} = \dfrac{1}{a^n}$ 　　$a^0 = 1$

$a^{\frac{1}{2}} = \sqrt{a}$ 　　$a^{\frac{1}{4}} = \sqrt{\sqrt{a}}$

$(a^n)(a^{-m}) = \dfrac{a^n}{a^m} = a^{n-m}$   [Ex.: $(a^5)(a^{-2}) = a^3$]

$\dfrac{a^n}{b^n} = \left(\dfrac{a}{b}\right)^n$

## Quadratic Formula [Appendix A–4]

Equation with unknown $x$, in the form
$$ax^2 + bx + c = 0,$$
has solutions
$$x = \frac{-b \pm \sqrt{b^2 - 4ac}}{2a}.$$

## Logarithms [Appendix A–8; Table p. A-11]

If $y = 10^x$, 　then 　$x = \log_{10} y = \log y$.
If $y = e^x$, 　then 　$x = \log_e y = \ln y$.

$\log(ab) = \log a + \log b$

$\log\left(\dfrac{a}{b}\right) = \log a - \log b$

$\log a^n = n \log a$

## Binomial Expansion [Appendix A–5]

$(1 + x)^n = 1 + nx + \dfrac{n(n-1)}{2 \cdot 1} x^2 + \dfrac{n(n-1)(n-2)}{3 \cdot 2 \cdot 1} x^3 + \cdots$   [for $x^2 < 1$]

$\approx 1 + nx$   if $x \ll 1$

[Example: $(1 + 0.01)^3 \approx 1.03$]

[Example: $\dfrac{1}{\sqrt{0.99}} = \dfrac{1}{\sqrt{1 - 0.01}} = (1 - 0.01)^{-\frac{1}{2}} \approx 1 - (-\frac{1}{2})(0.01) \approx 1.005$]

## Fractions

$\dfrac{a}{b} = \dfrac{c}{d}$ is the same as $ad = bc$

$\dfrac{\left(\dfrac{a}{b}\right)}{\left(\dfrac{c}{d}\right)} = \dfrac{ad}{bc}$

## Trigonometric Formulas [Appendix A–7]

$\sin\theta = \dfrac{\text{opp}}{\text{hyp}}$

$\cos\theta = \dfrac{\text{adj}}{\text{hyp}}$

$\tan\theta = \dfrac{\text{opp}}{\text{adj}}$

$\text{adj}^2 + \text{opp}^2 = \text{hyp}^2$   (Pythagorean theorem)

$\tan\theta = \dfrac{\sin\theta}{\cos\theta}$

$\sin^2\theta + \cos^2\theta = 1$

$\sin 2\theta = 2\sin\theta \cos\theta$

$\cos 2\theta = (\cos^2\theta - \sin^2\theta) = (1 - 2\sin^2\theta) = (2\cos^2\theta - 1)$

$\sin(180° - \theta) = \sin\theta$
$\sin(90° - \theta) = \cos\theta$
$\cos(90° - \theta) = \sin\theta$
$\sin\frac{1}{2}\theta = \sqrt{(1 - \cos\theta)/2}$ 　　$\cos\frac{1}{2}\theta = \sqrt{(1 + \cos\theta)/2}$
$\sin\theta \approx \theta$   [for small $\theta \lesssim 0.2$ rad]
$\cos\theta \approx 1 - \dfrac{\theta^2}{2}$   [for small $\theta \lesssim 0.2$ rad]
$\sin(A \pm B) = \sin A \cos B \pm \cos A \sin B$
$\cos(A \pm B) = \cos A \cos B \mp \sin A \sin B$

$\cos(180° - \theta) = -\cos\theta$

For any triangle:
$c^2 = a^2 + b^2 - 2ab\cos\gamma$   (law of cosines)
$\dfrac{\sin\alpha}{a} = \dfrac{\sin\beta}{b} = \dfrac{\sin\gamma}{c}$   (law of sines)

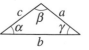

# Periodic Table of the Elements§

[Masses averaged over natural abundances]

**Transition Elements**

Key:
Symbol — Cl 17 — Atomic Number
Atomic Mass§ (averaged) — 35.453
Electron Configuration (outer shells only) — $3p^5$

(averaged over abundances of $^1_1H$, $^2_1H$)

| Group I | Group II | | | | | | | | | | | Group III | Group IV | Group V | Group VI | Group VII | Group VIII |
|---|---|---|---|---|---|---|---|---|---|---|---|---|---|---|---|---|---|
| H 1<br>1.00794<br>$1s^1$ | | | | | | | | | | | | | | | | | He 2<br>4.002602<br>$1s^2$ |
| Li 3<br>6.941<br>$2s^1$ | Be 4<br>9.012182<br>$2s^2$ | | | | | | | | | | | B 5<br>10.811<br>$2p^1$ | C 6<br>12.0107<br>$2p^2$ | N 7<br>14.0067<br>$2p^3$ | O 8<br>15.9994<br>$2p^4$ | F 9<br>18.9984032<br>$2p^5$ | Ne 10<br>20.1797<br>$2p^6$ |
| Na 11<br>22.98976928<br>$3s^1$ | Mg 12<br>24.3050<br>$3s^2$ | | | | | | | | | | | Al 13<br>26.9815386<br>$3p^1$ | Si 14<br>28.0855<br>$3p^2$ | P 15<br>30.973762<br>$3p^3$ | S 16<br>32.065<br>$3p^4$ | Cl 17<br>35.453<br>$3p^5$ | Ar 18<br>39.948<br>$3p^6$ |
| K 19<br>39.0983<br>$4s^1$ | Ca 20<br>40.078<br>$4s^2$ | Sc 21<br>44.955912<br>$3d^14s^2$ | Ti 22<br>47.867<br>$3d^24s^2$ | V 23<br>50.9415<br>$3d^34s^2$ | Cr 24<br>51.9961<br>$3d^54s^1$ | Mn 25<br>54.938045<br>$3d^54s^2$ | Fe 26<br>55.845<br>$3d^64s^2$ | Co 27<br>58.933195<br>$3d^74s^2$ | Ni 28<br>58.6934<br>$3d^84s^2$ | Cu 29<br>63.546<br>$3d^{10}4s^1$ | Zn 30<br>65.409<br>$3d^{10}4s^2$ | Ga 31<br>69.723<br>$4p^1$ | Ge 32<br>72.64<br>$4p^2$ | As 33<br>74.92160<br>$4p^3$ | Se 34<br>78.96<br>$4p^4$ | Br 35<br>79.904<br>$4p^5$ | Kr 36<br>83.798<br>$4p^6$ |
| Rb 37<br>85.4678<br>$5s^1$ | Sr 38<br>87.62<br>$5s^2$ | Y 39<br>88.90585<br>$4d^15s^2$ | Zr 40<br>91.224<br>$4d^25s^2$ | Nb 41<br>92.90638<br>$4d^45s^1$ | Mo 42<br>95.94<br>$4d^55s^1$ | Tc 43<br>(98)<br>$4d^55s^2$ | Ru 44<br>101.07<br>$4d^75s^1$ | Rh 45<br>102.90550<br>$4d^85s^1$ | Pd 46<br>106.42<br>$4d^{10}5s^0$ | Ag 47<br>107.8682<br>$4d^{10}5s^1$ | Cd 48<br>112.411<br>$4d^{10}5s^2$ | In 49<br>114.818<br>$5p^1$ | Sn 50<br>118.710<br>$5p^2$ | Sb 51<br>121.760<br>$5p^3$ | Te 52<br>127.60<br>$5p^4$ | I 53<br>126.90447<br>$5p^5$ | Xe 54<br>131.293<br>$5p^6$ |
| Cs 55<br>132.9054519<br>$6s^1$ | Ba 56<br>137.327<br>$6s^2$ | 57–71† | Hf 72<br>178.49<br>$5d^26s^2$ | Ta 73<br>180.94788<br>$5d^36s^2$ | W 74<br>183.84<br>$5d^46s^2$ | Re 75<br>186.207<br>$5d^56s^2$ | Os 76<br>190.23<br>$5d^66s^2$ | Ir 77<br>192.217<br>$5d^76s^2$ | Pt 78<br>195.084<br>$5d^96s^1$ | Au 79<br>196.966569<br>$5d^{10}6s^1$ | Hg 80<br>200.59<br>$5d^{10}6s^2$ | Tl 81<br>204.3833<br>$6p^1$ | Pb 82<br>207.2<br>$6p^2$ | Bi 83<br>208.98040<br>$6p^3$ | Po 84<br>(209)<br>$6p^4$ | At 85<br>(210)<br>$6p^5$ | Rn 86<br>(222)<br>$6p^6$ |
| Fr 87<br>(223)<br>$7s^1$ | Ra 88<br>(226)<br>$7s^2$ | 89–103‡ | Rf 104<br>(263)<br>$6d^27s^2$ | Db 105<br>(268)<br>$6d^37s^2$ | Sg 106<br>(271)<br>$6d^47s^2$ | Bh 107<br>(270)<br>$6d^57s^2$ | Hs 108<br>(277)<br>$6d^67s^2$ | Mt 109<br>(278)<br>$6d^77s^2$ | Ds 110<br>(281)<br>$6d^97s^1$ | Rg 111<br>(281)<br>$6d^{10}7s^1$ | Cn 112<br>(285)<br>$6d^{10}7s^2$ | 113¶<br>(286) | Fl 114<br>(289)<br>$6d^{10}7s^27p^2$ | 115¶<br>(289) | Lv 116<br>(293)<br>$6d^{10}7s^27p^4$ | 117¶<br>(294) | 118¶<br>(294) |

¶Preliminary evidence (unconfirmed) has been reported.

**†Lanthanide Series**

| La 57<br>138.90547<br>$5d^16s^2$ | Ce 58<br>140.116<br>$4f^15d^16s^2$ | Pr 59<br>140.90765<br>$4f^35d^06s^2$ | Nd 60<br>144.242<br>$4f^45d^06s^2$ | Pm 61<br>(145)<br>$4f^55d^06s^2$ | Sm 62<br>150.36<br>$4f^65d^06s^2$ | Eu 63<br>151.964<br>$4f^75d^06s^2$ | Gd 64<br>157.25<br>$4f^75d^16s^2$ | Tb 65<br>158.92535<br>$4f^95d^06s^2$ | Dy 66<br>162.500<br>$4f^{10}5d^06s^2$ | Ho 67<br>164.93032<br>$4f^{11}5d^06s^2$ | Er 68<br>167.259<br>$4f^{12}5d^06s^2$ | Tm 69<br>168.93421<br>$4f^{13}5d^06s^2$ | Yb 70<br>173.04<br>$4f^{14}5d^06s^2$ | Lu 71<br>174.967<br>$4f^{14}5d^16s^2$ |
|---|---|---|---|---|---|---|---|---|---|---|---|---|---|---|

**‡Actinide Series**

| Ac 89<br>(227)<br>$6d^17s^2$ | Th 90<br>232.03806<br>$6d^27s^2$ | Pa 91<br>231.03588<br>$5f^26d^17s^2$ | U 92<br>238.0289<br>$5f^36d^17s^2$ | Np 93<br>(237)<br>$5f^46d^17s^2$ | Pu 94<br>(244)<br>$5f^66d^07s^2$ | Am 95<br>(243)<br>$5f^76d^07s^2$ | Cm 96<br>(247)<br>$5f^76d^17s^2$ | Bk 97<br>(247)<br>$5f^96d^07s^2$ | Cf 98<br>(251)<br>$5f^{10}6d^07s^2$ | Es 99<br>(252)<br>$5f^{11}6d^07s^2$ | Fm 100<br>(257)<br>$5f^{12}6d^07s^2$ | Md 101<br>(258)<br>$5f^{13}6d^07s^2$ | No 102<br>(259)<br>$5f^{14}6d^07s^2$ | Lr 103<br>(262)<br>$5f^{14}6d^17s^2$ |
|---|---|---|---|---|---|---|---|---|---|---|---|---|---|---|

§ Atomic mass values averaged over isotopes in percentages they occur on Earth's surface; individual isotopes are found in Appendix B. For many unstable elements, mass of the longest-lived known isotope is given in parentheses. 2012 revisions.

08/21/15 3:18 PM

RENTAL

41793399900001942271